ELECTRONIC AND STRUCTURAL PROPERTIES OF AMORPHOUS SEMICONDUCTORS

Front Row (Left to Right), Professor Jonscher, Mrs. Steventon and daughter, Dr. Friedman, Mrs. Friedman, Professor Fritzsche, Dr. Shirley Emin, Dr. D. Emin, Mrs. Dingle, Professor Sir N. Mott, Professor Spear, Dr. Hilda Spear, Professor Dingle, Professor Grigorovici, Dr. Owen, Dr. Mort, Dr. Roberts, Dr. Davis, Dr. Le Comber, Dr. Main, Mrs. Millar, Dr. Robertson, Dr. Clark.

2nd Row (Left to Right), Mr. Bertrand, Mr. Razzetti, Dr. Kress, Mr. Madan, Professor Mullazzi, Mrs. Razzetti, Dr. Belu, Miss Brown, Mrs. Tyldesley, Dr. Magrabi, Mrs. Kastner, Mrs. Grover, Mrs. Hodgson, Dr. Maria Sabathy, Mr. Sabathy, Dr. Murgatroyd, Mr. Hodgson, Dr. J. Marshall, Mr. Fisher, Mr. Flynn, Dr. Hassan, Dr. Schlueter.

3rd Row (Left to Right), Mr. Wey, Mr. Steventon, Mr. Lahmers, Dr. Sharma, Dr. Bensoussan, Mr. Sussman, Dr. Hernandez, Mr. Gray, Professor Butcher, Mr. Kastner, Dr. Grover, Dr. Dohler, Mr. Babcenco, Dr. Dozier, Mr. Moore, Mr. Morys, Mr. Hall.

4th Row (Left to Right), Dr. Loveland, Dr. Leroux-Hugon, Dr. Kuindersma, Dr. Cracknell, Dr. El Mously, Mr. Litterst, Dr. Schneider, Mr. Vaughan, Mr. Kobliska, Mr. Frost, Mr. Petursson, Dr. Siggia, Miss Cross, Mr. Wallis, Mr. Allan, Dr. Passari, Mr. Tejeda.

Back Row (Left to Right), Dr. Evans, Dr. Sarges, Mr. Anderson, Mr. Turner, Mr. El-Fouly, Dr. Stagni, Mr. Greaves, Dr. Blom, Mr. Joiner, Mr. Abou Chacra, Dr. Rowland, Mr. Lewis, Captain Campbell, Mr. Last, Mr. Temple, Mr. Van Hay, Mr. Grandqvist, Dr. Kus, Dr. Ingleby, Mr. Farooq.

ELECTRONIC AND STRUCTURAL PROPERTIES OF AMORPHOUS SEMICONDUCTORS

Proceedings of the Thirteenth Session of the Scottish Universities Summer School in Physics, 1972

Edited by

P. G. LE COMBER

Carnegie Laboratory of Physics,
University of Dundee, Dundee, Scotland

and

J. MORT

Xerox Corporation, Rochester Corporate Research Center,
Rochester, New York, U.S.A.

1973

ACADEMIC PRESS · LONDON AND NEW YORK

A Subsidiary of Harcourt Brace Jovanovich, Publishers

ACADEMIC PRESS INC. (LONDON) LTD.
24/28 Oval Road
London NW1

United States Edition published by
ACADEMIC PRESS INC.
111 Fifth Avenue
New York, New York 10003

Library of Congress Catalog Card Number: 73–7034
ISBN: 0–12–440550–9

PRINTED IN GREAT BRITAIN BY
ROYSTAN PRINTERS LIMITED
Spencer Court, 7 Chalcot Road
London NW1

List of Participants

ORGANISING COMMITTEE

SPEAR, PROFESSOR W. E. *University of Dundee, Scotland* (*Director*).

LE COMBER, DR. P. G. *University of Dundee, Dundee, Scotland* (*Secretary and Joint Editor*).

ROBERTSON, DR. J. *University of Edinburgh, Edinburgh, Scotland* (*Treasurer*).

CLARK, DR. R. C. *University of Aberdeen, Aberdeen, Scotland* (*Local Organiser*).

MORT, DR. J. *Xerox Corporation, Rochester, New York, U.S.A.* (*Joint Editor*).

LOVELAND, DR. R. J. *University of Dundee, Dundee, Scotland* (*Steward*).

GOVERNING COMMITTEE

DINGLE, PROFESSOR R. B. *University of St. Andrews, St. Andrews, Scotland* (*Chairman*).

FINLAYSON, DR. D. M. *University of St. Andrews, St. Andrews, Scotland* (*Secretary-Treasurer*).

PARTICIPANTS PRESENTING LECTURES OR SEMINARS

APLING, DR. A. J. *University of Bristol, Bristol, England.*

AUSTIN, DR. I. G. *University of Sheffield, Sheffield, England.*

DAVIS, DR. E. A. *University of Cambridge, Cambridge, England.*

EMIN, DR. D. *Sandia Corporation, Albuqueruqe, New Mexico, U.S.A.*

FRIEDMAN, DR. L. *University of Cambridge, Cambridge, England.*

FRITZSCHE, PROFESSOR H. *University of Chicago, Chicago, U.S.A.*

GOODMAN, PROFESSOR C. H. L., *S.T.L. Ltd., Harlow, Essex, England.*

GREENWOOD, DR. D. *University of Bristol, Bristol, England.*

GRIGOROVICI, PROFESSOR R. *Institute of Physics, Bucharest, Rumania.*

JONSCHER, PROFESSOR A. K. *Chelsea College of Science and Technology, University of London, London, England.*

LE COMBER, DR. P. G. *University of Dundee, Dundee, Scotland.*

MAIN, DR. C. *University of Edinburgh, Edinburgh, Scotland.*

MORT, DR. J. *Xerox Corporation, Rochester, New York, U.S.A.*

MOTT, PROFESSOR SIR NEVILL. *University of Cambridge, Cambridge, England.*

OWEN, DR. A. E. *University of Edinburgh, Edinburgh, Scotland.*

ROBERTS, DR. G. G. *New University of Ulster, Coleraine, Northern Ireland.*

PARTICIPANTS

ABOU-CHACRA, MR. R. *University of Birmingham, Birmingham, England.*

ALLEN, MR. F. R. *University of Cambridge, Cambridge, England.*

ANDERSON, MR. D. A. *University of Dundee, Dundee, Scotland.*

BABCENCO, MR. A. *University of Liège, Liège, Belgium.*

BABCENCO, MRS. I. *University of Liège, Liège, Belgium.*

BELU, DR. A. *Institute of Physics, Bucharest, Rumania.*

BENSOUSSAN, DR. M. *CNET, Bagneux, France.*

BERTRAND, MR. P. *Laboratoire de Physico-Chemie et de Physique de l'Etat Solide, Heverlee, Belgium.*

BLOM, DR. F. A. P. *Eindhoven University of Technology, Eindhoven, Netherlands.*

BROWN, MISS F. E. *Westfield College, London, England.*

BUTCHER, PROFESSOR P. N. *University of Warwick, Coventry, England.*

CAMPBELL, CAPTAIN W. S. *AF Flight Dynamics Laboratory, Ohio, U.S.A.*

CRACKNELL, DR. A. P. *University of Dundee, Dundee, Scotland.*

CROSS, MISS J. A. *University of Cambridge, Cambridge, England.*

DOHLER, DR. G. *Max-Planck-Institut fur Festkorparforschung, Stuttgart, Germany.*

DOZIER, DR. A. W. *Vanderbilt University, Nashville, U.S.A.*

EL-FOULY, MR. M. H. *University of Dundee, Scotland.*

EL MOUSLY, DR. M. K. *Ain Shams University, Cairo, Egypt.*

EVANS, DR. A. G. R. *Strathclyde University, Glasgow, Scotland.*

FAROOQ, MR. M. U. *Birkbeck College, London, England.*

FISHER, MR. F. D. *University of Edinburgh, Edinburgh, Scotland.*

FLYNN, MR. B. W. *University of Edinburgh, Edinburgh, Scotland.*

FROST, MR. M. S. *Chelsea College of Science and Technology, University of London, London, England.*

GRANQVIST, MR. C. G. *Chalmers University of Technology, Gothenburg, Sweden.*

GRAY, MR. H. R. *University of Maine, Orono, U.S.A.*

GREAVES, MR. G. N. *University of Cambridge, Cambridge, England.*

GROVER, DR. M. K. *Brown University, Providence, U.S.A.*

HALL, MR. R. B. *Brunel University, Uxbridge, England.*

HASSAN, DR. A. R. *Instituto di Fisica dell Universita, Rome, Italy.*

HERNANDEZ, DR. J. P. *University of North Carolina, Chapel Hill, U.S.A.*

HODGSON, MR. J. M. *Polytechnic of Central London, London, England.*

INGLEBY, DR. M. *The Polytechnic, Huddersfield, England.*

JOINER, MR. B. A. *University of Texas, Austin, U.S.A.*

KASTNER, MR. M. *James Franck Institute, University of Chicago, Chicago, U.S.A.*

KOBLISKA, MR. R. J. *James Franck Institute, University of Chicago, Chicago, U.S.A.*

KRESS, DR. W. *Technische Universität München, Reakforgelande, Germany.*

KUINDERSMA, DR. P. I. *The University, Groningen, Netherlands.*

KUS, MR. F. *McMaster University, Ontario, Canada.*

LAHMERS, MR. H. *The Polytechnic, Huddersfield, England.*

LAST, MR. B. J. *University of Birmingham, Birmingham, England.*

LEROUX-HUGON, DR. P. *CNRS, Bellevue, France.*

LEWIS, MR. S. R. *University of Reading, Reading, England.*

LITTERST, DR. F. J. *Technische Universität München, Munchen, West Germany.*

MADAN, MR. A. *University of Dundee, Dundee, Scotland.*

MAGRABI, DR. C. *University of Sheffield, Sheffield, England.*

MOORE, MR. A. G. *University of Sheffield, Sheffield, England.*

MORETTINI, DR. L. *CNR–Lamel, Bologna, Italy.*

MORYS, MR. P. L. E. *University of Warwick, Coventry, England.*

MULAZZI, PROFESSOR E. *University of Parma, Parma, Italy.*

MURGATROYD, DR. P. N. *Loughborough University, Loughborough, England.*

NOBILI, PROFESSOR D. *CNR, Bologna, Italy.*

PASSARI, DR. L. *University of Ferrara, Italy.*

PETURSSON, MR. J. *University of Edinburgh, Edinburgh, Scotland.*

RAZZETTI, MR. C. *University of Parma, Parma, Italy.*

ROWLAND, DR. S. C. *Andrews University, Berrien Springs, U.S.A.*

SABATHY, MR. W. *Institut fur Hochfrequenztechnik, Wien, Austria.*

SARGES, DR. K. H. *Philips Forschungslaboratorium GmbH, Hamburg, Germany.*

SCHLUETER, DR. M. *Ecole Polytechnic Federale Lausanne, Lausanne, Switzerland.*

SCHNEIDER, DR. B. *Philips Forschungslaboratorium, Hamburg, Germany.*

SHARMA, DR. S. K. *Universitat Karlsruhe, Karlsruhe, West Germany.*

SIGGIA, DR. E. D. *Universite de Paris, Orsay, France.*

STAGNI, DR. L. *Universita di Roma, Rome, Italy.*

STEVENTON, MR. A. G. *P.O. Research Department, London, England.*

SUSSMAN, MR. R. S. *University of Sheffield, Sheffield, England.*

TEJEDA, MR. J. *Max-Planck-Institut, Stuttgart, Germany.*

TEMPLE, MR. B. K. *Pilkington Brothers Research and Development Laboratories, Ormskirk, England.*

TURNER, MR. R. E. *Wilkinson Sword (Research) Ltd., Slough, England.*

TYLDESLEY, MRS. D. *S.T.L. Ltd., Harlow, England.*

VAN HAY, MR. J. C. *ICTP, Miramare, Italy.*

VAUGHAN, MR. J. G. *Vanderbilt University, Nashville, U.S.A.*

WALLIS, MR. R. H. *University of Cambridge, Cambridge, England.*

WEY, MR. H. Y. *University of Chicago, Chicago, U.S.A.*

VISITORS

OVSHINSKY, MR. S. R. AND MRS. I. *Energy Conversion Devices Inc., Troy, Michigan, U.S.A.*

TANNHAUSER, PROFESSOR D. S. *Technion-Israel Institute of Technology, Haifa, Israel.*

Director's Preface

The Thirteenth Scottish Universities Summer School in Physics was held at the University of Aberdeen from 31st July to 18th August, 1972. The subject, electronic and structural properties of amorphous semiconductors, was one of particular current interest and we were fortunate that a distinguished team of lecturers, internationally known for their contributions to the field, accepted our invitation to participate. Since this School occurred in the year between the 4th and 5th International Conferences on Non-Crystalline Solids, lecturers and seminarists took the opportunity to discuss new ideas and as yet unpublished results. I hope therefore that the Proceedings will make a useful contribution both as an introduction to the subject and as a guide to the most recent developments.

This School was largely financed by a grant from the Scottish Universities. In addition, we were most grateful to the Science Research Council for three Senior Visiting Fellowships and to the Xerox Corporation and Rank-Xerox Limited for their contribution towards student bursaries.

Most participants and accompanying wives and children stayed in Dunbar Hall, situated in a delightful part of Old Aberdeen. We are greatly indebted to the Warden and his staff for the excellent domestic arrangements and to Professor R. V. Jones for allowing us to use the facilities of the Physics building.

Finally, it is a pleasure to acknowledge the contribution of all those who were involved in the planning and organisation of the School. Special thanks are due to Roger Clark who was responsible for all local arrangements and outings; to Peter Le Comber and Mrs. E. M. Miller who dealt with the correspondence and much of the organisation; to our Treasurer, John Robertson and our Steward, Ron Loveland for their efficient handling of financial and domestic problems, and to Mrs. H. D. Spear for her help with the ladies' programme.

W. E. Spear,
Professor of Physics,
University of Dundee.

Editors' Note

Each of the chapters in this book covers the material delivered by the lecturers at the Summer School and has been prepared by them with the minimum of editing. The articles by Dr. Apling, Dr. Friedman, Dr. Le Comber *et al.,* Dr. Austin *et al.*, Dr. Roberts, Dr. Main *et al.,* and Professor Goodman were presented as seminars. In order to achieve some uniformity within the volume, we have tried to adopt a consistent notation for those quantities that occur in more than one chapter. This does mean that on a few occasions the notation used differs from previously published work by the authors. For these alterations and all other changes from the contributions as received, the Editors accept full responsibility.

P. G. LE COMBER.
J. MORT.

Contents

I

SURVEYS OF THE PRESENT STATE OF THE SUBJECT

1. ELECTRONS IN NON-CRYSTALLINE MATERIALS

N. F. MOTT

Cavendish Laboratory, University of Cambridge, Cambridge, England

1.1. INTRODUCTION

The aim of the article is to survey certain aspects of electron transport in non-crystalline materials. As for crystalline materials, there are two situations to consider. In the first the density of states $g(\mathcal{E})$ is finite when the energy $\mathcal{E}$ is equal to the Fermi energy $\mathcal{E}_F$ and the current is determined by the behaviour of electrons with energies near $\mathcal{E}_F$. In the second $g(\mathcal{E}_F)$ is zero or small and current is carried by electrons or holes excited into conduction or valence bands. In the first class of material, which includes metals, electron correlation is always important and particularly so for narrow bands, where the presence of disorder also leads to important results. This article therefore begins with a review of the effect of correlation in crystalline materials, including the metal–insulator transition. Anderson localization and its effect on narrow bands is next introduced. Impurity conduction is then discussed, as providing the most direct example of localization and its consequences. The final section surveys some problems in non-crystalline semiconductors, particularly the relation between band gaps, mobility edges and structure.

It is only since about 1960 that the theory of electrons in non-crystalline materials has become an important part of solid state physics. This is perhaps curious, because one non-crystalline material, soda glass, has a property known to everybody long before science began; it is transparent. In the language of solid state physics, it has a band gap in the density of states; occupied states are separated from empty states by a range of forbidden energy values. For crystalline materials the theory of conduction bands, valence bands and band gaps goes back to the work of Bloch and Wilson in the early 1930's, and is linked with the behaviour of electrons in a periodic field; glass is not crystalline and so the field in which the electrons move is not periodic and yet the material is transparent. The "chemical" explanation, that in glass the electrons are firmly bound in silicon–oxygen bonds, is satisfactory as far as it goes, but it does not enable us to lay down exact conditions for a band gap, and we are still far from being able to do this.†

The study of the electrical properties of glasses begins with the work of the Leningrad school under Kolomiets[(2)] on the chalcogenide glasses. These appear to behave, whether in the glassy or liquid states, as *intrinsic* semiconductors with a band gap of less than 2 eV. A very important property established by this school, which differentiates them sharply from crystalline materials of similar composition, is that their conductivity depends little on purity or stoichiometry; for them the familiar shallow donors and acceptors of crystalline semiconductors do not seem to exist. This property has the result that the conductivity increases by many orders of magnitude when they crystallise, which is certainly the reason why chalcogenide glass switches show a memory effect.[(3,4)] In chemical terms the likely reason for this behaviour is well known. The structure of a glass adjusts itself so that each atom is surrounded by a number of nearest neighbours such that all its outer electrons are taken up in bonds. Thus in chalcogenide glasses arsenic atoms will be surrounded by three neighbours, germanium by four and so on. Doubtless the same is true in soda and phosphate glasses. The colours in the glass of medieval cathedrals were obtained by adding compounds of silver or of a transition metal, in which the $3d$ electrons do not enter into bonds; transitions within the d shell, which absorb visible light, are therefore possible. The vanadium phosphate glasses present an interesting intermediate case, in which apparently the V^{4+} ion in the state $3d^1$ can form a bond so that the electron is tightly bound, or with a different configuration of the

† See Section 1.13 and the correspondence by Ziman[(1)] and others in "Physics Today".

surroundings does not do so, so that it can hop from V^{4+} to V^{5+} sites with hardly any activation energy. This is probably the basis of a switching process (due to Drake and Scanlan[5]).

The translation of these ideas into the notation of mathematical physics, so that band structures can be calculated and the condition for a band gap established, is far from being achieved. Treatments which start from the tight-binding approximation and make use of the chemical concepts of bonding and anti-bonding orbitals may give a more direct description of the facts than other methods, despite the much greater success that some of them have had for crystalline materials. For instance, starting from the pseudopotential concept, it has not yet proved possible to show that a band gap exists.[1] On the other hand the slow convergence of the tight binding method is notorious (see Dresselhaus and Dresselhaus[6]). In Section 1.13 we review the band gap problem, stressing the experimental facts to be explained. Indeed throughout this article, we shall for each problem review the experimental situation, to see whether it gives any guide to the kind of mathematical approach most likely to be useful.

Impurity conduction in doped and compensated semiconductors is a phenomenon of quite a different kind which has proved of very great value in understanding the behaviour of electrons in a non-crystalline field, in this case the field of randomly distributed donors. It provides a direct experimental proof of the Anderson[7] localization theorem, which is basic to our understanding of many electrical properties of non-crystalline materials. In impurity conduction in compensated *n*-type material, electrons move from one donor centre to another, the donor centres being at random positions in the lattice and can be represented by potential wells with random depths. Impurity conduction, therefore, provides in contrast to glasses a situation in which the potential energy of the electron is known, and therefore is very useful for testing theories. This is why we include a description of it in this article.

In the early text books on electrons in crystalline materials, such as Sommerfeld and Bethe[8] and Mott and Jones,[9] two approximations are used: that of nearly free electrons and the "tight binding" method. The former has been brought back into the centre of the picture for band calculations by the invention of pseudopotentials, and for non-crystalline materials is the basis of Ziman's theory[10,11] of the resistivity of liquid metals. The latter method is very useful for certain problems in non-crystalline materials, notably for impurity conduction and for the discussion of "states in the gap". Anderson's localization theorem is the extension of the tight-binding method to a non-crystalline situation. We shall therefore summarise in the next sections certain useful formulae. We also distinguish—as in crystalline materials—between two kinds of problem. One is the movement of an electron in the conduction band of an insulator or semiconductor. In such problems it is a good approximation to treat the electrons as moving in a fixed potential determined by the crystal,[12] and problems of conductivity, trapping and so on are "one-electron" problems. In the other kind of problem, electron states at $T = 0$ are full up to a "Fermi energy" $\mathcal{E}_F$, *not* in a forbidden gap. This is the situation in liquid metals and in impurity conduction. In these problems the electron–electron interaction (e^2/r_{12}) always plays a role, though in crystals a surprisingly wide range of phenomena, such as a linear specific heat and Pauli paramagnetism, were explained without it, before Landau's work showed how to reconcile it with the existence of a sharp Fermi limit between occupied and unoccupied states. We shall in this article describe the effect of the term e^2/r_{12} for non-crystalline materials. It can be important, because it is precisely under the same conditions, namely narrow bands treated by the tight binding approximation, that both disorder and the interaction energy e^2/r_{12} will have their greatest effects. For the crystalline case, the position is set out in the next section.

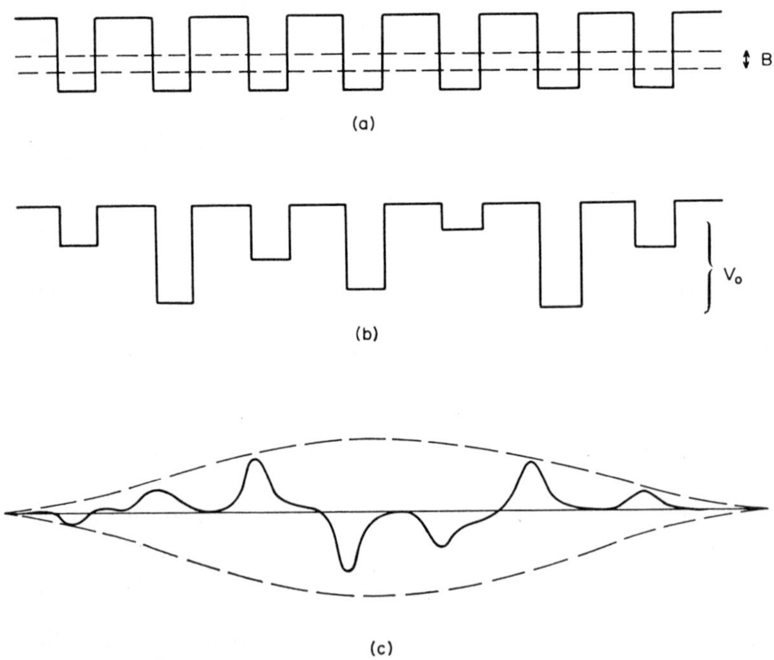

FIG. 1.1. (a) An array of potential wells appropriate to the tight-binding approximation; B is the band-width. (b) The potential energy when a random potential V_i is introduced on each well i (cellular disorder). The V_i lie in the range $-\frac{1}{2}V_0$ to $\frac{1}{2}V_0$. (c) Weakly localized wave functions for an electron acted on by the potential of FIG. 1.1(b).

1.2. The Tight-binding Approximation and the Effect of Correlation; the Metal–Insulator Transition

The tight binding approximation takes its simplest form for a potential energy which consists of a simple cubic array of non-overlapping wells $V(r)$ as in Fig. 1.1(a), in each of which the electron can exist in a $1s$ state of energy $\mathcal{E}_0$ with wave functions $\phi_i(r)$, i denoting the well (i). Then the (un-normalised) tight binding wave function is

$$\psi = \sum_i \exp(i\mathbf{k}\mathbf{a}_i)\,\phi_i(r).$$

This represents a wave moving through the lattice with wave number $\mathbf{k}$. The energy of such a state is

$$\mathcal{E}_0 - 2I(\cos k_x a + \cos k_y a + \cos k_z a), \tag{1.1}$$

where a is the lattice constant and I is the overlap energy integral

$$I = -\int \phi_i{}^* H\phi_{i+1}\,\mathrm{d}^3x \tag{1.2}$$

and H is the Hamiltonian. The band-width B is then given by

$$B = 2zI,$$

where z is the co-ordination number, 6 in our case.

An estimate of I is useful for Coulomb wells for which $V(r) = -e^2/r$, so that ϕ is the hydrogen function

$$\phi = (\alpha^3/\pi)^{\frac{1}{2}} \exp(-\alpha r), \qquad \alpha = me^2/\hbar^2 \tag{1.3}$$

If αa is in the range 2 to 4, it is a fair approximation to write[75]

$$\left.\begin{aligned} I &= 2\,e^{-a\alpha}(\alpha^3/\pi)\int (e^2/r)\,\phi\,\mathrm{d}^3x \\ &= 8\,e^2\,\alpha\,\mathrm{e}^{-a\alpha} \end{aligned}\right\} \tag{1.4}$$

The effective mass can be obtained by expanding eqn (1.1) for small values of $k(k^2 = k_x{}^2 + k_y{}^2 + k_z{}^2)$, giving

$$\mathcal{E}_0 = -6I + Ik^2a^2;$$

equating the second term on the right-hand side to $\hbar^2 k^2/2m$, we obtain

$$m = \hbar^2/2Ia^2. \tag{1.5}$$

The density of states for the simple cubic model has been calculated and is reproduced in various text books (e.g. Ref. 9, p. 85).

If a narrow band described by the tight binding approximation is partly occupied up to a Fermi energy $\mathcal{E}_F$, interaction between electrons can be important, and can lead to antiferromagnetism and a splitting of the band, and if there is exactly one electron per atom it can produce insulating rather than metallic behaviour. This effect is most simply described in the spirit of the tight binding approximations by considering the interaction between a pair of electrons only when they are on the same atom (necessarily with anti-parallel spins if the functions ϕ are non-degenerate as for *s*-states). This interaction energy may be written

$$U = \iint (e^2/r_{12})\, |\phi(r_1)|^2\, |\phi(r_2)|^2\, \mathrm{d}^3x_1\, \mathrm{d}^3x_2, \tag{1.6}$$

the integrations being over a well. If the functions ϕ are the 1*s* states given by eqn (1.3), the integration in eqn (1.6) can be carried out and is equal to[13]

$$U = \tfrac{5}{8}\, e^2\alpha. \tag{1.7}$$

This term was introduced by Hubbard,[14] who wrote down the well-known Hamiltonian

$$\Sigma\Sigma\, I(c_{i\uparrow}\, c_{j\uparrow}{}^{+} + c_{i\downarrow}\, c_{j\downarrow}{}^{+}) + \Sigma\, U\, n_{i\uparrow}\, n_{i\downarrow}.$$

The c_i are creation and annihilation operators and $n_{i\uparrow}$, $n_{i\downarrow}$ are the numbers of electrons on atom i with the two spin directions shown by the arrows. This Hamiltonian has been extensively used to discuss the properties of partly filled narrow bands when the correlation term [eqn (1.6)] is important, particularly magnetic properties and the metal-insulator transition (cf. Refs 15, 16, 17, 18, 19, 20 and many others).

We can summarise the properties of a crystalline solid described by this Hamiltonian as follows:

(i) *The insulating state*

If $U \geqslant B$, where B is the band width equal to $2zI$, and if there is one electron per atom, the material is an antiferromagnetic insulator. The wave function on each atom takes the form.

$$\phi_i + A \sum_j \phi_j, \tag{1.8}$$

where the summation is over the z nearest neighbours j, and A is a constant to be determined. The second term represents the extent to which the wave function overlaps onto neighbouring atoms. We determine A by minimising the total energy. For a ferromagnetic arrangement there can be no overlap

($A = 0$), but for an antiferromagnetic arrangement the energy per atom is $2AzI + zA^2 U$, which has its minimum when

$$A = -I/U. \tag{1.9}$$

The energy is then

$$-zI^2/U = -B^2/4zU. \tag{1.10}$$

This is the difference between the energies of the ferromagnetic and antiferromagnetic states. Assuming that half this energy disappears on disordering, the Néel temperature T_N should be given by

$$kT_N \ln 2 = B^2/8zU. \tag{1.11}$$

For transition metal compounds, $B \sim 1$–2 eV, $U \sim 10$–20 eV and $z \sim 6$, so that the right-hand side of eqn (1.11) is $\sim 1/240 - 1/60$ eV, which shows why values of the Néel temperature near or below room temperature are usual.†

Still assuming $U \gg B$, if the number of electrons per atom differs from unity, the material is *always metallic*. Or put in another way, if an extra electron is added, it can move from atom to atom without activation. In the same way, if an electron is taken away, it will produce a "hole" which can move without activation. This theorem will be used later in discussing impurity conduction to give experimental evidence for Anderson localization. It is due to Nagaoka.[22] We call the sites on which there is an extra electron, or where one is absent, "current carriers"; at zero temperature they can move with wave number k through the lattice, just like electrons or holes in normal semiconductors; though if the energy is large enough to excite antiferromagnetic spin waves they will be strongly scattered and lose energy rapidly.[23,24] Moreover, the carriers interact with the lattice in such a way as to produce ferromagnetism; a single carrier will produce a "spin polaron" and a comparatively small number should give a ferromagnetic state. The argument is as follows: we consider the case of an extra electron in the conduction band. Its wave function, if the spins were rigidly fixed, would have a small amplitude of order A (eqn (1.8)) on nearest neighbour atoms, because if A were zero the exclusion principle would prevent the electron from penetrating. Assuming that the integral I (eqn(1.2)) between next nearest neighbours is negligibly small, the "half band-width" for the electrons is of order zAI. But if the carrier turns the spin direction of the surrounding electrons over in a direction antiparallel to itself, the half band width does not contain the factor A. For a single carrier, therefore, it will polarize the spins in a direction antiparallel to itself over a volume R, which is obtained by minimising the energy

$$\hbar^2\pi^2/mR^2 + (4\pi R^3/3a^3)\, B^2/4zU. \tag{1.12}$$

† The treatment here is equivalent to that of Anderson. (Ref. 21).

The second term is the energy required to line up the spins over a volume $4\pi R^3/3$ in a direction antiparallel to that of the carrier; the first is the kinetic energy of an electron moving freely in a sphere of radius R, which it can do if not impeded by atoms where it cannot go because the spins are parallel to its spin. Minimising eqn (1.12) gives for the radius of the spin polaron

$$R = (\pi h^2 a^3 U/4m B^2)^{1/5}.$$

Spin polarons are usually discussed for problems such as that of an electron in the 6s-band of a magnetic semiconductor such as EuTe; the Hund's rule coupling between carrier and moment then produces the polaron.

Various discussions of the effective mass of a spin polaron have been given.[25,26,27] If the radius is large, its effective mass is large.† If it extends over next nearest neighbours only, it probably does not differ greatly from the undistorted band-mass m given by eqn (1.5).

If there are n excess carriers per atom, the ferromagnetic coupling between the spin polarons will lead to ferromagnetism; to produce ferromagnetism, energy equal to $B^2/4zU$ must be expended against the antiferromagnetic coupling, and we gain energy $-\frac{1}{2}nB$ per atom. There is also a term due to the increased Fermi energy, of order $n^{5/2}$, which we neglect if n is small. Thus the energy is

$$-\tfrac{1}{2}nB + B^2/4zU$$

and the condition for ferromagnetism is

$$n > \sim B/2zU. \tag{1.13}$$

Fig. 1.2 (from Nagaoka[22]) shows how the ferromagnetic region is separated from the non-ferromagnetic region.

(ii) *Behaviour as $U/B \to 1$ and the metal–insulator transition*

If $U/B \geqslant 1$, the energy required to take an electron from one atom to a distant one is, to a first approximation, equal to U, so the material should behave like an intrinsic semiconductor with $N \exp(-\frac{1}{2}U/kT)$ carriers per unit volume of both sign. But the Hubbard term U, which we shall write U_{scr}, is *not* the same as that which appears in the formula (1.11) for antiferromagnetism, and is always smaller. This is because U_{scr} is the energy required to form two *charged* centres,[27,28] and energy is gained by polarising the medium. The difference between U and U_{scr} is probably not important for an array of hydrogen atoms or a doped semiconductor, but is of major

† We believe that the hopping process characteristic of dielectric polarons does not occur for spin polarons.

importance for transition metal oxides where U_{scr} may be of order 1–2 eV and $U > 10$ eV. If polarons are formed by the carriers, the energy may be reduced still further (see article by Emin in this volume).

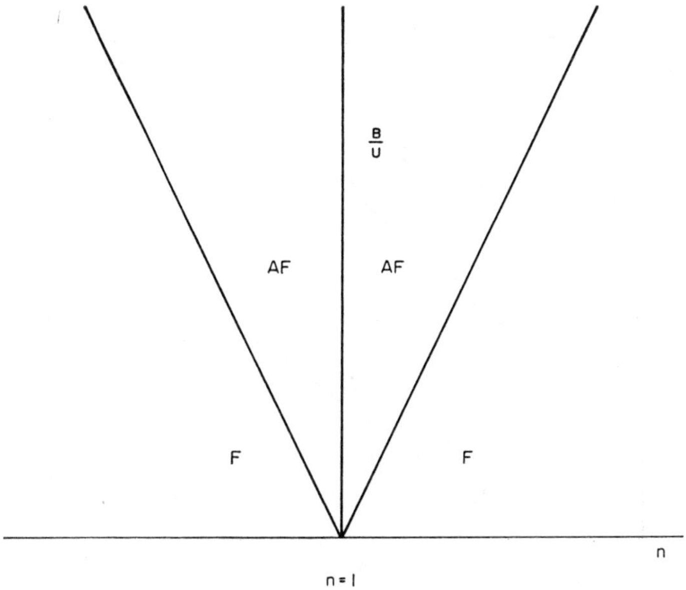

FIG. 1.2. Nagaoka's separation into antiferromagnetic (AF) and ferromagnetic (F) regions for non-integral values of n (the number of electrons per atom). (From Ref. 22).

If U/B decreases, for instance in a doped semiconductor by increasing the concentration, both carriers have band-widths of order B, so that the gap becomes

$$U - 2(\tfrac{1}{2}B) = U - B. \quad (1.14)$$

It is common in the literature to draw two "Hubbard bands" as in Fig. 1.3(a) as a function of k. We note however that these are not strictly one-electron pictures; the lower band refers to holes, the upper to electrons. Also except near the extremities of the bands $\Delta k/k$ is ~ 1, so k is a poor quantum number.[23,24] For the former reason we prefer the representation of Fig. 1.3(b).

The gap will become zero when $U = B$, so the material becomes metallic at this point, at any rate according to the Hubbard Hamiltonian. We have to ask what value of B should be taken. We have seen that if $U/B \gg 1$, the carrier has difficulty in penetrating to a nearest neighbour where the spin direction is the same as its own. But when $U/B \sim 1$, the value of A in eqn (1.9) is $1/2z$, and, since z neighbours contribute, the value of $\Sigma|A|^2$ is $1/4z$. So the nearest neighbours are still an obstacle. The analysis given by Mott[29] suggests that the "spin polaron" formed by the carrier turns round the moments only of nearest neighbours, and also that a band-width of order B is not far from the truth.

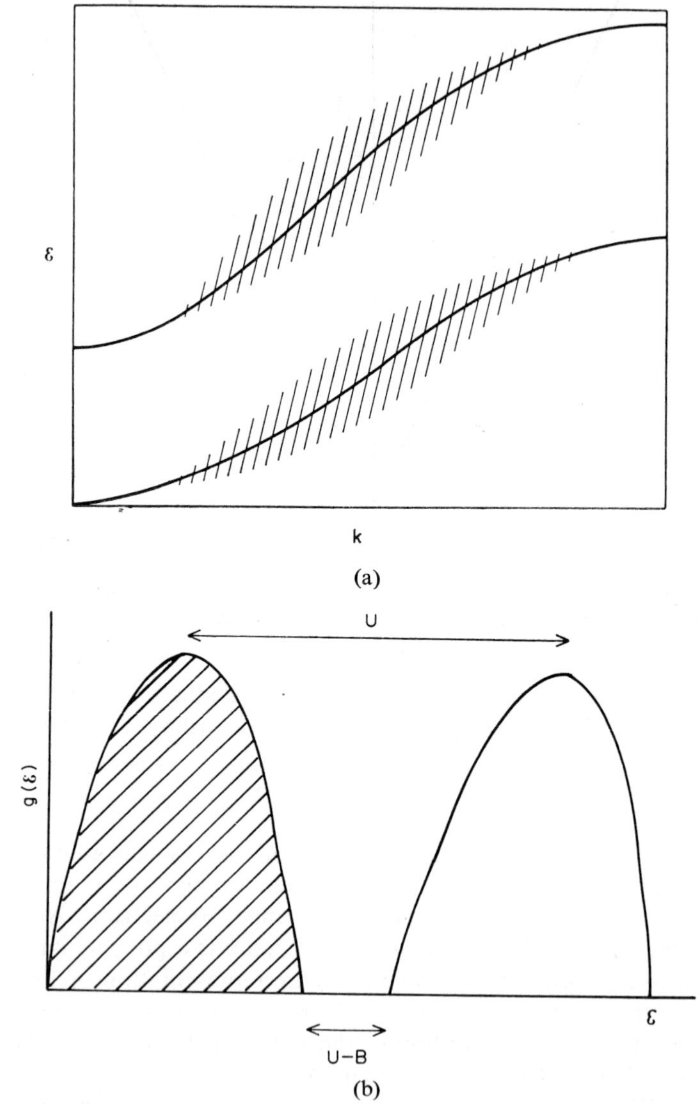

FIG. 1.3. (a) The plot of energy $\mathcal{E}$ against wave-number k for two Hubbard bands; the shaded region denotes large values of $\Delta k/k$. (b) Density of states $g(\mathcal{E})$ for two Hubbard bands; occupied states shaded.

If we insert the formulae eqn (1.4) and eqn (1.7) for B and U, we find from eqn (1.14) for the metal–insulator transition

$$n^{\frac{1}{3}} a_H = 0{\cdot}27 \qquad a_H = \hbar^2\kappa/me^2 \tag{1.15}$$

which is empirically correct for doped semiconductors.

The Hubbard Hamiltonian includes only the interaction between a pair of carriers when they are on the *same* atom. The treatment given by the present author[30] depended *essentially* on the coulomb nature of the interaction, and started from the principle that a small number of carriers could not exist, because a coulomb attraction between them would produce bound pairs. The condition for a metal insulator transition was taken to be the condition that the field $e^2/\kappa r_{12}$ between electron and hole is sufficiently screened to give no bound states. It is remarkable that this condition gives approximately the same formula as eqn (1.15).

The Hubbard Hamiltonian predicts a second order transition at $T = 0$ as B/U is varied, while the author's earlier model suggested that the transition must be first order. The work of Kohn[31] suggested a series of second order transitions and a mechanism[27,32] by which $\kappa \to \infty$ at the transition. Other descriptions have been given and the nature of the transition remains controversial.

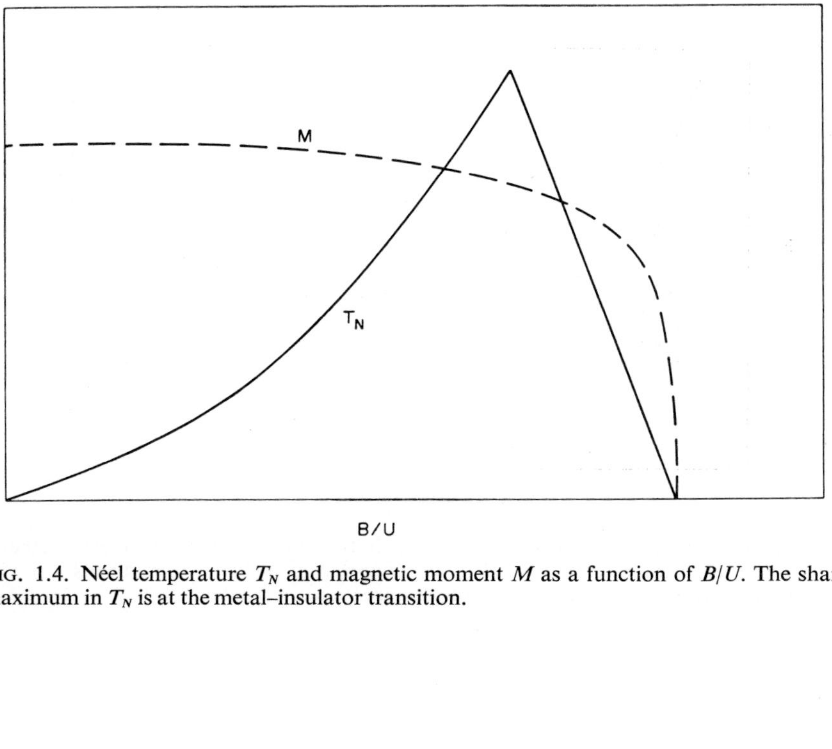

FIG. 1.4. Néel temperature T_N and magnetic moment M as a function of B/U. The sharp maximum in T_N is at the metal–insulator transition.

After the metal–insulator transition, with increasing values of B/U, we expect an antiferromagnetic metal. But considerations similar to eqn (1.13) show that with $B/U \sim 1$, the antiferromagnetism will disappear when $n \sim 1/2z$. We expect *paramagnetism* (rather than ferromagnetism) in this case, because the "electrons" and "holes" have an opposite effect on the magnetic properties of the spins. General considerations[26,15,16] suggest that the moments disappear as $\sqrt{(x - x_0)}$, where x is the parameter which is varying and leads to the variation of B/U. The Néel temperature and moment are expected to appear as in Fig. 1.4. This behaviour of the Néel temperature is that shown in Goodenough's books and papers.[33,34,35]

There are thus *two* transitions which occur as B/U changes:

(i) from insulator to an antiferromagnetic metal.

(ii) from an antiferromagnetic metal to a normal metal. The antiferromagnetic metal is frequently not observed, probably for the following reason. At the metal-insulator transition, extra negative energy proportional to $(x - x_0)^{5/2}$ appears;[26] x is B/U or any other parameter linked to it. At the concentration where moments appear, the energy is of the form given by Landau

$$AM^2 + BM^4 + \dots$$

and A must change sign, so the energy varies as $(x - x_0)^2$. Probably moments reach their full value rather rapidly, so a concave region is likely in the curve

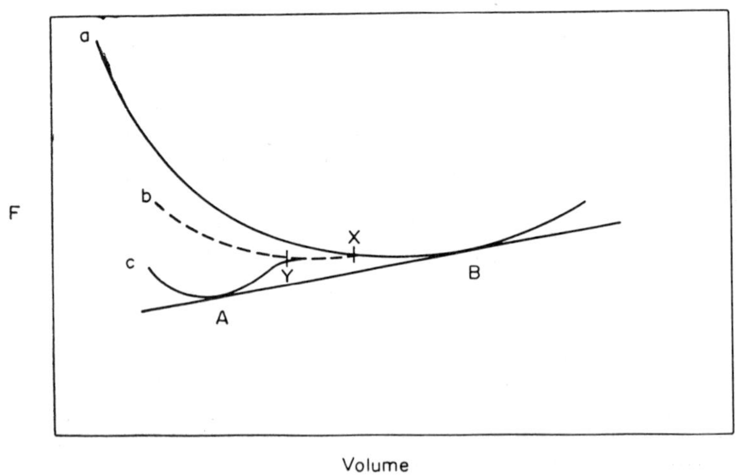

FIG. 1.5. Free-energy F as a function of volume in the neighbourhood of a metal–insulator transition. (a) is without current carriers which appear at X, but with moments, (b) with current carriers and (c) takes account of the disappearance of moments at Y.

plotting free-energy F against x. If the change in U/B is due, for instance, to a change in volume (V), an unstable region AB with a first-order phase change is then expected. This *may* cover the whole of the region in which the antiferromagnetic region lies, as shown in Fig. 1.5. Antiferromagnetic metallic oxides do exist however (see Ref. 26).

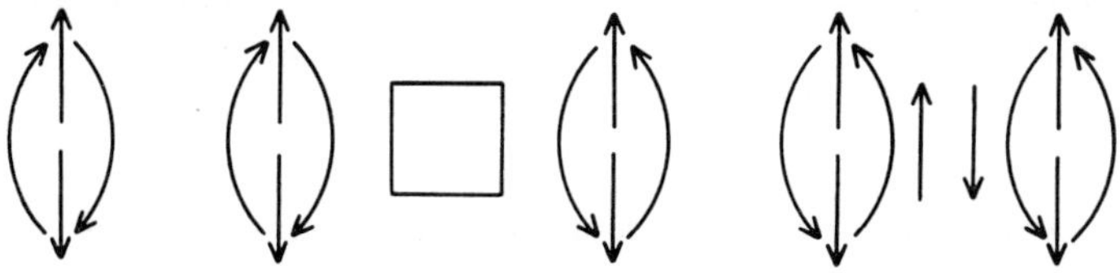

FIG. 1.6. Representing the occupation of centres and behaviour of moments in a Brinkman–Rice highly correlated gas. (a) is a hole-like carrier, (b) an electron-like carrier.

The "normal" metal near the metal–insulator transition has abnormal properties, as first pointed out by Brinkman and Rice(18). These authors use a Hamiltonian in which a (small) proportion c of the centres are doubly occupied, an equal proportion c is empty and $1 - 2c$ contain a single electron, which is resonating between its two positions. The situation is illustrated in Fig. 1.6. Since there is no antiferromagnetic (or other) lattice, then according to Luttinger's(36) theorem the correlation U cannot alter the (spherical) form of the Fermi surface or the volume included by it. The only carriers of current are the doubly and unoccupied sites, and they can be treated as having mass m where $\hbar^2/2ma^2$ is the overlap energy integral [eqn (1.2)]. So a field E applied for a time δt should produce a current

$$J = 2cNe^2 E\delta t/m. \tag{1.16}$$

But we can also treat the current by considering the *spherical* Fermi surface with *enhanced* mass m^*; thus

$$J = Ne^2 E\delta t/m^*. \tag{1.17}$$

Equating these two quantities, we find

$$m^*/m = 1/2c, \tag{1.18}$$

which is Brinkman and Rice's result.

This large effective mass is important in trying to understand the thermal properties of a highly correlated electron gas, such as for instance V_2O_3 under pressure (for a review and references see Refs 19, 26). In Section 1.12 of this article, we discuss the effect of the mass enhancement on Anderson localization.

The highly correlated electron gas has many properties that have not been thoroughly explored. Thus while the formula

$$\sigma = S_F e^2 L/12\,\pi^3\,\hbar \tag{1.19}$$

remains valid for the conductivity, where S_F is the Fermi surface area and L the mean free path, one would think that L ought to be calculated by putting $\delta t = \tau$ in eqn (1.16), so that

$$\sigma = 2c\,Ne^2\,\tau/m,$$

where τ is calculated for the current carriers. Comparing these equations, we see that

$$L = 2c\,\tau/(mk_F/\hbar) = \tau/m^* k_F.$$

McWhan *et al.*[37] have shown that in V_2O_3 doped with Cr_2O_3 the effect of Cr on the mean free path is too great to be explained by a cross section of order a^2 for each chromium atom. We think that $\tau/(mk_F/\hbar)$ is the quantity that should be related to $1/N_{\rm imp}\,a^2$; there is no velocity in the problem related to $h/m^* k_F$. Rice and Brinkman,[38] however, express an opposite conclusion.

There is also the problem of the thermopower discussed in Sections 1.9 and 1.12; for a highly correlated gas it should be small and change sign if the number of carriers is altered.

Finally, we think that the kind of transition discussed here may be one of many. Ramirez *et al.*[39] and Kimball,[40] for instance, have discussed one applicable to local moments from which electrons can be excited into an *s*-band.

1.3. Anderson Localization

Anderson's paper,[7] published in 1958, on the "absence of diffusion in certain random lattices" is the basis of much of our understanding of the behaviour of electrons in non-crystalline materials. Though originally written with spin-diffusion in view, it is directly applicable to impurity conduction. Miller and Abrahams[41] were the first in a published paper to treat the process of hopping from one localized state to another; curiously enough they did not refer explicitly to Anderson's work, though the localization concept was implicitly assumed. As far as I know, Twose's[42] unpublished Ph.D. thesis was the first to use the concept explicitly in this connection (see also Mott and Twose[43]).

Anderson's localization theorem has always seemed to the author essential if we are to understand even qualitatively the observed facts about impurity conduction which have been known for at least ten years. This point is made in Section 1.11. Some theorists, either ignorant of or not impressed by the experimental evidence, have found the theorem nonetheless difficult to accept. To see why this is so, we must first state what Anderson's theorem says. It is a statement about the behaviour of a single electron or other particle, and can therefore be treated by the one-electron Schrodinger equation. A crystalline array of sites is envisaged, with co-ordination number z and on each site there is a potential well leading to a single bound s-state. In the spirit of the tight binding approximation, interaction between nearest neighbour sites only is included. Anderson writes a_j for the probability amplitude that the particle is on site j, and for the time-dependence of a_j

$$(\hbar/i)\, \mathrm{d}a_j/\mathrm{d}t = \mathcal{E}_j a_j + \sum_i I_{ij} a_i, \tag{1.20}$$

The I_{ij}, defined for pairs by eqn (1.2), are the same for all pairs, but the essential point is that the energies $\mathcal{E}_j$ are distributed in a random way over a range that we shall call V_0. Alternatively we can write down the Schrödinger equation as

$$\nabla^2 \psi + (2m/\hbar^2)(\mathcal{E} - V)\psi = 0, \tag{1.21}$$

where V is the potential energy shown in Fig. 1.1(b), the depths of the wells being distributed at random over a range V_0. Anderson's theorem then says that if the parameter P defined by

$$P = V_0/B$$

is greater than some critical value P_0, a particle placed at time $t = 0$ on a given well will not, at zero temperature, diffuse away. It does not of course stay on its original well, but the chance as $t \to \infty$ that it has diffused a distance r decays exponentially as $e^{-\alpha r}$, where α depends on V_0/B. Another way to express the condition is to suppose that the electron is initially placed in a *large* volume of diameter Δx, so that the energy can be defined within a range

$$\Delta\mathcal{E} = Ia/\Delta x$$

by the uncertainty principle. Then, if the Anderson condition is satisfied, the electron will not move a distance comparable with Δx.

In terms of a solution of the Schrödinger eqn (1.20) or (1.21), the absence of diffusion means that the solutions are "localized" in the sense illustrated in Fig. 1.1(c). Each eigensolution has a quantized energy value and the envelope falls off exponentially as $e^{-\alpha r}$. Anderson,[44] Abram and Edwards[45] and Freed[46] have recently shown that, near the limit for localization, α behaves like $(P - P_0)^{0\cdot 6}$.

The intuitive objection that some people have experienced to Anderson localization arises perhaps for the following reason. It is felt that, starting with a given localized state, at a sufficient distance there must exist another state with just exactly the right energy to resonate with the wave function, which should therefore not be localized but have a series of peaks and valleys. This argument is wrong; it is of course true that, among the ensemble of all possible potentials of type shown by Fig. 1.1(b), there are some for which wave functions exist of this type; therefore it is not true to say that *all* wave functions are localized when one considers the whole ensemble (which must include the case when $V_0 = 0$). As emphasized particularly by Edwards,(47) localization can be properly defined only in terms of a quantity for which we can take an ensemble average, such as a Green's function or a product of Green's functions. The present author has taken the conductivity $\sigma(\mathcal{E})$ as an appropriate quantity. For this we use the formula

$$\sigma(\mathcal{E}) = (2\pi e^2 \hbar^3/m^2)\, |D(\mathcal{E},\mathcal{E}')|^2_{av}\, \{g(\mathcal{E})\}^2, \tag{1.22}$$

which is the conductivity of an electron gas if all states are filled up to the Fermi energy $\mathcal{E}_F$; at a finite temperature

$$\sigma = -\int \sigma(\mathcal{E}) \frac{\partial f}{\partial \mathcal{E}}\, d\mathcal{E}. \tag{1.23}$$

Here f is the Fermi distribution function and

$$D(\mathcal{E},\mathcal{E}') = \int \psi_{\mathcal{E}}^* \frac{\partial}{\partial \mathcal{E}} \psi_{\mathcal{E}'}\, d^3x,$$

and the average is over all pairs of states with energy $\mathcal{E}$. This can be shown as follows, following Mott and Davis(27) (Chapter 2).

We calculate $\sigma(\omega)$, the conductivity at frequency ω, and take for the d.c. conductivity the limit as $\omega \to 0$. Fig. 1.7 shows the states of a degenerate gas filled up to energy $\mathcal{E}_F$. $\sigma(\omega)$ for small values of $\hbar\omega$ is the product of:

(a) The number of electrons with energies within $\hbar\omega$ of the Fermi limit; this is $g(\mathcal{E}_F)\,\hbar\omega$.

(b) The number of states into which they can jump; this introduces another factor $g(\mathcal{E}_F)$.

(c) The transition probability; this depends on the square of the matrix element of x, where eEx is the potential energy of the electron in the field E. We find it useful to write

$$\int \psi_{\mathcal{E}'}^* x \psi_{\mathcal{E}}\, d^3x = (\hbar/m\omega) \int \psi_{\mathcal{E}'}^* (\partial/\partial x) \psi_{\mathcal{E}}\, d^3x.$$

(d) The energy $\hbar\omega$ absorbed.

It will be seen that ω cancels out and on making $\omega \to 0$ the conductivity is given by eqn (1.22).

If states are localized, they will not (except accidentally) be degenerate and all the wave functions are real. Thus the diagonal terms $D_{\mathcal{E}\mathcal{E}}$ vanish.

There are no non-diagonal terms, because two localized wave functions with the same energy cannot overlap. If they did they would split into states of odd and even parity.

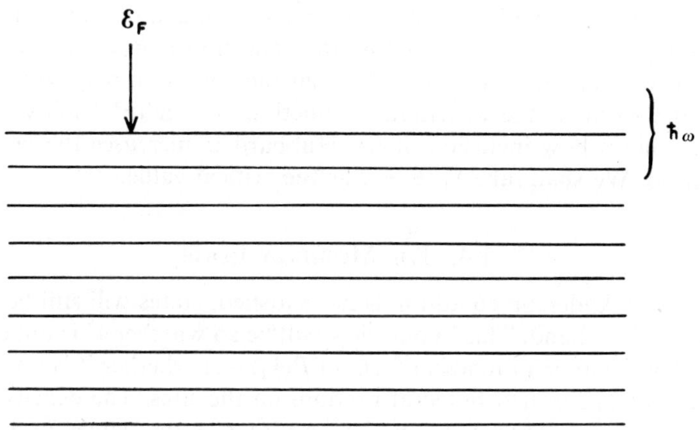

FIG. 1.7. Illustrating the states of a degenerate gas and the allowed transitions for a field with frequency ω.

There is now a large literature on Anderson localization, attempting to disprove it or to prove it with a greater or less degree of rigour. This will not be reviewed here. Some of the latest and most detailed treatments are those by Economou and Cohen[48,49,50] and Freed.[46]

We summarise, then, the situation when states throughout a band are localized. A single electron, with energy anywhere in the band and placed in any volume, cannot diffuse away. It can move only at a finite temperature through interaction with phonons (Section 1.5). A degenerate gas for which the Fermi energy lies in a range for which states are localized shows zero conductivity. Such a gas has been called by Anderson[51] a "Fermi glass".

A Fermi glass has linear specific heat, Pauli paramagnetism, and so on, just as if the states were non-localized (though correlation produces new effects, see Section 1.2). Moreover $\sigma(\omega)$ does not vanish, but behaves approximately at $T = 0$ as

$$\sigma(\omega) \sim \omega^2\{\ln(\omega_0/\omega)\}^4$$

We turn now to the critical value of the localization parameter. More recent work has considerably reduced Anderson's original value. Thus if states are distributed at random over a range V_0, with sharp upper and lower limits, and $z = 6$, Anderson found $(V_0/B)_{crit} = 5{\cdot}5$; but Edwards and Thouless[52] give as a result of numerical calculations a value nearer 2. Schönhammer[53] using† a probability density for the random energies V_i

$$P(V_i) = (2/\pi B^2)(B^2 - V_i^2)^{\frac{1}{2}} \tag{1.24}$$

gets 2.2. Economou and Cohen[50] using a Lorentzian distribution function for the energies, const/$(\mathcal{E}^2 + \Gamma^2)$, find the condition $\Gamma/B = \frac{1}{2}$ for localization —a much smaller value. This suggests that the condition may be quite sensitive to the form of the distribution function. Kikuchi[54] finds $V_0/B \simeq 4$, and also shows how inclusion of the Hubbard U increases the tendency to localization. We shall take $V_0/B = 2$ as the critical value.

1.4. The Mobility Edge

If the critical Anderson condition is not satisfied, states will still be localized in the tail of the band. This in our view will be so whether V_i is unbounded— as for a Gaussian or Lorentzian form of $P(V_i)$—or whether it is not, *provided* the values of V_i are distributed at random on the sites. The density of states in the two cases is shown in Fig. 1.8. As long as the values of V_i are dis-

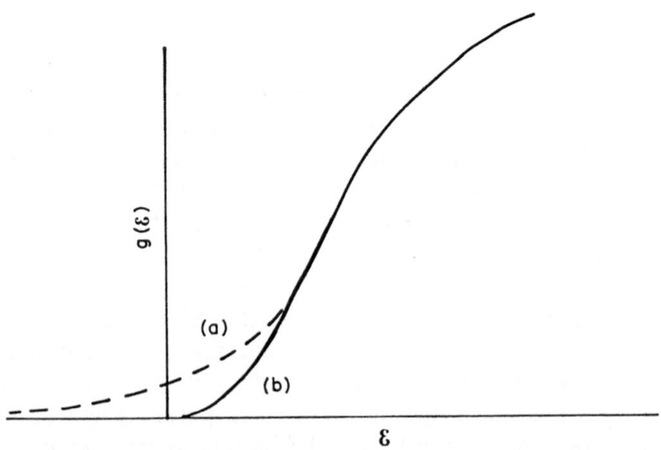

Fig. 1.8. Showing the "tail" at the bottom of the Anderson band. (a) An unlimited range of values of V_i. (b) A limited range.

† Note that his B and V_i are half ours.

tributed at random, a few large clusters will exist at which the energy of the electron is $V_{\min} - \frac{1}{2}B$; outside of such clusters the wave functions will decay exponentially, and, since large clusters will be rare, Anderson localization will occur. In fact we think that any potential for which there is a random element in *either* the depths *or* positions of wells must show a localized tail.

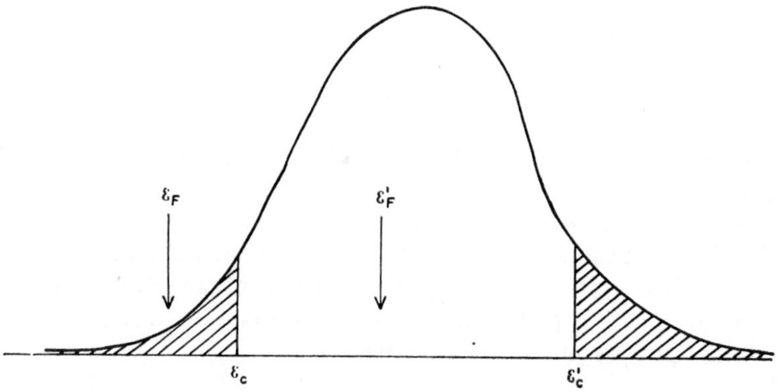

FIG. 1.9. Density of states in an Anderson band, showing the energies of the two "mobility edges". Localized states are shaded. $\mathcal{E}_F$, $\mathcal{E}_F'$, are two possible positions of the Fermi energy.

The density of states in infinite tails of this kind was one of the first problems in the field of this book to be investigated theoretically[55,56] and more recently.[57] In one dimension these authors find that $g(\mathcal{E})$, for large negative $\mathcal{E}$, decays exponentially as $|\mathcal{E}| \exp\{-\frac{4}{3}|2\mathcal{E}|^{3/2}\}$, with similar but more complicated formulae in three dimensions.

The present author first pointed out that, as long as one used a precise definition of localization such as $\langle\sigma\rangle = 0$, there must be a sharp energy, denoted by $\mathcal{E}_c$, separating energies where states are localized from energy where they are extended. This is shown in Fig. 1.9. The energy $\mathcal{E}_c$ is defined by

$$\begin{aligned} \sigma(\mathcal{E}) &= 0 \qquad \mathcal{E} < \mathcal{E}_c \\ &\neq 0 \qquad \mathcal{E} > \mathcal{E}_c \end{aligned}$$

The quantity $\mathcal{E}_c$ is called the "mobility edge". The calculation of its position has proved quite difficult, but has been solved for a Lorentizian form of $P(V_i)$ by Economou and Cohen.[50] Thus Schönhammer,[53] using formulae due to these authors, taking a distribution function

$$P(V_i) = (2/\pi V_0)\sqrt{(V_0^2 - V_i^2)}, \tag{1.25}$$

puts the mobility edge at

$$\mathcal{E}_c = \sqrt{(B_0{}^2 + V_0{}^2 + V_0{}^4/4B_0{}^2)},$$

where B_0 is the *half* band-width when V_0 vanishes. When V_0 is small, this gives

$$\mathcal{E}_c = B_0 + V_0{}^2/2B_0 - 3V_0{}^6/16\,B_0{}^5 \dots .$$

The half band width in this approximation is

$$\sqrt{(B^2 + V_0{}^2)} \simeq B_0 + \tfrac{1}{2}V_0{}^2/B_0,$$

so one always gets a mobility edge. However, as he points out, the true half band width must be $B_0 + V_0$.

For values of $\mathcal{E}$ above the mobility edge, conduction can occur without thermal activation; estimates of the mobility are given in Section 1.6.

For values below the mobility edge an electron can move only by exchanging energy with a phonon, and the motion is thermally activated. We deal with this process in the next section.

1.5. Thermally Activated Hopping

If the Anderson condition is satisfied everywhere in a band, an electron can only move from site to site by thermally activated hopping. This means that it moves from one localized state to another with the emission or absorption of a phonon. Processes in which the electron gains energy will be rate-determining, both for conduction and diffusion. If two sites are at a distance R from each other, and $\alpha R \gg 1$, then the probability that an electron jumps from one to the other with absorption of phonon energy will be of the form

$$\nu \exp(-2\alpha R - W/kT), \tag{1.26}$$

where W is the difference in the energies of the two states; ν is a factor depending on the phonon frequencies, which will be discussed later in this Section. If localization is strong ($\alpha R \gg 1$ for nearest neighbour sites), hopping processes to nearest neighbour sites will determine the diffusion coefficient and the mobility. The diffusion coefficient could be written

$$D = \tfrac{1}{6}\nu R^2 \exp(-2\alpha R - W/kT), \tag{1.27}$$

and the mobility, by Einstein's relation,

$$\mu = eD/kT.$$

But since for each hopping process both R and W vary, conduction and diffusion will be determined by the most difficult steps, and evaluation of these quantities involve considerations of percolation theory[58] which will not be considered here.

If we are dealing with a degenerate electron gas, the conductivity σ rather than the mobility is the relevant quantity. Here we argue that only electrons within a range $\sim kT$ of the Fermi energy make a significant contribution to σ. The jump probabilities in the direction of the field and in the opposite direction will be proportional to

$$\exp\{-W \pm eRE/kT\},$$

so we write for the current

$$e^2 g(\mathcal{E})\, kT\, \nu R \exp\{-2\alpha R - W/kT\} \sinh(eER/kT) \tag{1.28}$$

or, unless E is very strong, for the conductivity

$$\sigma = e^2 g(\mathcal{E})\, \nu R^2 \exp\{-2\alpha R - W/kT\}. \tag{1.29}$$

The same considerations about averaging over R and W apply.

The conditions $\alpha R \gg 1$ will not apply near a mobility edge, or when the Anderson condition is only just satisfied. If αR is not large, the factor $e^{-2\alpha R}$ may be put equal to unity, and if α is small the number ϕ of states for which this is so is given by

$$\phi \sim (\alpha R_0)^{-3}, \tag{1.30}$$

where R_0 is the distance to the nearest centre. A conductivity of the form

$$\sigma = e^2 g(\mathcal{E})\, \nu R_0{}^2 \phi\, e^{-W/kT} \tag{1.31}$$

is expected.

At sufficiently low temperatures the phenomenon of "variable range hopping" is expected in all cases.[59] This is because the factor

$$\exp\{-2\alpha R - W/kT\} \tag{1.32}$$

will not have its maximum value for nearest neighbour sites for which $R = R_0$. The most probable transitions will be those to sites for which W is small; within a distance R there will be $(4\pi/3)(R/R_0)^3$ such sites, and the average spacing between their energies will be $3/4\pi g(\mathcal{E})R^3$, where $g(\mathcal{E})$ is the density of states. Thus the most probable hopping distance is that for which the quantity

$$2\alpha R + [(4\pi/3)\, g(\mathcal{E})\, R^3\, kT]^{-1}$$

has its smallest value. This gives

$$R = \{9/\alpha\, g(\mathcal{E})\, 8\pi\, kT\}^{\frac{1}{4}}$$

and the well known law

$$\sigma = \sigma_0 \exp\left(-A/T^{\frac{1}{4}}\right). \tag{1.33}$$

The theoretical basis of this law has been extensively discussed in the literature; the averaging, so important in the nearest neighbour case, makes little difference for variable range hopping.[60,58] If $g(\mathcal{E})$ is not constant near the Fermi energy, other powers of T are possible.[61]

Another form of variable-range hopping is the well-known formula for the a.c. conductivity introduced by Pollak and Geballe.[62] These authors consider Debye loss due to hopping between pairs of centres at a distance R from each other and with energy difference W. At frequency ω, the contribution to $\sigma(\omega)$ by any pair of centres would be

$$\frac{1}{kT}\frac{e^2 R^2}{1+\omega^2\tau^2}\frac{1}{\tau}\exp\left(-\frac{W}{kT}\right).$$

Significant contributions came from pairs such that $W \sim kT$ and $\omega\tau \sim 1$, so that to a first approximation $\sigma(\omega) \sim \omega$. Since $1/\tau$ contains the factor $e^{-\alpha R}$ and there are $R^2\, dR$ centres in an integration over all space, $\sigma(\omega)$ is also proportional to R^4 where

$$\alpha R \simeq \ln(\omega_0/\omega),$$

where ω_0 is a constant. In practice the factor $\{\ln(\omega_0/\omega)\}^4$ makes the conductivity behave like $\omega^{0\cdot 8}$, with little dependence on T. However, most forms of Debye loss, whether due to ions or electrons, will give a value of $\sigma(\omega)$ very much of this form, and there is now some uncertainty about whether the $\omega^{0\cdot 8}$ term observed in some glasses is due to ionic movement (see articles by Jonscher and by Davis in this volume).

We have finally to discuss the factor ν. If localization is strong, so that the calculation refers to hops between pairs of states with strongly localized (real) wave functions ψ_a, ψ_b, then for an electron resonating between the pair we may write, if ψ_b has energy W above ψ_a,

$$\Psi = \psi_a + A\psi_b, \tag{1.34}$$

and minimise the energy

$$\int \Psi^* V \Psi\, d^3x.$$

This gives

$$A^2 W + 2AI = 0,$$

where

$$I = \int \psi_a V \psi_b \, d^3x;$$

so

$$A = -I/W.$$

The excited state has wave function

$$\psi_b - A\psi_a.$$

The factor $(I/W)^2$ will therefore be present in the transition probability.

The term ν is calculated by Miller and Abrahams[(41)] for the case when W is less than the maximum phonon energy. If more than one phonon is involved, the transition probabilities will certainly be small, and must be estimated by methods such as that used by Jortner and coworkers[(63,64)] for multiphonon transitions in molecules.

The factor $\nu e^{-2\alpha R}$ cannot be greater than the phonon frequency ω. Thus for electrons just below the mobility edge we expect, unless the temperature is low, a mobility of the form

$$\tfrac{1}{6} e\omega\phi R^2/kT \qquad (\omega \sim 10^{12}\,\mathrm{s}^{-1}),$$

which is of order $0{\cdot}1\ \mathrm{cm}^2/\mathrm{V.sec}$ if $\phi \sim 1$.

Finally we point out that, in a Fermi glass, states at the Fermi level are necessarily singly occupied, though states of lower energy can be doubly occupied.[29]

We turn now to the situation when a "mobility edge" $\mathcal{E}_c$ exists, and the Fermi energy lies in the range of localized states. While at very low temperatures the hopping processes described above are expected, at higher temperatures the main part of the current may be carried by electrons excited into extended states above the mobility edge. We need therefore to estimate the mobility there, and this is attempted in the next section. It is found to be in the range

$$\tfrac{1}{6} e\omega_{\mathrm{el}} R^2/kT, \qquad (\omega_{\mathrm{el}} \sim 10^{15}\,\mathrm{s}^{-1}), \tag{1.35}$$

so that, again apart from the factor ϕ, a discontinuity of order 10^3 is expected, though the factor ϕ may smooth this out somewhat. Two types of experiment may be planned to show this.

(a) A degenerate gas may have its density or random field varied so that $\mathcal{E}_F$ crosses $\mathcal{E}_c$. When this happens we go from unactivated conduction to conduction obeying the $T^{\frac{1}{4}}$ law expressed in eqn (1.33). This is described in the next Section.

(b) In semiconductors in which there is a range of localized states leading up to a mobility edge $\mathcal{E}_c$ in the conduction (or valence band), we expect a conductivity of the form

$$C \exp \{(\mathcal{E}_F - \mathcal{E}_A + W)/kT\} + D \exp \{-(\mathcal{E}_F - \mathcal{E}_c)/kT\}. \tag{1.36}$$

Here $\mathcal{E}_A$ is the energy which corresponds to the predominant hopping path through the material, W is the associated hopping energy, D as we shall see in the next Section is $\sim 200\ \Omega^{-1}\ \mathrm{cm}^{-1}$ and $C/D \sim 10^{-3}$. The $\ln \sigma - 1/T$ curve should show a kink. An example is shown for amorphous Si in the article by Le Comber and Spear in this volume (Chapter 10.)

Here we may add a note on the likely effect of a strong field E on these two terms and the Poole–Frenkel formula discussed by Jonscher in this volume. We think that the effect on the mobility of the second term is not likely to be large, and a major effect will be on the number of carriers. Since an electron and hole attract each other, the energy required to separate them from the exciton to the free state must be reduced by $\beta E^{\frac{1}{2}}$ ($\beta = e^{3/2}/\kappa^{\frac{1}{2}}$). If n is the number of carriers, and assuming the rate An^2 of recombination to be independent of field, it follows from the principle of detailed balancing that

$$n^2 \propto \exp (\beta E^{\frac{1}{2}}/kT),$$

so that

$$\sigma \propto \exp (\tfrac{1}{2}\beta E^{\frac{1}{2}}/kT). \tag{1.37}$$

The term $\frac{1}{2}$ in this formula will *not* appear if the Fermi energy is "pinned" by localized states in the gap (see Section 1.13).

1.6. A Minimum Metallic Conductivivty

If V_0/B is small, the mean free path in the Anderson model of Fig. 1.1(b) can be calculated using the Born approximation. Mott and Davis(27) (p. 27) give for a half-full simple cubic tight-binding band

$$\sigma = (e^2/3\hbar a)\, L/a \tag{1.38}$$

where

$$L/a = (4\pi/z^2)\,(B/V_0)^2.$$

The Born approximation and the assumption of single scattering break down when $L \sim a$; this happens when $V_0/B = (4\pi)^{\frac{1}{2}}/z \simeq 0{\cdot}59$. The conductivity is then

$$\sigma = e^2/3\hbar a. \tag{1.39}$$

Equation (1.38) could also be obtained by starting from the formula

$$\sigma = S_F\, e^2\, L/12\pi^3\, \hbar,$$

and putting S_F, the Fermi surface area, equal to $4\pi k_F{}^2 \simeq 4\pi^3/a^2$.

In the range

$$0{\cdot}59 < V_0/B < 2,$$

before localization occurs, we have to use the Kubo–Greenwood formula [eqn (1.22)] together with the assumption of random phases on each atom or each bond.(65,66,27,67,68) The conductivity comes out to be

$$\sigma = 0{\cdot}092(B/V_0)^2\, e^2/\hbar a.$$

Localization occurs when $B/V_0 \simeq 1/2$, so the smallest value that the conductivity can have before an activation energy occurs is

$$\sigma_{\min} = 0{\cdot}026\, e^2/\hbar a. \tag{1.40}$$

This quantity we call the "minimum metallic conductivity". It is usually in the range 200–300 $\Omega^{-1}\,\text{cm}^{-1}$, though for doped semiconductors, with $a >$ 100 A, it is about 50 times smaller. It may also be sensitive to co-ordination number [50], and be somewhat (2 to 4 times) larger for d functions, since their angular dependence will ensure that the number of centres for which there is overlap will be smaller (see 1.12).

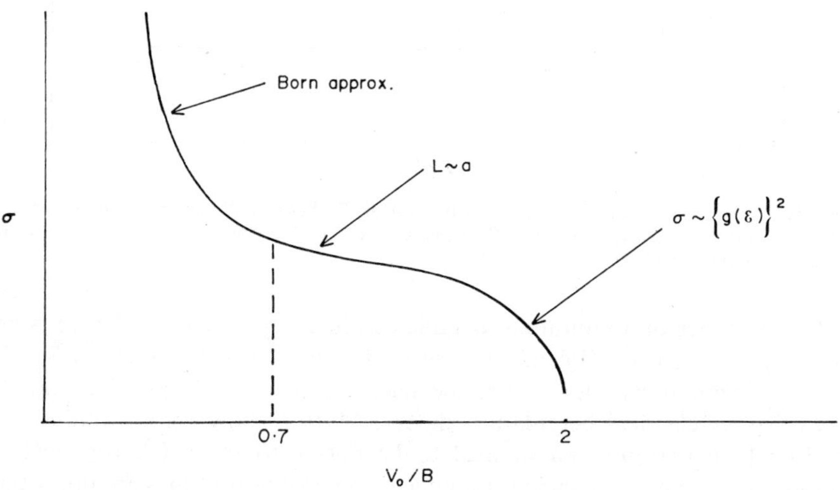

FIG. 1.10. A plot of conductivity σ for the Anderson band as a function of V_0/B, at zero temperature.

A change by a factor ~ 13 is thus expected in σ between the value when first $L \sim a$ and the onset of localization. A plot of σ against V_0/B (for constant a) is shown in Fig. 1.10.

Figure 1.11 shows the expected behaviour of the resistivity of a degenerate gas, as a function of temperature, as the situation at $\mathcal{E}_F$ goes from weak localization to extended states. This could occur *either* by variation of V_0/B through the Anderson critical value, *or* as $\mathcal{E}_F$ goes through $\mathcal{E}_c$, for instance in a doped and compensated semiconductor as the compensation is changed. Two examples of this behaviour are discussed in this article, and others are discussed by Mott and Davis[27] (pp. 83, 186).

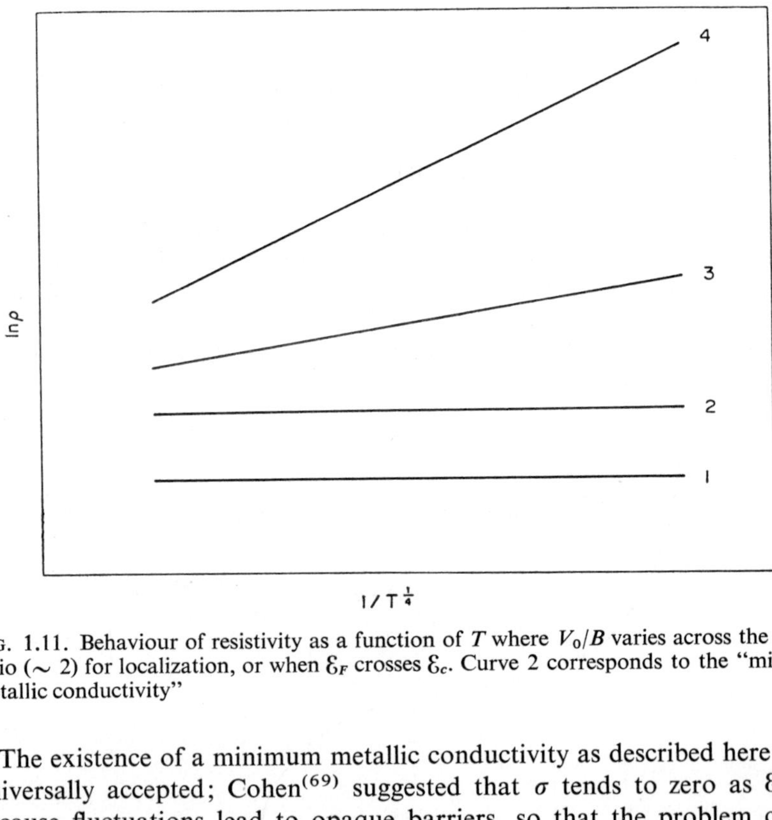

FIG. 1.11. Behaviour of resistivity as a function of T where V_0/B varies across the critical ratio ($\sim$ 2) for localization, or when $\mathcal{E}_F$ crosses $\mathcal{E}_c$. Curve 2 corresponds to the "minimum metallic conductivity"

The existence of a minimum metallic conductivity as described here is not universally accepted; Cohen[69] suggested that σ tends to zero as $\mathcal{E} \to \mathcal{E}_c$, because fluctuations lead to opaque barriers, so that the problem can be described by classical percolation theory. Mott[70] however maintains that statistical fluctuations cannot lead to barriers through which the electron cannot tunnel, and we consider that classical percolation edges are important only in unhomogeneous materials (1.13.3) and possibly liquids near a critical point.

1.7. Pseudogaps

There is a class of problem similar to Anderson localization, in which a full valence band and an empty conduction band overlap slightly, giving a "pseudogap" in the density of states as shown in Fig. 1.12. Examples are mercury at high temperatures and low densities[71,72] (for a review see Mott[73]) and the liquid tellurium alloys investigated by Warren[74] and others. For materials of this kind the present author[65] has introduced the factor

$$\mathscr{g} = g(\mathscr{E}_F)/g(\mathscr{E}_F)_{\text{free}} \tag{1.41}$$

and assumed that, if the *same* pseudopotential is responsible for the gap and for scattering, a value of $\mathscr{g}$ substantially less than unity implies that $L \sim a$. As we shall see below, a test of whether $\mathscr{g} < 1$ with $L \sim a$ is whether the Hall coefficient deviates from the classical formula; probably $\mathscr{g} \sim \frac{1}{2}$ before $L \sim a$ can *necessarily* be assumed.[73,75] The formula

$$\sigma = S_F e^2 L \mathscr{g}^2 / 12\pi^3 \hbar \tag{1.42}$$

is valid in general if L is the distance in which wave functions lose phase memory. If $L > a$, we may write $L = L_{\text{Ziman}}/\mathscr{g}^2$, so $\mathscr{g}$ cancels out, but if $L \sim a$ this is not possible, so

$$\sigma = S_F e^2 a \mathscr{g}^2 / 12\pi^3 \hbar. \tag{1.43}$$

S_F is here the Fermi surface area.

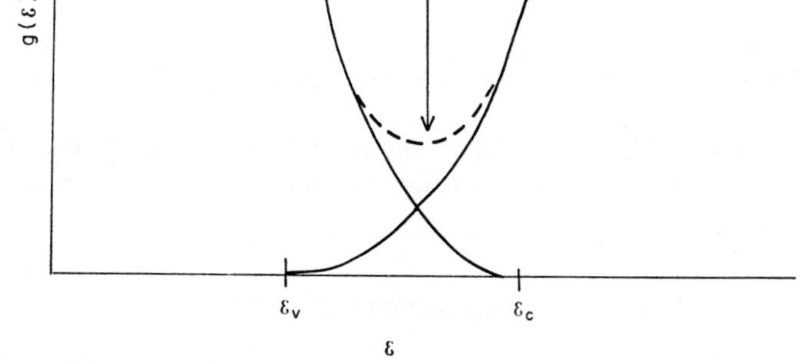

Fig. 1.12. Two overlapping bands in a non-crystalline material producing a "pseudogap". The dotted line shows the total density of states. States between $\mathscr{E}_v$ and $\mathscr{E}_c$ are localized.

Both bands should have localized states in the tails, so, if the pseudogap is deep, states will be localized there. The present author has assumed that the ratio (~ 12) between the conductivity when $L \sim a$ and the conductivity just before localization sets in applies to this case too, and that the formula (1.40) for the minimum metallic conductivity applies also. There is some uncertainty about the value of a, which is the distance between localized states; we do not think it should differ greatly from the lattice parameter. It follows that localization sets in when

$$g \simeq 1/\sqrt{12} \simeq 1/3{\cdot}5. \tag{1.44}$$

Applications of this formula have been made to mercury, liquid tellurium alloys and to evaporated Mg–Bi films;[(76)] other references are given by Mott and Davis.[(27)] One of the most striking predictions is that the Knight shift is proportional to $\sqrt{\sigma}$. There is considerable empirical evidence, too, that $g^2 \simeq 1/12$ is the criterion for localization.

Here we should emphasize that the pseudogap model does not differ essentially from that of two overlapping crystalline bands with $L \sim a$; we think it logically preferable since electrons move into holes extremely rapidly. But if we use normal semimetal theory, then σ is still proportional to S_F and hence to $g(\mathcal{E})^2$ if electrons and holes have the same mean free path.

We may possibly use eqn (1.44) to give the range of localized states at the edge of a conduction or valence band, in a chalcogenide glass for example. If so, and *if* $L \sim a$ throughout the band, one might guess that the mobility edge would appear when

$$g(\mathcal{E}_c) \sim \langle g(\mathcal{E}) \rangle / 3{\cdot}5,$$

where $\langle g(\mathcal{E}) \rangle$ is the mean density of states.

1.8. Electrons in Extended States Near the Mobility Edge

If the Fermi energy lies in a region of strongly localized states, or a region where $g(\mathcal{E}) = 0$, the current may be carried by electrons excited to the mobility edge. In this case we expect

$$\sigma = \sigma(\mathcal{E}_c) \exp(-\Delta\mathcal{E}/kT)$$

where

$$\Delta\mathcal{E} = \mathcal{E}_c - \mathcal{E}_F.$$

The mobility is here a quantity of interest. The number of electrons above the mobility edge is

$$g(\mathcal{E}_c)\, kT \exp(-\Delta\mathcal{E}/kT),$$

so

$$\sigma(\mathcal{E}_c) = e\, g(\mathcal{E}_c)\, \mu kT. \tag{1.45}$$

If $g(\mathcal{E}_c) \sim \frac{1}{3} \langle g(\mathcal{E})\rangle$, we find

$$\mu = 0{\cdot}075\, (e/\hbar a)/\langle g(\mathcal{E})\rangle\, kT.$$

If we write $\langle g(\mathcal{E})\rangle = 1/a^3 B$, this gives

$$\mu = 0{\cdot}075\, ea^2\, B/\hbar kT. \tag{1.46}$$

Cohen[77] first described this process as a quantum mechanical diffusive or Brownian motion, the electron jumping from site to site with an atomic frequency ν_{el}. If we compare eqn (1.46) with Cohen's formula

$$\mu = \tfrac{1}{6}\, \nu_{\mathrm{el}}\, ea^2/kT,$$

we find

$$\nu_{\mathrm{el}} \sim B/2\hbar.$$

If the band width B is of order 5 eV, we see that $\nu_{\mathrm{el}} \sim 10^{15}\ \mathrm{s}^{-1}$.

1.9. THERMOPOWER

Cutler[78] has shown that for the thermopower α we may use the "metallic" form

$$\alpha = \frac{\pi^2}{3} \frac{k^2 T}{e} \frac{\mathrm{d} \ln \sigma}{\mathrm{d}\mathcal{E}}, \tag{1.47}$$

as long as current is carried by electrons at the Fermi energy. This has several consequences.

(a) In the case of variable range hopping, α is of the form

$$\alpha = AT + BT^{3/4} \tag{1.48}$$

(b) In the case of hopping to nearest neighbours

$$\alpha = A + BT \tag{1.49}$$

(c) In the range $L \sim a$, α is proportional to $T\, \mathrm{d} \ln g(\mathcal{E})/\mathrm{d}\mathcal{E}$. Thus the sign of α depends on whether $g(\mathcal{E})$ is increasing or decreasing.

1.10. Hall Effect

For the theory of the Hall coefficient, either for the case of hopping or for the case $L \sim a$, we refer to Friedman[68] and the articles by Emin and by Friedman in this volume; see also Refs 79, 80, 81. Here we note that for a free carrier when $L \sim a$,

$$\mu_H \sim ev_{el}\, a^2/I \qquad (1.50)$$

compared with the conductivity mobility

$$\mu_D \sim ev_{el}\, a^2/kT.$$

Thus

$$\mu_H \ll \mu_D.$$

The sign of R_H will normally be negative.

In a Fermi glass, if there is a pseudogap, Friedman's analysis gives, if $L \sim a$

$$R_H = C/nec\, g \qquad (C \sim 0{\cdot}7). \qquad (1.51)$$

It should be emphasized that this is only so if $L \sim a$. If $L > a$, the empirical evidence[73] is that $R_H = 1/\text{nec}$, even if there is a pseudogap.

If the main part of the current according to formula (1.36) is carried by electrons in localized states, then the form of the Hall mobility has not yet been calculated, though it is probably of the form $\mu_0 \exp(-\gamma \Delta W/kT)$, where ΔW is the hopping energy and $\gamma < 1$; for polaron hopping under certain assumptions $\gamma = \frac{1}{3}$, as shown in the article by Emin in this volume. The recent observation of an activation energy in μ_H in certain chalcogenide glasses is to be expected if conduction is due to localized electrons at the band edge, and thus described by the first term in eqn (1.36); this may be an alternative explanation to that advanced by Emin. We note however that Nagels *et al.*[82] in As_2Se_3–As_2S_3 mixtures report the same activation energy in σ and R_H, which suggests motion in extended states.

1.11. Impurity-conduction

In extrinsic *n*-type semiconductors current is usually due to electrons excited from donors into the conduction band. At low temperatures, however, another form of conduction is predominant, which was first observed by Hung and Gliessman[83] in 1950. Its nature is now fully understood. It occurs *only* if the samples are compensated, that is to say if they contain some minority centres, in this case acceptors. Figure 1.13 shows the process involved. All the acceptors are occupied; their charges produce a random field giving a scatter of order $e^2/\kappa R$ in the energies of the donors; κ is here the dielectric constant and R the distance between centres. It is supposed in this argument that the compensation $K(= N_A/N_D)$ is not small, perhaps of order $\frac{1}{2}$. Since some of the donors are empty, because they have given up an electron to an acceptor, this hopping process will not be prevented by the correlation term $U = \langle e^2/\kappa r_{12}\rangle$. So the conductivity should depend on hopping processes of the kind shown in the diagram, for which the probability will be of the form given by eqn (1.29), namely

$$\nu \exp(-2\alpha R - \Delta W/kT). \tag{1.52}$$

Miller and Abrahams[(41)] and Twose[(42)] were the first to give theories of impurity conduction along these lines. Miller and Abrahams did not explicitly mention Anderson's work on localization; they assumed that states on each centre were localized, or rather that the whole process could be considered in terms of hops between pairs,described by the wave functions of eqn (1.34). The first mention of Anderson's work in this connection is contained in Twose's thesis (cf. Mott and Twose[(43)]). But Anderson localization becomes an essential concept when we consider hopping to more distant sites. We have pointed out in Section 1.2 that electrons in a crystalline array of wells will *always* be metallic if the number n of electrons per atom is non-integral. In a doped semiconductor, compensation ensures that $n < 1$. The random field ensures a term V_0 as described in Section 1.3. In addition the donors are distributed at random (a point not included in the Anderson model but with less effect on localization, (Section 1.3). Now if Anderson's theorem were *not* true, all wave functions would be extended, and a doped and compensated semiconductor would have a conductivity tending to a *finite* value as $T \to 0$. If

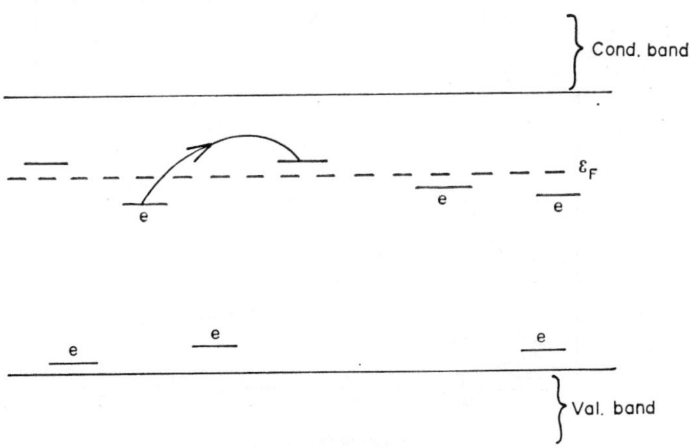

FIG. 1.13. Energy levels in a doped and compensated semiconductor, illustrating impurity-conduction.

Anderson localization is a valid theory, σ should tend to zero, ultimately as $\exp(-A/T^{\frac{1}{4}})$. There has never been any suggestion that lightly doped and compensated semiconductors are in fact metallic. though experiments at very low temperatures to observe the $T^{\frac{1}{4}}$ behaviour would be of interest. We believe then that the non-metallic behaviour of doped and compensated semiconductors is very strong evidence that the Anderson localization concept is a valid one.

We turn now to the properties of doped and compensated semiconductors in different ranges of composition. For concentrations n of majority carriers below that given by the relationship (1.15), viz

$$n^{\frac{1}{3}} a_H \simeq 0{\cdot}27 \qquad (a_H = \hbar^2\kappa/me^2), \tag{1.53}$$

the materials are non-metallic.[41] Conduction occurs at low temperatures through impurity conduction if compensation is present, with an activation energy $\mathcal{E}_3$; we expect $T^{\frac{1}{4}}$ behaviour at very low temperatures. The magnetic

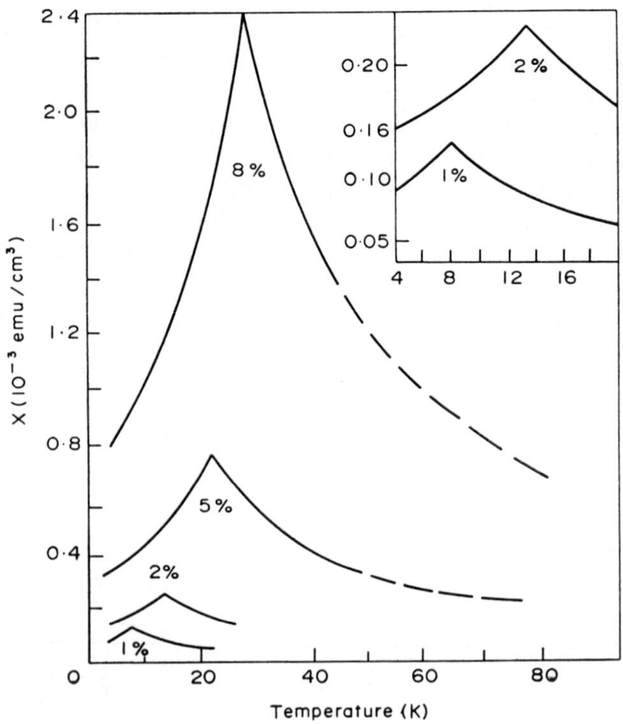

FIG. 1.14. Susceptibility in emu/cm³ versus T of AuFe alloys for various concentrations (in atomic per cent) of Fe (After Cannella *et al*, Ref. 85).

properties expected are those of an "amorphous antiferromagnetic"; each centre will have a fixed spin, as near as possible antiparallel to the spins in other directions. A similar problem arises in a "spin glass"—that is a metal such as copper containing transition metal atoms in solution, where the concentration of such atoms is great enough for the RKKY interaction† to exceed the Kondo energy. The properties of such a system are imperfectly understood; one problem is whether it has a "Néel temperature", namely a temperature at which M_z vanishes for all ions at the same time.(84) There seems to be evidence(85) that this is so in AuFe, shown in Fig. 1.14.

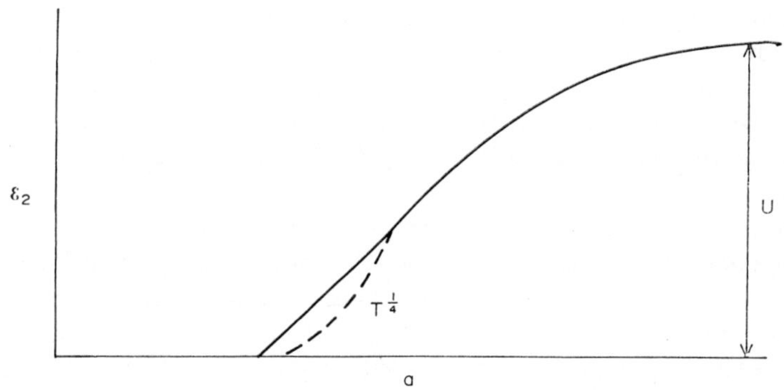

FIG. 1.15. Behaviour of ε_2 as a function of distance a between centres (schematic). The dotted line shows how the apparent value ε_2 may diminish when hopping replaces excitation to a mobility edge.

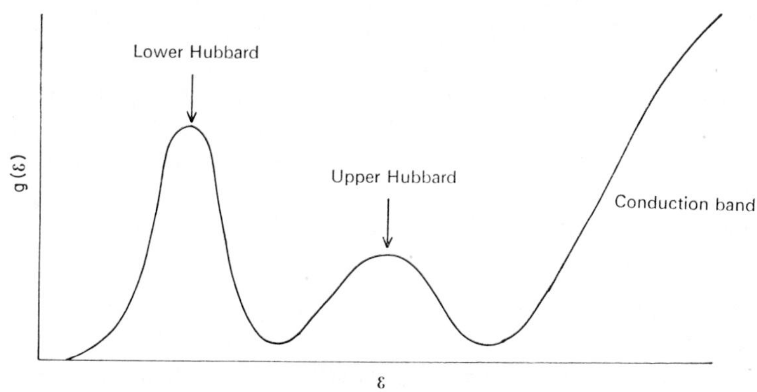

FIG. 1.16. Two Hubbard impurity bands and conduction band in doped and compensated semiconductors.

† The initials stand for the Ruderman–Kittel–Kasuya–Yosida interaction between moments in a metal, which falls off as $r^{-3} \cos 2kr$.

As the concentration is increased, we distinguish between the energy ε_1 required to free an electron and that (ε_2) required to remove an electron onto a distant site. The latter is the energy $U-B$ discussed in Section 1.2. For Ge:P its behaviour as a function of distance between the centres, from the measurements of Davis and Compton,[86] is shown by Mott and Davis[27] (p. 167). It is shown schematically in Fig. 1.15. We describe this behaviour in terms of two "Hubbard bands" as shown in Fig. 1.16. The form of the metal-insulator transition depends on the compensation. If this is considerable, $\mathcal{E}_F$ will lie in the lower Hubbard band, and metallic or non-metallic behaviour will depend on whether there is Anderson localization or not.

Experiments by Allen and Adkins[87] have shown that $T^{\frac{1}{4}}$ behaviour does occur at low temperatures when localization is weak; their results on *n*-type Ge with $K = 0{\cdot}20$ are shown in Fig. 1.17. When localization disappears, we expect that σ should be given by formula (1.40) for the minimum metallic conductivity, in this case $10\ \Omega^{-1}\ \mathrm{cm}^{-1}$. Mott[70] has discussed the evidence in favour of this behaviour.

If there is no compensation, we expect the behaviour to be similar; with increasing concentration ε_2 will drop, a $T^{\frac{1}{4}}$ region is expected and the metal-insulator transition occurs when states at $\mathcal{E}_F$ are non-localized in the Anderson sense. Experiments to see if this is so are needed.

Firm evidence on the question of whether there is a "metallic" range in which fixed moments still exist on centres is lacking, though the e.s.r. work of Morigaki and Onda[88] suggests that there is not (see Ref. 75). If the moments have disappeared, we expect *two* densities of states, one for current carriers and one (dotted line) the total density of states (Fig. 1.18). Evidence for the pseudogap for current carriers is provided by the work of Wolf *et al.*[89] on tunnelling into doped silicon, and for an enhanced susceptibility by the work of Ue and Maekawa,[90] shown in Fig. 1.19. The results of Yamanouchi *et al.*[91] show that the Hall coefficient obeys the classical formula through the whole metallic range; there is thus no Friedman anomaly (eqn 1.51) for the highly correlated gas. These results also strongly suggest $T^{\frac{1}{4}}$ behaviour on the insulator side.

For zero compensation, then, as the concentration increases, we expect the following behaviour:

(a) Excitations to the upper Hubbard band, the activation energy ε_2 decreasing as in Fig. 1.15, $T^{\frac{1}{4}}$ behaviour just before the metal insulator transition. Magnetic moments will remain at low concentrations, their strength gradually decreasing as we approach the transition.

(b) After the transition a range in which the gas is highly correlated, but in which the Hall effect is normal and in which there are no fixed moments. Probably $L \sim a$ in most of this range.

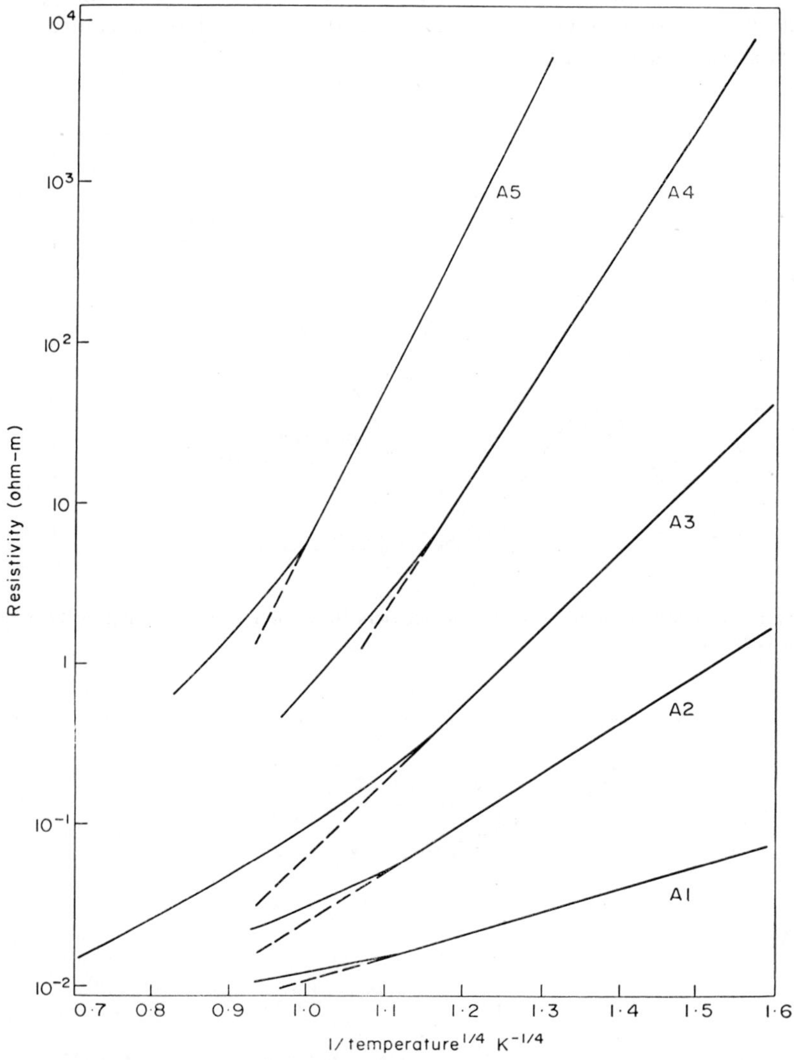

FIG. 1.17. Conductivity as a function of $T^{\frac{1}{4}}$ in *n*-type germanium (After Allen and Adkins Ref. 87). The doping levels are as follows:

Sample	N_D (cm^{-3})	K	$\mathcal{E}$ (meV)
A1	$4{\cdot}7\times10^{17}$	0·29	0·05
A2	$3{\cdot}5\times10^{17}$	0·23	0·10
A3	$1{\cdot}5\times10^{17}$	0·12	0·14
A4	$2{\cdot}2\times10^{17}$	0·25	0·21
A5	$1{\cdot}8\times10^{17}$	0·24	0·36
A6	$1{\cdot}3\times10^{17}$	0·26	0·57

(c) The range in which impurity and conduction bands merge. The resistivity can be calculated as due to scattering by charged centres as in the work of Krieger *et al.*[92]

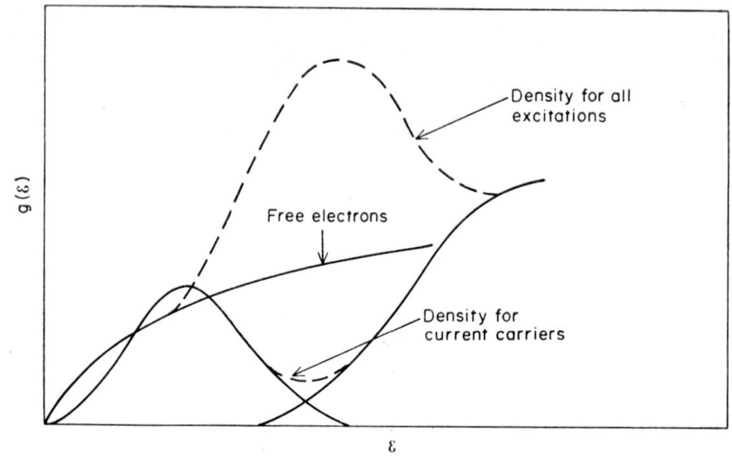

FIG. 1.18. Density of states for current carriers and total density in highly correlated impurity band.

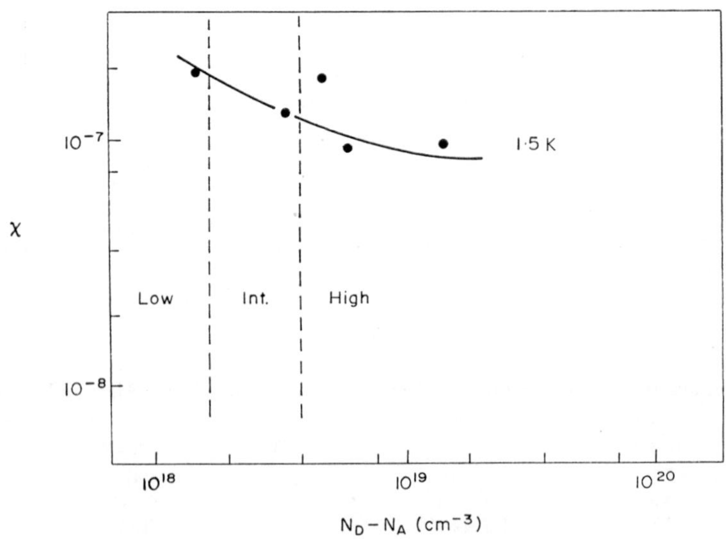

FIG. 1.19. Concentration dependence of the paramagnetic susceptibility at 1·3K of Si:P in the metallic range. (After Ue and Maekawa, Ref. 90).

1.12. Vanadium Monoxide (VO_x)

This material contains a high concentration of vacancies (~15%). It is not antiferromagnetic. We show in Figs. 1.20 and 1.21 the resistivity plotted against $1/T^{\frac{1}{4}}$ and the thermopower. The results are due to Banus and Reed[93] and Banus, Reed and Strauss.[94] The $T^{\frac{1}{4}}$ behaviour suggests that we have here a degenerate electron gas and that the random potential due to the

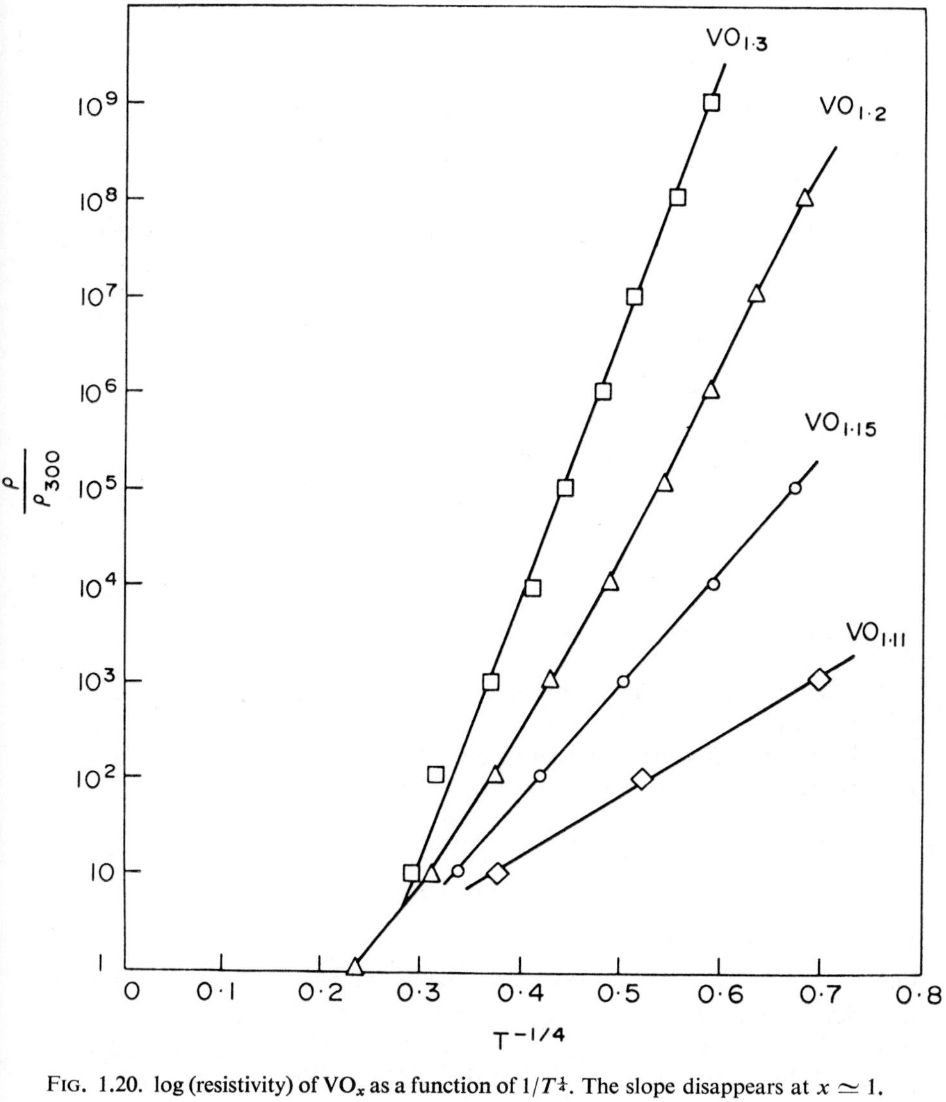

FIG. 1.20. log (resistivity) of VO_x as a function of $1/T^{\frac{1}{4}}$. The slope disappears at $x \simeq 1$.

vacancies can produce Anderson localization. The disappearance of the slope as x decreases is a typical example of the Anderson transition, as is the value of the conductivity ($\sim 1000\Omega^{-1}\,cm^{-1}$) just after the slope disappears, somewhat larger that $0.026\,e^2/ha$ given by eqn (1.40), perhaps because of the d-like nature of the wave functions (see Section 1.6).

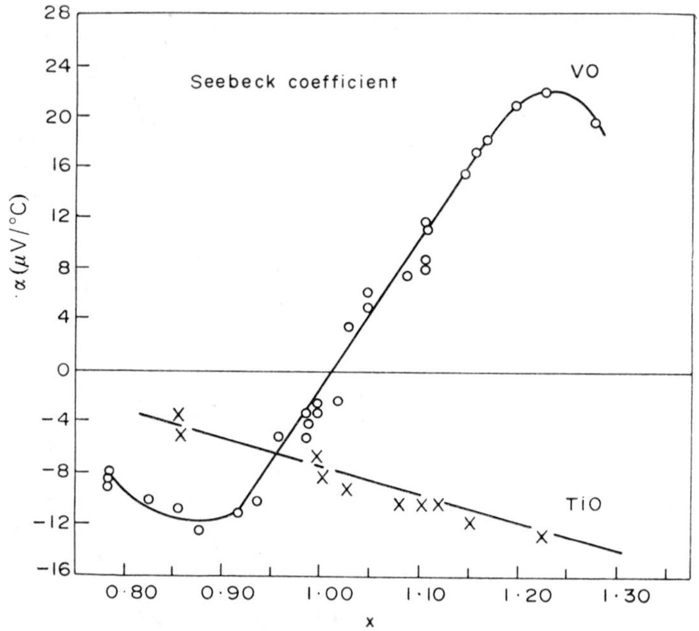

FIG. 1.21. Seebeck coefficient α of VO_x and TiO_x as a function of x at room temperature.

This material is included in this article because of the behaviour of thermo-power shown in Fig. 1.21. This is strong evidence that the gas is "highly correlated", α being proportional to $d \ln g(\mathcal{E})/d\mathcal{E}$ and $g(\mathcal{E})$ being the density of current-carrying excitation shown by the full line in Fig. 1.18.

1.13. NON-CRYSTALLINE SEMICONDUCTORS

1.13.1. INTRODUCTION

Our analysis up to this point has dealt with the Anderson model of "cellular" disorder, and its application to impurity-conduction in doped and compensated semiconductors. One of our conclusions is that for the Anderson potential (Fig. 1.1(b)), however small the random energy V_0, there is always a range of localized states at the extremities of the band, though if V_0 is small this range will be small too. In this section we examine the kind of model to be applied to real non-crystalline semiconductors and insulators such as silicon dioxide and chalcogenide glasses, amorphous silicon and germanium, liquid rare gases and water. The problems we have to ask ourselves are:

(a) What is the form of the density of states, and under what conditions can a band-gap exist?

(b) Is there always a range of localized states at the extremities of a band, and how wide is it?

(c) Is there a finite density of (localized) states at the Fermi level, and if so is it due to specific deviations from the bonded structure of an ideal glass ("dangling bonds")?

As is now well known, we have also to differentiate between the properties of materials, particularly evaporated Si and Ge, which contain voids and those which do not. Although we do not think that *statistical* fluctuations in composition will lead to potential fluctuations opaque to tunnelling (Ref. 70), charges on voids may well do so, and in our view it is only in this case—or for glasses in which two components have seggregated—that the "percolation edge" described in Professor Fritzsche's article exists.

Our answer to the first question—is there a gap in the density of states?—is that if no limits are set to the fluctuations in some parameter which determines the disorder, for instance to the thermal fluctuations in the density of a liquid, then in principle the conduction and valence bands will have infinite tails, so there is no real gap; but the fact that water and liquid rare gases transmit light shows that for them the density of states in the gap is entirely negligible. If limits are set on the fluctuations, a gap can certainly exist, as was first shown for the one-dimensional case by Borland[95] and in three dimensions by Taylor.[96] The situation for amorphous semiconductors such as silicon and germanium, which presents considerable problems, will be discussed later in this section.

On the second question—whether there is always a range of localized states at the bottom of a band—we believe that if the values of any parameter such as interatomic distance, bond angle and so on are statistically independent, even though they fluctuate in a limited range, there will always be a range of localized states if the energy of the bottom of the conduction band depends on these parameters, for the same reasons as for the Anderson model. The range may however be small compared with kT and in this case will have no appreciable effect on the mobility; some examples will be given.

1.13.2. RANDOM ARRAY OF POTENTIAL WELLS

In this Section we consider a random array of potential wells, all of the same depth. The problem is relevant to the impurity-band of an uncompensated semiconductor. It has been treated by Gaspard and Cyrot–Lackmann,(97) by Katz and Rice(98) and by Roth(99) (by the tight-binding method). A characteristic plot of density of states against energy is shown in Fig. 1.22. All calculations show an unsymmetrical band with a low energy tail. This is because the ground state for two charged donors, when close together, will give a localized state of energy $-4W$ compared with the energy $-W$ of a single donor, while the antibonding state of odd parity turns into a p state with energy $-W$. In an early paper Lifshitz(100) suggested that these bonding and antibonding states might produce an "erosion minimum". More recent work(97) shows that this occurs if at all only for small values of the number of centres ($\alpha a \gg 1$), and it is probably then not observable because of the large effect of correlation (Section 1.2).

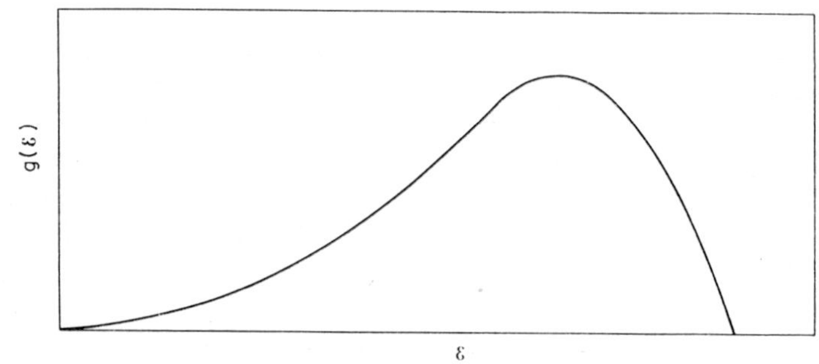

FIG. 1.22. Density of states for wells at random in space (lateral disorder).

The condition for Anderson localization in this case has been estimated roughly by Mott(28) as follows.† Following Lifshitz, we pair each atom with its nearest neighbour, which we suppose to be at a distance r_1 from it, given by

$$(4\pi/3)\, r_1{}^3 = 1/N,$$

where N is the number of centres per unit volume. The energy of an electron resonating between any two centres in a pair is $I_0 \exp(-\alpha r_1)$ and, since r_1 varies from pair to pair, we take this quantity as V_0 in the Anderson formula (a *very* rough approximation). The mean distance between pairs is

$$r_2 = 1/(\tfrac{1}{2}N)^{1/3},$$

† The estimate here is revised, following the revised Anderson localization criterion.

so we suppose the band width B to be

$$2zI_0 \exp(-\alpha r_2),$$

and, putting in the Anderson criterion $V_0 = 2B$, we have

$$\exp(-\alpha r_1) = 4z \exp(-\alpha r_2).$$

Thus

$$\alpha(r_2 - r_1) = \ln 4z \simeq 3{\cdot}2,$$

whence

$$\alpha N^{-\frac{1}{3}} \left\{ \left(\frac{1}{2}\right)^{\frac{1}{3}} - \left(\frac{3}{4\pi}\right)^{\frac{1}{3}} \right\} \simeq 3{\cdot}2 \tag{1.54}$$

or, if we write $a_H = \alpha^{-1}$, $N^{\frac{1}{3}} a_H = 0{\cdot}05$. This is a value much below that predicted for the metal–insulator transition (eqn (1.15)), and suggests that overlap between the two Hubbard bands may be smaller when localization occurs.

This calculation shows that for *s*-orbitals a range of localized states depends on *large* fluctuations in the distance between atoms, so that for the conduction band in liquid rare gases, for example, the range of localized states should be very small. There is ample evidence of this, for example, from the high mobilities of electrons injected into liquid argon. We have seen that for carriers with energies at the "mobility edge", mobilities of order 1 to 10cm^2/V sec are expected, while in liquid argon values as high as 500 cm^2/V sec have been measured (Miller *et al.*[101] and other references in Mott and Davis[27] (p. 78)), which can be accounted for as due to scattering by fluctuations in density. Another example is vitreous silica (SiO_2); the wave function in the conduction band should be 3*s*-like on the O^{2-} ions, and a long mean free path is not surprising. Fairly high mobilities have been measured for electrons injected into the conduction band; thus Williams,[102] by measuring the drift distance for injected electrons before trapping and assuming trapping by charged centres with cross sections $1{\cdot}3 \times 10^{-12}$ cm^2, estimates a microscopic conductivity mobility 17–34 cm^2/V sec. Perhaps more conclusive is the work of Goodman,[103] who has measured directly the Hall effect for electrons photoemitted into thermally grown layers of silicon dioxide from 3·5–6·6 μm thick. The values of the Hall mobility measured were in the range 6·8–62·5 cm^2/V sec with a mean of 29 cm^2/V sec. There is thus *no* large difference between conductivity and Hall mobility, as there should be according to Friedman (Section 1.10) for electrons with energies at the mobility edge (see also Mort *et al*[127]).

In a random array of atoms in *p* or *d* states, on the other hand, the position is quite different, and the present author has suggested that a range of localized states ought to exist, even if the fluctuation in the interatomic distance between nearest neighbours is neglected. This is because at the bottom of a band random *orientations* of the wave functions lead to random fluctuations of the overlap integral I. In such cases, where the band width B and the random field V are determined by the same quantity, the considerations of Section 1.7 suggest that $g(\mathcal{E}_c)$ should be *about* 1/10 of the mean density of states in the band. An example of this kind of behaviour may be provided

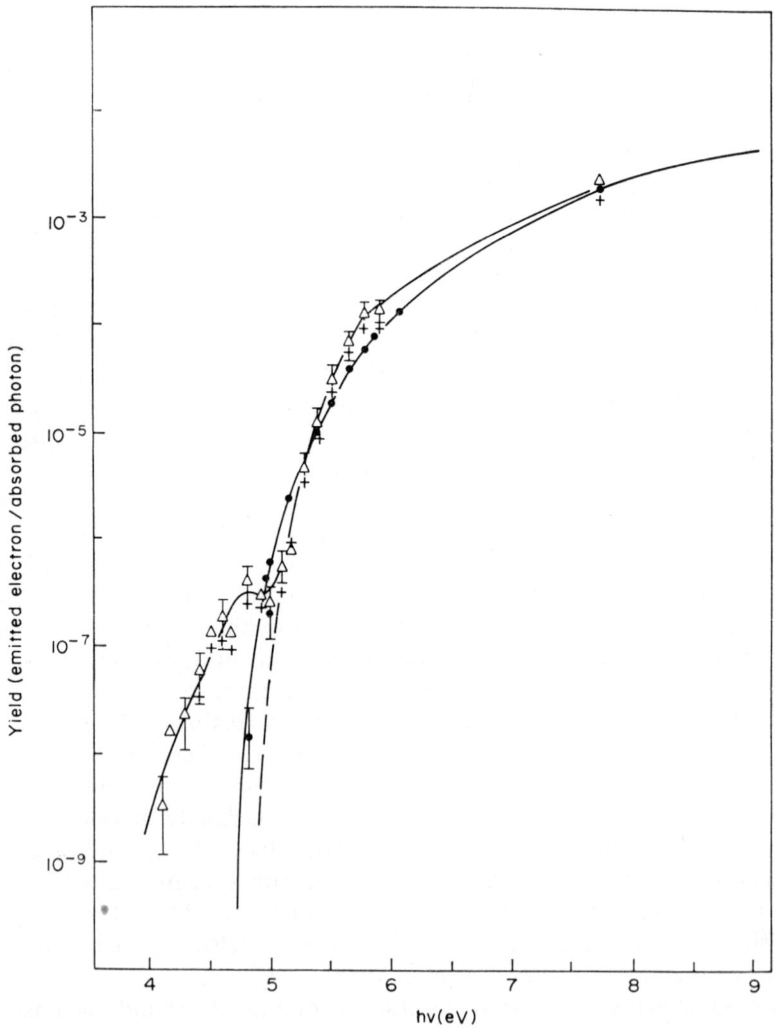

FIG. 1.23. Quantum yield of crystalline vacuum-cleaved trigonal —Te(●) and amorphous Te films ((△, +) Laude *et al.*, Ref. 104).

by the observations of Laude, Willis and Fitton[(104)] who compare photoemission from amorphous tellurium, which takes place from the second (d-like) conduction band, with that from vacuum cleaved crystalline Te. The yield (emitted electrons per absorbed photon) for both cases is shown in Fig. 1.23; these authors get good agreement by assuming a jump in the mobility in the amorphous case by $\sim 10^3$ at the edge.

1.13.3. SILICON, GERMANIUM AND CHALCOGENIDE GLASSES

Next we turn to amorphous silicon, germanium and the chalcogenide glasses. For these the evidence for a range of localized states leading up to a mobility edge is the following:

(a) For the conduction band of amorphous silicon, the work of Le Comber and Spear[(105,106)] suggests that the temperature-dependence of the electron drift mobility is controlled by shallow trapping levels (localized states) with energies located about 0·2 eV below the mobility edge.

(b) The existence of small, negative Hall coefficients in cases when the thermopower is p-type.[(107,108,109)] This aspect is discussed in Emin's article. The anomalous sign is expected whether electrons are in extended or localized states.

(c) The comparatively large and constant values of σ_0 when the conductivity varies with T according to the equation $\sigma_0 \exp(-\mathcal{E}/kT)$, as it does for chalcogenide glasses.

In discussing real materials, we must first reemphasize that in much work on evaporated amorphous silicon and germanium, the specimens are known to contain voids, and surface states (dangling bonds) on the surface of these voids are responsible for e.s.r. signals. These surfaces will probably be charged, and so random fields varying over large distances will exist, giving rise to conducting channels for carriers in the conduction and valence bands. If conductivity is due to such carriers, percolation theory would be appropriate for treating the current, as described in the article by Fritzsche in this volume† (Chapter 2). Probably however in such materials, particularly germanium, conduction is normally due to "states in the gap" and of $T^{\frac{1}{4}}$ form. The random fields have then a different effect, as we shall see. We shall

† The model of Fritzsche with long-range fields has an interesting consequence for the a.c. conductivity. Suppose $f (f \lesssim 1)$ is the volume fraction of the material that is in pockets with energies several multiples of kT below the percolation limit. Then a field E will produce a polarization $E/4\pi$ and thus a mean current $\omega f E/4\pi$, as long as $\sigma > \omega$, where ω is the frequency. If $\sigma > \omega$, the current will drop rapidly. Thus the conductivity is $f_T \omega/4\pi$ where f_T is the volume for which the inequality $\sigma > \omega$ applies. f_T should drop rapidly with decreasing T, unlike the Pollak–Geballe term or any contribution from ionic movement. As far as we know, such a temperature-dependent term has not been observed.

for the moment suppose that there are no voids, and that the long-range fields described by Fritzsche are absent.

We consider first amorphous silicon and similar materials in the "perfect" glassy state, such as that described by Polk[110] in which all bonds are satisfied at the expense of bond bending by $\sim 20\%$ (cf article by Grigorovici in this volume). Weaire and Thorpe,[111] Thorpe and Weaire[112] and Heine[113] treat this system in the following way. They take bonding and antibonding orbitals ψ_B, ψ_A for each pair of atoms, the wave functions in valence and conduction bands being of the form

$$\Sigma a_i \psi_{Bi}, \qquad \Sigma b_i \psi_{Ai},$$

the summation being over all pairs (i). a_i, b_i are coefficients which are not defined, though in our view they may well vary in a random way from atom to atom. The band widths are solely due to interactions on a given atom, and this leads to two flat bands as shown in Fig. 1.24, the energy being independent of the phases of a_i. Heine shows rigorously that with this model a gap can exist, whatever the values of the a_i.

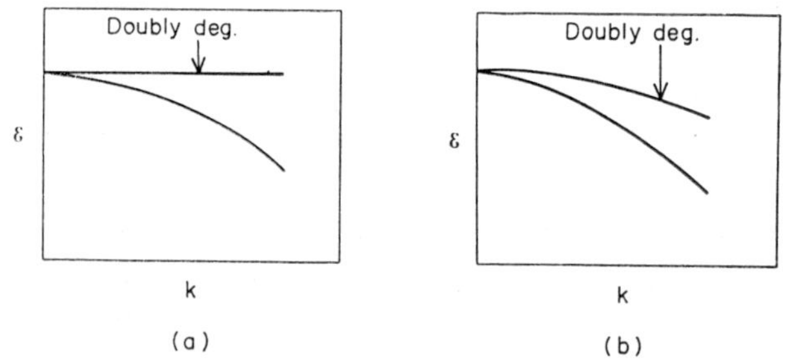

FIG. 1.24. Valence bands for (a) amorphous (after Thorpe and Weaire, Ref. 112) and (b) crystalline germanium.

Bringing in nearest and next nearest neighbours leads in the crystal to the well-known valence band diagram of Fig. 1.24(b).[6] It is not at all clear what happens in the amorphous material, since the interactions are large. Ziman[114] in a speculative paper states that the fields due to next nearest neighbours are insufficient to produce localization. The contrary view seems plausible too; since the nearest-neighbour approximation gives infinite effective mass, the conditions are exceptionally favourable for localization. It is true that the distances of next nearest neighbours will not vary much, but the orientations will, and according to Ref. 6 the third nearest neighbours give large overlap integrals, and the orientation of second nearest neighbours gives a random term too. On the other hand the experimental evidence does seem to favour the absence of strong tails in the valence and conduction bands. Thus Pierce and Spicer[115] show by photoemission that in "perfect" amorphous silicon the valence band edge is sharp and the gap little changed from the pure material. So we are inclined to think that bringing more distant neighbours into the tight binding approximation does not produce tails, but *may* produce a range of localized states. There will also be short-range fluctuations in the effective potential due to fluctuations in bond angle and bond length, which are bound to produce a small "tail" with localization, but probably too weak to be observed in photoemission.

We turn next to the existence of an energy gap. For materials such as liquid rare gases, for which the tight-binding approximation is appropriate, the problem is trivial. As shown by various investigators (see Fletcher[116]) quite early in the development of the subject, if limits are set on the fluctuations in the distance between atoms, or in the density, a gap can exist. If not, in general there will be localized states in the gap. Thus thermal fluctuation will in principle produce such levels—but the intensity must be negligible, since for example hot water is transparent. The analysis which shows the existence of a gap when fluctuations are limited is as follows: Consider an atom with states of energy $\mathcal{E}$, $\mathcal{E}'$. On the tight binding approximation these are spread out into bands of width

$$B = 2\sum_j I_{ij}, \qquad B' = 2\sum_j I_{ij}' \tag{1.55}$$

where I_{ij}, I_{ij}' are the transfer integrals to a neighbouring atom j. If two atoms come very near together, B and B' can be large and $\frac{1}{2}(B + B')$ may exceed $\mathcal{E}' - \mathcal{E}$; if so, some localized states will exist in the gap. But if limits are set to the fluctuation of the interatomic distance, so that $\frac{1}{2}(B + B')$ is always less than $\mathcal{E}' - \mathcal{E}$, then a band-gap will exist and the material will be transparent to light of low enough frequency. This treatment is similar to that of Taylor.[96]

This suggests that the tight-binding approach could be the best way to treat amorphous semiconductors, but as we have seen for covalent materials like silicon and the chalcogenide glasses the matter is not so simple. The obvious approach would be to start with bonding and antibonding wave functions, using again the tight binding approximation. However, in crystals this has not proved a satisfactory method of obtaining rapidly convergent results for numerical computation. Approximations of the nearly free electron type, starting with plane waves and using pseudopotentials, are reviewed by Greenwood in this volume; though successful in giving the position of the gap they have not yet proved useful for discussing properties at the band edge.

The experimental situation can give us some further indication of what approximations could be useful. In the introduction we have mentioned one of the most important ways in which covalent glasses differ from crystals. They cannot be doped, and the band gap is not influenced by small changes in composition. The current explanation is that in a glass each atom surrounds itself by a number of atoms equal to its valency; thus Ge will surround itself by four atoms, Se by 2, As by 3 and so on. The first indication of this kind of explanation is due to Gubanov[117] in 1963, who states (English translation p. 213) that "due to a rearrangement of atoms surrounding an impurity atom, the energy spectrum is smoothed out so that free levels approach the filled band . . ." In the terms given here, the explanation is due to Mott.[28] A similar idea is put forward by Phillips,[118] who says that in materials like amorphous germanium a self-consistent arrangement of atoms which maximises band energies can eject states which otherwise fall in the band tail to above the band edges. The insulating property depends then strongly on the structure. Indeed there is much evidence that this is so. Thus crystalline and amorphous silicon with coordination number 4 are non-metals, but liquid silicon with coordination number 6 is a metal. In fact in many liquids, as the temperature is raised, the coordination number increases and we can trace a gradual transition from one state to another (the pseudogap model). Moreover, bombardment by heavy ions by destroying the coordinated structure can increase the conductivity of some amorphous materials by many orders of magnitude.

The experiments do suggest, then, that a satisfactory theory of band gaps ought to start from the concept of the chemical bond, and bonding and antibonding orbitals, as in the work of Thorpe and Weaire.[112] We are however convinced, following Phillips' arguments, that the question of a band-gap in such materials as Ge, where the gap is small, is very intimately connected with the structure. Phillips even suggests that configurations which would push a valence band state into the gap are absent because they are energetically unfavourable; but such an explanation would not apply to conduction band tails. We think the problem of "tails" is essentially unsolved.

1.13.4. STATES IN THE GAP

We believe that in view of recent experimental evidence, particularly that of Spicer[115] and his colleagues on photoemission, that "perfect" amorphous Si and Ge have sharp band edges, but that though an "intrinsic" range of localized states can exist near the edges, states in the gap are normally due to structural defects ("dangling bonds"). The original model of Cohen, Fritzsche and Ovshinsky[119] introduced states into the gap by postulating "tails" to conduction and valence bands which could overlap, as in Fig. 1.25. The tail from the conduction band would consist of donors and the tail from the valence band of acceptors. Thus the higher acceptor states would give up an electron to the lower donors, producing positively charged donors and negative acceptors. The overlapping bands will moreover lock the Fermi energy near the middle of the band. This model, together with the assumption of "mobility edges", is known as the CFO model. It predicts that at high temperatures the material will behave like an intrinsic semiconductor, but since $g(\mathcal{E}_F)$ is finite, at low temperatures current will be carried by electrons at the Fermi energy, and the $T^{\frac{1}{4}}$ law (Section 1.5) is to be expected. A finite value of $g(\mathcal{E}_F)$ for any specimen which shows the Pollak–Geballe form of the a.c. conductivity has to be assumed.†

Davis and Mott[121] (see also Ref. 27) in an attempt to reconcile the high value of $g(\mathcal{E}_F)$ deduced from the a.c. conductivity‡ and other measurements in

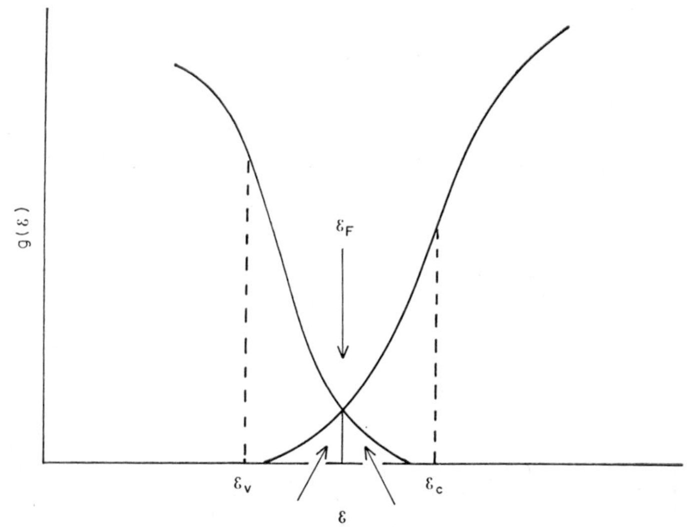

FIG. 1.25. Density of states in an amorphous semiconductor according to the CFO model; arrows show ranges of $\mathcal{E}$ where states are charged.

chalcogenide glasses with the low optical absorption coefficient in the infra-red, suggested that there might be a peak in the density of states in the middle of the gap, and we propose a model as to how this might occur. Assuming

† There is some possibility that this may be due to soft phonons in glasses (see Pollak and Pike;[120] see also articles by Fritzsche and by Davis in this volume).

‡ If the CFO tails are due to long range fluctuations in the potential, producing potential hills and valleys too opaque for tunnelling, the $T^{\frac{1}{4}}$ law would not occur and a percolation edge has to be defined as in Fritzsche's article in this volume. As stated earlier, we believe that such long-range fluctuations only exist if the material contains macroscopic voids with charged surface states or segregation into two components. Arguments that statistical fluctuations do *not* produce opaque barriers are given by the present author.[70]

there are no states in the gap when the co-ordination number is preserved, states can be introduced by "dangling bonds". Whether these are to be thought of as on the surface of *very* small voids or simply unsaturated bonds in the body of the glass is uncertain, and perhaps not essential to the argument, The important assumptions, at any rate for Si and Ge and in general materials without lone pair states, are as follows.

A dangling bond can provide a donor below the middle of the gap and an acceptor above it; but the disorder, and particularly random fields due to charged cavities, gives a range of energies for both, so that the two bands overlap, and some dangling bonds acquire a positive and some a negative charge. In this sense the model is like the CFO model. The density of states is illustrated in Fig. 1.26; $g(\mathcal{E}_F)$ is finite and the Fermi energy is "pinned".

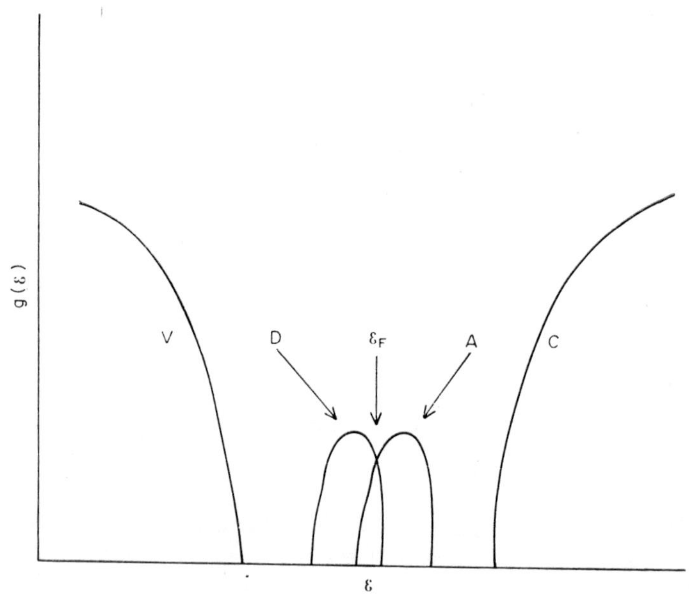

FIG. 1.26. Density of states in amorphous semiconductors in revised model, with dangling bonds at vacancies. *V* is the valence band, *C* the conduction band, *D* are deep donor states due to neutral dangling bonds, *A* acceptor states produced by double occupation.

The absorption line at 0·27 eV observed by Chittick[122] (see also Mott and Davis p. 302) and by Tauc *et al.*[123] in glow-discharge-deposited Ge may perhaps be ascribed to transitions between these states.

A very similar model has been proposed for surface states on crystalline silicon by various authors, who recognize that dangling bonds can act as donors or acceptors (traps) with higher energy, see for instance Chung,[124] who correlates the e.s.r. signal with the number of dangling bonds. The energy required to transfer an electron from the neutral band to another centre is the Hubbard energy U, and should be several eV; but it may be greatly reduced by the distortion of the bonds round the two centres when an electron is transferred. Band theory calculation such as that of Jones[(125)] for Si and Elices and Yndurain[(126)] for Ge will be appropriate only if the Hubbard term U is not large enough to split the band.

A rapid increase in conductivity on bombardment of amorphous germanium, silicon and III–V compounds by positive ions is observed by Olley and Yoffe;[(126)] this seems in accord with the model of dangling bonds or disturbed regions. We expect the conductivity to follow the $T^{\frac{1}{4}}$ law, and this is observed.

It is not, however, certain that a $T^{\frac{1}{4}}$ law or something similar *necessarily* implies variable range hopping. Various authors (for example Croitoru *et al.*[(129)]) have pointed out that if the product $g(\mathcal{E})\ \mu(\mathcal{E})$ is a minimum at $\mathcal{E}_F$ and increases rapidly with $\mathcal{E} - \mathcal{E}_F$, a conductivity of the form $\ln \sigma \alpha B T^{-1/n}$ may occur. In fact, with suitable assumptions it can be made to yield $T^{\frac{1}{4}}$, as was first shown by Vescan, Telnic and Popescu.[(130)] Suppose we write

$$\sigma = \int g(\mathcal{E})\ \mu(\mathcal{E})\ \mathrm{e}^{-\mathcal{E}/kT}\ \mathrm{d}\mathcal{E} \tag{1.56}$$

and assume that $\mu(\mathcal{E}) \sim \exp(\alpha\mathcal{E}^s)$ and that most of the integral occurs from the region where $\alpha\mathcal{E}^s - \mathcal{E}/kT$ is a maximum, so that

$$s\mathcal{E}^{s-1} - 1/kT = 0. \tag{1.57}$$

If $s = {}^{-\frac{1}{3}}$ this will give a $T^{\frac{1}{4}}$ behaviour.

Unfortunately, however, a derivation of this kind is only possible if μ is independent of T, or at any rate if it does not contain a hopping activation energy, as it should in the model given here. We think then that an alternative derivation of the $T^{\frac{1}{4}}$ law of this kind is possible only if defects in the gap are so dense that states are non-localized.

However, a minimum in $g(\mathcal{E})$ at the Fermi energy could lead to an equation of type (1.57), so that electrons at the Fermi level contribute little to the current, which is mainly carried by excited electrons with an energy given by eqn (1.57). If so the model should give for strong fields a variation of the number of carriers according to the Poole–Frenkel formula [eqn (1.37)], and hence

$$\sigma \propto E \exp(-\tfrac{1}{2}\beta E^{1/2}/kT). \tag{1.58}$$

A behaviour of this kind has been observed by Croitoru, Niklas and Stuke[(131)] for amorphous InSb.

This model can perhaps also be applied to the observations of Jonscher and Ansari[132] on strong field conduction in SiO, which shows low conductivity and Poole–Frenkel behaviour (cf. Hill[133]). The conductivity is thought to be hopping conduction between states in the gap; these authors distinguish between donors (neutral dangling bonds) and traps, which might be in our view dangling bonds for which the energy of an *extra* electron is low. For an electron hopping between the latter sites, there will be a temperature-dependent pseudo-mobility edge given by (1.57), and Poole–Frenkel behaviour for emission of electrons from dangling bonds to this "edge" is expected. We should expect a current of the form (1.58).

We now ask whether it is likely that the density of states at $\mathcal{E}_F$ shows a minimum, and whether there is evidence for it. If the broadening is due to random fields (rather than random environment), then we would expect the shape of both bands to be the same, so that $\mathcal{E}_F$ would lie in the middle at the minimum. Moreover, if diffusion is possible, positively and negatively charged centres would tend to diffuse away from positions of high energy, thus sharpening the minimum (Fritzsche, private communication). As regards experimental evidence, tunnelling is the most direct; but at a junction of the type Al/Al_2O_3/amorphous semiconductor, a minimum current at zero bias is observed[134] for so wide a range of amorphous semiconductors (Fritzsche, private communication) that it is doubtful whether internal, as opposed to surface states, are responsible. Sauvage, Mogab and Adler,[135] however, have observed tunnelling into amorphous silicon and the temperature variation obeys the $T^{\frac{1}{4}}$ law, showing that internal states are involved; they also find a minimum for zero bias, which seems to give evidence for a density of internal states with a minimum at $\mathcal{E}_F$.

The applicability of this model to chalcogenides, where the upper valence band is thought to be due to lone pair states, is more doubtful; it is thought that a dangling bond would produce a donor in a lone-pair† valence band. Since energy is needed to separate a positive and negative charge, a broken bond or end of a selenium chain must produce an acceptor, but there is no obvious reason why it should lie near the centre of the gap. In glasses with several components, a wide range of localized states at the band edges is probable and conduction may be by carriers in the localized states at the extremities of the bands. This is confirmed by the evidence for an activation energy in the Hall mobility described by Emin in this volume, which may be due to Anderson localization accentuated by polaron formation.

As emphasized by Fritzsche in this volume, the very good straight-line plot of $\ln \sigma$ against $1/T$ is evidence that the Fermi energy is pinned and these materials are not truly intrinsic. A possible model for this pinning by

† Lone pairs are *p* electrons not concerned in bonding.

states in the gap is as follows: Each dangling bond may produce, as we have seen, an acceptor. But the unpaired electron in the bond, because its orbital is more localized, will raise the energy of the lone pair state, so that in principle a low donor containing *two* electrons is produced below the acceptor produced by the dangling bond. Random environment could cause these to overlap (see also Kastner[138]).

Fritzsche (this volume) estimates from various evidence, a concentration $g(\mathcal{E}_F)$ in chalcogenide glasses of $3 \times 10^{18}\ \text{cm}^{-3}\ (\text{eV})^{-1}$, which is not out of line with the values deduced by Davis and Mott from the a.c. conduction. The absence of an e.s.r. signal and a rise in the susceptibility at low temperatures in specimens that do not contain Fe is a puzzle.[136,137] $g(\mathcal{E}_F)$ does not of course give the total number of free spins; if the bands of Fig. 1.26 are narrow and overlap considerably, the total density may be as low as $\sim 10^{17}\ \text{cm}^{-3}$. The discrepancy between the density of states $g(\mathcal{E}_F)$ in chalcogenides estimated in various ways and the absence of an e.s.r. signal or low-temperature paramagnetism seems to us the major contradiction in interpreting the evidence. A possible explanation is that the "dangling bonds" are not distributed at random but segregated on surfaces, and that they are near enough to interact and produce an "amorphous antiferromagnet" which would not give Curie–Weiss behaviour at low temperatures (compare the behaviour of Au–Fe observed by Cannella *et al.*[85]). At room temperature however an e.s.r. signal would be expected, and some other explanation, such as a spread in the g-values or short lifetime would be needed to explain its absence.

References

1. Ziman, J. M. *et al.*, "Physics Today", May 1972, p. 9.
2. Kolomiets, B. T., *Phys. Stat. Solidi* **7**, 359 and 713 (1964).
3. Ovshinsky, S. R., *Phys. Rev. Letters* **21**, 1450 (1968).
4. Fritzsche, H. and Ovshinsky, S. R., *J. Non-Cryst. Solids* **4**, 464 (1970).
5. Drake, C. F. and Scanlan, I. F., *J. Non-Cryst. Solids* **4**, 234 (1970).
6. Dresselhaus, C. and Dresselhaus, M. S., *Phys. Rev.* **160**, 649 (1967).
7. Anderson, P. W., *Phys. Rev.* **109**, 1492 (1958).
8. Sommerfeld, A. and Bethe, H. A., *Handb. d. Physik* **24/1**, 352 (1933).
9. Mott, N. F. and Jones, H., "Theory of the Properties of Metals and Alloys", Oxford, 1936.
10. Ziman, J. M., *Phil. Mag.* **6**, 1013 (1961).
11. Faber, T. E., "Liquid Metals". Cambridge, 1972.
12. Kohn, W., *Phys. Rev.* **105**, 509 (1967).
13. Slater, J. C., "Quantum Theory of Molecules and Solids". McGraw-Hill, 1963 Vol. 1.
14. Hubbard, J., *Proc. R. Soc.* **A277**, 237 (1964).
15. Cyrot, M., *J. de Physique* **33**, 125 (1972).
16. Cyrot, M., *Phil. Mag.* **25**, 1031 (1972).

17. Schrieffer, J. R., Evenson, W. E. and Wang, S. Q., *J. de Physique* **32,** C1–19 (1971).
18. Brinkman, W. F. and Rice, T. M., *Phys. Rev.* **B2,** 4302 (1970).
19. Rice, T. M., McWhan, D. B. and Brinkman, W. F., Proc. Tenth Intern. Conf. on Physics of Semiconductors, Cambridge (Mass.) p. 293 (1970).
20. Caron, L. G. and Kemeny, G., *Phys. Rev.* **B3,** 3007 (1971).
21. Anderson, P. W., *Phys. Rev.* **86,** 694 (1952).
22. Nagaoka, Y., *Phys. Rev.* **147,** 392 (1966).
23. Brinkman, W. F. and Rice, T. M., *Phys. Rev.* **B2,** 1324 (1970) and **B4,** 1566 (1971).
24. Ohata, N., *J. Phys. Soc. Japan* **29,** 1138 (1970).
25. Kasuya, T., *Solid State Commun.* **8,** 1635 (1970).
26. Mott, N. F. and Zinamon, Z., *Rep. Prog. in Physics* **33,** 881 (1970).
27. Mott, N. F. and Davis, E. A., "Electronic Processes in Non-Crystalline Materials". Oxford, 1971.
28. Mott, N. F., *Phil. Mag.* **20,** 1 (1969).
29. Mott, N. F., *Phil. Mag.* **24,** 935 (1971).
30. Mott, N. F., *Phil. Mag.* **6,** 287 (1961).
31. Kohn, W., *Phys. Rev. Lett.* **19,** 439 (1967).
32. Mott, N. F. and Davis, E. A., *Phil. Mag.* **17,** 1269 (1968).
33. Goodenough, J. B., "Magnetism and the Chemical Bond". Interscience, 1963
34. Goodenough, J. B., "Metallic oxides". In "Progress in Solid-State Chemistry" **5,** 145(1971).
35. Goodenough, J. B., *Phys. Rev.* **B5,** 2764 (1972).
36. Luttinger, J. M., *Phys. Rev.* **119,** 1153 (1960).
37. McWhan, D. B., Menth, A., Remeika, J. P., Brinkman, W. F. and Rice, T. M. *Phys. Rev.* **B7,** 1920 (1973).
38. Rice, T. M. and Brinkman, W. F., *Phys. Rev.* **B5,** 4350 (1972).
39. Ramirez, R., Falicov, L. M. and Kimball, J. C., *Phys. Rev.* **B2,** 3383 (1970).
40. Kimball, J. C., *Phys. Rev. Lett.* **20,** 117 (1972).
41. Miller, A. and Abrahams, F., *Phys. Rev.* **120,** 745 (1960).
42. Twose, W. D., Ph.D. Thesis, University of Cambridge (1959).
43. Mott, N. F. and Twose, W. D., *Adv. in Phys.* **10,** 107 (1961).
44. Anderson, P. W., *Proc. Nat. Acad. Sci. USA* **69,** 1097 (1972).
45. Abram, R. A. and Edwards, S. F., *J. Phys. C.* **5,** 1183 (1972).
46. Freed, K. F., *Phys. Rev.* **B5,** 4802 (1972).
47. Edwards, S. F., *Phil. Mag.* **6,** 617 (1961).
48. Economou, E. M. and Cohen, M. H., *Phys. Rev. Lett.* **25,** 1445 (1970).
49. Economou, E. M. and Cohen, M. H., *Mat. Res. Bull.* **5,** 577 (1970).
50. Economou, E. M. and Cohen, M. H., *Phys. Rev.* **B5,** 2931 (1972).
51. Anderson, P. W., *Comments on Solid State Physics* **2,** 193 (1970).
52. Edwards, J. T. and Thouless, D. J., *J. Phys. C.* **5,** 807 (1972).
53. Schönhammer, K., *Physics Lett.* **36A,** 181 (1971).
54. Kikuchi, M. J., *Phys. Soc. Japan* **29,** 296 (1970).
55. Halperin, B. I. and Lax, M., *Phys. Rev.* **148,** 722 (1966).
56. Zittartz, J. and Langer, J. S., *Phys. Rev.* **148,** 741 (1966).
57. Kirkpatrick, S., *J. Non-Cryst. Solids* **8-10,** 160 (1972).
58. Pollak, M., *J. Non-Cryst. Solids* **8-10,** 486 (1972).
59. Mott, N. F., *J. Non-Cryst. Solids* **1,** 1 (1968).
60. Ambegeokar, V., Halperin, B. I. and Langer, J. S., *Phys. Rev.* **B4,** 2612 (1971).

61. Hamilton, E. M., *Phil. Mag.* **26,** 1043 (1972).
62. Pollak, H. and Geballe, T. H., *Phys. Rev.* **122,** 1742 (1961).
63. Englman, R. and Jortner, J., *Molecular Physics* **18,** 145 (1970).
64. Nitzan, A. and Jortner, J., *J. Chem. Phys.* **56,** 2079 (1972).
65. Mott, N. F., *Adv. in Phys.* **16,** 49 (1967).
66. Mott, N. F., *Phil. Mag.* **19,** 835 (1970).
67. Hindley, N. F., *J. Non-Cryst. Solids* **5,** 17 and 31 (1970).
68. Friedman, L., *J. Non-Cryst. Solids* **6,** 329 (1971).
69. Cohen, M. H., *Physics Today* **24,** 26 (1971).
70. Mott, N. F., *Phil. Mag.* **26,** 1015 (1972).
71. Hensel, F. and Franck, E. U., *Rev. Mod. Phys.* **40,** 697 (1968).
72. Even, J. and Jortner, J., *Phys. Rev. Lett.* **28,** 31 (1972).
73. Mott, N. F., *Phil. Mag.* **26,** 505 (1972).
74. Warren, W. W., *Phys. Rev.* **B3,** 3708 (1971).
75. Mott, N. F., *Adv. in Phys.* **21** (1972) (to be published).
76. Ferrier, R. P. and Herrell, D. J., *Phil. Mag.* **19,** 853 (1969).
77. Cohen, M. H., *J. Non-Cryst. Solids* **4,** 391 (1970).
78. Cutler, M., *Phil. Mag.* **25,** 173 (1972).
79. Friedman, L. and Mott, N. F., *J. Non-Cryst. Solids* **7,** 103 (1972).
80. Acrivos, J. V., *Phil. Mag.* **25,** 753 (1972).
81. Holstein, T., *Phil. Mag.* **27,** 225 (1973).
82. Nagels, P., Callaerts, R. and Denayer, M., Procs 11th Intern. Conf. on Semiconductors (Warsaw), (1972).
83. Hung, C. S. and Gliessman, J. R., *Phys. Rev.* **79,** 726 (1950).
84. Mott, N. F. in "Cooperative Phenomena" (ed. M. Wagner), to be published (1973).
85. Cannella, V., Mydosh, J. A. and Budnick, J. I., *J. Appl. Phys.* **42,** 1689 (1971).
86. Davis, E. A. and Compton, W. D., *Phys. Rev.* **A140,** 2183 (1965).
87. Allen, F. and Adkins C. J., *Phil. Mag.* **26,** 1027 (1972).
88. Morigaki, K. and Maekawa, S., *J. Phys. Soc. Japan* **32,** 462 (1972).
89. Wolf, E. L., Losee, D. L., Cullen, D. E. and Dale Compton, *Phys. Rev. Lett.* **26,** 438 (1971).
90. Ue, H. and Maekawa, S., *Phys. Rev.* **B3,** 4232 (1971).
91. Yamanouchi, C., Mizuguchi, K. and Sasaki, W., *J. Phys. Soc. Japan* **22,** 859 (1967).
92. Krieger, J. B., Meeks, T. and Esposito, E., *Phys. Rev.* **B5,** 1499 (1972).
93. Banus, M. D. and Reed, T. B., "The Chemistry of Extended Defects in Non-metallic Solids" (ed. Le Roy Eyring and M. O'Keefe) North-Holland, p. 388 1970.
94. Banus, M. D., Reed, T. B. and Strauss, A. C. J., *Phys. Rev.* **B5,** 2764 (1972).
95. Borland, R. E., *Proc. Phys. Soc.,* **78,** 926 (1961).
96. Taylor, P. L., *Proc. Phys. Soc.* **88,** 753 (1966).
97. Gaspard, J. P. and Cyrot-Lackmann, F., *J. Phys. C.* (*Sol. State Phys.*) **5,** 3047 (1972).
98. Katz, I. and Rice, S. A., *J. Phys. C.* **5,** 1165 (1972).
99. Roth, L. M., *Phys. Rev. Lett.* **28,** 1570 (1972).
100. Lifshitz, I. M., *Adv. in Phys.* **13,** 483 (1964).
101. Miller, L. S., Howe, S. and Spear, W. E., *Phys. Rev.* **166,** 871 (1968).
102. Williams, R., *Phys. Rev.* **A140,** 569 (1965).

103. Goodman, A. M., *Phys. Rev.* **A164,** 1145 (1967).
104. Laude, L. D., Willis, R. F. and Fitton, B., *Phys. Rev. Lett.* **29,** 472 (1972).
105. LeComber, P. G. and Spear, W. E., *Phys. Rev. Lett.* **25,** 509 (1970). also LeComber, P. G., Madan, A. and Spear, W. E., *J. Non-Cryst. Solids,* **11,** 219 (1972).
106. LeComber, P. G., Madan, A. and Spear, W. E., Chapter 10 in this volume.
107. Male, J. C., *Br. J. Appl. Phys.* **18,** 1543 (1967).
108. Clark, A. H., *Phys. Rev.* **154,** 750 (1967).
109. Emin, D., Seager, C. H. and Quinn, R. K., *Phys. Rev. Lett.* **28,** 813 (1972).
110. Polk, D. E., *J. Non-Cryst. Solids* **5,** 365 (1971).
111. Weaire, D. and Thorpe, M. F., *Phys. Rev.* **B4,** 2508 (1971).
112. Thorpe, M. F. and Weaire, D., *Phys. Rev.* **B4,** 3518 (1971).
113. Heine, V., *J. Phys. C.* **4,** L221 (1971).
114. Ziman, J. M., *J. Phys. C.* **4,** 3129 (1971).
115. Pierce, D. T. and Spicer, W. E., *Phys. Rev.* **B5,** 3017 (1972).
116. Fletcher, N. H., *Proc. Phys. Soc.* **91,** 724 (1967).
117. Gubanov, A. I., "Quantum Electron Theory of Amorphous Conductors (1963)", translation by Consultants Bureau, New York 1965.
118. Phillips, J. C., *Phys. Stat. Sol.* **B44,** K1 (1971); Comments on Solid State Physics, **IV,** 9 (1971).
119. Cohen, M. H., Fritzsche, H. and Ovshinsky, S. R., *Phys. Rev. Lett.* **22,** 1065 (1969).
120. Pollak, M. and Pike, G. E., *Phys. Rev. Lett.* **28,** 1449 (1972).
121. Davis, E. A. and Mott, N. F., *Phil. Mag.* **22,** 903 (1970).
122. Chittick, R. C., *J. Non-Cryst. Solids* **32,** 475 (1970).
123. Tauc, J., Abraham, A., Zallen, R. and Slade, M., *J. Non-Cryst. Solids* **4,** 279 (1970).
124. Chung, M. F., *J. Phys. Chem. Solids* **32,** 475 (1971).
125. Jones, R. O., *J. Phys. C.* **5,** 1615 (1972).
126. Elices, M. and Yndarain, Y., *J. Phys. C.* **5,** L146 (1972).
127. Mort, J., Schmidlin, F. W., and Lakatos, A. I., *J. Appl. Phys.* **42,** 5761 (1971).
128. Olley, J. A. and Yoffe, A. D., *J. Non-Cryst. Solids* **8-10,** 850 (1972).
129. Croitoru, N., Popescu, C., Telnic, M. and Vescan, L., *Phys. Stat. Sol.* **48,** 429 (1971).
130. Vescan, L., Telnic, M. and Popescu, C., *Phys, Stat. Sol.* (in press).
131. Croitoru, N., Niklas, J. and Stuke, J., Proc. 11th Intern. Conf. on Semiconductors, (Warsaw). 536 (1972).
132. Jonscher, A. K. and Ansari, A. A., *Phil. Mag.* **23,** 181 (1971).
133. Hill, R., *Phil. Mag.* **23,** 59 (1971).
134. Koñák, J. and Stuke, J., *Phys. Stat. Sol.* (*a*) **9,** 333 (1972).
135. Sauvage, J. A., Mogab, C. J. and Adler, D., *Phil. Mag.* **25,** 1305 (1972).
136. Tauc, J. and Menth, A., *J. Non-Cryst. Solids* **8-10,** 569 (1972).
137. Di Salvo, F. J., Menth, A., Waszcsak, J. W. and Tauc. J., *Phys. Rev.* **B6,** 4574 (1972).
138. Kastner, M., *Phys.Rev. Lett.* **28,** 355 (1972).

2. A REVIEW OF SOME ELECTRONIC PROPERTIES OF AMORPHOUS SUBSTANCES

H. FRITZSCHE,

Department of Physics and James Franck Institute,
The University of Chicago, Chicago, Illinois.

2.1. EFFECTS OF HETEROGENEITY ON THE ELECTRONIC PROPERTIES OF AMORPHOUS SEMICONDUCTORS

2.1.1. INTRODUCTION

The electronic properties of amorphous semiconductors have been summarized and discussed recently by Mott and Davis.[1] They treat the amorphous solid as a rather ideal homogeneous, uniform material. In this lecture I wish to explore the possible effects of heterogeneities, voids, and potential fluctuations often encountered in real materials. First, the basic elements of the semiconductor model of amorphous materials will be discussed in terms of the experimental observations and of the chemical bond picture. This

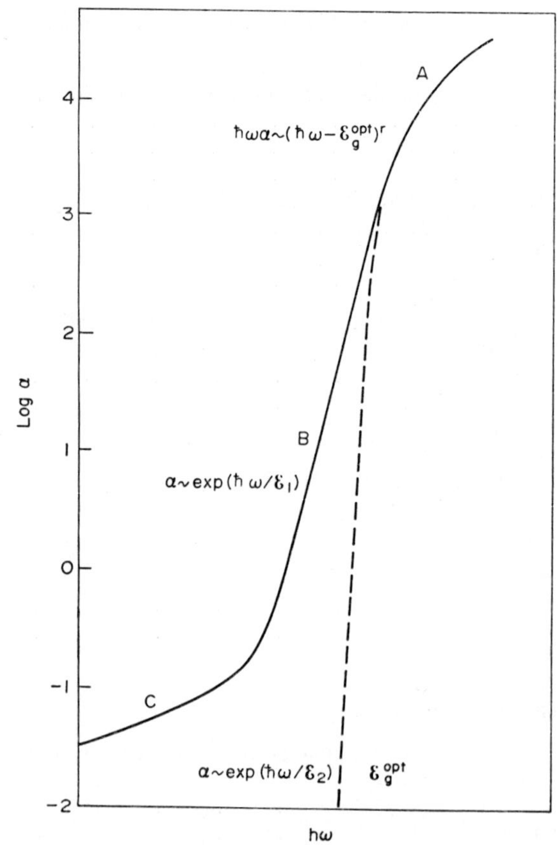

FIG. 2.1. Absorption constant α at the absorption edge. A = interband absorption with $\mathcal{E}_g^{opt} = \mathcal{E}_0$, the optical gap. B = exponential absorption. C = structure and composition dependent absorption region. (After Tauc and Menth, Ref. 2.)

leads to a classification of amorphous semiconductors according to the dominant character of their chemical bonds. Finally we discuss electronic transport in materials containing potential fluctuations and heterogeneities of sufficient magnitude so that one cannot use spatial averages of the transport parameters.

2.1.2. THE GAP

Amorphous semiconductors are transparent in the infrared spectral region. The onset of optical interband transitions defines an energy gap between the valence band and the conduction band. Fig. 2.1 shows the optical absorption of a-As_2S_3. The optical gap is usually obtained by fitting the high absorption region A to an expression describing interband absorption in amorphous materials for which k-conservation no longer holds.(2) It is proportional to the joint density of states and a matrix element which is assumed to be constant.

In this case the optical gap is $\mathcal{E}_0 \sim 2{\cdot}3$ eV. In other amorphous semiconductors the gap lies between 0·6 and 2·5 eV. At photon energies below $\mathcal{E}_0$ one finds an exponential absorption tail B. This so-called Urbach tail is attributed(3) to band edge exciton absorption modified by internal fields which are estimated to be of the order of 5×10^5 V/cm. Absorption region C is structure sensitive and apparently associated with absorption involving localized gap states.(2) The exponential absorption region B is not attributed to absorption from and to tail states because the logarithmic slope is so very much the same for almost all amorphous semiconductors and many crystalline compound semiconductors. In his contribution Dr. Davis will cover this subject in greater detail (see Chapter 13). For the purpose of this discussion we merely note how we obtain the optical gap $\mathcal{E}_0$.

Figure 2.2 shows that the electrical conductivity is thermally activated

$$\sigma = \sigma_0 \exp(-\Delta\mathcal{E}/kT). \tag{2.1}$$

Results for different materials are shown in this figure. This is no reason for concern as I am at present summarizing observations which are quite generally found in a large class of amorphous semiconductors and will discuss exceptions from these general rules later. One usually finds

$$\Delta\mathcal{E} \leqslant \tfrac{1}{2}\mathcal{E}_0. \tag{2.2}$$

so that one is led to conclude that the Fermi energy $\mathcal{E}_F$ is near the gap center. Does this mean that amorphous semiconductors are intrinsic and that the gap is free of states? Can we write

$$2\Delta\mathcal{E} = \text{band gap?} \tag{2.3}$$

The answer is NO. Consider the amorphous semiconductor model discussed by Professor Mott in his contribution (Chapter 1). If we integrate the expression

$$\sigma = e \int g(\mathcal{E})\, \mu(\mathcal{E})\, f(1 - f)\, d\mathcal{E} \tag{2.4}$$

assuming sharp mobility edges $\mu = 0$ for $\mathcal{E}_v < \mathcal{E} < \mathcal{E}_c$ we obtain

$$\sigma = e\, \mu_e\, g(\mathcal{E}_c)\, kT \exp\left[-\left(\frac{\mathcal{E}_c - \mathcal{E}_F}{kT}\right)\right] + e\, \mu_h\, g(\mathcal{E}_v)\, kT \exp\left[-\left(\frac{\mathcal{E}_F - \mathcal{E}_v}{kT}\right)\right] \tag{2.5}$$

where the two terms on the right-hand side represent the contributions from the mobile electrons and holes with densities n and p, respectively. Here

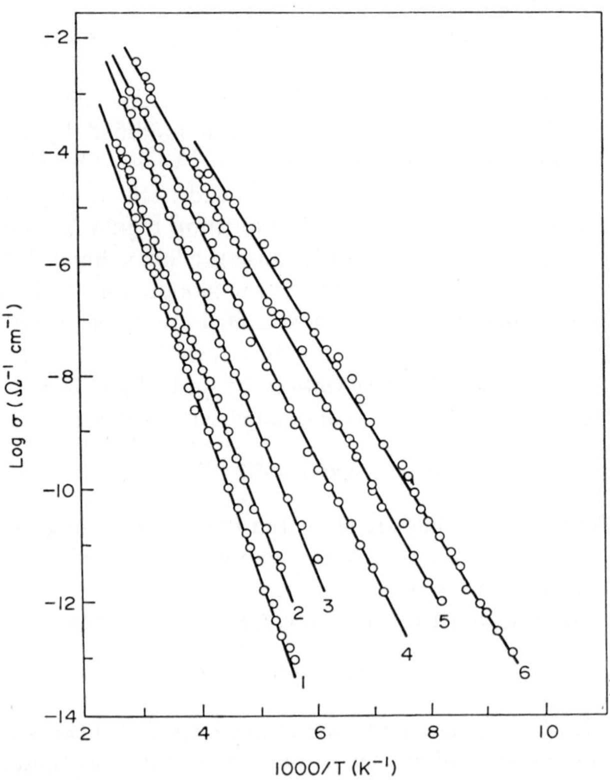

FIG. 2.2. Temperature dependence of conductivity of $Tl_2SeAs_2(Te_xSe_{1-x})_3$ according to Andriesh and Kolomiets (Ref. 28). The compositions of the samples labelled 1–6 correspond to $x = 0$, 0·2, 0·4, 0·6, 0·8, 1·0, repectively.

$g(\mathcal{E})$ are the densities of states and $f(\mathcal{E})$ is the Fermi function. In intrinsic crystalline semiconductors all electrons in the conduction band came from the valence band so that neutrality demands

$$n = p. \tag{2.6}$$

This condition determines $\mathcal{E}_F$ in terms of the conduction and valence band parameters so that

$$\sigma\ (\text{intrinsic}) = \text{const}\ \exp\left[-\left(\frac{\mathcal{E}_c - \mathcal{E}_v}{2kT}\right)\right]. \tag{2.7}$$

In amorphous semiconductors $n \neq p$ because of the much larger number of electrons and holes in localized tail states and gap states. A numerical example illustrates this. For a mobility gap $\mathcal{E}_c - \mathcal{E}_v = 1\text{eV}$ and $\mathcal{E}_F$ in the gap centre one obtains $n \sim 10^{10}\text{cm}^{-3}$ at 300K and $n \sim 10^{3}\text{cm}^{-3}$ at 150K. Figure 2.2 and many other amorphous semiconductors exhibit the "intrinsic" slope of the conductivity curve down to 150 K or lower. There are certainly more than 10^3 localized tail and gap states present. We therefore conclude that eqn (2.6) cannot be valid at these low temperatures. It is more likely that $\mathcal{E}_F$ is determined by the gap and tail states and that $n \neq p$. Either the first or the second term in eqn (2.5) dominates the conduction.

Then what is the meaning of $\Delta\mathcal{E}$? It is not equal to $(\mathcal{E}_c - \mathcal{E}_F)$ or to $(\mathcal{E}_F - \mathcal{E}_v)$ because of the temperature dependence of the electron energy states. This T-dependence arises from the thermal expansion of the material and the phonon contributions to the self-energy of electrons and holes. It has generally the form shown in Fig. 2.3. At room temperature the dependence is very nearly linear with $\mathcal{E}_c - \mathcal{E}_F$ (or $\mathcal{E}_F - \mathcal{E}_v$) $\simeq \Delta\mathcal{E} - \gamma T$, at low T it is temperature independent. From the T-dependence of the optical gap $\mathcal{E}_0$ one obtains a rough estimate of $\gamma \sim 3 \times 10^{-4}$ eV/K. As shown in Fig. 2.3, $\Delta\mathcal{E}$ is the $T = 0$ intercept of the tangent to the $\mathcal{E}$ versus T curve and therefore has no direct physical meaning. With the particular value of γ chosen here, $\Delta\mathcal{E}$ is 0·1eV larger than the relevant energy at 300 K and about 0·02 eV larger than $(\mathcal{E}_c - \mathcal{E}_F)$ or $(\mathcal{E}_F - \mathcal{E}_v)$ at $T = 0$. The term $\exp(\gamma/k)$ is contained in the pre-exponential factor σ_0. Except for the measurement of internal photoemission of electrons and holes from a metal electrode into an amorphous semiconductor, an experiment discussed in Dr. Mort's article (Chapter 15), I do not know of any experiment which yields the gap $\mathcal{E}_c - \mathcal{E}_v$.

2.1.3. IMPURITIES AND COMPOSITION DEPENDENCE OF CONDUCTION

The most striking difference between amorphous and crystalline semiconductors lies in the effect of impurities. The most interesting and technically useful feature of crystalline semiconductors is the possibility of changing their electrical properties by doping with donor and acceptor atoms, making them into electron conductors or hole conductors and forming *p–n* junctions. This is not possible with amorphous semiconductors. There appear to be no shallow donors near the conduction band or shallow acceptors near the valence band. The reason for this is the following. In crystals each impurity atom is forced by long-range order to have the coordination number of the host atom it substitutes. In amorphous materials, in contrast, each atom can satisfy its valence bonds by adjusting its nearest neighbour environment.

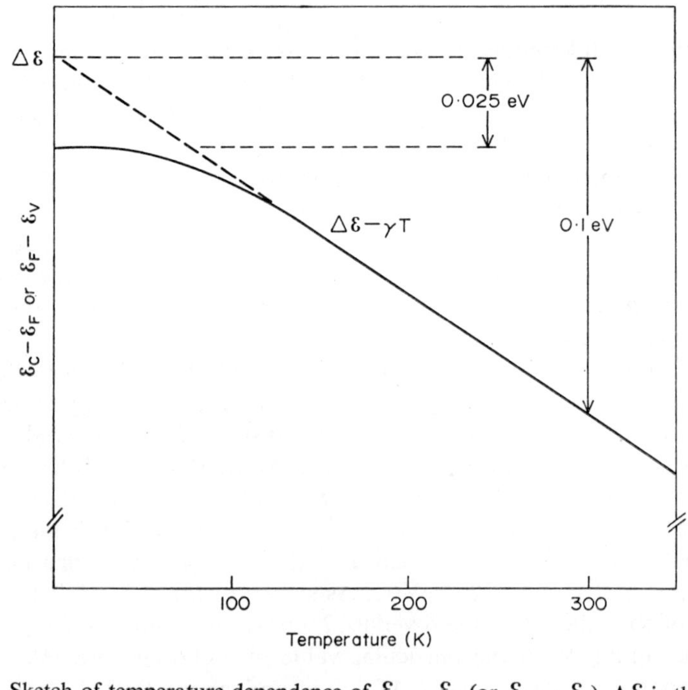

FIG. 2.3. Sketch of temperature dependence of $\mathcal{E}_c - \mathcal{E}_F$ (or $\mathcal{E}_F - \mathcal{E}_v$). $\Delta\mathcal{E}$ is the slope of the ln σ vs. $1/kT$ curve measured in the range $200 < T < 400$ K.

The possibility and the desire to satisfy valence bonds introduces a certain degree of chemical order into the disordered structure. Bonds do not form by chance or at random but strong bonds are favoured over weaker bonds with rather small statistical fluctuations. As an example Fig. 2.4 shows the heat capacity C_v of the $Ge_{17}Te_{83}$ eutectic mixture in its crystalline and its glassy form.[4] Below the glass transition temperature T_g, both forms have the same value of C_v. This shows that here C_v derives from the vibrational modes only. C_v of the glass increases at T_g because at that temperature additional degrees of freedom become available corresponding to configurational changes involving larger groups of atoms. Most significant however is the large peak in heat capacity at higher temperature below and above the eutectic melting temperature T_m. The area under the peak (26 cal/g), i.e. the excess enthalpy, is close to the heat of fusion of pure Te (33·7 cal/g). We conclude that there is a high degree of bond ordering in the glass distinctly different from the more random arrangement in the higher temperature liquid. Such excess enthalpy suggest a kind of order–disorder transformation occurring between 350–450°C. De Neufville[(4)] discovered this in a number of chalcogenide alloys which contain Ge.

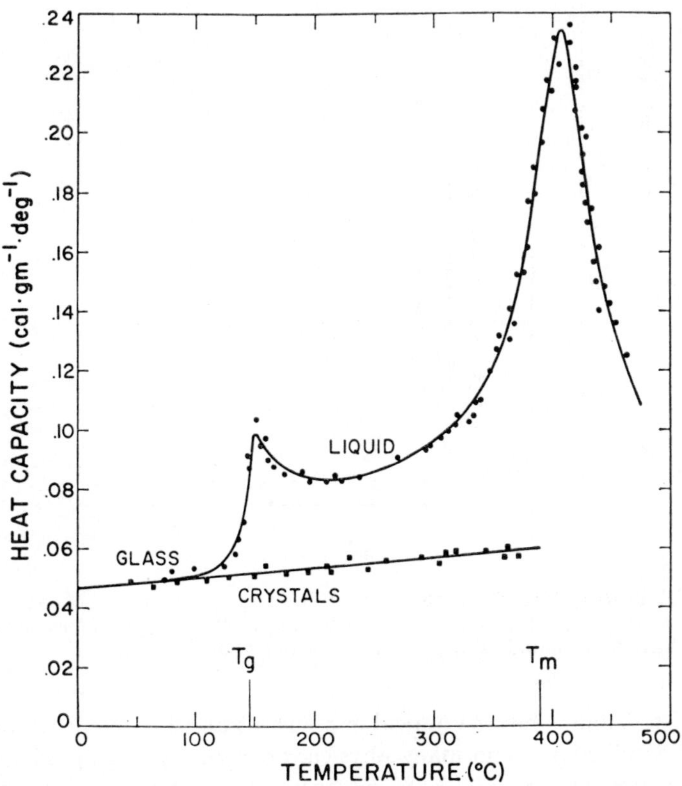

FIG. 2.4. Heat capacity of glassy, liquid, and crystalline $Ge_{17}Te_{83}$ vs. T. $Ge_{17}Te_{85}$ is the eutectic composition in the GeTe–Te system. The glass transition temperature is $T_g = 145$°C The eutectic melts at $T_m = 389$°C. (After de Neufville, Ref. 4.)

Figures 2.5 and 2.6 show the bonding chemistry at work in the Te–Tl system studied by Ferrier, Prado and Anseau.[5] These amorphous semiconducting films were deposited onto a cold substrate by co-evaporation from two independent sources. The conductivity shows a minimum at compound compositions Tl_2Te_n ($n = 1 - 7$) and at Tl_3Te_2 and Tl_5Te_3, whilst the activation energy $\Delta\mathcal{E}$ exhibits a maximum.

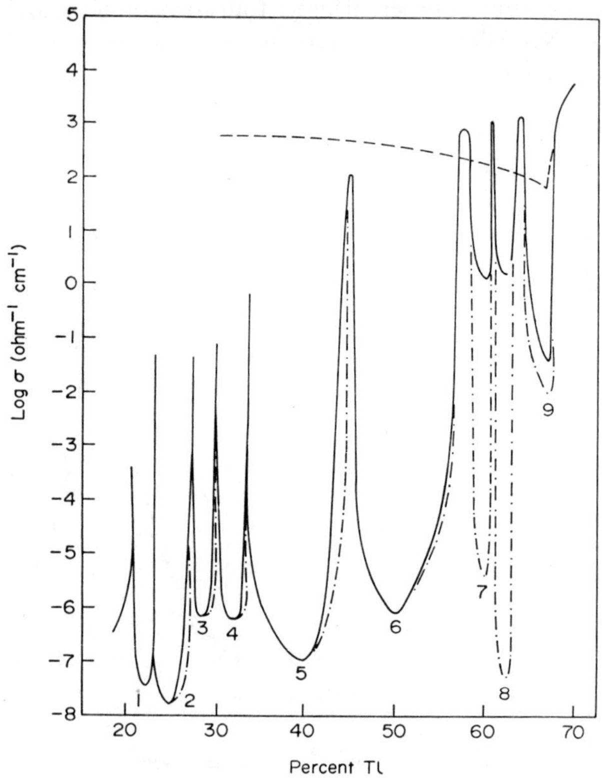

FIG. 2.5. Conductivity of co-evaporated films of Tl_xTe_{100-x} as a function of composition x. Solid curve is at 78 K as deposited. Dash-dotted curve: at 78 K after annealing. Dashed curve: conductivity of liquid at 425°C. (After Ferrier *et al.*, Ref. 5.)

Not only $\Delta\mathcal{E}$ and σ but the optical gap, glass transition temperature, hardness, heat conductivity, and many physical properties change gradually with the composition of the glass. Extrema occurring at certain compound compositions signify bond ordering in the alloy glass. Quite often these compositions do not exist in crystalline form. Addition of heavier elements often decrease $\Delta\mathcal{E}$, lighter elements increase $\Delta\mathcal{E}$ as shown[6] in Fig. 2.7. As these compositional changes occur, one still finds the relations expressed by eqns (2.1) and (2.2) are satisfied, which means that $\mathcal{E}_F$ remains near the gap center. There is no evidence of shallow donors or acceptors controlling the conductivity at low temperatures. In contrast, in crystalline semiconductors a minor deviation from stoichiometry causes a shift of $\mathcal{E}_F$ close to or into the valence or conduction band.

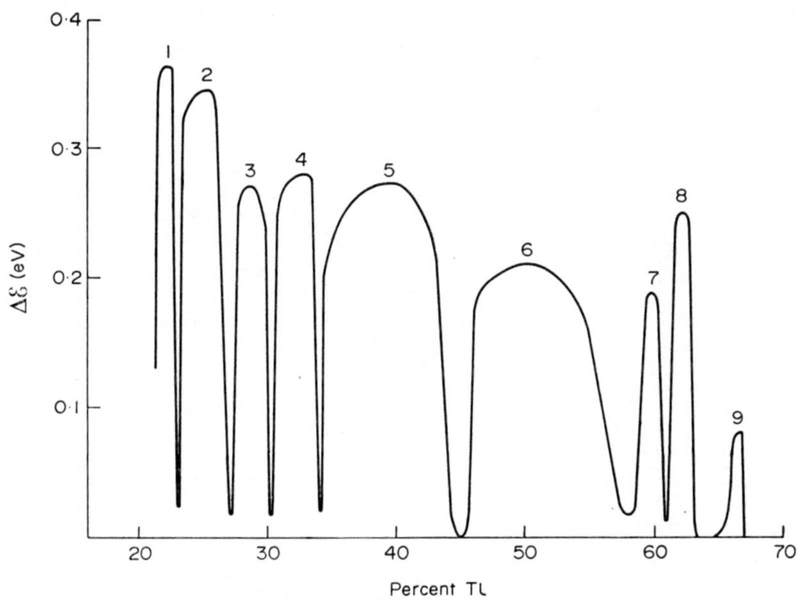

FIG. 2.6. Activation energy $\Delta\mathcal{E}$ of conductivity obtained from high temperature part of $\ln \sigma$ vs. $1/T$ curves. (After Ferrier *et al.*, Ref. 5.) Co-evaporated films of Tl_xTe_{100-x}.

2.1.4. CHEMICAL BOND DESCRIPTION

I showed several examples which should convince you that the basic electronic properties, the presence or absence of shallow donors and acceptors, the optical and electrical energy gap, are primarily determined by the chemical bonds between neighbours. I wish to demonstrate now, that the very different behaviour of large classes of amorphous semiconductors becomes plausible if one understands their chemical bonding. Let us contrast the origin of electronic states in two classes of semiconductors:

(a) tetrahedral semiconductors;

(b) semiconductors containing group VI elements in 2-fold coordination.

Figure 2.8 sketches the origin of states in Ge and Se. Ge in 4-fold coordination has hybridized sp^3 orbitals. On bonding these split into bonding (σ) and antibonding (σ^*) states which form in tetrahedral amorphous or crystalline semiconductors the valence and conduction band, respectively. In Se only four of the six *p*-states are utilized for bonding when Se is in 2-fold coordination. This leaves one non-bonding electron pair. In the solid these lone pair (LP) electrons form a band near the original *p* state energy. The bonding (σ) and antibonding (σ^*) bands are split symmetrically with respect to this reference energy.(7–9)

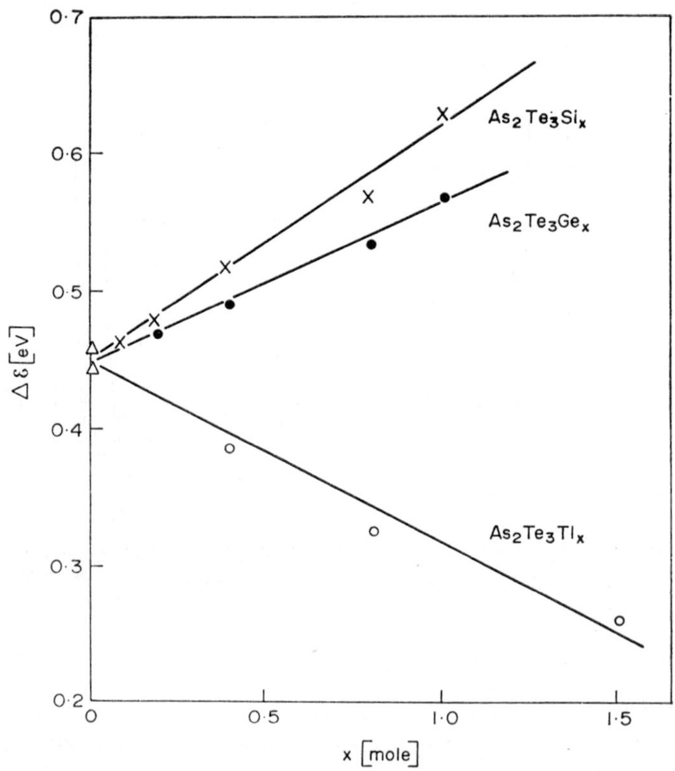

FIG. 2.7. Change of the conductivity activation energy $\Delta\mathcal{E}$ which occurs as Ge, Si, or Tl are alloyed to amorphous As_2Te_3. (After Stourac *et al.*, Ref. 6.)

This filled LP band is the highest filled band and thus the hole conduction band in any material which contains a considerable concentration of group VI elements in 2-fold coordination. The number of states in this LP valence band is equal to twice the number of group VI atoms in the material.

Kastner[9] explained that dangling bonds and impurity atoms form different states in tetrahedral and in LP semiconductors. When a dangling bond is formed, a full state is pulled out of the σ band and an unoccupied state from the σ^* band. In tetrahedral semiconductors these produce a donor near the gap center and an acceptor state above it separated from the donor by the correlation energy. In LP semiconductors the full state pulled from the σ band falls into the LP band, the unoccupied state from σ^* somewhat above it. We therefore see that in LP semiconductors dangling bonds do not give rise to midgap donors, in tetrahedral semiconductors they do.

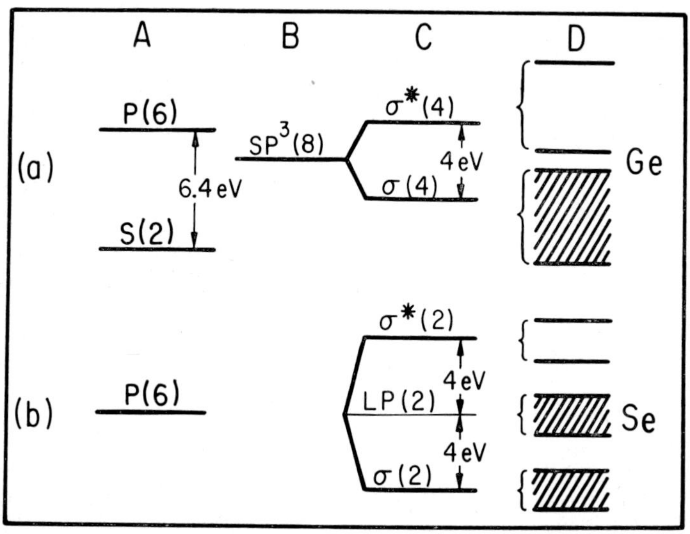

FIG. 2.8. Bonding in (a) Ge and (b) Se. (A) atomic states, (B) hybridized states, (C) molecular states, (D) states are broadened into bands in the solid. (After Kastner, Ref. 9.)

According to Kastner the effect of alloying is the following. An alloy contains different bonds; their relative number is determined by their strength and the composition. Each bond is characterized by its strength, the $\sigma - \sigma^*$ separation. LP states are associated only with group VI elements, which are found in 2-fold coordination. As a rough approximation the energetic centers of gravity of the particular bonds are thought to be shifted with respect to one another according to their relative ionization energy. Since all bonding states fall below or at best into the LP band, it is difficult to obtain donor states in the gap of a LP semiconductor. Localized donor states in the gap are formed by LP states whose energy is raised by the presence of nearby electropositive elements or by statistical fluctuations in the disordered material. These few examples have hopefully demonstrated the importance and usefulness of the chemical bond approach.

2.1.5. CLASSIFICATION

It is a great mistake to make general statements about amorphous semiconductors. Just as in crystalline semiconductors there are great differences between different materials. At the same time one can distinguish several large families with common properties, which are characterized by the nature

TABLE 2.1

Classification and examples of non-crystalline semiconductors

1. *Covalent Non-crystalline Solids*	
A. *Tetrahedral Amorphous Films*	
Si, Ge, SiC, InSb, GaAs, GaSb ...	
B. *Tetrahedral Glasses,* $A^{II} B^{IV} C_2^{V}$	
$CdGe_xAs_2$, $CdSi_xP_2$, $ZnSi_xP_2$, $CdSn_xAS_2$ ...	
C. *Lone Pair Semiconductors*	
(i) *elements and compounds*	
Se, S, Te, As_2Se_3, As_2S_3 ...	
(ii) *cross-linked networks*	
Ge–Sb–Se	Si–Ge–As–Te
Ge–As–Se	As_2Se_3–As_2Te
As–Se–Te	Tl_2Se–As_2Te_3
...	...
D. *Others*	
B, As, $(Cu_{1-x}Au_x)\,Te_2$ ...	
2. *Semiconducting Oxide Glasses*	
V_2O_5–P_2O_5	MnO–Al_2O_3–SiO_2
V_2O_5–P_2O_5–BaO	CoO–Al_2O_3–SiO_2
V_2O_5–GeO_2–BaO	FeO–Al_2O_3–SiO_2
V_2O_5–PbO–Fe_2O_3	TiO_2–B_2O_3–BaO
...	...
3. *Dielectric Films*	
SiO_x, Al_2O_3, ZrO_2, Ta_2O_3, Si_3N_4, BN ...	

of their chemical bonds. Table 2.1 groups the materials into three major categories. Anything said so far does not apply to the dielectric films listed as the last category. These are tight-binding insulators with band gaps of the order of 5 eV. They are electronic conductors and thus semiconductors on account of defects which act as donors or acceptors. They are not very sensitive to disorder and I doubt whether their bands have significant tails of localized states.

The second main group contains the oxide glasses, which have strong ionic bonds and are usually good insulators. Those listed are semiconducting because they contain transition metal ions in different valence states. Conduction is by charge transfer hopping between the transition metal ions. They are interesting but nothing I said applies to them. We are concerned here with the covalent semiconductors. They fall into very distinct groups. Starting from the bottom we have the lone pair semiconducting alloy glasses, then the chalcogenide elements and simple compounds.

The properties of the last subgroup are quite readily affected by minute amounts of impurities, one part per million in the case of Se. I believe *a*-Se has very few gap states and possibly even rather few tail states.

The tetrahedral glasses form a large family with a variety of interesting properties. They have optical gaps ranging from 0·5–2·3 eV. In contrast to the tetrahedral films the tetrahedral glasses can be obtained by fast cooling the melt. Some are *n*-type, some *p*-type, although $\mathcal{E}_F$ is always near the gap center. Some show ambipolar conduction such as $CdGeAs_2$.[10a]

The tetrahedral amorphous films can only be prepared by thin film deposition from the vapour phase or by electrolytic deposition because the melt has a metallic structure characterized by a higher coordination number. The properties of these films depend strongly on the deposition parameters and the annealing history. They show many interesting properties not shared by other amorphous semiconductors. Among these a large electron spin resonance, corresponding to 2×10^{20} spins/cc,[11] a negative magnetoresistance which is very sensitive to small amounts of impurities,[12] and a structure sensitive conduction which is associated with electrons hopping among localized states.[1]

Except for the very elegant and most informative experiments carried out on amorphous Si and Ge produced by glow discharge decomposition of SiH_4 and GeH_4 by Spear and LeComber,[13] the experiments with these tetrahedral films cannot be easily interpreted. This is because these films are structurally heterogeneous when they are deposited on a cold substrate by sputtering or evaporation. They contain a network of internal voids as shown in the electron microscope picture (Fig. 2.9) of Donovan and Heinemann.[14] This void structure disappears upon annealing and is not noticeable when the substrate temperature is held at about 200°C during deposition, but

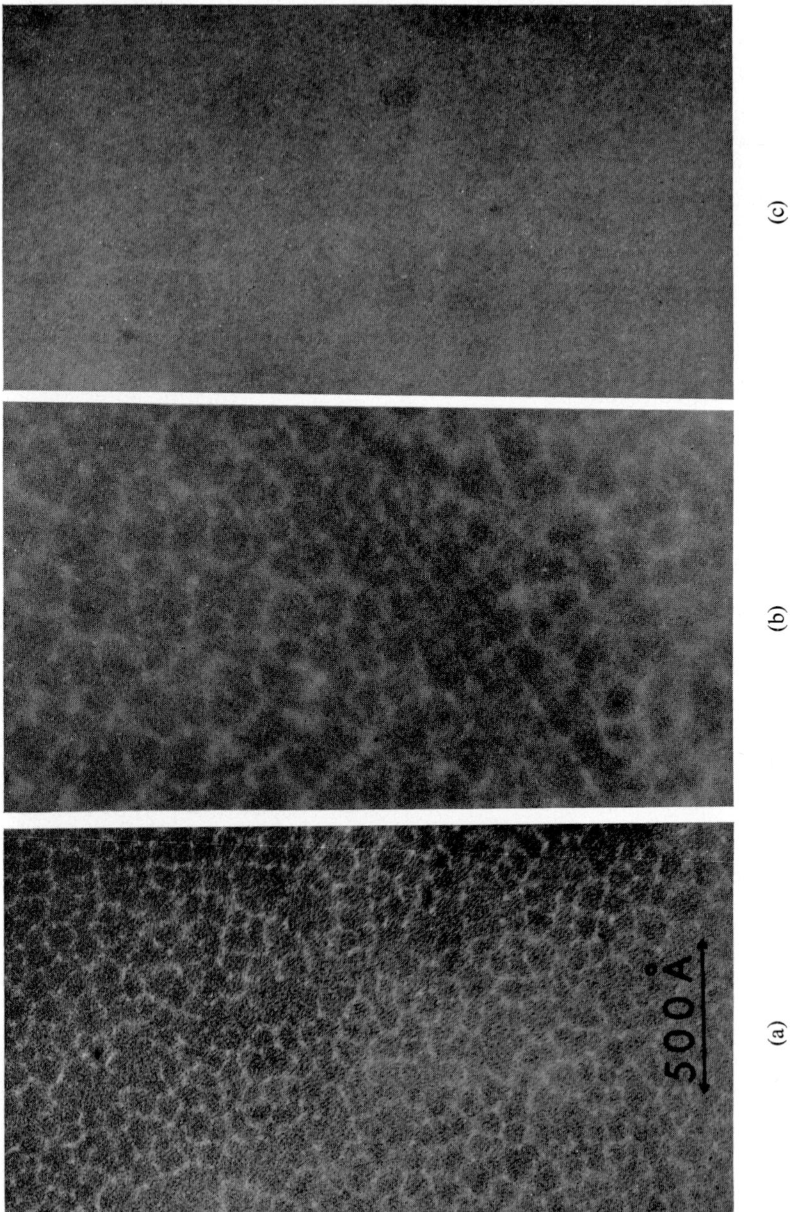

FIG. 2.9. Bright field electron micrograph of an *a*-Ge film formed at different substrate temperatures T_s. (a) $T_s = 25°C$, (b) $T_s = 150°C$, and (c) $T_s = 250°C$. The void network observed in the lower substrate temperature films is completely absent in the film prepared at 250°C. (After Donovan and Heinemann, Ref. 14.)

very few experiments have been carried out on void-free films. The presence of a connected void network in *a*-Ge can be observed by depositing it onto an oscillating quartz crystal and measuring the frequency shift corresponding to water adsorption in the film.(15) This method may be of interest to those lacking expertise or equipment for high resolution transmission electron-microscopy. Fig. 2.10 shows the frequency shift after exposing the film to a stream of wet and later dry nitrogen. As the *a*-Ge film is annealed at increasingly higher temperatures, the void network closes up and one obtains a monolayer surface absorption when the Ge film crystallizes. The amount of water absorbed in the virginal film at 20% humidity corresponds to 2×10^{20} H_2O/cm^3. Surprisingly, this water did not affect the electron spin resonance signal in any way.

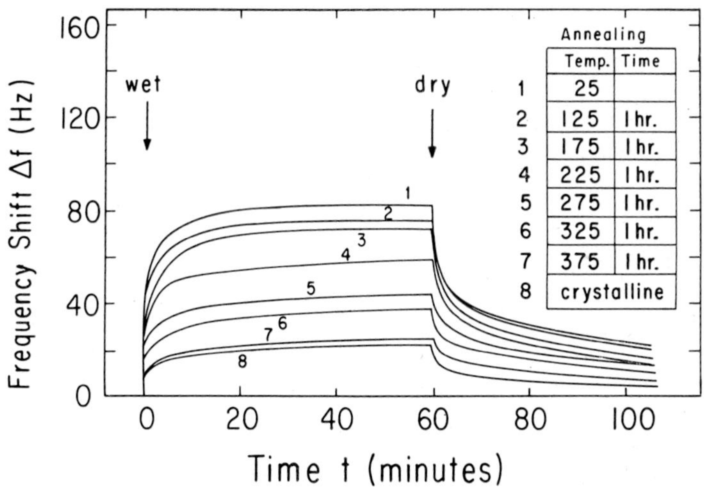

FIG. 2.10 Frequency shift of a 5 MHz quartz crystal carrying a 1 μm thick *a*-Ge film exposed to dry and wet (15% humidity) nitrogen. Film is porous after deposition at 300 K. Void structure disappears upon annealing. (After Fritzsche *et al.*, unpublished).

2.1.6. HETEROGENEITIES

The void structure and porosity of amorphous tetrahedral films are extreme examples of structural heterogeneity. Now we wish to consider cases which contain structural and compositional inhomogeneities whose own size or whose effect (by forming accumulation or depletion layers around it) extends over some hundred angstroms. These are dimensions which essentially prevent tunnelling and suggest a classical treatment of conduction in long-range potential fluctuations. First some examples of and possible reasons for such coarse heterogeneities.

(a) Since alloy glasses contain compositions which far exceed the limits of thermodynamic equilibrium, the metastable amorphous state is prone to phase separate into finely divided phases which may all be vitreous. This phase separation may not have reached completion so that the boundaries are soft and ill defined. This leads to fluctuations in the energy gap, and because of differences in electronegativity of the regions, to potential fluctuations.

(b) Figures 2.4 to 2.6 demonstrated the strong tendency of covalent glasses to form structural units. The size of these complexes and statistical deviations from them depend on the thermal history and the presence of structure modifying impurities.

(c) Most of the annealing effects in amorphous films occur below the glass transition temperature. At these low temperatures diffusion must be quite small. The annealing effects are thus indicative of a very loose structure. Annealing produces large changes in the electrical properties. In all likelihood this occurs inhomogeneously, starting at weak spots.

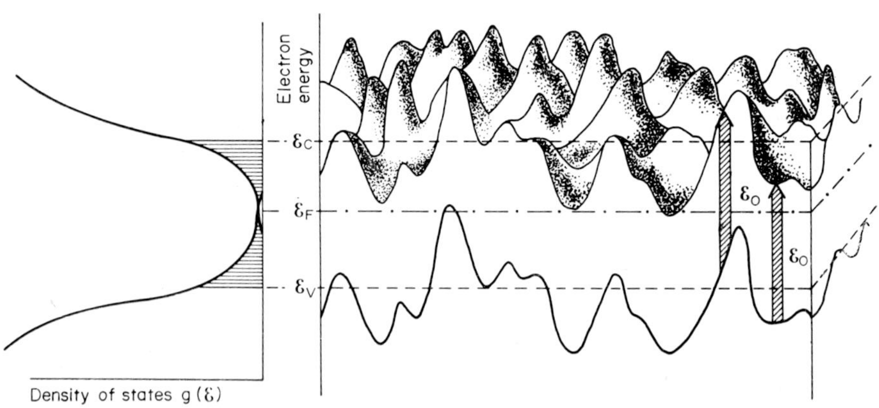

FIG. 2.11. Sketch of the symmetric part of long wavelength potential fluctuations. The short-range fluctuations, as well as variations which shift valence and conduction band states in opposite directions, are omitted for clarity. $\mathcal{E}_0$ corresponds to an average optical gap. $\mathcal{E}_c$ and $\mathcal{E}_v$ are percolation thresholds or mobility edges. The left hand side shows the density of states. The distribution of localized gap states between $\mathcal{E}_c$ and $\mathcal{E}_v$ is not necessarily smooth and monotonic. (After Fritzsche, Ref. 16.)

Fig. 2.11 shows a model of long-range potential fluctuations associated with screened Coulomb potentials and heterogeneities.[16] Short-range potential variations over distances of interatomic separations are omitted. Only Coulomb potentials which affect both bands equally are shown here. Fluctuations in density and composition will in addition cause spatial varia-

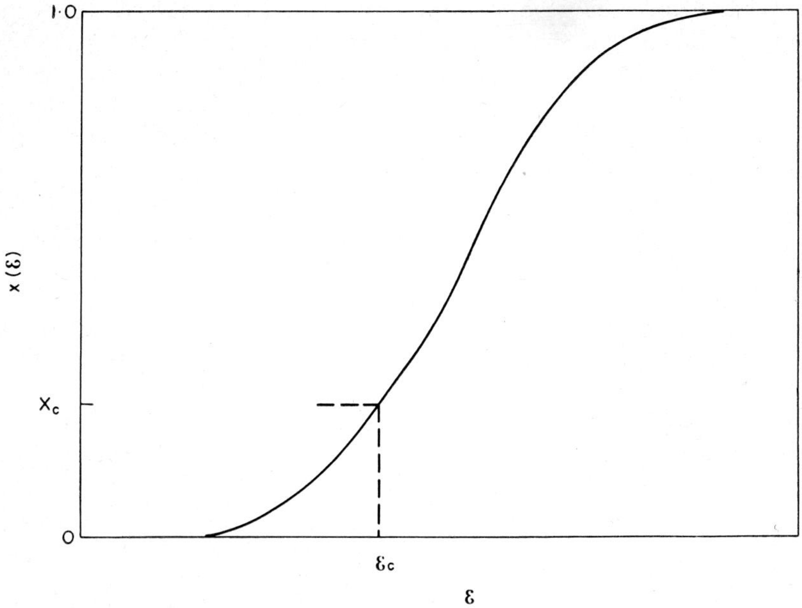

FIG. 2.12. The classically allowed volume fraction $x(\mathcal{E})$ as a function of energy for electrons near the bottom of the conduction band in a heterogeneous material.

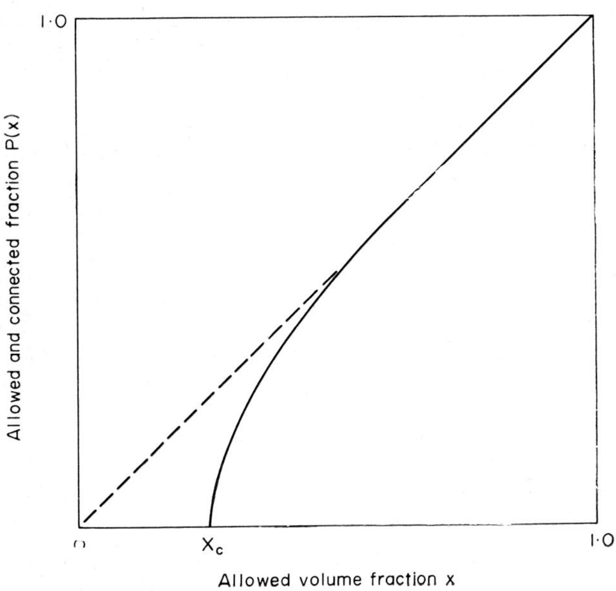

FIG. 2.13. Volume fraction $P(x)$ which is not only allowed but also connecting the electrodes.

tions in the local band gap. We have omitted and averaged over potential fluctuations that extend over distances which electrons can easily bridge by tunnelling. We limit ourselves here to fluctuations over a few hundred angstrom, so that the potential contour presents the classical turning points of the charge carriers. The vertical arrows marked $\mathcal{E}_0$ represent optical interband transitions. Since they occur in a volume element of the size of a coherence length, they are not much affected by potential fluctuations which raise or lower both the initial and the final states. The internal electric fields might of course dissociate the excitons and thus give rise to the exponential absorption tail.[3]

Let us consider the electronic transport properties which are expected to be much more strongly affected by the potential fluctuations. In crystalline semiconductors, the large mobility of the carriers and the long-range order permit us to use spatial averages of the transport parameters, i.e. one can obtain the conductivity by adding conductivity slices $\sigma(\mathcal{E})\,d\mathcal{E}$ which are well defined at each electron energy $\mathcal{E}$: we therefore write

$$\sigma = \int \sigma(\mathcal{E})\,d\mathcal{E} \tag{2.8}$$

for an amorphous semiconductor described by the model sketched in Fig. 2.11. Now we note that a constant energy slice cuts through regions of greatly different conductivities. Electrons are excluded from some regions and allowed in others. In a low energy slice the allowed regions form isolated pockets. At a higher energy allowed pockets are co-existing with allowed channel regions which connect one side of the sample with the other. At an even higher energy the slice contains allowed regions almost everywhere except for a few isolated forbidden islands. We distinguish at each energy the fraction $x(\mathcal{E})$ of volume allowed and, what is even more important for d.c. conduction, the fraction $P(x)$ of volume in extended channels, that is the volume allowed and not isolated in islands.

For any given potential one can figure out the fraction $x(\mathcal{E})$ of allowed volume for each energy slice. For the one shown in Fig. 2.11 this might look like Fig. 2.12. In order to derive an expression for d.c. conduction one then has to determine $P(x)$ the fraction of the total volume which is allowed and in connected channels. When the allowed volume encompasses most volume the chance is very small that some is isolated. Thus $P(x)$ is equal to x for large x. As the allowed volume fraction x decreases it becomes increasingly disjointed and disconnected. Since $P(x)$ counts only channels and not islands, $P(x)$ falls below x. Finally it vanishes rather abruptly when the allowed volume fraction becomes too small to yield any channels. This behaviour is sketched in Fig. 2.13.

The result of the $P(x)$ curve is called percolation theory.[17] It is useful for a number of very different problems.[18] It yields for the critical allowed fraction x_c of first channel formation $x_c = 0{\cdot}25$ in three dimensions. $x_c = 0{\cdot}25$ corresponds to a certain $\mathcal{E}_c$ which would mark the onset of d.c. conduction in our model of Fig. 2.12. One therefore might consider $\mathcal{E}_c$ the percolation edge replacing the mobility edge in these heterogeneous systems.

How does the average mobility depend on energy $\mathcal{E}$? If we assume an energy independent mobility μ_0 in the allowed regions one might set

$$\mu = \mu_0 P(x) \quad \text{where} \quad x = x(\mathcal{E}). \tag{2.9}$$

This yields a mobility shoulder like curve II in Fig. 2.14 instead of the abrupt mobility edge (curve I) assumed to exist without these large scale potential fluctuations.

This was our understanding until about a year ago. At that time, however, Last and Thouless[19] took a piece of conducting paper and measured its conductance G while punching into it an increasing number of holes. Surprisingly they found that G/G_0 was not equal to $P(x)$ but much smaller. Kirkpatrick[20]

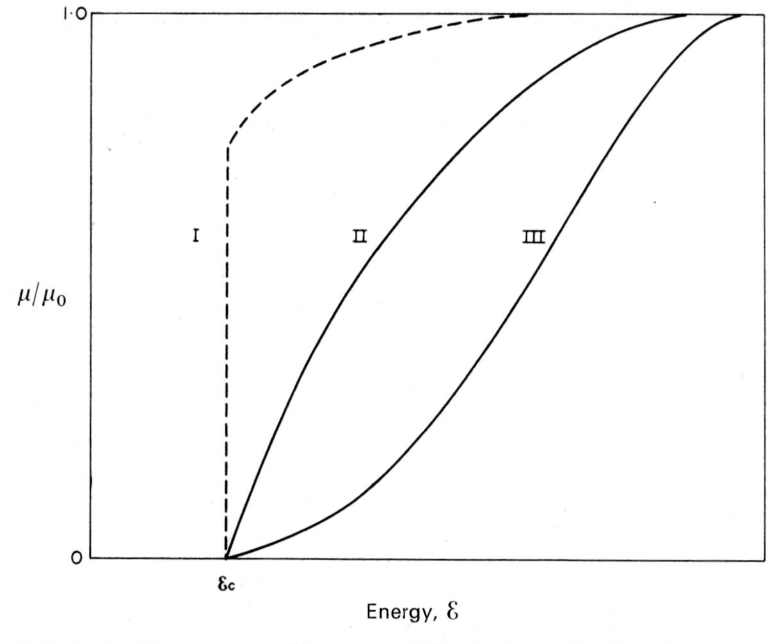

FIG. 2.14. Ratio of average mobility μ to mobility in classically allowed regions μ_0 as a function of $\mathcal{E}$. Curve I is the sharp mobility edge proposed by Mott for homogeneous material. Curve II is obtained from eqn (2.9) of text. Curve III is predicted by percolation theory for heterogeneous material.

repeated the experiment of Thouless and Last in three dimensions with the help of a computer. He confirmed their findings and obtained the data points shown in Fig. 2.15. Above the percolation threshold the mobility starts off as

$$\mu = \mu_0(x - x_c)^{1 \cdot 6}. \tag{2.10}$$

On account of this the mobility rises much more gradually with energy above the percolation threshold, as shown by curve III in Fig. 2.14. Curve II is incorrect for the following reason. Although the fraction of channel volume increases abruptly above the percolation threshold, the first channels are very constricted and convoluted in shape. They do not contribute much to conduction. In homogeneous materials the mobility edge may be as sharp as shown by curve I in Fig. 2.14. In sufficiently heterogeneous materials the mobility edge is softened by the exclusion of certain regions from conduction near the percolation threshold. In the latter case curve III in Fig. 2.14 represents the spatial average of the mobility. The local mobility μ_0 might of course be itself a function of $\mathcal{E}$. This classical picture of a heterogeneous solid leads to the following conclusions:

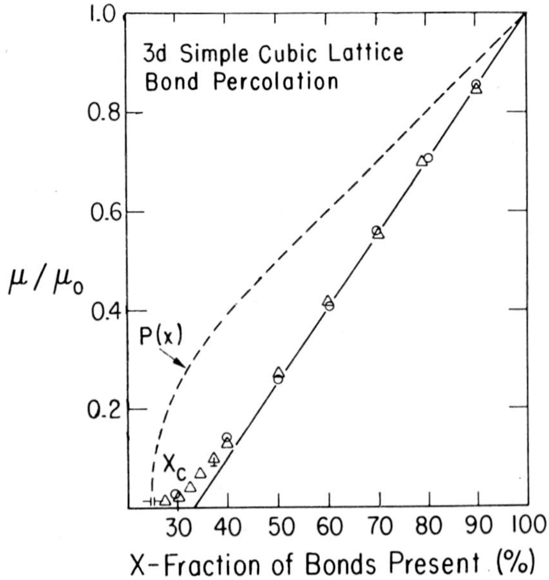

FIG. 2.15. The data points represent the average mobility relative to that in the classically allowed region as a function of the fraction x of the volume allowed. The full line represents the result of the effective-medium theory of conduction in mixtures. The dotted curve is the percolation probability $P(x)$, which is the fraction of material that is both allowed and connected to contacts. (After Kirkpatrick, Ref. 20.)

(1) Electrons are localized predominantly where there are few holes and holes are localized at places different from electrons.

(2) The optical gap $\mathcal{E}_0$ bears no direct relationship with the percolation gap $\mathcal{E}_c - \mathcal{E}_v$.

(3) If $\mathcal{E}_0$ is about equal to the average energy gap, then $\mathcal{E}_c - \mathcal{E}_v < \mathcal{E}_0$ because $x_c < 0{\cdot}5$.[21] The percolation threshold for electrons $\mathcal{E}_c$ lies below the centre line of the conduction band fluctuations, $\mathcal{E}_v$ lies above the centre line of the valence band fluctuations.

(4) The optical matrix element from localized to channel states and between channel states should be smaller than that between extended band states because of the spatial relationship between these different states.

(5) Regions allowed for electrons near $\mathcal{E}_c$ do not overlap in space with regions allowed for holes near $\mathcal{E}_v$. Consequently, recombination either requires hopping or tunnelling to overcome the spatial separation or more energetic electrons $\mathcal{E} > \mathcal{E}_c$ and energetic holes $\mathcal{E} < \mathcal{E}_v$.

One might erroneously think that a soft mobility onset might be in conflict with the observation of a constant activation energy $\Delta\mathcal{E}$. This is not be case. If in eqn 2.4 one has $\mu(\mathcal{E}) = \mu_0\,(\mathcal{E} - \mathcal{E}_c)^m$ then a temperature dependence $(kT)^m$ will move into the prefactor σ_0. In addition the term A in the expression for the thermopower

$$S = \frac{k}{e}\left(\frac{\Delta\mathcal{E}}{kT} + A - \frac{\gamma}{k}\right) \tag{2.11}$$

will be increased by an additive term m.

Dielectric properties, both the dielectric constant and the loss as a function of frequency, are strongly affected by inhomogeneities because in mixtures having different resistivities space charge polarizations occur. Several simple geometries have been analysed by Volger[22] and van Beek.[23] It might be interesting to explore whether the differences in conductivities measured by conventional methods and by contactless induction methods[24,25] are caused by heterogeneities. There also is a wealth of information on so-called granular metals[26,27] which is relevant for understanding the electrical, optical, and magnetic properties of heterogeneous materials. These materials are made with a good control over the size and density of the metal grains by simultaneous sputtering or evaporation of a metal and a dielectric such as SiO_2.

If large scale potential fluctuations are important one clearly has to reexamine a number of transport properties. In doing so one should note that the position of the percolation thresholds $\mathcal{E}_c$ and $\mathcal{E}_v$ depend self-consistently on the occupation of the localized states. They will therefore be changed with the electron distribution at high injection or photoexcitation levels or by high electric fields.

2.2. Contacts, Surfaces, Space Charges, Field Effect and Injection into Amorphous Semiconductors

2.2.1. INTRODUCTION

It is a surprising fact that most papers reporting electrical transport properties of amorphous semiconductors do not mention problems with contacts or special effects associated with surfaces although many experiments use thin films. Those of us who lived through the early years of crystalline semiconductor research remember the know-how needed to make ohmic contacts and the care taken by four probe measurements and surface preparation to obtain results which reflect the true bulk properties of the semiconducting crystals.(29) As a matter of fact, were it not for the useful rectifying and injecting effects of semiconductor contacts discovered in the early 40's, which led to a large variety of devices, this field might never have attracted much attention.(30)

In this section I wish to discuss what is known about contacts and surfaces of amorphous semiconductors and explore why they behave differently from those of crystalline materials.

In this brief review I am unfortunately unable to discuss the pioneering and comprehensive work on contacts by Simmons *et al.*(30a)

2.2.2. SPACE CHARGE LAYERS

There is no reason why there should not be space charge regions in amorphous semiconductors adjacent to contacts or surfaces. When a metal is placed in contact with an amorphous semiconductor the different electronegativity of the materials will cause a transfer of charge until the Fermi levels of the two materials are equalized. When an appreciable density of surface or interface states ($\sim 10^{14}$ cm^{-2} eV^{-1}) is present, the metal and the semiconductor come into equilibrium with the interface layer essentially independently, but again a space charge region in the semiconductor screens the surface potential.

The extra charges in the semiconductor reside in localized and extended states. The form of the screening potential $U(x)$ is obtained by solving Poisson's equation

$$\frac{d^2 U}{dx^2} = -\rho/\kappa\epsilon_0 \tag{2.12}$$

where κ = dielectric constant and $\epsilon_0 = 8{\cdot}86 \times 10^{-14}$ F/cm. The charge density $\rho(x)$ in turn depends on the local potential $U(x)$ as

$$\rho(x) = -e \int g(\mathcal{E}) \left[f(\mathcal{E} + eU(x)) - f(\mathcal{E}) \right] d\mathcal{E} \tag{2.13}$$

so that we have to assume a distribution of states $g(\mathcal{E})$ before we are able to solve eqn (2.12). It is interesting to see what happens in a semiconductor whose band gap is free of localized states. Only the free electrons (n) and holes (p) are then available for screening and one obtains the screening potential of an intrinsic semiconductor[29]

$$\tanh(v/4) = \tanh(v_s/4) \exp(-x/\lambda) \tag{2.14}$$

where

$$v = eU/kT \tag{2.15}$$

and v_s is the value of v at the interface. The screening length λ is determined by the number density of screening charges $\mathcal{N}_s$ through the relation

$$\lambda = (\kappa \epsilon_0 kT/e^2 \mathcal{N}_s)^{\frac{1}{2}}. \tag{2.16}$$

For the intrinsic case considered here $\mathcal{N}_s = n + p$. For a semiconductor with a band gap of 1 eV and a dielectric constant $\kappa = 16$, the free carrier densities are about 3×10^{10} cm^{-3} (at 300 K) which yields $\lambda \sim 15$ µm. Such a large screening length would strongly influence the properties of thin film samples whose thickness is of this magnitude or less. Since such effects have not been observed we conclude that λ must be smaller than given by eqn (2.16) and that screening is largely determined by localized states.

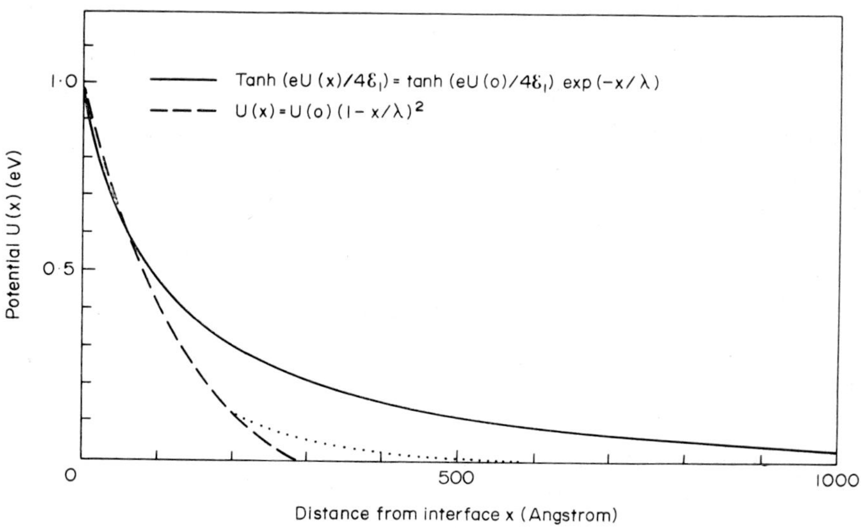

FIG. 2.16. Potential $U(x)$ in space charge layer near contact. The dotted line represents an exponential tail not described by eqn (2.19). It is found for large x when $\mathcal{E}_F$ is within the band of midgap states. (After Fritzsche, Ref. 31.)

Figure 2.16 compares the screening potential obtained from two models.[31] In both cases an interface potential $U(0) = 1$ eV was assumed and $\kappa = 16$. In the first model we assumed that tails of localized states drop exponentially toward $\mathcal{E}_F$ as

$$g(\mathcal{E}_v) \exp[-(\mathcal{E} - \mathcal{E}_v)/\mathcal{E}_1] \quad \text{and} \quad g(\mathcal{E}_c) \exp[-(\mathcal{E}_c - \mathcal{E})/\mathcal{E}_1]. \qquad (2.17)$$

The calculation of the barrier potential is similar to that for an intrinsic semiconductor so that $U(x)$ is obtained from eqn (2.14) with

$$v(x) = eU(x)/\mathcal{E}_1 \qquad (2.18)$$

The screening length λ is given by eqn (2.16) with the effective density of screening charges $\mathcal{N}_s = g(\mathcal{E}_F)\,kT$ in this case. As a numerical example we take a material having a mobility gap $\mathcal{E}_c - \mathcal{E}_v \sim 1$ eV and $\mathcal{E}_1 = 0{\cdot}14$ eV and obtain $g(\mathcal{E}_F) = 10^{18}$ cm^{-3} eV^{-1} and $\lambda = 300$ Å.

As a last example let us consider the screening potential which results from the hump of midgap states as proposed by Davis and Mott.[32] A hump of midgap states of height $g(\mathcal{E}_F)$ and halfwidth δ yields a constant space charge near the interface and thus a parabolic Schottky barrier

$$U(x) = U(0)\,(1 - x/\lambda)^2 \qquad (2.19)$$

with λ given by eqn (2.16) with $\mathcal{N}_s = g(\mathcal{E}_F)\,\delta$. Fig. 2.16 shows as a numerical example $U(x)$ for $g(\mathcal{E}_F) = 5 \times 10^{17}$ eV^{-1} cm^{-3} and halfwidth $\delta = 0{\cdot}05$ eV so that again $\lambda = 300$ Å is obtained. As expected the screening is more effective the closer the localized states are concentrated near $\mathcal{E}_F$. We now turn to the experimental evidence for such space charge layers and try to explain why they do not give rise to rectifying contacts.

2.2.3. EXPERIMENTAL EVIDENCE

The space charge region is either one of depletion or of accumulation of the carriers dominating the conduction. The width λ of the space charge region may be estimated by looking for deviations from the linear relation between conductance and the film thickness of the amorphous semiconductor. As far as I know no deviations have been noticed down to thicknesses of 1000 Å within the accuracy of reproducibility of film depositions.

Another method[33] uses the different absorption depth of light of different photon energies as a probe to measure the width λ. As a monitor one needs an effect which can distinguish light absorbed in the space charge region from light absorbed in the field free bulk region. Such an effect is the photovoltage. It is reasonable to assume that only those electron-hole pairs which are photoexcited in the field region of the space charge layer contribute to the photovoltaic effect. Let us see whether the experimental results justify such an interpretation.

Contact Photovoltage

Figure 2.17 shows the open circuit d.c. photovoltage[(34)] of a chalcogenide alloy glass $Ge_{16}As_{35}Te_{28}S_{21}$ which was sputtered as a 5 μm thick film onto a conducting SnO_2 electrode. The material has a conductivity

$$\sigma = 3\times 10^{-8}\ \text{ohm}^{-1}\ \text{cm}^{-1}$$

at 300 K and an optical gap $\mathcal{E}_0 = 1{\cdot}18$ eV. The upper half of Fig. 2.17 shows the spectral response of the photovoltage of a sample with a top electrode of Sb. The lower part of the figure shows the same for a sample with Au as a top electrode. The samples were illuminated through the SnO_2 bottom electrode with different light intensities produced by neutral density filters. The spectral response looks so complicated because one is dealing here with three photovoltaic sources which are connected in series. For $h\nu > 1{\cdot}5$ eV the semiconductor is essentially opaque so that only the bottom electrode is illuminated. The small negative voltage originates from the space charge layer in the SnO_2. It is absent when the bottom electrode is a semitransparent metal. The positive contribution between $1{\cdot}4 < h\nu < 2{\cdot}8$ eV in Fig. 2.17 is the barrier photovoltage of the bottom semiconductor contact. For $h\nu < 1{\cdot}5$eV the absorption coefficient is sufficiently small so that light penetrates the semiconductor and produces an additional photovoltage at the top contact. The fact that this voltage has the opposite sign for Sb and Au demonstrates that we are dealing here with a contact photovoltage rather than a Dember voltage which depends only on the difference in mobilities of electrons and holes. From the sign of the photovoltage measured with different contact materials we conclude that at contacts with Au, Al, SnO_2, Bi, Pb, In, Sn, and nichrome, the bands bend downward (negative space charge). The Sb contact of Fig. 2.17 (upper half) gave rise to a positive space charge. Other Sb contacts and also Te often yield a neutral or flat band condition. This is qualitatively what one would expect because the first group of electrode materials is more electropositive than the amorphous semiconductor, whereas Sb and Te are not.

Figure 2.18(a) and (b) shows the spectral response of the photovoltage of a sample contacted with one Sb and one Ni/Cr contact and with two nichrome (Ni/Cr) contacts, respectively.[(33)] The direction of illumination is from the top as shown. The photon energies at which the product of the absorption coefficient α and the sample thickness d is equal to 1 and 5, respectively, are indicated by arrows on the top scale of the figures. The absence of any photovoltage for $h\nu > 1{\cdot}7$ eV in Fig. 2.18(a) shows that we achieved a flat band condition at this Sb contact. In this case then, the photovoltage peak

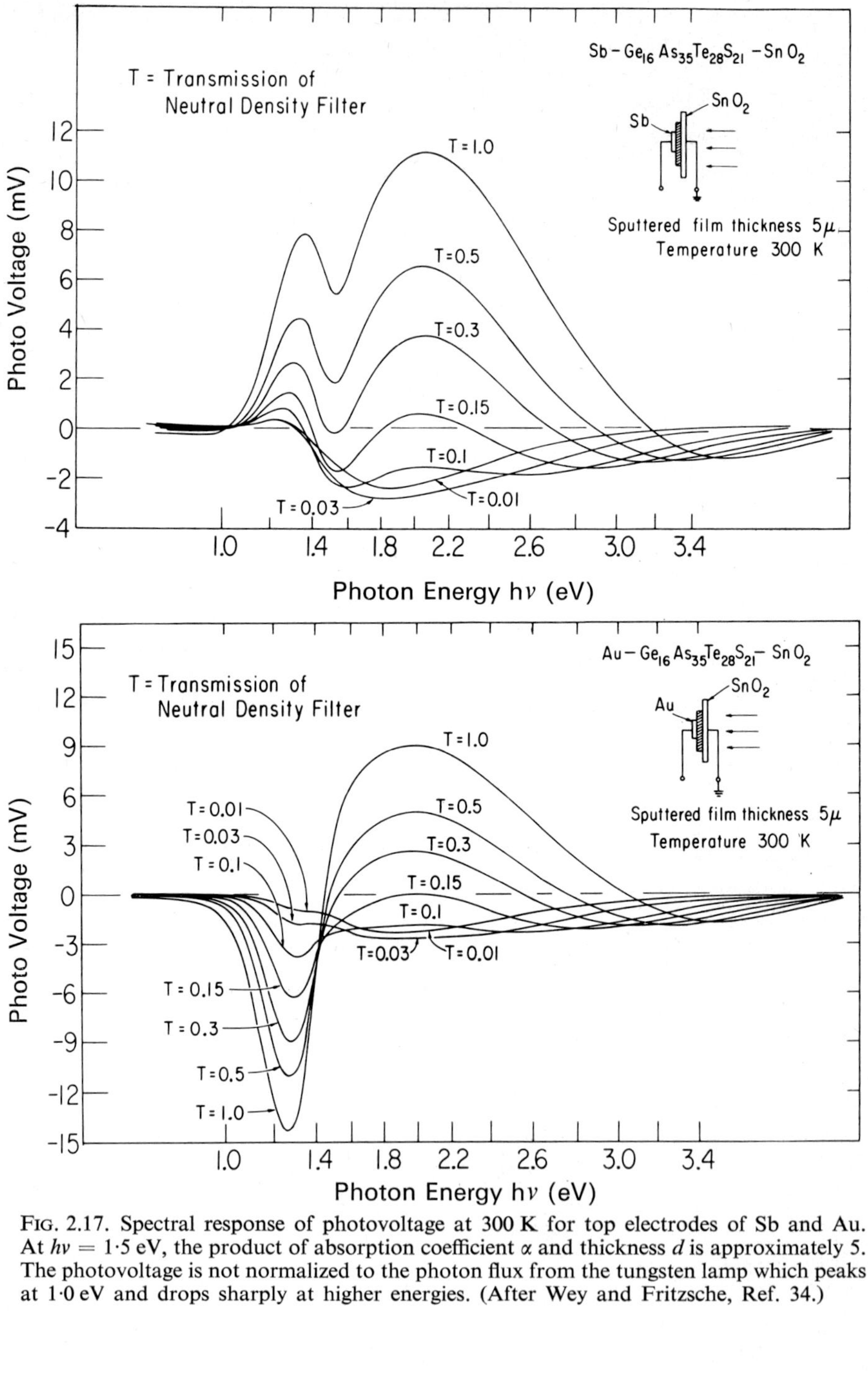

FIG. 2.17. Spectral response of photovoltage at 300 K for top electrodes of Sb and Au. At $h\nu = 1{\cdot}5$ eV, the product of absorption coefficient α and thickness d is approximately 5. The photovoltage is not normalized to the photon flux from the tungsten lamp which peaks at 1·0 eV and drops sharply at higher energies. (After Wey and Fritzsche, Ref. 34.)

is entirely due to the bottom Ni/Cr contact. Because this sample contains only one space charge layer we used this sample to determine the width of this layer by measuring the space charge capacitance as a function of frequency. This will be discussed later.

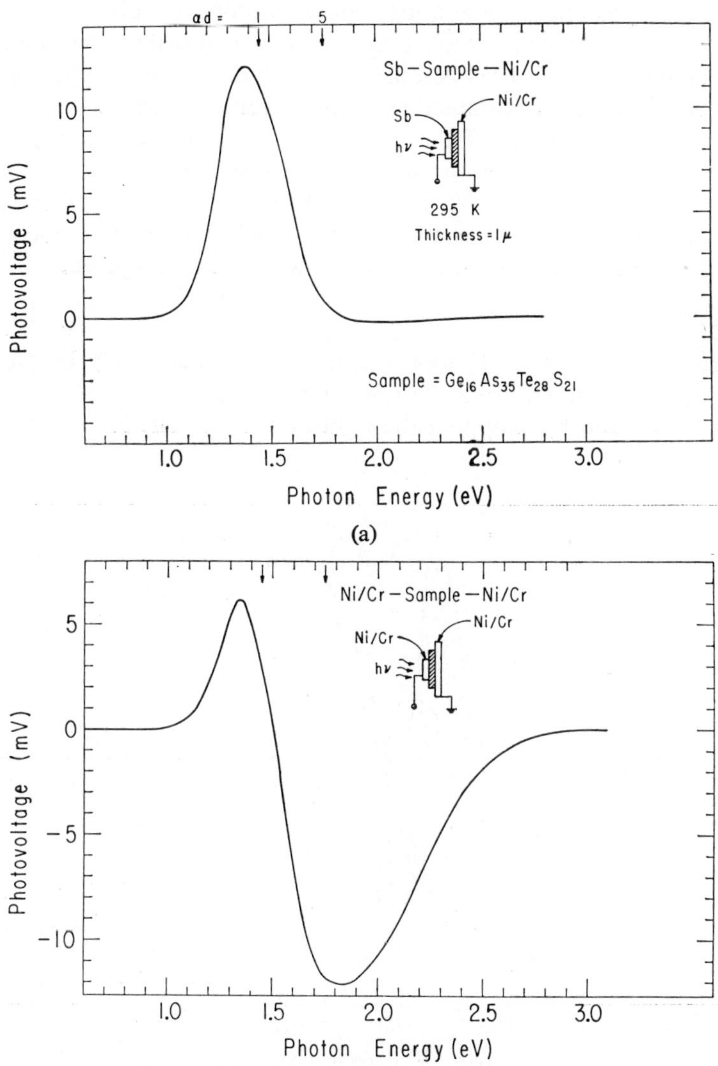

FIG. 2.18. Spectral response of photovoltage (a) for top electrode of Sb which produces a flat band contact and (b) for Ni/Cr contact which produces a negative space charge layer. (After Wey and Fritzsche, Ref. 33.)

Figure 2.18(b) shows the photovoltage of two Ni/Cr contacts. The sign for the upper contact photovoltage is negative because its internal field points in a direction opposite to that of the lower contact. At large *hv* only the top photovoltage is seen; at small *hv* one observes the resultant of the top and bottom photovoltage. The fact that the positive sign of the bottom photovoltage predominates although more photons are absorbed near the top shows that the contacts were different. This is possible even though the same contacting metal is used because the interface is different if the sequence of deposition of the semiconductor and the metal is reversed.

A quantitative analysis of the spectral response of the photovoltage is easier if one need not worry about the light loss and reflection due to the semitransparent electrode. We therefore measured the photovoltage of a free surface and made the semiconductor thick enough that the bottom electrode remained unexposed in the spectral range investigated.

Surface Photovoltage

The surface photovoltage V_s was measured[33] capacitatively by placing a semitransparent Au electrode 25 μm apart from the free surface of the semiconductor and by chopping the light at 150 Hz. The insert in Fig. 2.19

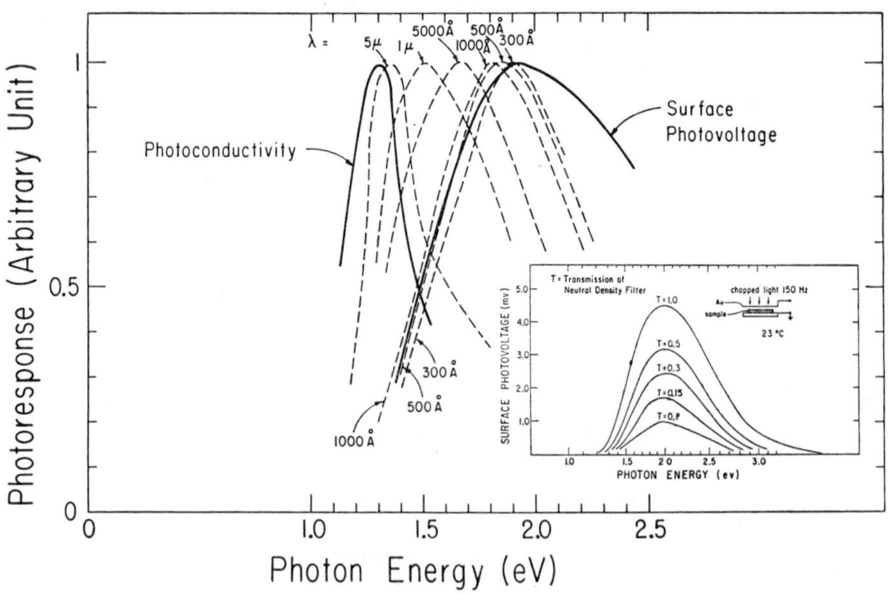

FIG. 2.19. Spectral photoresponse of photoconductivity and of surface photovoltage (full curves) compared with dashed curves calculated from eqns (2.20) and (2.21). The thickness of the photoconductivity sample was about 6 μm. (After Wey and Fritzsche, Ref. 33.)

shows the spectral response of V_s for various light intensities at 300 K. The sign of the photovoltage corresponds to a negative space charge region adjacent to the surface. If we assume that only those photons contribute to V_s which create electron–hole pairs in the space charge region of width λ then V_s can be written

$$V_s = C\eta\mathcal{N}_0 \tag{2.20}$$

$$\eta = 1 - \exp[-\alpha\lambda] \tag{2.21}$$

where $\mathcal{N}_0(h\nu)$ is the incident photon flux. Here we assume that each absorbed photon produces an electron–hole pair and that those generated within λ contribute equally to V_s regardless of the photon energy. The factor $\eta(h\nu)$ increases with $h\nu$ because of the increase in the absorption coefficient α. One expects η to saturate when all photons are absorbed in the region λ. Since we know $\mathcal{N}_0$ and α as a function of $h\nu$ we can calculate V_s for various space charge widths λ. Such a family of curves is shown in Fig. 2.19 by the dashed curves. The peak values are normalized to unity. The heavy lines in Fig. 2.19 represent the measured spectral response of the surface photovoltage and

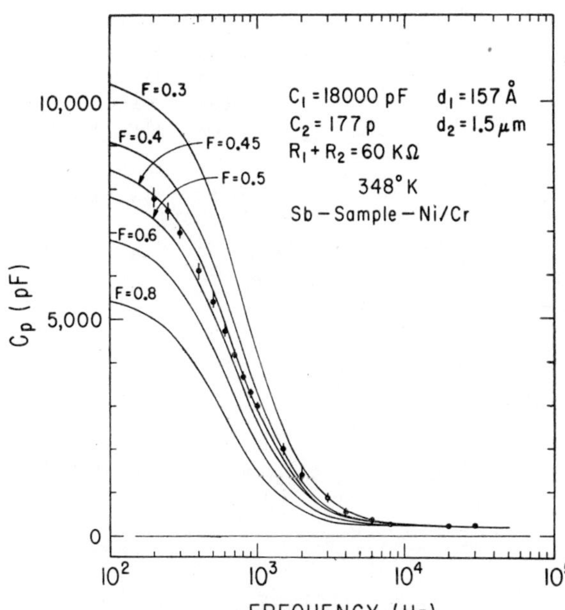

FIG. 2.20. Parallel capacitance C_p as a function of frequency for various resistance ratios $F = R_2/R_1$. Best fit to data points with $F = 0{\cdot}45$. Width of space charge region is $d_1 = \lambda$, electrode separation is d_2. (After Wey and Fritzsche, Ref. 33.)

of the photoconductivity, for which the same arguments apply if λ is replaced by the sample thickness d. A modestly good fit to the experimental curves is obtained with λ between 300 and 500 Å for V_s and $d = 6$ μm for the photoconductivity.

This simple model does not take account of the decrease of the internal field from its maximum value at the surface to zero at $x = \lambda$. Furthermore the model predicts V_s to increase linearly with light intensity which is not borne out by experiment, as shown in the insert of Fig. 2.19. Nevertheless, the comparison of the spectral response of the surface photovoltage with that of the photoconductivity suggests that the effective space charge layer is much less than the thickness. It appears to be of the order of a few hundred angstrom.

Capacitance

The frequency dependence of the capacitance was measured[33] on the sample whose photovoltage is shown in Fig. 2.18(a). According to the photovoltage measurements the bands are flat adjacent to the Sb electrode and bend

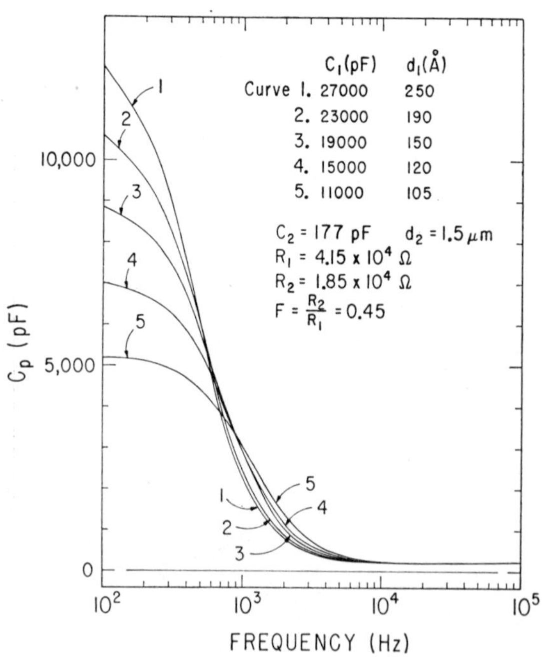

FIG. 2.21. Parallel capacitance C_p as a function of frequency for various screening lengths λ. Best fit to data points shown in Fig. 2.20 is obtained with $\lambda = d_1 = 160$ Å. (After Wey and Fritzsche, Ref. 33.)

downward at the Ni/Cr electrode. The parallel capacitance C_p measured at 75°C over the frequency range between 100 Hz and 5 HMz is shown in Figs. 2.20 and 2.21. The capacitance at low frequencies is about 40 times higher than that at high frequencies. This behaviour can be analyzed by assuming the model shown in Fig. 2.22. The space charge region of thickness λ is represented by a capacitance C_1 and resistance R_1. The bulk region of thickness d_2 has a capacitance C_2 and resistance R_2. This parallel series combination is represented by the equivalent circuit of the parallel capacitance C_p and resistance R_p. It is clear from Fig. 2.22 that the space charge capacitance can only be measured if the dielectric relaxation time

$$\tau_{\text{rel}} = \kappa\epsilon_0/\sigma$$

in the space charge region is different from that in the bulk. We believe that the dielectric constant κ is the same in the space charge region and the bulk. Thus we require that the conductivity σ_1 be smaller than σ_2. If that is so one observes at high frequencies the geometric capacitance of the sample sandwich $C_p = C_2$. At low frequencies C_p rises to a value which depends on

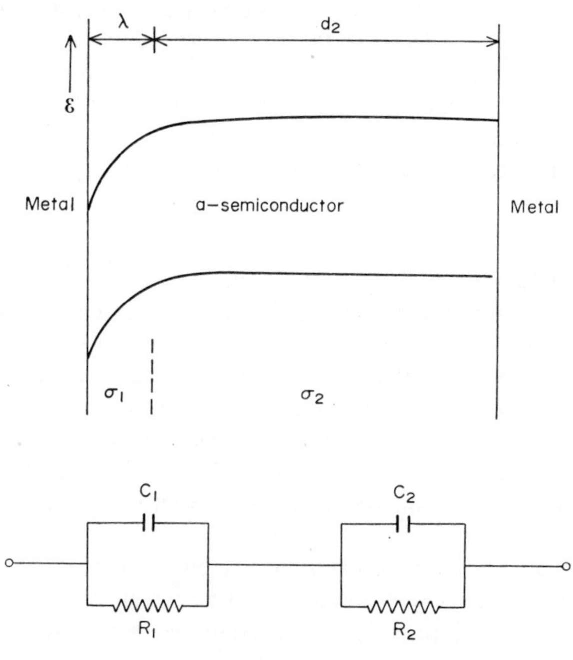

FIG. 2.22. Equivalent network used for amorphous semiconductor with one space charge layer of thickness λ and conductivity σ_1. The bulk material of thickness d_2 has conductivity σ_2.

the relative magnitudes of the thicknesses and relaxation times of the space charge region and the bulk. The full expression of C_p in terms of resistances and capacitances in Fig. 2.22 is

$$C_p = \frac{R_1{}^2 C_1 + R_2{}^2 C_2 + \omega^2 R_1{}^2 R_2{}^2 C_1 C_2 (C_1 + C_2)}{(R_1 + R_2)^2 + \omega^2 R_1{}^2 R_2{}^2 (C_1 + C_2)^2}. \qquad (2.22)$$

Only the geometric capacitance C_2 and the total resistance $R = R_1 + R_2$ can be measured independently. Equation (2.22) still contains the thickness of the space charge region λ and the ratio of the conductivities σ_1/σ_2 as the two unknown parameters. These can be determined by fitting the experimental curve by calculating C_p for a variety of these parameters and finding the best fit curve. This is shown in Figs. 2.20 and 2.21 which present C_p as a function of frequency for various resistance ratios $F = R_2/R_1$ and different values for the space charge thickness λ. With the additional measurements of the sample thickness $= 1{\cdot}5\ \mu$m and the total bulk resistance = 60 K ohms and contact area of 0·02 cm^2 we obtain for the best fit curve at 75°C a space charge width of 160 Å, a bulk resistivity of $2{\cdot}3 \times 10^6$ ohm cm, and a resistivity in the space charge region of $5{\cdot}3 \times 10^8$ ohm cm.

These capacity measurements were carried out at 75°C rather than at room temperature in order to decrease the dielectric relaxation time so that the frequency at which C_p begins to deviate from the geometric capacitance lies in the middle of the frequency range accessible to our capacitance bridge. At room temperature the deviation of C_p occurs at such low frequencies that very little of the rise of C_p could be measured.

2.2.4. RECTIFYING PROPERTY

It is interesting to note that all samples discussed above, which showed a sizable contact photovoltage, exhibit ohmic 1–V characteristics up to fields of about 10^4 V/cm. From the I–V curves alone one would have concluded that there are no space charge regions and that the conductance is determined by the bulk conductivity. A negative space charge region should be blocking to holes which are the dominant current carriers in the chalcogenide alloys. Why do we not find contacts to amorphous semiconductors more often rectifying or blocking? To answer this question let us look at what gives rise to rectification in crystalline semiconductors.

If the material is an n-type crystalline semiconductor as shown in Fig. 2.23(a), the bulk conductance is determined by the electron density n_0 which is large because $\mathcal{E}_F$ is close to $\mathcal{E}_c$. The reverse bias saturation current or the contact conductance is exponentially limited by the barrier height because the minority carrier density p_0 is exceedingly small. Consequently the barrier resistance is large compared to the bulk resistance and a high rectification ratio is observed. Let us now consider a contact to an amorphous semiconductor as sketched in Fig. 2.23(b). For simplicity we again assume predominant electron conduction. The main difference between this and the previous case is that now the Fermi level $\mathcal{E}_F$ is close to the gap centre. This decreases n_0 and increases the bulk resistance by a factor of order

$$\exp (\mathcal{E}_c - \mathcal{E}_v)/2\,kT.$$

At the same time the barrier resistance is lowered because holes can now participate. As a consequence the electron current in the bulk region has no greater conductance than the hole current in the barrier region except for the difference in electron and hole mobilities. Rectification is poor or does not occur. Generation and recombination provide charge continuity and equalize the currents near the onset of the space charge region. These are essentially the same reasons why contacts to intrinsic crystalline semiconductors fail to be rectifying.

A barrier region which is only a few hundred Angstrom wide should actually not be presented by a smooth potential as shown in Fig. 2.23(b). The number of localized states which would yield such a small screening length are only 100 Å apart on the average. Therefore the fluctuating local potential can no

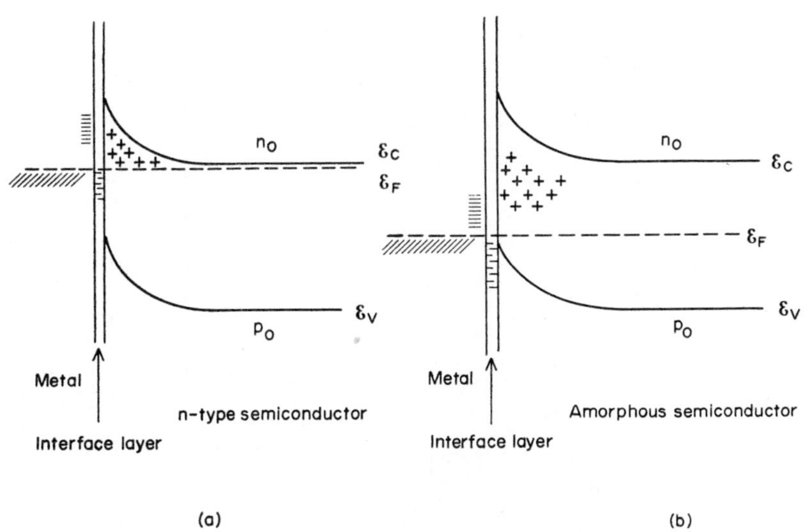

FIG. 2.23. Potential distribution near metal–semiconductor contacts. The sign of the charge in the interface layer is chosen arbitrarily. (a) n-type crystalline semiconductor. The resistance in the bulk is low on account of the large value of n_0. In comparison, the Schottky barrier presents a large contact resistance. (b) amorphous semiconductor with $\mathcal{E}_F$ near gap centre. Ambipolar current flow allows the barrier region to have a negligible resistance compared to that of bulk.

longer be neglected. This might permit carriers to pass the barrier at certain spots. Moreover, in writing eqn (2.13) we assumed that the density of localized and extended states $g(\mathcal{E})$ remains unchanged by the electrostatic potential $U(x)$. This is permissible only as long as variations in $U(x)$ over distances equal to the electron wavelength are small compared to kT/e. This assumption will not hold near the interface.[35]

Rectifying effects can be observed as soon as $\mathcal{E}_F$ is not longer close to the gap centre. This is the case in *a*-Ge and *a*-Si studied by Grigorovici *et al.*,[36] Henisch and Vendura,[37] and Jayadeviah *et al.*[38] Very high contact resistances are usually encountered in wide gap (larger than 1·5 eV) amorphous semiconductors such as Se, As_2Se_3, and As_2S_3. They have fewer localized states and wide space charge regions. Figure 2.24 illustrates a few points which are often observed at contacts to semi-insulating materials.[39,40] Shown here are I–V curves plotted logarithmically of three different contact com-

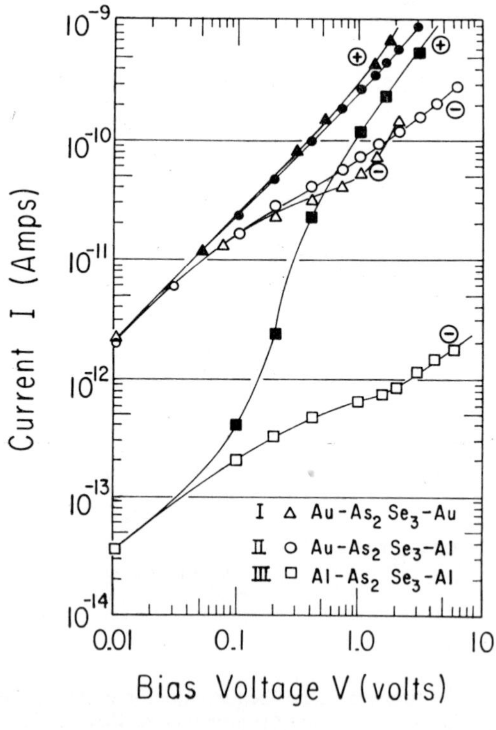

FIG. 2.24. Current voltage characteristics at 300 K of a 2 μm thick evaporated As_2Se_3 film sandwiched between different evaporated metal electrodes. The code name of the diodes lists the materials in sequence of deposition. The polarity is that of the top with respect to the bottom electrode. (After Carruthers, Ref. 39.)

binations to a 2 μm thick evaporated As_2Se_3 film.[39] The first metal in the code name for the diode shown in Fig. 2.24 was evaporated first, then As_2Se_3 with the second metal electrode on top. The voltage is that of the top electrode with respect to the bottom electrode. One often observes that

(i) when the bottom electrode is a nonoxidizing metal, the low voltage ohmic region corresponds to the bulk resistance of the amorphous material (diodes I and II).

(ii) when the bottom electrode is allowed to oxidize before it is covered with the amorphous film, it gives rise to a larger contact resistance (diode III).

(iii) at larger biases ($V \sim 0{\cdot}5$ volt) the oxidized contact passes electrons from the metal into the amorphous semiconductor but not in reverse. This is often called a hole blocking contact.

(iv) the choice of metal for the top electrode is not so critical because oxidation is less likely to occur. The top electrode might differ in its behaviour from the bottom electrode even when the materials are the same (diode I), because the interfaces are prepared quite differently.

2.2.5. FIELD EFFECT

The surface potential and hence the space charge adjacent to the surface or interface can be changed by applying an electric field perpendicular to the surface. This can be done conveniently in the conventional insulated gate field effect geometry where the amorphous semiconductor is used as one plate of a parallel plate capacitor with mylar, mica, Al_2O_3, or SiO_2 as an insulator. The dielectric strength of a typical insulator permits about $\Delta\mathcal{N}_t = 2 \times 10^{12}$ cm^{-2} carriers to be deposited per unit area in the surface region. If no gap states or surface states were available then these extra carriers would go into the conduction band (or valence band for opposite polarity). This would substantially increase the conductivity in the space charge region. A film of thickness $d = 500$ Å for example, would receive on the average an extra carrier density of $\Delta\mathcal{N}_t/d = 4 \times 10^{17}$ cm^{-3}. This is an increase by a factor 10^7 over the equilibrium carrier density $n = 3 \times 10^{10}$ cm^{-3} of a semiconductor having a gap $\mathcal{E}_c - \mathcal{E}_v = 1$eV at 300°K.

Egerton[41] studied the field effect in a number of chalcogenide alloy glasses and found typically a conductance change of only a factor 2 as shown in Fig. 2.25. The smallness of the conductance change implies that most of the field-induced charges went into gap and surface states and produced a shift of the Fermi level of only $\Delta\mathcal{E}_F \sim 0{\cdot}02$ eV according to

$$\frac{\Delta\sigma}{\sigma} = \frac{\Delta\mathcal{E}_F}{kT}. \tag{2.23}$$

A density of gap states of about

$$g(\mathcal{E}_F) = \Delta\mathcal{N}_t/d\Delta\mathcal{E}_F = 2\times 10^{19}\,\mathrm{cm}^{-3}\,\mathrm{eV}^{-1} \tag{2.24}$$

would explain the results or a surface state density of

$$\Delta\mathcal{N}_t/\Delta\mathcal{E}_F = 10^{14}\,\mathrm{cm}^{-2}\,\mathrm{eV}^{-1}.$$

Although these values may be reduced by a factor 2 if we assume surface and bulk states to be both present. Interface state densities of order 5×10^{13} $\mathrm{cm}^{-2}\,\mathrm{eV}^{-1}$ are quite common in crystalline semiconductors.

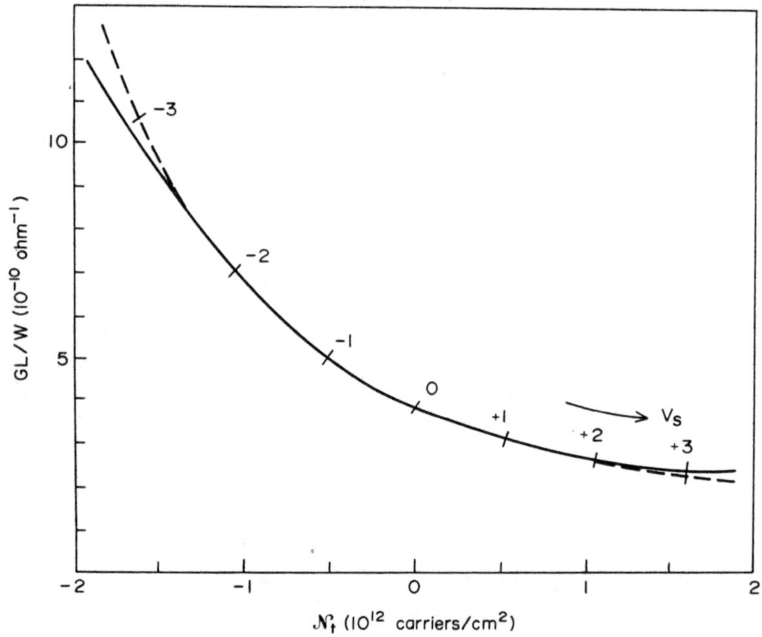

FIG. 2.25. Field induced change of conductance G of a 450 Å thick film of chalcogenide alloy glass measured with 50 Hz modulation. W = film width. GL/W is plotted against $\Delta\mathcal{N}_t$, the surface density of electrons deposited. Dotted line shows fit of eqn (2.27) to experimental curve assuming no surface states. (After Egerton, Ref. 41.)

Of all amorphous semiconductors studied so far there is only one material which showed a large field effect. This is amorphous Si produced by radio frequency glow discharge of SiH_4. Spear and LeComber[(13)] were able to map out the distribution of gap states through most of the gap. The density varies between 10^{17} and $10^{19}\,\mathrm{cm}^{-3}\,\mathrm{eV}^{-1}$. These results will be discussed in the section on localized states (Section 2.3).

Equation (2.23) allows a rough estimate of the conductance change only when the film thickness is small compared to the screening length λ. In general, the rapid change of potential within λ requires integration over the coordinate x normal to the interface.[15] The solution of Poisson's equation yields for the shift of $\mathcal{E}_F$ relative to the conduction band edge

$$\Delta\mathcal{E}_F(x) = \frac{\Delta\mathcal{N}}{g(\mathcal{E}_F)\,\lambda} \exp(-x/\lambda) \tag{2.25}$$

where $\Delta\mathcal{N}$ is the number of field-induced electrons per unit area which are not trapped in surface states. The gap state density $g(\mathcal{E}_F)$ is assumed to be constant. With $\Delta\mathcal{E}_F(x)$ governing the change in free carriers one obtains

$$\frac{\Delta\sigma(x)}{\sigma_0} = \frac{b}{1+b}[\exp(\Delta\mathcal{E}_F/kT) - 1] + \frac{1}{1+b}[\exp(-\Delta\mathcal{E}_F/kT) - 1]. \tag{2.26}$$

The parameter b is the ratio of electron to hole mobility when $\Delta\mathcal{E}_F = 0$. The field-induced conductance change is obtained by integrating eqn (2.26) over the sample thickness d

$$\frac{\Delta G}{G_0} = \frac{\lambda}{d}\left[\frac{b}{1+b}F(-v_s) + \frac{b}{1+b}F(v_s)\right] \tag{2.27}$$

where

$$F(v_s) = \int_0^{v_s} (e^{-v} - 1)\,\mathrm{d}V/V = \sum_{m=1}^{\infty} \frac{(-v_s)^m}{m.m!} \tag{2.28}$$

and $v_s = \Delta\mathcal{E}_F(0)/kT$. A fit of eqn (2.28) to the experimental curve is shown in Fig. 2.25. Egerton[41] finds with $b = 0{\cdot}2$ that the holes are more mobile than electrons. The field effect turns out to be one of the few experiments which yields information about the relative magnitude of the mobilities.

2.2.6. RECOMBINATION INJECTION

Recently van Roosbroeck and collaborators[42,43] have drawn attention to the fact that electronic transport exhibits a number of novel features when the dielectric relaxation time τ_{rel} becomes long compared with the excess carrier recombination time τ_0. Van Roosbroeck's theory has cast some doubts on the conventional interpretation of several transport phenomena in amorphous semiconductors and suggests a theory of switching. It therefore seems important to present here the major elements of this theory and to explore whether amorphous semiconductors indeed fall into this so-called relaxation time regime.

Let us consider minority carrier injection in (a) the conventional lifetime case $\tau_{rel} < \tau_0$ and (b) the relaxation time case $\tau_{rel} > \tau_0$. At equilibrium the electron and hole distributions are characterized by one Fermi level $\mathcal{E}_F{}^0$. A pulse of extra free carriers Δn injected into the conduction band can be described by a quasi-Fermi level $\mathcal{E}_F{}^n$ which is merely a different way of keeping count of the new free electron density

$$n = n_0 + \Delta n = N_c \exp[-(\mathcal{E}_c - \mathcal{E}_F{}^n)/kT] \tag{2.29}$$

$\mathcal{E}_F{}^n$ does not reveal any information about the occupancy of localized states or of the hole density, which is described by a hole quasi-Fermi level $\mathcal{E}_F{}^p$ analogous to eqn (2.29). Equilibrium is restored by two simultaneously acting processes: relaxation and recombination. The field associated with the extra minority carrier charge $-e\Delta n$ attracts majority carriers Δp from neighbouring volume elements until the charge is neutralized. The characteristic time for this process is the relaxation time $\tau_{rel} = \kappa\epsilon_0/\sigma$. The end result of this process is neutrality rather than equilibrium. The recombination-generation process characterized by a lifetime τ_0 acts locally to restore the law of mass action, i.e. it causes $\mathcal{E}_F{}^n$ and $\mathcal{E}_F{}^p$ to coalesce into one Fermi level which, however, may differ in energy from the equilibrium $\mathcal{E}_F{}^0$. The end result of the recombination-generation process is that the product of electron and hole concentrations recovers its equilibrium value

$$np = n_0 p_0. \tag{2.30}$$

Only the combined action of relaxation and recombination restores equilibrium. Van Roosbroeck *et al.* points out that transport phenomena involving nonequilibrium carriers depend strongly on which of the two processes is rate limiting. Equilibration of a volume element is shown in Fig. 2.26(a) and (b) for the conventional lifetime regime ($\tau_{rel} < \tau_0$) and the new relaxation regime ($\tau_{rel} > \tau_0$), respectively. In ordinary crystalline semiconductors space charge neutrality is restored almost immediately, $\tau_{rel} \sim 10^{-10}$ sec. The charge of an excess minority carrier density Δn is thus compensated by a lateral influx of Δp majority carriers. The two quasi-Fermi levels split and slowly relax to equilibrium within the minority carrier lifetime which may be of order milliseconds (see Fig. 2.26(a)). In the relaxation regime, on the other hand, the *np* product is quickly restored by decreasing the local majority carrier density as shown in Fig. 2.26(b). The space charge is thereby enhanced. It relaxes slowly with τ_{rel}.

Because of the persistance of space charge the analysis of injection in the relaxation regime is rather complex and the results differ greatly from those in the conventional regime. One finds, for example, a majority carrier depletion region adjacent to a minority carrier injecting contact followed by

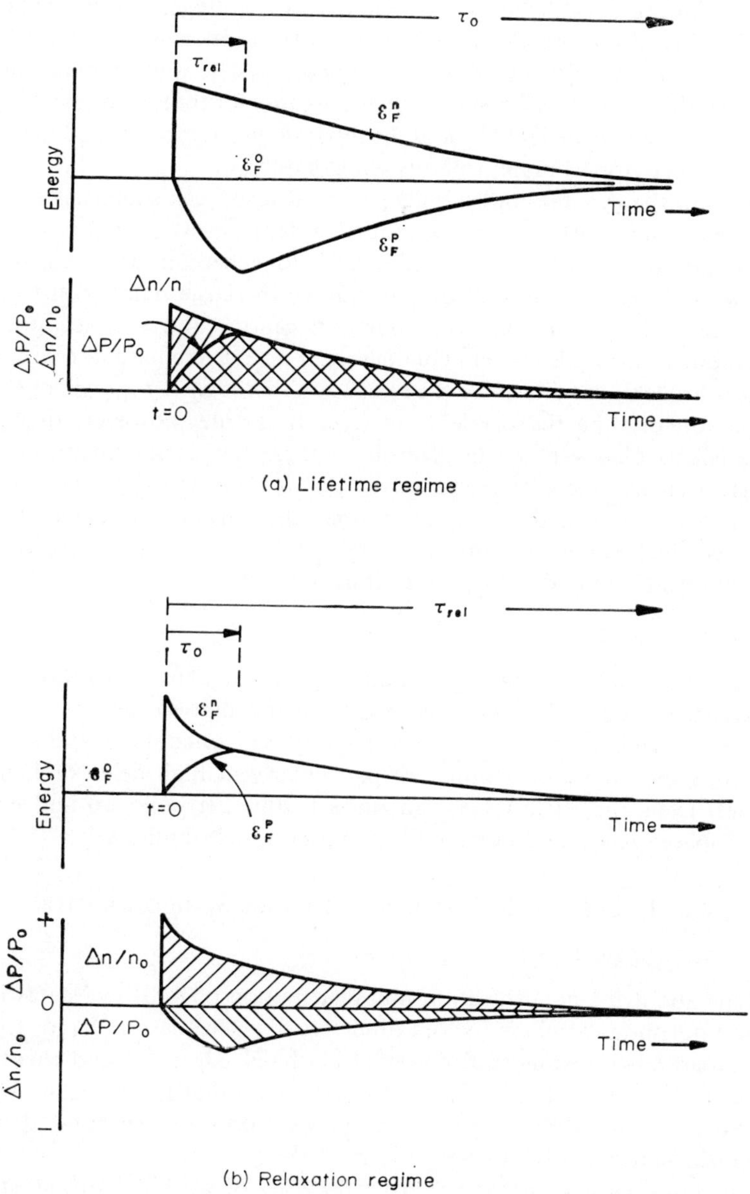

FIG. 2.26. Response of quasi Fermi levels $\mathcal{E}_F{}^n$ and $\mathcal{E}_F{}^p$ and of the relative carrier concentrations to a pulse of excess minority carriers (electrons), (a) in the lifetime regime $\tau_0 > \tau_{rel}$ and (b) in the relaxation regime $\tau_0 < \tau_{rel}$.

a narrow recombination front. The majority carrier depletion decreases the current, causing a sublinear $V^{\frac{1}{2}}$ dependence on voltage. Experimental evidence for this was found[44] in crystalline GaAs diodes which contained large and equal concentrations of donors and acceptors. It was claimed, furthermore, that this recombination injection is the cause for threshold switching,[43] the negative Hall coefficient,[45] and the intrinsic-like temperature dependence of the conductivity[46] in amorphous semiconductors.

We doubt the general applicability of van Roosbroeck's analysis to amorphous semiconductors.[47] Photoconductive decay measurements show that the equilibration time τ_0 is considerably longer than assumed by van Roosbroeck *et al.* The reason for this may be the large number of trapping levels which do not act as recombination centres and perhaps potential fluctuations in amorphous semiconductors. For a semiconductor with traps van Roosbroeck's condition for the relaxation case is $\tau_0 < f\tau_{\text{rel}}$ where $f \ll 1$ is the fraction of time the carriers are free. It is true, however, that space charge effects play a much larger role in these low conductivity materials than is often assumed. It seems advisable to familiarize ourself with van Roosbroeck's ideas and to be aware that the various characteristic time constants which govern the injection and equilibration kinetics are very different in crystalline and amorphous semiconductors.

2.2.7. CONCLUSIONS

The apparent ease of making conducting contacts to amorphous semiconductors should not deceive us into believing that we understand these contacts. Space charge layers are present adjacent to contacts and surfaces. They are likely to affect conduction particularly at high fields or high frequencies. Long dielectric relaxation times favour the build up and persistance of space charge and nonequilibrium carrier distributions.

2.3. Localized States in Amorphous Semiconductors

2.3.1. INTRODUCTION

Opinions about the concentration and distribution of localized states in the gap of amorphous semiconductors differ greatly. On the one hand, there is some experimental evidence for very sharp band edges,[48] and theoretical arguments[49,50] have been presented which claim that there are hardly any localized gap states. Other experiments suggests, on the other hand, densities of localized states as high as $5 \times 10^{19}\ \text{cm}^{-3}\ \text{eV}^{-1}$.

We first ask: how can that be? After 30 years of working with crystalline semiconductors we should know how to determine the density and energy of impurity levels. The problem arises because the two most informative experiments, Hall effect and resistivity measurements, disappoint us in amor-

TABLE 2.2

Density of gap states near Fermi energy
(Units of $cm^{-3} eV^{-1}$ except where otherwise stated)

Material classes	Tetrahedral amorphous films		Tetrahedral glasses	Chalcogenides	
	Evaporated	SiH_4 decomposed		Elements, compounds	Alloys
Examples	Ge, Si, . . .	Si	$CdGe_xAs_2$, . . .	Se, As_2Se_3, . . .	Ge–Te–As, . . .
Absorption	$<10^{17}$	—	$<10^{17}$	$<10^{17}$	—
Photoemission	$<3\times10^{19}$	—	—	$<3\times10^{19}$	—
Field effect	$>10^{19}$	10^{17}–10^{19}	—	—	$\sim10^{19}$
Tunnelling	$\sim10^{20}$	—	—	—	—
Screening length	—	—	—	—	5×10^{18}
A.C. conductivity	—	—	—	3×10^{17}–5×10^{19}	5×10^{19}
Conduction near $\mathcal{E}_F$	3×10^{18}–10^{20}	—	—	—	—
Glow curves	—	—	—	10^{18}–10^{19}	10^{18}–10^{19}
ESR[a]	3×10^{20} cm^{-3}	—	—	$<10^{14}$ cm^{-3}	$<10^{14}$ cm^{-3}
Paramagnetism[a]	—	—	10^{19}–10^{20} cm^{-3}	$<3\times10^{16}$ cm^{-3}	—

[a] Numbers refer to concentrations of paramagnetic centres.

phous semiconductors. They do not reveal an extrinsic region at low temperatures which is governed by the localized states and from which their density and distribution can be ascertained. Amorphous Ge, Si, and III–V compounds are exceptions but these materials present other problems. For the large majority of amorphous semiconductors the Hall effect and resistivity remain "intrinsic" to the lowest temperatures measured. Oddly enough this is to be expected when there are either extremely few or a large number of gap states.

We wish to examine here the experiments which yield information about the density of localized states. We find significant discrepancies between experiments. Part of the conflict is resolved if one abandons the attempt to describe structurally and chemically very different substances with one unifying model. We use instead the classification of amorphous semiconductors, discussed in (Section 2.1) and in Ref. 4, in a few major categories which are distinguished by their structure and physical properties. In other cases conflicting evidence forces us to reexamine the validity of the interpretation of some experiments. Let us first define the problem and then discuss the experimental evidence which is summarized in Table 2.2.

2.3.2. THE PROBLEM

When a semiconductor becomes disordered the identity of the energy band is sufficiently preserved that one finds about $5 \times 10^{22}\ \mathrm{cm}^{-3}$ states in a band as in a crystal. As sketched in Fig. 2.27, this yields an average density of

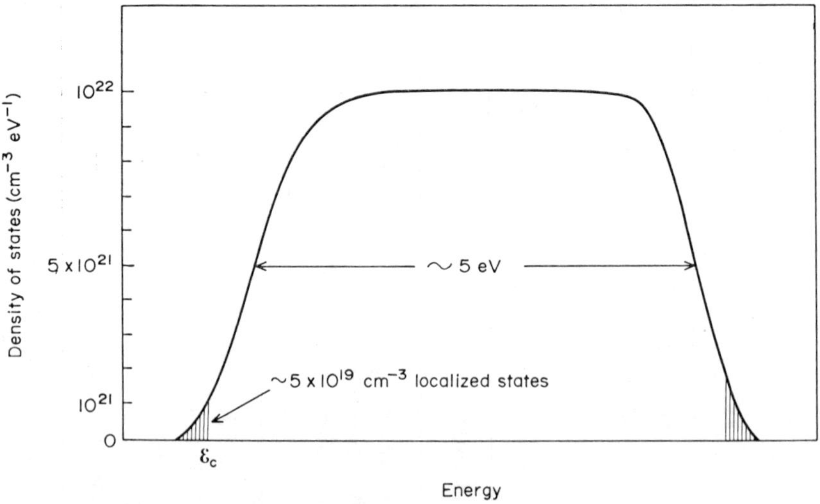

FIG. 2.27. Sketch of density of states of a conduction band in a typical amorphous semiconductor. States at the band edges are localized.

band states of $g(\mathcal{E}) = 10^{22}\,\text{cm}^{-3}\,\text{eV}^{-1}$ for a typical bandwidth of 5 eV. According to Anderson, Mott, and Cohen,[1] the states become localized near the top and bottom of the band when their density falls below about $10^{21}\,\text{eV}^{-1}\,\text{cm}^{-3}$. An estimated concentration of $5 \times 10^{19}\,\text{cm}^{-3}$ localized tail states are expected when the width of the tails is about 0·1eV. This value should depend on the disorder parameter and hence on the material but it is important to realize that the number is of order 0·1 percent of all band states. It is convenient to distinguish these *tail states* caused by the lack of long-range order from *gap states* which might originate from dangling bonds, impurities, and other defects in and deviations from a random network structure.

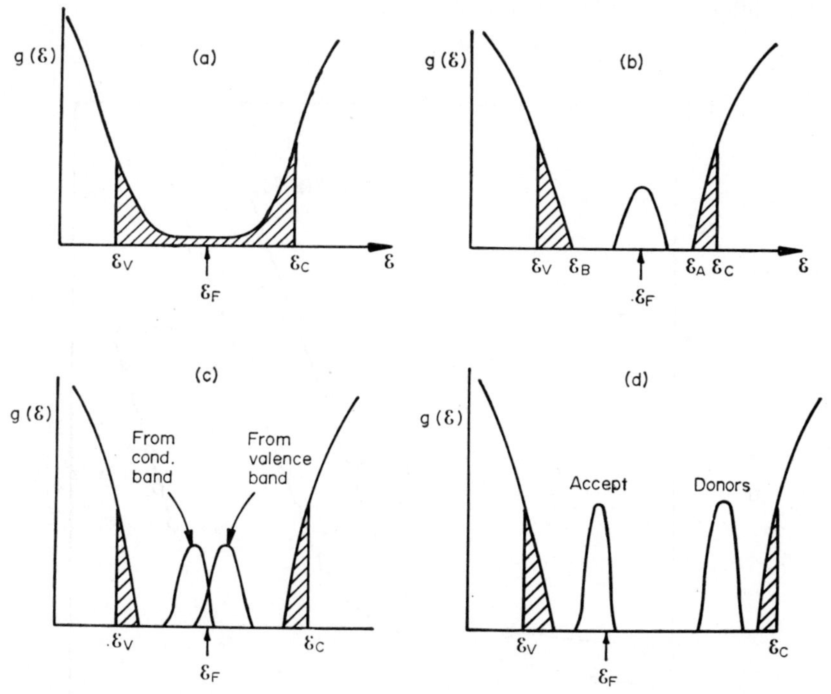

FIG. 2.28. (a) CFO model (Ref. 52), (b) Davis and Mott model (Ref. 32), (c) Mott model (Ref. 53), (d) Marshall and Owen model (Ref. 54).

Figure 2.28 shows sketches of several model density of states which have been proposed[32,52–54] to explain one or the other experiments in different materials. Common to all are the critical energies (mobility edges) $\mathcal{E}_c$ and $\mathcal{E}_v$ which separate the localized tail states from band states. The Fermi level $\mathcal{E}_F$ is close to the gap centre.

2.3.3. EXPERIMENTAL EVIDENCE

Optical Absorption

Figure 2.29 shows as an example the spectral dependence of the absorption in some glasses measured by Tauc *et al.*[2] The absorption in region *C* (see Fig. 2.1) occurs at energies which correspond to excitations between localized gap states and extended band states. This low absorption region is structure and impurity sensitive. It has been explored in only a few materials because it requires large homogeneous samples and great care to eliminate light scattering. Its magnitude is about 10^4–10^5 times smaller than the interband absorption in region *A*. If the transition matrix elements for the two processes are the same then absorption coefficients $\alpha < 1\text{cm}^{-1}$ imply that the density of gap states must be less than about $10^{17}\ \text{cm}^{-3}\ \text{eV}^{-1}$.

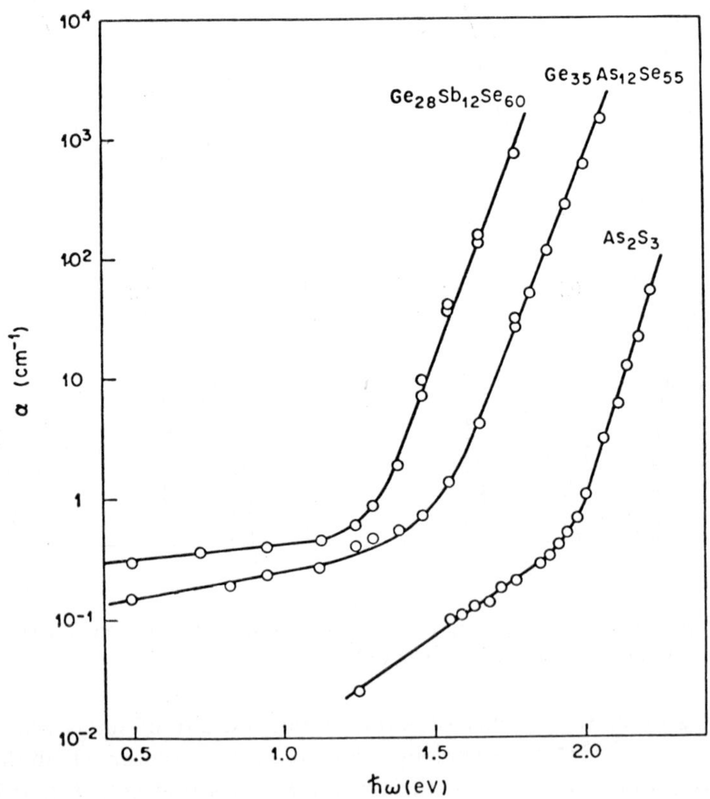

FIG. 2.29. Optical absorption α below the band gap in amorphous As_2S_3, $Ge_{28}Sb_{12}Se_{60}$, and $Ge_{35}As_{12}Se_{55}$. (After Tauc and Menth, Ref. 2.)

At photon energies less than about half the optical gap optical absorption should drop to much lower values because the spatial separation makes transitions between localized states very improbable. Unfortunately, absorption measurements have not been extended low enough to verify the presence of an absorption edge near $0{\cdot}5\,(\mathcal{E}_c - \mathcal{E}_v)$. The model illustrated in Fig. 2.28(c) predicts an edge in the low absorption tail near $0{\cdot}8\,(\mathcal{E}_c - \mathcal{E}_v)$. No such structure has been discovered.

The estimate of gap state densities from α requires the knowledge of the transition matrix elements. The present uncertainty in this parameter allows us to increase the upper limit noted in Table 2.2 to $5 \times 10^{17}\ \mathrm{cm}^{-3}\ \mathrm{eV}^{-1}$.

The exponential absorption tail between $10^2 < \alpha < 10^4\ \mathrm{cm}^{-1}$ is not considered as evidence for tail states. This Urbach tail is observed in most amorphous and many crystalline semiconductors with about the same slope.[55] According to Dow and Redfield[3] it arises from internal field broadening of exciton absorption.

Photoemission

The sensitivity of absorption measurements to localized gap states can be enhanced by increasing the sample thickness as long as homogeneity is assured. This is not possible in photoemission experiments which probe only a thin layer adjacent to the surface, the 50–100 Å thick escape depth of the photoemitted electrons. This limits the detectability to about $3 \times 10^{19}\ \mathrm{cm}^{-3}\ \mathrm{eV}^{-1}$ or to a surface state density of about $10^{13}\ \mathrm{cm}^{-2}\ \mathrm{eV}^{-1}$. Careful comparison of the shape of the valence band density of states between amorphous and crystalline Ge, Si, or Se has not revealed convincing evidence for a tail of states as postulated by the models shown in Fig. 2.28.

Field Effect

As discussed earlier[(13)] the field effect is an ideal method for probing the density of gap states because the field-induced excess charges distributed over the gap states shift the Fermi level. This shift in turn is monitored by measuring the conductance along the thin semiconductor film. Spear and LeComber[(13)] observed conductance changes of several orders of magnitude in *a*-Si produced by SiH_4 decomposition. They mapped out the density of gap states for these special *a*-Si films for several substrate temperatures as shown in Fig. 2.30. The density of states have been increased by a correction factor 5 compared to those reported earlier to account for polarization effects in the glass insulator which were discovered and eliminated in the meantime.[(56)] These are the first measurements of the gap state distribution in an amorphous semiconductor. Spear *et al.* did not assume localized interface states. If they are present, which is very likely, then the gap state density deduced from this experiment would be smaller.

The very small field effect in evaporated or sputtered *a*-Ge, *a*-Si, or in chalcogenide alloys shows that the gap state density must be appreciably higher in these materials. The value $g(\mathcal{E}_F) \sim 10^{19}\ \mathrm{cm}^{-3}\ \mathrm{eV}^{-1}$ quoted in Table 2.2 may be reduced somewhat if one finds evidence for a substantial density of surface states.

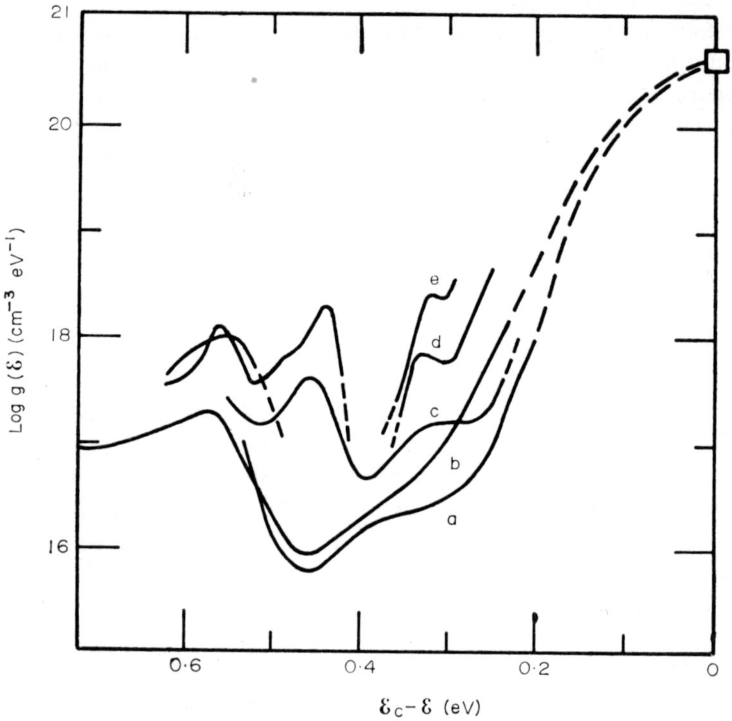

FIG. 2.30. Density of states in *a*-Si produced by decomposition of SiH_4 on substrates held at various temperatures: (a) 630 K, (b) 530 K, (c) and (d) 470 K, and (e) 400 K. (After Spear and LeComber, Ref. 13.)

Surprisingly, attempts to measure the field effect in Se, As_2Se_3, and As_2S_3 have failed so far. We suspect that these materials have very low densities of gap states, $g(\mathcal{E}_F) < 3 \times 10^{16}\ \mathrm{cm}^{-3}\ \mathrm{eV}^{-1}$, which should yield a large field effect. Unfortunately, their very low conductivity make low field measurements very difficult in these materials.

Tunnelling

By measuring the conductance for electron tunnelling from a metal (Al) through an insulator (Al_2O_3) into an amorphous semiconductor, one should be able to observe strong structure in $g(\mathcal{E})$ such as shown in Fig. 2.28(b)–(d). Although the localized gap states do not contribute much to the d.c. conduction, they are in equilibrium with the conducting bands *via* the recombination-generation kinetics so that they can act as tunnelling states. Moreover, one hoped to observe the mobility edges as a sharp increase in tunnelling conductance at a bias voltage which shifts the Fermi level of the metal to either $\mathcal{E}_c$ or $\mathcal{E}_v$ of the semiconductor.

The results[57] shown in Fig. 2.31 do not reflect any of the expected features. Adler *et al.*[58] deduce from the temperature dependence of the zero bias conductance, which follows approximately $\exp - (T_0/T)^{1/4}$, a

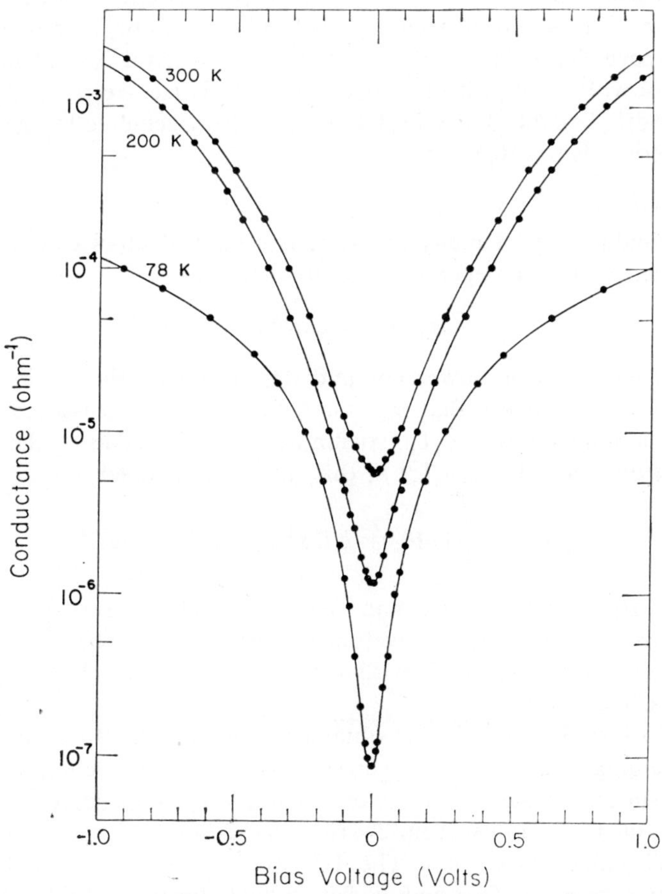

FIG. 2.31. Differential conductance dI/dV as a function of bias voltage of a sandwich made of Al–Al_2O_3–amorpous Ge, and a top electrode. (After Osmun and Fritzsche, Ref. 57.)

value $g(\mathcal{E}_F) \sim 10^{20}\,\mathrm{cm}^{-3}\,\mathrm{eV}^{-1}$ for *a*-Ge and *a*-Si. We doubt that an interpretation of Fig. 2.31 in terms of $g(\mathcal{E})$ is correct. The bias dependence of the tunnelling conductance implies that the density of gap states has a minimum exactly at the Fermi energy and that its distribution is nearly symmetric above and below $\mathcal{E}_F$. This appears to be very unlikely. Moreover, the shape of the tunnelling conductance is nearly independent of preparation condition and the same for amorphous Ge,[59] Si,[59,60] and III–V compounds.[61]

Screening Length

The field effect and tunnelling are strongly affected by surface states. A true bulk effect is the width of the space charge layer adjacent to a contact or surface. Measurements of the space charge capacitance are difficult to carry out because of the small difference in dielectric relaxation time between the bulk and space charge layer. Wey *et al.* used the spectral dependence of the surface photovoltage to probe the space charge layer (see Section 2.2.3). They found a width $\lambda \sim$ 200–300 Å in a sputtered chalcogenide alloy glass. This value corresponds to $g(\mathcal{E}_F) \sim 5 \times 10^{18}\,\mathrm{cm}^{-3}\,\mathrm{eV}^{-1}$.

AC Conductivity

The a.c. conductivity of many amorphous semiconductors exhibits over a large frequency range a dependence on frequency as

$$\sigma_{\mathrm{ac}}(\omega) = \mathrm{const}\,\omega^s \tag{2.31}$$

with $0{\cdot}7 < s < 1{\cdot}0$. This power law and the proportionality of σ_{ac} with T is often taken as evidence that σ_{ac} is caused by thermally assisted hopping of carriers among states near $\mathcal{E}_F$. Austin and Mott[62] adapted the hopping theory of Pollak and Geballe[63] to this case and obtained

$$\sigma_{\mathrm{ac}}(\omega) = \frac{\pi}{3}\,[g(\mathcal{E}_F)]^2\,kT\,e^2\,\alpha^{-5}\,\omega\,[\ln(\nu_p/\omega)]^4 \tag{2.32}$$

where $\nu_p \sim 10^{13}\,\mathrm{sec}^{-1}$ is a phonon frequency and α describes the radial decay of the localized state wavefunction $\exp(-\alpha r)$. The last factor in eqn (2.32) decreases very slowly with ω so that eqn (2.32) yields the observed frequency dependence.

Table 2.3 lists the values $g(\mathcal{E}_F)$ obtained from eqn (2.32) for various materials assuming $\alpha^{-1} = 8$ Å. A larger value of α^{-1} appears unreasonable for deep gap states, a smaller value would make $g(\mathcal{E}_F)$ even larger. The $g(\mathcal{E}_F)$ values obtained here range from 3×10^{17} in As_2Se_3 to $5 \times 10^{19}\,\mathrm{cm}^{-3}\,\mathrm{eV}^{-1}$ in a chalcogenide alloy glass. The $1{\cdot}5 \times 10^{18}\,\mathrm{cm}^{-3}\,\mathrm{eV}^{-1}$ value for Se is particularly surprising. This high density would effectively pin $\mathcal{E}_F$. This contradicts the observations by Lakatos and Tabak[64] that as few as $10^{16}\,\mathrm{cm}^{-3}$ Cl and As impurities can drastically change the electrical properties of Se. Mort and Lakatos[65] showed that $\mathcal{E}_F$ in Se adjusts to the Fermi level of acontacting metal electrode. Furthermore, Schottmiller *et al.*[66] observed excess electron lifetimes in Se as long as milliseconds.

TABLE 2.3

Density of states $g(\mathcal{E}_F)$ from AC conductivity data and eqn (2.32) of text for $\alpha^{-1} = 8$ Å.

Material	σ_{DC} (ohm^{-1} cm^{-1}) at 300 K	σ_{AC} (ohm^{-1} cm^{-1}) at 300 K $\omega = 10^6$ Hz	$g(\mathcal{E}_F)$ (eV^{-1} cm^{-3}) for $\alpha^{-1} = 8$ Å	Reference
As_2Te_3	5×10^{-4}	2×10^{-6}	5×10^{19}	*a*
Te_2AsSi	5×10^{-5}	2×10^{-6}	5×10^{19}	*a*
$Te_{53}As_{32}Si_{10}Ge_5$	5×10^{-6}	$1{\cdot}6\times10^{-6}$	$4{\cdot}5\times10^{19}$	*a*
As_2Se_3	10^{-12}	8×10^{-10}	10^{18}	*b*
As_2Se_3	6×10^{-13}	8×10^{-11}	3×10^{17}	*c*
As_2Se_3	10^{-12}	8×10^{-9}	3×10^{18}	*d*
As_2Se_3 + 0·2% Ag	5×10^{-11}	8×10^{-9}	3×10^{18}	*b*
As_2S_3	3×10^{-16}	2×10^{-10}	5×10^{17}	*b*
As_2S_3	5×10^{-15}	$1{\cdot}2\times10^{-9}$	$1{\cdot}2\times10^{18}$	*d*
As_2S_3 + 0·2% Ag	3×10^{-14}	8×10^{-10}	10^{18}	*b*
Se	$2{\cdot}8\times10^{-14}$	$1{\cdot}8\times10^{-9}$	$1{\cdot}5\times10^{18}$	*d*

[a] Rockstad, H. K., *Solid State Comm.* **9**, 2233 (1971).
[b] Owen, A. E. and Robertson, J. M., *J. Non-Cryst. Solids* **2**, 40 (1970).
[c] Crevecoeur, C. and deWit, H. J., *Solid State Comm.* **9**, 445 (1971).
[d] Lakatos, A. I. and Abkowitz, M., *Phys. Rev.* **B3**, 1791 (1971).

It appears unfair at this time to discredit the hopping process underlying eqn (2.32) on grounds of difficulties encountered with the $g(\mathcal{E}_F)$ values it predicts. There are sufficient uncertainties with the choice of the decay parameter α and simplifying assumptions in the derivation of eqn (2.32) to raise or lower the values quoted in Table 2.3 by a factor of 10.

However, it is important to draw attention to the fact that the frequency dependence expressed by eqn (2.31) is not a unique identification of the hopping processes. On the contrary, it results quite generally from any relaxation mechanism in the material which involves a wide and flat distribution $n(\tau)$ of relaxation times τ since[16,67,68]

$$\sigma(\omega) = \text{const}\,\omega \int n(\tau) \frac{1}{\omega\tau + 1/\omega\tau}\, d\tau. \tag{2.33}$$

As a consequence, the order of magnitude, temperature and frequency dependence of σ_{ac} observed here is very much the same as that of a large number of insulating noncrystalline solids,[16,67] e.g. fused quartz, alkali glasses, porcelain, plexiglass, marble, pyrex, celluloid, bakelite, and anodic oxide films. We therefore have to use $g(\mathcal{E}_F)$ values derived from eqn (2.32) with proper reservation. A comparison of σ_{ac} with the measured $g(\mathcal{E})$ values in a-Si produced by SiH_4 decomposition would be most useful.

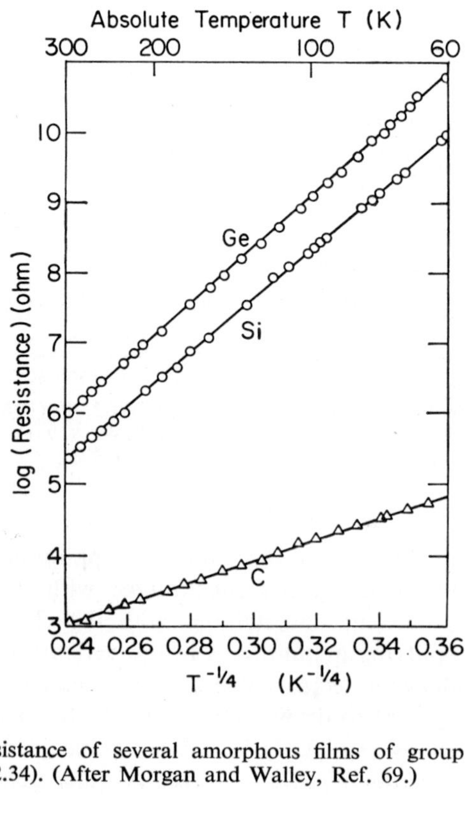

FIG. 2.32. D.C. resistance of several amorphous films of group IV elements plotted according to eqn (2.34). (After Morgan and Walley, Ref. 69.)

DC Tunnelling near $\mathcal{E}_F$

Figure 2.32 shows that the d.c. resistivity of evaporated amorphous Ge, Si, and C follows[69] the relation

$$\sigma = \text{const} \exp [-(T_0/T)^{1/4}] \tag{2.34}$$

derived by Mott[53,70] for thermally assisted hopping of carriers near the Fermi level. The slope yields T_0 which in turn is related to $g(\mathcal{E}_F)$ as

$$kT_0 = 16\alpha^3/g(\mathcal{E}_F) \tag{2.35}$$

where α is the radial decay parameter of the localized state wavefunction. For Ge and Si $T_0 = 7{\cdot}1 \times 10^7$ K and for carbon $T_0 = 1{\cdot}4 \times 10^6$ K. This yields with $\alpha^{-1} = 10$ Å, $g(\mathcal{E}_F) = 2{\cdot}5 \times 10^{18}$ cm^{-3} eV^{-1} for Ge and Si and $1{\cdot}2 \times 10^{20}$ cm^{-3} eV^{-1} for C. Unfortunately, as discussed in Section 2.1, these measurements have not yet been carried out on amorphous films which are free of microvoids.[51] Moreover, it is surprising that the chalcogenide glasses, which contain a substantial density of localized states according to Table 2.2, do not show this conduction process near $\mathcal{E}_F$.

Glow Curves and Space Charge Limited Currents

Further evidence for localized gap states can be extracted from measurements of thermostimulated currents, thermostimulated depolarization[71] and space charge limited currents. These phenomena are complicated by the fact that one deals with distributions rather than discrete states. Quantitative estimates presently quoted have to be treated with great caution.

2.3.4. MAGNETIC PROPERTIES

What can one predict about the spin state of the localized states and hence the magnetic properties of amorphous semiconductors? The tail states of the valence band are expected to be doubly occupied and this corresponds to paired electrons. The conduction band tail states are spinless when unoccupied. On the other hand, localized states near the gap centre should be paramagnetic. These may be charged states produced by charge transfer from high lying valence band states to lower conduction band tail states or dangling bonds. If the localization volume of midgap centres is about atomic size, then the doubly occupied state is shifted upward with respect to the singly occupied state by an appreciably Coulomb repulsion energy.

Singly occupied gap states should give rise to a Curie paramagnetism

$$X_c = \frac{N\mu^2}{3\,kT}\,. \tag{2.36}$$

The number of paramagnetic states should also be measurable by electron spin resonance (ESR) unless their relaxation time is too short or the distribution of g-values too wide. The gyromagnetic ratio g may differ from that of a free electron because of spin–orbit interactions.

Table 2.2 summarizes the results. Evaporated and sputtered Ge and Si exhibit a very large ESR signal[72] corresponding to $1\text{–}3 \times 10^{20}\ \text{cm}^{-3}$ spins. The g-values are 2·022 for Ge and 2·006 for Si, very close to those of spins observed at freshly cleaved (III) crystalline surfaces. These ESR signals decrease with annealing and are therefore considered to be associated with the microvoids which are characteristically found in films of tetrahedral semiconductors deposited at room temperature. Such large ESR signals have also been observed in sputtered amorphous SiC but not in *a*-Si films produced by SiH_4 decomposition.[73]

No spin resonance signal was observed in amorphous Se, As_2Se_3, As_2S_3 as well as a number of chalcogenide alloy glasses, both sputtered and quenched from the melt.[74] The detection sensitivity was about 10^{14} spins/cm^3. Very short spin relaxation times or a wide spread in g-values may be the reason for the absence of an ESR signal. For that reason, the measurement of the Curie paramagnetism is a more reliable test for the presence of paramagnetic centres. A search for these in Se, As_2Se_3, and As_2S_3 has so far brought to light only some Fe impurities. After these were removed no paramagnetic centres were observable[75] with a detection sensitivity of 3×10^{16} centres/cm^3. A large number of paramagnetic centres were found[76] in the so-called tetrahedral glasses. In Cd $Ge_x As_2$ the concentration of centres scaled linearly with x reaching a total concentration of about $10^{20}\ \text{cm}^{-3}$ at $x = 1$. Such a large number of centres are difficult to accommodate in a 1eV wide gap. One again suspects magnetic impurities. Amorphous Ge, Si and chalcogenide alloy glasses should show a sizable paramagnetic term. Unfortunately, they have not yet been studied.

2.3.5. SUMMARY AND CONCLUSIONS

According to Table 2.2 there appears to be a conflict between the upper limit to the density of gap states deduced from optical absorption and the density of midgap gap states estimated from the screening length, field effect, and tunnelling experiments. Furthermore, the number of unpaired spins in chalcogenide semiconductors is far below the number we expect in view of $g(\mathcal{E}_F)$ obtained from AC conductivity and field effect experiments. The tetrahedral amorphous films and glasses on the other hand, exhibit spin densities which are too large to be accounted for by the present density of state models.

First a remark concerning optical absorption. When the oscillator strength for transitions between localized and extended states is distributed over energies of the order of the bandwidth (~ 5 eV), then α yields the average density of gap states rather than $g(\mathcal{E}_F)$. A hump of midgap states, 0·1 eV wide and $10^{18}\,\mathrm{cm}^{-3}\,\mathrm{eV}^{-1}$ in magnitude, for example, yields about the same α as an average $g(\mathcal{E}) = 10^{17}\,\mathrm{cm}^{-3}\,\mathrm{eV}^{-1}$. Moreover, potential fluctuations will cause the oscillator strength to be shifted from lower to higher energies[16] where it falls into the interband absorption. These factors may explain why localized gap states are not easily observed optically.

Let us now summarize our conclusions for each of the material classes.

(i) *Evaporated and sputtered* Ge *and* Si. Most measurements were carried out on films containing a network of microvoids. The large density of unpaired spins is believed to be associated with internal surface states of these microvoids. The rather high conductivity of these films and the presence of Mott's $\ln \sigma$ versus $T^{-1/4}$ hopping conduction near $\mathcal{E}_F$ suggest that $g(\mathcal{E}_F)$ is indeed high, perhaps 5×10^{18}–$5 \times 10^{19}\,\mathrm{cm}^{-3}\,\mathrm{eV}^{-1}$ depending on the deposition temperature or state of annealing. Mott's $T^{-1/4}$ hopping conduction was recently[77] observed in void free Si with $T_0 = 2 \times 10^8$ K. This yields $g(\mathcal{E}_F) = 10^{18}\,\mathrm{cm}^{-3}\,\mathrm{eV}^{-1}$ with $\alpha^{-1} = 10$ Å and $7 \times 10^{19}\,\mathrm{cm}^{-3}\,\mathrm{eV}^{-1}$ with $\alpha^{-1} = 2{\cdot}5$ Å.

(ii) SiH_4 *decomposed* Si. Field effect measurements[13] allow a direct determination of $g(\mathcal{E})$. The densities obtained should be reduced somewhat when the presence of surface states is established. This material should be used to test the interpretation of several of the other experiments, particularly that of the AC conductivity. The AC conductivity could be a valuable tool in materials whose field effect is screened by surface states. Moreover, the question of the spin of gap states could be answered by measuring ESR and the paramagnetic term of the susceptibility.

(iii) *Tetrahedral glasses*. These materials are most interesting. There seems to be no void structure, yet the paramagnetic term is surprisingly large. It scales with the concentration of the group IV element which makes it unlikely but not impossible that accidental magnetic impurities are responsible for the paramagnetic terms. Unfortunately, too few of the other experiments have been carried out on these materials. The paramagnetic term is absent in pure $CdGeAs_2$ according to recent experiments at Bell Laboratories.

(iv) *Chalcogenide elements and alloys*. These materials seem to have very few localized gap states according to optical and paramagnetic measurements. From the effects of doping[64,75] we estimate $g(\mathcal{E}_F) \sim 10^{16}\,\mathrm{cm}^{-3}\,\mathrm{eV}^{-1}$. The AC conductivity might have to be explained by a different mechanism. The glow curves do not permit quantitative analysis at present. The density of gap states is probably higher in evaporated and sputtered films than in annealed bulk samples. One also should be aware of the fact that defect states can be introduced in As_2Se_3 and As_2S_3 by light. These photochromic effects will be discussed later.

(v) *Chalcogenide alloys*. The small magnitude of the field effect and the measurement of the screening length suggest $g(\mathcal{E}_F) \sim 5 \times 10^{18}\ \text{cm}^{-3}\ \text{eV}^{-1}$. A measurement of the paramagnetic term of the susceptibility would be interesting. The density of gap states will depend of course on the alloy composition and on the method of preparation.

2.4. Tunnelling, Anomalous Heat Capacity and other Unsolved Problems in Amorphous Semiconductors

2.4.1. Introduction

The fascinating aspect of a new field is that more often than not a new experiment yields completely unexpected results. This may be frustrating for the orderly minded but it is interesting and challenging to find new models and concepts appropriate for describing the new phenomena.

In this Section I wish to discuss several such experiments. They deal with electron tunnelling into amorphous semiconductors and the anomalous specific heat and heat conduction in noncrystalline materials discovered by Zeller and Pohl. At the end I wish to remark on some other experimental results which have not yet been explained satisfactorily.

2.4.2. Tunnelling

The major interest in electron tunnelling started with Ivar Giaever's elegant experiment on tunnelling into superconductors.[(78,79)] At the same time interband tunnelling in semiconductors[(79,80)] yielded a wealth of information on band structure effects in crystalline semiconductors and the interaction of the tunnelling electron with the various elementary excitations in the solid.

We carried out tunnel experiments with amorphous semiconductors[(57)] hoping to obtain the density distribution of gap states and the energy of the mobility edge. Tunnelling experiments are not expected to yield the density of states when the initial and final tunnel states are Bloch states.[(79)] This, however, is not the case in amorphous semiconductors and we expected to observe the gap state distribution, particularly the hump of midgap states proposed by Davis and Mott[(32)] and the position of the Fermi energy relative to the mobility edges.

A tunnelling experiment is quite simple. A tunnel junction is made by first evaporating a strip of aluminium on a glass substrate, then forming a 20–40 Å aluminium oxide layer by exposing the Al to an oxygen plasma or to air and finally evaporating several 0·2 cm wide cross strips of the amorphous semiconductor to be investigated over the oxidized Al. A metal strip is deposited on top of the semiconductor in order to reduce the resistance of the top electrode. One junction (metal-insulator-metal) along the strip is always made without the semiconductor to test the tunnelling properties of the Al_2O_3 layer. Another junction (metal-semiconductor-metal) is made to monitor the properties and the resistance drop in the amorphous semiconductor, which was found to be negligible compared to that of the oxide

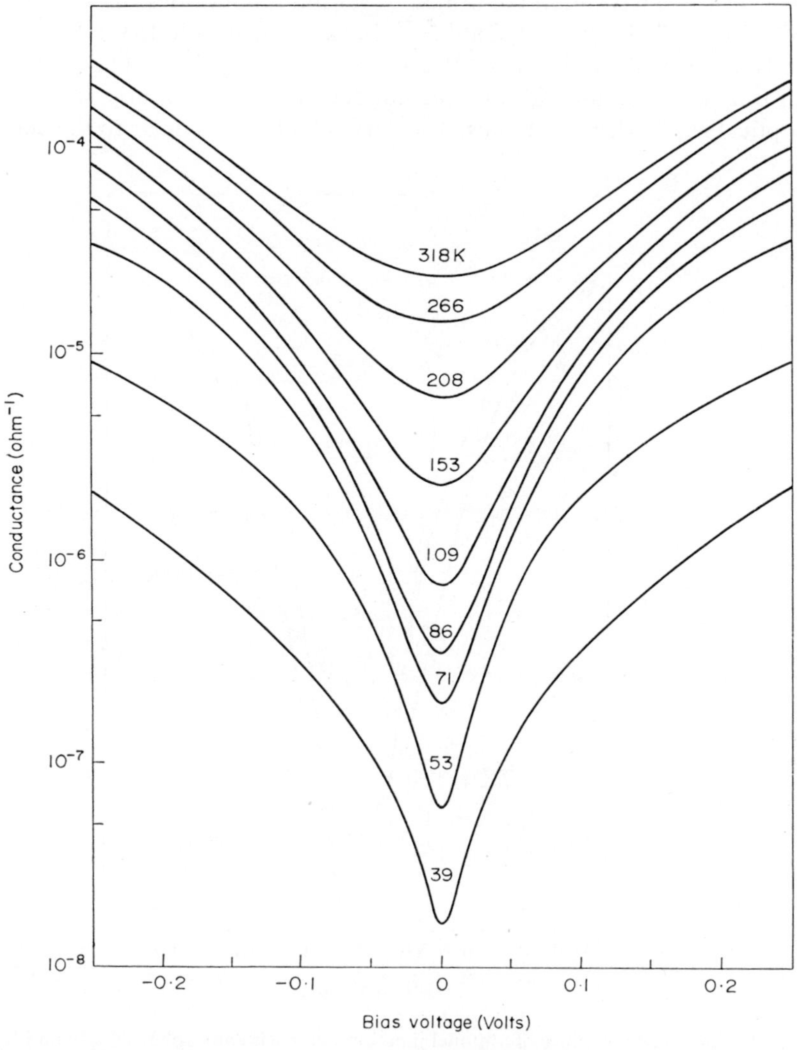

FIG. 2.33. Differential conductance dI/dV as a function of bias voltage of an Al–Al_2O_3–amorphous Ge tunnelling junction. (After Osmun, Ref. 81.)

layer. The current I or the differential conductance $G = dI/dV$ is then measured as a function of bias.

The tunnel conductance obtained[57] with (300 K) evaporated Ge is shown in Fig. 2.31. Figure 2.33 shows measurements in the small bias region extended to lower temperatures.[81] Most striking is (i) the symmetry of the curves with respect to sign reversal of the bias, (ii) the lack of any structure, and (iii) the fact that minimum conductance occurs always at zero bias. These general features have been confirmed in several laboratories on amorphous Ge,[59] Si,[59,60] InSb, and GaSb.[61] Measurements on the III–V compounds by Konak and Stuke[61] are shown in Fig. 2.34. These authors point out that the half width of the conductance curves nearly scales with the optical gap in these materials. The curves look very similar to the density

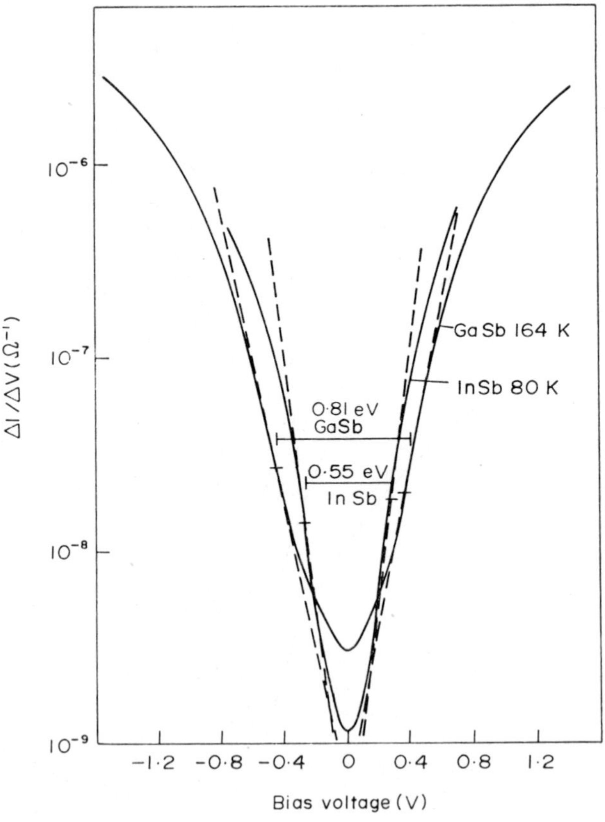

FIG. 2.34. Tunnel conductance of a tunnel junction made with amorphous GaSb and InSb. The half widths of the conductance minima are compared with the respective optical gaps. (After Konak and Stuke, Ref. 61.)

of states curve of the CFO model.[52] It is unlikely, however, that there is a direct relationship. The CFO model was never meant to apply to amorphous elements and simple tetrahedral amorphous semiconductors. Furthermore, there is evidence that $\mathcal{E}_F$ in these materials lies closer to the conduction band. It would be very fortuitous indeed if the Fermi level were exactly at the minimum of $g(\mathcal{E})$ in several different materials, independent of preparation conditions. The non-monotonic density of states distribution of SiH_4 decomposed Si obtained by Spear and LeComber[13] is closer to what we expect in these materials.

On first sight one is surprised that tunnelling is temperature dependent. However, at these temperatures we are dealing with a rather wide distribution of carriers and the tunnelling transfer function $P(\mathcal{E})$ will certainly depend on energy. We tried to attribute the temperature dependence to that of the Fermi functions $f_M(\mathcal{E}, T)$ and $f_{SC}(\mathcal{E}, T)$ in the metal and the semiconductor, respectively, by writing

$$G(V, T) = \int P(\mathcal{E})\,[f_M(\mathcal{E}, T) - f_{SC}(\mathcal{E} - eV, T)]\,d\mathcal{E}. \tag{2.37}$$

The symmetry of $G(V, T)$ requires that $P(\mathcal{E})$ is symmetric above and below $\mathcal{E}_F$. Setting therefore

$$P(\mathcal{E}) = \text{const}\exp\left[-a_1 - a_2\,|\mathcal{E} - \mathcal{E}_F|\right] \tag{2.38}$$

one can solve[82] eqn (2.37) and obtain for the temperature dependence

$$\frac{G(V, T)}{G(V, 0)} = \frac{a_2\,kT}{\sin(\pi\,a_2\,kT)}. \tag{2.39}$$

The coefficient a_2 governs the steepness of the conductance increase with bias, its physical meaning depends on the tunnelling process. These calculations predict that the T-dependence is independent of bias and that it becomes less strong at low T in contrast to the observations. Phonon-assisted tunnelling, on the other hand, yields a stronger T-dependence as the temperature is lowered. Sauvage, Mogab, and Adler[58] suggest that an electron must tunnel a certain distance into the semiconductor to find a localized state. This distance is about the average separation of the number of states $g(\mathcal{E})\,kT$ available at T in $d\mathcal{E}$ at $\mathcal{E}$. With phonon-assistance electrons can tunnel to states of different energy and thus take advantage of gap states lying closer to the interface. At low T, on the other hand, phonons become scarce and tunnelling on the average must take place to states further away from the oxide-semiconductor interface. These are the arguments very similar to those on which the derivation[70] of Mott's $\ln\sigma$ versus $T^{-1/4}$ law is based. Surprisingly one finds a good fit to the data by plotting $\ln G(0, T)$ versus $T^{-1/4}$. Sauvage, Mogab and Adler analyze their data in this manner and obtain $g(\mathcal{E}_F)$ as high as 10^{20} cm^{-3} eV^{-1}. This analysis accepts that $g(\mathcal{E})$ has the general shape of $G(V)$ which we believe is very unlikely.

It is interesting to note that deep and symmetric tunnelling conductance minima centred at zero bias have been known for some time prior to experiments on tunnelling into amorphous semiconductors. These so-called giant resistance anomalies are observed(83) with transition metal oxides as one of the tunnelling electrodes. Figure 2.35 shows an example(84) in which a thin layer of TiO_2 was deposited between the oxide and the top metal electrode. A few tens of Angstrom are sufficient. Without this interface layer one finds a nearly bias independent tunnelling conductance. These phenomena have not been explained. It might be that tunnelling in amorphous semiconductors occurs to interface states rather than bulk states.(57)

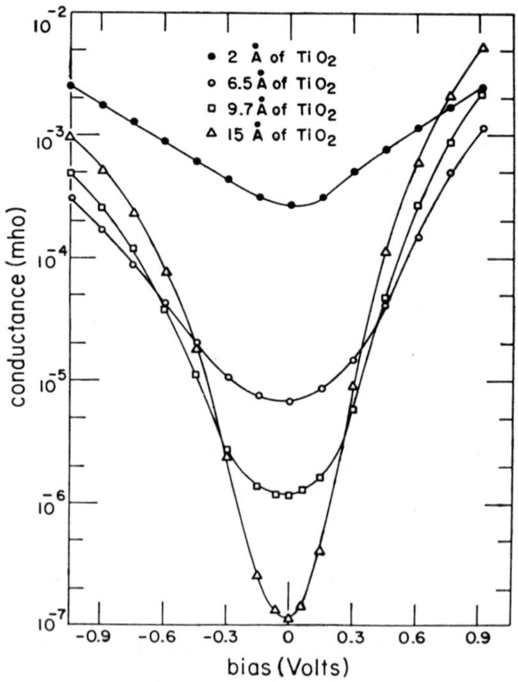

FIG. 2.35. Effect of thin layers of titanium oxide in $Al–Al_2O_3–Ag$ tunnel junctions. The curves were taken at 4·2 K. (After Nielson, Ref. 84.)

An asymmetric tunnel characteristic was first observed by Osmun(81) in a tunnel junction made with a flash evaporated chalcogenide alloy glass $Tl_2SeAs_2Te_3$. His results are shown in Figs. 2.36 and 2.37. About the same time Mott suggested(85) that the polaron formation in a semiconductor should give rise to such an asymmetry. Electrons can easily tunnel out of their polaron cloud and thus out of the semiconductor but the bias has to provide the polaron binding energy before electrons can tunnel from the metal into unoccupied polaron states. The asymmetry noticeable in Figs. 2.36 and 2.37 amounts to about 0·070 eV and has the correct polarity. These ideas were first proposed by Hall, Racette, and Ehrenreich[86] to explain a zero bias conductance minimum in the characteristic of *p*–*n* tunnel junctions made of III–V compounds.

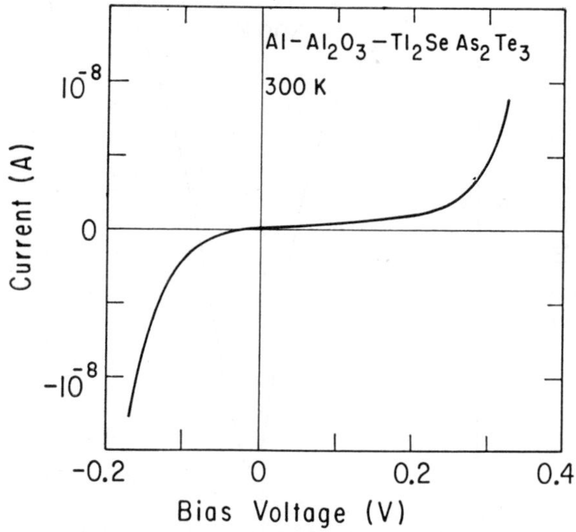

FIG. 2.36. I–V characteristic of a tunnel junction made with a flash evaporated chalcogenide alloy $Tl_2SeAs_2Te_3$. (After Osmun, Ref. 81.)

It appears that tunnelling studies do not yield an easy answer to the question of the localized state distribution but instead they pose new and interesting problems.

2.4.3. HEAT CONDUCTIVITY AND SPECIFIC HEAT

The measurements by Zeller and Pohl[87] of the heat conductivity and specific heat are the most exciting new experiments in the field of noncrystalline solids. They have very important and interesting consequences. Figure 2.38 contrasts the thermal conductivity k of crystalline and fused quartz. In crystalline quartz k rises rapidly as T is lowered because the phonon mean free path λ increases as the number of phonons available for scattering decreases. It reaches a peak near 10 K and then decreases as T^3. The decrease is caused by the fact that λ reaches a constant value when it becomes equal to the sample size and k then follows the T^3-dependence of the specific heat C_v:

$$k = \frac{1}{3} C_v \lambda v \tag{2.40}$$

where v = sound velocity. In fused quartz, in contrast, k is a monotonically decreasing function of temperature.

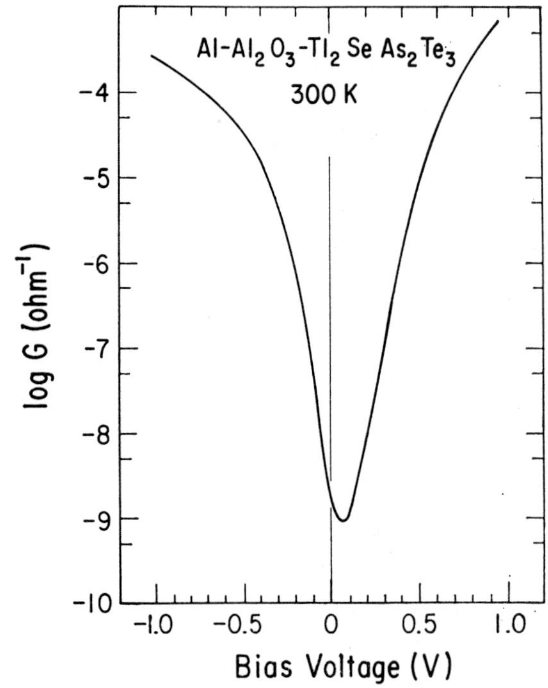

FIG. 2.37. Differential conductance of tunnel junction shown in Fig. 2.36. (After Osmun, Ref. 81.)

What is surprising is that all noncrystalline solids studied so far, vitreous Se, GeO_2, polymers, oxide glasses, have nearly the same heat conductivity. A few examples are shown in Fig. 2.39. Characteristic is the low temperature slope and the plateau near 10 K. Exceptions to this rule are nylon, teflon and polystyrene, but these materials are partially crystallized.

One might expect the same $\mathcal{K}$ if this were the smallest possible value with λ of the order of the interatomic distance. This is not the case, however. Figure 2.38 shows $\mathcal{K}$ of a KCl crystal doped with CN. This crystal has at low temperatures a smaller thermal conductivity than the amorphous materials.

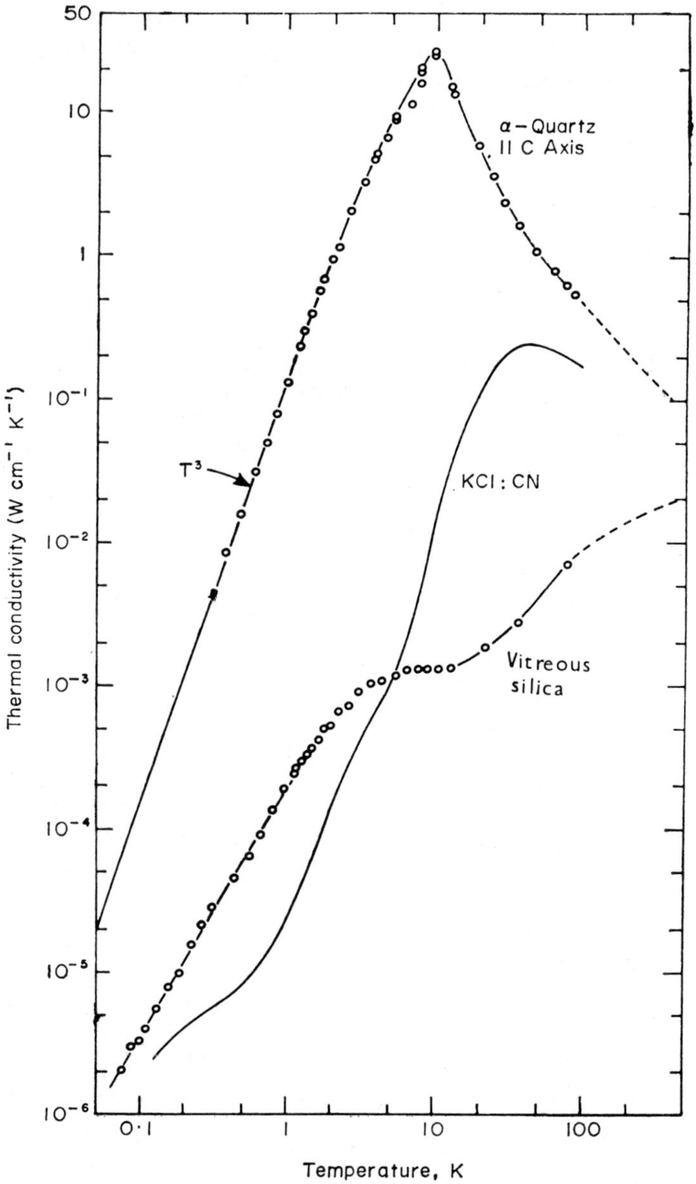

FIG. 2.38. Thermal conductivity of crystalline and vitreous SiO_2 and crystalline KCl containing $4{\cdot}9 \times 10^{19}$ cm^{-3} CN. (After Zeller and Pohl, Ref. 87.)

Moreover, λ is of interatomic size only at high temperatures (for both crystalline and amorphous materials). Since v and C_v in eqn 2.40 are known one can find λ from k. As shown in Fig. 2.40 one finds that λ of amorphous materials rises to rather large values, 10–100 μm at low T. Why then is the heat conductivity of "all" noncrystalline solids the same?

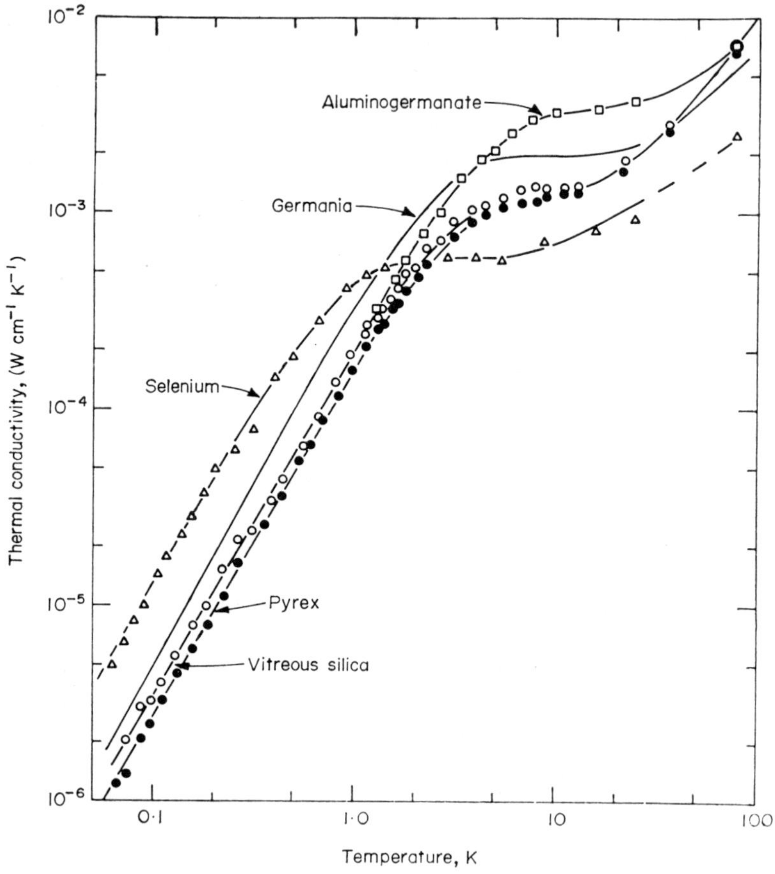

FIG. 2.39. Thermal conductivity of vitreous SiO_2, GeO_2, Se, and Pyrex 7740. (After Zeller and Pohl, Ref. 87.)

The dotted lines in Fig. 2.40 are estimates of λ from Rayleigh scattering at interstitials and vacancies, assuming that each atom in the amorphous solid is slightly displaced and acts as a scatterer. This yields $\lambda \propto \omega^4$ and a sharp rise of λ as the temperature is lowered because the dominant phonon frequency effective in heat conduction decreases with T. The frequency and wavelength of the dominant phonon are shown on the upper scales in Fig. 2.40. Below 10 K the ω^4 rise of λ seems to be interrupted by another scattering

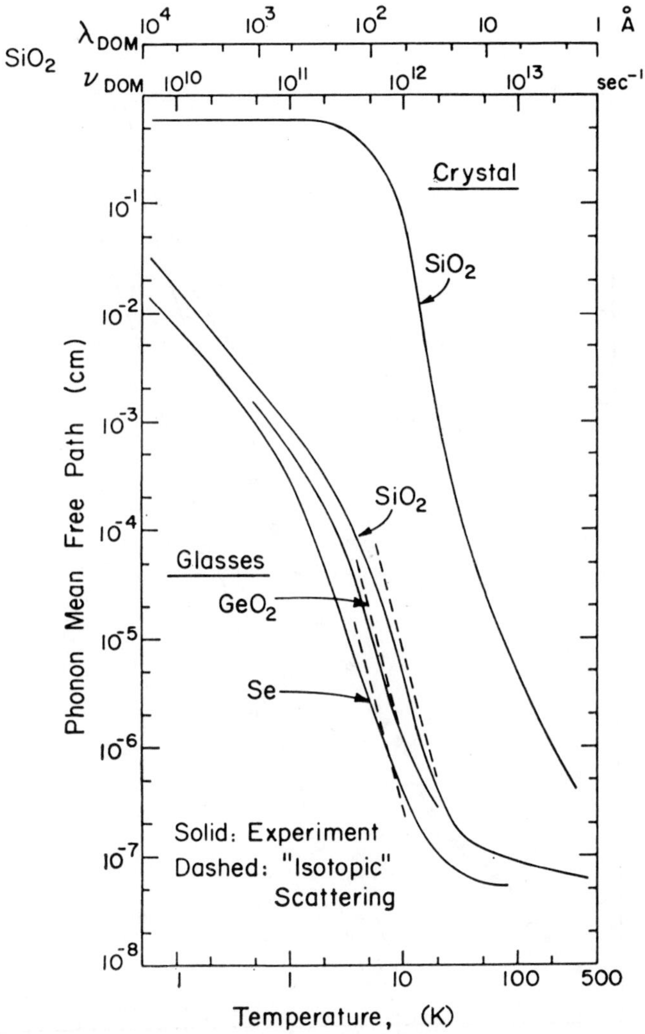

FIG. 2.40. Average phonon mean free path λ calculated from eqn (2.40) for quartz and for vitreous SiO_2, GeO_2, and Se. The dashed lines are λ values calculated from Raighley scattering. At the top are indicated the wavelengths and frequencies of the dominant phonons effective in thermal conduction at the corresponding temperatures in quartz. (After Zeller and Pohl, Ref. 87.)

mechanism. The approximate size of the new scattering modes may be estimated by the wavelength of the dominant phonons below 10 K.

The new modes should show up as an excess contribution to the specific heat C_v. Figure2.41 shows that this is indeed the case. At low temperatures C_v of the vitreous materials lies appreciably above that of crystalline SiO_2.

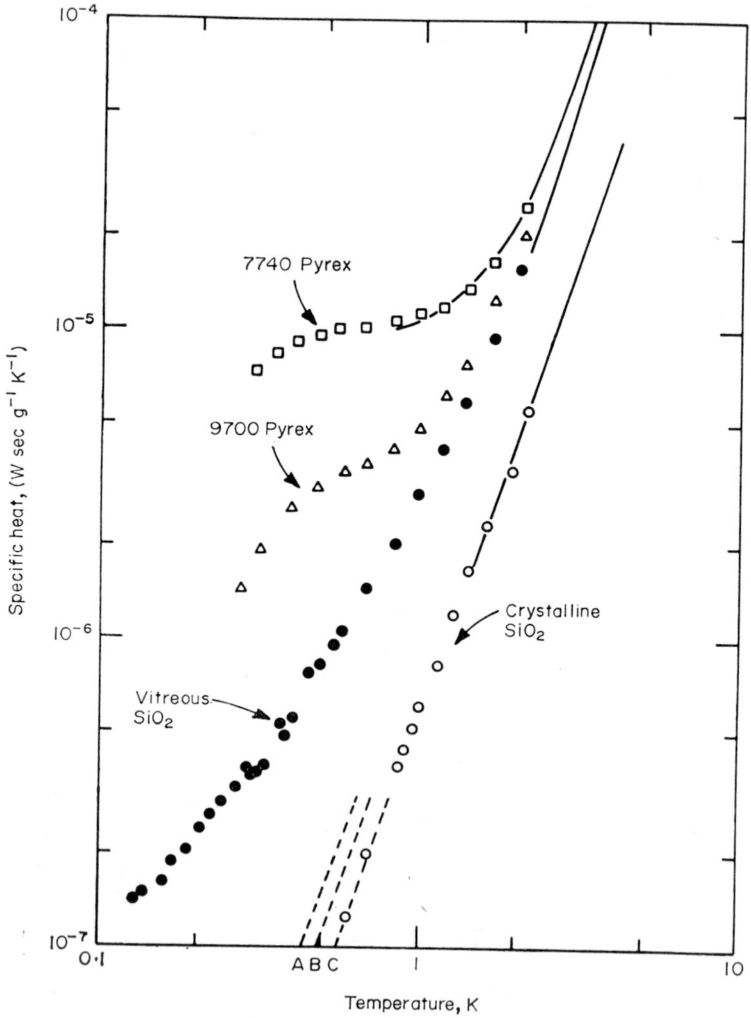

FIG. 2.41. Specific heat of vitreous and crystalline SiO_2 and of Pyrex. (After Zeller and Pohl, Ref. 87.)

Moreover, C_v deviates strongly from the normal T^3 dependence. Zeller and Pohl[87] find that the excess term in C_v is linear in T.

$$C_v = AT + BT^3. \tag{2.41}$$

Surprisingly, the anomalous term A is of the same order of magnitude in many noncrystalline solids: for vitreous SiO_2, GeO_2 and Se, they found A to be 10, 16, and 7 erg/gK^2, respectively. These values are comparable in magnitude with the electronic specific heat of copper ($A = 120$ erg/gK^2). The factor B is about twice as big as expected from the elastic constants or the phonon spectrum. Hence, the B term is also anomalous.

A specific heat term which is linear in temperature can be derived from a density of states spectrum of excitations which is independent of energy. In these cases one finds a density of order 10^{21} cm^{-3} eV^{-1} between

$$0 < \mathcal{E} < 2 \times 10^{-3} \text{ eV}.$$

This is shown in Fig. 2.42 together with the phonon density of states $g(\mathcal{E}) \propto \mathcal{E}^2$. We do not know how the spectrum of the new excitations continues at larger

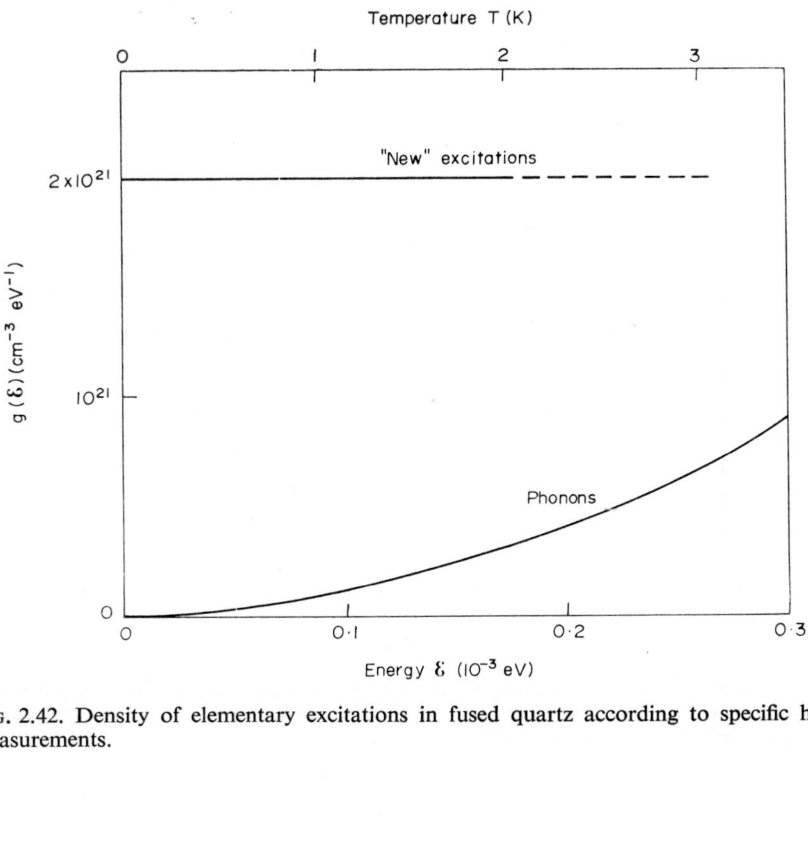

FIG. 2.42. Density of elementary excitations in fused quartz according to specific heat measurements.

energies because above 10 K the second term in eqn (2.41) dominates the first term.

Such a large density of low energy excitations may be very important for any kind of interaction with electrons or with phonons. In perturbation treatments the strength of such interactions is proportional to the square of the matrix element divided by an energy denominator. Since there is a finite density of zero energy excitation, the interaction can become very strong unless the matrix element vanishes. One might explore whether these interactions favour polaron formation in noncrystalline solids.

The next problem is to find the origin of these low energy excitations. One should search for an explanation which applies to noncrystalline solids in general because this linear specific heat term was observed in many different classes of noncrystalline materials. Anderson, Halperin, and Varma[88] and Phillips[88a] suggest the following. In any glass system there are many structural configurations, which differ by the positions of groups of atoms, but which have nearly the same energy. Energy barriers separate the various configurational states. At the glass transition temperature T_g, the thermal energy is sufficiently large so that the system can attain approximately $3\mathcal{N}$ different con-

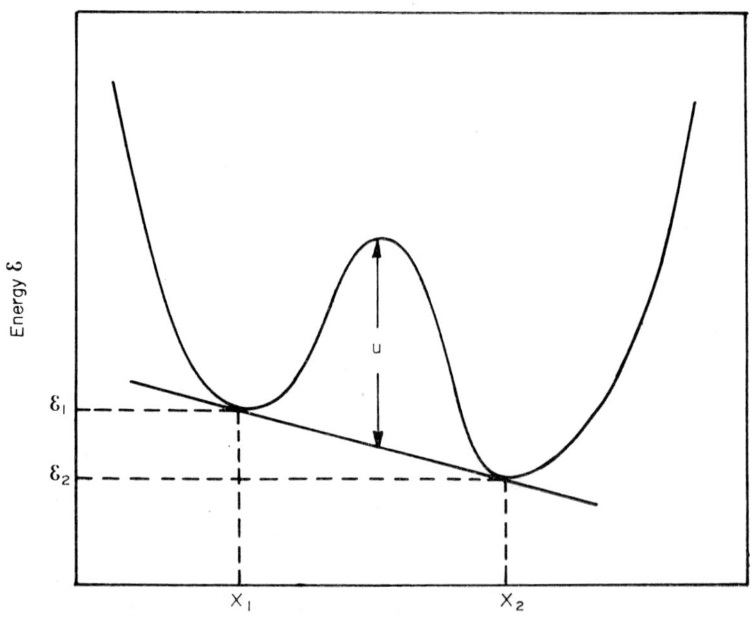

FIG. 2.43. Energy of a noncrystalline solid as a function of a configurational coordinate x measuring positions along a line connecting two nearby local minima of $\mathcal{E}$. (After Anderson, Halperin and Varma, Ref. 88.)

figurational states. Above T_g the number of degrees of freedom is therefore greatly enhanced. From the value of T_g, $400 < T_g < 600$ K, one estimates the magnitudes of the potential barriers to be about $\langle U \rangle \sim 0.1$ eV. The distribution of U has a low energy tail. The number of very low barriers is sufficiently small to assure overall rigidity of the glass structure for $T < T_g$.

Let us consider two of such configurational states separated by a potential barrier U along a configurational coordinate connecting two local minima in the $3\mathcal{N}$ dimensional configuration space as shown in Fig. 2.43. Transitions between configurational states can also occur by tunnelling. Of course the tunnelling probability is small because one is dealing here with groups of several atoms so that the de Broglie wavelength is very much smaller than that of electrons. The tunnelling distance must therefore be of order one Angstrom for tunnelling to take place within the duration of the experiment ($<10^3$ sec). The energies of these configurational states will be random quantities, and the energy differences $\Delta\mathcal{E}$ will smoothly cover a spectrum of values. It is important that the density of levels per unit $\Delta\mathcal{E}$, with energy difference $\Delta\mathcal{E}$ and with tunnelling times in the acceptable range, is non-zero and continuous in the vicinity of $\Delta\mathcal{E} = 0$ if a linear specific heat term is to be obtained. Configurational transitions important for the linear specific heat are those with small $\Delta\mathcal{E}$ *and* having low potential barriers U. An interesting prediction of this model is that the observed specific heat should increase roughly logarithmically with the length of observation time.

2.4.4. OTHER PROBLEMS

AC Conductivity

I mentioned earlier that the frequency dependence of the AC conductivity

$$\sigma_{ac} = \text{const}\, \omega^s \tag{2.31}$$

with $0.7 < s < 1.0$ is observed in a variety of noncrystalline systems. In many insulating and dielectric materials it is impossible to associate this loss mechanism with thermally assisted hopping of electrons in localized states near the Fermi energy. The general occurrence and similar behaviour of the AC loss in many glass systems suggests that it might instead be associated with the spectrum of low energy excitations discovered by Zeller and Pohl. Regardless of their origin, if these excitations correspond to different polarizabilities so that they are coupled to the electro-magnetic field, then the observed frequency dependence of the AC conductivity can be derived(68) from the spectrum of low energy configurational excitations.

Infrared Absorption Peak

Figure 2.44 shows an absorption peak near $h\nu = 0{\cdot}23$ eV which was first observed by Chittick(89) in *a*-Ge prepared by decomposition of GeH_4. Later, this absorption was found by Tauc *et al.*[90] also in evaporated, sputtered and in electrolytically deposited amorphous germanium. The absorption band at $h\nu = 0{\cdot}09$ eV seen in Fig. 2.44 is probably due to an oxygen-defect complex and the slow rise of α at higher photon energies is the tail end of interband absorption. The origin of the 0·23 eV absorption band has not yet been explained. Its magnitude is $\alpha = 60$, 900 and 1100 cm^{-1} in sputtered, electrolytic, and GeH_4 decomposed Ge, respectively.[90] If it is caused by a molecular vibration, the impurity must be present in concentrations of 1% or more. Furthermore, it must be very light (perhaps hydrogen) to account for its high frequency. The possibility of an electronic transition has been considered by Tauc *et al.*[90] However, the narrow line width of the absorption suggests a transition between localized states rather than a transition between a defect state and a band of states. Such an infrared absorption peak has not been observed in amorphous Si.

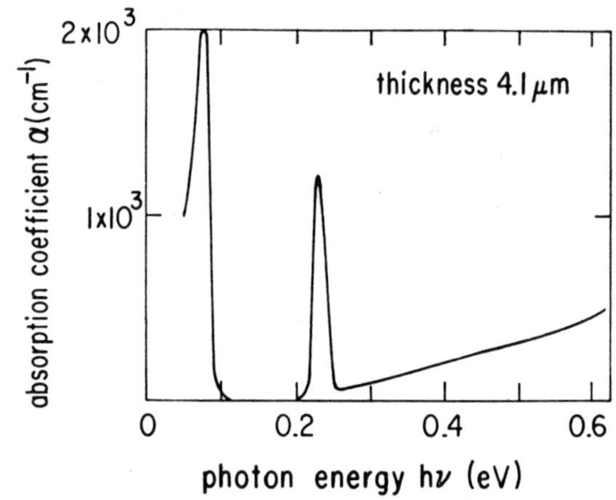

FIG. 2.44. Infrared absorption of amorphous Ge prepared by glow discharge decomposition of GeH_4. (After Chittick, Ref. 89.)

Exponential Absorption Edge

The onset of interband absorption in amorphous semiconductors is characterized by a region for α between 1 and 10^4 cm^{-1} in which the absorption coefficient increases exponentially as

$$\alpha = \text{const} \exp h\nu/\mathcal{E}_1. \tag{2.42}$$

The slope $\mathcal{E}_1$ lies between 0·05 and 0·08 eV for almost all amorphous semiconductors.[2] In a few cases (vitreous Se) it increases with temperature. In some materials ($CdGeAs_2$) the value of $\mathcal{E}_1$ is the same in the crystalline and the vitreous state.

Dow and Redfield[3] attribute this exponential absorption tail to a dissociation of excitons by internal fields present in the material. The internal field lowers the Coulomb potential of the exciton allowing the electron to tunnel out into an adjacent potential minimum. $\mathcal{E}_1$ is proportional to the magnitude of the internal fields. The question why this magnitude is so similar in semiconductors belonging to different material classes and containing very different concentrations of localized states remains unresolved.

Acknowledgements

The support by the Air Force Office of Scientific Research (AFSC), United States Air Force, under contract F44620–71–C–0025, and the general support of the Materials Research Laboratory by the National Science Foundation in the work described in this article, is gratefully acknowledged.

References

1. Mott, N. F. and Davis, E. A., "Electronic Processes in Non-Crystalline Materials", Oxford, Clarendon Press, 1971.
2. Tauc, J. and Menth, A., *J. Non-Cryst. Solids,* **8-10,** 569 (1972).
3. Dow, J. D. and Redfield, D., *Phys. Rev.* **B5,** 594 (1972).
4. De Neufville, J., *J. Non-Cryst. Solids* **8-10,** 85 (1972).
5. Ferrier, R. P., Prado, J. M. and Anseau, M. R., *J. Non-Cryst. Solids* **8-10,** 798 (1972).
6. Stourac, L. A., Abraham, A., Hruby, A. and Zavetova, M., *J. Non-Cryst. Solids* **8-10,** 353 (1972).
7. Mooser, E. and Pearson, W. B., *Phys. Rev.* **101,** 1608 (1956).
8. Mooser, E. and Pearson, W. B., *Progress in Semiconductors* **5,** 104 (1960).
9. Kastner, M., *Phys. Rev. Letters* **28,** 355 (1972).
10. Tauc, J., Stourac, L., Vorlicek, V. and Zavetova, M., Proc. of the IX Intl. Conf. on the Physics of Semiconductors, Moscow, p. 1251 (1968).
10a. Callaerts, R., Denayer, M., Hashmi, F. H. and Nagels, P., Discussions of the Faraday Soc. **50,** 27 (1970).
11. Brodsky, M. H., Title, R. S., Weiser, K. and Pettit, G. D., *Phys. Rev.* **B1,** 2632 (1970).
12. Mell, H. and Stuke, J., *J. Non-Cryst. Solids* **4,** 304 (1970).
13. Spear, W. E. and LeComber, P. G., *J. Non-Cryst. Solids* **8-10,** 727 (1972). See also their article in this volume (Chapter 10).
14. Donovan, T. and Heinemann, K., *Phys. Rev. Letters* **27,** 1794 (1971).
15. Kastner, M. and Fritzsche, H., *Mat. Res. Bull.* **5,** 631 (1970).
16. Fritzsche, H., *J. Non-Cryst. Solids* **6,** 49 (1971).
17. Zallen, R. and Scher, H., *Phys. Rev.* **B4,** 4471 (1971).
18. Shante, V. K. S. and Kirkpatrick, S., *Adv. in Phys.* **20,** 325 (1971).
19. Last, B. J. and Thouless, D. J., *Phys. Rev. Letters* **27,** 1719 (1971).
20. Kirkpatrick, S., *Phys. Rev. Letters* **27,** 1722 (1971).
21. Shklovski, B. I., *JETF* **61,** 2033 (1971).
22. Volger, J., *Progress in Semiconductors* **4,** 205 (1970).

23. Van Beek, L. K. H., *Progr. in Dielectrics* **7,** 69 (1967).
24. Pozdnev, V. P., *Soviet Physics Solid State* **4,** 693 (1962).
25. Chaberski, A. Z., *J. Appl. Phys.* **42,** 940 (1971).
26. Gittleman, J. I., Goldstein, Y. and Bozowski, S., *Phys. Rev.* **B5,** 3609 (1972).
27. Sheng, P. and Abeles, B., *Phys. Rev. Letters* **28,** 34 (1972).
28. Andriesh, A. M. and Kolomiets, B. T., *Soviet Physics Solid State* **6,** 2652 (1965).
29. Many, A., Goldstein, Y. and Grover, N. B., "Semiconductor Surfaces". North-Holland Publ. Co., 1965.
30. Henisch, H. K., "Rectifying Semiconductor Contacts". Oxford, Clarendon Press, 1957.

30a. Simmons, J. G. and Taylor, G. W., *Phys. Rev.* **B6,** 4793, 4804 (1972); Simmons, J. G. and Nadkarni, G. S., *Phys. Rev.* **B6,** 4815 (1972); Simmons, J. G. and Wei, L. S., *Sol. State Electronics* **16,** 43, 53 (1973).

31. Fritzsche, H., Chapter 5 *In* "Amorphous and Liquid Semiconductors", (ed. J. Tauc), Plenum Press, 1973.
32. Davis, E. A. and Mott, N. F., *Phil Mag.* **22,** 903 (1970).
33. Wey, H. Y. and Fritzsche, H., Intl. Conf. on the Physics of Semiconductors, Warsaw, Poland 1972, and to be published.
34. Wey, H. Y. and Fritzsche, H., *J. Non-Cryst. Solids* **8-10,** 336 (1972).
35. Garrett, G. G. B. and Brattain, W. H., *Phys. Rev.* **99,** 376 (1955).
36. Grigorovici, R., Croitoru, N., Devenyi, A. and Teleman, E., Proc. of the VII Intl. Conf. on the Phys. of Semiconductors. (Dunod, Paris), p. 423 (1964).
37. Henisch, H. K. and Vendura, G. J., *Appl. Phys. Letters* **19,** 363 (1971).
38. Jayadevaiah, T. S. and Busmundrud, O., *Electronics Letters* **8,** 75 (1972).
39. Carruthers, T. (to be published).
40. Polanco, J. I., Roberts, G. G. and Maers, M. B., *Phil. Mag.* **25,** 117 (1972).
41. Egerton, R. F., *Appl. Phys. Letters* **19,** 203 (1971).
42. van Roosbroeck, W. and Casey, Jr., H. C., *Phys. Rev.* **B5,** 2154 (1972).
43. van Roosbroeck, W., *Phys. Rev. Letters* **28,** 1120 (1972).
44. Queisser, H. J., Casey, Jr., H. C. and van Roosbroeck, W., *Phys. Rev. Letters* **26,** 551 (1971).
45. Queisser, H. J., *J. Appl. Phys.* **42,** 5567 (1971).
46. Queisser, H. J. *In* "Proceedings of the European Solid State Device Research Conference", Lancaster, England, Sept. 1972, (Institute of Physics, Bristol, to be published).
47. Fagen, E. A., to be published.
48. Donovan, T. M., Spicer, W. E., Bennett, J. M. and Ashley, E. J., *Phys. Rev.* **B2,** 397 (1970).
49. Weaire, D., *J. Non-Cryst. Solids* **6,** 181 (1971).
50. Van Vechten, J. A., *Solid State Comm.* **11,** 7 (1972).
51. Fritzsche, H., "Annual Review of Materials Science", (ed. R. A. Huggins). Annual Reviews Inc., Palo Alto, Cal., **2,** 697 (1972); See also reference 31.
52. Cohen, M. H., Fritzsche, H. and Ovshinsky, S. R., *Phys. Rev. Letters* **22,** 1065 (1969).
53. Mott, N. F., *J. Non-Cryst. Solids* **8-10,** 1 (1972).
54. Marshall, J. M. and Owen, A. E., *Phil. Mag.* **24,** 1281 (1971).
55. Tauc, J., Chapter 4 *In* "Amorphous and Liquid Semiconductors", (ed. J. Tauc), Plenum Press, 1973.
56. Spear, W. E., private communication.
57. Osmun, J. W. and Fritzsche, H., *Appl. Phys. Letters* **16,** 87 (1970).

58. Sauvage, J. A., Mogab, C. J. and Adler, D., to be published.
59. Smith, C. W., and Clark A. H., *Thin Solid Films* **9,** 207 (1972).
60. Sauvage, J. A. and Mogab, C. J., *J. Non-Cryst. Solids* **8-10,** 607 (1972).
61. Konak, C. and Stuke, J., *phys. stat. sol.(a)* **9,** 333 (1972).
62. Austin, I. G. and Mott, N. F., *Adv. in Phys.* **18,** 41 (1969).
63. Pollak, M. and Geballe, T. H., *Phys. Rev.* **122,** 1742 (1961).
64. Lakatos, A. I. and Tabak, M. D. *Bull. Am. Phys. Soc. Series II,* **17,** 322 (1972).
65. Mort, J. and Lakatos, A. I., *J. Non-Cryst. Solids* **4,** 117 (1970).
66. Schottmiller, J., Tabak, M., Lucovsky, G. and Ward A., *J. Non-Cryst. Solids* **4,** 80 (1970).
67. Owen, A. E., *Progress in Ceramic Science* **3,** 77 (1963).
68. Pollak, M. and Pike, G. E., *Phys. Rev. Letters* **28,** 1449 (1972).
69. Morgan, M. and Walley, P. A., *Phil. Mag.* **23,** 661 (1971).
70. Mott, N. F., *Phil. Mag.* **19,** 835 (1969).
71. Kolomiets, B. T., Ljubin, V. M. and Averjanov, V. L., *Mat. Res. Bull.* **5,** 655 (1970).
72. Brodsky, M. H. and Title, R. S., *Phys. Rev. Letters* **23,** 581 (1969).
73. Brodsky, M. H., *J. Vacuum Sci. and Tech.* **8,** 125 (1971).
74. Agarwal, S. C., *Phys. Rev.* (in print).
75. Tauc, J., Di Salvo, F. J., Peterson, G. E. and Wood, S. L., Proc. of Symposium on Amorphous Magnetism, Detroit, Michigan, Aug. 1972.
76. Matyas, M., *J. Non-Cryst. Solids* **8-10,** 592 (1972).
77. Lewis, A., *Phys. Rev. Letters* **29,** 1555 (1972).
78. Giaever, I., *Phys. Rev. Letters* **5,** 147, 464 (1960).
79. Duke, C. B., "Tunneling in Solids", Solid State Physics, Supplement 10, Academic Press, N.Y. 1969.
80. Esaki, L., *Phys. Rev.* **109,** 603 (1957).
81. Osmun, J. W., to be published.
82. Stratton, R., *J. Phys. Chem. Solids* **23,** 1177 (1962).
83. Shen, L. Y. L. and Rowell, J. M., *Solid State Comm.* **5,** 189 (1967).
84. Nielsen, P., *Phys. Rev.* **B2,** 3819 (1970).
85. Mott, N. F., private communication (1972).
86. Hall, R. N., Racette, J. H. and Ehrenreich, H. *Phys. Rev. Letters* **4,** 456 (1966).
87. Zeller, R. C. and Pohl, R. O., *Phys. Rev.* **B4,** 2029 (1971).
88. Anderson, P. W., Halperin, B. I. and Varma, C. M., *Phil. Mag.* **25,** 1 (1972).
88a. Phillips, W. A., *J. Low Temp. Phys.* **7,** 351 (1972).
89. Chittick, R. C., *J. Non-Cryst. Solids* **3,** 255 (1970).
90. Tauc, J., Abraham, A., Zallen, R. and Slade, M., *J. Non-Cryst. Solids* **4.** 279 (1970).

3. THE ELECTRONIC DENSITY OF STATES IN AMORPHOUS MATERIALS

D. A. GREENWOOD

H. H. Wills Physics Laboratory,
University of Bristol,
Bristol.

3.1. INTRODUCTION

This contribution surveys attempts which have been made to reconcile the physicist's and chemist's approaches to the problem of understanding, and calculating, the density of one-electron states in an amorphous semi-conductor. The theory of potential scattering from a single centre is briefly reviewed, and the concepts of a scattering resonance and a Wigner delay time are introduced. Perturbation methods based on the Green function are described, and the difficulty of applying such methods to the open structure of an amorphous semi-conductor pointed out. The Friedel sum for a scatterer is related to the Wigner delay time, and an alternative approach to the density of states of a condensed system, based on the Friedel sum, is described. The generalization of this argument to clusters of scatterers, suggested by McGill and Klima and based on a formalism of

Lloyd, is set out. The method allows short-range order to be included. Numerical calculations are presented. In the final sections, the results obtained with the Weaire model Hamiltonian are discussed, together with Ziman's conjectures on the nature of electronic states in the valence band of an amorphous semi-conductor.

Most of our understanding of the physics of electrons in metals and semi-conductors is based on the concept of energy bands. These we see as arising from the periodicity of the crystal lattice, and the mathematical structure of our theories is based on Bloch's theorem. Faced with a liquid metal, or, worse still, an amorphous semi-conductor, our theories are inapplicable. Nevertheless, liquid metals are metallic, and amorphous semi-conductors exist. To a chemist, who thinks in terms of hybridized atomic orbitals and localized bonds, an amorphous semi-conductor is far less mysterious, and, indeed, is rather to be expected. Our object, in these lectures, is to reconcile the conventional point of view of the solid state physicist with that of the chemist.

We shall assume that it is sufficient to work in the usual one-electron approximation, in which we construct a self-consistent potential $V(\mathbf{x})$ for the valence electrons of a metal or semi-conductor. If $V(\mathbf{x})$ is a periodic potential, there are the various familiar methods for computing the energy eigenvalues and eigenstates of the Schrödinger equation†

$$\{-\nabla^2 + V(\mathbf{x})\}\psi_n = \mathcal{E}_n\psi_n, \tag{3.1}$$

and hence obtaining the band-structure.

If $V(\mathbf{x})$ is not periodic, but corresponds to a disordered condensed system, we can, first of all, ask for the density of states per unit volume per unit energy range

$$g(\mathcal{E}) = \frac{2}{\Omega}\sum_n \delta(\mathcal{E} - \mathcal{E}_n), \tag{3.2}$$

or the integrated density of states

$$N(\mathcal{E}) = \int_{-\infty}^{\mathcal{E}} g(\mathcal{E})\, d\mathcal{E}. \tag{3.3}$$

Here Ω is the volume of the system, and we have included a factor of 2 for electron spin.

† We use atomic units, in which the unit of length is the Bohr radius $a_0 = \hbar^2/me^2$, and the unit of energy the Rydberg, 1 Ry $= e^2/2a_0 = 13{\cdot}6\ eV$. In these units, $\hbar = 2m = e^2/2 = 1$.

In fact, the positions of the ions in a disordered system will not be precisely known, so that what we can actually calculate will be an ensemble average of $N(\mathcal{E})$. And, for the same reason, the question of the nature of the eigenstates in a disordered system is a difficult one. We shall touch on it later, but our primary concern will be the calculation of $N(\mathcal{E})$, as the simplest quantity to calculate, and a necessary, even if not sufficient, ingredient in the calculation of many experimentally observable quantities. In particular, a gap in the density of states has immediate consequences in optical absorption studies. Our primary concern will be with amorphous Group IV semi-conductors, but we shall also refer occasionally to liquid metals for comparison and contrast.

We consider an "ideal" amorphous semi-conductor, following the model of Grigorovici,[1] in which each ion is in a locally ordered, nearly perfect, tetrahedral environment. This is, then, a rather open structure with a co-ordination number of four, in contrast to the "random close-packing" of a liquid metal for which the coordination number is typically between ten and twelve. We no longer have a periodic lattice, but the chemist's picture of a covalently bonded solid suggests that it is the local configuration of neighbouring ions which determines the electron energies and electron density around a particular ion. Essentially, the cluster scattering approach developed at Bristol which will be described in this article explores the extent to which this is true.

3.2. Construction of the Muffin-tin Potential

To a first approximation, the potential $V(\mathbf{x})$ may be taken to be a sum of spherically symmetrical, non-overlapping potentials $u(|\mathbf{x} - \mathbf{X}_i|)$ centred on the ion sites $\mathbf{X}_i$, with a constant interstitial potential V_M:

$$V(\mathbf{x}) = V_M + \sum_i u(|\mathbf{x} - \mathbf{X}_i|). \tag{3.4}$$

This approximation is familiar in the standard APW or KKR methods of calculating band structure.

For constructing the muffin-tin potential, $u(r)$, the recipe suggested by Mattheiss[2] can equally well be followed in the amorphous case. Essentially, one constructs a potential by superposing the electron charge distributions of neutral atoms (as calculated, for example, by Herman and Skillman[3]), placed at the ion sites, and using Poisson's equation. A spherical average is taken, and a contribution for electron exchange added. The method is described in detail in the book by Loucks.[4] Experience of band structure calculations suggests that this can be a reliable technique, though the results may be somewhat sensitive to the approximation used for the exchange term.[5] Possibly the best approach is to use the Slater free-electron expression

$$u_{ex}(r) = -3\{3\rho(r)/\pi\}^{1/3}, \tag{3.5}$$

multiplied by a parameter ($\sim 0{\cdot}8$) which is determined by fitting the band structure to experiment.

We assume that in the amorphous material the muffin-tin potential can be taken to be the same at each site; this is a good approximation for the Grigorovici model in which each site has a tetrahedral environment. It should be noted that the muffin-tin potential has the local environment built into it to some extent—a different environment would lead to a different $u(r)$.

In the open structure of a semi-conductor, the interstitial constant potential region fills some two-thirds of the volume, even when the muffin-tins touch. It is clear that this cannot be a very satisfactory approximation, nor is the assumption of spherically symmetric potentials likely to be an accurate one. Keeping V_M as a parameter helps, in practice. Keller [6] has avoided this arbitrariness by packing the interstitial region with yet more muffin-tins. The restriction to spherically symmetric potentials can also be removed, at least in principle. However, we will assume that the essential physics can be understood using the potential given by eqn (3.4).

3.3. Scattering by a Single Centre

Consider the solution of the Schrödinger equation with the potential of eqn (3.4). At energies above V_M, which we can conveniently take as our energy zero, electrons can be thought of as propagating freely in the interstitial region, and being scattered by the muffin-tin potentials. We shall review briefly the description of scattering by a single scattering centre with a potential $u(r)$, where $u(r) = 0$ for $r > r_M$.[7]

We exploit the spherical symmetry of the scattering potential by splitting up the incident wave e^{ikz}, at energy $\mathcal{E} = k^2$, into its angular momentum components, which will be conserved:

$$e^{ikz} = e^{ikr\cos\theta} = \sum_l (2l+1) i^l j_l(kr) P_l(\cos\theta) \underset{r\to\infty}{\sim} \sum_l (2l+1)\frac{1}{2ikr} \times [e^{ikr} - (-1)^l e^{-ikr}]\, P_l(\cos\theta), \tag{3.6}$$

using the asymptotic form for the spherical Bessel function. In the presence of the scattering centre the incoming spherical waves e^{-ikr} cannot be affected, but the potential alters the outgoing waves by a phase factor, which we write as $S_l = e^{2i\eta_l}$. These are elements of the "S-matrix". Since the outgoing flux must equal the incoming flux, for each angular momentum component, only a phase factor, with η_l real, can be introduced. Thus the scattering solution must be of the form

$$\sum_l (2l+1)\frac{1}{2ikr}\left[e^{2i\eta_l}e^{ikr} - (-1)^l e^{-ikr}\right] P_l(\cos\theta) \tag{3.7}$$

in the asymptotic region.

We can rearrange this, using eqn (3.6), as

$$e^{ikz} + \sum_l (2l+1)\frac{1}{2ikr}(e^{2i\eta_l} - 1)\, e^{ikr} P_l(\cos\theta)$$

$$= e^{ikz} + \frac{1}{k}\sum_l (2l+1)\, e^{i\eta_l}\sin\eta_l \frac{e^{ikr}}{r} P_l(\cos\theta)$$

$$= e^{ikz} - \sum_l (2l+1) T_l P_l(\cos\theta)\frac{e^{ikr}}{r}$$

(where the last line defines $T_l = -\dfrac{1}{k}e^{i\eta_l}\sin\eta_l$)

$$= \text{incident wave} + \text{scattered wave.} \tag{3.8}$$

Thus the scattering properties of the potential are expressed entirely in terms of the phase shifts η_l. We can remove the "modulo π" ambiguity by putting $\eta_l(\mathcal{E}) = 0$ at $\mathcal{E} = 0$. The alternative convention is to take $\eta_l(\infty) = 0$ (at sufficiently high energies the incident electron will not be affected by the scattering potential), but we can make the reduction since Levinson's theorem[8] states that $\eta_l(0) - \eta_l(\infty) = n_l\pi$, where n_l is the number of bound states of angular momentum l in the potential. We shall not be concerned with states below the muffin-tin zero, so our choice is convenient.

In the empty space region, $r > r_M$, the radial part $R_l(r)$ of solutions of the Schrödinger equation is of the form

$$a_l\{\cos\eta_l\, j_l(kr) - \sin\eta_l\, n_l(kr)\}.$$

(This goes over to give eqn (3.7) in the asymptotic region with a suitable choice of a_l.)

At $r = r_M$ the radial wave function $R_l(r)$ and its derivative must be continuous, so that

$$\left[\frac{1}{R_l}\frac{dR_l}{dr}\right]_{r=r_M} = \frac{k[\cos\eta_l\, j_l' - \sin\eta_l\, n_l']_{r=r_M}}{[\cos\eta_l\, j_l - \sin\eta_l\, n_l]_{r=r_M}}. \tag{3.9}$$

This equation we can solve for $\tan \eta_l$. The left-hand side of eqn (3.9) can be found by numerical integration of the radial Schrödinger equation out from the origin, given the potential $u(r)$. Conversely, the logarithmic derivative of the wave-function at the muffin-tin radius is expressible in terms of the phase shifts.

With the scattering of each partial wave is associated a *Wigner delay time*.[9] For simplicity, consider an incident spherical "wave-packet" formed by the superposition of waves k, k':

$$\psi_{\text{in}} = \frac{1}{r}(e^{-i(kr-\varepsilon t)} + e^{-i(k'r-\varepsilon' t)}).$$

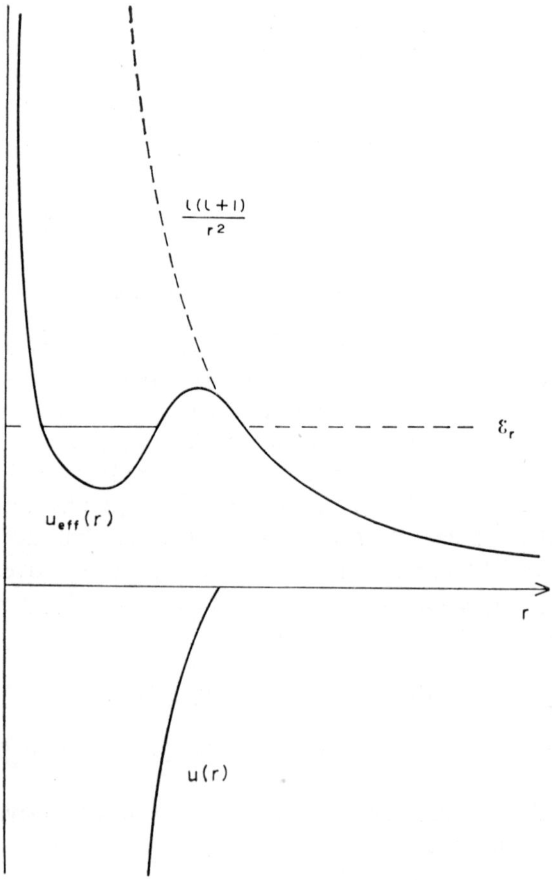

FIG. 3.1. Effective radial potential, giving rise to a metastable state and a scattering resonance.

The group velocity of the superposed wave is obtained from the condition that the two components remain in phase:

$$kr - \mathcal{E}t = k'r - \mathcal{E}'t$$

or

$$r = \frac{\mathcal{E} - \mathcal{E}'}{k - k'}t = \frac{d\mathcal{E}}{dk}t,$$

i.e. the group velocity is $d\mathcal{E}/dk$.

The outgoing wave is

$$\psi_{\text{out}} = \frac{1}{r}(e^{i(kr - \mathcal{E}t) + 2i\eta} + e^{i(k'r - \mathcal{E}'t) + 2i\eta'}),$$

and the components are in phase when

$$kr - \mathcal{E}t + 2\eta = k'r - \mathcal{E}'t + 2\eta'$$

or

$$r = \frac{d\mathcal{E}}{dk}t - 2\frac{d\eta}{dk} = \frac{d\mathcal{E}}{dk}\left(t - 2\frac{d\eta}{d\mathcal{E}}\right).$$

Thus the outgoing wave-packet is retarded by a time

$$\tau_w = 2\frac{d\eta}{d\mathcal{E}}$$

relative to the time in the absence of the scattering potential.

The Wigner delay time τ_w can be very long if $\eta(\mathcal{E})$ is a very rapidly varying function of energy, as in the neighbourhood of a *scattering resonance*. When an attractive potential $u(r)$ is combined with the centrifugal potential $l(l + 1)/r^2$ the resulting effective potential in the radial Schrödinger equation can have the form shown in Fig. 3.1. In this potential there may be quasi-bound states, which we can represent by a complex energy $\mathcal{E}_r - i\Gamma$, where Γ determines the life-time of the metastable state. At energies near $\mathcal{E}_r$, an incident electron is trapped into the metastable state, and the Wigner delay time is correspondingly long, of the order of the life-time of the metastable state. Assuming a Lorentzian form for the Wigner delay time for energies in the neighbourhood of $\mathcal{E}_r$,

$$\tau_w = 2\frac{d\eta}{d\mathcal{E}} = \frac{2\Gamma}{\Gamma^2 + (\mathcal{E}_r - \mathcal{E})^2}, \tag{3.10}$$

will give

$$\eta(\mathcal{E}) = \eta_0 + \tan^{-1}\frac{\Gamma}{\mathcal{E}_r - \mathcal{E}}, \tag{3.11}$$

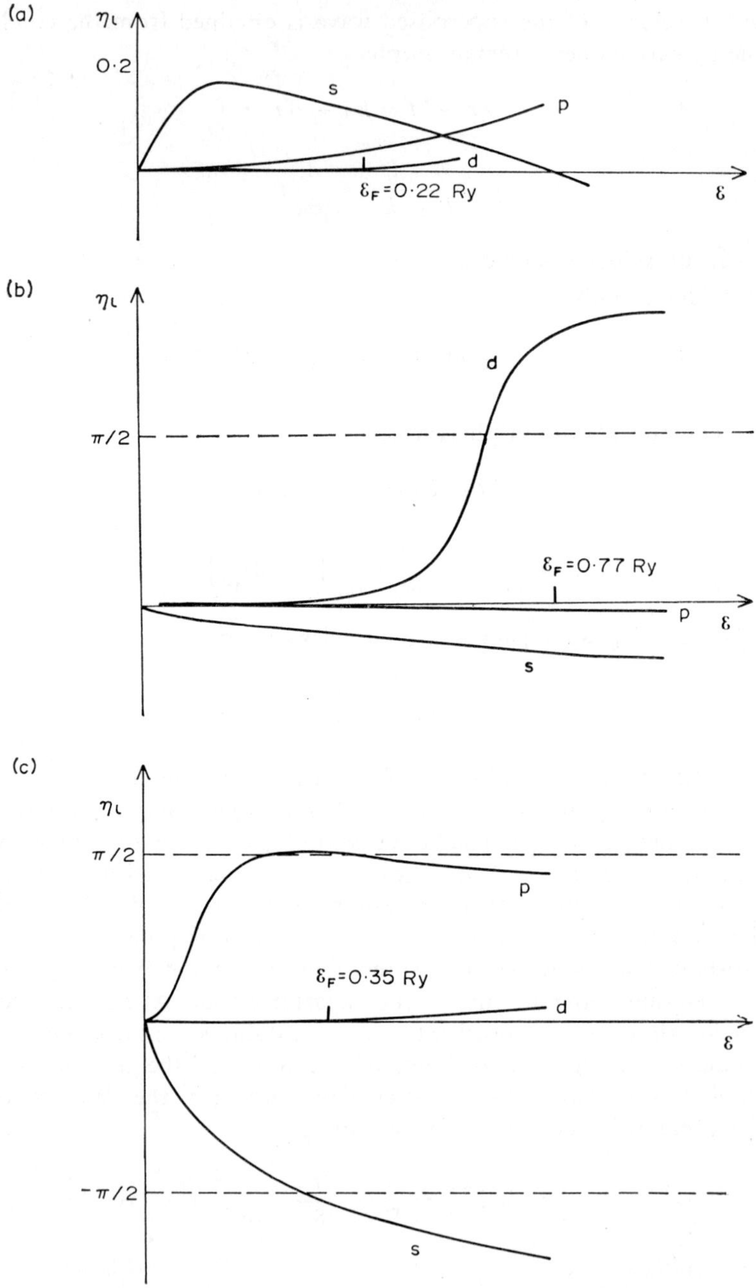

FIG. 3.2. Typical "muffin-tin" phase shifts for (a) sodium (b) iron (c) germanium. (From unpublished calculations of A. Jain, R. Evans, and P. V. Smith, University of Bristol.)

where η_0 is a contribution which varies slowly with energy. It can be shown that this form is, indeed, correct. From eqn (3.11), we see that the phase shift rises steeply in passing through a resonance. The cross-section for scattering of the lth partial wave, which is proportional to $|T_l|^2$, is also of Lorentzian form, with maximum at the resonance energy.

Clearly, we can construct potentials for which the Wigner delay time is arbitrarily long. However, the magnitude of a negative delay time is severely restricted by causality arguments: semi-classically, the scattered wave-packet cannot be advanced in space by a distance greater than $2a$, where a is the range of the scattering potential, so that

$$\tau_w = 2\frac{\mathrm{d}\eta}{\mathrm{d}\mathcal{E}} > -\frac{2a}{(\partial\mathcal{E}/\partial k)} = -\frac{a}{k}.$$

We shall refer to a large negative value of $\mathrm{d}\eta_l/\mathrm{d}\mathcal{E}$ as an anti-resonance; in such a case, an electron of angular momentum l does not penetrate into the interior of the scattering potential.

For the muffin-tin potentials and energies with which we are concerned, only the $l = 0$, 1, 2 phase shifts are significant. (A classical particle with angular momentum $> 2kr_M$ would not see the potential.) Typical phase shifts, calculated using eqn (3.9), for a simple free-electron metal, a transition metal, and a semi-conductor are shown in Fig. 3.2.

In the case of sodium the phase shifts are all small up to the Fermi energy. In contrast, in a transition metal, the *d*-phase shift is important, and there is a *d*-resonance in the region of the Fermi energy: in forming the muffin-tin potential, the overlapping of the neutral atoms lifts the atomic *d*-states above the muffin-tin zero, but they survive as a resonance. In the semi-conductor both the *s*- and *p*-phase shifts are large at the Fermi energy. Thus the nature of the atomic core is very directly reflected in the single-site scattering phase shifts.

3.4. Pseudo-potentials and Perturbation Theory

Returning to the multiple scattering problem, we can see that the eigenvalues of the Schrödinger equation for the assembly of non-overlapping potentials are, in fact, determined by the positions of the muffin-tins and the muffin-tin phase shifts. In constructing an eigenstate, the logarithmic derivatives of the solutions to the Schrödinger equation inside the muffin-tins have to be matched, at the surface of each sphere, to the solution of the Schrödinger equation in the interstial region. These logarithmic derivatives are expressed, by eqn (3.9), in terms of the phase shifts. If these are known as a function of energy then the eigenstates in the interstial region may, in principle, be constructed. It follows that, if we are only interested in the density of states, any "pseudo-potential" which has the same scattering properties as the original muffin-tin potential may be used.[10] In particular, it may be possible to define a pseudo-potential which is sufficiently weak to be treated by perturbation theory. Such pseudo-potentials will, in general, be energy dependent and angular momentum dependent. It is worth remarking here that the pseudo-potentials which have been used so successfully in the theory of simple metals[11] are not usually of "muffin-tin" form, but overlap, and so do not quite fit in with the definition we have made. However, the basic idea behind them is the same.

The calculation of the density of states for a disordered system by perturbation theory has acquired an extensive and sophisticated literature.[12] The density of states can be expressed in terms of the Green function operator:

$$G(\mathcal{E}) = \frac{1}{\mathcal{E} + i\varepsilon - H}, \tag{3.12}$$

where $H = -\nabla^2 + V$, and the small imaginary part $i\varepsilon$ is introduced to remove singularities when $\mathcal{E}$ is an eigenvalue of H.

Using the identity

$$\frac{1}{x + i\varepsilon} = P\left(\frac{1}{x}\right) - i\pi\delta(x), \tag{3.13}$$

we can write for the density of states

$$\begin{aligned} g(\mathcal{E}) &= \frac{2}{\Omega}\sum_n \delta(\mathcal{E} - \mathcal{E}_n) \\ &= -\frac{2}{\pi}\frac{1}{\Omega}\,\mathrm{Im}\sum_n (n|G(\mathcal{E})|n). \end{aligned} \tag{3.14}$$

The sum over the eigenstates n gives just the trace of the operator $G(\mathcal{E})$, and the trace is the same in any representation. In a perturbation theory it is natural to use free-electron plane wave states $(\Omega)^{-\frac{1}{2}}e^{i\mathbf{k}\cdot\mathbf{r}}$ as a basis. Then

$$g(\mathcal{E}) = -\frac{2}{\pi}\frac{1}{\Omega}\,\mathrm{Im}\sum_{\mathbf{k}} (\mathbf{k}|G(\mathcal{E})|\mathbf{k}). \tag{3.15}$$

In the limit $\Omega \to \infty$,

$$\frac{1}{\Omega}\sum_{\mathbf{k}} \to \frac{1}{(2\pi)^3}\int \mathrm{d}^3k.$$

We can easily check that in the absence of scattering centres, with $H = H_0 = -\nabla^2$, we obtain the free-electron density of states,

$$g_0(\mathcal{E}) = \begin{cases} 0, & \mathcal{E} < 0. \\ \dfrac{1}{2\pi^2}\mathcal{E}^{\frac{1}{2}}, & \mathcal{E} > 0. \end{cases} \tag{3.16}$$

If $H = H_0 + V$, we can iterate the operator identity

$$\frac{1}{\mathcal{E} + i\varepsilon - (H_0 + V)} = \frac{1}{\mathcal{E} + i\varepsilon - H_0} + \frac{1}{\mathcal{E} + i\varepsilon - H_0} V \frac{1}{\mathcal{E} + i\varepsilon - (H_0 + V)}$$

or

$$G = G_0 + G_0 V G, \tag{3.17}$$

to give an expansion which we can write, formally,

$$G = G_0 + G_0 V G_0 + G_0 V G_0 V G_0 + \ldots$$

or, in full, in the momentum representation,

$$(\mathbf{k}|G(\mathcal{E})|\mathbf{k}) = \frac{1}{\mathcal{E} + i\varepsilon - k^2} + \frac{1}{\mathcal{E} + i\varepsilon - k^2}(\mathbf{k}|V|\mathbf{k})\frac{1}{\mathcal{E} + i\varepsilon - k^2}$$
$$+ \sum_{\mathbf{k}'} \frac{1}{\mathcal{E} + i\varepsilon - k^2}(\mathbf{k}|V|\mathbf{k}')\frac{1}{\mathcal{E} + i\varepsilon - k'^2}(\mathbf{k}'|V|\mathbf{k})\frac{1}{\mathcal{E} + i\varepsilon - k^2} + \ldots \tag{3.18}$$

In the expanded form, it is possible to take an ensemble average over the position of the scattering centres.

It is assumed then that the series can be resummed to give

$$(\mathbf{k}\,|G(\mathcal{E})|\,\mathbf{k}) = \frac{1}{\mathcal{E} + i\varepsilon - k^2 - \Sigma(\mathcal{E}, k)}, \tag{3.19}$$

where $\Sigma(\mathcal{E}, k) = A - iB$ is a complex *self-energy*: a particle initially in a state $\mathbf{k}$ will be both shifted in energy by an amount A, and scattered into other states so that it decays exponentially in time like $\exp(-Bt)$, in the presence of the scattering centres. If we now assume that $\Sigma(\mathcal{E}, k)$ can be expanded in powers of the potential, we can compare the Taylor expansion of eqn (3.19) with eqn (3.17) to get the coefficients in the expansion of $\Sigma(\mathcal{E}, k)$. The procedure is not mathematically rigorous, but the self-energy assumption is physically plausible. If a weak pseudo-potential can be used to describe the scattering then, to second order in

$$V = \sum_i u_{ps},$$

we find easily enough

$$\Sigma(\mathcal{E}, k) = \frac{1}{(2\pi)^3} \frac{\Omega}{\mathcal{N}} \int d^3q \frac{|u_{ps}(q)|^2 a(q)}{\mathcal{E} - |\mathbf{k} + \mathbf{q}|^2 + i\varepsilon}, \tag{3.20}$$

where

$$a(q) = \frac{1}{\mathcal{N}} \left\langle \left| \sum_i \exp(i\mathbf{q} \cdot \mathbf{X}_i) \right|^2 \right\rangle$$

is the structure factor. Here $\mathcal{N}$ is the number of scattering centres in the volume Ω.

The brackets denote an ensemble average over the positions of the scattering centres, and the zero of energy has been conveniently redefined to give $\langle(\mathbf{k}|V|\mathbf{k})\rangle = 0$.

The density of states is then to be found from eqn (3.15)

$$g(\mathcal{E}) = -\frac{2}{\pi} \operatorname{Im} \frac{1}{(2\pi)^3} \int d^3k \frac{1}{\mathcal{E} - k^2 - \Sigma(\mathcal{E}, k)}. \tag{3.21}$$

To this order, only the radial distribution function, via the structure factor $a(q)$, enters into the ensemble average over the positions of the scatterers. The expression can be improved somewhat by making it self-consistent, and modifying it to allow for the energy dependence of a practical pseudo-potential (which gives difficulties in the basic expression for $g(\mathcal{E})$ as Lloyd[(13)] pointed out). The result of such calculations by Chan and Ballentine[(14)] for liquid mercury are shown in Fig. 3.3. Mercury is a quite good free-electron metal, but the resistivity measurements show that the scattering of conduction electrons is strong compared, say, with sodium. There is some structure in the density of states, but the curve is still, really, close to the free-electron parabola.

The perturbation series can be rearranged and manipulated, so that, for example, the scattering at an ion is correctly described by a "T-matrix", rather than in the Born approximation to which our expression eqn (3.20) is restricted. It is, furthermore, possible to rearrange the expression for the *integrated* density of states so that only "on-energy-shell" elements of the T-matrix appear, providing that we have a non-overlapping muffin-tin potential. The on-energy-shell elements correspond to the T_l defined in eqn (3.8), and are expressible in terms of the phase shifts. We have argued on general grounds that such a result must be possible. The paper by Lloyd[13] in which this formalism is derived is the basis of the cluster scattering calculations I shall describe.

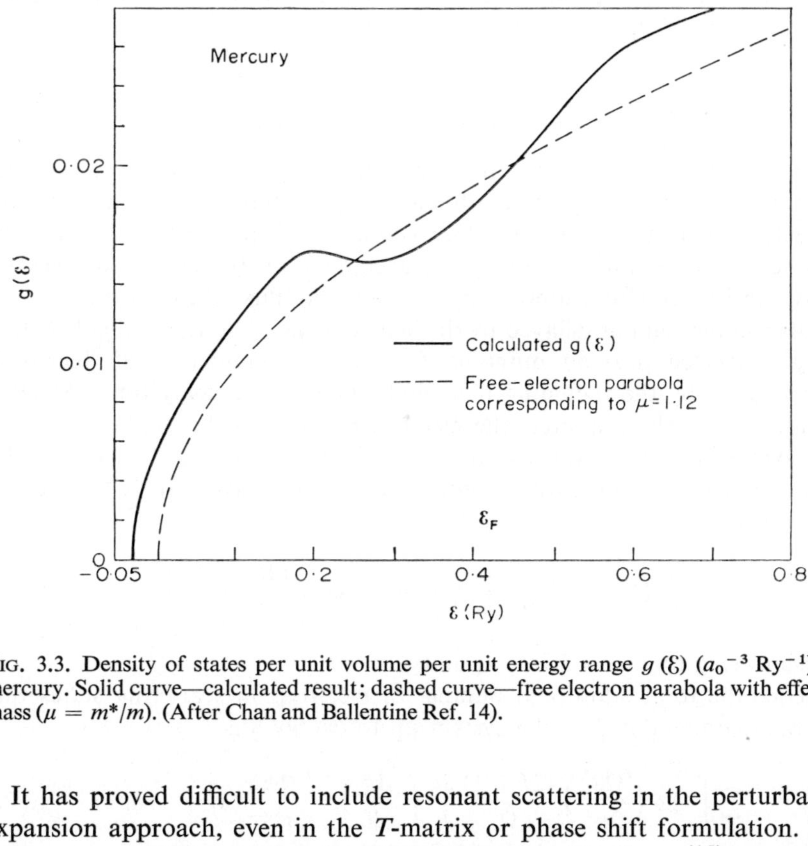

FIG. 3.3. Density of states per unit volume per unit energy range $g(\mathcal{E})$ (a_0^{-3} Ry^{-1}) for mercury. Solid curve—calculated result; dashed curve—free electron parabola with effective mass ($\mu = m^*/m$). (After Chan and Ballentine Ref. 14).

It has proved difficult to include resonant scattering in the perturbation expansion approach, even in the T-matrix or phase shift formulation. The coherent potential approximation (CPA) introduced by Soven[15] is a further development in the multiple scattering theory, and appears better able to handle scattering resonances. In the CPA, each scatterer is considered to be immersed in a medium described by an effective Hamiltonian, such that the single particle scattering due to the difference between the true Hamiltonian and the effective Hamiltonian is, on average, zero. The application of this idea to liquid metals has been considered by Gyorffy.[16]

However, the difficulty in the application of either perturbation expansions, or CPA, to the amorphous semi-conductor is clear. Any essential improvement on a formula like eqn (3.20) involves some approximations for the three-particle, and higher order, correlation functions, which are not at all well known, either theoretically or experimentally. We have already argued that, in an amorphous semi-conductor, the local tetrahedral environment of a scatterer is likely to be of dominant importance. The pair correlation function is not sufficient to determine this. The cluster scattering approach which we describe in the next sections attempts to include the local structure from the start of the calculation.

3.5. The Friedel Sum Density of States

Consider, now, a single scattering centre in a volume Ω. Suppose the one-electron states in this volume are filled up to an energy $\mathcal{E}$. With no scatterer present, the density of states per unit volume is given by $g(\mathcal{E}) = \mathcal{E}^{\frac{1}{2}}/2\pi^2$. The presence of the scatterer leads to an accumulation (or deficit) of electron density in its neighbourhood, since an incoming flux of electrons, of given angular momentum, is delayed by the Wigner delay time $\tau_w = 2\mathrm{d}\eta_l/\mathrm{d}\mathcal{E}$ before being converted into an outgoing flux. The accumulation of density is simply given by the incoming flux, multiplied by the delay time. At large distances from the scatterer, the electron density will be unaltered. Using the asymptotic form given by eqn (3.7), the nett incoming flux of angular momentum l associated with a normalized plane wave $(\Omega)^{-\frac{1}{2}}e^{i\mathbf{k}\cdot\mathbf{r}}$ is easily found to be

$$\frac{1}{\Omega}(2l+1)^2\frac{2k}{4k^2}\int P_l^{\,2}(\cos\theta)\,\mathrm{d}\omega = \frac{(2l+1)}{\Omega}\frac{2\pi}{k}, \qquad (3.22)$$

(remembering that $m = \frac{1}{2}$ in our units).

Hence the total additional number of electrons in scattering states that can be accommodated in the system up to energy $\mathcal{E}$ is

$$\sum_l \int_0^{\mathcal{E}} \mathrm{d}\mathcal{E}\left(\frac{\Omega\mathcal{E}^{\frac{1}{2}}}{2\pi^2}\right)\frac{(2l+1)}{\Omega}\frac{2\pi}{k}\left(2\frac{\mathrm{d}\eta_l}{\mathrm{d}\mathcal{E}}\right) = \frac{2}{\pi}\sum_l (2l+1)\,\eta_l(\mathcal{E}). \qquad (3.23)$$

This is the *Friedel sum*. Other elementary derivations of the result can be given,[17] but this one shows the physics behind the result rather clearly.

It follows that the integrated density of states, $N(\mathcal{E})$, is given by

$$N(\mathcal{E}) = N_0(\mathcal{E}) + \frac{1}{\Omega}\frac{2}{\pi}\sum_l (2l+1)\,\eta_l(\mathcal{E}),$$

where $N_0(\mathcal{E})$ is the free-electron contribution, and

$$g(\mathcal{E}) = g_0(\mathcal{E}) + \frac{1}{\Omega}\frac{2}{\pi}\sum_l (2l+1)\frac{\mathrm{d}\eta_l}{\mathrm{d}\mathcal{E}}. \qquad (3.24)$$

If we have $\mathcal{N}$ scatterers in the volume Ω, and assume that interference between the scatterers can be neglected, then

$$g(\mathcal{E}) = g_0(\mathcal{E}) + \frac{\mathcal{N}}{\Omega}\frac{2}{\pi}\sum_l (2l+1)\frac{d\eta_l}{d\mathcal{E}}. \tag{3.25}$$

This is a crude "single-site" approximation. Nevertheless if a scattering resonance is important, as in a transition metal, this formula immediately picks out the resonant contribution to the density of states, and is the basis of the calculations of Anderson and McMillan[18] for liquid iron, and Schwartz and Ehrenreich[19] for copper.

The formula can be rewritten in terms of the matrix elements $S_L = e^{2i\eta_l}$ of the "S-matrix" which converts the incoming wave into the outgoing wave:

$$\begin{aligned} N(\mathcal{E}) &= N_0(\mathcal{E}) + \frac{\mathcal{N}}{\Omega}\frac{1}{\pi}\,\text{Im}\sum_{l,m}\log S_L(\mathcal{E}) \\ &= N_0(\mathcal{E}) + \frac{\mathcal{N}}{\Omega}\frac{1}{\pi}\,\text{Im}\,\text{tr}\{\log S(\mathcal{E})\}, \end{aligned} \tag{3.26}$$

where the trace is over angular momentum states $L = (l, m)$. In this form it remains valid, even if the scattering centre is not spherically symmetric.

3.6. Cluster Expansion

The basic idea of the cluster scattering method of McGill and Klima[20] is now very simply expressed: we divide up the amorphous material into clusters of ions, and apply the Friedel sum expression with the cluster as scattering element, instead of the single ion. Within a cluster, the scattering of an electron between the ions of the cluster is correctly treated. Interference between different clusters of atoms is neglected in the calculations. Clusters can be chosen to represent typical configurations, α, found in the amorphous structure, and added in the Friedel sum with appropriate weights p_α, so that we have, approximately,

$$\langle N(\mathcal{E})\rangle = \frac{\mathcal{E}^{3/2}}{3\pi^2} + \sum_\alpha p_\alpha \frac{1}{\pi\omega_\alpha}\,\text{Im}\,\text{tr}\,\{\log S^\alpha(\mathcal{E})\}, \tag{3.27}$$

where ω_α is the volume occupied by the cluster. It can be shown that eqn (3.27) is the first term of a formally correct expansion. Higher-order terms in this expansion represent the effect of interference between clusters of scatterers. But the computational problem of estimating the size of such terms, and the convergence of the expansion, is not one to be contemplated: the arguments for eqn (3.27) have to be physical ones. We expect the short-range order to be significant, and much of this is included, even in a quite small cluster. Since there is no long-range order, there will be partial cancellation of the randomized multiple scattering effects between clusters, which should tend to reduce the importance of these corrections.

The expression in curly brackets in eqn (3.27) which represents the scattering properties of the cluster can, by our general arguments, be rewritten in terms of the phase shifts and positions of the muffin-tins making up the cluster. For computational purposes it is convenient to use the expression obtained by Lloyd.[21] A mixed position-angular momentum representation is used, and the trace replaced by the evaluation of a determinant, using the identity

$$\operatorname{tr}\{\log M\} = \log\{\det \|M\|\}. \tag{3.28}$$

Lloyd showed that

$$\operatorname{Im}\operatorname{tr}\{\log S(\mathcal{E})\} = -2\operatorname{Im}\log\{\det\|\delta_{LL'}\,\delta_{ij} + G_{LL'}(\mathbf{X}_i - \mathbf{X}_j)\,k_{L'}{}^{j}(\mathcal{E})\|\}, \tag{3.29}$$

where $k_L{}^j(\mathcal{E}) = -\tan\eta_l{}^j(\mathcal{E})/k$ are elements of the "reaction matrix" describing the scattering at the site $\mathbf{X}_j$, and depend on the muffin-tin phase shifts,

$$G_{LL'}(\mathbf{R}) = \begin{cases} ik\sum\limits_{L''} 4\pi i^{l''} h_{l''}{}^{+}(kR)\, Y_{L''}(\hat{\mathbf{R}})\, C_{LL'}^{L''}, & |\mathbf{R}| > 0, \\ ik\delta_{LL'}, & |\mathbf{R}| = 0, \end{cases} \tag{3.30}$$

is the "propagator" which describes the motion of the electrons between sites $\mathbf{X}_i$ and $\mathbf{X}_j$ at energy $\mathcal{E}$, and

$$C_{LL'}^{L''} = \int Y_L(\Omega)\, Y_{L'}{}^{*}(\Omega)\, Y_{L''}(\Omega)\, d\Omega$$

are integrals over spherical harmonics (known as Gaunt numbers).

The subscripts on the determinant run over all sets $(L, i) = (l, m; i)$. Thus, if *s*-, *p*-, and *d*-phase shifts have to be included for each of the N atoms in a cluster, we shall have to evaluate a $9N \times 9N$ determinant. Note that the atoms of a cluster need not be all the same. The size of a cluster for which the determinant can be evaluated is clearly limited by the computational problem, and for the largest clusters dealt with at Bristol it has taken several hours of computer time to produce the density of states curve.

3.7. Numerical Results for Amorphous Semiconductors

The calculations of McGill and Klima[20] for carbon, silicon and germanium were restricted to clusters of eight atoms, in the staggered and eclipsed configurations shown in Fig. 3.4. (In the Grigorovici model of an amorphous semi-conductors all intermediate rotations between these configurations will also appear.) The phase shifts were adjusted slightly so that a reasonable band-structure for the crystalline material could be obtained, bearing in mind the limitations of the muffin-tin model for open structures. For carbon, *s*- and *p*-phase shifts sufficed; for silicon and germanium *d*-phase shifts had to be included also. Typical results, for silicon, are shown in Fig. 3.5, which shows the effect of increasing cluster size. The single atom cluster result, which corresponds to the single Friedel sum of eqn (3.25) shows a peak in the density of states, corresponding to the sharp rise in the *p*-phase shift (Fig. 3.2). As the size of the cluster increases, a dip appears, and the density of states falls below the free-electron density of states. This is shown clearly in Fig. 3.6, where the rather similar results for both staggered and eclipsed configurations of eight atoms are given. Averaging over configurations tends to smooth out any structure appearing

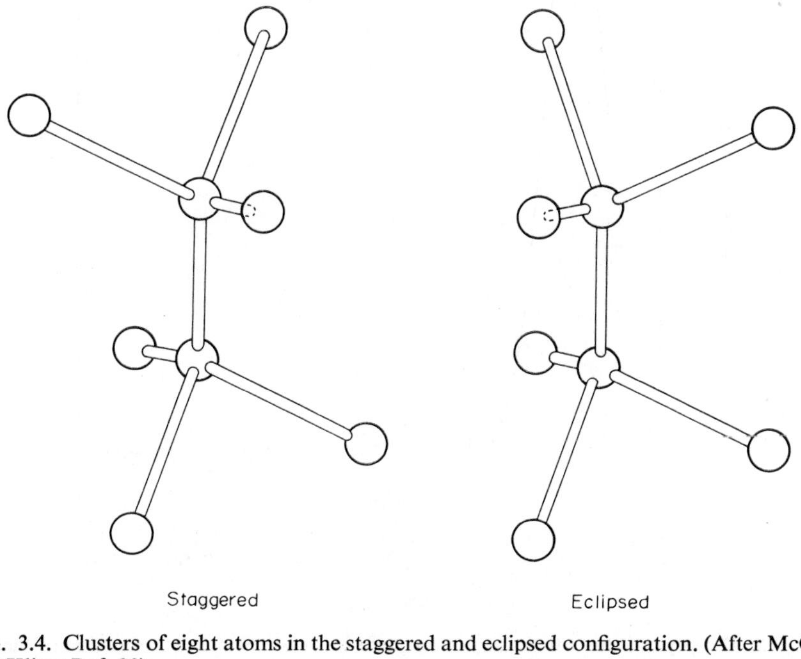

FIG. 3.4. Clusters of eight atoms in the staggered and eclipsed configuration. (After McGill and Klima Ref. 20)

in the results. We have, in fact, a *pseudo-gap* appearing in the density of states. The pseudo-gap is of about the same width as the energy-gap in a perfect crystal, but appears at a rather higher energy.

Calculations for larger clusters, of up to thirty carbon atoms, have been made by Keller.[22] As the number of atoms in a cluster increases, the pseudo-gap tends to shift to lower energies, where the energy gap is in the perfect crystal (Fig. 3.7). For clusters corresponding to an "ideal" amorphous structure, there is no sort of structure in the pseudo-gap. It is

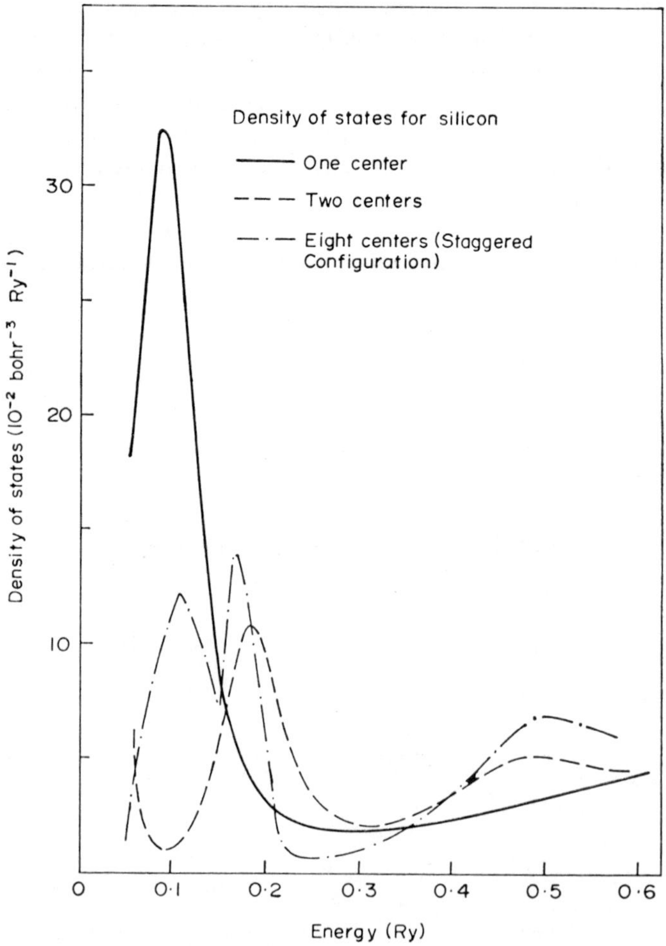

FIG. 3.5. Density of states for clusters of one, two, and eight Si sites as a function of energy. (After McGill and Klima Ref. 20)

interesting that, when calculations were made for a "highly disordered" cluster, in which the angles of some of the bonds were much smaller (15%) than in the "normal" tetrahedral configuration, states appeared in the pseudo-gap. Such "geometrical disorder" tends to change the next nearest neighbour distance for many atoms in the cluster, and it is clear that the "directional bonding" is important in producing a gap.

In fact, of course, even a cluster of thirty atoms is small, and all of the atoms in it are, more or less, on the surface. Keller also attempted to

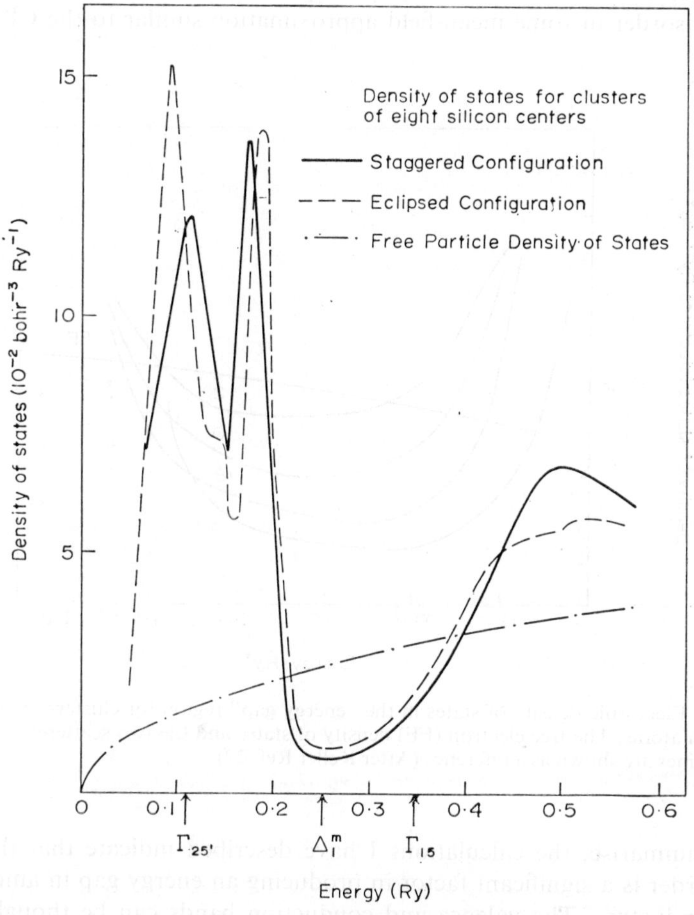

FIG. 3.6. Density of states for clusters of eight Si sites in the staggered and eclipsed configurations, and the free-particle density of states as a function of energy. The locations of relevant energy levels in the band structure are shown along the energy axis. (After McGill and Klima Ref. 20)

estimate the effect of inter-cluster scattering by regarding it as a boundary condition on the cluster, and varying the boundary conditions. For example, if the cluster is imagined repeated periodically, a band-structure calculation becomes possible, and a true band-gap will appear. In Keller's calculations this band gap was not found to be sensitive to the nature of the cluster, or to the existence of "broken bonds" in the periodic structure. However, this periodic arrangement does not properly represent the correlations between clusters in a real amorphous material; these correlations must, surely, be included in the calculation if a true energy-gap rather than a pseudo-gap is to be found. It may, perhaps, be possible to treat the "long-range" disorder in some mean-field approximation similar to the CPA.

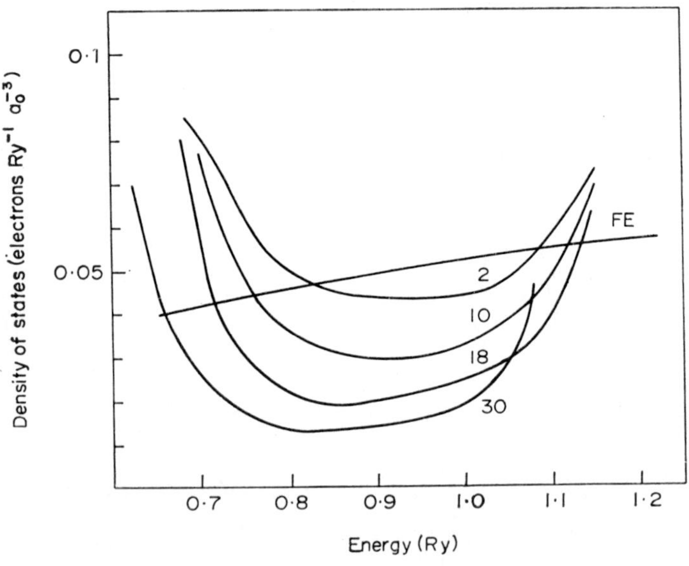

FIG. 3.7. Electronic density of states in the "energy gap" region for clusters of 10, 18 and 30 carbon atoms. The free electron (FE) density of states and the two scatterers (2) density of states lines are shown as a reference (After Keller Ref. 22)

To summarise, the calculations I have described indicate that the short range order is a significant factor in producing an energy gap in amorphous semi-conductors. The valence and conduction bands can be thought of as resulting from resonances in the scattering by a cluster. In the gap, an "anti-resonance" makes it extremely difficult for an electron to propagate through the crystal. At energies in the gap both *p* and *s* muffin-tin phase shifts are large, near $\pi/2$ and $-\pi/2$ respectively, and the cluster combines the individual scatterings into an "anti-resonance". Scattering properties of a cluster have in fact been investigated more closely by Keller and Smith.(23)

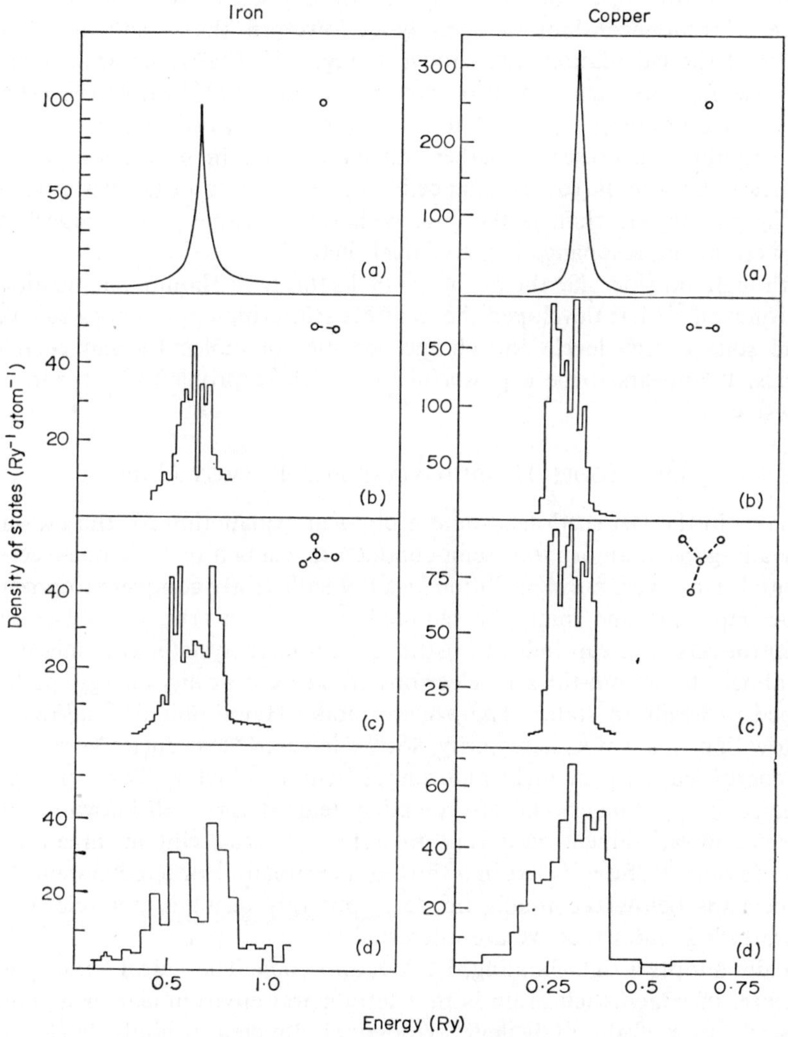

FIG. 3.8. The electronic density of states for iron and copper. (a) Single atom; (b) Two scatterers; (c) Four scatterers; (d) Band structure results for solid iron (Wood 1962) and solid copper (Burdick 1963). (After Keller and Jones Ref. 24)

3.8. Other Applications of the Cluster Method

We have remarked on the difficulty of including a scattering resonance in a perturbation expansion approach. In liquid transition metals there is a *d*-resonance in the neighbourhood of the Fermi energy. The cluster method seems ideally suited to handle this. Keller and Jones[24] have made calculations for iron, and for copper. In copper there is a somewhat sharper *d*-resonance than in iron, lying lower in the conduction band. Results of the calculations are shown in Fig. 3.8. The single atom Friedel sum gives far too narrow a peak, compared with the width of the *d*-band in the corresponding crystal, but it is immediately broadened out when two scatterers are present. Further calculations are in progress at Bristol. It appears that the narrow resonance in copper leads to rather better results with larger clusters than is the case with iron; as one would expect, the sharpness of the resonance is the critical factor.

Although outside the theme of these lectures, it should be mentioned that Johnson[25] has developed the multiple scattering approach to calculate bound state energy levels and charge densities of molecules and chemical radicals. It appears to be a powerful tool in these problems of theoretical chemistry.

3.9. Model Hamiltonians and Energy Gaps

We have, in the last sections, found a physical explanation for the existence of pseudo-gaps in amorphous semi-conductors, using a realistic one-electron potential. However, our description was not sufficiently complete to produce a true gap, and one might be tempted to ask if a true gap, above the muffin-tin zero V_M, can indeed exist in a disordered system. One might feel that above the muffin-tin zero electrons could exist at any energy, perhaps trapped in localized states. The Weaire model Hamiltonian[26] allows one to show that this is not, necessarily, so: we have an "existence theorem" for true energy gaps in an ideal amorphous semi-conductor. That such gaps can exist in one-dimensional disordered systems is very well known,[27] but scattering in one dimension is very different from scattering in three dimensions. Taylor[28] showed that in a three-dimensional disordered system there can be gaps below the muffin-tin zero, but this result is not relevant to the scattering states that we are interested in.

Weaire adopts a tight-binding, LCAO, approach. The "ideal" amorphous structure, in which each atom is in a tetrahedral environment, is assumed. At each site α, four equivalent orthogonal directed orbitals $|\alpha\, i\rangle$, with $i = 1, 2, 3, 4$, are constructed, along the bonding directions to nearest neighbours. They are to be thought of as the familiar sp^3 hybridized orbitals of theoretical chemistry, but we do not need to write them down explicitly in these terms. Orbitals on different sites are assumed to be orthogonal also. In the LCAO approximation, all states $|\psi\rangle$ are then linear combinations of these orbitals:

$$|\psi\rangle = \sum_{\alpha,i} a_{\alpha i}|\alpha i\rangle. \tag{3.31}$$

The only matrix elements of the Hamiltonian which are considered in this model are those between orbitals on the same site:

$$\langle\alpha i|H|\alpha j\rangle = V_1, \qquad i \neq j, \tag{3.32a}$$

and those between orbitals of neighbouring atoms which lie along the same "bond" (Fig. 3.9).

$$\langle\alpha i|H|\beta j\rangle = V_2, \quad \text{for} \quad \alpha i, \beta j \tag{3.32b}$$

making a bond pair.

We set $\langle\alpha i|H|\alpha i\rangle = 0$ for convenience; this is just a choice of energy zero.

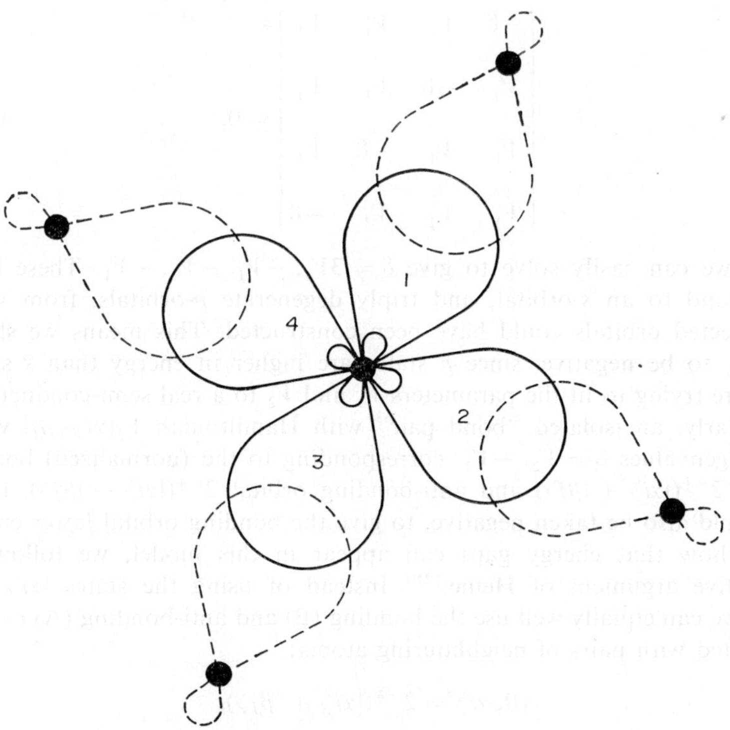

FIG. 3.9. Directed orbitals in the Weaire model (schematic)

V_1 and V_2 are taken as constants for all the atoms of the network. In Dirac notation, we can summarise this by writing

$$H = \sum_{\alpha, i \neq j} V_1 |\alpha i\rangle\langle\alpha j| + \sum_{\text{bonds}} V_2 |\alpha i\rangle\langle\beta j|, \tag{3.33}$$

but this gives no additional information.

The Hamiltonian for a tetrahedrally coordinated lattice, like the diamond or wurzite lattice, will also be of this form; the difference between an amorphous structure and a crystal structure, or between different crystal structures, will appear in the enumeration of the sum over bonds part of the Hamiltonian.

To illustrate the formalism, consider a single isolated atom, for which the Hamiltonian reduces to

$$H_{\text{atom}} = \sum_{i \neq j} V_1 |i\rangle\langle j|.$$

The Schrödinger equation, $H|\psi\rangle = \mathcal{E}|\psi\rangle$, gives the secular equation for the eigenvalues:

$$\begin{vmatrix} -\mathcal{E} & V_1 & V_1 & V_1 \\ V_1 & -\mathcal{E} & V_1 & V_1 \\ V_1 & V_1 & -\mathcal{E} & V_1 \\ V_1 & V_1 & V_1 & -\mathcal{E} \end{vmatrix} = 0, \tag{3.34}$$

which we can easily solve to give $\mathcal{E} = 3V_1, -V_1, -V_1, -V_1$. These levels correspond to an *s*-orbital, and triply degenerate *p*-orbitals, from which the directed orbitals could have been constructed. This means we should take V_1 to be negative, since *p* states are higher in energy than *s* states, if we are trying to fit the parameters V_1 and V_2 to a real semi-conductor.

Similarly, an isolated "bond pair" with Hamiltonian $V_2|\alpha i\rangle\langle\beta j|$ would have eigenvalues $\mathcal{E} = V_2, -V_2$, corresponding to the (normalized) bonding orbital $2^{-\frac{1}{2}}(|\alpha i\rangle + |\beta j\rangle)$ and anti-bonding orbital $2^{-\frac{1}{2}}(|\alpha i\rangle - |\beta j\rangle)$. Hence V_2 should also be taken negative, to give the bonding orbital lower energy.

To show that energy gaps can appear in this model, we follow the instructive argument of Heine.[(29)] Instead of using the states $|\alpha i\rangle$ as a basis, we can equally well use the bonding (B) and anti-bonding (A) orbitals associated with pairs of neighbouring atoms:

$$\begin{aligned} |B, \alpha i\rangle &= 2^{-\frac{1}{2}}(|\alpha i\rangle + |\beta j\rangle), \\ |A, \alpha i\rangle &= 2^{-\frac{1}{2}}(|\alpha i\rangle - |\beta j\rangle), \end{aligned} \tag{3.35}$$

where $|\beta j\rangle$ is the state pairing with $|\alpha i\rangle$. These are also orthogonal and normalized.

The only non-vanishing matrix elements of H between the new basis states are then easily found:

$$\langle B, \alpha i|H|B, \alpha i\rangle = V_2, \qquad \langle A, \alpha i|H|A, \alpha i\rangle = -V_2,$$
$$\langle B, \alpha i|H|B, \alpha i'\rangle = \pm \langle B, \alpha i|H|A, \alpha i'\rangle$$
$$= \pm \langle A, \alpha i|H|A, \alpha i'\rangle = \tfrac{1}{2}V_1, \tag{3.36}$$

where the signs are determined by the relative phases of the anti-bonding orbitals.

If we assume that the valence band can be constructed out of bonding orbitals alone, then an eigenstate of the valence band can be written

$$|\psi\rangle = \sum_{\text{bonds}} b_{\alpha i}|B, \alpha i\rangle \tag{3.37}$$

The normalization integral is

$$\langle\psi|\psi\rangle = \sum_{\text{bonds}} |b_{\alpha i}|^2 = \tfrac{1}{2} \sum_{\alpha, i} |b_{\alpha i}|^2, \tag{3.38}$$

since in the latter sum each bond is counted twice.

The expectation value of the energy for this state is

$$\mathcal{E} = \frac{\langle\psi|H|\psi\rangle}{\langle\psi|\psi\rangle} = \frac{\Sigma b_{\alpha' i'}{}^* b_{\alpha i}\langle B, \alpha' i'|H|B, \alpha i\rangle}{\langle\psi|\psi\rangle} \tag{3.39}$$

where the sum runs over all pairs of bonds. We get contributions V_2 from terms $\alpha i = \alpha' i'$, and $\frac{1}{2}V_1$ from terms $\alpha = \alpha'$, $i \neq i'$, so that

$$\mathcal{E} = V_2 + \tfrac{1}{2}V_1 \left[\sum_{\alpha} \sum_{i \neq i'} b_{\alpha i'}{}^* b_{\alpha i}\right] \Big/ \tfrac{1}{2} \sum_{\alpha} \sum_{i} |b_{\alpha i}|^2 \tag{3.40a}$$

$$= V_2 + V_1 \left[\sum_{\alpha} \left\{ \Big|\sum_{i} b_{\alpha i}\Big|^2 - \sum_{i} |b_{\alpha i}|^2 \right\}\right] \Big/ \sum_{\alpha} \sum_{i} |b_{\alpha i}|^2 \tag{3.40b}$$

$$= V_2 + V_1 \left[\sum_{\alpha} \left\{ 3 \sum_{i} |b_{\alpha i}|^2 - \sum_{i > i'} |b_{\alpha i} - b_{\alpha i'}|^2 \right\}\right] \Big/ \sum_{\alpha} \sum_{i} |b_{\alpha i}|^2. \tag{3.40c}$$

where the algebraic identities are readily checked.

From eqn (3.40b) we see that the minimum value of the coefficient of V_1 is -1 (when each $\sum_i b_{\alpha i} = 0$) and from eqn (3.40c) the maximum value is 3 (when every $b_{\alpha i} = b_{\alpha i'}$). Hence $\mathcal{E}$ must lie between $V_2 + 3V_1$ and $V_2 - V_1$.

The minimum $V_2 + 3V_1$, is reached when all bonds are combined with the same sign, so that each bond has an interaction energy $\frac{1}{2}V_1$ with the three bonds at either end, and the maximum when, at each atom, there are two bonds with coefficients of positive sign in the expansion, and two with negative sign, giving four anti-linking pairs and two linking pairs for each bond. We shall discuss this description of the valence band in more detail later.

We can, similarly, assume that the conduction band is to be constructed wholly from the anti-bonding states, and find limits $-V_2 + 3V_1$, $-V_2 + V_1$ for the conduction band.

In this approximation, there is no overlap between the bands provided that $|V_2| > 2|V_1|$. If we now allow an admixture of anti-bonding orbitals into the valence band, and bonding orbitals into the conduction band, it can be shown with the help of a little matrix algebra that the two bands are pushed further apart. Thus a true energy gap is possible in the Weaire model, of magnitude at least $2|V_2 - 2V_1|$, when the condition $|V_2| > 2|V|_1$ is satisfied.

It is clear from Heine's argument that, in the case $|V_2| > 2|V_1|$, there are four states per atom (counting spin degeneracy) below the gap, and four above the gap. Thus all bonds are saturated in the valence band, in accord with the chemist's picture of covalent bonding.

The Weaire model is sufficiently simple for its properties to have been explored in some detail.[30] It can be shown rigorously that the valence band and conduction band are indeed essentially constructed out of the bonding and anti-bonding orbitals. The band structure, and density of states, have been worked out for the regular diamond lattice. The translational symmetry reduces the problem to the diagonalization of an 8×8 matrix (since the unit cell of the diamond lattice contains two atoms, and there are four states associated with each atom). An analytic solution may be obtained. In this lattice the bounds we have found for the energy bands are, in fact, achieved. With appropriately chosen values for V_1 and V_2 ($V_1 = -2{\cdot}5$ eV, $V_2 = -6{\cdot}75$ eV for germanium), the results may be compared (see Fig. 3.10) with those of a realistic band-structure calculation.[31] The restrictions of the Weaire model as a description of a real amorphous semi-conductor are apparent: the valence band is reproduced quite well, but the conduction band is not.

This result is not really surprising. The limitations of the tight-binding method as *a priori* description of electrons in extended states above the muffin-tin zero have often been stressed,[32] though the method can provide a useful interpolation scheme.[33] The Weaire model, for all its interest, just does not provide enough parameters to describe a real semi-conductor. Furthermore, some degree of variation in V_1 and V_2 would lead to a tailing of states into the gap, which the experimental evidence may, perhaps, preclude. Phillips[34] has suggested that a carefully prepared amorphous structure relaxes in such a way as to exclude states appearing in the gap, in order to maximise bond energies, but to investigate this suggestion quantitatively would be difficult.

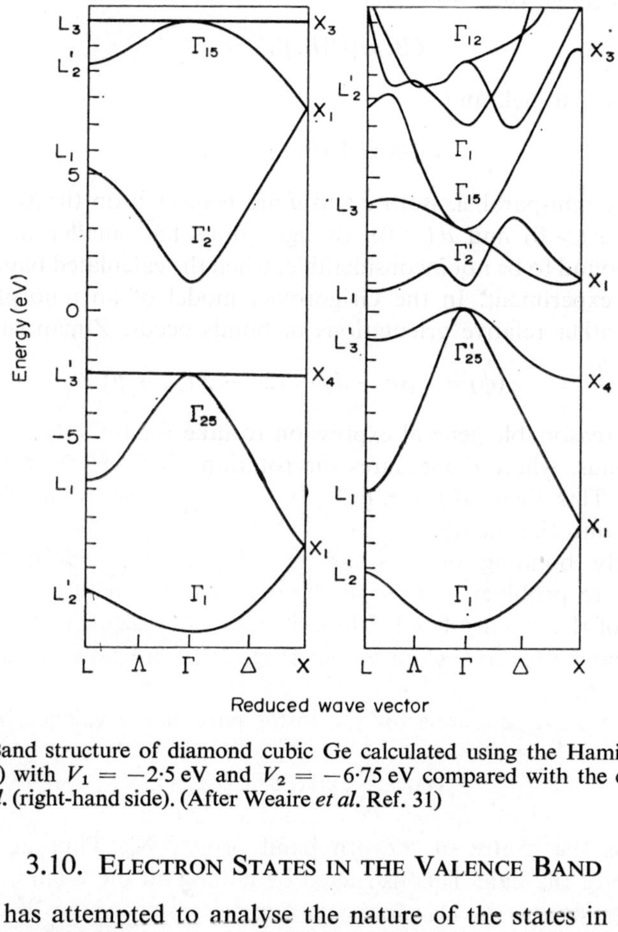

FIG. 3.10. Band structure of diamond cubic Ge calculated using the Hamiltonian given by *eqn* (3·33) with $V_1 = -2{\cdot}5$ eV and $V_2 = -6{\cdot}75$ eV compared with the calculation of Herman *et al.* (right-hand side). (After Weaire *et al.* Ref. 31)

3.10. Electron States in the Valence Band

Ziman[35] has attempted to analyse the nature of the states in the valence band, using a more realistic Hamiltonian. Wave functions in the valence band are expanded in terms of bond orbitals, as in eqn (3.37), but more interaction terms are included in the discussion. This description was in fact introduced by Hall[36] in a discussion of the valence band in diamond structures. Hall's bond orbitals are not constructed out of atomic orbitals, but rather defined as "best" orthogonalised one-electron orbitals in a self-consistent one-electron description. In the diamond lattice, all configurations are of the "staggered" type (Fig. 3.4). Hall considers not only the matrix elements between bonds hinging on the same atom:

$$\langle B, \alpha i|H|B, \alpha' i'\rangle = b, \qquad i \neq i',$$

(so that b corresponds to $\frac{1}{2}V_1$ in eqn (3.36), but also matrix elements between bonds on neighbouring atoms α, β:

$$\langle B, \alpha i|H|B, \beta j\rangle = c$$

if i and j are parallel, and

$$\langle B, \alpha i|H|B, \beta j\rangle = d$$

if i and j are non-parallel. Both c and d are neglected on the Weaire model but in fact $c\,(> 0)$ and $d\,(< 0)$, though somewhat smaller in magnitude than b, are found to be not inconsiderable, when the calculated band-structure is fitted to experiment. In the Grigorovici model of an amorphous semiconductor, other relative orientations of bonds occur. Ziman suggests that

$$g(\phi) = \tfrac{1}{3}\{(c + d) + (2c - 2d)\cos\phi\} \tag{3.41}$$

might be a reasonable general expression to take for the interaction energy between bonds, where ϕ measures the rotation about $\alpha\beta$ from the parallel orientation. This gives $g(0) = c$, $g(2\pi/3) = d$, and does not introduce further parameters into the theory.

Since only bonding orbitals are included in the wave-function, the band-structure problem reduces to the diagonalization of a 4×4 matrix. The width of the valence band is found to be $8|b + 2d|$. The flat sub-bands which appear with the Weaire model (Fig. 3.10) are now spread out to a width $8|c - d|$.

The Bloch wave-functions for the lower part of the valence band are of the form

$$\psi_{\mathbf{k}}(\mathbf{r}) = \sum_p \exp(i\mathbf{k}.\mathbf{R}_p)\,\mathbf{X}_p(\mathbf{r}), \tag{3.42}$$

where $\mathbf{R}_p$ is the centre of the pth band orbital $\mathbf{X}_p$. Thus at $\mathbf{k} = 0$ the lowest state of the band is constructed by adding all the bond orbitals with equal amplitudes, as we also found in our discussion of the Weaire model. Other states near the bottom of the band are obtained by modulating this function with plane wave phase factors.

Ziman suggests that such states must be nearly as good eigenfunctions for the disordered crystal, near the bottom of the valence band. The wave vector $\mathbf{k}$ is not, of course, a good quantum number for the disordered system but might be a "nearly good" quantum number, for sufficiently small $\mathbf{k}$. Clearly, such states are in no sense localized.

The top of the valence band is a little more difficult to discuss. In the perfect crystal the bands are triply degenerate at $\mathbf{k} = 0$ at the top of the band. As we should now expect, the corresponding wave functions are such that at each atom there are two orbitals with positive sign and two with negative sign, so that maximum advantage is taken of the "anti-linking" matrix elements. At a particular site, we can split the four bonds into pairs in just three different ways, corresponding to the triple degeneracy. Bonds with the same sign run in strings through the crystal, and locally appear as in Fig. 3.11 where the black bonds are associated with positive signs.

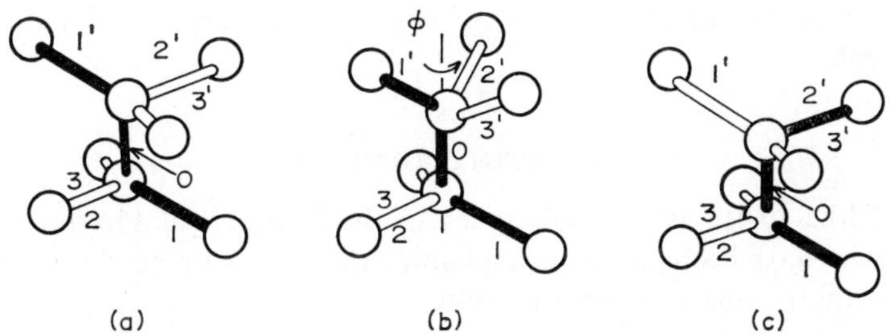

FIG. 3.11. Local assignments of black and white bonds: (a) in ideal, "staggered" case, (b) best choice for case of configuration rotated through ϕ, (c) badly perturbed case. (After Ziman, Ref. 35)

Ziman suggests that, in the amorphous crystal, states at the top of the valence band will attempt a "best fit" to the arrangement on the perfect crystal. Topological disorder makes this difficult to achieve, but something like strings of "black" bonds running through the amorphous structure can be constructed and optimized, and it can be argued that the triple degeneracy at the top of the band is preserved. Again, it is clear that these states are extended. Near the top of the band, the states can be modulated with a phase factor $\exp(i\mathbf{k}.\mathbf{R}_p)$ as in the perfect crystal, to give approximate eigenstates and a "nearly good" quantum number $\mathbf{k}$. In this picture we have "hole" states at the top of the valence band. The initial restriction to bonding orbitals rather removes any possibility of localized states above the band edge. Ziman does not consider a discussion of the conduction band in terms of anti-bonding orbitals to be realistic.

It should perhaps be emphasized that the *true* eigenstates of an amorphous system are not open to calculation: the amorphous structure is only statistically specified. Nevertheless, approximate eigenstates constructed in the way Ziman suggests might perhaps serve as a basis for more detailed computational work.

We have made no attempt in this section to discuss the way in which departures from the "ideal" amorphous structure affect the electronic structure and, in particular, the band edges. This is a difficult problem, but might perhaps be open to investigation on the simple Weaire model.

3.11. Conclusions

One is expected to draw conclusions from a review. The clear conclusion is that much remains to be done to understand fully the electronic structure of amorphous systems. We have not, as yet, a mode of discussion that is simple enough to be theoretically tractable, and at the same time realistic enough to allow calculations to be made which can be compared with experiment.

Aknowledgements

I should like to thank my colleagues in the Theoretical Physics group at Bristol, who have been patient in explaining their work to me. Dr. P. V. Smith criticized the manuscript, very usefully.

References

1. Grigorovici, R. and Manaila, R., *J. Non-Cryst. Solids* **1,** 371 (1969). See also the lectures by Professor Grigorovici in this volume (Chapter 5).
2. Mattheiss, L. F., *Phys. Rev.* **134,** A970 (1964).
3. Herman, F. and Skillman, S., "Atomic Structure Calculations". Prentice-Hall, New Jersey, 1963.
4. Loucks, T. L., "Augmented Plane Wave Method". Benjamin, New York, 1967.
5. See, for just one example, Boring, M. and Snow, E. C., *Phys. Rev.* **B5,** 1221 (1972).
6. Keller, J., *J. Phys. C: Solid St. Phys.* **4,** L85 (1971).
7. Messiah, A., "Quantum Mechanics". North-Holland, Amsterdam, 1962, or any other standard text.
8. Levinson's theorem may be made plausible along the lines of the discussion of the Friedel sum given in reference 17, switching on the potential adiabatically.
9. Wigner, E. P., *Phys. Rev.* **98,** 145 (1955).
10. Lloyd, P., *Proc. Phys. Soc.* **86,** 825 (1965).
11. Harrison, W. A., "Pseudopotentials in the Theory of Metals". Benjamin, New York, 1966.
12. For a review see Edwards, S. F. *Adv. Phys.* **16,** 359 (1967).
13. Lloyd, P., *Proc. Phys. Soc.* **90,** 207 (1967).
14. Chan, T. and Ballentine, L. E., *Can. J. Phys.* **50,** 813 (1972).
15. Soven, P., *Phys. Rev.* **156,** 809 (1967).
16. Gyorffy, B. L., *Phys. Rev.* **B1,** 3290 (1970).
17. See Ziman, J. M., "Principles of the Theory of Solids". Cambridge, 1964, where the Friedel sum is discussed from a different point of view.

18. Anderson, P. W. and McMillan, W. L., *Proc. Int. Sch. Phys. Enrico Fermi* **37,** 50 (1967).
19. Schwartz, L. and Ehrenreich, H., *Ann. Phys.* **64,** 100 (1971).
20. McGill, T. C. and Klima, J., *Phys. Rev.* **B5,** 1517 (1972).
21. For a full review of multiple scattering theory in condensed materials, see Lloyd, P. and Smith, P. V., *Adv. Phys.* **21,** 69 (1972).
22. Keller, J., *J. Phys. C: Solid St. Phys.* **4,** 3143 (1971).
23. Keller, J. and Smith, P. V., *J. Phys. C: Solid St. Phys.* **5,** 1109 (1972).
24. Keller, J. and Jones, R., *J. Phys. F: Metal Phys.* **1,** L33 (1971).
25. Johnson, K. H., *J. Chem. Phys.* **45,** 3085 (1966); *Int. J. Quantum Chem.* **1S,** 361 (1967).
26. Weaire, D., *Phys. Rev. Lett.* **26,** 1541 (1971).
27. Borland, R. E. *Proc. Phys. Soc.* **78,** 926 (1961).
28. Taylor, P. L., *Proc. Phys. Soc.* **88,** 753 (1966).
29. Heine, V., *J. Phys. C: Solid St. Phys.* **4,** L221 (1971).
30. Weaire, D. and Thorpe, M. F., *Phys. Rev.* **B4,** 2508 (1971). Thorpe, M. F. and Weaire, *D., Phys. Rev.* **B4,** 3518 (1971).
31. Weaire, D., Thorpe, M. F. and Heine, V., *J. Non-Cryst. Solids* **8–10,** 128 (1972).
32. Ziman, J. M., *Solid State Phys.* **26,** 1 (1971), (eds. F. Ehrenreich, F. Seitz and D. Turnbull, Academic Press).
33. Dresselhaus, G. and Dresselhaus, M. S., *Phys. Rev.* **160,** 649 (1967).
34. Phillips, J. C., *Comments on Solid St. Phys.* **4,** 9 (1971).
35. Ziman, J. M., *J. Phys. C: Solid St. Phys.* **4,** 3129 (1971).
36. Hall, G. G., *Phil. Mag.* **43,** 338 (1952); *Phil. Mag.* **3,** 429 (1958).

II

PREPARATION AND STRUCTURAL PROPERTIES

4. PREPARATION OF AMORPHOUS MATERIALS AND THE GLASS TRANSFORMATION

A. E. OWEN

School of Engineering Science,
University of Edinburgh,
Edinburgh.

4.1. The Preparation of Amorphous Solids

4.1.1. Introduction

Amorphous solids can readily be prepared from solid, liquid or vapour phases by processes which essentially cause an excess of free energy (or chemical potential) to be more or less temporarily "frozen-in" to the system. This widespread origin of amorphous solids is illustrated schematically in Fig. 4.1 which has been adapted from a diagram used by Roy [(1)] to show the relationship between different amorphous solids and their parent phases. Various features in this figure will be explained in later paragraphs. All amorphous solids are, therefore, thermodynamically unstable and they tend to relax towards a stable or metastable phase. The time scale on which these changes occur varies enormously and is usually a sensitive function of temperature. In the common silicate glasses which are the basic materials of the glass industry, for example, the time constant of this relaxation process is hundreds of years at room temperature, but of the order of a few minutes at the glass transformation temperature, T_g (500–700°C in typical cases). For practical purposes, therefore, silicate glasses are "stable" solids at or

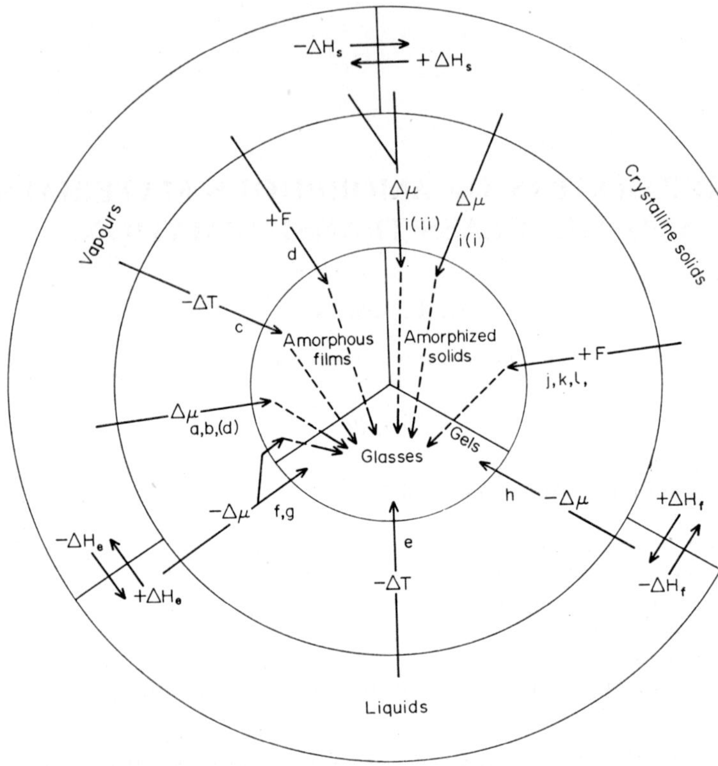

FIG. 4.1. Schematic illustration of the origin and inter-relationships of amorphous solids (adapted from Ref. 1). For explanations see text.

near room temperature. Many other glasses or amorphous solids are much less "stable" especially some such as amorphous Ge, which are of particular interest in the context of electronic conduction. In some cases the relative lack of stability is simply a consequence of a low glass transformation temperature T_g; in other cases—typically, some vapour-deposited amorphous solids—it is not possible to identify a T_g. These thermodynamic and kinetic aspects of glass formation will be discussed later.[2] It is worth recognising at the outset, however, that a certain amount of confusion has arisen in the literature through the use of the terms "glass", "vitreous solid", "amorphous" and "non-crystalline solid (N.C.S.)" as synonyms. The arguments can rapidly become semantic and in the present paper we shall adopt the view that "glasses" or "vitreous solids" are materials which are prepared by cooling a liquid and which, therefore, are structurally continuous with the parent (liquid) phase. Thus in principle glasses have a specifiable thermodynamic relationship with their parent phase.[(2)] "Amorphous" or "non-crystalline solid" will be used as general terms which include glasses as a special case.

The various techniques available for the preparation of amorphous solids will be described, the emphasis being on those materials of special interest because of their electronic properties. The object is to make a general survey and not give experimental details; the latter information can be found in the references. Some schemes for the classification of amorphous solids will also be reviewed.

4.1.2. PREPARATION METHODS

There are at least 12 different methods for the preparation of amorphous solids. A brief description of each is given below, beginning with techniques which start from the vapour phase and finishing with those that start from the solid. All of them have been, or can be, used to produce amorphous solids of interest in the present context. Methods for producing amorphous organic polymers are not explicitly included.

(a) *Vapour phase-hydrolysis*

The best known example is the reaction between silicon tetrahalide vapour and water vapour to produce silica:

$$SiCl_4 + 2H_2O \rightarrow SiO_2 + 4HCl.$$

The reaction is carried out at very high temperature (in an oxy-hydrogen flame) and is used commercially to produce the so-called synthetic grade of high purity fused silica.[(3)] It is probably also appropriate to include under this heading reactions such as the thermal decomposition (at 700–900°) of boron triethyl in vacuum to produce amorphous B_2O_3.[(4)] Many other materials can be decomposed in the vapour phase to give thin oxide films (e.g. $VOCl_3 \rightarrow VO_2$)[(5)] although the product may not always be amorphous.

(b) *Glow-discharge decomposition*

In this method decomposition takes place at a comparatively low temperature, the energy being provided from a radio-frequency discharge.[(6,7)] Amorphous films of silicon and germanium can readily be prepared from their hydrides (SiH_4 and GeH_4). The growth rate is typically 2 to 12 Å/s at a pressure of about 0·1 torr. Compounds can also be prepared. A mixture of silane (SiH_4) and nitrous oxide, for example, gives SiO_2; silane and ammonia gives Si_3N_4. Boron triethyl can be decomposed to B_2O_3 and triethyl-silicate [$(Et)_3SiO_4$] to SiO_2.

(c) *Thermal evaporation*

This is probably the most widely used method for the production of amorphous solids in thin-film form. It would be impractical to quote even a fraction of the literature, so widespread is its use, but references 8, 9 and 10 give general information on experimental details. Thermal evaporation is normally carried out in a vacuum of 1 micron or better and the source of material is usually heated by an electric filament or by electron-beam bombardment. Flash evaporation[11] is a variant of the process in which small amounts of the material to be evaporated are fed at a suitable rate onto an electrically heated "hearth" which is already at a sufficiently high temperature so that on contact the material evaporates almost instantaneously. This is intended to overcome the problem of differential evaporation of constituents when attempting to deposit a compound or "alloy" (see below).

Most of the materials of interest as amorphous semiconductors can be prepared by thermal evaporation, e.g. Si†, Ge,† III–V compounds, Se and probably all of the chalcogenide family of materials, SiO, SiO_x and SiO_2, and many others. To the author's knowledge, however, there have been no reports of transition-metal oxide glasses prepared by thermal evaporation.

The major disadvantage of thermal evaporation as a preparative method is lack of control over purity and, in the case of compounds or alloys, composition. There are many experimental variables and it is often difficult to exercise close control over most of them.
The following list gives some examples but this is probably far from complete:

(i) Filament temperature—a function of filament current *and* filament geometry.

(ii) The source material—its size, shape, etc.

(iii) Source-substrate separation.

(iv) Substrate temperature.

(v) Substrate preparation and cleaning (affecting adhesion).

(vi) Residual gas pressure, including the unwanted presence of hydrocarbons from any diffusion pump oil.

(vii) Impurities from the filament or evaporation crucible, or other parts of the vacuum system including desorbed water vapour and other gases, metal hydrides and other volatile reaction products, etc.

† Note, however, that there are significant differences in the electronic properties of amorphous Si and Ge prepared by thermal evaporation and by glow discharge decomposition of silane. (See the article by LeComber, Madan and Spear in Chapter 10 of this volume for a discussion of the electronic properties of amorphous Si prepared by both methods).

As mentioned before, a particular difficulty with compounds and alloys is the differential evaporation of the various constituents. Mueller and Wood[12] have recently investigated this problem in connection with the deposition of thin-films of Sb_2Se_3. They point out that mass spectrometry shows that the equilibrium vapour species over molten Sb_2Se_3 is mainly SbSe. Thus one might at least expect to obtain a reasonably consistent film of composition SbSe. But the result quoted is appropriate to the condition of dynamic equilibrium between melt and vapour and this is not the situation in conventional vacuum evaporation where the vapour is immediately removed from above the molten source. Instead one is concerned with the dissociation vapour pressure of the element of highest vapour pressure, i.e. in Mueller and Wood's case with Se vapour over a solid or molten source which has an initial composition Sb_2Se_3 but which changes with time. As the evaporation proceeds the source is gradually depleted of Se and, therefore, the composition of the deposit changes with time and thickness, This is illustrated in Fig. 4.2

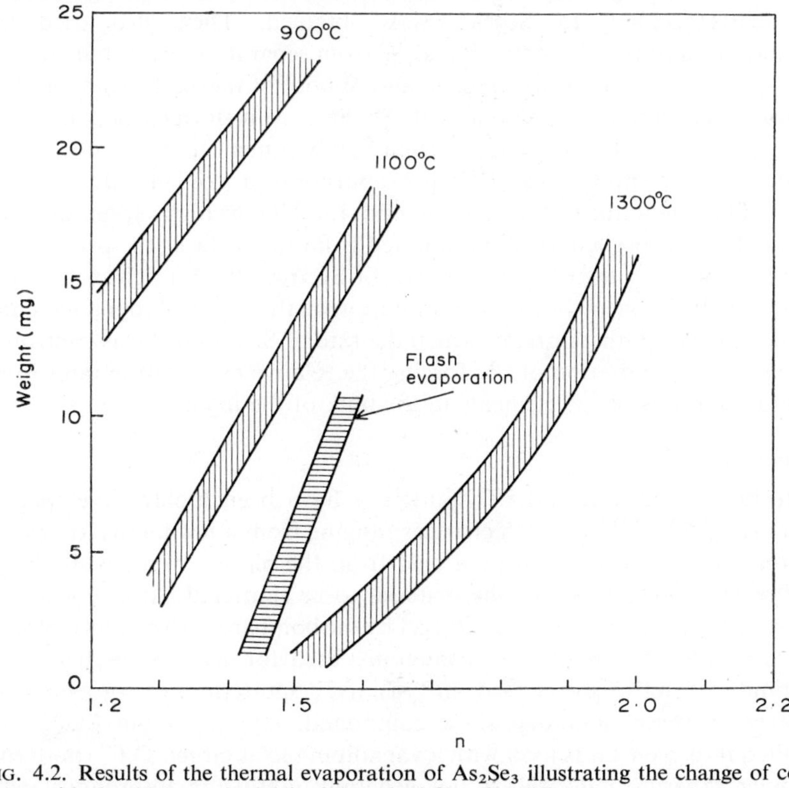

FIG. 4.2. Results of the thermal evaporation of As_2Se_3 illustrating the change of composition with the thickness (weight) of deposited film, for different source temperatures. The composition is expressed by the index n in $AsSe_n$.

with some results from the author's laboratory for the thermal evaporation of As_2Se_3 (the source material actually had a composition $As_2Se_{3\cdot018}$) for three different source temperature—900°C, 1100°C and 1300°C. These results should not, of course, be taken to do more than indicate general features obtained in the particular system used for the experiments. As emphasised earlier, it is difficult enough to reproduce conditions in a particular evaporator and one should not expect that the results shown in Fig. 4.2 would be quantitatively reproduced in different systems. They show, however, that the stoichiometry of the deposit changes significantly as the weight increases, i.e. as the thickness of film increases. Thus, the stoichiometry of the film is graded with thickness. Figure 4.2 also shows that flash evaporation leads to a considerable improvement in the stoichiometry of the deposit but it does not completely solve the problem. Moreover, flash evaporation can be erratic and sometimes widely divergent compositions are obtained. In their investigation of Sb_2Se_3 Mueller and Wood found that with a source temperature of 720°C and starting composition $SbSe_{0\cdot9}$ a deposit with an *average* composition corresponding to Sb_2Se_3 was obtained. They also tried the co-evaporation of the elements Sb and Se from separate sources, but without much success. According to Mueller and Wood[(12)] the best technique was to induce the *surface evaporation* of Sb_2Se_3. A melt-quenched ingot of Sb_2Se_3 was evaporated by electron-beam bombardment from a water-cooled hearth at a slow enough rate so that evaporation proceeded only from the surface of the ingot, the rest remaining solid. Initially the high vapour pressure element (Se) was preferentially evaporated from the surface; the process was then rate limited by the evaporation rate of Sb from the Sb-rich surface and the rate of diffusion of Se from the interior of the ingot. A dynamic equilibrium was set-up, therefore, in which the rate of Se removal was controlled by the rate of evaporation of Sb, forcing the relative evaporation rates to be equal to the ratios of the elements in the bulk of the ingot.

(d) *Sputtering*

The techniques of d.c. and r.f. sputtering have been highly developed in recent years.[(9, 10, 13)] Ions with energies ranging from a few tenths to several electron volts are accelerated by a field from the plasma of a gas discharge and directed onto a target of the material to be sputtered. Atoms *or groups of atoms* are "knocked-out" of the target by the bombarding ions and collected on a substrate. Argon is most commonly used for the gas discharge but sometimes a reactive gas is used, in "reactive" sputtering, to combine with the target material and deposit a compound, e.g. vanadium oxides are deposited if oxygen gas is used with a vanadium metal target. D.C. sputtering is simpler experimentally but is not generally useful for amorphous semiconductors as many of the materials of interest are not sufficiently conducting.

A positive surface charge builds up on the target (cathode) which repels the bombarding ions. Most frequently, therefore, r.f. techniques are employed to sputter amorphous semiconductors and dielectrics. D.C. sputtering can be useful, however, especially when it is realised that it is not necessary for the target to be amorphous. Some of the complex chalcogenide materials, for instance, are sufficiently conducting when *crystalline* to be used as targets in d.c. sputtering.

As in the case of thermal evaporation, virtually all of the materials of interest as amorphous semiconductors can be depositied as thin films by sputtering techniques. Some of the transition-metal oxide glasses have been r.f. sputtered in the author's laboratory, using an argon-oxygen mixture.

Sputtering has the major advantage that sputtering yields do not vary widely from element to element—in contrast to melting points, vapour pressures and hence evaporation rates. Compositional changes *do* occur, however, and some results obtained by Bosnell and Thomas[14] on glasses of nominal composition $Ge_{10}Si_{12}Te_{48}As_{30}$ and $Ge_{15}Te_{81}Al_4$ are shown in Table 4.1.

TABLE 4.1

$Ge_{10}Si_{12}Te_{48}As_{30}$				
	Ge	Si	Te	As
Target, wt%	7·8	3·7	64·8	23·7
0·1 μm film, wt%	7·9	1·65	69·0	21·4
2·9 μm film, wt%	8·9	1·95	65·0	24·1
$Ge_{15}Te_{81}Al_4$				
	Ge	Te	Al	
Target, wt%	8·7	82·7	8·6	
0·4 μm film, wt%	8·8	91·0	0·20	
1·6 μm film, wt%	9·0	91·0	0·21	

In the four-component glass there is a small gain in Te and a commensurate loss in Si, which is consistent with the relatively low sputtering yield of the latter.[14] Notice also that there are changes in composition with the thickness of the deposited film. In the three-component material the amorphous films are very deficient in Al which is known to be extremely difficult to sputter.[14] The data shown in Table 4.1 were obtained with d.c. sputtering; results from the author's laboratory using r.f. sputtering have also shown small compositional changes. There is in addition the likelihood of contamination of the deposited film by the sputtering gas.

(e) *Glass formation by cooling a liquid*

The formation of a glass by cooling a liquid requires the prevention of the nucleation and growth processes responsible for crystallisation. Both the rate of *homogeneous* nucleation and the subsequent rate of crystal growth go through a maximum as the temperature is lowered below the thermodynamic melting point.[15] Thus if the temperature of a liquid can be lowered below these maxima quickly enough so as to avoid nucleation or the growth of detectable crystals, further cooling will make crystallisation *less* likely and the liquid will eventually solidify as a glass. In this sense, glass formation is a kinetic process but there are features of the phenomenon which can be described within the framework of thermodynamics. The thermodynamic and kinetic aspects of the problem will be discussed more fully later.[2]

Whether or not a particular liquid can be cooled to form a glass will obviously depend upon the rates of the atomic and/or molecular transport

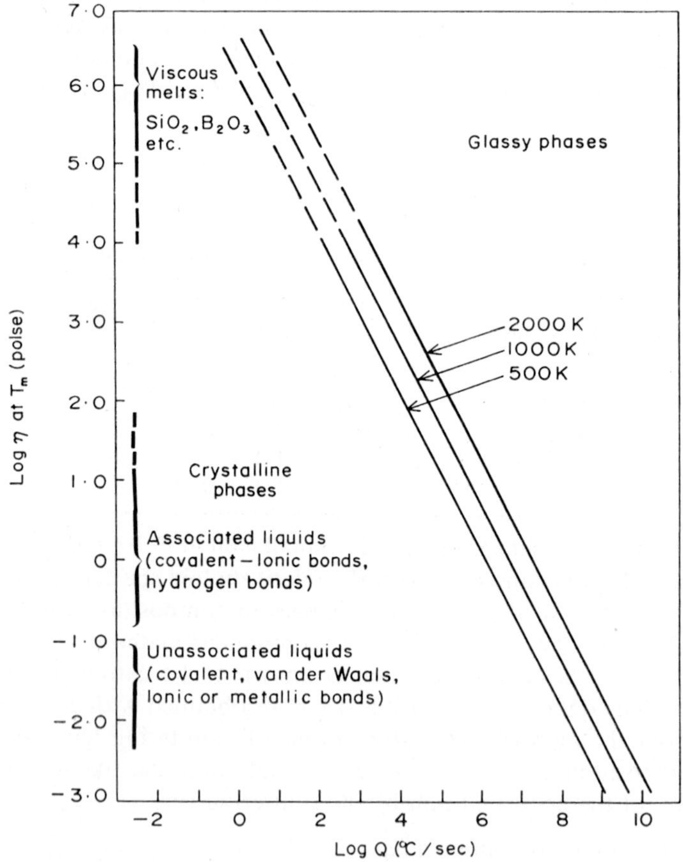

FIG. 4.3. The logarithm of melt viscosity versus critical cooling rate at various melting temperatures. (from Ref. 16).

processes involved in nucleation and crystal growth. Viscosity is often regarded as the determining factor and while it is true that many liquids which can be readily cooled to form glasses are comparatively viscous at their melting points (e.g. $10^5 - 10^8$ poise), this is not always an infallible guide. It is known, for example, that in the homologous series of primary alcohols the lowest members (e.g. methyl) crystallise very rapidly on cooling, as do the higher members with long chains, while intermediate members (e.g. propyl, butyl) tend to form glasses. Viscosity, on the other hand, increases steadily with increasing chain length. There are clearly steric factors involved which are not simply related to the flow processes. Nevertheless it is possible to roughly relate the glass forming tendency of a melt with a transport process such as viscosity. The result of one such attempt is shown in Fig. 4.3, taken from Sarjeant and Roy.[(16)] The critical cooling rate of Fig. 4.3 is the rate which must be exceeded if a melt of a given viscosity is to be cooled from a specified temperature to form a glass. Note that the critical cooling rate is a *minimum* condition imposed on the rate.

In principle it should be possible to cool any liquid to form a glass—it should merely (!) be a matter of achieving rapid enough cooling rates. Experimentally, surprisingly large quench rates are in fact accessible. The most rapid technique is the "splat cooling method" pioneered by Duwez and his colleagues[(17)] in which it has been estimated that quench rates of up to 10^7 deg/s can be reached.[(16)]

This compares with rates of a few degrees per second which are sufficient for the readily prepared glasses such as silicates and some chalcogenide glasses. The experimental range of quench rates is listed in Table 4.2, taken from Sarjeant and Roy.[(16)] The slowest of these correspond to what is normally called "annealing". In the splat cooling technique developed by Duwez,[(17)] very small liquid droplets are ejected at high velocity on to a cold metal substrate. Chen and Miller[(18)] have described a variant in which the liquid droplet is ejected into rotating rollers so that it is squeezed out into a thin ($\sim 10-100\ \mu$m) glassy "ribbon" of reasonably uniform thickness. The "strip furnace" mentioned in Table 4.2 consists simply of a metallic (e.g. Pt) strip filament, heated electrically, on which a small droplet is melted; quenching is achieved by quickly turning off the current.[(16)] It should be noted that the quench rates listed in Table 4.2 cover twelve orders of magnitude; the difference between the rates of ordinary air quenching and the extremely rapid splat cooling techniques is $10^6 - 10^7$ orders.

There are some materials, such as liquid metals and molten alkali halides, which are beyond the glass-forming capability of even the most rapid splat-cooling techniques.[(16)] The elementary semiconductors Si and Ge, and the related III–V compounds are also apparently in this category. There is, nevertheless, a very wide range of materials which can be prepared as glasses by the techniques listed in Table 4.2, even if only in small quantities. This includes most of the chalcogenide group of materials, diamond lattice-type semiconductors typified by $CdGeAs_2$, the transition-metal oxide glasses and even some "pure" transition-metal oxides (e.g. V_2O_5, WO_3),[19] and many others. Perhaps the most notable achievement has been the preparation of glassy metallic alloys such as Pb–Si[17] and Au–Si–Ge[20] alloys.

TABLE 4.2. Experimental cooling rates[16]

Annealing	Large telescope mirrors	10^{-5} deg/s
	Optical glasses	$\sim 3 \times 10^{-4}$ deg/s
	Ordinary glasses	10^{-3}–10^{-2} deg/s
Air quenching—typical experimental batches (10's of grams)		1–10 deg/s
Small drops (1–100 mg) into liquid Hg		10^2–10^3 deg/s
Strip furnace (mg batches)		10^3–10^4 deg/s
Splat cooling		10^5–10^7 deg/s
Evaporation, sputtering		?

It is pertinent to comment briefly at this stage on the structure of glasses and other amorphous solids. As glasses are continous with the liquid phase it is not surprising to find that they have a structure which is essentially like that of the liquid, with the same type of local or short-range-order but no long-range order. Glasses are often spoken of, in fact, as having a "frozen-in liquid-like configuration". Materials which can be cooled from the liquid state to form glasses can also be prepared in an amorphous form by deposition from the vapour phase (the cooling rate necessary for a particular material is a *minimum* condition). So far as is known these amorphous solids also have structures similar to their *liquid* phases and would show all the features typical of the glass transformation.[2] On the other hand an appreciable number of materials, of which amorphous Si and Ge are typical, can *only* be prepared in an amorphous form by deposition from the vapour-phase. Generally speaking amorphous solids of this type have a local atomic configuration which is typical of the crystalline form rather than the liquid. Amorphous Ge, for example, has a tetrahedral local co-ordination as in the crystal[21, 36] whereas liquid Ge has eightfold co-ordination. On heating, amorphous solids of this kind crystallise before anything recognisable as a glass transformation[2] occurs. This situation is not clear-cut, however, and there are a few examples of materials which can only be prepared in an amorphous form by vapour-deposition methods, but which have a liquid-like structure (e.g. Bi, Ga).[22]

In glasses formed from two or more components the possibility of compositional inhomogeniety (i.e. phase separation) on a microscopic or macroscopic scale must always be reckoned with. Indeed there is a school of thought which regards all "good" glasses (which tend to have complex compositions) as inhomogeneous, on a microscopic scale at least. Phase separation may occur via stable liquid–liquid immiscibility, by metastable liquid–liquid immiscibility[23] below the solidus or by spinodal decomposition.[24] The last is supposed to be characterized by a structure in which all phases are interconnected. Liquid–liquid phase separation requires homogeneous nucleation and growth of one phase from another, and hence should be characterised by a structure of droplets in a continuous medium of the other phase. Haller[25] has pointed out, however, that a precipitated droplet structure, formed through stable or metastable liquid–liquid immiscibility, could grow until the droplets merge giving a final structure with interconnected phases, much as expected from spinodal decomposition. The visual appearance is not, therefore, a conclusive indication of the mode of phase separation. The complex chalcogenide glasses, which often contain three or more elements, are quite likely to be inhomogeneous and there is experimental evidence for this in several cases.[26, 27] The degree of phase separation will depend very critically, however, on the method of preparation and subsequent heat treatment.

(f) *Chemical reaction and precipitation from solution*

Amorphous As_2S_3 may be precipitated from a solution of As_2O_3 in a dilute hydrochloric acid solution by passing H_2S gas and, similarly, amorphous Sb_2S_3 from an acid solution of $SbCl_3$.[28] In both cases the amorphous precipitates show typical glass transformation behaviour on heating with T_g's very close to those observed for melt-quenched samples.

(g) *Electrolytic deposition from solution*

Comparatively thick ($\sim 30\ \mu m$) amorphous films of Ge can be deposited on a copper cathode by the electrolysis of $GeCl_4$ in glycol.[29]

(h) *Desiccating gels*

The best known example is amorphous silica (SiO_2) prepared by the low temperature dehydration of silicic acid solutions. Complete dehydration is achieved by baking at high temperatures ($\sim 1000°C$). Some amorphous aluminates and alumina silicates can also be prepared in the same way.

(i) *Reaction amorphized solids*

(i) *Oxidation*. Amorphous thin films of SiO_2 can be prepared by the anodic oxidation of Si.[30] Aqueous electrolytes can be used but tend to give porous films; this can be overcome by using non-aqueous electrolytes of suitable dielectric constant.[31a] By making appropriate additions to the electrolyte amorphous SiO_2–P_2O_5 or SiO_2–B_2O_3 films can also be prepared by anodic oxidation. Essentially the same reaction can be induced by the thermal oxidation of Si with dry oxygen or steam at high temperatures.[31b] The properties of amorphous SiO_2 films prepared by the different methods can be quite different however. Amorphous oxides of other elements and compounds (e.g. Al, Ge, III–V compounds) can also be prepared by similar techniques.

(ii) *Loss of a volatile constituent.* Almost the reverse of the above is the decomposition and loss of a volatile component from a solid. Some complex kaolinite-type minerals, for example, are reported to become amorphous on dehydration, decarbonation or deoxygenation (loss of H_2O, CO_2 or O_2, respectively).[1] This process is similar to the formation of a gel from an aqueous solution (h).

(j) *Irradiation*

(i) *Neutron bombardment.* Crystalline quartz,[32] cristobalite and tridymite[33] can be transformed by heavy dosages ($\sim 10^{20}$ neutrons/cm^2) of fast neutrons into an amorphous form of SiO_2 with a density of 2·26 g/cm^3. Although less than that of quartz this is still greater than that of "ordinary" melt-quenched fused silica (density 2·205 g/cm^3). If the completely irradiated amorphous SiO_2 is annealed at $\sim$ 1100°C, however, its density decreases further to 2·205 g/cm^3 which is identical with that of fused silica (the refractive indices are also the same). Incompletely irradiated material reverts on annealing to crystalline quartz. Beryl, a mineral which is structurally related to quartz, can also be transformed into an amorphous form by neutron irradiation.

(ii) *Ion bombardment.* Amorphous surface layers can be prepared on some solids by ion bombardment. Amorphous layers of up to 600 Å can, for example, be formed on single crystal Ge by 100 keV oxygen ions ($\sim 10^{15}$ ions/cm^2).[34] This can often happen inadvertently if ion-bombardment is used for surface cleaning.

(k) *Shock-wave transformation*

Quartz is transformed into amorphous SiO_2 when struck by explosively driven metal plates developing a shock pressure of about 600 kilobars and a temperature, in the shock front, of $\lesssim$ 1200°C.[35] The material produced seems to be identical with ordinary fused silica (e.g. densities and refractive indices are practically the same). The mechanism of transformation is not known but it is thought unlikely that it caused by melting in the high temperature front of the shock wave. The mineral albite ($NaAlSi_3O_8$) has also been prepared as a glassy phase by shock-wave transformation.

(l) *Shear-amorphized layers*

Amorphous surfaces can be produced on quartz, cristobalite and possibly other crystals by grinding or comminution.[1]

Summary

In methods (a) to (d), amorphous films are produced from the vapour phase although, of course, (c) and (d) start from a solid and may proceed through a liquid phase. Methods (e) to (h) start from a liquid (or solution) and in (i) to (l) the amorphous phase is produced directly from a solid.

In (j), (k) and (l) an excess of free energy is "forced" into the crystalline material; (i) is probably best viewed as an addition of chemical potential to the system, i.e.

$$\text{Cryst. solid} \xrightarrow{+\Delta\mu} \text{Amorph. solid} + \text{vapour; [i(i)]}$$

$$\text{Cryst. solid} + \text{vapour} \xrightarrow{+\Delta\mu} \text{amorph. solid; [i(ii)]}$$

$$\text{Cryst. solid} \xrightarrow{+F} \text{amorph. solid; (j), (k), (l).}$$

These paths are shown in Fig. 4.1. The dotted lines are meant to indicate that the "amorphized solid" *may* be in a state identical to that defined as "vitreous" or "glassy".[2]

Method (e) is the traditional method for glass formation induced by a rapid decrease in the temperature of a melt to "freeze-in" the glassy configuration. Methods (f), (g) and (h) also start from a liquid (solution) but are probably best regarded as proceeding via a change in chemical potential, i.e.

$$\text{Liquid} \xrightarrow{-\Delta T} \text{Amorph. solid; (e)}$$

$$\text{Liquid} \xrightarrow{\Delta\mu} \text{amorph. solid; (f), (g), (h).}$$

In the example quoted for method (g) the product (amorphous Ge) is an amorphous film which cannot readily be identified with a glass; there seems to be no reason, however, why this should always be so.

In methods (a) to (d) a vapour phase is, in effect, the starting point. In (a) and (b) a chemical reaction is induced and hence the amorphous film is formed via a change in chemical potential. Method (c) is like an extension of (e)—it involves a rapid decrease in temperature. Sputtering (method (d)) is probably best viewed as an extension of (j)–(l); free energy is "pumped" into the system through bombardment with energetic ions. Reactive sputtering is akin to methods (a) and (b). Thus:

$$\text{Vapour} \xrightarrow{\Delta\mu} \text{Amorph. solid; (a) ,(b), [(d)]}$$

$$\text{Vapour} \xrightarrow{-\Delta T} \text{Amorph. solid; (c)}$$

$$\text{Vapour} \xrightarrow{+F} \text{Amorph. solid; (d).}$$

Again, the dotted lines in Fig. 4.1. are meant to indicate that the product may be identical to a "glass".[(2)]

As Roy[(1)] points out, various organic polymers may be prepared in an amorphous form by processes which could be represented:

$$\text{Liquid} \xrightarrow{\Delta\mu} \text{Amorph. solid}$$

$$\text{Liquid + Vapour} \xrightarrow{\Delta\mu} \text{Amorph. solid}$$

$$\text{Vapour} \xrightarrow{\Delta\mu} \text{Amorph. solid}$$

4.1.3. THE STRUCTURE AND CLASSIFICATION OF AMORPHOUS SOLIDS

Perhaps the broadest scheme of classification for solids (or other materials) is in terms of bond types—i.e. metallic, ionic, covalent and molecular solids. This is of very little help in the present context since amorphous solids can be formed in all four categories, although as a class *covalent* materials probably provide the majority of the well-known glasses and other amorphous solids. There is no doubt, of course, that in solid state physics, structure and in particular, structural symmetry—has provided the most fruitful basis of classification. Structural symmetry is also the key to the detailed theoretical analysis of band structures and hence to all the varied electronic and optical properties which follow from that. Amorphous solids cannot, however, be categorized in terms of structural symmetries in the usual way. Although, as is well known, they have a high degree of local, or short range, order (SRO) over the first few interatomic distances, amorphous solids have no long range order (LRO), and hence no readily definable translational symmetry.† Diffraction data on amorphous solids is best represented by a radial distribution function (r.d.f.)[(36)] which could conceivably be used as a "fingerprint" of the structure (e.g. by integrating over a certain interval). Unfortunately it does not appear to be a very sensitive function (often because of experimental uncertainties) and different structural models—ranging from the "microcrystalline" to random-network—often seem to fit equally well to experimental r.d.f.'s.[(36, 37)] Moreover, it may well be misleading to even talk

† The structure of amorphous solids is discussed in detail in the article by Grigorovici.[(36)]

of *the* structure of a given amorphous solid as the degree of order could easily vary spatially, from region to region. Figure 4.4 for example, illustrates schematically an extreme case of a two-dimensional ordered (crystalline) arrangement of triangles merging continuously into the so-called random-network arrangement. An experimental r.d.f. would average over all such spatial variations. In addition, amorphous solids of two or more components may also have compositional fluctuations ranging from macroscopic phase separation to more subtle local ordering.[38]

FIG. 4.4. A two-dimensional network of triangles illustrating schematically a variation in local order from a periodic array to a random network.

In the absence of adequate analytical tools for the quantitative study of aperiodic lattices one must resort to more empirical methods. One such scheme, depending on structural inferences from either the origin or behaviour of amorphous solids, has been proposed by Roy.[1] It is known that in many cases the basic structural features are carried over from the parent materials and the preparation process and Roy makes the plausible supposition that this is generally true of amorphous solids. Figure 4.1 (adapted from Ref. 1) and the summary to Section 4.1.2 illustrate the main features of this classification. It emphasizes what Roy refers to as the "polymorphism" of amorphous solids. Silica (SiO_2), for example, can be prepared in an amorphous form by any of the basic methods starting from vapour, liquid or solid; Roy argues[1] that it should not be expected that the various forms of amorphous SiO_2 are "the same" any more than the several crystalline modications of SiO_2 are the same.

Useful as such empirical classification schemes may be there is no doubt that the absence of some suitably definitive analytical method of quantifying the structure is a major obstacle to more quantitative theories of properties of amorphous solids. Very recently, however, two methods have been proposed for the description of non-periodic structures which, although not fully developed at the moment, may offer some solution to this problem. Stevels[39] has introduced the idea of a Repeatability Number (RN) defined by the following steps. One first defines a coefficient of linear repetition for arbitrary directions i – CLR_i:

$$\mathrm{CLR}_i = \frac{\text{number of atoms in direction } i \text{ in } n \text{ steps}}{n}$$

The directions *i could* be chosen as cartesian coordinates x, y and z, but could also, for example, be chosen to correspond with bond directions from a given atom. The weighted average of all CLR's (limited to regions within one phase) gives the Average Coefficient of Repetition, ACR;

$$\mathrm{ACR} = \frac{\sum_i c_i \, \mathrm{CRL}_i}{\sum_i c_i}$$

where the coefficients c_i represent weighting for different directions. From this is obtained a Coefficient of Repetition in Space, CRS;

$$\mathrm{CRS} = \frac{\sum_j N_j \cdot \mathrm{ACR}}{\sum_j N_j}$$

averaged over all N_j atoms in a given volume. The CRS is a function of distance r but in a perfect crystal would be unity, independent of r. In a "real" crystal the CRS would decrease slowly with r but in glasses or amorphous solids would deviate from unity after a few interatomic distances, the deviation occurring more quickly with r the more rapidly the material is quenched. This is illustrated in Fig. 4.5(a) (from Ref. (39)). Stevels defines the Repeatability Number over a distance r_0 (or volume V_0), $(\mathrm{RN})_{r_0}$:

$$(RN)_{r_0} = \frac{1}{V_0}\int_0^{V_0} \mathrm{CRS}\,(V)\,\mathrm{d}V = \frac{3}{4\pi r_0^{\,3}}\int_0^{r_0} \mathrm{CRS}(r)\,4\pi r^2\,\mathrm{d}r$$

So defined, $(\mathrm{RN})_{r_0}$ would be a direct measure of the degree of order in the structure in a limited region defined by r_0 or V_0. In a material in which the

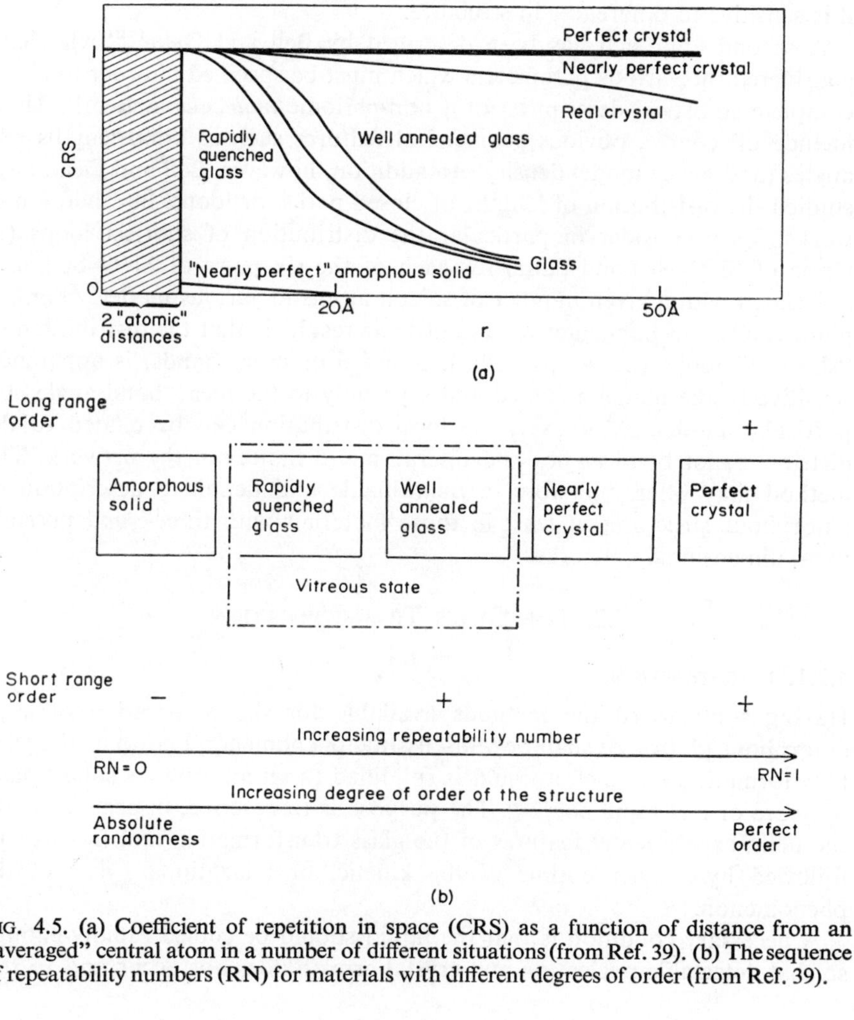

FIG. 4.5. (a) Coefficient of repetition in space (CRS) as a function of distance from an "averaged" central atom in a number of different situations (from Ref. 39). (b) The sequence of repeatability numbers (RN) for materials with different degrees of order (from Ref. 39).

degree of order is the same everywhere one could define a Repeatability Number, *RN*, which is a measure of the overall degree of order. If, as is quite likely, the degree of order varies, it would be necessary to define a weighted average of all the local $(RN)_{r_0}$'s. Figure 4.5 (b) illustrates schematically the way in which RN varies according to Stevels, but this is slightly incorrect in that the left-hand end corresponding to "absolute randomness" (RN = 0) should presumably be linked with the "ideal gas" and not an amorphous solid. There have been no attempts to as yet evaluate RN or $(RN)_{r0}$, and local variations in the degree of order would certainly cause difficulties. It would nevertheless be worthwhile to calculate RN for certain typical models—e.g. the random network and "microcrystalline" models—in order to see whether it is sensitive to differences in structure.

A second approach has been described by Bell and Dean[(40)] who have considered the various parameters which must be specified in order to give a complete statistical description of a non-periodic *molecular* network. These include of, course, obvious geometrical features such as bond lengths and angles (and hence model density). In addition, however, Bell and Dean have studied the distribution of lengths of closed paths, or loops, in random networks. They consider in particular the distribution of shortest loops (in terms of Si–O–Si bond pairs) for each of the six pairs of Si–O–Si bonds associated with a given number of silicon atoms in various models of amorphous SiO_2. An important feature of their results is that the distribution of "shortest" loops (i.e. loops with 4, 5 and 6 or more bonds) is apparently sensitive to the model assumed and especially to the mean bond angle of a particular model. Moreover, the loop distribution can be related to the distribution of bond angles and interatomic distances in the network. The method may offer, therefore, a quantifiable and definitive description of amorphous structures at least in those materials with three-, and possibly two-, dimensional networks.

4.2. The Glass Transformation

4.2.1. Introduction

Having summarised the methods available for the preparation of solid amorphous phases we shall now discuss the phenomenon known as the glass transformation in which a liquid is solidified to an amorphous solid (glass) by more or less rapid cooling. The purpose is to describe, in general terms, the main experimental features of the glass transformation and this will be followed by a brief outline of the kinetic, or relaxational, view of the phenomenon.

A necessary condition is that the liquid should be cooled rapidly enough so that detectable nucleation and crystal growth does not occur and in this sense glass formation is clearly a kinetic process. Several authors have treated this aspect of the problem.[15, 16, 41] It is possible to make rough estimates of the cooling rate which must be exceeded (the critical cooling rate, R_c) in order to encourage glass formation in particular liquids; it will obviously be a function of diffusion and flow processes in the liquid. Sarjeant and Roy[16] make an empirical estimate of R_c by first measuring the velocity of crystallisation in a small melt of NaCl subjected to very rapid cooling (10^4–10^5 deg/s). They then assume that for NaCl the spontaneous nucleation and growth rates become extremely low at 10° of supercooling and that the minimum detectable crystalline ordering occurs when the crystallite size is $\sim 10^{-7}$ cm. Thus, R_c is the rate which will not allow the growth of a crystallite of that size in a 10° interval. For NaCl Sarjeant and Roy find this to be $1{\cdot}2 \times 10^9$ deg/s. They further suggest that R_c can be written

$$R_c = Z\frac{T_m}{\tau}$$

where Z is a constant, T_m the thermodynamic melting point and τ the appropriate jump rate of "structural units" which, for NaCl, is equated with the Restrahlen frequency. Hence, from known parameters, $Z \sim 2{\cdot}0 \times 10^{-6}$. To relate R_c to viscosity η, the simplest general relationship is assumed, viz

$$\tau = \frac{V\eta}{\mathcal{N}_A k T_m}$$

where V is the molar volume and $\mathcal{N}_A$, Avogadro's number. Thus,

$$R_c = 2{\cdot}0 \times 10^{-6}\frac{T_m^{\,2} R}{V\eta}$$

where R is the gas constant. It is this relationship which is shown schematically in Fig. 4.3. and it is roughly in accord with experience.

But this says nothing about the processes which occur when, having prevented crystallisation, the super-cooled liquid solidifies to a glass, i.e. the phenomenon associated with the glass transformation temperature T_g. This is a subject of some controversy and not a little confusion—a situation which is due in part at least to the difficulty of characterising and quantifying the glass transition over a wide range of conditions. There are, in fact, no generally accepted models or theories of the glass transformation. Nevertheless, there are several reasons quite apart from intrinsic interest, why it is desirable in the present context to have as clear an understanding of the process as possible:

(i) Although amorphous solids can be prepared from vapour and solid phases, as well as from liquids, most (*but not all*) show typical glass transformation behaviour during subsequent thermal cycling, regardless of their method of preparation. Amorphous materials produced by widely different methods are often compared—an exercise which needs to be done with some caution.

(ii) Switching is often related to glass formation.

(iia) Threshold switching behaviour is associated with a very "stable"† glass whilst memory switching is most reliably observed in a relatively less "stable" glass.[42] Thus the composition $As_{30}Te_{48}Si_{12}Ge_{10}$ is a typical threshold switching material and it is at the centre of a large "glass-forming region" in the quaternary system As–Te–Si–Ge.‡ By contrast the composition $Te_{81}Ge_{15}As_4$ is on the edge of the "glass-forming region", and hence is less "stable" as a glass; it is more suitable therefore for memory-switching. According to current ideas memory switching is intimately connected with the kinetics of nucleation and crystal growth for switching ON, and with revitrification to switch OFF, and is less, perhaps, a problem of the glass transformation as such.

The question may arise, however: does crystallisation and revitrification occur via a process glass $\leftrightarrow$ liquid $\leftrightarrow$ crystal or directly glass $\leftrightarrow$ crystal?

(iib) The threshold voltage V_{TH} in chalcogenide glass switches decreases with temperature and there have been several reports which show that it extrapolates to zero at a temperature equal or close to the glass transformation temperature, T_g.[43] Moreover, in a series of related glasses V_{TH} sometimes increases linearly with T_g. It has been suggested, therefore, that the critical conditions at threshold are somehow linked with the glass transformation.

(iii) In the recently proposed optical applications of certain amorphous semiconductors e.g., reversible optical imaging, structural changes associated with T_g may also be important.[44] Mass transport, flow properties and visco-elastic effects are all affected by the glass transformation and such processes could play a part in optical image formation or, indeed, in electric field-induced switching. In macro-molecular glasses the degree of polymerisation affects the transformation temperature and in complex glasses phase separation processes are accelerated in the neighbourhood of T_g.

† As implied before the term "stable" is not used in its strict thermodynamic sense.

‡ The concept of a "glass-forming region" is not precise. It depends very much on the method of preparation and generally speaking the more rapid the method of cooling the larger the glass-forming region. In the As–Te–Si–Ge system, for example, it could be extended to embrace the *whole* system if suitable vapour deposition techniques were used.

4.2.2. EXPERIMENTAL SITUATION

It is assumed that the liquid has been cooled rapidly enough to prevent crystallisation. The subsequent behaviour of, for example, volume as a function of temperature is illustrated in Fig. 4.6. (adapted from Ref. 45) for vitreous selenium. This is typical, in its general form, for all glasses. The experimental characteristics of the glass transformation process are as follows:

(i) First order extensive thermodynamic variables such as volume V, enthalpy H and entropy S are continuous with temperature (and pressure) but show a change of slope on the transition liquid → glass. The *range* of temperature over which this transition takes place is called the glass transformation temperature T_g.

(ii) Although first order variables are continuous, there are discontinuities in second order thermodynamic quantities such as expansion coefficient α, specific heat C_p and compressibility κ (see bottom right insert to Fig. 4.6). For this reason the glass transformation has sometimes been regarded as a second order thermodynamic transition.[46] This is probably not correct, or at best, is a gross oversimplification of the situation.

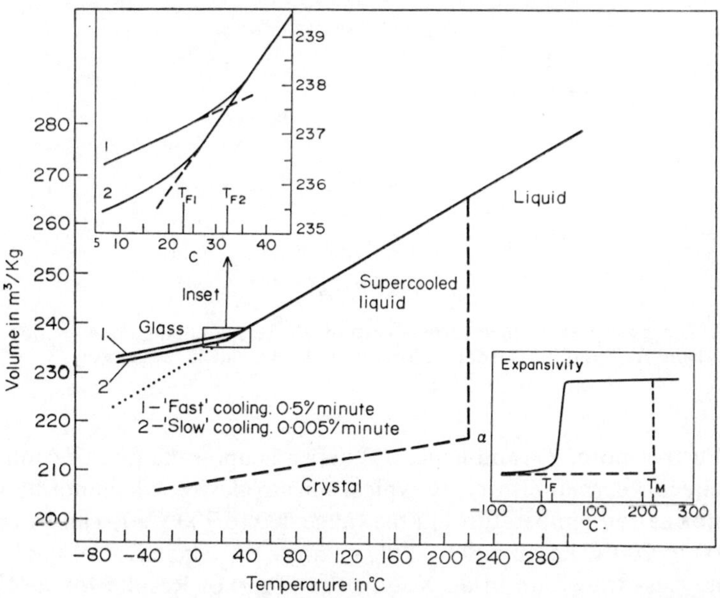

FIG. 4.6. The volume–temperature relationship of selenium at the liquid–crystal and liquid–glass transition. The upper insert shows in more detail the effect of different cooling rates, and the lower one illustrates the discontinuity in the expansion coefficient (a second order property). The data is taken from Ref. 45.

(Note. This figure is taken from "Semiconducting Glasses" Part 1. *Contemp. Phys.* 11, 227 (1970), A. E. Owen).

(iii) It is important to realise that T_g is not an accurately defined temperature. In the first place, the glass transition is not sharp but occurs over a range of temperature. It is sometimes convenient to *define* as a specific temperature the point at which the extrapolated liquid and glass lines cross (see top left insert to Fig. 4.6). In glass technology this temperature is known as the fictive temperature T_F.[47] It is the temperature at which the glass would be in (metastable) equilibrium if it could be brought to that temperature instantaneously.

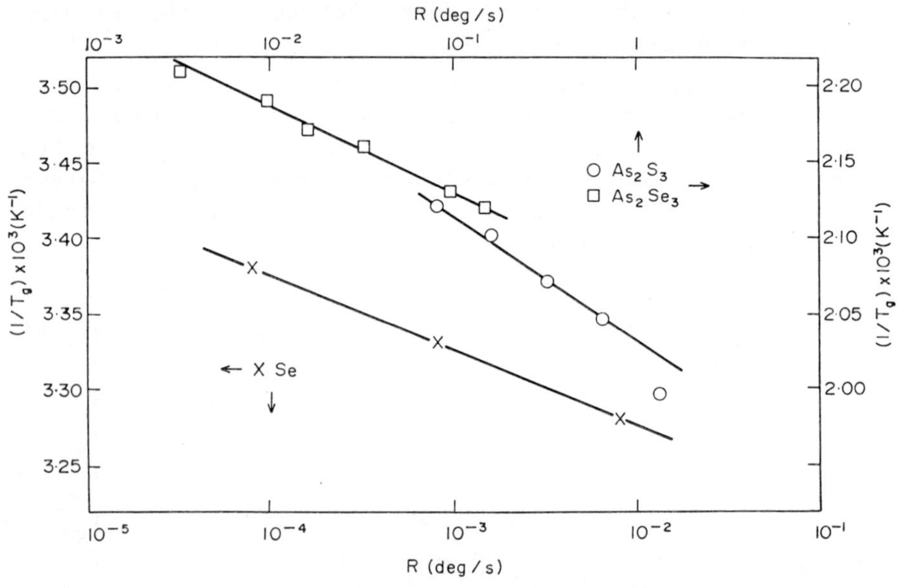

FIG. 4.7. The glass transformation temperature T_g, versus cooling rate, R, for Se (×), As_2S_3 (○) and As_2Se_3 (□). The data is from Refs. 45, 49, and 48 respectively.

(iv) Furthermore, T_g (and hence T_F) depends upon the rate of cooling and subsequent thermal history. In typical silicate glasses, which normally have transformation temperatures in the range 400 to 700°C, T_g can be changed by 100 to 200°C i.e., by 20 to 30% or more.[46] The effect of cooling rate on the glass transition in Se is shown in Fig. 4.6. Results for Se[45], and As_2Se_3[48] and As_2S_3[49] are plotted in Fig. 4.7; they can be expressed empirically in the form,

$$R = R_0 \exp\left[-\frac{1}{C}\left(\frac{1}{T_g} - \frac{1}{T_m}\right)\right]$$

where R is the cooling rate and R_0 and C are constants. Expressed in this way, R_0 is the cooling rate at which $T_g = T_m$. The constants obtained from the data in Fig. 4.7 are listed in Table 4.3.

TABLE 4.3.

	R_0(deg/s)	C(deg^{-1})
Se	10^{23}	$2{\cdot}2 \times 10^{-5}$
As_2Se_3	$4{\cdot}5 \times 10^8$	$2{\cdot}6 \times 10^{-5}$
As_2S_3	7×10^3	$3{\cdot}6 \times 10^{-5}$

There are two features of the data in Fig. 4.7 and Table 4.3 worthy of particular mention. Firstly, for these materials at any rate, T_g changes by 10 to 20° for a two-order of magnitude change in cooling rate. Recalling (Table 4.2) that cooling rates may easily vary by 8 to 10 or more orders of magnitude, it is clear that very substantial differences in T_g might be expected between, say, an evaporated "glassy" film and the same material cooled relatively slowly from the melt. Secondly, the magnitudes of R_0 in Table 4.3 should be noted. For Se it is 10^{23} deg/s and this is certainly not attainable by any known technique. The value for As_2Se_3 (4.5×10^8 deg/s) is also beyond the range of quenching techniques although vapour-deposition could, presumably, give an "effective" rate in that region. For As_2S_3, however, R_0 is 7×10^3 deg/s and this could certainly be achieved or exceeded. Assuming that the extrapolation is valid, the implication is that in As_2S_3, T_g could be made to occur at *or even above* T_m. At first sight this may not seem plausible and it would be impossible, of course, to observe the "glass transformation" in the liquid state by methods implicit in Fig. 4.6. This apparent anomaly does, however, illustrate a most important characteristic of the phenomenon and it will be returned to later.

(v) A glass is thermodynamically unstable and if held at a temperature below T_g it will tend to approach the (meta) stable equilibrium of the supercooled liquid at that temperature. Consider in Fig. 4.8 for example, two glasses prepared by different rates of cooling and, therefore, with different T_g's—T_{g_1} (slow cooling) and T_{g_3} (fast cooling).† If they are both held at a temperature T', the volumes of both will tend towards that of the supercooled liquid at that temperature, i.e., V'. The T_g's will also change, of course. This process is known as "stabilisation". The rate at

† The T_g's are being used here, and subsequently, as specific temperatures; strictly one should use the fictive temperature T_F.

which stabilisation occurs will depend very sensitively on temperature. Within a few degrees of T_g it will normally take place on a time scale of minutes or hours. Far removed from T_g it will take years and, as mentioned before, silicate glasses at room temperatures are "stable" for practical purposes. Stabilisation involves a decrease in V (and H and S) and T_g. It is also possible to increase V and T_g by heat treatment above T_g. Suppose that, in Fig. 4.8, for example, we heat-treat the two glasses with T_{g1} and T_{g3}, at an intermediate temperature T_{g2} (i.e., a T_g corresponding to an "intermediate" rate of cooling) and then cool them rapidly to T'. The volume of glass 3 will decrease to V_2 and its transformation temperature will also decrease to T_{g2}. On the other hand the volume of glass 1 will *increase* to V_2 and its transformation temperature will also increase to T_{g2}. The *rates* at which these two processes occur will be different; the change $T_{g3} \rightarrow T_{g2}$ (and $V_3 \rightarrow V_2$) will be more rapid than $T_{g1} \rightarrow T_{g2}$ (and $V_1 \rightarrow V_2$).[47] From this and the previous paragraphs it should be clear therefore, that the volume and other physical properties, of a glass *and* T_g are sensitive to the total thermal history, i.e., to rate of cooling from the liquid as well as to subsequent heat treatment in the neighbourhood of T_g.

(vi) If we reheat a glass in order, say, to determine T_g by a D.T.A. experiment, for example, anomalies will occur unless the *rate of heating is the same as the original rate of cooling*. This can also be illustrated by Fig. 4.8.

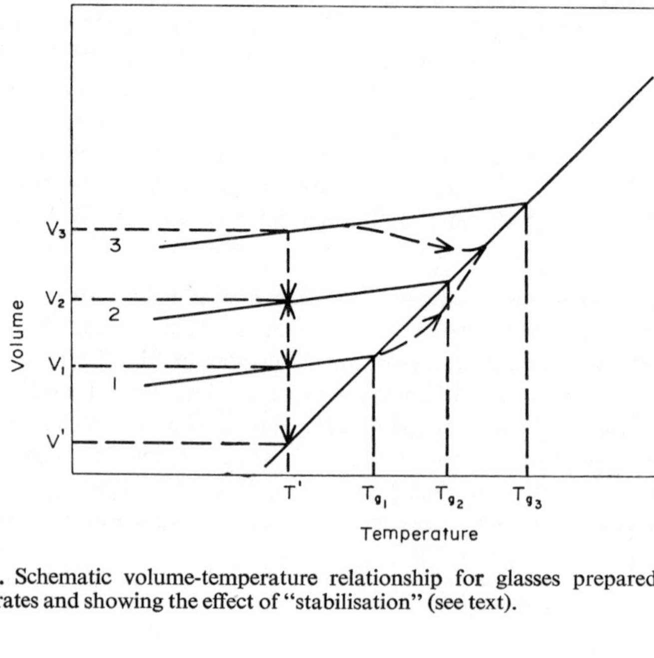

FIG. 4.8. Schematic volume-temperature relationship for glasses prepared by different cooling rates and showing the effect of "stabilisation" (see text).

Suppose glass 1 (prepared by *slow* cooling) with $T_g = T_{g_1}$ is reheated at the (faster) rate corresponding to glass 2 (intermediate rate of cooling). Its volume will tend to undershoot the supercooled liquid line (shown by the dotted curve in Fig. 4.8) because it is being reheated too rapidly to come into equilibrium. On the other hand if glass 3 (fast cooling) is reheated at the intermediate rate its volume will tend to decrease before T_{g_3} is reached (dotted curve in Fig. 4.8) because its rate of heating is slow enough to allow volume changes (corresponding to stabilisation) to take place at a temperature lower than its own T_g. The derivative curves (α, C_p) will be as shown in Fig. 4.9 and if we were using these experiments to determine T_g say, the measured value would be closer to T_{g_2} rather than the values corresponding to the original rates of cooling. *To measure T_g by a heating experiment the rate of heating must be the same as the original rate at which the glass was cooled.* It should also be noted that if the rate of heating is less than the rate of cooling the derivative properties will undershoot their normal curves [(3) in Fig. 4.9]; conversely, if the heating rate is faster there will be some overshoot (curve (1)).[47]

(vii) It is possible to define and observe a glass transformation pressure P_g. The derivative (dV/dP) is negative, of course, and shows a change of slope but no discontinuity at P_g. The experimental V–P–T relationship for vitreous selenium is shown in Fig. 4.10. It is worth noting from this figure that T_g *increases* with pressure.

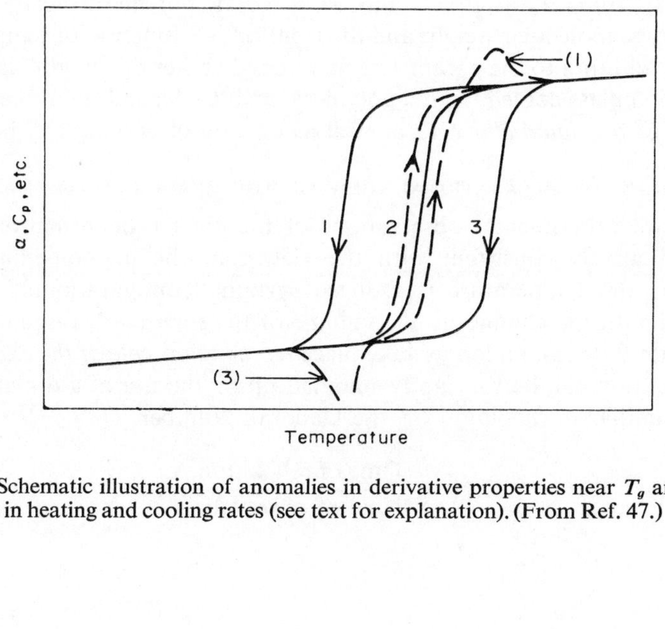

FIG. 4.9. Schematic illustration of anomalies in derivative properties near T_g arising from differences in heating and cooling rates (see text for explanation). (From Ref. 47.)

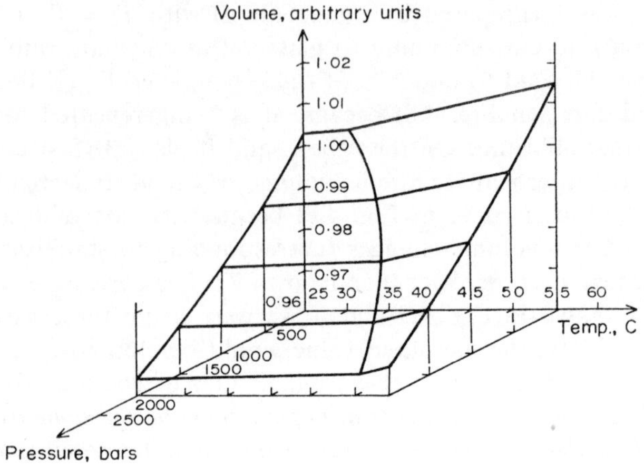

FIG. 4.10. The volume–pressure–temperature relationship for vitreous selenium. (From Ref. 50.)

(viii) In polymeric materials, which would include Se and possibly other chalcogenide glasses, a glass transformation molecular weight M_g can also be defined.(50) The three quantities T_g, P_g and M_g are empirically related:(50)

$$T_g = T_g^{\infty}(0) - \left(\frac{\partial T_g}{\partial(1/M)}\right)\frac{1}{M_g} + \left(\frac{\partial T_g}{\partial P}\right)P_g$$

where $T_g^{\infty}(0) = T_g$ at $P = 0$ and $M = \infty$. In polymeric *liquids* like Se the average molecular weight and distribution is a function of temperature. Thus in addition to the parameters mentioned earlier one might also expect the T_g of a glass cooled from a polymeric melt to depend upon the thermal history of *the liquid above* T_m as well as on rate of cooling etc. below T_m.

4.2.3. KINETIC OR RELAXATIONAL VIEW OF THE GLASS TRANSFORMATION

Probably all experimental observations of the glass transformations are at least qualitatively consistent with the view that the phenomenon occurs because as the temperature is reduced certain "configurational" changes associated with the changes in the volume of the supercooled liquid become so slow that they can no longer take place *on the time scale of the experiment.*

The situation can be very aptly summed up by the use of a dimensionless number familiar in rheology, viz. the Deborah Number, D.N.:(51)

$$\text{D.N.} = \frac{\text{Time of relaxation, } \tau}{\text{Time of observation}}.$$

The difference between "solids" and fluids is then defined by differences in D.N. If the time of observation is large, *or* τ is small, D.N. is small and the consequences of the relaxation process (whatever they may be, e.g., plastic flow) can be observed. On the other hand if D.N. is large the material behaves as a solid. We can look at this in terms of the V–T diagram for a glass (e.g. Fig. 4.6.) and consider the relaxation time for the structure to adapt to a change in temperature. In the liquid, the relaxation time will be perhaps 10^{-18}–10^{-13} sec. (depending on the type of liquid) and hence D.N. is very small for normal experimental times. As the temperature is reduced to T_g the relaxation time approaches a few minutes, *i.e.*, *T_g is defined as the temperature where D.N.* = 1. In the glassy state ($T < T_g$) τ increases rapidly with decreasing T, the structure cannot change to its appropriate equilibrium configuration and D.N. becomes very large. The simple concept involved in the definition of D.N. can help us to resolve the apparent anomally mentioned in Section 4.2.2 paragraph (iv)—i.e., the *possibility* that in As_2S_3, T_g could be equal or greater than T_m. In a normal glass-formation experiment the time of observation is kept fixed, and D.N. is changed by varying τ through its temperature dependence. An exactly equivalent experiment, however, would be to change D.N. by keeping the temperature, and hence τ, fixed and varying the time of observation. This would be most conveniently done by making measurements as a function of frequency, e.g., to measure the specific heat C_p or compressibility κ, of a liquid, by ultra-sonic absorption.[52] The expected results of such an experiment are shown schematically in Fig. 4.11. At low

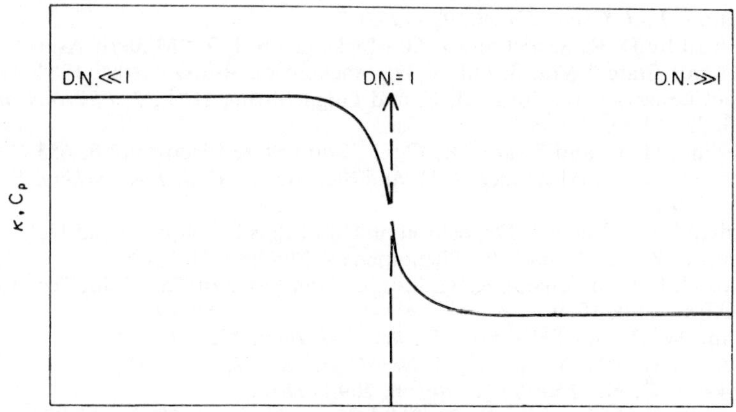

FIG. 4.11. Schematic diagram showing the expected behaviour of derivative properties (specific heat C_p, or compressibility κ) measured as a function of frequency in a liquid. The change of DN is also shown, when DN > 1 the liquid behaves as a "glass" (see text for explanation).

frequencies, the time of observation ($=1/2\pi f$) is long compared with τ and D.N. is very small. When $\tau = (1/2\pi f)$, D.N. $= 1$ and if the frequency is increased further a typical sigmoid-type relaxation curve is obtained. At high frequencies D.N. becomes very large. The particular temperature at which the experiment is performed could be regarded as a glass transformation temperature corresponding to the observation time ($1/2\pi f_{\mathrm{mid}}$), where f_{mid} is the frequency of the mid-point in the relaxation curve. This is completely consistent with the kinetic view of the liquid $\leftrightarrow$ glass transition. As in all relaxation processes, the sharp decrease in κ or C_p as the frequency increases through f_{mid} is caused by the inability of the system to respond to the external stimulus. In this situation, however, the liquid is at all times in internal thermal equilibrium, i.e., the "glassy" state corresponding to $f > f_{\mathrm{mid}}$ is an *equilibrium* glass.[52] It should also be realized that several different mechanical and electrical relaxation processes are possible in liquids and glasses;[52, 53] only when the relaxation involves configurational or structural changes is it realistic to equate it with a glass transformation. In the field of organic polymers, however, there are often structural relaxation processes connected with, for example, side groups; these processes usually freeze-in at temperatures below the main glass transformation and are referred to as secondary glass transitions.

References

1. Roy, R., *J. Non-Cryst. Sol.* **3,** 33 (1970).
2. See Section 4.2 of this article.
3. Hetherington, G and Jack, K. H., *Phys. Chem. Glasses* **3,** 129 (1962). Dalton, J., *J. Chem. Ed.* **40,** 99 (1963).
4. Quoted by D. R. Secrist and J. D. Mackenzie p. 149. "Modern Aspects of the Vitreous State." Vol. 3. (ed. J. D. Mackenzie), Butterworths, 1964.
5. MacChesney, J. B., Potter, J. F. and Guggenheim, H. J., *J. Electrochem. Soc.* **115,** 52 (1968).
6. Sterling, H. F. and Swann, R. C. G., Solid-State Electronics **8,** 653 (1965).
7. Chittick, R. C., Alexander, J. H. and Sterling, H. F. *J. Electrochem. Soc.* **116,** 77 (1969).
8. Holland, L., "Vacuum Deposition of Thin Films", Chapman and Hall, 1963.
9. Chopra, K. L., "Thin Film Phenomena." McGraw Hill, 1969.
10. Maissel, L. I. and Glang, R., (Editors). "Handbook of Thin Film Technology". McGraw Hill, 1970.
11. Beam, W. R. and Takahashi, T., *Rev. Sci. Instr.* **35,** 1623 (1964).
12. Mueller, R. and Woods, C., *J. Non-Cryst. Sol.* **7,** 301 (1972).
13. Jackson, G. N., *Thin Solid Films* **5,** 209 (1970).
14. Bosnell, J. and Thomas, C. B., *Solid-State Electronics* **15,** 1261 (1972).
15. McMillan, P. W., "Glass-Ceramics". Academic Press, 1964.
16. Sarjeant, P. T. and Roy, R., *Mat. Res. Bull.* **3,** 265 (1968).

17a. Duwez, P. and Willens, R. H., *Trans. Met. Soc. A.I.M.E.* **227,** 362 (1963).
b Duwez, P., Willens, R. H. and Crewdson, R. C., *J. Appl. Phys.* **36,** 2267 (1965).

18. Chen, H. S. and Miller, C. E., *Rev. Sci. Instr.* **41,** 1237 (1970).

19. Sarjeant, P. T. and Roy, R., *J. Am. Ceram. Soc.* **50,** 500 (1967).
20. Chen, H. S. and Turnbull, D., *Appl. Phys. Lett.* **10,** 284 (1967).
21. Grigorovici, R., *Mat. Res. Bull.* **3,** 13 (1968).
22. Richter, H., *J. Vac. Sci. Tech.* **6,** 855 (1969).
23. Roy, R., *J. Am. Ceram. Soc.* **43,** 670 (1960). Also: Symp. Nucleation and Crystallisation in Glasses. (ed. M. K. Reser) p.39 Am. Ceram. Soc., (1964).
24. Cahn, J. W. and Charles, R. J., *Phys. Chem. Glasses* **6,** 181 (1965).
25. Haller, W., *J. Chem. Phys.* **42,** 686 (1965).
26. Bunton, G. V., *J. Non-Cryst. Sol.* **6,** 72 (1972).
27. Moynihan, C. T., Macedo, P. B., Aggarwal, I. D., and Schnaus, U. E., *J. Non-Cryst. Sol.* **6,** 322 (1972).
28. Onodera, N., Suga, H. and Seki, S., *J. Non-Cryst. Sol.* **1,** 331 (1969).
29. Tauc, J., Abraham, A., Zallen, R. and Slade, M., *J. Non-Cryst. Sol.* **4,** 279 (1970).
30. Young, L., "Anodic Oxide Films." Academic Press, 1961.
31.a Schmidt, P. F. and Owen, A. E., *J. Electrochem. Soc.* **111,** 682 (1964).
b Schmidt, P. F., O'Keefe, T. W., Oroshnik, J. and Owen, A. E., *J. Electrochem. Soc.* **112,** 800 (1965).
32. Primak, W., *Phys. Rev.* **110,** 1240 (1958).
33. Wittels, M. and Sherrill, F. A., *Phys. Rev.* **93,** 1117 (1954).
34. Parsons, J. R., *Phil. Mag.* **12,** 1159 (1965).
35. De Carli, P. S. and Jamieson, J. C., *J. Chem. Phys.* **31,** 1675 (1959).
36. Grigorovici, R., Chapter 5 of this volume.
37. Shevchik, N. J., "The Structure of Tetrahedrally Coordinated Amorphous Semiconductors." Technical Report No. ARPA-44 (1972).
38. Porai-Koshits, E. A., Proc. VII Int. Congr. on Glass (Brussels), Paper 28 (1965).
39. Stevels, J. M., *J. Non-Cryst. Sol.* **6,** 307 (1971).
40. Bell, R. J. and Dean, P., *Phil. Mag.* **25,** 1381 (1972).
41. Turnbull, D., "Modern Aspects of the Vitreous State." Vol. 1. p. 38. (ed. J. D. Mackenzie), Butterworths, 1960.
42. Fritszche, H., *I.B.M. J. Res. Dev.* **13,** 515 (1969).
43.a Bosnell, J. R. and Thomas, C. B., *Solid State Electronics* **15,** 1261 (1972).
b Gupta, Y. P., *J. Non-Cryst. Sol.* **3,** 148 (1970).
44. Feinleib, J., Isawa, S., Moss, S. C., de Neufville, J. F. and Ovshinsky, S. R., *J. Non-Cryst. Sol.* **8-10,** 909 (1972).
See also: Fritzsche, H., This volume Chapter 19.
45. Dzhalilov, S. U. and Rzaev, K. I., *Phys. Stat. Sol.* **20,** 261 (1967).
46. Stevels, J. M., "Handbuch der Physik." Vol. 13. p. 510 (ed. S. Flugge), Springer-Verlag, 1962.
47. Jones, G. O., "Glass". Methuen, 1956.
48. Kolomiets, B. T., Lyubin, V. M. and Shilo, V. P., *Inorg. Mat. U.S.S.R.* **7,** 751 (1971).
49. Onodera, N., Suga, H. and Seki, S., *J. Non-Cryst. Sol.* **1,** 331, (1969).
50. Eisenberg, A., *J. Phys. Chem.* **67,** 1333 (1963).
51. Stevels, J. M., *J. Non-Cryst. Sol.* **6,** 307 (1971).
52. Litovitz, T. A., "Non-Crystalline Solids". p. 252. (ed. V. D. Frechette), Wiley, 1960.
53. Hopkins, I. L. and Kurkjian, C. R., "Physical Acoustics". Vol. II Part B. p. 91. (ed. Warren P. Mason), Academic Press, 1965.

19. Sarkar, [illegible] and [illegible] W. R., J. Am. Ceram. Soc. [illegible] (1967).
20. Cohen, M. H. and Turnbull, D., [illegible] (1960).
21. [illegible] (1966).
22. [illegible]
23. [illegible] J. Am. Ceram. Soc. [illegible] Also, "Sodium [illegible]" [illegible] Crystallization in Glasses [illegible] J. Am. Ceram. Soc. (1964).
24. [illegible] Phys. Chem. Glasses [illegible] (1964).
25. [illegible] (1962).
26. [illegible] (1963).
27. [illegible] and [illegible] [illegible] (1962).
28. [illegible] J. Am. Ceram. Soc. [illegible]
29. [illegible] and [illegible] J. [illegible] (1970).
30. [illegible]
31. [illegible]
32. [illegible]
33. [illegible]
34. [illegible]
35. [illegible]
36. [illegible]
37. [illegible]
38. [illegible]
39. [illegible]
40. [illegible]
41. [illegible]
42. [illegible] (1963).
43. [illegible] (1972).
44. [illegible]
45. [illegible] and [illegible] (1971).
46. [illegible]
47. [illegible] J. Non-Cryst. Solids [illegible] (1971).
48. [illegible]
49. [illegible] and [illegible] J. Non-Cryst. Solids 1, [illegible] (1969).
50. [illegible] (1963).
51. [illegible] (1949).
52. [illegible] (1969).
53. [illegible] [illegible] and [illegible] Academic Press, 1964.

5. THE STRUCTURE OF AMORPHOUS SEMICONDUCTORS

R. Grigorovici

Institute of Physics,
Bucharest, Rumania.

5.1. Introduction

The work described in this article has been divided into four main Sections dealing with the basic amorphous structure, with defects in this structure, with the type of short range order found in the amorphous state and finally with a discussion of the continuous network model. Under the first heading the methods aimed at investigating the basic structure of amorphous semiconductors are discussed. Most emphasis is given to X-ray, electron and neutron diffraction and to the extended X-ray absorption fine structure. Infra-red and Raman spectra, nuclear magnetic resonance and thermal effects are briefly considered in this context.

The methods of investigating defects in the basic structure of amorphous semiconductors are discussed in Section 5.3. These include electron microscopy, local chemical analysis by an electron microprobe technique, infra-red dark-field microscopy, X-ray and electron small angle scattering, electron spin resonance and paramagnetic susceptibility measurements.

Section 5.4 is devoted to the short-range order in amorphous semiconducting elements and alloys, especially as far as the nearest-neighbour configuration is concerned. The merits and short-comings of micro-crystal models are critically reviewed in the light of the accessible experimental data. The final Section reviews the progress achieved during recent years in the building of continuous network models. A critical discussion of this alternative approach to the structural properties of a number of amorphous semiconductors is given.

5.2. Methods of Investigating the Basic Structure of Amorphous Solids

5.2.1. X-ray diffraction

Amorphous solids yield an X-ray diffraction pattern similar to that of a microcrystalline powder, i.e. broad rings vanishing rapidly with increasing diffraction angle (Fig. 5.1). The angular distribution $I(s)$ of the scattered X-ray intensity is determined by the scattering factors f_i of the i types of atoms composing the solid and by the randomly oriented interatomic distances x_{ij} between all pairs of i and j atoms present within one crystallite containing $\mathcal{N}$ atoms.[1,2] $I(s)$ is given by the Debye formula

$$I_{\mathcal{N}}(s) = \sum_i f_i^2 + \sum_{i}^{\mathcal{N}} {}_1 \sum_{j \neq i} f_i f_j \frac{\sin 2\pi s x_{ij}}{2\pi s x_{ij}}; \qquad s = \frac{2 \sin \theta}{\lambda}. \tag{5.1}$$

In an elemental solid eqn (5.1) becomes

$$I_{\mathcal{N}}(s) = \phi(s)\, \mathcal{N} f^2 = \mathcal{N} f^2 \left[1 + \frac{2}{\mathcal{N}} \sum_n \frac{\sin 2\pi s x_n}{2\pi s x_n}\right], \tag{5.2}$$

where $\phi(s)$ is the so-called interference function and $n_{max} = \mathcal{N}(\mathcal{N}-1)/2$. Obviously the sharpness of the diffraction rings will depend on the size of the crystallites, the rings being the broader, the smaller the crystallites.

A vectorial description of the short-range order in a homogeneous elemental amorphous solid will not be attempted. Consequently, we use either the probability $P(r)$ of finding an atom on a sphere of radius r or the radial distribution function (RDF) $4\pi r^2 \rho_a(r)$, where $\rho_a(r)$ is the atomic density related to $P(r)$ and the average atomic density ρ_0 by $\rho_a(r) = \rho_0 P(r)$. In this case

$$\phi(s) = \left\{1 + \frac{2\rho_0}{s} \int_0^\infty r[P(r) - 1] \sin(2\pi sr)\, dr\right\}$$

$$= 1 + \int_0^\infty 4\pi r^2 [\rho_a(r) - \rho_0] \frac{\sin 2\pi sr}{2\pi sr}\, dr. \tag{5.3}$$

The RDF gives a one-dimensional description of the atomic distribution. If appropriately normalized, the RDF has a number of peaks at certain values of r and the area below each peak is equal to the number of atoms contained in the respective coordination shell (see, for instance, Fig. 5.11). The width of the peaks describes the radial fluctuations of the corresponding interatomic distances. This description is appropriate for an amorphous solid consisting of a continuous network of interconnected atoms.

If the amorphous solid (a-solid) contains more than one type of atom, one has to take into account that it is the electron cloud surrounding each nucleus which scatters the X-rays. Mostly one assigns therefore each atom i a given number of effective electrons K_i so that eqn (5.3) can be rewritten as

$$\phi(s) = 1 + \mathcal{N} f_e^2 \int_0^\infty 4\pi r^2 \left[\sum_i \sum_j c_j K_i K_j \rho_{ij}(r)\right.$$

$$\left. - \left(\sum_i c_i K_i\right)^2 \rho_0\right] \frac{\sin 2\pi sr}{2\pi sr}\, dr. \tag{5.4}$$

Here f_e is the free electron scattering factor; c_j is the concentration of atoms of type j; ρ_{ij} is the density distribution of the atoms of type j about an atom of type i; ρ_0 is the mean atom density. $4\pi r^2 \sum_i \sum_j c_j K_i K_j \rho_{ij}(r)$ is the so-called atomo-electronic ERDF which is the RDF of the atom density expressed in electronic units. For other RDF's see references 3–5.

The practical problem is to get from the experimental $\phi(s)$ the RDF or ERDF. This is done by a Fourier transformation of the form

$$4\pi r^2 \rho_a(r) = 4\pi r^2 \rho_0 + 8\pi r \int_0^\infty s[\phi(s) - 1] \sin(2\pi sr)\, ds; \phi(s) = \frac{I(s)}{\mathcal{N} f^2}; \quad (5.5)$$

or

$$4\pi r^2 \sum_{i\,1}^{m} \sum_{j\,1}^{m} c_j K_i K_j \rho_{ij}(r) = 4\pi r^2 \left(\sum_{i\,1}^{m} c_i K_i\right)^2 \rho_0$$

$$+ 8\pi r \int_0^\infty s \frac{I(s) - I(0) - \mathcal{N} \sum_i c_i f_i^2}{\mathcal{N} f_e^2} \sin(2\pi sr)\, ds, \quad (5.6)$$

respectively. $I(0)$ is the not observable central scattered intensity peak; $\mathcal{N} f^2$ and $\mathcal{N} f_e^2$ are obtained by adapting the corrected experimental $I(s)$ curve to a normalized $\mathcal{N} f^2$ or $\mathcal{N} f_e^2$ curve at high diffraction angles.

In practice the $\mathcal{N} f_e^2$ curve is calculated by using tabulated atomic scattering factors,[6] taking into account absorption of X-rays in the sample, geometrical and polarization corrections in order to correct the $I(s)$ curve and adding to $\mathcal{N} fe^2$ the incoherently scattered radiation, if it was not eliminated by using monochromatic radiation. Then the experimental curve is fitted to the calculated one by different procedures, all aimed at making both curves coincide at high diffraction angles (Fig. 5.1). The integration in the subsequent Fourier transformation eqn (5.5) or eqn (5.6) is necessarily limited to the highest possible diffraction angle $2\theta_{max} = \pi$ and therefore to $s_{max} = 2/\lambda$. As a consequence a δ-shaped coordination sphere of $\mathcal{N}$ atoms shows up in the RDF not as a δ-shaped peak, but as a

$$\sin 2\pi s_{max}(r - a)/2\pi s_{max}(r - a)$$

function, whose maximum lies at $r = a$, is $2 s_{max} \mathcal{N}$ high, $1/s_{max}$ wide at the base and covers an area of $1{\cdot}18\,\mathcal{N}$. The secondary peaks surrounding the central one can simulate non-existing coordination spheres, especially by overlapping with other secondary peaks. Supplementary broadening of the peaks may result from overlapping of neighbouring maxima (Fig. 5.2a) or from static and dynamic (thermal) fluctuations in r (Fig. 5.2b). In this

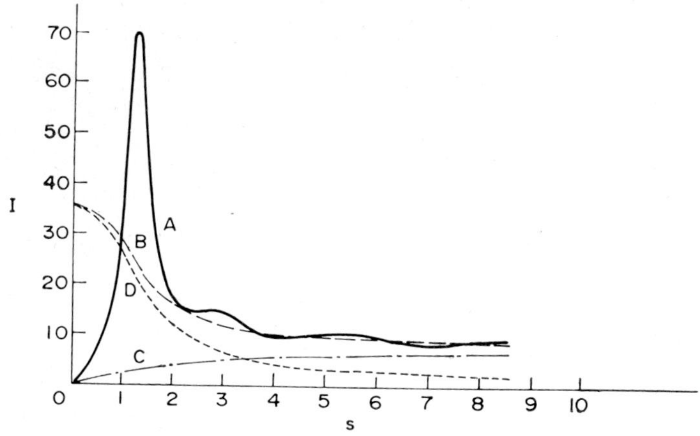

FIG. 5.1. Experimental X-ray scattering curve of amorphous sample (A); total independent scattering curve (B); incoherent scattering (C); independent coherent scattering $\mathcal{N}f^2$ (D). (After Ref. 1.)

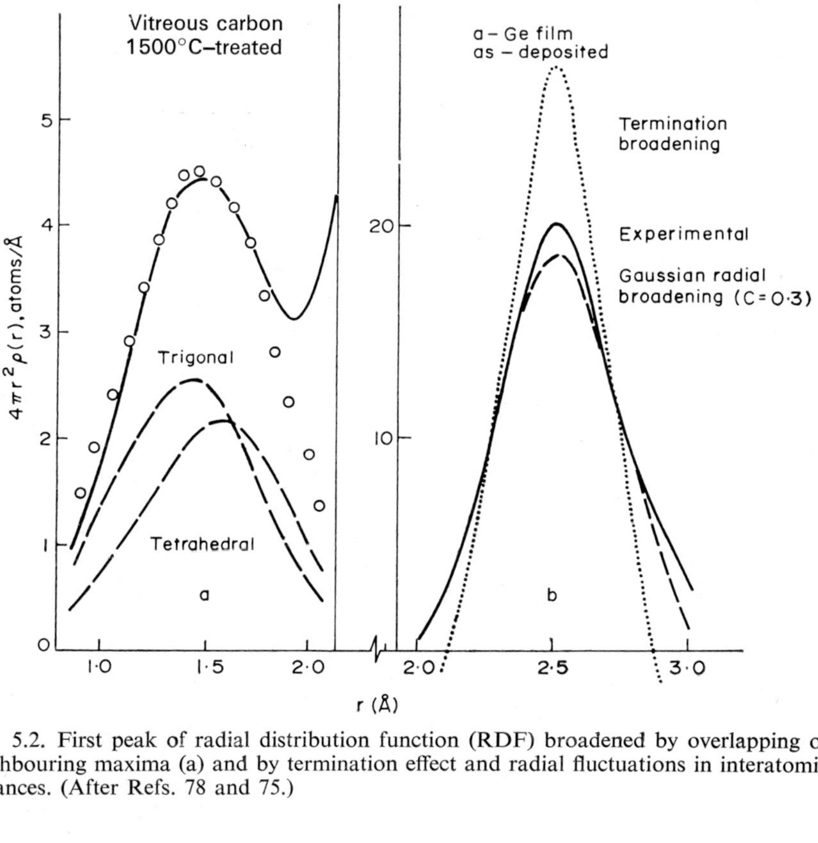

FIG. 5.2. First peak of radial distribution function (RDF) broadened by overlapping of neighbouring maxima (a) and by termination effect and radial fluctuations in interatomic distances. (After Refs. 78 and 75.)

latter case the real shape of the peak results from a convolution product of all the partial broadening functions. Therefore the real fluctuations in r are obtained only by a deconvolution procedure or by comparing the experimental shape to that calculated from s_{max} and specific broadening functions, mostly Gaussian ones (see Fig. 5.2b).

Incorrect normalizing and correction factors and especially experimental errors at high diffraction angles introduce periodic ripples without physical significance into the RDF, because the transformed function

$$s[\phi(s) - 1] = s\left[\frac{I(s) - \mathcal{N}f^2}{\mathcal{N}f^2}\right]$$

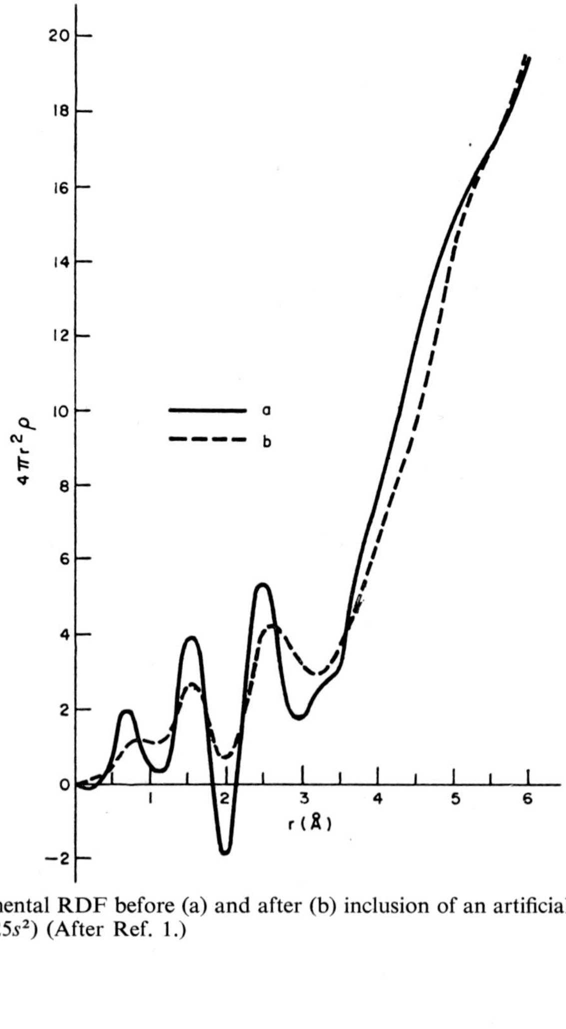

FIG. 5.3. Experimental RDF before (a) and after (b) inclusion of an artificial temperature factor exp (—0·025s^2) (After Ref. 1.)

is obviously very sensitive to small errors in $I(s)$ at high values of s, where $I(s)$ is very near $\mathcal{N}f^2$ and the difference between them is multiplied by s.

One can detect these artifacts and even get rid of them in a number of ways:

(i) by multiplying $I(s)$ by an appropriate artificial temperature factor e^{-Bs^2}; false peaks disappear (Fig. 5.3), but the main peaks are broadened and resolution is lost;

(ii) by extending the upper limits of the integrals of eqn (5.5) and (5.6) to ∞ by an iteration procedure.[7] By performing the Fourier transformation with increasing s_{max}, the general trends of the improvement of the initial

$$G_i(r) = (\text{RDF} - 4\pi r^2 \rho_0)/r = 4\pi r\,(\rho - \rho_0)$$

function are found. A guess of the correct $G_c(r)$ is made and a new Fourier transformation gives a corrected $I'(s)$ now extending to $s = \infty$ and differing from $I(s)$ below s_{max}. A Fourier transformation of $I'(s)$ limited to s_{max} yields another $G_c'(r)$ which includes termination errors. These are defined by $\Delta G(r) = G_c' - G_c$ and used to correct $G_i\,(r)$, giving $G_i'(r)$ which can be compared with $G_c\,(r)$ indicating the changes required in G_c. The procedure

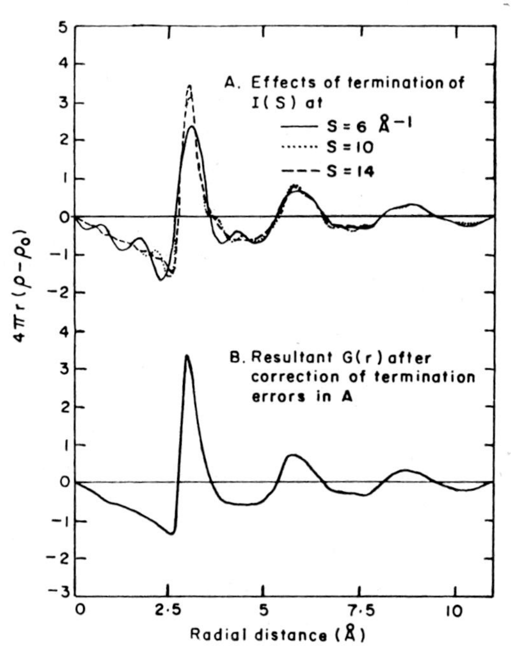

FIG. 5.4. Correction of termination effects by iteration procedure. (After Ref. 7.)

is then repeated until $I(s)$ fits arbitrarily well the experimental values below s_{max} and extends beyond s_{max} to ∞.

If correctly calculated, the final RDF or $G(r)$ must be practically free of ripples below the first peak (Fig. 5.4). The areas below the peaks of the RDF will be equal to $\mathcal{N}$ and their widths will correspond to real fluctuations in r only. $G(r)$ starts from $r = 0$ as a straight line of slope $-4\pi\rho_0$, so providing an independent value of ρ_0; it also yields correct values of the coordination sphere radii.

(iii) by continuing the $i(s) = \phi(s) - 1$ function analytically if it has the form of a damped sine wave. Caution is needed, however, because if two discrete sets of not too different interatomic distances are involved in the structure, the amplitude of $i(s)$ may first decrease and then increase again (see, for instance, Fig. 5.13).

Thus, in elements, and in particular in amorphous semiconductors (a-SC) where the first peak of the RDF can usually be easily separated from the rest of the curve, the information conveyed by the RDF is restricted primarily to the positions and the radial fluctuations within the successive coordination spheres as well as to the corresponding coordination numbers. If however the whole RDF agrees with that of a given structural model, radial fluctuations can be expressed by fluctuations in bond angles and in dihedral angles. In a tetrahedrally interconnected network model of a-Ge, for instance, the broadening of the second peak of the RDF is due to fluctuations in both bond lengths and bond angles.

In amorphous compounds and alloys covalent radii differ too little from each other to allow for a splitting of the peaks of the RDF into partial pair functions. Consequently peaks in these materials are generally broadened more than in elements. The atoms involved in the first peak can be inferred by analogy with the related crystals. The area (expressed in electronic units, i.e. electrons^2/Å) and shape of the first peak can be used to find out to what extent the short-range order of the crystal has been conserved in the a-SC in the first coordination sphere.

In order to obtain a complete picture of the radial distribution in an alloy, the partial interference functions

$$\phi_{ij}(s) = 1 + \int_0^\infty 4\pi r^2 \rho_0 [\rho_{ij}/c_j \rho_0 - 1] \frac{\sin 2\pi sr}{2\pi sr} \, dr \tag{5.7}$$

must be known. In a binary alloy there are three such functions and therefore at least three $I(s)$ functions must be recorded under conditions in which the structure remains the same, but the scattering factors f_i differ from each other as much as possible from one measurement to another. If ϕ_{ij} is assumed to be independent of composition, three different compositions and the same radiation can be used to investigate the same system.[8] In this case the three partial RDF's ($4\pi r^2\rho_{11}(r)$; $4\pi r^2\rho_{12}(r)$; and $4\pi r^2\rho_{22}(r)$) can be obtained. The fourth RDF is $4\pi r^2\rho_{21}(r) = (c_1/c_2)4\pi r^2\rho_{12}(r)$, where c_1 and c_2 are the concentrations of the two elements. The area corresponding to a given atom pair will be

$$A_{ij} = c_i K_i K_j n_{ij}, \tag{5.8}$$

where n_{ij} is the number of j-atoms surrounding an i-atom. If the three A_{ij} overlap, as in most a-SC,

$$\begin{aligned} A_{\text{total}} &= \textstyle\sum_i \sum_j c_i K_i K_j n_{ij} \\ &= c_1 K_1{}^2 n_{11} + c_2 K_2{}^2 n_{22} + c_1 K_1 K_2 n_{12} + c_2 K_1 K_2 n_{21}. \end{aligned} \tag{5.9}$$

In general, however, other radiations, like electrons and neutrons, have to be used in conjunction with X-rays in order to solve the problem.

5.2.2. ELECTRON AND NEUTRON DIFFRACTION

Electrons are scattered much stronger than X-rays. Therefore the use of electron diffraction is restricted to very thin films (10^2–10^3 Å) only. The angular dependence of $\phi(s)$ is steeper than for X-rays and the dependence of f on the atomic number Z follows at low diffraction angles a $Z^{1/3}$ law rather than a Z^2 law as with X-rays. It is therefore easier to detect light atoms in a mixture with heavier ones by electron than by X-ray diffraction. Recording an electron diffraction pattern takes much less time than with X-rays, but inelastically scattered electrons must be eliminated and corrections for multiple scattering have to be applied.

Neutrons are scattered by the atomic nucleus. The scattered intensity is spherically symmetrical. The samples must be rather big compared with those used with both electrons and X-rays. As a whole the scattering cross-section b for neutrons increases slowly with the mass number A (about as $A^{1/3}$), but suffers strong fluctuations from nucleus to nucleus in a seemingly random manner due to resonance and magnetic scattering. The recording time is very long.

By combining X-ray and neutron diffraction so as to make, for instance, $f_1 > f_2$, but $b_1 < b_2$, peaks of the RDF to which 1–1 pairs contribute mostly can be recognized by being enhanced in the X-ray compared with the neutron RDF.

As an example, studying the eutectic $Ge_{0.17}Te_{0.83}$ alloy by X-ray diffraction alone leaves many possibilities open for the way Ge and Te atoms are linked directly together:

(a) All atoms are 2-fold coordinated, either at random ($2r$) or so that each Ge atom is surrounded by two Te atoms ($2d$);

(b) All Ge atoms are surrounded by three Te atoms; 1/6 of the Te atoms are linked to one Ge and two Te atoms, another 2/6 to one Ge and one Te atom, the rest to two Te atoms (3*r*). Or the alloy is phase-separated into 3-fold coordinated Ge Te and 2-fold coordinated Te (3*d*).

(c) Each Ge atom is 4-fold, each Te atom 2-fold coordinated, either so that no Ge–Ge bonds exist (4*d*) or at random (4*r*) or it is phase-separated into amorphous $GeTe_2$ and *a*-Te (4*p*).

It has been shown recently[(9, 9a)] that X-ray diffraction data favour the 3- and 4-fold coordinated models, but do not exclude 2-fold coordinated Ge atoms at least in Te-rich alloys, while neutron diffraction which enhances Ge–Ge pairs decides in favour of models *b* or *c*.

If a complete solution of the short-range order of a binary alloy is looked for, either three radiations like X-rays, electrons and neutrons have to be used or the isotopic composition of the samples has to be changed so as to change as much as possible the scattering cross-section *b* for neutrons from one measurement to the other. No such complete investigation has yet been performed on *a*-SC.

5.2.3. THE EXTENDED X-RAY ABSORPTION FINE STRUCTURE (EXAFS)

The *K* absorption edge for X-rays is due to the ejection of photoelectrons from the deepest electronic orbit. These electrons are scattered by the atoms surrounding the ejecting atom and the photoelectric cross-section will depend on the energy of the incident photon because of the variation with energy of the dipole transition matrix element between the initial electron state (*s*-state) and its final state. Therefore the X-ray absorption coefficient will show, beyond the absorption edge proper, a characteristic energy dependence which reflects the arrangement of the atoms surrounding the atom which absorbs the X-ray photon.

A recent treatment of the already old problem[(10)] shows that the oscillatory part of the EXAFS $\chi(k)$ depends only on the number of atoms $\mathcal{N}_j$ in a given coordination sphere of radius r_j:

$$\chi(k) = -kf(k)\textstyle\sum_j[\mathcal{N}_j \exp(-\gamma r_j)/r_j^{\,2}] \times \exp(-\sigma_j^{\,2}k^2/2)\sin[2kr_j + 2\eta(k)]. \qquad (5.10)$$

Here $k = 2\pi/\lambda$; $f(k)$ is the electronic scattering factor; γ describes the decay of the final photoelectron wave function ψ_f; σ_j is the mean square amplitude of the relative displacements of the atoms in the *j*th shell around r_j and $\eta(k)$ is the phase shift of the photoelectron caused by the potential of the absorbing atom.

This formula is in excellent agreement with the experimental $\chi(k)$ obtained on crystalline metals and SC. The dependence of $\chi(k)$ only on $\mathcal{N}_j$ and its character of a sum of damped waves suggests the possibility of obtaining a damped RDF $D(r)$ by its Fourier inversion

$$D(r) = -\left(\frac{2}{\pi}\right)^{\frac{1}{2}} \int_0^\infty \frac{\chi(k)}{kf(k)} \sin\left[2kr + 2\eta(k)\right] \mathrm{d}k$$

$$= \Sigma_j \left[\mathcal{N}_j \frac{\exp(-\gamma r_j)}{r_j^2 \sigma_j}\right] \exp\left[-\frac{2(r - r_j)^2}{\sigma_j^2}\right] + \Delta(r), \qquad (5.11)$$

where $\Delta(r)$ can be neglected. $D(r)$ has the form of a sum of Gaussian peaks whose areas are not proportional to $\mathcal{N}_j$ because of the damping factor γ. However, the values of γ and $\eta(k)$ can be determined by recording the EXAFS of the related crystal for which the RDF is well known and therefore the $\mathcal{N}_j$ of the first few coordination spheres can be evaluated also in the *a*-SC.

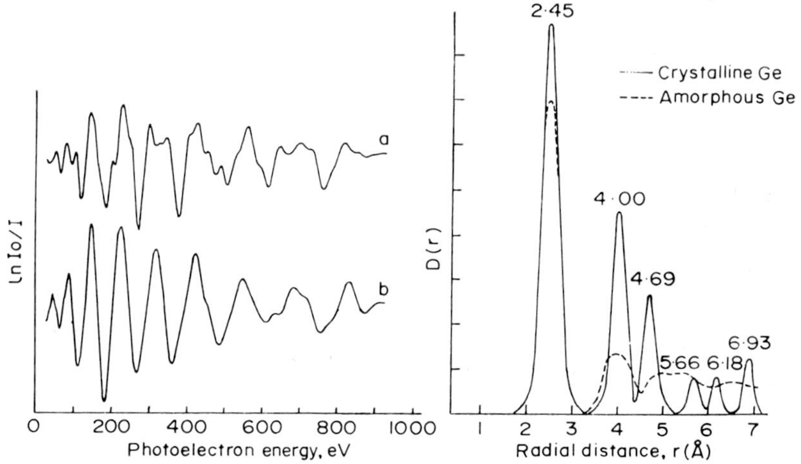

FIG. 5.5. Extended X-ray absorption fine structure in crystalline (a) and amorphous (b) Ge and resulting radial distribution functions *D*(*r*). (After Ref. 11.)

Results obtained on crystalline and *a*-Ge films[11] are shown in Fig. 5.5. Naturally the RDF of *a*-Ge (Fig. 5.12 and 5.19) and the *D*(*r*) of Fig. 5.5 differ by the areas of the successive peaks, but both display the characteristic narrowness of the first peak compared with the broadening of the second one and the near disappearance of the third one.

The EXAFS offers interesting prospects in the structural analysis of alloys because now each type of atom has its own absorption edge and a different surrounding. Indeed, the EXAFS of *a*-GeSe clearly shows[12] the difference in the surrounding of Ge and Se atoms, respectively. Each Se atom seems to bind covalently two Ge atoms at a distance of 2·35 Å, while each Ge atom is surrounded in a distorted tetrahedral configuration by four Se atoms. Both surroundings differ, however, from that in crystalline GeSe which has a distorted rock-salt structure.

5.2.4. INFRA-RED AND RAMAN SPECTRA

The IR and Raman spectra of crystalline solids reveal characteristic frequencies associated with different modes of atomic vibrations; the actual assignments being based on selection rules and calculated frequency values. These in turn involve force constants, atomic masses and geometrical configurations between nearest and higher order neighbours. If it is supposed that in *a*-SC's atoms are linked together by the same type of chemical bonds and in geometrical configurations similar to those found in different existing crystal polymorphs, a comparison between the IR and Raman spectra of the *a*-SC and its different related crystal polymorphs can yield information about the configurations present in the amorphous phase and their statistical weight.

TABLE 5.1. IR and Raman modes (cm^{-1}) in amorphous, trigonal and monoclinic selenium (fundamental modes). (After Ref. 13.)

a-Se		Trigonal Se		Monoclinic Se		Assignments	
IR	Raman	IR	Raman	IR	Raman	Se_8-ring	Se-chain
—	50	—	—	—	50	E_2	—
—	80	—	—	—	84	E_2	—
95	—	102	—	92–97	—	E_1	A_2
—	112	—	—	—	114	A_1	—
120	—	—	—	116–122	—	B_2	—
—	—	—	—	128	—	E_3	—
135	138	144	143	—	—	—	E
—	235	225	233	—	—	—	E
—	—	226	237	—	—	—	A_1
—	250	—	—	—	249	A_1	—
—	—	—	—	—	254	E_1	—
254	—	—	—	254	—	E_1	—

Detailed examples are given in Chapter 14 by Dr. Mort. Only one point will be stressed here by discussing the earlier evidence for the simultaneous presence of trigonal chains and 8-fold rings in the *a*-Se structure based on the identification of vibrational frequencies in the IR and Raman spectra of trigonal, monoclinic and *a*-Se[13] (Table 5.1). If one looks, however, at the expressions used in calculations of vibrational spectra of chain or ring molecules, one finds that only force constants related to 3rd and lower-order neighbours need to be taken into consideration in order to get an excellent fit with experiment. Conversely, identification of vibration frequencies and state of polarization in IR and Raman spectra yield information about the short-range order not farther than 3rd order neighbours. As Se chains and rings differ only by the fact that in the chains the dihedral angle has the same sign in all bonds, while it changes its signs alternatively in the octagonal ring, the presence of both types of frequencies in the spectrum of *a*-Se only proves that both types of successions of dihedral angles are present in this phase, but not that the chains are trigonal over more than three atoms or that the rings really close and are indeed 8-fold.

Recently an attempt has been made[14] to calculate the Raman spectrum of the *a*-Si random network model of Henderson and Herman[15] (see Section 5.5.6) using the band stretching and bending force constants of the normal Si crystal. The degree of agreement reached with experiment[16] is shown in Fig. 5.6 which also contains the calculated Raman spectrum of the high pressure polymorph Si^{III} whose structure includes distorted hexagonal and pentagonal rings like the network model.

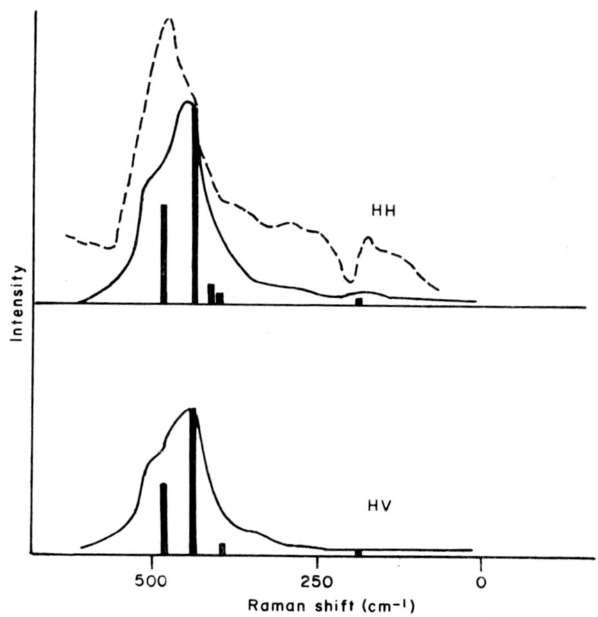

FIG. 5.6. Raman spectrum of *a*–Si (calculated: full curve; experiment: broken curve) and of Si^{III} (vertical bars). *HH*—incident and scattered light polarised horizontally. *HV*—incident light horizontally polarised and scattered light vertically polarised. (After Refs 14 and 16.)

5.2.5. NUCLEAR MAGNETIC RESONANCE (NMR)

The theoretical NMR frequency $\omega_0 = \gamma_N H_0$ can be shifted to a different value due to the interaction between the electron cloud of the atom and the nucleus of the considered atom. This is the chemical shift whose magnitude yields some information about the type of chemical bond existing between the considered atom and the surrounding ones. A much stronger shift appears due to the interaction between the nucleus and the unpaired conduction electrons near the Fermi level in metals and the charge carriers near the band edges in heavily doped SC's, respectively. This so-called Knight shift is a sensitive indicator of any change in the Fermi level connected with changes in the short-range order. The NMR line width is related to the site-to-site variation in the chemical shift interaction in the amorphous solid. In a liquid, however, due to the atomic motion, these variations can average to zero, narrowing the line, if significant changes in configuration happen

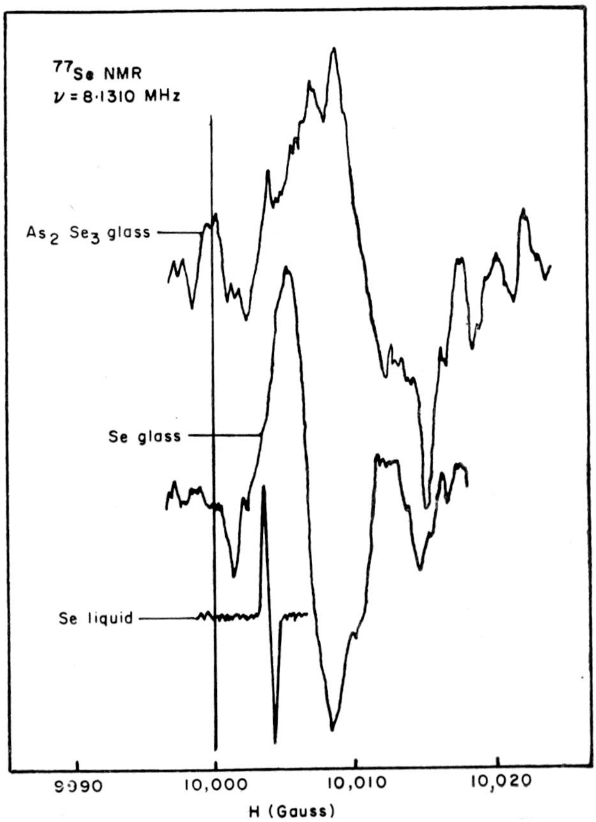

FIG. 5.7. ^{77}Se nuclear magnetic resonance in different disordered structures. (After Ref. 20.)

often enough, i.e. if the correlation time τ is short compared with the time associated with the reciprocal of the line width in the solid.

While NMR has been used extensively to investigate the structure of oxide glasses[17] and liquid metals,[18] *a*-SC's have been studied only recently.[19,20] Using NMR in ^{205}Tl and ^{77}Se, glassy and liquid $Tl_2SeAs_2Te_3$, As_2Se_3 and Se have been investigated. Fig. 5.7 and Table 5.2 show that the As–Se bond can be differentiated from the Se–Se bond by its chemical shift, that the As–Se bond remains unbroken in the melt for at least 10^{-4} seconds (there is no change in the resonance width due to melting), while the narrowing of the resonance in molten Se shows that the Se–Se bonds are broken every 10^{-6}—10^{-7} seconds. In $Tl_2SeAs_2Te_3$ the frequent breaking of Se–Tl bonds sets in only at ~420 K, i.e. markedly above the glass transition temperature $T_g = 359$ K.

TABLE 5.2. NMR ^{77}Se signal in disordered structures (after Ref. 20)

Material	Chemical shift (%)	Width of signal
H_2SeO_3 solution	0	0·1 *G*
Liquid and glassy As_2Se_3	0·111	6·2 *G*
Glassy Se	0·064	3·3 *G*
Liquid Se	0·042	0·1 *G*

5.2.6. THERMAL EFFECTS

Thermal effects can yield some information about the validity of a given structural model. Knowing, for instance, how much the local configurations differ statistically from those in an ideal crystal, e.g. by increases and fluctuations in bond lengths, by fluctuations in bond and dihedral angles and also knowing the increase in free energy due to these deformations, the heat of crystallization can be evaluated and compared with the experimental value.

Indeed, in the continuous network model of *a*–Ge proposed by Grigorovici and Mănăilă,[21] the static radial fluctuations in bond length were evaluated to be ±0·12 Å, the bond angle fluctuation to be ±8° and the number of eclipsed bonds to be 40 per cent of the total number of bonds. Elimination of the static fluctuations and of all eclipsed bonds by crystallization yields a decrease in free energy of 3·08 kcal/mole[22] which compares well with the experimental heat of crystallization of *a*–Ge of 2·75 kcal/mole.[23] Similar results were obtained[24] using Shevchik and Paul's random network model[25] as a starting point. In contrast, the *a*–Se model of Kaplow *et al.*,[26] obtained by distorting monoclinic α–Se locally just as much as needed in order to match the experimental RDF of *a*–Se, yields a static bond length fluctuation of $\pm 0{\cdot}20$ Å and a bond angle fluctuation of $\pm 11°$. If β–Se or trigonal Se were used as a starting point, the fluctuations were significantly higher. The lowest calculated heat of crystallization is 9·3 kcal/mole which compares unfavourably with experimental values of 1·1 to 1·4 kcal/mole.[27]

The value of the glass transition temperature T_g may also serve to decide between different structural models. For example, in Ge_xTe_{1-x} alloys T_g shows a sharp peak at $x = 0{\cdot}33$ ($GeTe_2$) as a function of composition.[28] This favours, for $x < 0{\cdot}33$, a short-range order of the SiO_2 type in which every Ge atom is surrounded by four Te atoms, while the Te atoms are 2-fold coordinated, i.e. the 4*d* model discussed in Section 5.2.2.

5.3. Methods of Investigating Defects in the Basic Structure of Amorphous Semiconductors

Macroscopically *a*–SC's appear to be isotropic, homogeneous and continuous. Microscopically they may be, however,

(i) anisotropic, i.e. contain oriented domains (crystallites) separated by high-angle grain boundaries or disordered regions;

(ii) chemically inhomogeneous if they are an alloy or compound;

(iii) structurally inhomogeneous, if they contain different types of ordered regions or voids;

(iv) discontinuous, if part of the covalent bonds are broken (dangling bonds).

5.3.1. Electron Microscopy

Normal electron microscopy provides information about the geometrical aspects of thin *a*–SC films like continuity, homogeneity, surface roughness, and about the evolution of crystallization processes.

While evaporated *a*–C films used as substrates in electron microscopy are known to be very smooth and homogeneous and so also are evaporated and sputtered *a*–Se and *a*–chalcogenide films used in imaging and switching devices, *a*–Ge and *a*–Si films are smooth too,[29,30] but sometimes far from homogeneous or even continuous. From this point of view *a*–Ge and *a*–Si films have been found to be either continuous and homogeneous[31] or to contain a certain number of spherical voids[32,39] or even cracks,[34] depending on the substrate, deposition conditions and annealing. In any case, however, polycrystalline and even epitaxial films have a much rougher surface and scatter therefore short wave-length light much more than *a*–films.

Dark-field microscopy should be able to show if an *a*–SC film consists of very small randomly oriented crystallites or not. By eliminating the central

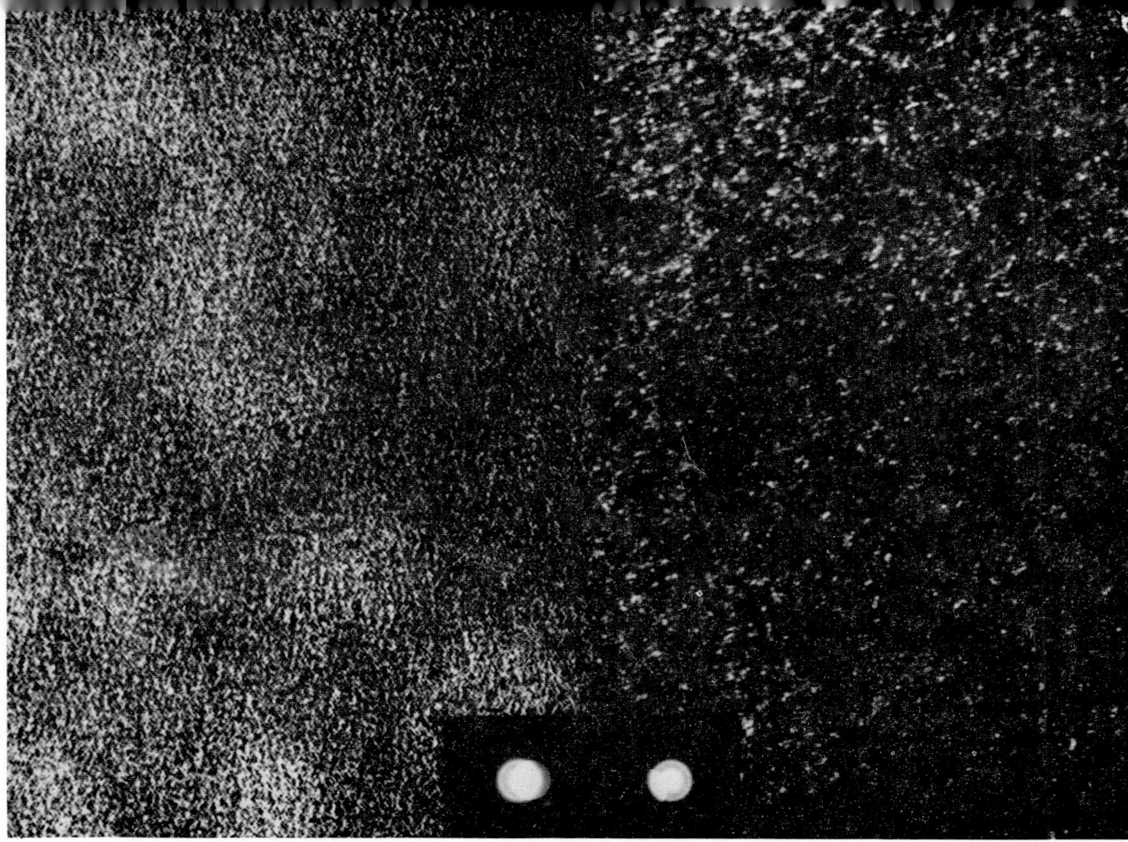

(a) | 100 Å | (b)

FIG. 5.8. Dark-field electron micrograph of *a*-Ge film (after Ref. 36). (a)—selected area of diffraction pattern includes central peak; the image shows lattice fringes. (b)—selected area of diffraction pattern includes only diffraction rings; the image shows bright spots only.

maximum of the intermediate diffraction pattern formed by the object and letting the image be formed only by the electrons which cross a certain part of the diffraction rings, the image will be in general dark and contain bright areas only in regions where, by chance, the diffracted electrons were directed towards the selected area of the diffraction pattern (Fig. 5.8b). It is a general feature of *a*–SC films that under the above conditions they do show images which contain discrete bright regions about 15 Å in diameter.[35]

If a part of the central peak and one of the diffraction rings (e.g. the 111 ring of *a*–Ge) are used to form the image, lattice fringes are observed in these small areas.[31,36] Their distance corresponds to the interplanar distance in the 111 direction of crystalline Ge, i.e. 3·3 Å (Fig. 5.8a).

All these observations were interpreted by Rudee[36] as showing that *a*–Ge and *a*–Si consist of crystallites about 15 Å in diameter. The whole problem will be discussed again in Section 5.4. Here it will be noted that the above conclusion is now under extensive discussion. Chaudhari *et al.*[33] and Shevchik[37] have shown independently that random networks of tetrahedrally coordinated atoms contain enough short-range order to extend coherence between electrons scattered by atoms as far apart as 15–20 Å, yielding peaks in the scattered intensity distribution sharp enough to generate bright regions of this size in the dark-field image. Howie *et al.*[38] have, however recently shown that the appearance of lattice fringes is incompatible with a random network model of *a*–Ge and that such models, if extended enough in space, indeed generate narrow diffraction peaks, but that the corresponding bright spots in the image should be only 3 Å in diameter.

Another argument against the microcrystal structural model stems from the observation[39] that by heating, for instance, an *a*–$Ge_{10}Te_{90}$ film above the glass transition temperature T_g, where it behaves like a supercooled liquid, the bright spots disappear, only to appear again in other places when cooling the film back below T_g, and this in spite of the fact that in other parts of the image crystallization can be seen to have already started. These facts show that the bright spots are not small crystallites increasing in size above the glass transition temperature.

5.3.2. LOCAL CHEMICAL ANALYSIS

Local chemical analysis can be performed by an electron microprobe technique. An electron beam is focussed on an area of about 1 μm^2 of the sample and the characteristic X-ray spectrum is recorded. It gives quantitative information about the local atomic composition at the surface of the sample irrespective of chemical binding.

The presence of chemical inhomogeneities in *a*–SC's may form an important argument in favour of the existence of local fluctuations in the position and width of the energy gap within the *a*–SC which are often held responsible for certain unusual optical and transport phenomena in SC glasses.

5.3.3. INFRA-RED DARK-FIELD MICROSCOPY

Infra-red dark-field microscopy reveals inhomogeneities which, at a scale of the order of microns, produce variations in the refractive index in a thick enough sample. These are mostly of chemical nature, but can also be due to density fluctuations. The principle of the method is similar to that already discussed in Section 5.3.1, with the exception that mostly the whole diffraction pattern is used, rather than just the central peak.[40] This has proved to be an excellent tool for the control of the homogeneity of chalcogenide glasses intended for optical applications.

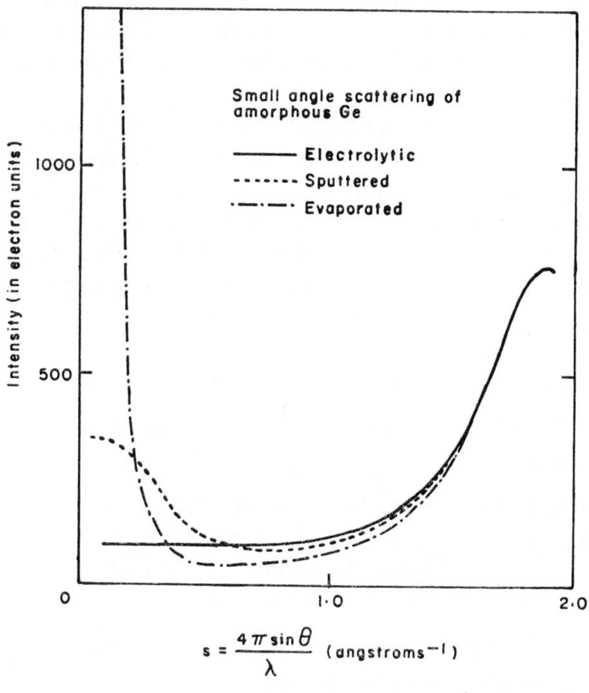

FIG. 5.9. Small-angle scattering of X-rays in amorphous Ge films. (After Ref. 42.)

5.3.4. SMALL-ANGLE SCATTERING

Small-angle scattering of electrons[41] and of X-rays[42] was detected in *a*–Si and *a*–Ge films (Fig. 5.9) and revealed the presence of inhomogeneities consisting of particles having an effective radius *R* and differing much in density from the surrounding material. The theory of small-angle scattering[1,43] yields the following expression for the angular dependence of the scattered X-ray intensity:

$$I(kR) = M\mathcal{N}_e^{\,2}\,\phi^2(kR)\,\{1 + P[5\sin(2\,kR)/(2\,kR) - 6\,\phi(2\,kR)]\}, \qquad (5.12)$$

with

$$k = 2\pi\theta/\lambda; \qquad \phi(kR) = \frac{3}{(kR)^3}\,[\sin(kR) - (kR)\cos(kR)].$$

Here M is the number of particles; $\mathcal{N}_e$ the number of electrons per particle; θ the scattering angle and $0 < P < 1$ a parameter characterizing the closeness of packing ($P = 0$ means a low packing density). If the angular resolution is good enough and the particles are of uniform size, the maxima in $\phi^2(kR)$ can be detected and used to determine R. If the resolution is insufficient, $\phi^2(x) \approx \exp(-x^2/5)$ and the slope of a $\log I$ versus θ^2 plot can be used to determine the mean particle size.

The conclusion drawn from this type of measurement on *a*–Si and *a*–Ge films was[44] that non-annealed vacuum deposited films contain an important number (a few volume per cent) of voids 10 to 30 Å in diameter which tend to disappear during annealing, thus increasing the density of the film. Obviously the initial density of the as-deposited films also depends on the deposition conditions like, for example, the rate of deposition as shown by X-ray glancing angle measurements on sputtered *a*–Ge films.[45]

5.3.5. ELECTRON SPIN RESONANCE (ESR)

ESR in SC's is linked to unpaired localized electrons or to free carriers. An ESR line is characterized by:

(i) the Landé factor g which is determined by the contributions of the magnetic moments of the spin and that of the orbit to the total magnetic moment of the electron and therefore allows a calculation of the degree of localization of the resonating unpaired electrons;

(ii) the absorption strength given by the area below the line which permits the number of resonating electrons to be calculated;

(iii) the line form and width which depend on the type and strength of the interactions leading to relaxation;

(iv) the fine and hyperfine structure of the line which is linked with the interaction of the resonating electrons with the crystal field and the nuclear magnetic moment, respectively.

ESR measurements on *a*–Si films[46,47] revealed the existence of a line whose *g*-factor, line form and width matched exactly that of the surface states in Si crystals (Table 5.3). The density of unpaired electrons reached 10^{20} cm^{-3} in as-deposited films and diminished by one to two orders of magnitude with annealing, i.e. in the same ratio as the conductivity decreased.[48] This density is assumed to represent the number of dangling bonds per cm^3 either within the high-angle grain boundaries between the crystallites, or within the continuous network, or—even more probable because of the similarity with the ESR response of surface states—with dangling bonds at the inner surfaces of voids. Similar densities and a similar dependence on annealing were found in the acceptor density in *a*–Ge deduced from barrier capacity measurements in *a*–Ge/*n*–Ge heterojunctions.[49] ESR measurements on different *a*–SC's mostly reveal the presence of unpaired electrons of relatively low densities ($<10^{16}$ cm^{-3}) in glasses[50] and relatively high densities ($<10^{20}$ cm^{-3}) in non-annealed vacuum-deposited elemental *a*–SC films (Table 5.3). Paramagnetic susceptibility measurements[51,52] lead to similar conclusions.

TABLE 5.3. ESR in amorphous semiconductors (after Ref. 47)

a-semicond.	*g*	Width (Gauss)	Surface states in crystals of	*g*	Width (Gauss)
a-Si	2·0055	4·7	Si	2·0055	7–8
a-Ge	2·021	39	Ge	2·023	50
a-SiC	2·003	6	SiC	2·0027	5·5

5.4. THE SHORT-RANGE ORDER IN AMORPHOUS ELEMENTS AND ALLOYS

In the context of this discussion the short-range order in *a*–SC's will be characterized chiefly by the nearest neighbour configuration, though sometimes next-nearest neighbours will also be taken into consideration. The first coordination number $\mathcal{N}_1$ will be used as a criterion of classification. The importance of this criterion is emphasized by the validity of Joffe and Regel's rule[53] which states that a molten or amorphous material maintains the SC properties it displays when in a crystalline form if it conserves in the disordered state the short-range order it had in the crystal. In this context the short-range order is defined by the first coordination number $\mathcal{N}_1$ and the first interatomic distance r_1.

5.4.1. TWO-FOLD COORDINATION (CHAIN AND RING STRUCTURES)

a–Se, both quenched from the melt and vacuum-deposited, shows a very narrow first peak in its RDF (Fig. 5.10) at an interatomic distance intermediate between that of trigonal Se (helicoidal chain structure) and mono-

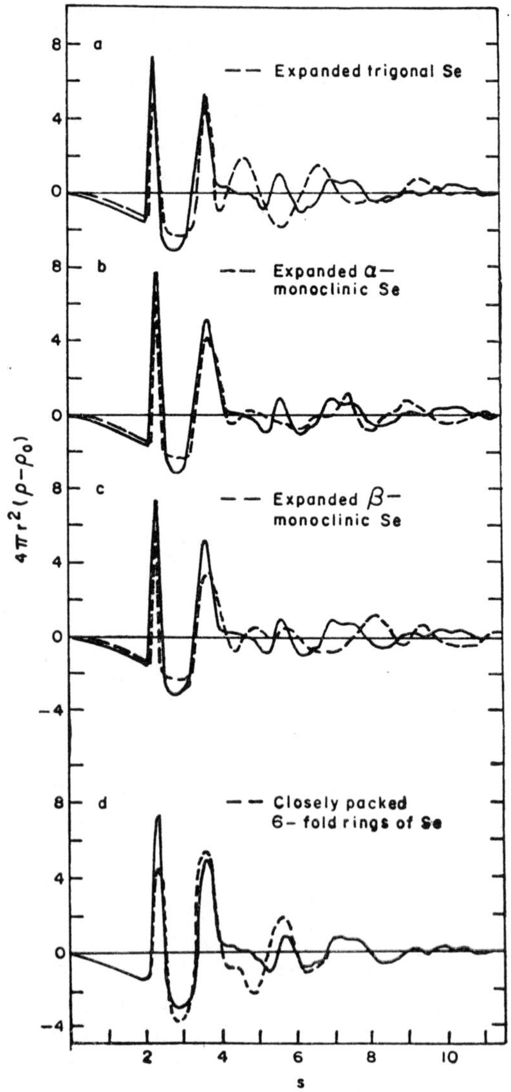

FIG. 5.10. Differential RDF of amorphous Se film (full curve) compared with RDFs of trigonal (a), α-monoclinic (b), β-monoclinic (c) and hypothetical packed 6-fold ring Se crystals. (After Refs. 26 and 102.)

clinic Se (8-fold ring structure). The width of the first peak shows that fluctuations in r_1 are smaller than in crystalline monoclinic Se at the same temperature. $\mathcal{N}_1$ is exactly 2, as in all three crystalline polymorphs of Se. The next peaks are broadened, thus showing that fluctuations in bond angles prevail over those in bond lengths. They also do not fit those of any known crystalline polymorph of Se. Nevertheless it is mostly stated[26,54–56] that *a*–Se consists of a mixture of trigonal chains and 8-fold rings. Vacuum-deposited *a*–Se films and cast glassy Se differ in the details of their RDF's. Films deposited on substrates held at low temperature display a simpler RDF, but annealing brings it near the RDF of cast *a*–Se.

5.4.2. THREE-FOLD COORDINATION (LAYER STRUCTURES)

This coordination is found in *a*–As and *a*–Sb[54,56–59] with r_1 0·3 to 0·7 per cent smaller than in the corresponding crystals (Fig. 5.11). The higher-order

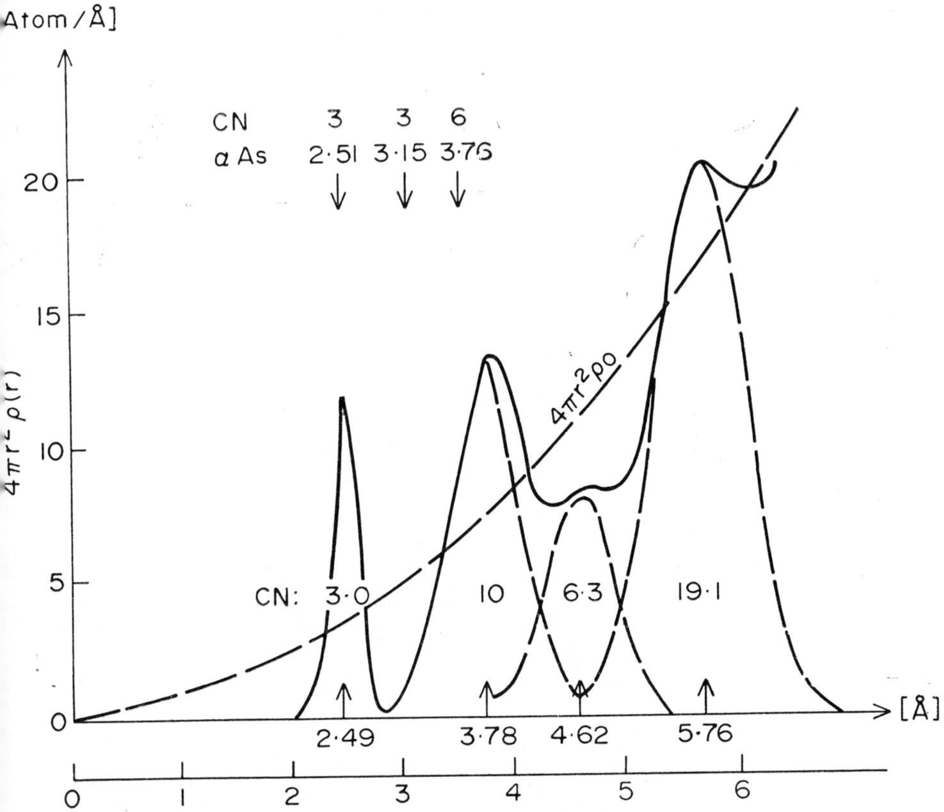

FIG. 5.11. RDF of amorphous As. (After Ref. 64.) CN—coordination numbers in α-As crystal; arrows down: peaks in α-As crystal; arrows up: peaks in *a*-As.

peaks of the RDF do not fit the RDF of any known crystalline polymorph of these elements.

5.4.3 FOUR-FOLD COORDINATION (TETRAHEDRAL STRUCTURES)

This coordination is found in *a*–Si and *a*–Ge, in *a*–$A^{IV}B^{IV}$, *a*–$A^{III}B^{V}$, *a*–$A^{II}B^{IV}C_2{}^{V}$ compounds and the $CdGe_xAs_2$ ($0 < x < 1{\cdot}2$) system. In the elements Si[41,60,61] and Ge[11,25,59,60,62,63] r_1 is only slightly increased (by 0·01—0·04 Å) over the crystal values, while the width of the peak after deconvolution of the end-of-series and thermal broadenings indicates a static standard deviation in r_1 between 0·06 and 0·12 Å. In general the second peak of the RDF (Fig. 5.12) which corresponds to the edge of the elementary tetrahedron, i.e. to the distance between atoms lying at its corners, is relatively strongly broadened thus showing that, like in *a*–Se, angular distortions

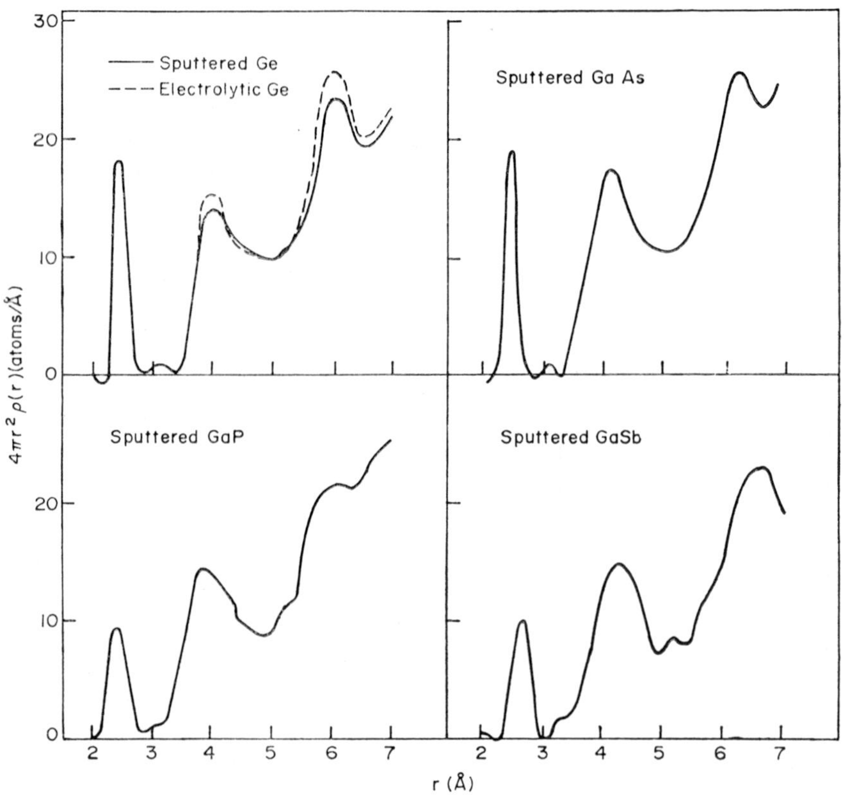

FIG. 5.12. RDFs of amorphous Ge, GaAs, GaP and GaSb films. (After Ref. 68.)

largely exceed radial distortions in the elementary tetrahedra of 4-fold co-ordinated *a*–SC's. The third peak, which would be as high as the second one if *a*–Si and *a*–Ge consisted of small normal crystallites, (see Fig. 5.18a) is reduced to a shoulder or a strongly shifted and diminished peak. All these features are present indiscriminately in evaporated, sputtered, electrolytically deposited films and in heavily ion implanted and amorphized crystals[39, 65,66].

It may be mentioned here that liquid Ge, Si and InSb present a completely different diffraction pattern and RDF from their amorphous counterparts. In liquid Ge, for instance[67] (Fig. 5.22), $\mathcal{N}_1 = 7{\cdot}1$; $r_1 = 2{\cdot}8$ Å and the short-range order extends to much shorter distances than in solid *a*–Ge. Correspondingly, liquid Ge has lost its semiconducting properties and behaves like a molten metal.

In $A^{III} B^{V}$ and $A^{II} B^{IV} C_2^{V}$ compounds electron transfer from the group V elements to the group II and III atoms must take place in order to provide the latter with the necessary four electrons involved in the four covalent bonds. This gives in GaAs, for example, the Ga atoms a negative charge, the As atoms a positive one and the Ga–As bond a partially ionic character. In spite of that, the RDF of *a*–GaAs films is identical in all its details with that of *a*–Ge, while *a*–GaP and *a*–GaSb films show the same general features[68] but all the peaks of their RDF are broadened compared with those in *a*–GaAs (Fig. 5.12). This suggests that in *a*–$A^{III} B^{V}$ compounds not only III–V but also III–III and V–V "wrong" pairs are contributing to the first peak of the RDF in spite of both atoms carrying electric charges of the same sign. But while Ga^- and As^- ions have practically the same covalent radius, namely that of an Ge atom, P^- and Sb^- differ in size from Ga^- and therefore all peaks of the RDF, including the first one, are broadened. Raman spectra confirm the presence of "wrong" pairs in *a*–$A^{III} B^{V}$ compounds.[69] Obviously "wrong" bonds must be admitted to be weaker than normal ones, without being longer, however.

a–$A^{II} B_x^{IV} C_2^{V}$ alloys with $0 < x < 1{\cdot}2$ can be obtained not only as thin vacuum-deposited films, but also by quenching their melt if $r_A \geqslant 1{\cdot}10\, r_B$, e.g. in $CdGe_xAs_2$.[51,70–72] Both $CdAs_2$ crystals[73] and $A^{II} B^{IV} C_2^{V}$ crystals[74,75] have tetrahedrally coordinated lattices in which the local tetrahedron is more or less radially and angularly distorted. Obviously one electron from each of two As atoms has to be transferred to one Cd atom to provide the latter with the necessary four valence electrons and achieve a prevailing sp^3-type bonding. Adding Ge atoms does not change anything from this point of view, only that in $CdAs_2$ each Cd atom is surrounded by four As atoms and each As atom by two Cd and two As atoms, while in $CdGeAs_2$ each As atom is surrounded by two Cd and two Ge atoms, and each Cd or Ge atom by four As atoms. This time, however, electrostatic repulsion between Cd^{--} ions seems to impede the formation of wrong Cd–Cd pairs. Indeed, an analysis of the shape of the first peak of the RDF in *a*–$CdGeP_2$ showed[75] that in the *a*-phase each *P* atom is also surrounded by two Cd and two Ge atoms situated at different distances, as in the crystal. Therefore it must be supposed that both the high tetragonal distortion in the $CdGeAs_2$ and $CdGeP_2$ crystals and the relatively easy vitrification of their melt have the same origin, namely the difference in the length and in the ionic character of the Cd–As and Ge–As bonds. In alloys with increasing quantities of Ge or Si, these atoms are supposed[73] to occupy, for $x < 0{\cdot}2$, the empty space of the $CdAs_2$ lattice and for $x > 0{\cdot}2$, the empty centres of the As tetrahedra.

5.4.4. SIX-FOLD COORDINATION (ICOSAHEDRAL STRUCTURES)

The RDF of chemically deposited *a*–B films[76] has its first two peaks of 6·6 and 17·5 atoms area lying at 1·8 and 3·1 Å, respectively. This compares favourably with the first two coordination spheres of α-rhombohedral boron, one of the three polymorphs of this element ($\mathcal{N}_1 = 6$; $r_1 = 1{\cdot}71$ to $2{\cdot}03$ Å; $\mathcal{N}_2 = 17$; $r_2 = 2{\cdot}9$ to $3{\cdot}3$ Å). In all three polymorphs the essential structural units are nearly regular icosahedra containing twelve 6-fold coordinated B atoms. Five of the bonds lie within the icosahedron, the sixth helps to link the icosahedra directly or indirectly, via supplementary 4-fold coordinated B atoms, to each other. The resulting elementary cells are huge; that of tetragonal boron, for instance, contains as many as 50 atoms because the icosahedron itself, due to its 5-fold symmetry axes, is incompatible with the translational symmetry of a crystal lattice. On the other hand the electrical, magnetic and optical properties of crystalline boron are very similar to those of other elemental *a*–SC's. It was therefore suggested[77] that local order of this sort is sufficient to confer in a crystal the characteristics of an amorphous material in spite of the long-range order presented by the elementary cells as a whole which contain a certain number of such odd building units.

5.4.5. MIXED 3–4 COORDINATION

This mixture is found in *a*–carbon films and in vitreous C, where $\mathcal{N}_1$ lies between 3 and 4 and the first peak of the RDF not only lies somewhere between the graphite and the diamond r_1, but also is so broad that it must be supposed to consist of two merged peaks[78] (Fig. 5.2a). This is strongly supported by a careful analysis of the electron diffraction pattern of *a*–C films (Fig. 5.13) which can be matched by an intensity function calculated with the aid of the Debye formula eqn (5.1) only if two first- and two second-neighbour distances equalling those in diamond and graphite ($r_{1d} = 1{\cdot}54$; $r_{2d} = 2{\cdot}53$; $r_{1g} = 1{\cdot}41$; $r_{2g} = 2{\cdot}44$ Å) are used in the calculation with nearly equal statistical weights.[79] If the first peak of the RDF is therefore split into two narrower ones, their width exceeds $1/s_{max}$ only slightly but significantly. The corresponding fluctuations of about 0·11 Å are still higher than the thermal fluctuations in the corresponding crystals and must therefore be attributed to static fluctuations.

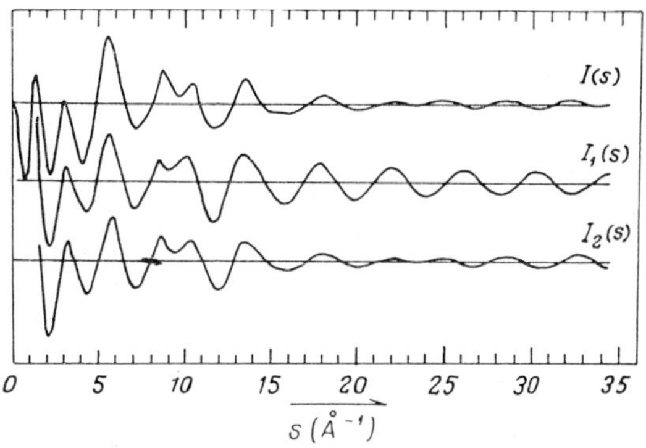

FIG. 5.13. Interference function of amorphous carbon films. (After Ref. 79.) $I(s)$, experimental curve, $I_1(s)$, calculated curve based on only two interatomic distances ($r_1 = 1{\cdot}50$ and $r_2 = 2{\cdot}54$Å); $I_2(s)$, calculated curve based on four interatomic distances ($r_{1d} = 1{\cdot}54$; $r_{1g} = 1{\cdot}41$; $r_{2d} = 2{\cdot}53$ and $r_{2g} = 2{\cdot}44$ Å).

In accordance with these structural characteristics, as-deposited *a*–C films display semiconducting properties. Annealing, without inducing crystallization, brings the short-range order nearer that of crystalline graphite and, correspondingly, most but not all physical properties approach those of polycrystalline graphite. Similarly, in vitreous C annealing reduces the statistical weight of the diamond-type bonds in the first coordination sphere.[78]

a–InSe is probably the only case in which, due to the very different lengths of the In–In (2·60 Å) and In–Se (3·15 Å) bonds, the first peak of the RDF is clearly split into two and partial pair functions can be obtained.[80] The resulting short-range order (every Se atom surrounded by 3·46 In atoms; every In atom by 3·46 Se atoms and 0·95 In atoms) compares fairly well with the InSe crystal, where the corresponding numbers are 3 and 3+1.

In *a*–Si_3N_4 films the Si:4N; N: 3Si configuration of the Si_3N_4 crystal is also maintained.[81]

5.4.6. MIXED 2–3–4 COORDINATION

These more complex configurations appear in *a*–$A_2^{III}B_3^{VI}$ films and splat-cooled alloys as well as in *a*–$A_2^{V}B_3^{VI}$ films and glasses with addition of different elements, especially Ge, Si and I. Of special interest are Ge-

and Si-doped AsTe alloys used in memory and threshold switches. If the As content is zero or low enough, the same materials can be considered to be $A^{IV} B^{VI}$ alloys with small C^{V} additions. Some examples will best illustrate the similarities and dissimilarities of the short-range order with that in the related crystals.

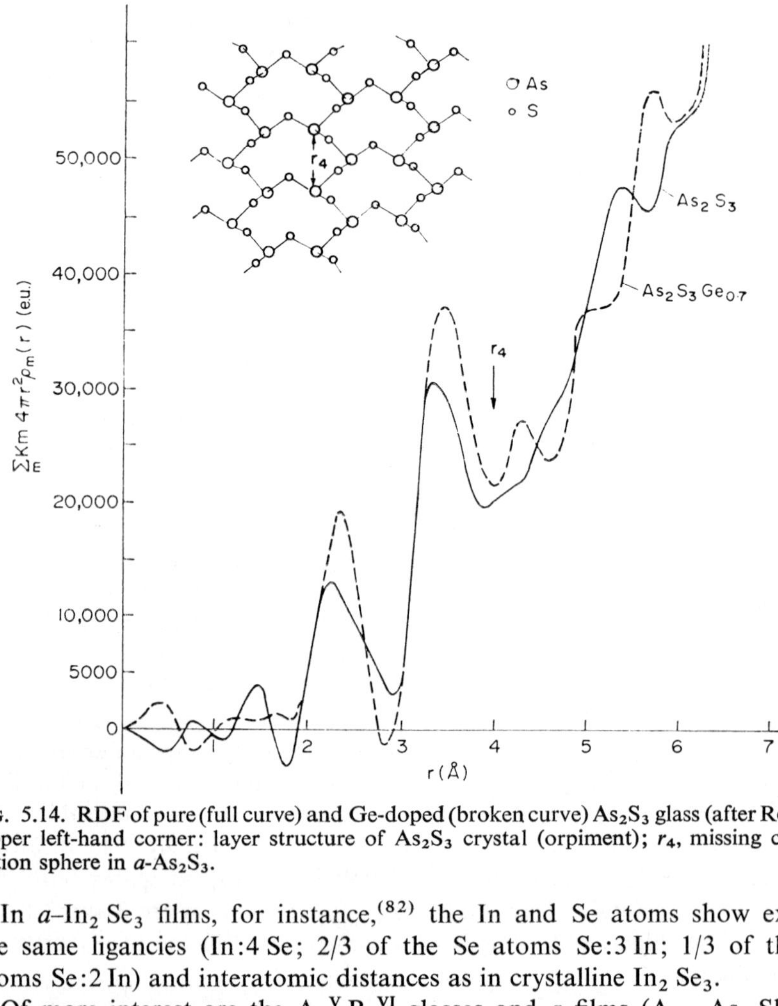

FIG. 5.14. RDF of pure (full curve) and Ge-doped (broken curve) As_2S_3 glass (after Ref. 86). Upper left-hand corner: layer structure of As_2S_3 crystal (orpiment); r_4, missing coordination sphere in a-As_2S_3.

In a–In_2Se_3 films, for instance,[82] the In and Se atoms show exactly the same ligancies (In:4 Se; 2/3 of the Se atoms Se:3 In; 1/3 of the Se atoms Se:2 In) and interatomic distances as in crystalline In_2Se_3.

Of more interest are the $A_2^{V} B_3^{VI}$ glasses and a–films (A = As, Sb, Bi; B = S, Se, Te).[83,84] We shall discuss here in more detail a–As_2S_3 and a–As_2Se_3 which were shown a long time ago[85] to have a short-range order simliar to that of the analogous crystal—orpiment. Orpiment has a layer structure (Fig. 5.14), each layer consisting of merged puckered 12-fold rings. Within the ring each S atom is linked to two As atoms, while each As atom is linked to two S atoms within one ring and to a third S atom in the neighbouring ring. Again, however, in the RDF of a–As_2S_3 one of the peaks of the orpiment RDF is absent[86] (Fig. 5.14), namely that corresponding to the distance r_4 between second-nearest atoms of the same type within the ring, as if the layers were heavily distorted. On the other hand the lowest angle diffraction peak (Fig. 5.15) corresponding to a slightly increased inter-layer distance (5·1 Å in a–As_2S_3 compared with 4·8 Å in orpiment) seems to show, in contrast, that the parallelism of the layers is well conserved in the glass.

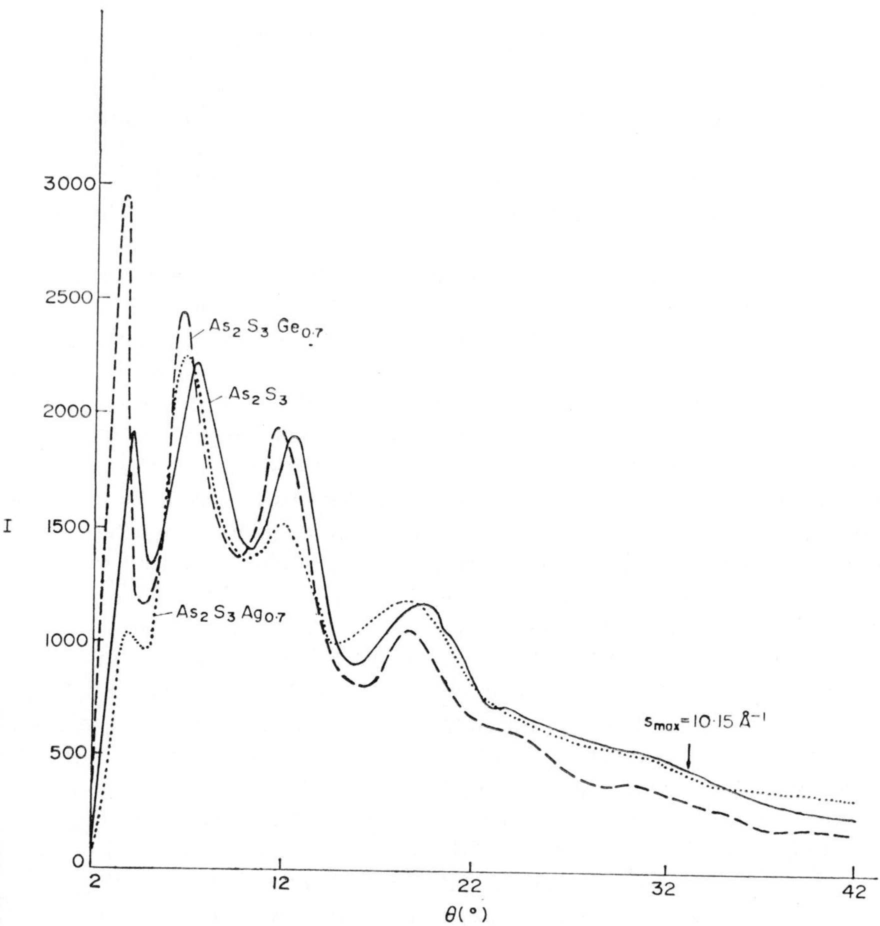

FIG. 5.15. Scattered X-ray intensity in pure (full curve), Ge-doped (broken curve) and Ag-doped (dotted curve) As_2S_3 glass. (After Ref. 86.)

This contradiction becomes even more obvious if one introduces different impurities into the glass. While the addition of Ag does not modify the position of the above-mentioned peak, but rapidly destroys both the inter- and intra-layer structure and soon leads to partial crystallization,[87] addition of Ge eases vitrification, enhances the inter-layer diffraction peak, at least for not too high concentrations, and shifts it to smaller angles, i.e. pushes the hypothetical layers apart, conserving, however, the periodicity. Thermal properties of Ge–doped As_2Se_3 glasses[88,89] point to a strengthening of the chemical bonds between the layers, also suggesting the intercalation of Ge atoms between the layers. Excess S and As atoms have been shown[90] to form supplementary "wrong" S–S and As–As bonds, thus increasing the size and reducing or increasing the connectivity of the rings.

Much less can be concluded about Ge–doped As_2Te_3 glasses. The first two peaks of the RDF in a $Ge_{11}As_{45}Te_{40}$ glass, for example,[84] lie at 2·50 and 4·02 Å, respectively. Obviously the first peak contains all possible inter-atomic distances ranging from Ge–Ge = 2·43 Å to Te–Te = 2·86 Å, so that little can be inferred about the short-range order. Alloys of this class form the standard material of threshold switches and are therefore difficult to crystallize.

Of interest are also $A^{IV}B^{VI}$ compounds and alloys. In technologically important *a*-SiO both Si and O atoms seem to be 2-fold coordinated, but local segregation into Si and SiO_2 cannot be ruled out completely.[91] *a*–Ge_xTe_{1-x} and *a*–Ge_xSe_{1-x} alloys in the form of glasses or films, eventually containing small quantities of elements like P, As, S, Si or I in order to ease vitrification, have been investigated quite thoroughly.[84,92–98] The most important results have been discussed in Section 5.2.2. The essential feature is that the atoms seem to be coordinated as predicted by covalent bonding (Ge:4; Te:2) and that this coordination differs completely from that in crystalline GeTe and GeSe which have a slightly distorted rock-salt structure, every Ge and Te atom being linked to six Te and Ge atoms, respectively. It may also be mentioned that the $GeTe_6$ alloy in which some Te is substituted by As ($\sim$5%) is the easiest to vitrify and that this composition corresponds to a eutectic. Alloys of this type crystallize relatively easily by heating or under strong illumination, separating into two phases, namely Te and GeTe. They are therefore the standard material for memory switches and imaging. It is generally believed that in this alloy trigonal Te chains about three atoms long are linked together at their ends by Ge atoms.

Te chains and interposed Au and Ag atoms are supposed to form the best description of the structure of splat-cooled *a*–$(Au, Ag)_{30}Te_{70}$ alloys.[99]

Other *a*–alloys whose structure has been analysed in terms of the analogous crystal structures are *a*–$MgBi_x$ and *a*–$MgSb_x$ films for which the RDF agrees with the hypothesis that they consist of a mixture of small crystallites of Mg_3X_2 and of X, X standing for Mg, Bi or Sb, respectively, depending on which of them is in excess.[100]

5.4.7. MOLECULAR STRUCTURES

A curious structural effect has been detected in a–Te_xTl_2 films[101] which display sharp minima in their electrical conductivity when x is an integer comprised between 1 and 7 (Fig. 5.16). This hints to the possibility of a–SC's presenting a molecular type short-range order, an idea which has not been given much thought hitherto.

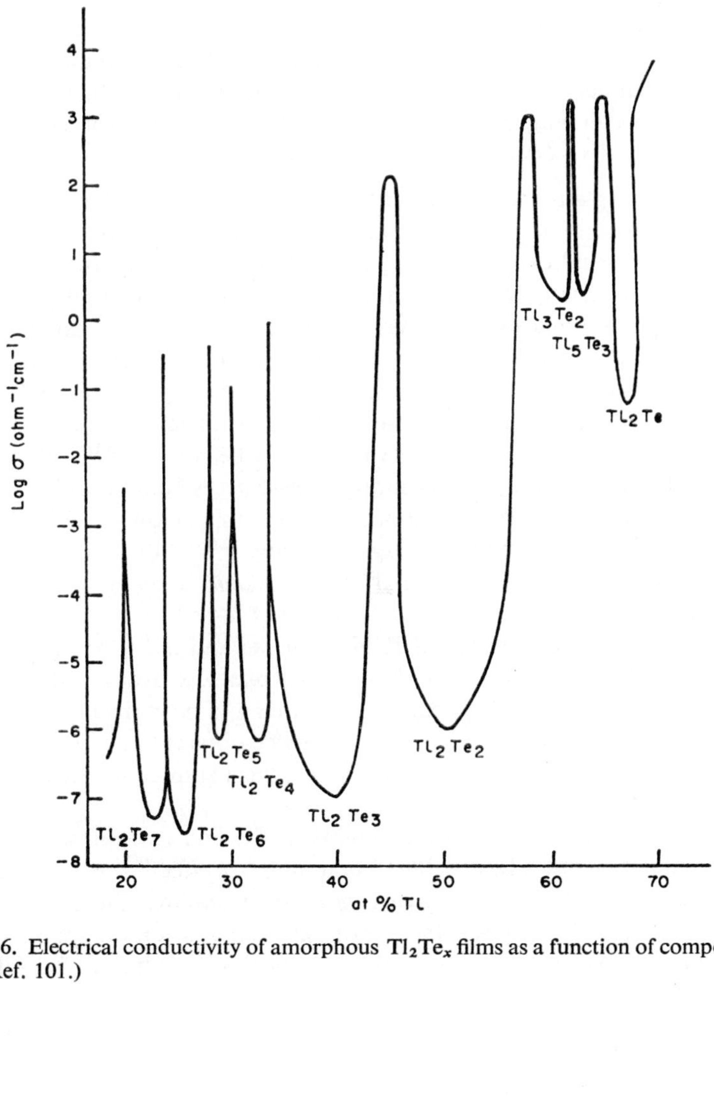

FIG. 5.16. Electrical conductivity of amorphous Tl_2Te_x films as a function of composition. (After Ref. 101.)

5.4.8. CONCLUSIONS

(i) *a*–SC's mostly display a nearest-neighbour short-range order based on predominantly covalent bonds like their analogous crystals. Sometimes covalent bonding prevails in an *a*–SC even if the corresponding crystal has a structure based on other types of bonds.

(ii) Small radial and strong bond-angle distortions are found in the local arrangements.

(iii) The higher-order neighbours are mostly in a configuration which differs from that found in any existing crystal.

There are two ways in which these facts can be explained:

(a) *The microcrystal or cluster approach* in which the *a*-SC is supposed to consist of randomly oriented crystallites or atom clusters linked together in a well defined manner and separated from each other by an ill-defined disordered region. The broadening of the diffraction rings is thus interpreted as a size effect. The necessary crystal sizes are of the order of 10 to 15Å.

In order to explain the existing differences between the I(*s*) and RDF of *a*–SC's and those of the related crystals, the crystallites must be supposed to belong to unusual or non-existing polymorphs, to have particular shapes, to be heavily distorted, or to contain a very high number of multi-vacancies.

For example, no mixture whatever of 8-fold rings and trigonal chains can reproduce the RDF of vacuum-deposited *a*–Se films, because of the peaks at 5·7 and 7·2 Å which are absent in all crystalline forms of Se (see Fig. 5.10 a–c). However, an appropriate periodic arrangement of 6-fold ring molecules, like those present in Se vapour, has been shown[102] to give the required RDF (Fig. 5.10d). Other more or less successful molecular arrangements consist of mixtures of trigonal chains with not only 8-fold but also bigger rings[54,56] or of coplanar zig-zag chains.[103] Still another solution consists in distorting locally trigonal or monoclinic Se in such a way as to obtain a good fit with the experimental RDF of *a*–Se.[7]

To reproduce the experimental RDF of *a*–As and *a*–Sb one must either shape the crystallites into randomly oriented platelets, rotate the flat tetrahedra forming one atom layer against each other and push the layers apart[58] or shift the layers parallel to each other in three different configurations and add a certain quantity of hypothetical crystallites similar to those of black phosphorus.[54]

The abnormal ratio between the heights of the first two diffraction peaks in *a*–Ge and *a*–Si cannot be attributed to a pure size effect[41] (see Fig. 5.17e). It can again be explained by giving normal crystallites the form of thin platelets.[60] But to have these platelets oriented at random, the space between them, which considerably exceeds the occupied space, has to be filled by

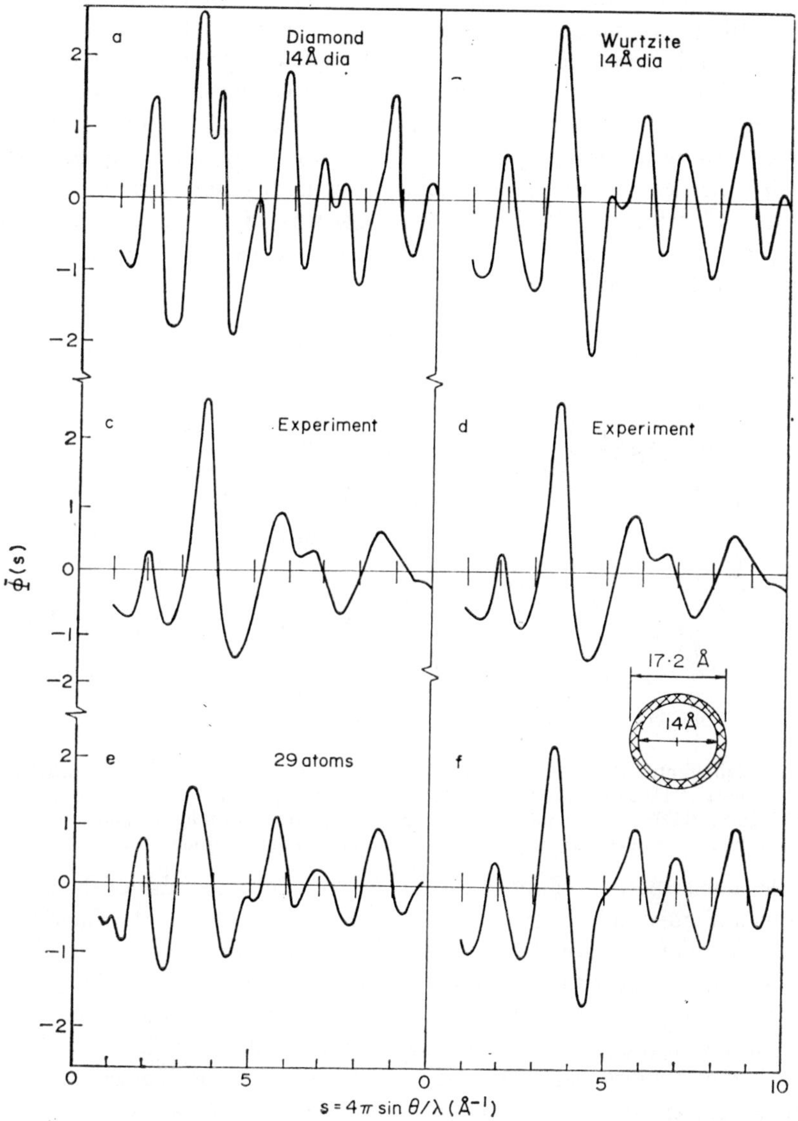

FIG. 5.17. Interference function of amorphous Si films. *c* and *d*, experiment. (After Ref. 41); *a* and *e*, calculated for diamond-type crystallites, 14 Å in diameter (after Ref. 31) and containing only 29 atoms (after Ref. 41); *b* and *f*, calculated for wurtzite-type crystallites, 14 Å in diameter, surrounded by randomly and preferentially oriented crystallites. (After Ref. 38.)

a very disordered matrix of unknown structure. Better results were obtained by mixing normal diamond-type crystallites with so-called amorphons, i.e. tetrahedrally interconnected atom clusters(61,62) in which advantage is taken of the small misfit between the tetrahedral angle of 109° 28′ and the pentagonal angle of 108°. Indeed, instead of linking neighbouring tetrahedra exclusively in the staggered configuration which gives rise to the diamond lattice consisting of merged puckered 6-fold rings only, they may be linked to each other exclusively in the eclipsed configuration, giving rise to a dodecahedral arrangement (Fig. 5.18d) consisting of merged planar 5-fold rings (Fig. 5.18c). Such arrangements arise in fact by multiple twinning in polycrystalline Ge films.(104)

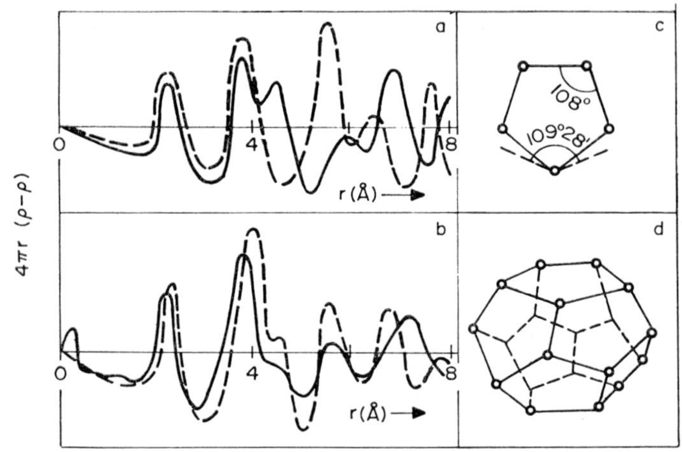

FIG. 5.18. Differential RDFs of *a*-Si (after Ref. 61) and the amorphon structure (after Ref. 62). *a*, calculated RDFs of Si crystal (full curve) and of amorphonic structure (broken curve) *b*, experimental RDF of *a*-Si (full curve) and of 3/2 mixture of Si microcrystals and amorphonic clusters (broken curve); *c* and *d*, 5-fold ring and pentagonal dodecahedron.

While such structures cannot be extended to infinity because of the above-mentioned angular misfit, they may easily extend over 10–15 Å. Their RDF differs from that of normal crystallites by the absence of those very peaks which are much reduced in height in the RDF of *a*–Ge and *a*–Si (Fig. 5.18a). Therefore an appropriate mixture of normal crystallites and amorphons fits remarkably well the experimental RDF of *a*–Ge and *a*–Si (Fig. 5.18b).

Another possibility has been indicated by Rudee and Howie, the most recent advocates of the microcrystal approach. They first showed(31) that both the I(s) and the RDF of wurtzite type crystals gives a better, if not perfect fit with those of *a*–Si and *a*–Ge (Fig. 5.17b). This is easy to understand if one considers that in the wurtzite crystal one quarter of the bonds are eclipsed. Later they improved the fit (Fig. 5.17f) by assuming a certain preferential relative orientation of neighbouring wurtzite crystallites.(38)

In spite of these partial successes, the microcrystal approach is unsatisfactory because of the following considerations:[105]

(i) To account for the low number of dangling bonds in *a*–SC's (less than one in a thousand bonds), the disordered matrix between the crystallites must be supposed to consist of mainly interconnected atoms.[36] If the relative orientation of neighbouring crystallites is more or less aleatory, this matrix must have the character of a random network. On the other hand, in a cubic crystallite containing 1000 atoms, which is more than in a crystallite of 15 Å edge, more than half the atoms lie at the surface, so that most atoms would anyhow be included in a random network.

(ii) There is no thermodynamic explanation why the crystallites should be heavily distorted, belong to non-equilibrium phases or to never encountered lattices or to noncrystalline spatial arrangements.

(iii) Crystallization processes in *a*–SC's have the characteristics of reconstructive phase transformations rather than that of a recrystallization process.[21,32]

(b) *The continuous network approach*[106] in which the *a*–SC is considered to consist of a network of atoms linked together by the same types of chemical bonds as in the correspondent crystals, but lacking long-range order. Such a network has no discontinuities, no high-angle boundaries between ordered regions and therefore eliminates from the very beginning the first objection raised above against the microcrystal approach. A continuous network model in which all covalent bonds are satisfied is called *ideal* and with such models we shall deal in the following section.

5.5. Modelling of the Ideal Structure of Amorphous Elements and Binary Alloys

The building of continuous ideal network structural models (CNM) of *a*–SC's raises a series of problems which must be discussed first in general terms.

5.5.1. The Destruction of Long-Range Order

In order to destroy the long-range order in *a*–SC's and, at the same time, conserve the short-range order, one has to mix together intimately two or more different nearest- or higher-order neighbour configurations compatible with the geometry of the covalent bonds found in some crystalline polymorphs. Angular misfits will always appear and have to be compensated by letting the interatomic distances and the bond and dihedral angles fluctuate to some extent about some mean values.[21,106] If the distribution functions of these magnitudes are predetermined before building the model (and especially if the dihedral angle distribution is a uniform one) such models are called random network models (RNM).[107] The intimate mixture of different configurations generates local groupings like 5- or 7-fold rings which do not allow for the conservation of long-range order, though the short-range order essentially retains its original symmetry.

5.5.2. SIMULATION

Modelling is supposed to simulate as accurately as possible:

(i) the *geometrical* characteristics of the appropriate covalent bonds, like their symmetry and directional character and the values of the covalent radii;

(ii) the *energetic* characteristics of the covalent bonds, like the force constants for stretching, bending, rotation about the bonds, etc;

(iii) the *formation process* of *a*–SC's like deposition procedure and rate, quenching of a melt, implantation of ions;

(iv) *thermodynamic* aspects like lowest free energy configuration, annealing, crystallization.

5.5.3. MODEL BUILDING UNITS AND PROCEDURES

Models can be built *by hand* using *building units* which simulate the chemical bonds and a *building procedure* which simulates the formation process of the *a*–SC.

The traditional form of building unit is the usual ball-and-spoke unit of the organic chemist. The spokes simulate well the directivity of the covalent bond and its opposition against bending. It allows for increases in bond length and for rotation about the bond, but does not oppose them; it does not allow for compression. It allows for an easy location of individual units, for an evaluation of the density and homogeneity of the model as well as of fluctuations in bond lengths, bond angles and dihedral angles. Springs replacing the spokes can simulate better the directional and energetic characteristics of the covalent bonds, but have begun to be used only very recently. Rigid Voronoi polyhedra have also been used in modelling. They are limited by plane faces which lie perpendicularly at mid-distance between neighbouring atoms in the crystals. The biggest faces correspond to the strongest bonds and have to be brought into contact or nearly so in a CNM. Simulation of the covalent bond is in general worse than with ball-and-spoke type building units.

Two building procedures have been used to date. One may start with a central atom, satisfy all its bonds by a first layer of atoms, then continue with a next "correlation" layer which satisfies all the bonds left unsatisfied during the first stage, and go on like that. One may satisfy all bonds by allowing for some rearrangements in the previous layers and one will get an ideal CNM. Or one may impose some restrictions to the rearrangements, resulting in a certain number of unsatisfied bonds and therefore a non-ideal CNM. This procedure simulates a deposition process.

The second procedure starts with a completely disordered array of atoms of a convenient density. Then the atoms are allowed to move in small steps so as to arrange themselves with the minimum of distortions into the wanted short-range order. This procedure simulates a quenching process.

Obviously modelling can be done *by computer*. Unfortunately, the better the bond simulation, the higher the number of operations necessary to build the model. So far the inclusion of energetic characteristics of the chemical bond into the calculation has proved to be prohibitive in CNM building by computer.

5.5.4. CHECKING AND USING THE MODEL

In order to check the validity of a given model, the following criteria have to be used:

(i) it has to reproduce the experimental diffraction pattern and RDF, to correspond to the experimental electron microscope image, to give the appropriate values of the density, heat of crystallization, etc.;

(ii) it has to explain the metastability of the structure, the kinetics and effects of annealing, the crystallization process, the phonon spectrum and the behaviour of the *a*–SC under pressure and uniaxial stress, etc.;

(iii) it should allow for a complete statistical description of the structure by appropriate distribution functions (radial and angular distributions, homogeneity parameters).

In the following, particular attempts to build CNM's of *a*–SC's will be discussed in some detail.

5.5.5. AMORPHOUS CARBON

The first CNM of an *a*–SC to be built was that of *a*–C.[79] Building units were of the ball-and-spoke type with either the tetrahedral symmetry and bond length of the C atom in the diamond lattice (1.54 Å) or the trigonal symmetry and shorter bond length of the C atom in graphite (1·41 Å). A very good agreement with both the diffracted electron intensity function and the RDF was obtained. The angular misfits were used to explain qualitatively the fluctuations in interatomic distances which were found experimentally to exceed the thermal fluctuations.

5.5.6. AMORPHOUS SILICON AND GERMANIUM

The first ideal CNM was built[21] by using Voronoi polyhedra with tetrahedral symmetry (Fig. 5.19), which admitted only two different third-order neighbour configurations: the staggered and the eclipsed one. As a conse-

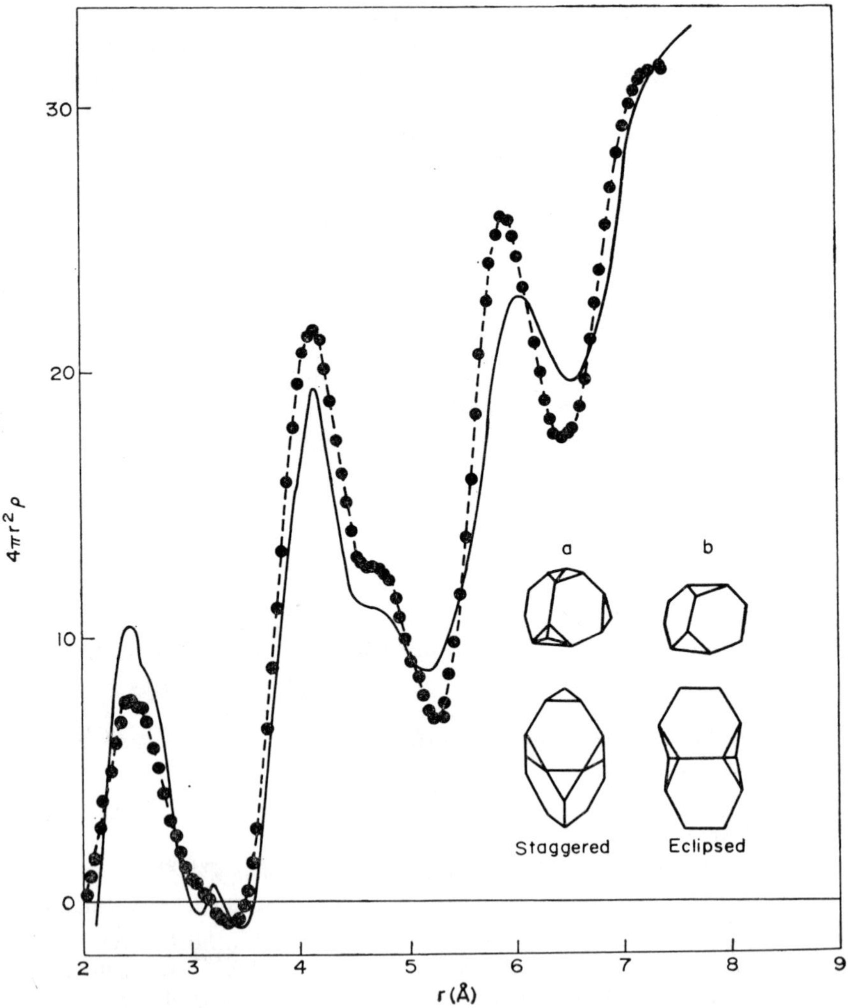

FIG. 5.19. RDF of *a*-Ge. Full curve, experiment (after Ref. 60) dotted curve, continuous network model of Grigorovici and Mănăilă (Refs. 108, 109); lower right-hand corner, exact (a) and simplified (b) Voronoi polyhedron in staggered and eclipsed configuration.

TABLE 5.4. Configurations in successive correlation layers of Grigorovici and Månăilă's continuous network model of amorphous germanium

Correlation layer I		II		III		IV		V	Statistical weight $\times 10^6$	
s/e	$\mathcal{N}$	s/e	$\mathcal{N}$	s/e	$\mathcal{N}$	s/e	$\mathcal{N}$	$\mathcal{N}$		
4/0	4	12/0	12	36/0	24	60/0	42	64	16	
						54/6	42	66	61	
						48/12	42	68	92	
						42/18	42	70	61	
						36/24	42	72	16	
				33/3	24	60/0	43	65	489	
						54/6	43	67	489	
				30/6	24	60/0	44	66	1 465	
				27/9	24	60/0	45	67	977	
				24/12	24	60/0	46	68	244	
		9/3	12	36/0	25	64/0	43	66	489	
						57/7	43	66	489	
						58/6	43	68	1 465	
						51/13	43	68	1 465	
						52/12	43	70	1 465	62 500
						45/19	43	70	1 465	
						46/18	43	72	489	
						39/25	43	72	489	
				33/3	25	64/0	44	67	3 906	
						57/7	44	67	3 906	
		6/6	12	32/6	26	52/12	43	68	5 860	
						46/18	43	70	11 719	
						40/24	43	72	5 860	
		3/9	12	24/18	27	45/21	42	67	7 813	
						39/27	42	69	7 813	
		0/12	12	12/36	28	28/36	40	60	3 906	
3/1	4	12/0	12	30/6	24	60/0	42	64	6 945	
						54/6	42	66	6 945	
				27/9	24	60/0	43	65	6 945	
						54/6	43	67	6 945	
						55/7	43	63	41 667	
				24/12	24	50/14	44	66	41 667	250 000
				21/15	24	45/21	45	68	13 889	
		9/3	12	30/6	25	56/8	43	66	31 250	
						49/15	43	66	31 250	
				27/9	25	56/8	44	67	31 250	
						49/15	44	67	31 250	
2/2	4	10/3	12	26/8	24	54/14	43	64	93 748	
				23/11	24	46/26	44	61	187 494	375 000
				20/14	24	38/38	45	58	93 748	
1/3	4	6/9	12	18/15	24	42/24	42	64	125 000	250 000
				15/18	24	27/45	43	65	125 000	
0/4	4	0/18	12	0/36	24	0/54	36	52	62 500	62 500
									1 000 000	1 000 000

quence, a finite number of possible arrangements resulted for a given number of correlation layers (Table 5.4), each having a certain statistical weight and being characterized by the ratio of staggered (s) and eclipsed (e) bonds and by the number $\mathcal{N}$ of atoms in the layer. The model could be increased in size indefinitely. It consisted essentially of 6-fold puckered or boat-like rings and of 5-fold planar rings which all merged into a continuous network.

Good agreement with experiment was obtained in the general shape of the RDF[108,109] (Fig. 5.19), though the first peak was broadened rather too much and the second too little, showing that due to the rigidity of the building units bond lengths were artificially increased and suffered exaggerated fluctuations, while fluctuations in bond angles were under-estimated. Another success was the calculation of the heat of crystallization[22] by taking into account the gain in energy when the static distortions in bond lengths, bond angles and dihedral angles disappeared during crystallization (see Section 5.2.6).

It could also be shown that clusters of about 100 atoms in which eclipsed bonds prevailed corresponded to a lower free energy than crystallites of the same size (Fig. 5.20), thus explaining why small vacuum-deposited nuclei of Ge and Si prefer to be amorphous rather than crystalline.

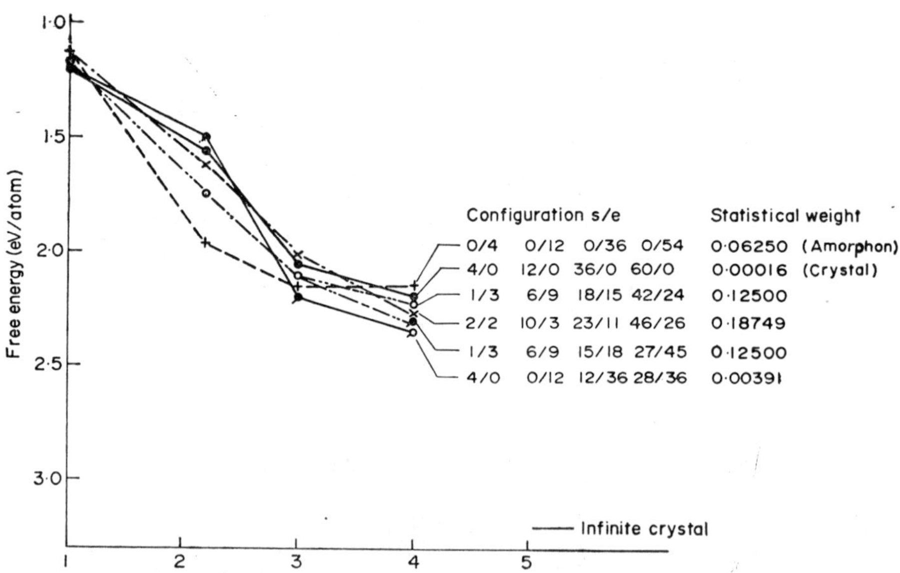

FIG. 5.20. Free energy of various tetrahedrally bound atom clusters of increasing size[21] (see Table 5.4). *s*/*e*—ratio between numbers of staggered and eclipsed bonds.

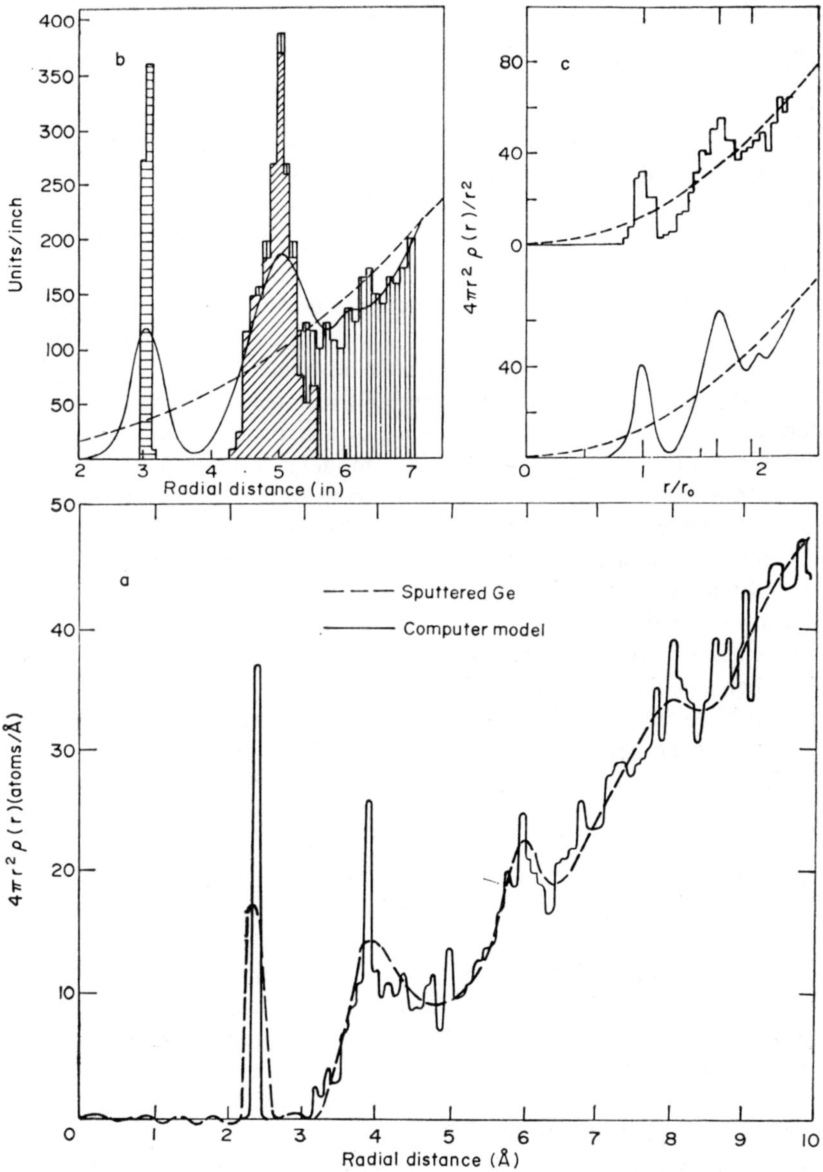

FIG. 5.21. RDFs (histograms) of random network models of *a*-Si and *a*-Ge. *a*, computer-built non-ideal network of Shevchik and Paul;[25] *b*, hand-built ideal network of Polk;[110] *c*, computer-built ideal network of Henderson and Herman;[15] full curve in *b* and lower curve in *c*, experiment on *a*-Si.[41]

The second ideal CNM was built by using the same building procedure as before, but more flexible and freely rotatable ball-and-spoke units.[110] An ideal CNM resulted which also consisted of 5- and 6-fold rings and probably contained also 7-fold ones. However, their shape varied rather significantly, angular fluctuations were high ($\pm 10°$) and dihedral angles were distributed broadly over all values between 0° and 60°, which served to attribute to this model the character of randomness (RNM). The general agreement with the experimental RDF was better than with the previous model, except for the extreme narrowness of the first peak and the appearance of another narrow peak superposed over the broader second peak, both artifacts due to the arbitrary suppression of interatomic bond length fluctuations (Fig. 5.21b). The density proved to be only a few per cent below that of crystalline Ge in agreement with experimental data on highly annealed samples containing about one in a hundred thousand unsatisfied bonds.[111]

Later RNM of *a*–Ge and *a*–Si were built in much the same way by computer, but also allowing for the presence of unsatisfied bonds when a newly arriving atom found at the surface of the model a configuration which could not accommodate it ideally without breaking bonds already formed.[25,68] The results were very similar to those of the hand-built RNM (Fig. 5.21a) and furnished a new basis for an evaluation of the heat of crystallization[24] in accordance with experimental data on *a*–Ge, though the energy gain due to rotation of neighbouring tetrahedra into the staggered configuration was not taken into consideration.

Another calculation[15] started with a randomly packed computer array of appropriate density. Then the atoms were allowed to arrange themselves stepwise into the best possible tetrahedral environment. The resulting RDF (Fig. 5.21c) fitted the experimental RDF of *a*–Si very well.

5.5.7. LIQUID GERMANIUM

Modelling of a RNM by hand using ball-and-spoke units was also tried for liquid Ge.[67] The short-range order of the nearest neighbours was taken to be white tin-like: four nearest neighbours in a flattened tetrahedral configuration, two next-nearest neighbours only slightly (13 %) further apart (Fig. 5.22). Due to this quasi-cubical symmetry an arrangement lacking long-range order could be achieved only if a large number of second-nearest neighbours were eliminated. The resulting quasi-vacancies gave the model a liquid- rather than a solid-like character. The form of the resulting RDF depended on the predetermined width of the bond angle distribution—broader for model 1, narrower for model 2. The fit with the experimental RDF is satisfactory. The white tin-like short-range order correlates well with the metallic properties of liquid germanium.

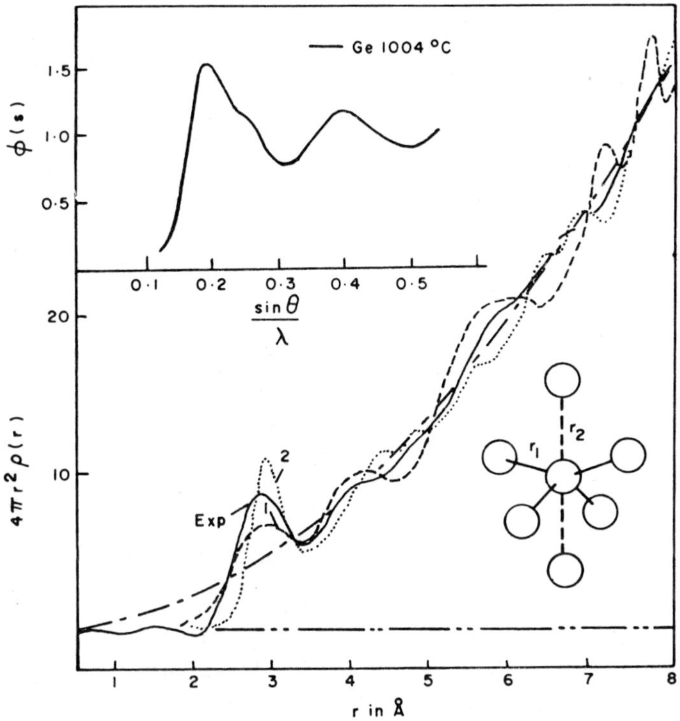

FIG. 5.22. Interference function $\phi(s)$ and RDF of liquid Ge (after Ref. 67). Full curves, experiment; curves 1 and 2, RDFs of random network models based on a white tin-like building unit shown in lower right-hand corner; $r_2/r_1 = 1{\cdot}13$.

5.5.8. AMORPHOUS $A^{III}B^{V}$ COMPOUNDS

The RDF of *a*–GaP, *a*–GaAs and *a*–GaSb show striking similarities with those of *a*–Ge. However, as already stated, while in *a*–GaAs the width of the peaks is the same as in *a*–Ge, those of *a*–GaP and *a*–GaSb are significantly broadened (see Section 5.4.3 and Fig. 5.12). A CNM of these compounds is therefore expected to be very similar to that of *a*–Ge. The different broadening of the RDF peaks was interpreted[(68)] as being due to the presence of "wrong" A–A and B–B pairs. In the framework of any of the proposed CNM's, which all contain rings with an odd number of atoms, the existence of wrong pairs appears as a natural consequence.

5.5.9. AMORPHOUS SELENIUM

The difficulty with modelling *a*–Se consists essentially in the fact that the atoms are 2-fold coordinated by strong and short ($r_1 = 2{\cdot}32$ Å) covalent bonds. In all three crystalline polymorphs the bond angle is 105°, the dihedral angle 102°. In the trigonal lattice, which consists of parallel helicoidal chains, the dihedral angle has the same positive or negative sign along the whole chain. The chains are packed together in such a way as to link also each Se atom by longer ($r_2 = 3{\cdot}46$ Å) and therefore weaker (probably resonant) bonds to four atoms contained in neighbouring chains. Thus trigonal Se has a distorted 6-fold coordination which can be attributed to only partially filled $d^2\,sp^3$ orbitals.(112) This interpretation is also suggested by the way the r_2/r_1 ratio and the bond and dihedral angles vary in the series of heavy group VI elements (Table 5.5). The heaviest of all, polonium, has a simple cubic structure in which all six orbitals are equivalent.

TABLE 5.5. Structural constants of Group VI element crystals

Element	r_2/r_1	Bond angle	Dihedral angle
Se	1·49	105°	102°
Te	1·21	102°	100°
Po	1·00	90°	90°

In the two monoclinic polymorphs of Se (α– and β–Se) the Se atoms are again 2-fold coordinated, only now the dihedral angle alternates in sign along the chain, a fact which leads to its closing into an 8-fold puckered ring. These rings are slightly irregular so as to fit into a monoclinic lattice if packed together in two particular ways. Therefore in α– and β–Se the atoms are not equivalent as are the rings, the next-nearest neighbours are farther apart than 3·46 Å and the crystal is clearly molecular. Tellurium does not have monoclinic polymorphs because alternating the sign of the dihedral angle of 100° does not close rings for a bond angle of 102°.

All proposed models of *a*–Se are vague (see Section 5.4.1) and are essentially molecular. They are mostly characterized as unspecified mixtures of trigonal chains with either 6-fold, or 8-fold or even higher-number rings. Obviously, however, in order to obtain in such a mixture the characteristic peaks in the RDF at 5·7 and 7·2 Å an appropriate way of packing these structural units together is needed. This has not yet been proposed, except for the microcrystal model of Richter and Herre(102) based on 6-fold rings.

An adequate atomic building unit has been proposed(113) in the form of a modified Voronoi quasi-polyhedron (Fig. 5.23) where the biggest faces A correspond to the two strong covalent bonds, the smaller B faces correspond to the four weaker resonant bonds, and the rounding off of the edges of the exact Voronoi polyhedron allows for the building and packing together of closed rings. However, no CNM based on these building units has been devised so far.

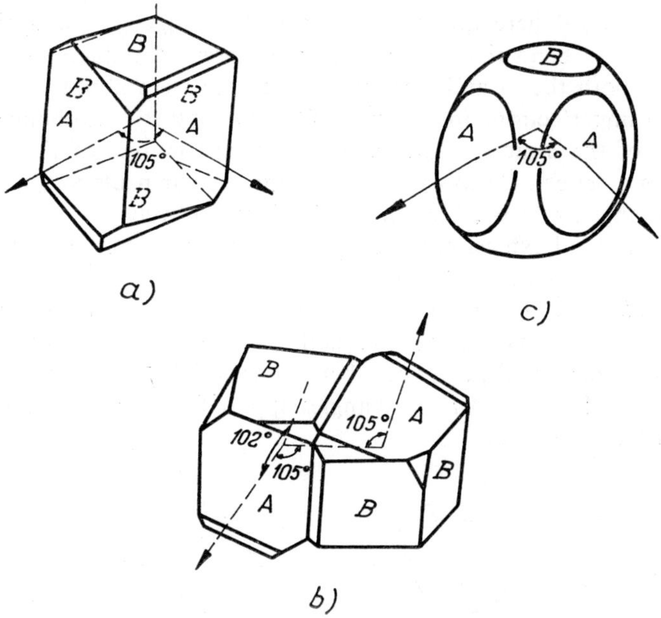

FIG. 5.23. Voronoi polyhedron of Se. *a*, exact polyhedron of trigonal Se; *b*, two polyhedra in normal configuration; *c*, simplified quasi-polyhedron allowing for the building and packing together of 8-fold rings. (After Ref. 113.)

5.5.10 COVALENT BOND SIMULATION BY SPRINGS

The use of springs instead of rods in order to better simulate the energetic characteristics of the covalent bond has been restricted so far to two-dimensional lattices[114] with the aim of investigating the behaviour of non-periodic networks under stress.

In crystals a uniform radial deformation changes only distances, but not angles; uniaxial deformation splits both angles and distances into two groups of values. A non-periodic continuous network may react in a different way. In order to approach this problem, units consisting of three coplanar and symmetrical springs were used to build both a periodic lattice consisting of merged hexagons only (*c*-model, Fig. 5.24a) and a continuous disordered network consisting of merged pentagons, hexagons and heptagons (*a*–model, Fig. 5.24b).

A comparison of the RDF of the *a*–model with the coordinations of the *c*-model (Fig. 5.24c) shows that only those spheres are conserved in the *a*-model which correspond to the interatomic distances within a zig-zag chain (peaks 1, 2, 4 and 6). This is similar to the rule established earlier by Richter and Breitling[60] for the tetrahedral configuration of which the models discussed here are two-dimensional analogues. The springs being much easier to bend than to extend or compress, the fluctuations in bond lengths in both the *c*- and *a*-model (Fig. 5.25a) were about equal and due chiefly to imperfections in the units. On the contrary, the bond angle distribution of the *a*-model differed from the narrow one of the *c*-model by displaying a series of peaks which flank the regular pentagon, hexagon and heptagon angles of 108°, 120° and 128° 34′, respectively.

Uniform radial extension left the bond length distribution of the *c*-model unchanged, but changed the bond angle distribution of the *a*-model drastically (Fig. 5.25b). The formation of only three peaks near the regular polygon angles was a consequence of the reduction in the fluctuations of bond angles within each individual polygon, a behaviour which can be explained by the fact that already highly deformed angles oppose further deformation stronger than less deformed ones.

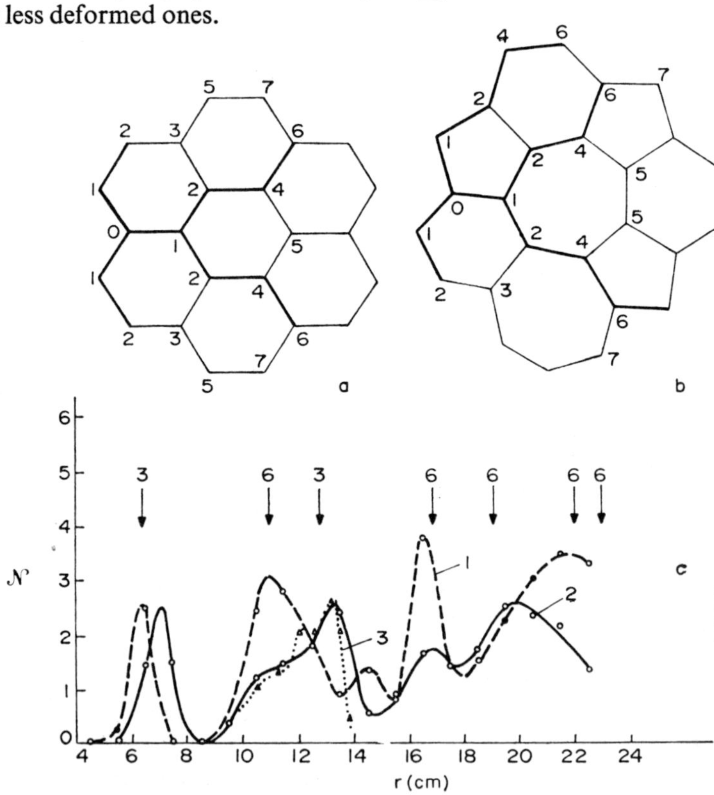

FIG. 5.24. Two-dimensional model using elastic springs (after Ref. 114). *a*, fragment of periodic (crystalline) network model; *b*, fragment of disordered (amorphous) network model; *c*, RDF of undistorted (broken curve) and uniaxially strained (full curve) amorphous model; dotted curve, calculated RDF (second peak) based on angular distortion (Fig. 5.25c) only; arrows, coordination numbers of crystalline model.

Uniaxial deformation of the *c*-model yielded the expected splitting of both the bond lengths and bond angles (Fig. 5.25c) into two groups. In the *a*-model, however, the width of the bond length distribution hardly increased, while the bond angle distribution became much wider and very complicated. As a consequence the first peak of the RDF of the uniaxially extended *a*-model only shifted to higher distances, while the second peak was strongly distorted in accordance with the new angular distribution (Fig. 5.24).

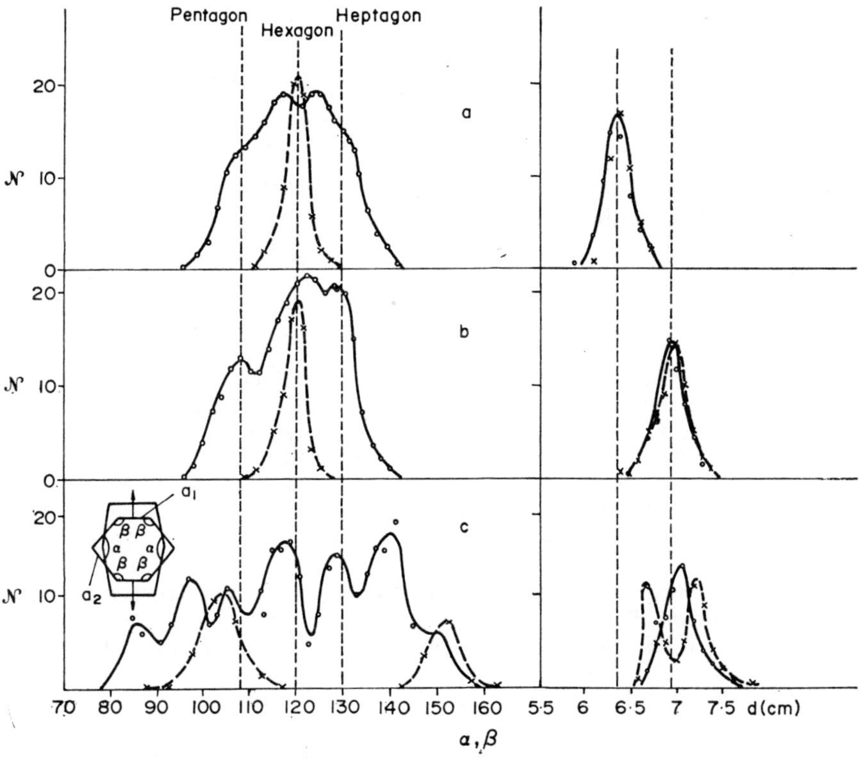

FIG. 5.25. Bond angle (left) and bond length (right) distribution in crystalline (broken curves) and amorphous (full curves) model of Fig. 5.24. *a*, unstrained models; *b*, models under uniform radial deformation; *c*, models under uniaxial strain.

It can be concluded from these model experiments that the behaviour of amorphous and crystalline SC's of similar short-range order under hydrostatic pressure or uniaxial deformation is to be expected to differ not only because of differences in their electronic band structure, but also because their short-range order suffers different geometrical changes. This difference in behaviour will be enhanced, for instance, in *a*–$A^{III}B^{V}$ compounds[24] by the existence in the non-periodic structure of "wrong" A–A and B–B pairs linked together by much weaker bonds than normal A–B pairs.

The use of elastic springs instead of flexible roads may also markedly modify the distribution functions of the structural parameters like bond lengths, bond angles and dihedral angles in continuous random networks.

5.5.11. CONCLUSIONS

Continuous network models and in particular random network models seem to give the best description of the structure of *a*–SC's so far. There are only few contradictions and many agreements with experimental data. Improvements can be expected from a better simulation of the covalent bonds, including their energetic characteristics. Another open question is if the distribution functions of bond lengths, bond angles and dihedral angles constitute, at least in tetrahedrally coordinated elements, a satisfactory statistical description of a continuous network model.

REFERENCES

1. Klug, H. P. and Alexander, L. E., "X-Ray Diffraction Procedure for Polycrystalline and Amorphous Materials." John Wiley, New York, 1954.
2. Furukawa, K., *In* "Reports on Progress in Physics." (ed. A. C. Stickland), Institute of Physics, London, Vol. **25,** p. 395 (1962).
3. Filipovitch, V. N., *J. Tehn. Fiz.* **25,** 1604 (1955).
4. Paalman, H. H. and Pings, C. J., *Rev. Modern Phys.* **35,** 389 (1963).
5. Leadbetter, A. J. and Wright, A. C., *J. Non-cryst. Solids* **7,** 23, (1972); *ibid* **7,** 141 (1972).
6. "International Tables for X-Ray Crystallography". Vol. III, Kynoch Press, Birmingham, 1962.
7. Kaplow, R., Strong, S. L. and Averbach, B. L., *Phys Rev.* **138A,** 1336 (1965).
8. Steeb, S. and Hezel, R., *Z. Physik* **191,** 398 (1966).
9. Betts, F., Bienenstock, A., Keating, D. T. and de Neufville, J. P., *J. Non-Cryst. Solids* **7,** 417 (1972).

9a. Nicotera, E., Corchia, M., De Giorgi, G., Villa, F. and Antonini, M., *J. Non-Cryst. Solids* **11,** 417 (1972).

10. Sayers, D. E., Lytle, F. W. and Stern, E. A., *In* "Advances in X-Ray Analysis" (ed. W. N. Mueller) Plenum Press, New York, Vol. **13,** p. 248 (1970).
11. Sayers, D. E., Stern, E. A. and Lytle, F. W., *Phys. Rev. Letters* **27,** 1204 (1971).
12. Sayers, D. E., Lytle, F. W. and Stern, E. A., *J. Non-Cryst. Solids* **8-10,** 401 (1972).
13. Lucovsky, G., Mooradian, A., Taylor, W., Wright, G. B. and Keezer, R. C., *Solid State Commun.* **5,** 113 (1967).
14. Weaire, D., Thorpe, M. F., Alben, R. and Goldstein, S., Proc. 11th Internat. Conf. Phys. Semicond., Warsaw (1972) p. 470.
15. Henderson, D. and Herman, F., *J. Non-Cryst. Solids* **8-10,** 359 (1972).

16. Smith, J. E., Brodsky, M. H., Crowder, B. L., Nathan, M. I. and Pinczuk, A., *Phys. Rev. Letters* **26,** 642 (1971).
17. Bray, P. J., *In* "Modern Aspects of the Vitreous State" (ed. J. D. Mackenzie), Butterworths, London, Vol. I (1960).
18. Ziman, J. M., *Advances in Physics* **16,** 421 (1967)
19. Bishop, S. G., Taylor, P. C. and Mitchell D. L. *J. Non-Cryst. Solids* **8-10,** 106 (1972).
20. Bishop, S. G. and Taylor, P. C. Proc. 11th Internat. Conf. Phys. Semicond., Warsaw (1972) p. 576.
21. Grigorovici, R. and Mănăilă, R., *J. Non-Cryst. Solids* **1,** 371 (1969).
22. Grigorovici, R. and Mănăilă, R., *Nature* (*London*) **226,** 143 (1970).
23. Chen, H. S. and Turnbull, D., *J. Appl. Phys.* **40,** 4214 (1969).
24. Connell, G. A. N. and Paul, W., *J. Non-Cryst. Solids* **8-10,** 215 (1972).
25. Shevchik, N. J. and Paul, W., *J. Non-Cryst. Solids* **8-10,** 381 (1972).
26. Kaplow, R., Rowe, T. A. and Averbach, B. L., *Phys. Rev.* **168,** 1068 (1968).
27. Natl. Bur. Standards, Circ. 500, Selected Values of Chemical Thermodynamical Properties Washington (1952).
28. de Neufville J. P., *J. Non-Cryst. Solids* **8-10,** 85 (1972).
29. Tauc, J., Abrahám A., Pajasová, L., Grigorovici, R. and Vancu, A., Proc. Internat. Conf. Phys. Non-Cryst. Solids, Delft (1964), p. 606.
30. Barna, A., Barna, P. B., Pócza, E. F., Croitoru, N., Devenyi, A. and Grigorovici, R., Proc. Colloq. Thin Films, Budapest (1965), p. 49.
31. Rudee, M. L. and Howie, A., *Phil. Mag.* **25,** 1001 (1972).
32. Barna, A., Barna, P. B. and Pócza, J. F., *J. Non-Cryst. Solids* **8-10,** 36 (1972).
33. Chaudhari, P., Graczyk, J. F. and Charbnau, H. P., *Phys. Rev. Letters* **29,** 425 (1972).
34. Donovan, T. M. and Heinemann, K., *Phys. Rev. Letters* **27,** 1794 (1971).
35. Rudee, M. L., *Phys. Stat. Solidi* (*b*) **46,** K 1 (1971).
36. Rudee, M. L., *Thin Solid Films* **12,** 207 (1972).
37. Shevchik. N. L., *Phys. Stat. Solidi* (*b*) **52,** K 121 (1972).
38. Howie, A., Krivanek, O. and Rudee, M. L., *Phil. Mag.* **27,** 235 (1973).
39. Chaudhari, P., Graczyk, J. F. and Herd, S. R., *Phys. Stat. Solidi* (*b*) **51,** 801 (1972).
40. Vasko, A., *Silikáty* (Czech.) **4,** 169 (1960).
41. Moss, S. C. and Graczyk, J. F., *Phys. Rev. Letters* **23,** 1167 (1969); Proc. 10th Internat. Conf. Phys. Semicond., Cambridge (Mass.) (1970), p. 658.
42. Shevchik, N. L. and Paul, W., *J. Non-Cryst. Solids* **8-10,** 381 (1972).
43. Guinier, A., Fournet, G., Walker, C. B. and Yudowitch, K. L., "Small Angle Scattering of X-Rays", John Wiley, New York, 1955.
44. Ehrenreich, H. and Turnbull, D., *Comments on Solid State Phys.* **3,** 75 (1970).
45. Renner, O., *Czech. J. Phys.*(*B*) **22,** 1007 (1972).
46. Ditina, Z. Z., Strakhov, L. P. and Helms, H. H., *Fiz. Tehn. Poluprov.* **2,** 1199 (1969).
47. Brodsky, M. H. and Title, R. S., *Phys. Rev. Letters* **23,** 581 (1969).
48. Title, R. S., Brodsky, M. H. and Crowder, B. L., Proc. 10th Internat. Conf. Phys. Semicond., Cambridge (Mass.) (1970) p. 794.
49. Grigorovici, R., Croitoru, N., Devenyi, A and Teleman, E., Proc. 7th Internat. Conf. Phys. Semicond., Paris (1964) p. 423.
50. Smith, R. E., *J. Non-Cryst. Solids* **8-10,** 598 (1972).

51. Červinka, L., Hrubý, A., Matyáš, M., Šimeček, T., Škácha, J., Štourač, L., Tauc, J., Vorliček, V. and Höschl. P., *J. Non-Cryst. Solids* **4**, 258 (1970).
52. di Salvo, F. J., Menth, A., Waszczak, J. V. and Tauc, J., (to be published).
53. Joffe, A. F. and Regel, A. R., *In* "Progress in Semiconductors" (ed. A. F. Gibson) Heywood, London, Vol. 4, 237 (1960).
54. Krebs, H. and Schultze-Gebhardt, F., *Acta Cryst.* **8**, 412 (1955).
55. Andreyevski, A. I., Nabitovitch, I. D. and Voloshchuk, Ya. V., *Kristallografiya* **5**, 369 (1960).
56. Krebs, H. and Steffen, R., *Z. Anorg. Chem.* **327**, 224 (1964).
57. Krebs, H., Schultze-Gebhardt, F. and Thees, R., *Z. Anorg. Chem.* **282**, 177 (1955).
58. Richter, H. and Gommel, G., *Z. Naturforsch.* **12a**, 996 (1957).
59. Breitling, G. *J., Non-Cryst. Solids* **8-10**, 395 (1972).
60. Richter, H. and Breitling, G., *Z. Naturforsch.* **13a**, 988 (1958).
61. Coleman, M. V. and Thomas, D. J. D., *Phys. Stat. Solidi* **24**, K 111 (1967).
62. Grigorovici, R. and Mănăilă, R., *Thin Solid Films* **1**, 343 (1968).
63. Light, T. B. and Wagner, C. N. J., *J. Appl. Cryst.* **1**, 199 (1968).
64. Krebs, H., *J. Non-Cryst. Solids* **1**, 455 (1969).
65. Pavlov, P. V. and Teitelbaum, D. I., *Dokl. Akad. Nauk SSSR* **175**, 823 (1967).
66. Crowder, B. L., Title, R. S., Brodsky, M. H. and Petit, G. D., *Appl. Phys. Letters* **16**, 205 (1970).
67. Isherwood, S. P., Orton, B. R. and Mănăilă, R. *J., Non-Cryst. Solids* **8-10**, 691 (1972).
68. Shevchik, N. L., PhD Thesis, Harvard University (1972).
69. Wihl, M., Cardona, M. and Tauc, J., *J. Non-Cryst. Solids* **8-10**, 172 (1972).
70. Goriunova, N. A., *Izv. Akad. Nauk SSSR, Seriya Neorg. Mater.* **1**, 885 (1965).
71. Vaipolin, A. A., Osmanov, E. O. and Rud, Yu.V., *Fiz. Tverd. Tela* **7**, 2266 (1965).
72. Hrubý, A. and Štourač, L., *Mater. Res. Bull.* **4**, 745 (1969).
73. Červinka, L., Hosemann, R. and Vogel, W., *J. Non-Cryst. Solids* **3**, 294 (1970).
74. Borschchevskii, A. S., Goriunova, N. A., Kesamanly, F. P. and Nasledov, D. N., *Phys. Stat. Solidi* **21**, 9 (1967).
75. Grigorovici, R. Mănăilă, R. and Vaipolin, A. A., *Acta Cryst.* **B24**, 535 (1968).
76. Badzian, A. R., *Mater. Res. Bull.* **2**, 987 (1967).
77. Zaremba, A., Commun. at Conf. Amorph. Liq. and Vitreous Semicond., Sofia (1972).
78. Noda, T. and Inagaki, M., *Bull. Chem. Soc. Japan* **37**, 1534 (1964).
79. Kakinoki, J., Katada, K., Hanawa, T. and Ino, T., *Acta Cryst.* **13**, 171 (1960).
80. Tatarinova, L. I. and Kazmazovskaya, T. S., *Kristallografiya* **6**, 668 (1961).
81. Coleman, M. V. and Thomas, D. J. D., *Phys. Status Solidi* **25**, 241 (1968).
82. Andreyevski, A. I., Nabitovitch, I. D. and Voloshchuk, Ya. V., *Kristallografiya* **7**, 865 (1962).
83. Andreyevski, A. I., Nabitovitch, I. D. and Voloshchuk, Ya. V., *Kristallografiya* **6**, 662 (1961).
84. Hilton, A. R., Jones, C. E. and Brau, M., *Phys. Chem. Glasses* **7**, 105 (1966).
85. Porai-Koshits, E. A. and Vaipolin, A. A., *Fiz. Tverd. Tela* **5**, 246 (1963).
86. Mănăilă, R. and Popescu, M., Commun. at Conf. Amorph. Liq. and Vitreous Semicond., Sofia (1972).

87. Andreichin, R. E., Nikiforova, M. S., Skordeva, E. R. and Yurikova, L. S., Commun. at Conf. Amorph. Liq. and Vitreous Semicond., Sofia (1972).
88. Kolomiets, B. T., Pajasová, L. and Štourač, L., *Fiz. Tverd. Tela* **7,** 1588 (1965).
89. Štourač, L., Kolomiets, B. T. and Shilo, V. P., *Czech. J. Phys.(B)* **18,** 92 (1968).
90. Tsushihashi, S. and Kawamoto, Y., *J. Non-Cryst. Solids* **5,** 286 (1971).
91. Yasaitis, J. A. and Kaplow, R., *J. Appl. Phys.* **43,** 995 (1972).
92. Luo, H. L. and Duwez, P., *Appl. Phys. Letters* **2,** 21 (1963).
93. Bienenstock, A., Betts, F. and Ovshinsky, S. R., *J. Non-Cryst. Solidi* **2,** 347 (1970).
94. Dove, D. B., Heritage, M. B., Chopra, K. L. and Bahl, S. K. *Appl. Phys. Letters* **16,** 138 (1970).
95. Betts, F., Bienenstock, A., Keating, D. T. and de Neufville, J. P., *J. Non-Cryst. Solids* **7,** 417 (1972).
96. Betts. F., Bienenstock, A. and Bates, C. W., *J. Non-Cryst. Solids* **8-10,** 364 (1972).
97. Fawcett, R. W., Wagner, C. N. J. and Dargill III, G. S., *J. Non-Cryst. Solids* **8-10,** 369 (1972).
98. Dove, D. B., Chang, J. and Molnar, B., *J. Non-Cryst. Solids* **8-10,** 376 (1972).
99. Duwez, P. and Tsuei, C. C., *J. Non-Cryst. Solids* **4,** 345 (1970).
100. Ferrier, R. P. and Herrell, D. J., *J. Non-Cryst. Solids* **4,** 338 (1970).
101. Ferrier, R. P., Prado, J. M. and Anseau, M. R., *J. Non-Cryst. Solids* **8-10,** 798 (1972).
102. Richter, H. and Herre, F., *Z. Naturforsch.* **13a,** 874 (1958).
103. Richter, H. and Breitling, G., *Z. Naturforsch.* **21a,** 1710 (1966).
104. Mader, S., *J. Vacuum Sci. Techn.* **8,** 247 (1971).
105. Grigorovici, R., *J. Non-Cryst. Solids* **1,** 303 (1969).
106. Zachariasen, W. H., *J. Amer. Chem. Soc.* **54,** 3841 (1932).
107. Turnbull, D. and Polk, D. E., *J. Non-Cryst. Solids* **8-10,** 19 (1972).
108. Grigorovici, R., "Comptes Rendus Congr. Internat. Couches Minces". (ed. M. R. Gerbier), Cannes, 1970, p. 481.
109. Grigorovici, R., *Thin Solid Films* **9,** 1 (1971).
110. Polk, D. E., *J. Non-Cryst. Solids* **5,** 365 (1971).
111. Donovan, T. M., Spicer, W. E., Bennett, J. M. and Ashley, E. J., *Phys. Rev.* **2B,** 397 (1970).
112. Pauling, L., "The Nature of the Chemical Bond." 3rd edition, Cornell Univ. Press, Ithaca, 1960.
113. Grigorovici, R., Abstr. Internat. Conf. Amorph. Liq. Semicond., Cambridge (1969), p. 60.
114. Grigorovici, R. and Belu, A., Proc. 11th Internat. Conf. Phys. Semicond., Warsaw (1972) p. 453.

6. THE STRUCTURE OF SOME SIMPLE ARSENIC CHALCOGENIDE GLASSES

A. J. APLING

School of Chemistry,
University of Bristol.

6.1. INTRODUCTION

Arsenic chalcogenides form glasses which promise to extend our knowledge of the glassy state and present a challenge to current methods of interpreting diffraction data. Two properties of major interest are the asymmetry of the fundamental structural unit and the ability to form glasses over a wide range of composition. Since diffraction data can only give an average picture of the structure, several techniques of direct data analysis and model fitting are needed in attempting to establish the effect of these properties on the glass structure.

Results are given for X-ray and neutron scattering experiments on glassy As_2S_3 and arsenic selenide glasses covering a range of stoichiometry. Strong evidence for a sheet structure in all these glasses is obtained from the scattered intensity pattern and in the arsenic selenide glasses the extent of this sheet structure is shown to be a function of arsenic content. Comparisons of experimental correlation functions with models based on the Orpiment structure show that the twelve membered ring structure of the crystal sheets is missing in the glasses. Calculations based on the nearest neighbour correlations indicate that these are all network glasses in which covalent bonding requirements are completely satisfied.

The most thoroughly studied two-component glasses are nearly all four coordinate, tetrahedrally bonded structures.[1-3] An exception is the study of B_2O_3 by Mozzi and Warren.[4] Previous studies of the arsenic chalcogenides,[5-8] however, indicate that the local coordination of chalcogenide atoms about arsenic atoms is threefold as in the crystal forms. It is therefore to be expected that the fundamental structural unit of these glasses is similar to that of the crystal, namely, a shallow pyramid with an arsenic atom at the apex and chalcogenide atoms at the corners of the triangular base. This basic asymmetry of the structural unit contrasts sharply with the tetrahedral glasses and is reflected in the typical crystal structure "Orpiment", which consists of stacks of two dimensional, covalently bonded, infinite sheets of corner linked pyramidal units.[9] The existence of sheet structures in the glasses is frequently mentioned in the literature, but more as an intuitive extrapolation from the crystal structure than a deduction from the diffraction data. An investigation of chalcogenide glass structure is, therefore, inevitably linked with the problem of whether diffraction experiments, which can only give an averaged picture of the structure, are capable of revealing asymmetric features such as sheet formations.

Arsenic chalcogenides have a further property of interest in that they form glasses over a wide range of compositions. Until the recent work of Tsuchihashi and Kawamoto,[10] on arsenic sulphide glasses, no diffraction study had been reported of the effect of stoichiometry on glass structure.

6.2. Theory

6.2.1. Fourier analysis of diffraction data

The presentation of scattering formalism in the literature displays a considerable diversity of notation and definition and for the time being it is necessary to precede analyses of experimental results by a summary of the formalism to be used. A full account of the theory used in this paper is given by Leadbetter and Wright.[11]

From an X-ray or neutron diffraction pattern, $I'(Q)$, we can obtain an intensity function, $i(Q)$, where Q is the magnitude of the scattering vector and is numerically equal to $4\pi \sin\theta/\lambda$ in the case of elastic scattering. This function is the structure dependant, coherent contribution to the measured intensity, normalized to absolute units with respect to some representative compositional unit, and it may be Fourier transformed to give a correlation function, $D'(r)$. For neutron diffraction this is computed as follows,

$$D^{\mathcal{N}}(r) = \frac{2}{\pi} \int_0^{Q_{\max}} Q\, i(Q)\, M(Q) \sin rQ \, dQ$$

and for X-rays

$$D^X(r) = \frac{2}{\pi}\int_0^{Q_{max}} \frac{Qi(Q)}{f_0^{\,2}(Q)} M(Q) \sin rQ \, dQ$$

$M(Q)$ is some arbitrary function used to minimise the spurious effects of terminating $I(Q)$ at Q_{max}. Traditionally this is the "Artificial Temperature Factor", but in this study the more convenient and well behaved "Resolution Function" of Lorch,[12] is used. The function $f_0(Q)$ is the so-called "Sharpening Factor", which is conveniently the square of the average scattering factor per electron.

$$f_0(Q) = \frac{\sum_i f_i(Q)}{\sum_i Z_i}$$

Other representations of the Fourier transform are easily computed by adding $D_0(r)$, the correlation function that would be obtained from a perfectly disordered substance of the same composition and density as the sample, to $D'(r)$. The linear distribution function may then be written as $D'(r) + D_0(r)$ and the familiar Radial Distribution Function as $r(D'[r] + D_0[r])$.

The correlation function for a glass containing two types of atoms, A and B, contains the three individual pair correlation functions, A–A, B–B and A–B, and only in rare circumstances can these be separated out. Writing this formally we have,

$$D'(r) = \sum_i \sum_j d_{ij}'(r)$$

where the double summation is over all i atoms in the compositional unit and all j atom types. The relationship between $d_{ij}'(r)$ and the "True" pair correlation function, $d_{ij}(r)$, is a convolution with the Fourier transform of the modifying function $M(Q)$ and, in the X-ray case, with the sharpening function, so that

$$d_{ij}'(r) = \int_0^{\infty} d_{ij}(u) \left[P_{ij}'(r-u) - P_{ij}'(r+u) \right] du$$

where

$$P_{ij}^{\mathcal{N}}(r) = \frac{\bar{b}_i \bar{b}_j}{\pi} \int_0^{Q_{max}} M(Q) \cos rQ \, dQ$$

and

$$P_{ij}^X(r) = \frac{1}{\pi}\int_0^{Q_{max}} M(Q) \frac{f_i(Q) f_j(Q)}{f_0^{\,2}(Q)} \cos rQ \, dQ$$

$\bar{b}$ is a neutron scattering length and $f(Q)$ an X-ray form factor. $D'(r)$ is seen to be a summation of the independent pair correlation functions, each of which is convoluted with its own "Peak Function", $P_{ij}'(r)$.

6.2.2. MODEL CALCULATIONS

Two types of model are used in this work as aids to the interpretation of intensity and correlation functions. The first is the Peak Function model used by Mozzi and Warren.[1] It utilises our ability to calculate the peak function, $P_{ij}'(r)$, in conjunction with a guess at the value of some commonly occurring interatomic separation in the glass. In terms of such interatomic separations, the complete correlation function can be written as,

$$D'(r) + D_0(r) = \sum_i \sum_j \frac{\mathcal{N}_{ij}}{2r_{ij}} P_{ij}'(r)$$

where $\mathcal{N}_{ij}$ is the number of atomic separations of magnitude r_{ij} per unit of composition. This complete calculation is, of course, out of the question, but the experimental functions of most glasses have first peaks, corresponding to the first coordination sphere, which are well separated from the rest of the pattern so that calculations of the contribution to the correlation function of the first coordination sphere alone should be directly comparable with experiment.

The second model, described by Leadbetter and Wright,[11] and known as the "Quasi-Crystalline" model, represents an attempt to produce a one-parameter description of a glass in terms of a related crystal structure and a correlation length, L, beyond which the crystalline correlations are destroyed. It is expressed by writing the intensity as the Fourier transform of the individual correlation functions.

$$Qi(Q) = \sum_{i \neq j} \sum c_i c_j \int_0^L d_{ij}(r)\, F(r) \sin Qr \, dr$$

The functions $d_{ij}(r)$ are obtained from the positional parameters for some known crystal structure and $F(r)$ is some suitable function such that,

$$F(0) = 1$$

$$F(r \geqslant L) = 0$$

The calculated intensities can be directly compared with experiment or Fourier transformed under exactly the same conditions as the experimental data.

6.3. Results and Discussion

6.3.1. ARSENIC SULPHIDE

Arsenic sulphide, of composition As_2S_3, may be regarded as the parent glass of the arsenic chalcogenides. The material is available commercially in pure form and samples are easily prepared by cooling from the melt.

Figure 6.1 shows the total diffracted intensity of X-rays and neutrons obtained from As_2S_3 glass. A distinctive feature of both patterns is the sharpness of the first diffraction maximum which corresponds in position to the

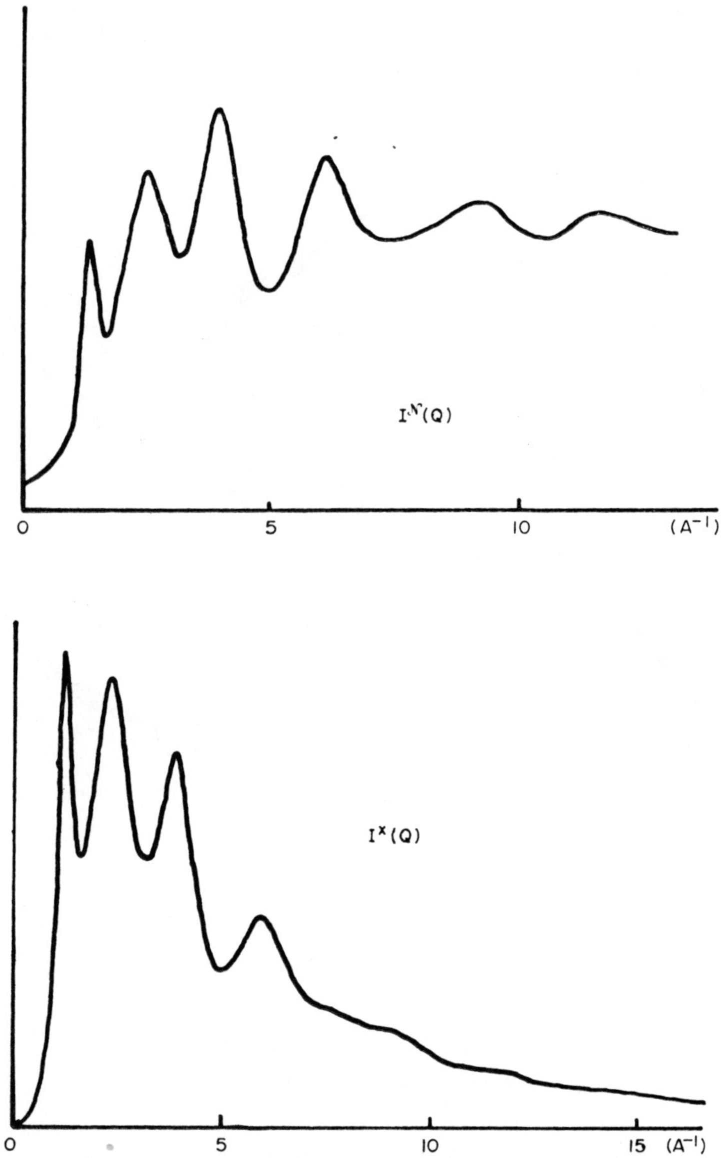

FIG. 6.1. X-ray and Neutron diffraction patterns for As_2S_3.

(020) Orpiment reflection. The strong crystalline reflection arises from the stacking of parallel sheets of atoms along the b_0 axis with two sheets per unit cell. (See Fig. 6.2). This suggests that the strong diffraction feature of the glass pattern is due to some similarly layered structure. To obtain an estimate of the extent of this possible layer structure, the Scherrer equation for line broadening as a function of crystallite size was applied to the width of the

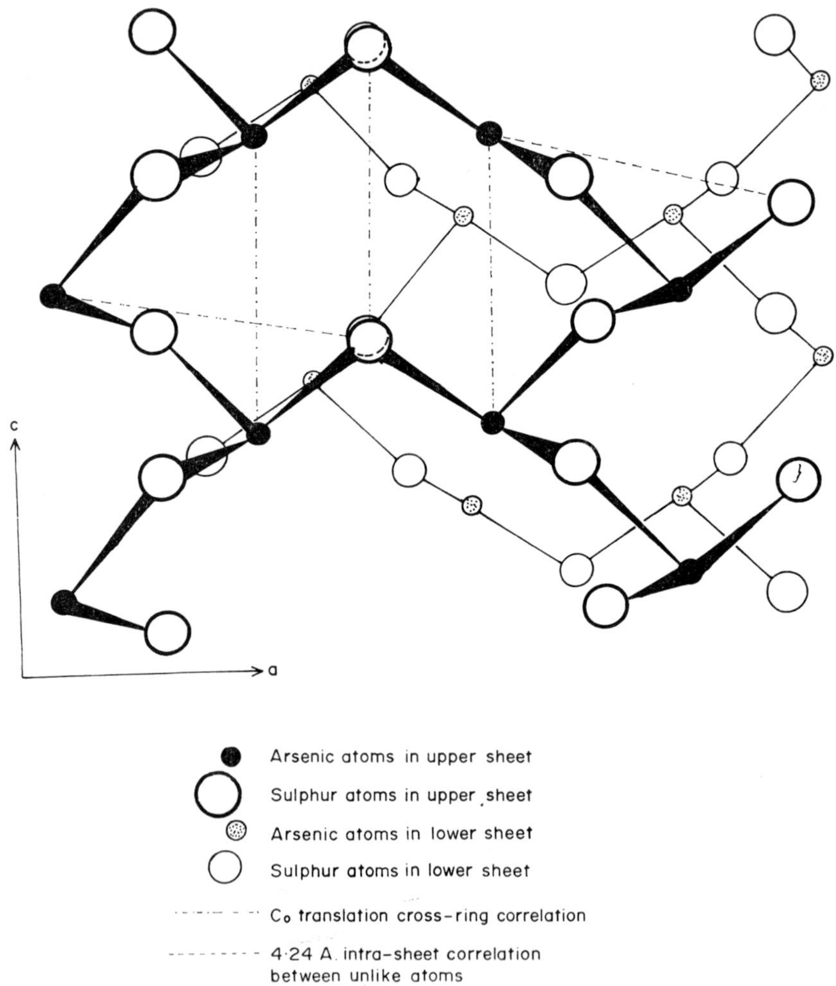

FIG. 6.2. The sheet structure of Orpiment viewed along the b_0 axis. The relative orientation of two adjacent sheets is shown, atoms in the upper sheet being linked by directional bonds.

first diffraction maximum. The equation can be written as

$$D = \frac{K\lambda}{\beta \cos \theta}$$

where K is a constant of about 0·9, λ is the radiation wavelength, θ is the Bragg angle of reflection, β is the line half width in radians and D is the effec-

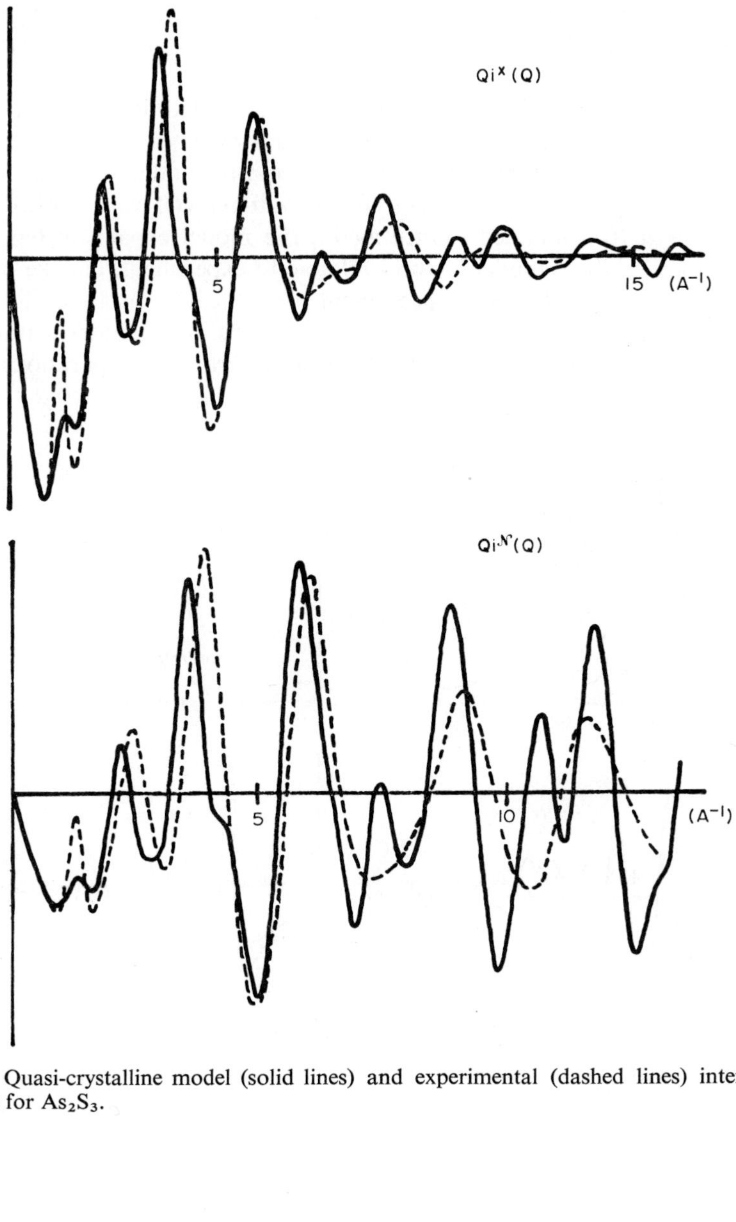

FIG. 6.3. Quasi-crystalline model (solid lines) and experimental (dashed lines) intensity functions for As_2S_3.

tive crystallite thickness perpendicular to the reflecting planes. On this basis, the first As_2S_3 glass peak corresponds to $D = 15A$ which is a high value for a glass correlation length.

Figure 6.3 shows a comparison of the experimental intensities with Quasi-Crystalline model calculations based on the Orpiment structure. The model, which has a characteristic correlation length of 12A, has been isotropically expanded to match the glass density which explains the shift of model peaks to lower Q values. The fit is otherwise good except for the first experimental maximum, which in both X-ray and neutron cases occurs at a lower value of Q and is relatively more pronounced than for the model. This suggests that the layers are more widely separated in the glass than in the crystal and the layer structure persists over a greater radial distance than the intra-layer atomic correlations. Figure 6.4 shows the experimental and model data after Fourier inversion. (For the model, the individual $i_{ij}(Q)$ can be calculated and hence the $d_{ij}'(r)$ obtained.) Once again, the model expansion has resulted in a general shift of model features relative to experiment, but the overall fit is good. The exception is the prominent model feature at about 4·4A which has no counterpart in the experimental patterns. This model feature arises almost entirely from intra-sheet, like-atom interatomic separations charac-

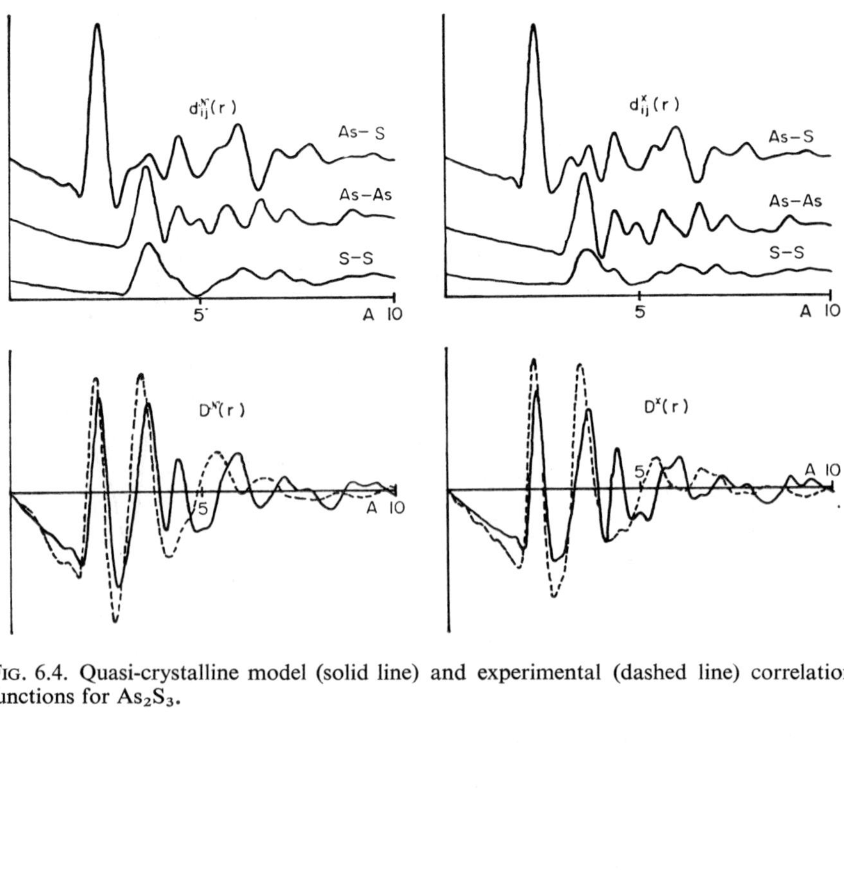

FIG. 6.4. Quasi-crystalline model (solid line) and experimental (dashed line) correlation functions for As_2S_3.

teristic of the c_0 translation, together with some cross-ring correlations between unlike atoms. These correlations are shown on the Orpiment structure of Fig. 6.2. It is therefore clear that the regular, twelve membered ring strucsure of the crystal is completely absent from the glass.

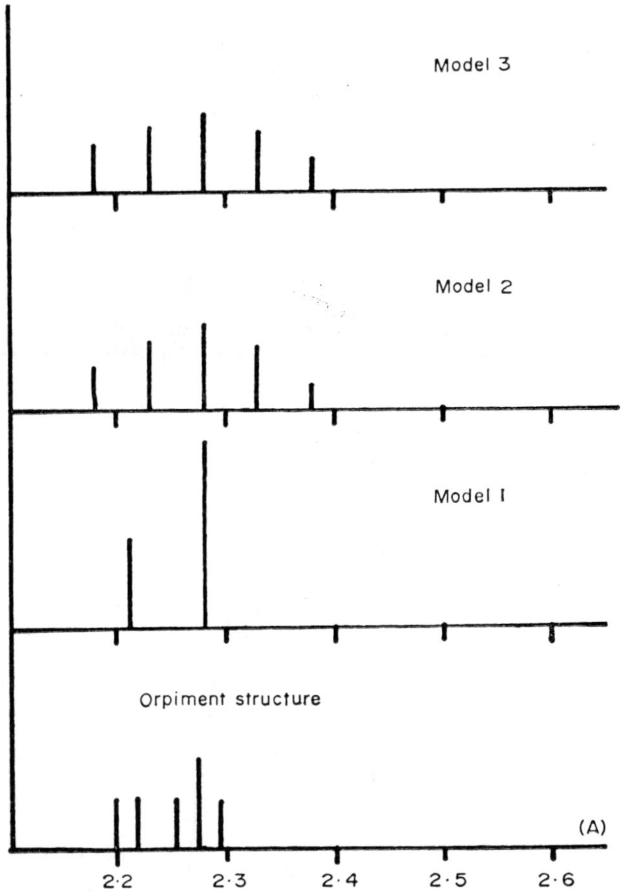

FIG. 6.5. Arsenic–Sulphur bond length distributions for the crystal structure and Peak Function models.

Figure 6.5 shows the distribution of arsenic–sulphur bond lengths in Orpiment, together with three simple model distributions. The results of peak function calculations, based on these distributions and assuming the threefold coordination of AsS_3 pyramidal units, are compared with experiment in Fig. 6.6. The best fit, obtained with the broadest distribution of bond lengths, could be even further improved by additional broadening, but

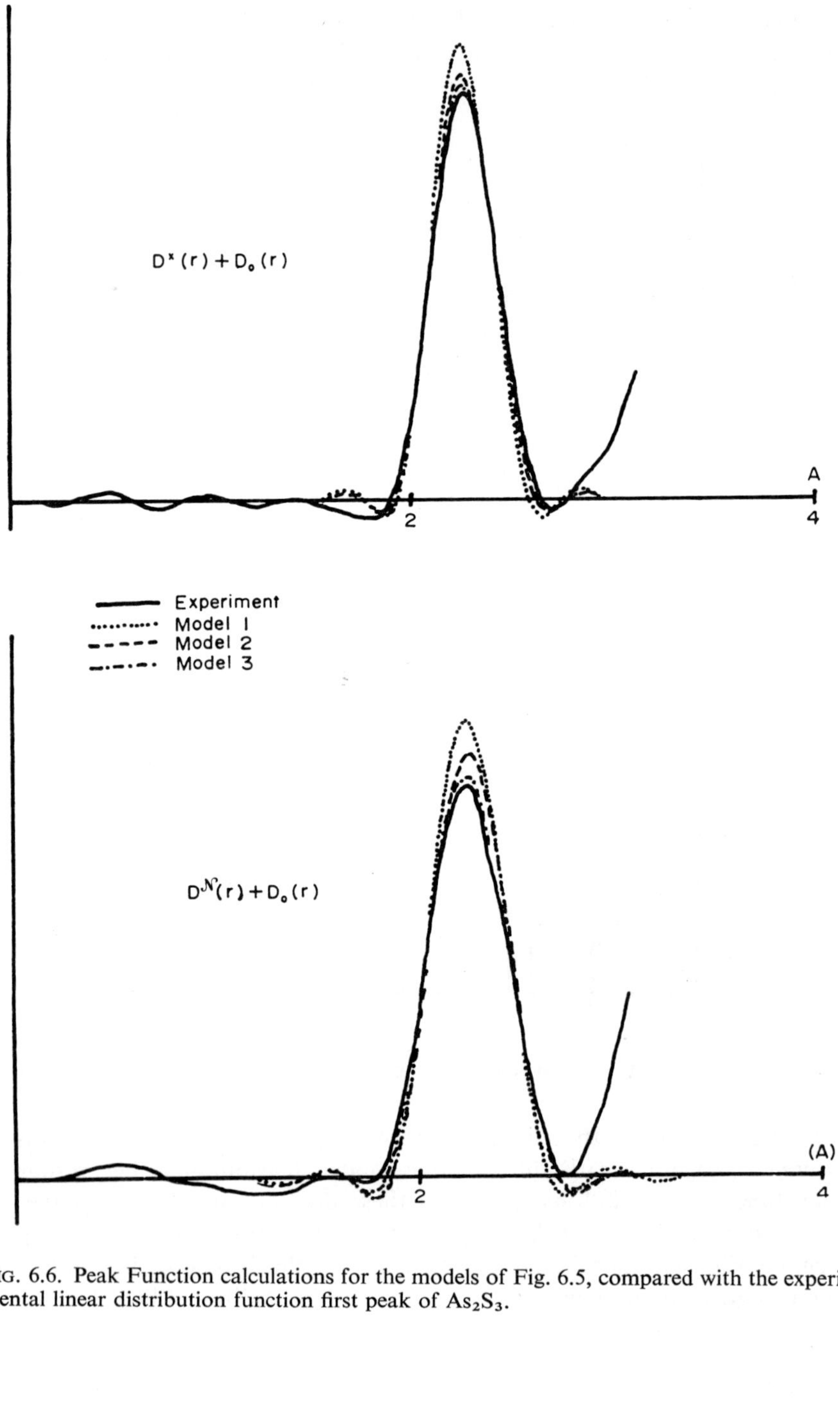

FIG. 6.6. Peak Function calculations for the models of Fig. 6.5, compared with the experimental linear distribution function first peak of As_2S_3.

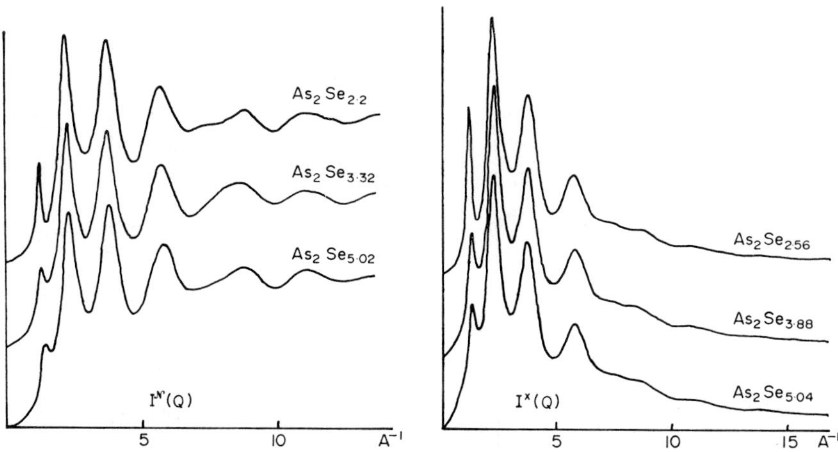

FIG. 6.7. Neutron and X-ray diffraction patterns for arsenic–selenium glasses.

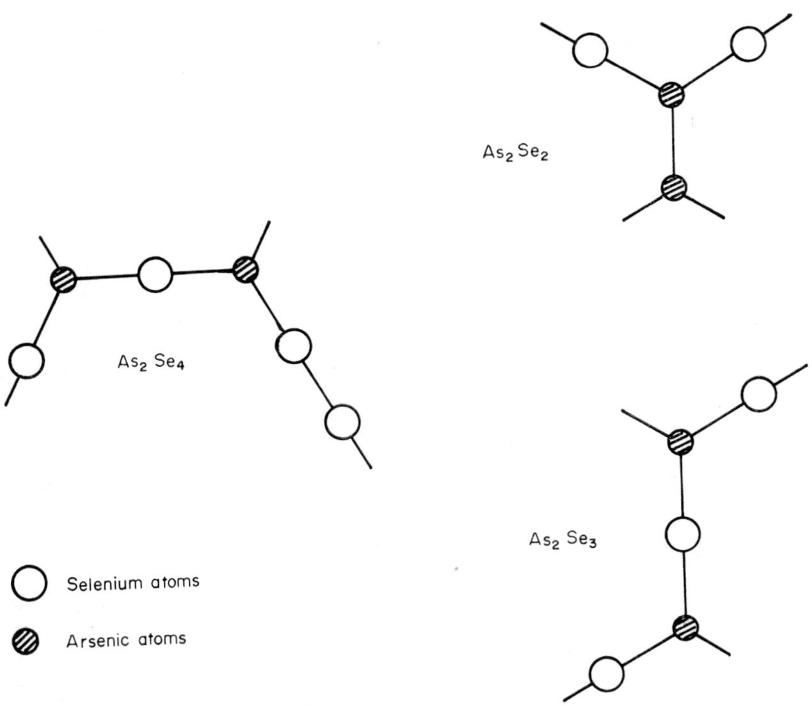

FIG. 6.8. Nearest neighbour models of arsenic selenium glasses with excess elements included.

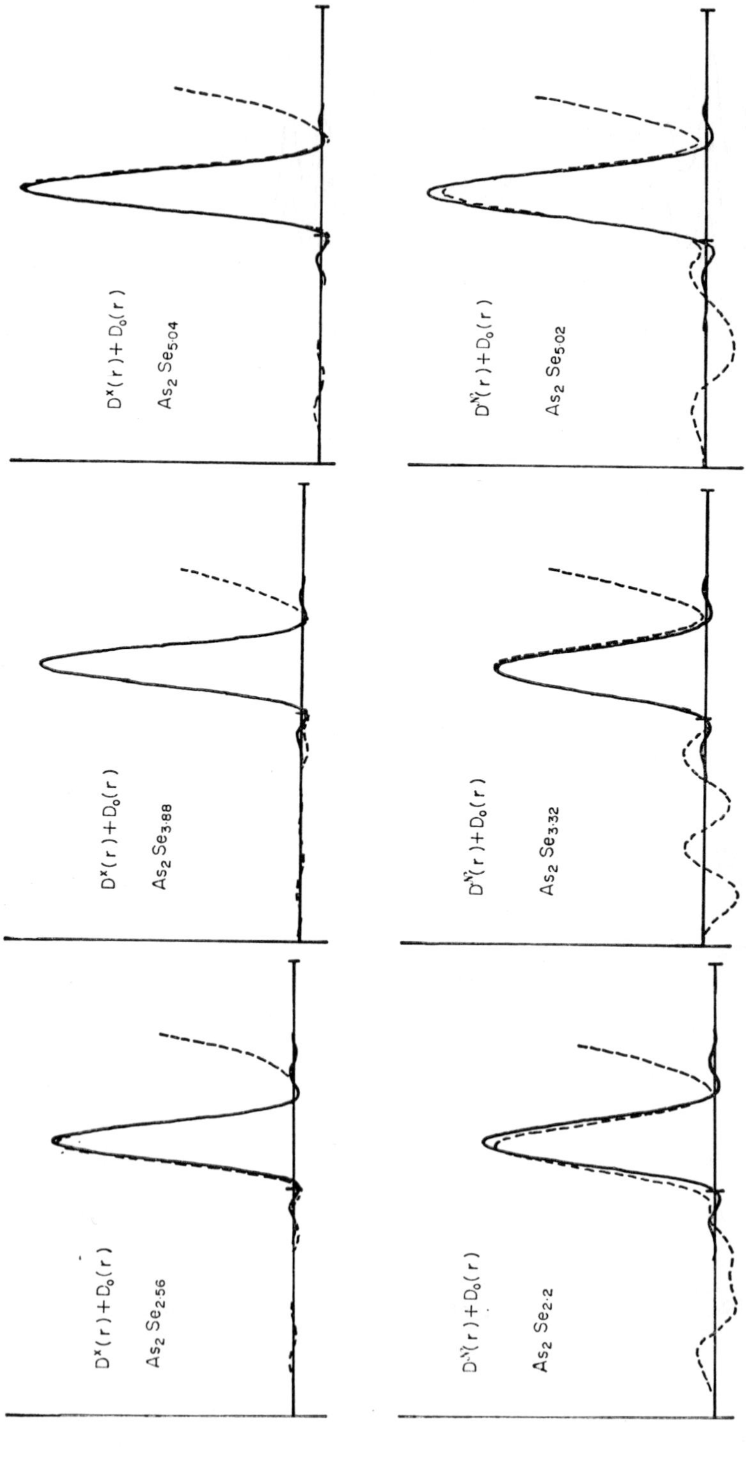

FIG. 6.9. Experimental linear distribution function first peaks for arsenic–selenium glasses (dashed lines) compared with peak function model calculations in which the excess elements have been included.

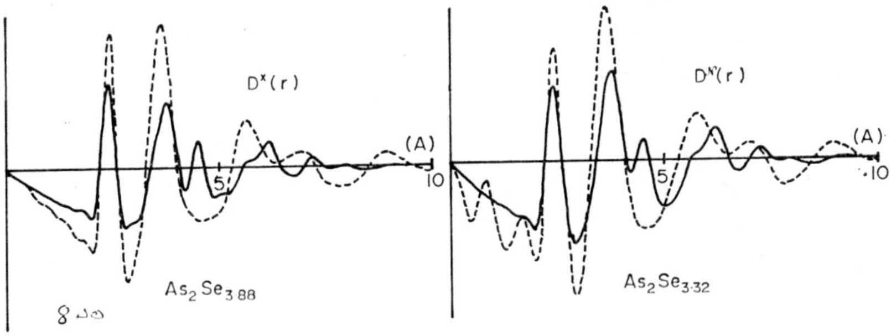

FIG. 6.10. Correlation functions from Quasi-crystalline model calculations based on an expanded Orpiment structure (solid lines) compared with the experimental results for some arsenic–selenium glasses (dashed lines).

is sufficient to demonstrate that the range of arsenic–sulphur bond lengths is greater in the glass than in the crystal.

6.3.2. ARSENIC SELENIDE GLASSES

The arsenic selenide glasses were prepared by fusion of the pure elements. While samples for the neutron and X-ray experiments were not exactly similar in composition, in each case three materials were studied, one close to the As_2Se_3 composition and one each containing excess arsenic and selenium. The coherent intensities obtained from both types of experiment are given in Fig. 6.7 and the overall similarity to the arsenic sulphide data is immediately obvious. However, not only are the first diffraction maxima unusually sharp, but their relative intensity is a function of composition, increasing with arsenic concentration. The half-height width of the first diffraction maxima for the arsenic rich compositions indicates a correlation extending over about 20A. By analogy with arsenic sulphide, a layered structure is suggested and the positions of the first diffraction maxima indicate layer separations ranging from 5·1A in the selenium rich glasses to 5·6A for the arsenic rich glasses.

In order to incorporate the excess elements into a model of the nearest neighbour correlations a simple bonding picture was adopted in which the glass structure consists of a network formed with complete satisfaction of the two and threefold covalent bonding requirements of selenium and arsenic. The effect of an element in excess of the As_2Se_3 stoichiometry is to produce bonds between like atoms so that a further model criterion was required such that the number of like atom bonds was a minimum. Figure 6.8 illustrates the sort of models that can be constructed on this basis. From the point of view of covalent bonding requirements, these models can be linked together in any proportion to form continuous networks of various average com-

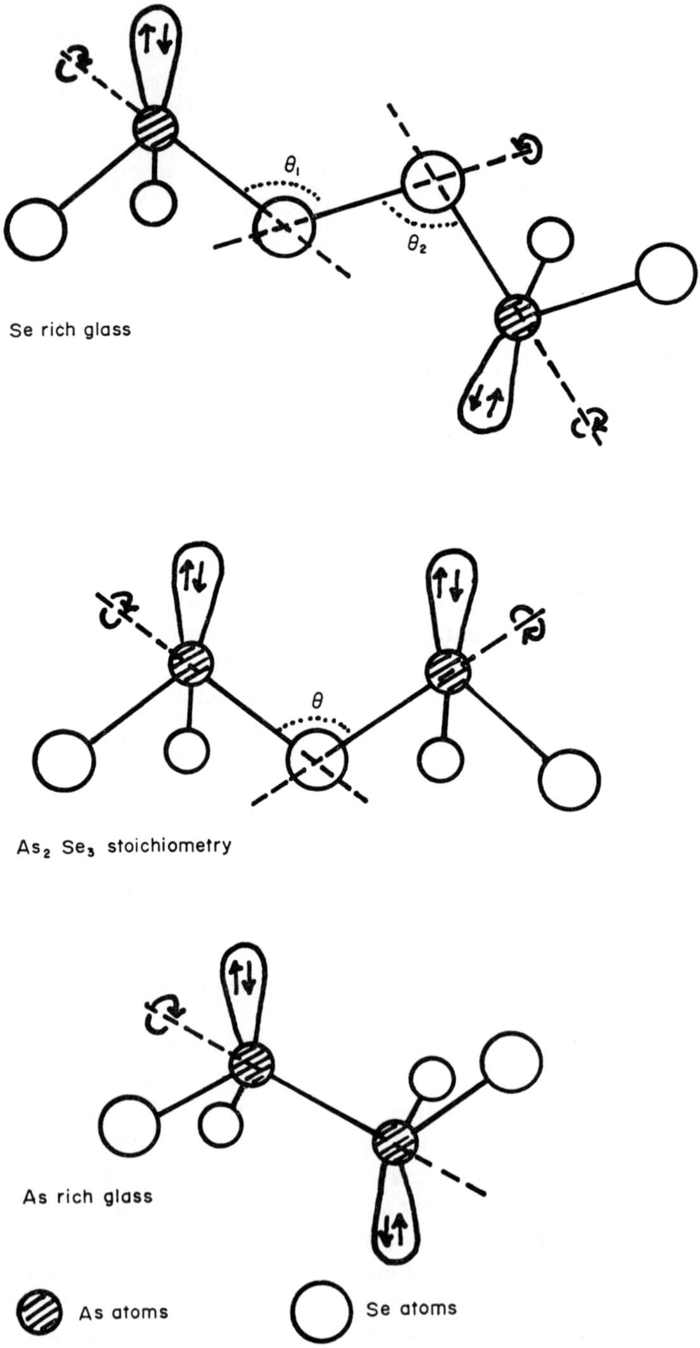

FIG. 6.11. Near neighbour correlations in arsenic–selenium glasses as a function of composition.

positions by linking the atoms with unassigned bonds to unlike atoms. On the basis of crystal chemistry data and the values of normal covalent radii, the selenium–selenium and arsenic–arsenic bond lengths were set at 2.32A and 2·42A respectively. Figure 6.9 shows the fit of experimental data with peak function calculations for these models. The shape fit is good and in all cases the area match is better than 3%.

Quasi-crystalline calculations, based on a suitably expanded Orpiment structure, are compared with the experimental correlation functions for the glasses closest to As_2Se_3 stoichiometry in Fig. 6.10. As in the case of arsenic sulphide, the cross ring correlations are absent from the experimental data. The intra-sheet correlation is therefore very different from the crystal, although some degree of correlation is indicated out to at least 10A.

The picture of arsenic selenide glasses as covalently bonded networks suggests a mechanism for the dependence of layer structure on composition. The intensities of Fig. 6.7 show that the first diffraction maximum is enhanced by increasing arsenic concentration. Figure 6.11 illustrates how the glass forms from the melt in a way which leads to a more or less pronounced sheet structure. In selenium rich glasses, the selenium-selenium bonds introduce a high variability in the relative orientation of connected $AsSe_3$ units and tend to nullify the basic asymmetry of the covalent arsenic bonds. In glasses of As_2Se_3 stoichiometry, which consist solely of corner linked pyramidal units, there are less opportunities for orientational variation with the added possibility of steric hindrance effects between the lone pair electrons in the filled arsenic sp^3 hybrid orbitals. When arsenic–arsenic bonds are formed in the arsenic rich glasses the steric hindrance effect attains considerable importance and the restriction of rotation about the arsenic–arsenic bond means the this simple unit alone embodies a strong tendency to sheet formation.

References

1. Mozzi, R. L. and Warren, B. E., *J. Appl. Cryst.* **2,** 164 (1969).
2. Leadbetter, A. J. and Wright, A. C., *J. Non-Cryst. Solids* **7,** 156 (1972).
3. Wright, A. C., Ph. D. Thesis, Bristol University (1969).
4. Mozzi, R. L. and Warren, B. E., *J. Appl. Cryst.* **3,** 251 (1970).
5. Petz., J. I., Kruh, R. F. and Amstutz, G. C., *J. Chem. Phys.* **34,** 526 (1961).
6. Hopkins, T. E., Pasternak, R. A., Gould, E. S. and Herdon, J. R., *J. Phys. Chem.* **66,** 733 (1962).
7. Vaipolin, A. A. and Porai-Koshits, E. A., *Soviet Physics Solid State* **5,** 178 (1963).
8. Vaipolin, A. A. and Porai-Koshits, E. A., *ibid.* **5,** 497 (1963).
9. Wyckoff, R. W. G., "Crystal Structures" Vol. 1, Interscience, New York, 2nd Ed., (1963).
10. Tsuchihashi, S. and Kawamoto, Y., *J. Non-Cryst. Solids* **5,** 286 (1971).
11. Leadbetter, A. J. and Wright, A. C., *J. Non-Cryst. Solids* **7,** 23 (1972).
12. Lorch, E. A., Ph.D. Thesis, Birmingham University (1967).

III

ELECTRICAL TRANSPORT PHENOMENA

7. ASPECTS OF THE THEORY OF SMALL POLARONS IN DISORDERED MATERIALS

DAVID EMIN

Sandia Laboratories,
Albuquerque, New Mexico, 87115, *U.S.A.*

7.1. INTRODUCTION

The essential purpose of my lectures and those of some other lecturers at this school is to discuss some aspects of electrical transport in noncrystalline materials. In this endeavour we must continually bear in mind that the problems of this field are intrinsically much more difficult than the study of systems for which we may apply models that are characterized by assumed periodicity. Thus, the reliability of our conjectures and the applicability of our models will be, and should be, generally greeted with more scepticism and debate than are associated with studies of defined crystalline systems. With this preface let me proceed to qualitatively outline some of the difficulties which are encountered in attempting to understand electrical transport phenomena in solid noncrystalline materials.

To begin, let us consider what we mean by a crystalline material. Ideally, a crystal is a stoichiometric material characterized by the presence of a definite long-range order of the atomic constituents of the material. In such a situation we may encounter some impurities, but these are generally viewed, in some sense, as a small perturbation of the crystalline system, although the consequences of the presence of such dopants may be considerable as far as our uses for the doped material are concerned. Furthermore, a feature generally associated with the order of the material is the very low mobility of the atoms (or ions) of the crystal.

In a noncrystalline solid, such as a glass, the situation is considerably different. Essentially, by definition, such a material does not present itself as a well-ordered array of the atomic constituents. In particular, while there most certainly are some local juxtapositions of neighbouring atoms in the material which are preferred over others, there is no long-range order in the substance. Indeed, such a noncrystalline compound as a glass is in general not a stoichiometric compound. It is characterized as being inhomogeneous on a scale which is larger than atomic dimensions and (hopefully) smaller than sample dimensions. Thus one can conceive of spatial regions containing different atomic constituents. Hence such notions as those of strings, strands, and ribbons of nearly-ordered constituents (either elements or compounds) present themselves. While there generally is short-range order in these regions, there is presumably no long-range order (that is, they may not be true microcrystals). Thus we will be confronted in general by a system containing islands of a material which differs in its chemical composition from that which surrounds it. An additional example of inhomogeneity is the occurrence of regions which contain no material, the so-called voids. Of course, there may well be distinct boundaries between regions of differing chemical composition or even between regions possessing a similar chemical composition but a different local order. Furthermore, it is quite conceivable that at least a certain fraction of the atomic constituents of, let us say, a glass will be able to move (at least within a limited spatial regime) much more freely than they would in a crystal. Thus having enunciated this physical picture of a noncrystalline solid (such as a glass), it must be said that each of the noncrystalline aspects of such a solid need not present comparable difficulties when one attempts to understand the results of electrical transport measurements on these substances. Indeed, as a practical matter, one ignores some aspects of the problem and focuses one's attention upon creating a model for the effects of one, or a few, ingredients of the total problem.

Having perhaps created a terrifying image of the problems associated with studying noncrystalline (particularly glassy) materials, let me now proceed to make some preliminary observations which (while not in any sense completely resolving these problems) may perhaps provide some indication of one way in which the situation (at least in some circumstances) may be improved. To begin, let us note that even in a glass one constituent may exist in sufficient quantity to form a reasonably well-connected entity throughout the sample; such a situation is schematically illustrated by the shaded regions of Fig. 7.1. If the mobilities and conductivities associated with electronic charge carriers within this constituent are sufficiently high compared with those in the surrounding regions, it is conceivable that the carrier motion may in some sense be qualitatively dominated by this constituent (either an element or compound which presumably possesses short-range order). However, this requires that breaks in the continuity of the dominant regions have a relatively small effect on the resistivity of the sample. As shown by Lipskis, Sakalas, and Viscakus[1] for a model resistance network, such a situation can prevail even in the circumstance of spatially separated (totally unconnected) regions of a "dominant constituent" if the voltage drop associated with the dominant constituent is large compared to that in the intervening boundary regions. The presumed (at least partial) connectivity of a dominant constituent should make the prevalence of this situation even more likely (though certainly not necessary) in a real glass. The presence of regions of a glass in which few carriers move may at the very simplest level be viewed as providing an inert ingredient interjected within and about regions of the dominant constituent.

Finally, it should be observed that the inclusion of "inert" material or the presence of voids surely affects the measured electrical transport properties. This is illustrated by a calculation by Juretschke, Landauer, and

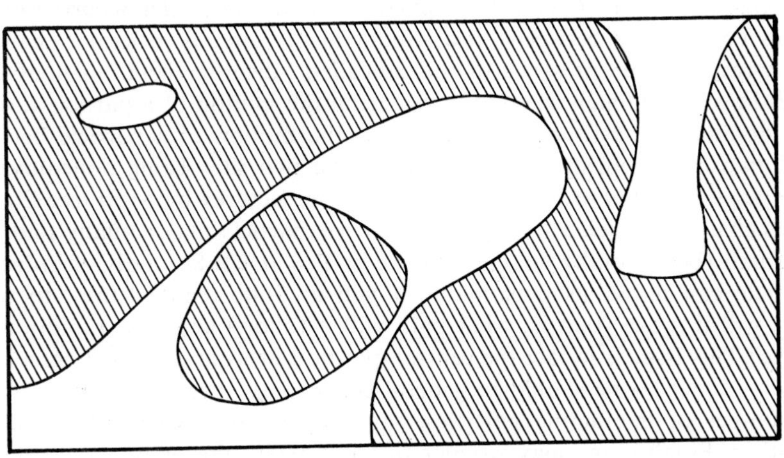

FIG. 7.1. A two-dimensional illustration depicting small regions of a glass characterized by relatively high carrier mobility and conductivity (dark area) and those in which relatively few carriers move (white areas). The regions may differ in their local order, proximity to "defects" or "impurities", or in their chemical composition.

Swanson[2] in which they report that for isolated (noninteracting) spherical voids in an otherwise continuous conductive medium, the measured conductivity, σ_m, and Hall mobility, μ_m, of the otherwise continuous medium are given respectively by

$$\sigma_m = \sigma_0 \frac{(1 - f)}{(1 + \frac{1}{2}f)}, \tag{7.1}$$

and

$$\mu_m = (\sigma_0 R_0) \frac{(1 - \frac{1}{4}f)}{(1 + \frac{1}{2}f)}, \tag{7.2}$$

where f is the fraction of missing conductive material and σ_0 and R_0 are, respectively, the conductivity and Hall constant of the conductive medium. It is important to note that the fractional change of the observed conductivity and Hall mobility from their intrinsic values is of the order of the fraction of the volume occupied by voids. Thus, the presence of regions which exclude the bulk of charge carriers will not necessarily drastically influence the electrical transport properties of a noncrystalline material; that is, the measurements will essentially relate to the transport properties of the "dominant constituent."

A final, and also far from trivial, problem is that of the carrier motion within the dominant constituent itself. Specifically, the absence of long-range order, and the possible presence of voids, boundaries, other constituents, traps, and impurities all contribute to imposing a "disorder field" upon the dominant constituent. Thus the immediate environment that a carrier experiences as it moves between "nearly-equivalent" sites is generally changing. As a result of the presence of this so-called "disorder field," a number of questions immediately arise in considering the motion of a particle through the dominant constituent. First, how is the motion of a particle between, say, two neighbouring sites affected by the two sites being nonequivalent. One aspect of this problem, discussed in the lectures by Mott (see Chapter 1), is that for sufficiently large disorder the particle may actually find itself localized (being totally unable to move at $T = 0$ K). Secondly, if in fact the particle can move, as presumably it will at finite temperatures, to which sites will it move (that is, what path will it follow) and how rapidly will it move? This issue is in fact the central question of what is termed percolation theory. Indeed, one should in general consider the percolation problem for a number of interacting particles, say for electrons which can repel one another. Finally, one may wonder to what extent the presence of a charge carrier alters the positions of (and distorts) the atoms in its immediate vicinity. It is the issue raised by this which gives rise to the notion of the "self-trapping" of a carrier. This phenomenon, the notion of self-trapping, will be the principal focus of much of the remainder of these lectures. That is, we shall essentially restrict our attention to the questions of polaron formation (the self-trapping of a carrier) and the transport properties of these polarons in ordered and disordered materials. Namely, for simplicity and definitiveness, we shall generally proceed by first considering these questions for a simplified model of an ordered system and then extend the results to disordered systems.

Specifically, the outline of these lectures will be as follows. The concept of a polaron is introduced and discussed in Section 7.2; in particular, the applicability of the polaron concept to non-polar materials via the utilization of a short-range electron–lattice interaction is considered. In Section 7.3 an unpublished study by T. Holstein of the situation of an electron interacting with a deformable continuum via a short-range electron–lattice interaction is presented. It is argued that in this circumstance a self-trapped carrier is always severely localized (a small polaron). To provide a basis for pursuing the problem of an excess carrier in a discrete system the Molecular-Crystal Model is introduced in Section 7.4. The principal features of a variational calculation of the energy spectrum of an electron in a deformable lattice are presented in Section 7.5. Specifically, the calculation generates criteria for the dynamic stability of both the situation of a carrier existing in a slightly deformed lattice, and that of a carrier which exists as a self-trapped unit (small-polaron) in the coupled electron–lattice system. The physical significance of the stability conditions is then discussed. Finally, it is suggested that the presence of disorder tends to constrain a carrier's motion, thereby making small-polaron formation generally easier in a disordered material than in its crystalline counterpart.

The topic of small-polaron transport is broached in Section 7.6, in which the results of systematic studies of small-polaron motion withn the Molecular-Crystal Model are succinctly reviewed. It is pointed out that the situation which is most likely to be of relevance to small-polaron motion in non-crystalline solids is that in which the small polaron is viewed as moving through the material via a succession of thermally-activated hops. The calculation of the small-polaron jump rate appropriate to a carrier in the Molecular-Crystal Model is outlined in Section 7.7. The role of vibrational dispersion in the calculation and the sharp distinction between small-polaron hopping in the low-temperature regime and the complementary high-temperature regime are emphasized in the discussion of this Section. In Section 7.8 the rate associated with a small-polaron hop between two lattice sites which are characterized by different local electronic energies is calculated. Again in this model of a disordered situation the distinction between the high- and low-temperature characteristics of the jump rate is apparent. In Section 7.9 the semiclassical theory of the small-polaron jump rate in the high-temperature regime is sketched. This approach to the calculation of the jump rate contributes to our understanding of the problem by providing a physical picture of the jump process in the high-temperature regime. The modification of the semiclassical theory that is required in order to incorporate the effects of various types of disorder is also outlined. In Section 7.10 it is pointed out that an implicit assumption inherent in the previously discussed hopping studies is that the atomic distortion associated with a hop relaxes sufficiently rapidly after the hop so as to permit treating each hop of a carrier as independent of its prior hops. Furthermore, the results of studying the question of possible correlated small-polaron hopping motion within the semiclassical theory for ordered and disordered systems are then reviewed.

The principal conclusions of the rather extensive theory of the Hall effect of the small polaron are set forth in Section 7.11. In particular, the magnitude, temperature dependence, and sign of the Hall mobility for both excess electrons and excess "holes" are considered. In Section 7.12 the thermoelectric power attributed to small-polaron hopping motion in the high-temperature regime is explored, emphasizing the influence of disorder on the small-polaron thermoelectric power. The high-frequency ($\omega > \omega_{\text{phonon}}$) conductivity of the small-polaron is the subject of a cursory discussion in Section 7.13. Section 7.14 is devoted to a compact review of the ramifications of small-polaron formation with respect to a number of experimentally observed quantities. In particular, the question of the relationship of the "optical gap" to the "transport gap" is considered and the predictions of the simplest version of small-polaron theory with regard to the d.c. conductivity, thermoelectric power, and Hall mobility of small polarons in a disordered material or in the "dominant constituent" of a glass are presented. Finally, the theory is favourably compared with a number of experiments on chalcogenide glass systems.

While much experimental and theoretical work remains to be done to elucidate the role of small polarons in disordered materials, the formation of small polarons in noncrystalline solids is certainly a plausible occurrence for which some presently available experimental evidence lends support. The point of view taken here is, therefore, that it behooves us to be aware of the implications of such a situation so that we may be open to the possibility of small polarons being formed in specific noncrystalline materials.

7.2. The Notion of a Small-Polaron

Having given a very cursory qualitative discussion of the specific problems associated with the disordered nature of noncrystalline solids, we now proceed to discuss the question of small-polaron formation. The motivation for this entire endeavour stems from the fact that the observed Hall mobilities of an extremely large number of noncrystalline substances are sufficiently low so as to raise the possibility that they are "low-mobility solids," in which the trap-free mobility associated with the conductivity of the material is low, (less than $\sim 1\text{cm}^2/\text{V-sec}$). Qualitatively one may say that if in fact the carriers in these materials do move sufficiently slowly, the circumstance of their being "self-trapped" becomes quite feasible. However, it should be cautioned at the onset that a low value of a measured Hall mobility is not of itself deemed a proof of the presence of small polarons (self-trapped carriers) in such low-mobility materials.

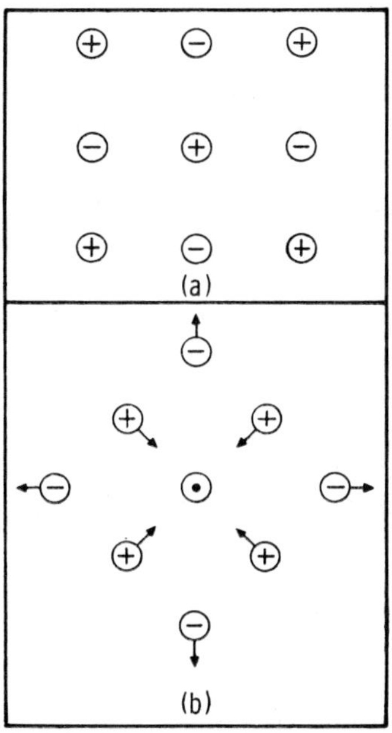

FIG. 7.2. Subfigure (a) depicts a portion of a lattice of ionic charges in which no excess charge carriers are present, while (b) illustrates the readjustment of the positions of the ions in response to an excess electron being placed on the central positive ion.

To begin our discussion let us develop a picture of what is meant by a small polaron in a crystal. Let us consider the regular two-dimensional ionic lattice illustrated in Fig. 7.2(a), in which the neighbouring sites are oppositely charged ions; this is a simplified analogue of an alkali halide crystal. If an electron is added to the system and is stationary at the site of a positive ion, the surrounding atoms will tend to be displaced in response to its presence. Such a situation is illustrated in Fig. 7.2(b). The potential well that the added electron sees as a result of the atomic displacements its presence induces, tends to hold it fixed at its place in the lattice. If this carrier-induced potential well is sufficiently deep, the carrier may occupy a bound state, being unable to move without an alteration of the positions of the surrounding atoms. If the increase in the strain energy associated with the displacement of the ions from their locations prior to the introduction of the excess electron is smaller than the decrease of the energy of the electron due to its being bound in its self-induced well, then the situation is energetically stable. The unit comprising the self-trapped carrier and its concomitant lattice distortion is termed a *polaron*. If the carrier is essentially confined to single site it is referred to as a *small-polaron*.

The name polaron arises from the early consideration of the possibility of self-trapping in ionic (polar) materials.[3] However, the principal arguments that we have employed are actually quite general and not at all restricted to a polar system. In fact, the best experimental evidence for the existence of self-trapped carriers is found in such materials as orthorhombic sulphur, studied by Gibbons and Spear,[4] realgar (As_4S_4),[5] MnO[6] and the so-called V_k centres (self-trapped holes) in the alkali halides.[7] In these materials the electron–lattice interaction is believed to possess either an entirely short-range character or at least a strong short-range component. Thus, for the purpose of these lectures we shall not discuss the usual "optical polaron" problem, in which one's concern is restricted to a study of the effects of the long-range polarization forces that are associated with the interaction between a carrier and the ions of a polar insulator. It may be commented that even in a real ionic material the presence of a dominant short-range component of the electron–lattice interaction can easily be envisaged and cannot be disregarded *a priori*.

7.3. Continuum Model

Let us now try to elucidate the physical nature of the state which results when a carrier interacts with the lattice via a short-range electron–lattice interaction. To begin, let us consider the adiabatic limit; that is, we assume that the vibrational motion of the atoms is so slow compared with the motion of an electron that one may assume that the electron adjusts to the instantaneous position of the atomic constituents. Operationally this means that we shall, at least for the moment, neglect the kinetic energy of the atoms of the material. Such a procedure has been carried out for a one-dimensional (linear chain) model by Holstein,[8] and by Toyozawa[9] for a three-dimensional continuum model in which the well induced by the electron is assumed to be a "three-dimensional square well." The following analysis of the nature of an electron placed in a deformable continuum is based on a generalization of these earlier works which was suggested by Holstein.[10]

To begin, the Hamiltonian for the coupled electron–continuum system in the adiabatic approximation may be written as

$$H_{\text{total}} = H_e + H_{\text{dis}}, \tag{7.3}$$

where

$$H_e = T_e - C_1 \, \Delta(\mathbf{r}) \tag{7.4}$$

and

$$H_{\text{dis}} = \tfrac{1}{2} S \int \Delta^2(\mathbf{r}') \, d^3 \mathbf{r}'. \tag{7.5}$$

Here the electronic Hamiltonian H_e is the sum of the electron kinetic-energy operator, T_e, and the potential energy associated with a dilation of the continuum, $\Delta(\mathbf{r})$; C_1 is a constant. The interaction term $-C_1 \, \Delta(\mathbf{r})$ is assumed (as is the usual but not universal practice) to be a linear function of the dilation. The distortional energy contribution [eqn (7.5)] from the carrier-induced distortion is simply the strain energy as given by Hooke's law; S is just a strain constant. The energy of this system is then simply

$$\mathcal{E}_{\text{total}} = \int \psi^*(\mathbf{r}) \, H_{\text{total}} \, \psi(\mathbf{r}) \, d^3 \mathbf{r} \tag{7.6}$$

$$= \mathcal{E}_{\text{el}} + \mathcal{E}_{\text{dis}}, \tag{7.7}$$

the sum of the electronic energy and the strain energy. Minimizing the total energy with respect to the dilation, one readily finds that

$$\Delta(\mathbf{r}) = \frac{C_1 \, |\psi(\mathbf{r})|^2}{S}. \tag{7.8}$$

Thus the distortion at a given position $\mathbf{r}$ is proportional to the electron–continuum coupling constant and to the probability that the carrier be on the site, and is inversely proportional to a constant which reflects the stiffness of the material, S. To find $\psi(\mathbf{r})$ one must solve the equation

$$\mathcal{E}_{\text{el}} \, \psi(\mathbf{r}) = -\frac{\hbar^2}{2m} \nabla^2 \psi(\mathbf{r}) - \frac{C_1{}^2 \, |\psi(\mathbf{r})|^2}{S} \, \psi(\mathbf{r}) \tag{7.9}$$

[analogous to Holstein's eqn (19) in reference 8]. If the electronic energy $\mathcal{E}_{\text{el}}$ is negative, the electron will have formed a bound electronic state. Furthermore, the occurrence of a stable self-trapped unit in this model requires that $\mathcal{E}_{\text{total}} = \mathcal{E}_{\text{el}} + \mathcal{E}_{\text{dis}}$ be negative. While Holstein solved the one-dimensional model and found a stable bound state with a finite radius of distortion, Toyozawa found from analysis of his three-dimensional model that there was no stable finite-radius bound state. Indeed, Toyozawa's analysis implied that, in the adiabatic approximation, the particle and its distortion would either be localized in a vanishingly small spatial region or the particle would be sufficiently spread out so as to have a negligible distortion: There is no such thing as a finite-radius continuum polaron in Toyozawa's model.

Let us now proceed to understand these results. Writing the energy of the system as the sum of the electron's kinetic energy, T_e, reduced potential energy due to the electron–continuum interaction, V_{int}, and the strain energy, $\mathcal{E}_{\text{dis}}$, we have

$$\mathcal{E}_{\text{total}} = T_e - V_{\text{int}} + \mathcal{E}_{\text{dis}}, \tag{7.10}$$

where

$$T_e = -\frac{\hbar^2}{2m} \int \psi^*(\mathbf{r}) \nabla^2 \psi(\mathbf{r}) \, d^3\mathbf{r}, \tag{7.11}$$

$$V_{\text{int}} = \frac{C_1^{\ 2}}{S} \int |\psi(\mathbf{r})|^4 \, d^3\mathbf{r}, \tag{7.12}$$

$$\mathcal{E}_{\text{dis}} = \tfrac{1}{2} V_{\text{int}}, \tag{7.13}$$

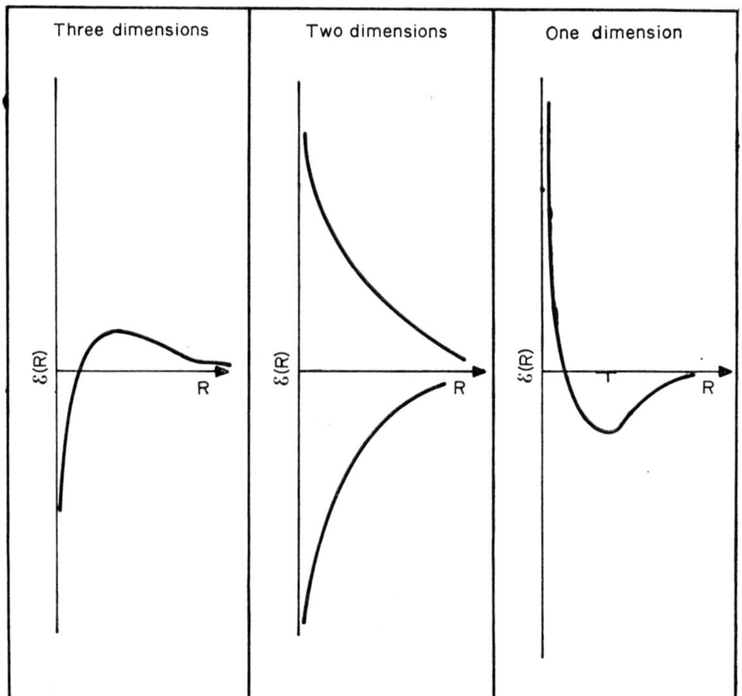

FIG. 7.3. The energy of the system comprising an electron and deformable continuum is plotted against the scaling factor R which is related to the size of the polaron. A finite-radius polaron in this system is only energetically stable when the system is restricted to one dimension.

and $\psi(\mathbf{r})$ is the assumed solution of eqn (7.9). Let us now consider how a change of spatial scale would affect the above three terms.

Specifically, if we replace $\psi(\mathbf{r})$ by $R^{-n/2}\,\psi(\mathbf{r}/R)$, where n is the dimensionality of the system (the prefactor ensures the normalization of the wave-function), we find

$$\mathcal{E}_{\text{total}}(R) = \left(\frac{T_e}{R^2} - \frac{V_{\text{int}}}{R^n}\right) + \frac{V_{\text{int}}}{2R^n} = \frac{T_e}{R^2} - \frac{V_{\text{int}}}{2R^n}, \tag{7.14}$$

where the bracketed term is the electronic energy contribution to the total energy. If in fact there is a stable (finite-radius) bound state of the carrier, then $\mathcal{E}_{\text{total}}(R)$ must have a minimum at $R = 1$. However, it is clear from Fig. 7.3, where the result of the three-dimensional model ($n = 3$) is plotted, that the only minima occur at $R = \infty$, corresponding to an unbound electron in an undistorted continuum, and at $R = 0$, corresponding to an electron self-trapped in an infinitely deep and infinitesimally narrow potential well. In the two-dimensional case illustrated in Fig. 7.3, a minimum occurs at $R = 0$ or at $R = \infty$ depending on whether the kinetic energy term, T_e, is respectively less than or greater than half the interaction term, $\frac{1}{2}V_{\text{int}}$. The one-dimensional case does, however, yield results which are qualitatively different; namely, in this case, as is illustrated in Fig. 7.3, a stable finite-radius minimum is found. This solitary minimum occurs at $R = 4T_e/V_{\text{int}} = 1$. The electronic energy is then given by [see the bracketed term of eqn (7.14)]

$$\mathcal{E}_{\text{el}} = T_e\left(\frac{V_{\text{int}}}{4T_e}\right)^2 - V_{\text{int}}\left(\frac{V_{\text{int}}}{4T_e}\right) \tag{7.15}$$

$$= -\tfrac{3}{16}(V_{\text{int}})^2/T_e = -\tfrac{3}{4}V_{\text{int}}; \tag{7.16}$$

since the electronic energy is negative, the particle may be said to be bound. The strain energy contribution is (of course) positive and is given by

$$\tfrac{1}{2}\frac{V_{\text{int}}}{R} = \tfrac{1}{8}(V_{\text{int}})^2/T_e. = \tfrac{1}{2}V_{\text{int}}. \tag{7.17}$$

Thus, the total energy at the minimum is found to be negative:

$$\mathcal{E}_{\text{total}} = -\tfrac{3}{16}\frac{(V_{\text{int}})^2}{T_e} + \tfrac{1}{8}\frac{(V_{\text{int}})^2}{T_e} = -\tfrac{1}{16}\frac{(V_{\text{int}})^2}{T_e} = -\tfrac{1}{4}V_{\text{int}}; \tag{7.18}$$

hence self-trapping is in this case energetically favoured. Furthermore, the ratio of the electronic energy to the strain energy is 3:2 in accord with Holstein's more detailed analysis of the one-dimensional model [specifically, examine his eqn (32) in reference 8].

However, since we shall, in the main, be interested in three-dimensional systems, we must now contend with the fact that the (three-dimensional) adiabatic continuum model does not permit a finite-radius self-trapped carrier. One certainly does not expect an electron to actually be localized at an infinitesimal point. We expect, however, that in a lattice an electron (and the distortion induced by it) can be confined essentially to a unit cell. Thus, in order to better understand the physical situation governing the formation of (or failure to form) a self-trapped entity, one must investigate a model that includes the discreteness which characterizes the microscopic nature of a solid.

7.4. The Molecular Crystal Model

The simple model upon which we shall base most of our remaining discussion is termed the Molecular-Crystal Model.[8] Specifically, we shall consider a single excess carrier placed in an ordered array of deformable molecules. Each molecule, in addition to its position **g**, is characterized by a configurational coordinate, $x_{\mathbf{g}}$, which represents the deviation of the molecule from its equilibrium configuration. For instance, $x_{\mathbf{g}}$ may be thought of as the molecular-distortion variable involved in the breathing (vibrational) mode of the molecule (such as the interatomic separation of the two atoms of a diatomic molecule).

Taking cognizance of only one such parameter per molecule, the Hamiltonian corresponding to the harmonic vibrations of the molecular-crystal lattice is simply

$$H_{\text{vib}} = \sum_{\mathbf{g}} \left\{ \frac{1}{2M} \left(\frac{\hbar}{i} \frac{\partial}{\partial x_{\mathbf{g}}} \right)^2 + \frac{M}{2} \omega_0^{\,2} x_{\mathbf{g}}^{\,2} + \frac{M}{2} \sum_{\substack{\mathbf{h} \\ \text{neighbours}}} \omega_1^{\,2} x_{\mathbf{g}} x_{\mathbf{g}+\mathbf{h}} \right\}, \tag{7.19}$$

where M is an appropriate reduced atomic mass, ω_0 is the harmonic-oscillator frequency associated with the configurational coordinate of the isolated molecule, and the final term couples neighbouring molecules and gives rise to dispersion of the vibrational frequencies. In particular, the relationship between a normal-mode frequency and its associated wavevector, **q**, is[11]

$$\omega_{\mathbf{q}}^{\,2} = \omega_0^{\,2} + \frac{\omega_1^{\,2}}{2} \sum_{\mathbf{h}} \cos\left(\mathbf{q}\cdot\mathbf{h}\right), \tag{7.20}$$

where $-\pi \leqslant \mathbf{q}\cdot\mathbf{h} \leqslant \pi$.

While the inclusion of vibrational dispersion is a crucial ingredient in a theory of small-polaron hopping motion, it plays no essential role in the present considerations. Thus, for simplicity the intermolecular vibrational coupling will for the moment be ignored; that is, it is assumed that $\omega_{\mathbf{q}} = \omega_0$. Finally, it is expedient to reexpress the vibrational portion of the Hamiltonian in terms of the phonon operators, $b_{\mathbf{q}}^*$ and $b_{\mathbf{q}}$, which respectively create and annihilate phonons of wavevectors $\mathbf{q}$, thereby obtaining[12]

$$H_{\text{vib}} = \sum_{\mathbf{g}} \hbar\omega_0 \,(b_{\mathbf{q}}^* \, b_{\mathbf{q}} + \tfrac{1}{2}). \tag{7.21}$$

That segment of the total Hamiltonian of the system which refers to the excess electron may be written as the sum of two terms. The first expression describes the electron in the molecular-crystal lattice when the individual molecules of the system are constrained to maintain their equilibrium configuration. Explicitly, this term of the Hamiltonian is

$$H_e = \frac{1}{2m}\,(\text{grad}_r)^2 + \sum_{\mathbf{g}} U(\mathbf{r} - \mathbf{g}), \tag{7.22}$$

where m and $\mathbf{r}$ are respectively the electron's mass and position vector and $U(\mathbf{r} - \mathbf{g})$ is the contribution of the $\mathbf{g}$th rigid molecule to the electron's potential energy. The remaining portion of the electronic Hamiltonian depicts the change in the electron's potential energy as a result of a deviation of the molecules of the system from their equilibrium configuration. In particular, the electron–lattice interaction energy is assumed to be a short-range linear function of the configuration coordinates. Generally, a linear electron-lattice interaction may be written in terms of $b_{\mathbf{q}}^*$ and $b_{\mathbf{q}}$ as[12]

$$H_{\text{int}} = \sum_{\mathbf{q}} [V_{\mathbf{q}}(\mathbf{r})\, b_{\mathbf{q}} + V_{\mathbf{q}}^*(\mathbf{r})\, b_{\mathbf{q}}^*]. \tag{7.23}$$

To relate this expression to the specific form of the electron–lattice interaction that has been utilized in the small-polaron transport studies, it is recalled that in these works the energy of an electron residing on the $\mathbf{g}$th molecule is, apart from an arbitrary constant, taken to possess the manifestly short-range form [see eqn (8) in reference 8]

$$\mathcal{E}(x_{\mathbf{g}}) = -Ax_{\mathbf{g}}, \tag{7.24}$$

where A is a constant. In terms of the phonon operators this energy may be rewritten as

$$\mathcal{E}(x_{\mathbf{g}}) = -A(\hbar/2M\mathscr{n}\omega_0)^{\frac{1}{2}} \sum_{\mathbf{q}} (\mathrm{e}^{i\mathbf{q}\cdot\mathbf{g}}\, b_{\mathbf{q}} + \mathrm{e}^{i\mathbf{q}\cdot\mathbf{g}}\, b_{\mathbf{q}}^*), \tag{7.25}$$

where $\mathscr{n}$ is the number of molecules in the lattice. Calculating the alteration of the energy of an electron at site $\mathbf{g}$ due to the presence of the electron–

lattice interaction, H_{int}, yields, from eqn (7.23),

$$\mathcal{E}(x_{\mathbf{g}}) = \sum_{\mathbf{q}} \langle \mathbf{g} | V_{\mathbf{q}}(\mathbf{r}) | \mathbf{g} \rangle b_{\mathbf{q}} + \langle \mathbf{g} | V_{\mathbf{q}}^*(\mathbf{r}) | \mathbf{g} \rangle b_{\mathbf{q}}^*, \tag{7.26}$$

where $|\mathbf{g}\rangle$ is the electronic eigenfunction of an excess electron at molecule $\mathbf{g}$ when the molecule is maintained in its equilibrium configuration. Comparing eqns (7.25) and (7.26), one may make the association

$$\langle \mathbf{g} | V_{\mathbf{q}}(\mathbf{r}) | \mathbf{g} \rangle = -A(\hbar/2M\mathcal{N}\omega_0)^{\frac{1}{2}} \, e^{i\mathbf{q}\cdot\mathbf{g}}. \tag{7.27}$$

Finally, considering the matrix elements of H_{int} between different molecular sites to be ignorable, Holstein's Molecular-Crystal Model is obtained. In other words, the Hamiltonian comprising the three terms H_{vib}, H_e, and H_{int} [eqns (7.21), (7.22), and (7.23), respectively], along with the subsidiary condition

$$\langle \mathbf{g}' | V_{\mathbf{q}}(\mathbf{r}) | \mathbf{g} \rangle = \delta_{\mathbf{g}',\mathbf{g}} [-A(\hbar/2M\mathcal{N}\omega_0)^{\frac{1}{2}} \, e^{i\mathbf{q}\cdot\mathbf{g}}], \tag{7.28}$$

serves as an adequate specification of the Molecular-Crystal Model.

7.5. Self-trapping in a Discrete Lattice: A Variational Approach

Having established the Hamiltonian associated with the Molecular-Crystal Model, it is now our task to investigate the eigenstates of this system. In the absence of a bona fide solution of this problem, we shall investigate a variational solution that may provide some insight into the question of the conditions which determine when a charge carrier will be self-trapped or untrapped. In the spirit of the tight-binding approximation the trial wavefunction of the variational calculation is written as an appropriate linear superposition of "local" wavefunctions. Here the local wavefunctions are each taken to be the product of a local electronic wavefunction and the vibrational wavefunction associated with the concomitant carrier-induced molecular deformation. The linear superposition compatible with a carrier of wavevector $\mathbf{k}$ existing in our periodic system is [compare eqn (2) in reference 13]

$$\psi_{\mathbf{k},\ldots\mathcal{N}_{\mathbf{q}}\ldots}(\mathbf{r}) = \frac{1}{\sqrt{n}} \sum_{\mathbf{g}} e^{i\mathbf{k}\cdot\mathbf{g}} \, \Phi_{\mathbf{g}}^{\mathbf{k}}(\mathbf{r}) \, \chi^{\mathbf{k}}_{\mathbf{g},\ldots\mathcal{N}_{\mathbf{q}}\ldots}(\ldots x_{\mathbf{g}} \ldots), \tag{7.29}$$

where $\Phi_{\mathbf{g}}^{\mathbf{k}}(\mathbf{r})$ is the local electronic wavefunction appropriate to a carrier of wavevector $\mathbf{k}$ situated about molecular site $\mathbf{g}$, and $\chi^{\mathbf{k}}_{\mathbf{g},\ldots\mathcal{N}_{\mathbf{q}}\ldots}(\ldots x_{\mathbf{g}} \ldots)$ is the wavefunction characterizing vibrational motion about the altered equilibrium configuration of the system's molecules that is commensurate with electron occupancy of the state $\Phi_{\mathbf{g}}^{\mathbf{k}}(\mathbf{r})$; the local vibrational state

$$\chi^{\mathbf{k}}_{\mathbf{g},\ \dots \mathcal{N}_{\mathbf{q}} \dots}(\dots x_{\mathbf{g}} \dots)$$

will in general be characterized by the number of phonons $\mathcal{N}_{\mathbf{q}}$ associated with each vibrational mode $\mathbf{q}$.

In order to elucidate the variational nature of this solution, the "local" vibrational wavefunction is expressed explicitly in terms of the variational parameters, the $f_{\mathbf{q}}(\mathbf{k})$'s:[13]

$$\chi^{\mathbf{k}}_{\mathbf{g},\ \dots \mathcal{N}_{\mathbf{q}} \dots} = \prod_{\mathbf{q}} \exp\left[f_{\mathbf{q}}(\mathbf{k})\, e^{i\mathbf{q}\cdot\mathbf{g}}\, b_{\mathbf{q}} - f_{\mathbf{q}}^{*}(\mathbf{k})\, e^{-i\mathbf{q}\cdot\mathbf{g}}\, b_{\mathbf{q}}^{*}\right] |\mathcal{N}_{\mathbf{q}}\rangle; \tag{7.30}$$

here $|\mathcal{N}_{\mathbf{q}}\rangle$ represents the undisplaced harmonic oscillator wavefunction associated with the $\mathcal{N}_{\mathbf{q}}$th excited state of vibrational mode $\mathbf{q}$ in a carrier-free lattice. The magnitude of $f_{\mathbf{q}}(\mathbf{k})$ determines the magnitude of the carrier-induced distortion and the q-dependence of $f_{\mathbf{q}}(\mathbf{k})$ determines the spatial extent of such a distortion. For example, while $f_{\mathbf{q}}(\mathbf{k}) = 0$ refers to an undistorted lattice and $f_{\mathbf{q}}(\mathbf{k}) = \text{constant} \neq 0$ corresponds to a distortion which is restricted to a single molecule, a situation in which $f_{\mathbf{q}}(\mathbf{k})$ is only substantial for $|\mathbf{q}|$ less than some value q_0 ($q_0 < \pi/a$, a being the lattice constant) is characterized by a local distortion of range $1/q_0$. Finally, it should be noted that, consistent with taking the electron–lattice interaction to possess only a linear term in the molecular-distortion parameters, the $x_{\mathbf{g}}$'s, the "local" vibrational wavefunction permits only a displacement of the equilibrium positions of the $x_{\mathbf{g}}$'s. Specifically, it does not include such higher order electron–lattice effects as a carrier-induced shift in the local vibrational frequencies.

The "local" electronic wavefunction will most generally depend on the nature of the local distortion of the molecular crystal. Thus local electronic and vibrational wavefunctions must, in principle, be determined self-consistently. While such a procedure appears to be a prohibitively formidable task for an arbitrary-ranged electron–lattice interaction, our study of the three-dimensional continuum situation suggests that the situation may be substantially simplified when one restrict's one's interest to the circumstance of a short-range electron–lattice interaction such as that encountered in the Molecular-Crystal Model. In particular, our previous study of the spatial extent of a "self-trapped" state (in the continuum model, see Section 7.3) indicates that the "self-trapped" carrier is always essentially localized at a single lattice site; it is a small polaron. Thus, for a "self-trapped" carrier it is not an unreasonable first approximation to take $\Phi_{\mathbf{g}}^{\mathbf{k}}(\mathbf{r}) = |\mathbf{g}\rangle$, where $|\mathbf{g}\rangle$ is an electronic wavefunction associated with site $\mathbf{g}$. Furthermore, in regard to the suitability of the trial wavefunction for describing an untrapped carrier, it is readily demonstrated that the wavefunction which emerges from a perturbation treatment of the electron–lattice interaction,[14] H_{int} [eqn (7.23)], is of the form of our trial wavefunction. Thus it is concluded that, although the trial wavefunction is not the most general that can be conceived, it is a reasonable selection that possesses the virtue of being in consonance with the simplifications which are commonly employed in studying both the "self-trapped" and untrapped situation.

By minimizing $\langle\psi_{\mathbf{k}}|H|\psi_{\mathbf{k}}\rangle$ with respect to $f_{\mathbf{q}}(\mathbf{k})$ one can calculate the eigenvalues and eigenfunctions which represent the dynamically stable states of our system. Since it is a straightforward, albeit tedious, enterprise to carry out this variational procedure, we shall now simply summarize and discuss our principal results.[15]

Specifically, the energy of the system which corresponds to minimizing $f_{\mathbf{q}}(\mathbf{k})$ is given by

$$\begin{aligned}\mathcal{E}_{\mathbf{k}} = -J \sum_{\substack{\mathbf{h}\\ \text{neighbours}}} \mathrm{e}^{-i\mathbf{k}\cdot\mathbf{h}}\, \mathrm{e}^{-S_{\mathbf{k}}(\mathbf{h})}\,[1+S_{\mathbf{k}}(\mathbf{h})] \\ - \mathcal{E}_b\left(\frac{1}{n}\right)\sum_{\mathbf{q}}\Bigg\{1+(J/\hbar\omega_0)\sum_{\substack{\mathbf{h}\\ \text{neighbours}}} \mathrm{e}^{-i\mathbf{k}\cdot\mathbf{h}}\,\mathrm{e}^{-S_{\mathbf{k}}(\mathbf{h})}\,[(2\mathcal{N}_{\mathbf{q}}+1) \\ - \mathcal{N}_{\mathbf{q}}\,\mathrm{e}^{-i\mathbf{q}\cdot\mathbf{h}} - (\mathcal{N}_{\mathbf{q}}+1)\,\mathrm{e}^{i\mathbf{q}\cdot\mathbf{h}}]\Bigg\}^{-1} + \sum_{\mathbf{q}} \hbar\omega_0\,(\mathcal{N}_{\mathbf{q}}+\tfrac{1}{2}),\end{aligned} \qquad (7.31)$$

where J is an electronic transfer integral, $\mathcal{E}_b = A^2/M\omega_0{}^2$ is the small-polaron binding energy in the limit of $J = 0$, and $S_{\mathbf{k}}(h)$ satisfies the relationship

$$\begin{aligned}S_{\mathbf{k}}(\mathbf{h}) = (\mathcal{E}_b/\hbar\omega_0)\left(\frac{1}{n}\right)\sum_{q}[(2\mathcal{N}_{\mathbf{q}}+1) - \mathcal{N}_{\mathbf{q}}\,\mathrm{e}^{-i\mathbf{q}\cdot\mathbf{h}} - (\mathcal{N}_{\mathbf{q}}+1)\,\mathrm{e}^{i\mathbf{q}\cdot\mathbf{h}}] \\ \times\Bigg\{1+(J/\hbar\omega_0)\sum_{\substack{h'\\ \text{neighbours}}} \mathrm{e}^{-i\mathbf{k}\cdot\mathbf{h}'}\,\mathrm{e}^{-S_{\mathbf{k}}(\mathbf{h}')} \\ \times[(2\mathcal{N}_{\mathbf{q}}+1) - \mathcal{N}_{\mathbf{q}}\,\mathrm{e}^{-i\mathbf{q}\cdot\mathbf{h}'} - (\mathcal{N}_{\mathbf{q}}+1)\,\mathrm{e}^{i\mathbf{q}\cdot\mathbf{h}'}]\Bigg\}^{-2}\end{aligned} \qquad (7.32)$$

[$\mathrm{e}^{-S_{\mathbf{k}}(\mathbf{h})}$ is the "vibrational-overlap" factor].

In proceeding to investigate the solutions of eqn (7.32), it is useful to first introduce a number of simplifications. To begin, it is recalled that up to this point of our discussion the phonon occupation numbers, the $\mathcal{N}_{\mathbf{q}}$'s, have been considered arbitrary parameters. However, a considerable computational simplification (which appears to possess no essential liability) is effected by restricting our attention to the situation in which the phonon distribution is the Planck distribution:

$$\mathcal{N}_{\mathbf{q}} = [\exp(\hbar\omega_0/kT) - 1]^{-1} \equiv \mathcal{N}. \tag{7.33}$$

Here k is the Boltzmann constant and T is the temperature. In addition, it is apparent from an examination of eqn (7.32) that the $S_{\mathbf{k}}(\mathbf{h})$'s associated with different nearest neighbour position vectors, $\mathbf{h}$, are coupled to one another. Rather than contend with the difficulties inherent in solving a set of coupled equations of the form of eqn (7.32), it is deemed sufficient for our purposes at this stage of our discussion to study a simpler mathematical situation, namely, that which arises when $\mathbf{k} = 0$.

It is first noted that for an isotropic cubic molecular crystal $S_{\mathbf{k}=0}(\mathbf{h})$ is identical for all values of $\mathbf{h}$. Thus $S_{\mathbf{k}=0}(\mathbf{h})$ may be obtained by investigating only one equation of the form of eqn (7.32) rather than a coupled set of such equations. In particular, replacing $\mathcal{N}_{\mathbf{q}}$ by $\mathcal{N}$ and utilizing the (above-mentioned) symmetry associated with the $\mathbf{k} = 0$ state, one finds that $S_{\mathbf{k}=0}(\mathbf{h}) \equiv S_0$ is the solution of the following equation:

$$S_0 = (\mathcal{E}_b/\hbar\omega_0)(2\mathcal{N} + 1)\left(\frac{1}{\mathscr{n}}\right)\sum_{\mathbf{q}}[1 - \cos q_x]$$
$$\times \Big\{1 + [(6J/\hbar\omega_0)(2\mathcal{N} + 1)\,e^{-S_0}]$$
$$\times \left[1 - \frac{1}{3}(\cos q_x + \cos q_y + \cos q_z)\right]\Big\}^{-2}. \tag{7.34}$$

Although $S_0(\mathbf{h})$ is a real quantity, it should be mentioned that $S_{\mathbf{k}}(\mathbf{h})$ generally possesses an imaginary part. It may also be noted that the solution of eqn (7.34) becomes trivial in the limit of $J \to 0$; namely, $(1/\mathscr{n})$ multiplied by the summed term yields unity and $S_0 = (\mathcal{E}_b/\hbar\omega_0)(2\mathcal{N} + 1)$. This is the result obtained by Holstein in the $J = 0$ limit.[16] Furthermore, in the limit of $(\mathcal{E}_b/\hbar\omega_0)(2\mathcal{N} + 1)$ approaching infinity, S_0 must increase and the denominator approach unity, once again yielding Holstein's formula:

$$\lim_{(\mathcal{E}_b/\hbar\omega_0)(2\mathcal{N}+1)\to\infty} S_0 = (\mathcal{E}_b/\hbar\omega_0)(2\mathcal{N} + 1). \tag{7.35}$$

Evaluating eqn (7.34) in order to obtain a plot of S_0 versus $(\mathcal{E}_b/6J)$ yields

Fig. 7.4. In this figure the two portions of the curve which correspond to a positive slope are drawn with solid lines and the negative-sloped region is shown as a dashed curve. The lower solid curve corresponds to the "weak-coupling" regime while the upper solid curve refers to what is essentially a small-polaron solution. The dashed section of the curve appears to be

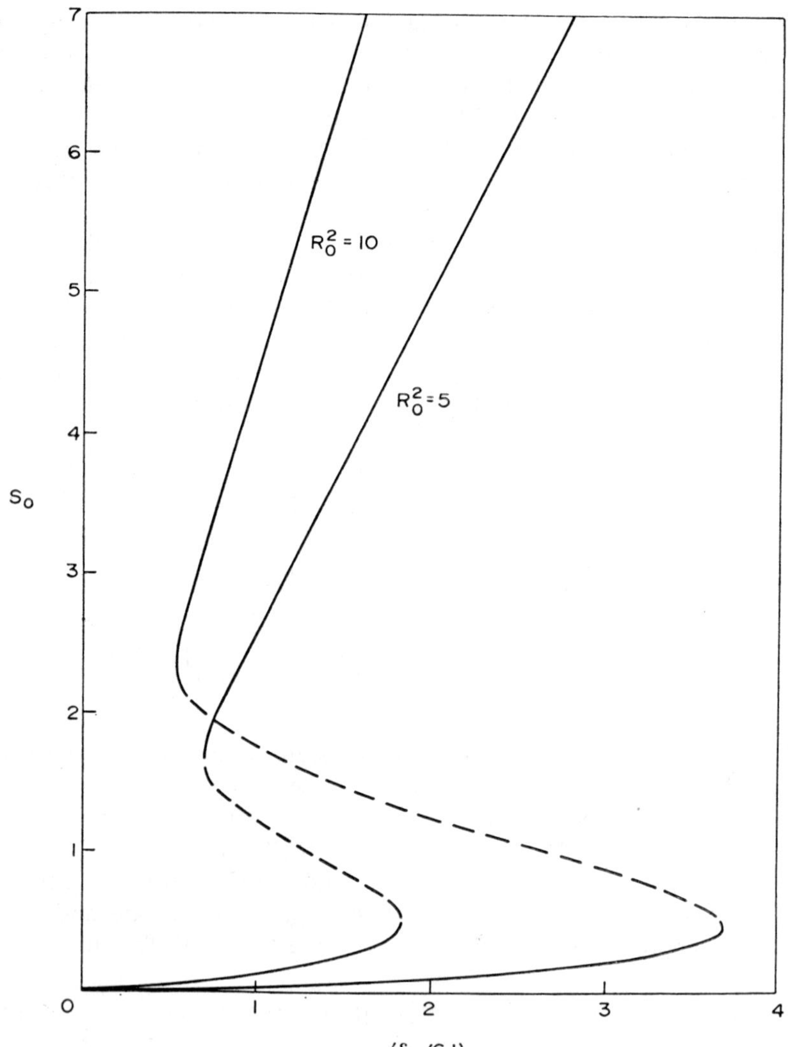

FIG. 7.4. S_0 is plotted as a function of $\mathcal{E}_b/6J$ for two values of $R_0^2 \equiv (6J/\hbar\omega_0)(2\mathcal{N}+1)$. This curve is the solution of eqn (7.34).

irrelevant for our purposes and hence will not be considered. Thus two qualitatively distinct types of states are found to be possible eigenstates of our system. Explicitly, the eigenstates of the system are either those of a weakly-coupled electron–lattice system or they are small-polaron-band states.

The weakly-coupled situation may be visualized as that characterized by an electron carrying a slight lattice deformation with itself as it moves through the lattice. The characteristic radius of the distortion R_p is

$$[(6J/\hbar\omega_0)(2\mathcal{N} + 1)]^{\frac{1}{2}} a,$$

where a is the lattice constant.

This radius is much larger than the lattice constant since $6J$, the rigid-lattice electronic band halfwidth, has been taken to be much greater than $\hbar\omega_0$. The energy of the $\mathbf{k} = 0$ weak-coupling state is simply

$$\begin{aligned} \mathcal{E}_{\mathbf{k}=0} &= -6J + \mathscr{n}h\omega_0 [\mathcal{N} + \tfrac{1}{2}] - \mathcal{E}_b/R_p{}^2 \\ &= -(6J + \mathcal{E}_b/R_p{}^2) + \mathscr{n}h\omega_0 (\mathcal{N} + \tfrac{1}{2}), \end{aligned} \tag{7.36}$$

where $R_p{}^2 \gg 1$.

The self-trapped (or small-polaron) solutions are characterized by the carrier-induced distortion being confined essentially to the site occupied by the excess electron. The energy of a $\mathbf{k} = 0$ small polaron obtained from this calculation is

$$\begin{aligned} \mathcal{E}_{\mathbf{k}=0} &= -6J\,\mathrm{e}^{-S_0} + \mathscr{n}h\omega_0 (\mathcal{N} + \tfrac{1}{2}) - \mathcal{E}_b \\ &= -(6J\,\mathrm{e}^{-S_0} + \mathcal{E}_b) + \mathscr{n}h\omega_0 (\mathcal{N} + \tfrac{1}{2}), \end{aligned} \tag{7.37}$$

where $S_0 = (\mathcal{E}_b/\hbar\omega_0)(2\mathcal{N} + 1) \geqslant 1$.

In the weak-coupling solution the principal contribution to the carrier-related energy is simply the *rigid-lattice* banding energy $-6J$. The small-polaron energy expression, on the other hand, clearly manifests the effects of the electron–lattice interaction in both its terms. The principal contribution to the electron-related energy in this case is, in general, the electron–lattice coupling energy term itself, $\mathcal{E}_b$, while in addition the small-polaron banding-energy contribution also manifests the electron–lattice interaction via its substantial reduction from the rigid-lattice value. Hence, it is clear that the energy expressions associated with the weak-coupling and small-polaron regimes are in fact qualitatively different from one another.

In agreement with the results of applying the adiabatic approximation to the continuum model (Section 7.3), we find that in the adiabatic limit [that is, letting ω_0 tend to zero while the stiffness $M\omega_0{}^2$ remains a constant—the infinite atomic mass limit] there are always two qualitatively distinct eigenstates of the system. However, it is interesting to note that as one relaxes this hypothetical limit, situations arise in which either the weakly-coupled states or the small-polaron states are dynamically unstable. In particular, one may expect that if the lattice is sufficiently "soft" ($M\omega_0^2$ being small enough) and the electron–lattice coupling strength, A, is sufficiently large, a carrier placed in the lattice will immediately distort the lattice so as to ultimately find itself in a self-trapped situation.

The variational calculation yields the result that the ($\mathbf{k} = 0$) weak-coupling state is dynamically stable only if (apart from a minor numerical factor ~ 1)

$$\mathcal{E}_b \equiv \frac{A^2}{2M\omega_0^2} < J\left(\frac{6J}{\hbar\omega_0}\right)(2\mathcal{N} + 1). \tag{7.38}$$

To understand this condition somewhat better physically we may rearrange its terms and take the square root of both sides to obtain (again ignoring numerical factors ≈ 1)

$$\frac{\hbar}{6J} < \frac{[h\omega_0(\mathcal{N} + \frac{1}{2})/M]^{\frac{1}{2}}}{(A/M)}. \tag{7.39}$$

This condition may be read as saying that the weakly-coupled states will be dynamically stable when the time that a carrier associated with a band of halfwidth $6J$ can be confined to a site ($\hbar/6J$) is less than the time it takes for the surrounding atoms to adjust to its presence. The latter time is essentially the velocity attributable to the configurational coordinate of an occupied molecular site of the crystal divided by the acceleration associated with the change in the configurational coordinate of a molecule when occupied by a carrier (simply A/M). Thus one may say that for a carrier to exist in a material (characterized by a short-range electron–lattice interaction) without being self-trapped, it must be able to move between sites sufficiently rapidly so as to preclude the atomic constituents from being able to respond to its presence. Therefore, mechanisms which impede the motion of a carrier tend to inhibit the possibility of having untrapped carriers in a material.

One might expect also that as $\mathcal{E}_b$ ($\mathcal{E}_b \equiv A^2/2M\omega_0^2$) is reduced, it will become impossible to form a dynamically stable small polaron. As before it is found that such a phenomenon is indicated by the variational solution (outside of the adiabatic limit). Namely, it is found that the small-polaron state is only dynamically stable provided that

$$\left(\frac{7}{3}\right)\left(\frac{\mathcal{E}_b}{h\omega_0}\right)\left(\frac{6J}{h\omega_0}\right)(2\mathcal{N} + 1)^2 \exp\left[-\left(\frac{\mathcal{E}_b}{h\omega_0}\right)(2\mathcal{N} + 1)\right] \leqslant 1. \tag{7.40}$$

In particular, the calculation indicates that the small polaron remains essentially confined to a single site throughout the entire domain in which it is dynamically stable. As in understanding the condition governing the stability of the weak-coupling situation, it is worthwhile to develop ancillary arguments, independent of the detailed variational calculation, to display the physical origin of the small-polaron condition. To begin, consider the situation where the carrier-induced distortion is restricted to a single site, i.e., the carrier is a small polaron. The distortion-related energy of the $\mathbf{k} = 0$ small-polaron state expressed in terms of the local distortion parameter, x, is the appropriate minimum of [eqn (16) in Ref. 13]

$$\mathcal{E}(x) = \tfrac{1}{2}M\omega_0^2 x^2 - Ax - 6J \exp\{-[x^2/(\hbar/M\omega_0)][\mathcal{N} + \tfrac{1}{2}]\}, \quad (7.41)$$

where the first term is the lattice distortion energy, the second term depicts the reduction of the carrier's energy as a result of a linear electron–lattice

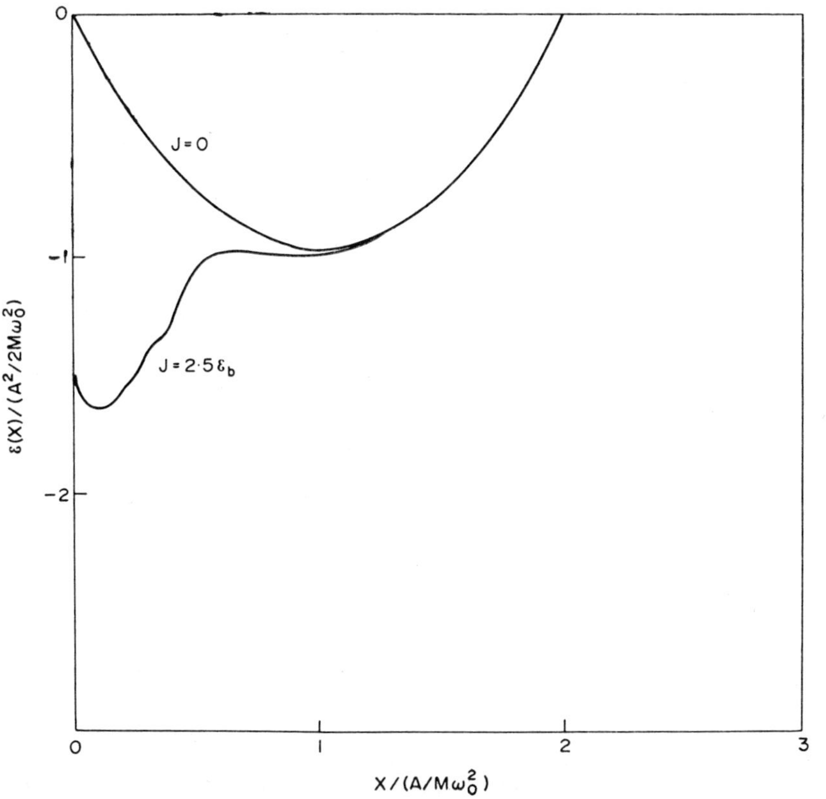

FIG. 7.5. $\mathcal{E}(x)$ in units of $\mathcal{E}_b \equiv A^2/2M\omega_0^2$ is plotted against x, the distortion parameter, measured in units of $A/2M\omega_0^2$. The small polaron solutions are stable when $\mathcal{E}(x)$ possesses a minimum at $x \approx 1$. With an increase of J to $J/\mathcal{E}_b \approx 2{\cdot}5$, the small-polaron minimum disappears.

interaction, and the final term is the lowering of the energy of the system associated with the ability of the small polaron to move through a periodic lattice (this term is the small-polaron band halfwidth). The small-polaron state will represent a dynamically stable state of the system only if $\mathcal{E}(x)$ possesses a minimum at a value of x which is consistent with our assumption of the carrier being a small polaron, namely, for $x \approx A/M\omega_0^2 \equiv [2\mathcal{E}_b/M\omega_0^2]^{\frac{1}{2}}$. Analysis of $\mathcal{E}(x)$ reveals that the minimum in $\mathcal{E}(x)$ at $x = A/M\omega_0^2$ remains relatively unshifted as J increases from zero to a critical value for which the minimum ceases to exist [see Fig. 7.5]. The critical condition for the presence of this minimum is essentially that given by the above inequality [eqn (7.40)]. The essence of this circumstance is that small-polaron formation becomes unstable when a slight decrease in x from its "minimum" at $x = A/M\omega_0^2$ yields a net reduction in $\mathcal{E}(x)$ due to the fact that the rate of lowering of the banding-energy term exceeds the rate at which the strain energy increases.

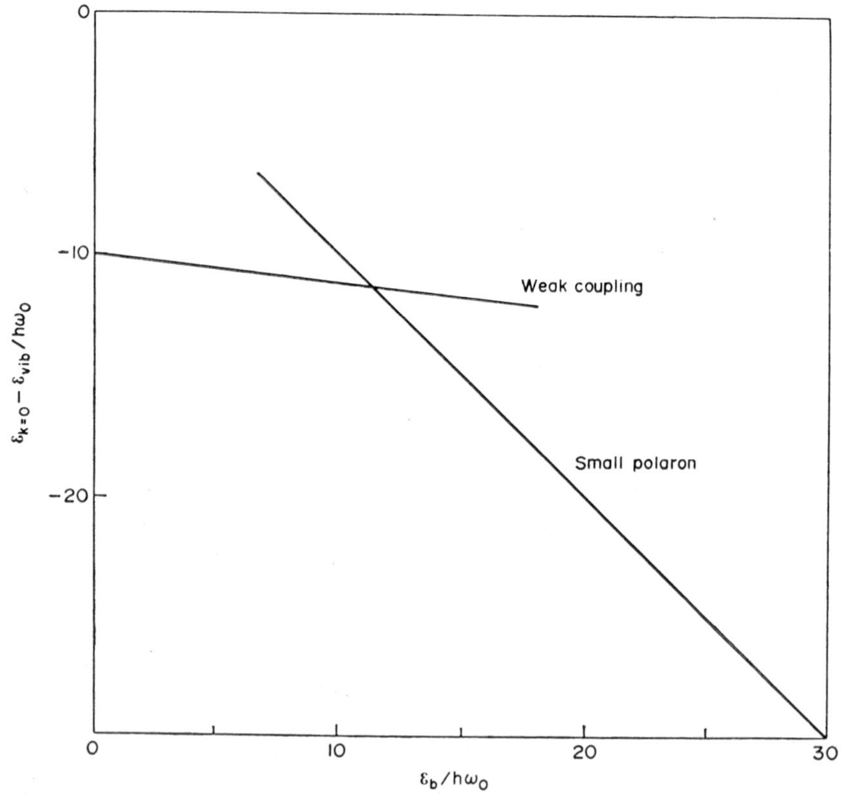

FIG. 7.6. The energy of the two dynamically stable states of the coupled electron–lattice system is plotted against the coupling constant ($\mathcal{E}_b/\omega\tau_0$) for $6J/\tau\omega_0 \approx 10$ and $\mathcal{N} \approx 0$. The two curves intersect at $\mathcal{E}_b \approx 6J$.

The energy of the dynamically stable $\mathbf{k} = 0$ states versus the coupling constant $(\mathcal{E}_b/\hbar\omega_0)$, is plotted in Fig. 7.6. One can observe that a decreasing J or increasing $(\mathcal{E}_b/\hbar\omega_0)$ favours small-polaron formation energetically and dynamically.

Before proceeding to speculate on the application of the physical ideas expressed here to the question of small-polaron formation in the glasses, a number of comments should be made concerning the detailed validity of the

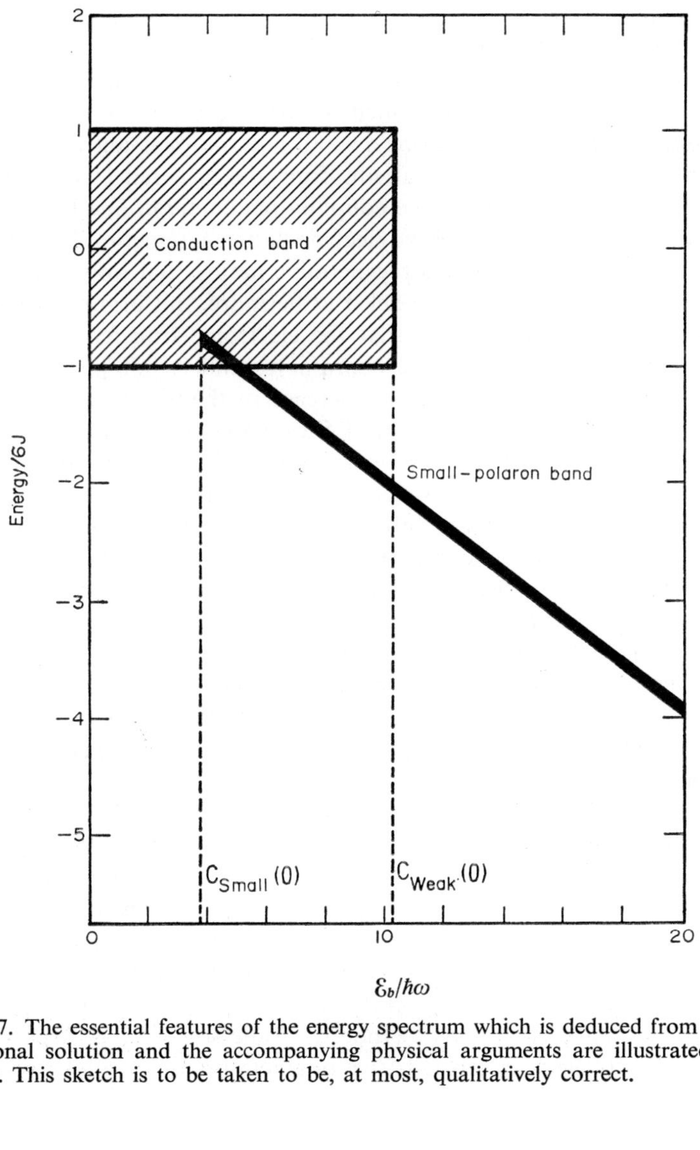

FIG. 7.7. The essential features of the energy spectrum which is deduced from the $\mathbf{k} = 0$ variational solution and the accompanying physical arguments are illustrated here for $\mathcal{N} = 0$. This sketch is to be taken to be, at most, qualitatively correct.

calculations. In particular, only the $\mathbf{k} = 0$ state has been considered here. The results for $\mathbf{k} \neq 0$ are qualitatively similar to the $\mathbf{k} = 0$ solution but cannot, in the opinion of the present author, be profitably pursued in detail by a variational scheme due to increasing doubt regarding the reliability of the $\mathbf{k}$-variation of qualities deduced from a variational theory. However, the situation is presumably as illustrated in Fig. 7.7: The width of the band of states in the weakly-coupled situation is essentially that corresponding to a rigid-lattice band ($\approx 12J$), whereas the small-polaron bandwidth is generally extremely narrow (compared, say, to kT) and is essentially $12J \exp[-(\mathcal{E}_b/\hbar\omega_0)(2\mathcal{N} + 1)]$.

To summarize, we have seen that in materials characterized by a short-range electron–lattice interaction, such as may be expected in covalently-bonded solids, a carrier either forms a small polaron or is untrapped. Such a situation may be thought of as arising from the fact that [see eqn (7.34)] as one increases the electron–lattice coupling parameter $\mathcal{E}_b$ one increases S_0, which in turn increases the "effective electron–lattice interaction" by reducing $J \exp(-S_0)$. It is because of this regenerative nature that only two distinct possibilities arise in this simple model. Furthermore, this argument suggests another which relates to the fact that the electronic transfer integral J is generally dependent on the atomic coordinates and is not just a constant (as assumed here for simplicity). An increase in the electron–lattice interaction manifests itself by, among other things, reducing the effective transfer integral, $J \exp(-S_0)$, not only by increasing S_0 but also by producing a deeper potential well for an excess electron, thereby yielding electronic wavefunctions which are presumably more localized and hence reducing the actual value of J. Thus as a carrier is slowed down its interaction with the lattice increases (the distortion which will be formed about it increases) and the particle is slowed down still further. Viewing "self-trapping" as arising from such "feedback" mechanisms provides some physical insight into understanding why two distinctly different situations arise in our treatment of an excess electron in a deformable lattice. Finally, it may be added that similar arguments can be applied to the effect of the electron–lattice interaction on localized states. Namely, the interaction of a particle in a localized state with the lattice becomes more pronounced as the spatial extent of the localized state is reduced. (Only phonons for which $|q|\, R \lesssim 1$ interact with a localized state of radius R). Since the potential well associated with the localized state is deepened by the carrier-induced distortions, the electron–lattice interaction acts to reduce the spatial extent of the localized state, which in turn acts to increase the effect of the electron–lattice interaction. Again the "feedback" aspect of the interaction of a carrier with the lattice displacements is apparent. Indeed, Anderson[17] has argued that while a sufficiently weak electron–lattice interaction will have little effect on localized states, a sufficiently strong electron–lattice interaction will produce very deep, severely localized, states.

It is speculated that the role of disorder in the solids to which the small-polaron concept may find applicability is that of providing a mechanism which tends to localize carriers; that is, the disorder may aid in small-polaron formation by providing a means to "slow down" or localize a carrier. Such a situation can be associated with both compositional and topological disorder and generally will manifest itself in differences of the local-site energies (the so-called Anderson disorder) and in variations of the transfer integrals associated with different pairs of sites. Of course, the microscopic inhomogeneity of a material may also contribute to small-polaron formation via local changes in the electron–lattice coupling strength and the material's stiffness. It should be clear that much more work needs to be done in order to understand small-polaron formation in noncrystalline as well as crystalline materials. In the foreseeable future the absence or presence of small-polarons in a given material must be decided on the basis of experimental results that will or will not manifest those properties which are expected to be hallmarks of the presence of small polarons. The remainder of this article will be devoted to elucidating the electrical transport phenomena expected to be manifestations of small polarons in noncrystalline solids.

7.6. An Introduction to Small-Polaron Hopping Motion

Having argued that it is plausible for the charge carriers in some disordered materials to form small polarons, it is now our task to investigate the ramifications of such a conjecture as far as experimentally determined quantities are concerned. In particular, let us now proceed to focus our attention on the transport properties that are associated with the motion of small polarons. Our attack on the problem of calculating the small-polaron transport properties will utilize the previously introduced Molecular-Crystal Model. It should be borne in mind that the Molecular-Crystal Model is concerned with a particular situation in which the carrier interacts only with a single optical mode. In general, however, one does not expect the short-range distortion involved in the formation of a small polaron to be associated with only a single optical mode. Rather, the carrier's interaction with all of the optical and acoustical branches should in principle be treated on an equal footing. The justification for studying the Molecular-Crystal Model, however, rests not only on the computational simplifications arising from its utilization but from the fact that this model appears to be adequate for developing an understanding of the essential physical phenomena involved in small-polaron transport. We shall first consider the question of small-polaron motion in a crystalline lattice and then extend our considerations to explicitly take account of some of the effects of charge carriers moving in a disordered environment.

To begin, let us write the wavefunction of an excess electron placed in the molecular crystal as a superposition of the local electronic wavefunctions associated with the carrier residing on any one of the deformable molecules which comprise the molecular crystal:[16]

$$\Psi(\mathbf{r}, \ldots x_{\mathbf{g}} \ldots) = \sum_{\mathbf{g}} a_{\mathbf{g}} (\ldots x_{\mathbf{g}} \ldots)\, \phi_{\mathbf{g}} (\mathbf{r}, x_{\mathbf{g}}). \tag{7.42}$$

Here $\mathbf{r}$ is the position vector of the electron, $\phi_{\mathbf{g}}(\mathbf{r}, x_{\mathbf{g}})$ is the wavefunction of a charge carrier situated on the molecule that is characterized by the position vector, $\mathbf{g}$, and the deformation parameter, $x_{\mathbf{g}}$. The amplitudes $a_{\mathbf{g}}$ that arise in this tight-binding scheme are generally functions of the molecular distortions, the $x_{\mathbf{g}}$'s. If one projects the molecular-crystal-model Hamiltonian on to the system's wavefunction, eqn (7.42), and then employs the approximations which are appropriate to the utilization of a tight-binding approach, one finds that the amplitudes of the tight-binding expansion satisfy the following equation:[16]

$$ih \frac{\partial a_{\mathbf{g}}}{\partial t} = [H_{\text{vib}} - Ax_{\mathbf{g}}]\, a_{\mathbf{g}} - J \sum_{\substack{\mathbf{h} \\ \text{neighbours}}} a_{\mathbf{g}+\mathbf{h}}, \tag{7.43}$$

where H_{vib} is given by eqn (7.19), and the simplifications of assuming (1) that the energy of a carrier on molecule $\mathbf{g}$ is simply $-Ax_{\mathbf{g}}$ and (2) that the electronic transfer integral $J(x_{\mathbf{g}}, x_{\mathbf{g}+\mathbf{h}})$ may be approximated by the constant J have been incorporated into our result.

Considering first the case of $J = 0$, eqn (7.43) is simply the Schrödinger equation which describes the vibratory atomic motion about the displaced equilibrium positions that are assumed when an excess carrier is placed on a site $\mathbf{g}$. For instance, in the absence of vibrational dispersion [$\omega_1 = 0$ in eqn (7.19)] the Hamiltonian in eqn (7.43) (with $J = 0$), $H_{\text{vib}} - Ax_{\mathbf{g}}$, is simply

$$\left[\frac{-\hbar^2}{2M} \frac{\partial^2}{\partial x_{\mathbf{g}}^2} + \tfrac{1}{2}M\omega_0^2 x_{\mathbf{g}}^2 - Ax_{\mathbf{g}}\right] + \sum_{\mathbf{g}' \neq \mathbf{g}} \left[\frac{-\hbar^2}{2M} \frac{\partial^2}{\partial x_{\mathbf{g}'}^2} + \tfrac{1}{2}M\omega_0^2 x_{\mathbf{g}'}^2\right], \tag{7.44}$$

which may be rewritten as

$$\left[\frac{-\hbar^2 \partial^2}{2M\partial x_{\mathbf{g}}^2} + \tfrac{1}{2}M\omega_0^2 \left(x_{\mathbf{g}} - \frac{A}{M\omega_0^2}\right)^2 - \frac{A^2}{2M\omega_0^2}\right] + \sum_{\mathbf{g}' \neq \mathbf{g}} \left[\frac{-\hbar^2 \partial^2}{2M\partial x_{\mathbf{g}'}^2} + \tfrac{1}{2}M\omega_0^2 x_{\mathbf{g}'}^2\right]. \tag{7.45}$$

Here the equilibrium position of the configuration coordinate associated with the **g**th site is displaced by an amount $A/M\omega_0^2$ while the energy of the coupled electron–lattice system is reduced by the amount $A^2/2M\omega_0^2$. In general, the eigenfunctions $a_{\mathbf{g}}(\ldots x_{\mathbf{g}} \ldots)$ in the limit of $J = 0$ are designated by the site index referring to the occupied site and the set of vibrational quantum numbers, $\ldots \mathcal{N}_{\mathbf{q}} \ldots$, which describe the vibratory state of the molecular-crystal lattice. Finally, the $J = 0$ restriction means that if the system is prepared with a particle placed on a specific site the carrier will never be able to move to another molecular site.

For nonzero J an excess carrier will be able to move from site to site within the crystal. The processes by which a charge carrier moves between adjacent sites may be divided into two classes. The first, the so-called "diagonal processes," are those in which the vibrational quantum numbers, the $\mathcal{N}_{\mathbf{q}}$'s, remain unaltered in a charge transfer between neighbouring sites; these processes correspond to a simple one lattice-spacing translation of the charge carrier and its associated lattice distortion without the emission or absorption of phonons. Consideration of these processes by themselves gives rise to small-polaron band motion. That is, just as in a non-deformable lattice, the exact equivalence of a carrier's occupation of any one of a number of sites yields Bloch-type eigenstates for the system, the eigenvalues of which (in the tight-binding scheme) form an energy band of width $12J \exp(-S)$ (for a simple cubic material). This small-polaron bandwidth is very much narrower than that of the band which occurs if the lattice is essentially non-deformable, as is seen from the fact that the factor e^{-S} is generally very much smaller than unity (typically $\sim 10^{-4}$–10^{-5}). The factor e^{-S} is actually just the overlap integral between the two wavefunctions which describe the vibrational state of the system when the carrier resides respectively on either of two adjacent sites in the crystal; in a nondeformable lattice the vibrational wavefunction is unaltered by the transfer of the carrier between neighbouring sites, and hence the overlap factor for a strictly rigid lattice is just unity. The second class of transfer processes, termed "non-diagonal" processes, pertains to the situation where the charge transfer is accompanied by the absorption and/or emission of phonons. In other words, in non-diagonal processes the vibrational quantum numbers are altered when a particle moves between adjacent sites in the material.

If one considers a small polaron moving in a perfectly ordered material, it is found that its description in terms of an essentially free carrier, which suffers occasional scatterings associated with the presence of the atomic vibratory motion, is no longer appropriate when the energy uncertainty associated with such a scattering event is comparable to the small-polaron bandwidth.[16] In this situation one adopts the complementary approach of considering a particle as residing at a site and occasionally hopping (with the emission and/or absorption of phonons) to a neighbouring site. The scattering approach corresponds to a situation in which the particle moves through the lattice primarily via diagonal transitions, while the hopping regime corresponds to the situation in which the carrier moves between sites predominantly by non-diagonal transitions. As a result of the extreme narrowness of the small-polaron band, $<10^{-4}$–10^{-5} eV, the hopping regime is assumed to be the only regime of practical interest in disordered materials, since the fluctuations of the local site energies are generally estimated to be larger than the small-polaron bandwidth. Indeed, even in models of a *perfectly* ordered small-polaron material the band-like motion is found to predominate only at temperatures well below room temperature.(12) Thus our attention will henceforth be restricted to the phonon-assisted hopping of a small polaron.

In thinking about the dc conductivity σ associated with small-polaron hopping motion it is convenient to write

$$\sigma = n_c e\mu, \tag{7.46}$$

where n_c is the carrier density, e is the carrier's charge, and μ is its mobility. Furthermore one can utilize the Einstein relationship to relate the mobility and the diffusion constant D, in say a cubic lattice structure:

$$\mu = \frac{e}{kT} D, \tag{7.47}$$

where T is the temperature and k is the Boltzmann constant. Anticipating the result of studying the hopping motion of a small polaron, we may write the diffusion constant as the product of the rate characterizing a jump of a carrier between neighbouring sites and the square of the lattice constant a (the jump distance):

$$D = Ra^2. \tag{7.48}$$

Thus the small-polaron dc conductivity in a crystalline system is simply related to the small-polaron jump rate R:

$$\sigma = (n_c e^2 a^2/kT) R. \tag{7.49}$$

In a disordered material the jump rate characterizing hops between pairs of adjacent sites is, in general, not identical for all pairs. Hence one must, at least in principle, consider the percolation problem in calculating the small-polaron dc conductivity. However, it should be pointed out that if the relative variation in the jump rate between various pairs of adjacent sites is sufficiently small, eqn (7.49) may be a reasonable approximation to the dc conductivity, R being some suitably defined average. The relationship between the jump rate and the dc conductivity is a question which we shall discuss in somewhat greater detail later. For the moment let us focus our attention on the calculation of the jump rate R.

7.7. THE SMALL-POLARON JUMP RATE

A natural way to begin one's studies of small-polaron hopping motion is to exploit the fact that a carrier does not move if $J = 0$ by carrying out a perturbation expansion of the jump rate in powers of J. While it is not obvious at the outset what the range of validity of such a procedure is, we shall see that ultimately we shall be able to develop criteria for its validity (in at least what we shall call the high-temperature semiclassical regime). The most direct way of obtaining the jump rate is via time-dependent perturbation theory. The somewhat more formal approach of obtaining the jump rate, from a calculation of the dc conductivity which starts from the Kubo formula(18) and is restricted to lowest order in J, can readily be shown to be completely equivalent to the perturbation approach. The details of the formulation of the calculation may be found in Section III of a paper by Holstein.(16) We shall begin our discussion by rewriting Holstein's general expression for the small-polaron jump rate [eqn (54) of his paper]:

$$R = \frac{J^2}{\hbar^2} e^{-2S} \int_{-\infty}^{\infty} dt \, \{e^{F_1(t)} \cos [F_2(t)] - 1\}, \tag{7.50}$$

where the additional steps of (1) extending the limits of integration to infinity and of (2) subtracting the contribution of the diagonal transition to the jump rate (see pages 354 and 359 of Holstein's article(16)) have been taken. In eqn (7.50) we have

$$S = \frac{1}{n} \sum_{\mathbf{q}} \frac{(A^2/2M\omega_{\mathbf{q}}^2)}{(\hbar\omega_{\mathbf{q}})} [1 - \cos \mathbf{q}\cdot\mathbf{h}] \coth (\beta\hbar\omega_{\mathbf{q}}/2), \tag{7.51a}$$

$$F_1(t) = \frac{2}{n} \sum_{\mathbf{q}} \frac{(A^2/2M\omega_{\mathbf{q}}^2)}{(\hbar\omega_{\mathbf{q}})} [1 - \cos \mathbf{q}\cdot\mathbf{h}] \coth (\beta\hbar\omega_{\mathbf{q}}/2) \cos (\omega_{\mathbf{q}} t), \tag{7.51b}$$

$$F_2(t) = \frac{2}{n} \sum_{\mathbf{q}} \frac{(A^2/2M\omega_{\mathbf{q}}^2)}{(\hbar\omega_{\mathbf{q}})} [1 - \cos \mathbf{q}\cdot\mathbf{h}] \sin (\omega_{\mathbf{q}} t), \tag{7.51c}$$

where $\mathbf{h}$ is a nearest-neighbour position vector, n is the number of molecules in the lattice and $\beta = 1/kT$. At this point let us direct our attention to the narrow vibrational-dispersion regime in which [see eqn (7.20)]

$$\omega_{\mathbf{q}} \approx \omega_0 + \tfrac{1}{2}\omega_b \sum_{\substack{\mathbf{h} \\ \text{neighbours}}} \cos(\mathbf{q}\cdot\mathbf{h}), \tag{7.52}$$

where $\omega_b \equiv \omega_1{}^2/2\omega_0$ and we consider the limit of ω_b/ω_0 being arbitrarily small. Then we may write[19]

$$S = (\mathcal{E}_b/\hbar\omega_0)\coth(\beta\hbar\omega_0/2), \tag{7.53a}$$

$$F_1(t) = 2S\left(\frac{1}{n}\right)\sum_{\mathbf{q}}(1 - \cos \mathbf{q}\cdot\mathbf{h})\cos(\omega_{\mathbf{q}}\, t), \tag{7.53b}$$

$$F_2(t) = 2(\mathcal{E}_b/\hbar\omega_0)\left(\frac{1}{n}\sum_{\mathbf{q}}(1 - \cos \mathbf{q}\cdot\mathbf{h})\sin(\omega_{\mathbf{q}}\, t), \tag{7.53c}$$

where $\mathcal{E}_b = A^2/2M\omega_0{}^2$.

A method for the evaluation of such **q**-summations has been developed previously [see references 20 and 11] and thus we can just quote the results of carrying out the summations (the details of these calculations are to be published by this author):[19]

$$F_1(t) = 2S\, A(\omega_b\, t)\cos[\omega_0\, t - \phi(\omega_b\, t)], \tag{7.54(a)}$$

$$F_2(t) = 2(\mathcal{E}_b/\hbar\omega_0)\, A(\omega_b\, t)\sin[\omega_0\, t - \phi(\omega_b\, t)]; \tag{7.54(b)}$$

in the weak-dispersion regime, $\omega_b/\omega_0 \ll 1$, the "relaxation amplitude" $A(\omega_b\, t)$ and the "relaxation phase-angle" $\phi(\omega_b\, t)$ vary very little with time over the vibrational period $2\pi\omega_0{}^{-1}$. The physical significance of the so-called "relaxation functions" has been discussed in detail elsewhere[11,20,21] and will be pursued further at a later point in these lectures (Section 7.10). However, for the present it suffices to observe that a local distortion created at a time $t = 0$ relaxes (dissipating the energy associated with its distortion to the remainder of the material) in a characteristic time which depends on the coupling between the vibrating motion of different molecules; this coupling is associated with a nonzero dispersion of the vibrational frequencies and is governed by the parameter ω_b. The relaxation amplitude $A(\omega_b\, t)$ may be obtained from the bottom line of Table 1 of reference 11 by utilizing eqn (78) of that paper; the relaxation amplitude is plotted in Fig. 7.13, which will be presented later.

Exploiting the relatively small variation of the relaxation functions during a vibrational period, one may reexpress the time integration of eqn (7.50) as a sum of integrals over a vibrational period:

$$\int_{-\infty}^{\infty} d(\omega_0 t) \rightarrow \sum_{n=-\infty}^{\infty} \int_{\pi(2n-1)}^{\pi(2n+1)} d(\omega_0 t). \tag{7.55}$$

Carrying out the integrations over a vibrational period, one obtains

$$R = (2\pi J^2/\hbar^2 \omega_0) e^{-2S} \sum_{n=-\infty}^{\infty} \{I_0[(2\mathcal{E}_b/\hbar\omega_0) A_n \operatorname{csch}(\beta\hbar\omega_0/2)] - 1\}, \tag{7.56}$$

where I_0 is a modified Bessel function and A_n is the relaxation amplitude after n vibrational periods have elasped, i.e., at a time $t \approx n(2n/\omega_0)$. One should observe that in the limit of zero vibrational dispersion $A_n = 1$ and the summation will in general not converge but yield an infinite result. Thus it is seen that the inclusion of vibrational dispersion is a necessary ingredient of a theory of small-polaron hopping motion. The jump rate R (for a simple cubic lattice) measured in units of $2\pi J^2/\hbar^2\omega_0$ is plotted versus $\hbar\omega_0/kT = \theta/T$ in Fig. 7.8 for $\mathcal{E}_b/\hbar\omega_0 = 10$ [the values used in the curves shown in reference 8]; the two curves refer to different choices of the dispersion parameter ω_b/ω_0: curve (a) $\omega t/\omega_6 = \omega_b/\omega_0 = 0{\cdot}1$,curve (b) $\omega_b/\omega_0 = 0{\cdot}01$.

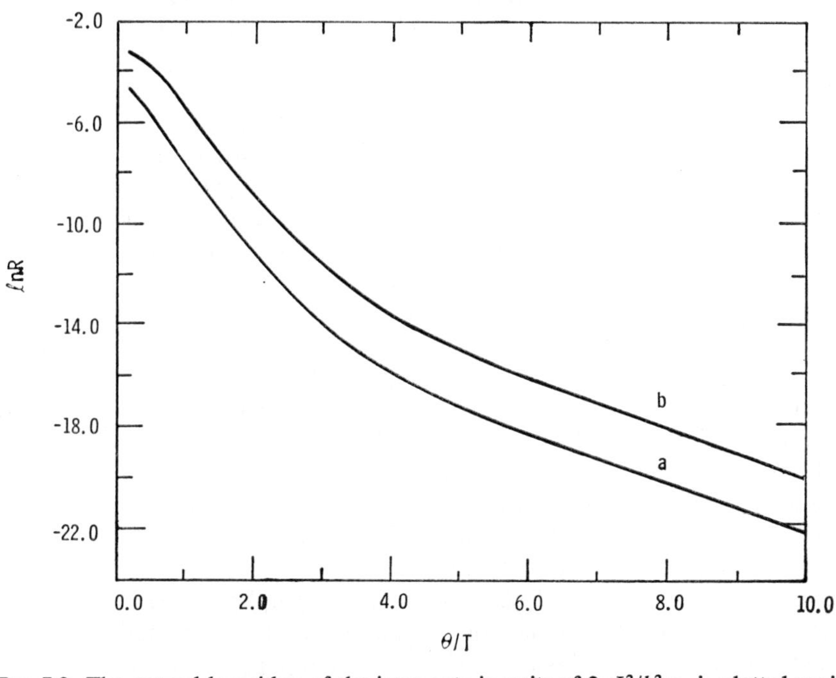

FIG. 7.8. The natural logarithm of the jump rate in units of $2\pi J^2/\hbar^2\omega_0$ is plotted against θ/T ($\theta \equiv \hbar\omega_0/k$) for $\mathcal{E}_b/\hbar\omega_0 = 10$ and two choices of the vibrational dispersion parameter: (a) $\omega_b/\omega_0 = 0{\cdot}1$ and (b) $\omega_b/\omega_0 = 0{\cdot}01$.

As a first step toward understanding these curves let us note their salient features. To begin, a comparison of the two curves shows that a decrease in the vibrational dispersion yields an increase in the jump rate, this effect becoming smaller with increasing temperatures. Each of the curves also manifests two distinct regimes: a low-temperature regime for $\theta/T > 3$ in which $\ln R \propto \theta/T$, and a high-temperature regime ($\theta/T \lesssim 3$) with a considerably greater temperature dependence.

To begin to understand the physical significance of these results let us write down the jump rate, eqn (7.56), in the low-temperature limit, for which the argument of the Bessel function is small:

$$\operatorname{csch} \beta\hbar\omega_0/2 \approx 2\exp[-\beta\hbar\omega_0/2] \ll 1.$$

Explicitly, in this limit the rate is

$$R = \left(\frac{2\pi}{\hbar^2\omega_0} J^2 \,\mathrm{e}^{-2S}\right)\left(\frac{2\mathcal{E}_b}{\hbar\omega_0}\right)^2 \mathrm{e}^{-\beta\hbar\omega_0} \sum_{n=-\infty}^{\infty} A_n{}^2. \tag{7.57}$$

This expression represents the two-phonon transition rate in which one phonon of energy $\sim\hbar\omega_0$ is emitted, and another is absorbed. The relevant Bose factor which arises in these two phonon processes is essentially $\approx\mathcal{N}(\mathcal{N}+1) \approx \mathrm{e}^{-\beta\hbar\omega_0}$, as seen in eqn (7.57). Furthermore the two-phonon character of the rate is manifested by the rate's proportionality to the square of $(2\mathcal{E}_b/\hbar\omega_0)$. It should be noted that the lowest-order phonon-assisted hopping process within the molecular-crystal model is just the two-phonon process. The necessity of the inclusion of vibrational dispersion in calculating a bona fide transition rate is seen by observing the n-summation; as already stated, in the absence of vibrational dispersion $A_n = 1$ and the sum yields an infinite result. In fact, for a one-dimensional chain, since $A_n{}^2 \propto n^{-1}$ for large n, the series also diverges. However, for a two- or three-dimensional system (where $A_n{}^2 \to n^{-2}$ and n^{-3}, respectively), the series does in fact converge. Two additional features of the result perhaps deserve comment. First, it should be noted that the transition rate is proportional to the square of the total transfer integral, $J\mathrm{e}^{-S}$, rather than just the electronic transfer integral J. Such a situation also arises in considering phonon-assisted hopping between impurity states, although in the cases of either a very weak electron–lattice coupling strength or a very shallow impurity level the effect may be small, i.e. $\mathrm{e}^{-S} \approx 1$. Finally, it should be emphasized that the details of the temperature dependence of the jump rate at low temperatures are artifacts of the details of one's model of the coupling of the carrier to the lattice vibrations of the system; our model considers only a narrow band of optical frequencies. However, the phenomenon of the phonon-assisted hopping processes being "frozen out" as one reduces the temperature is expected to be a feature which is independent of the details of the model.

Proceeding now to investigate the high-temperature regime, defined by the product $(2\mathcal{E}_b/\hbar\omega_0)\operatorname{csch}(\beta\hbar\omega_0/2)$ being large compared with unity, it is observed that for sufficiently small values of n, A_n is close to unity and the argument of the Bessel function in eqn (7.56) is therefore also much larger than unity. In this case one may insert the large-argument approximation for the Bessel function into the formula for the jump rate, note that the Bessel function is much larger than unity, and then combine the high-temperature Bessel-function approximation with the high-temperature form for e^{-2S} to obtain an expression for the nth term of the summation, for sufficiently small n. Of particular interest to us is the largest single term, the $n = 0$ term:

$$R(n=0) = \frac{J^2}{\hbar}\sqrt{\frac{\pi}{4\mathcal{E}_2 kT}}\, e^{-\mathcal{E}_2/kT}, \tag{7.58}$$

where $\mathcal{E}_2 \equiv \mathcal{E}_b/2$. It is just this expression which was found by Holstein to be the principal term contributing to the jump rate.[16]

The relative contribution of the $n = \pm 1$ terms compared with the $n = 0$ term is readily found to be[19]

$$\frac{1}{\sqrt{A_1}} \exp\{[A_1 - 1](2\mathcal{E}_b/\hbar\omega_0)(2kT/\hbar\omega_0)\}. \tag{7.59}$$

Noting from reference 11 that $A_1 \equiv A(t = 2\pi/\omega_0) \approx 1 - \frac{1}{2}\pi(\omega_b/\omega_0)^2$, one sees that the condition for the $n = \pm 1$ terms to be small compared with the $n = 0$ term is

$$(2\pi)^2 \left(\frac{\omega_b}{\omega_0}\right)^2 \frac{(\mathcal{E}_2 kT)}{(\hbar\omega_0)^2} \gg 1. \tag{7.60}$$

This condition, obtained initially by Holstein,[16] is indicative of a quantum-mechanical effect in that it is automatically satisfied in the classical limit $\hbar \to 0$. The physical significance of this condition will be discussed when we approach the jump-rate calculation from a semiclassical point of view (Section 7.9). Of course it is clear that, regardless of both the size of the electron–lattice coupling strength, as measured by $(2\mathcal{E}_b/\hbar\omega_0)$, and the temperature, the argument of the Bessel function will be small for sufficiently large n (small A_n). Hence, as in the low-temperature regime, our jump-rate calculation only yields a finite result for two- and three-dimensional systems.

The failure of these rate calculations to give a finite result for a one-dimensional model was previously discovered by de Wit via a computer evaluation of the original jump-rate integral.[22]

7.8. The Jump Rate in a Disordered System

While the calculation of the rate characterizing the jump of a small polaron between neighbouring sites directly yields the dc conductivity (associated with non-diagonal transitions) in an ordered system, such is not the case in a disordered material. In particular, the introduction of a spread in each of the material parameters which govern the small-polaron jump rate (namely A, ω_0, J, ω_b), in addition to the imposition of a spread in the local-site energies, generally will yield a myriad of jump rates which are functions of the particular initial and final sites involved in the hop. The calculation of the d.c. conductivity then involves a solution of a rather involved statistical problem. Of course a necessary ingredient of such a calculation is a knowledge of what the small-polaron jump rate will be between dissimilar sites. It has been a relatively standard practice to somewhat simplify the problem associated with the introduction of disorder by assuming that some of the local parameters do not vary between adjacent carrier positions. This procedure may gain some degree of justification in a number of materials (such as glasses) with the realization that these substances possess short-range order but no long-range order. Furthermore, the incorporation of local variations in ω_0 and A into our calculation has little effect on the primary characteristics of our results. Also, since in a three-dimensional system the lattice relaxation amplitude $A(t)$ is essentially determined by the transfer of vibrational energy from the pair of sites involved in a hop to only their (≈ 10) neighbours, it is a reasonable first approximation to represent this relaxation using a single parameter, ω_b, which may be assumed to be local in character. When restricting our attention to the calculation of the jump rate between a pair of sites, the issue of the relative importance of changes in J and in the local-site energies between different pairs of sites need not be considered, since J can be regarded as the J appropriate (as a scale factor) to the particular pair of sites under consideration. Nevertheless, although the relative variation in the J's and the local-site-energy variations are not completely unrelated, the J variations may often be of subsidiary importance. Thus we shall consider a situation in which the disorder is manifested solely through the removal of the equality of the local-site energies.

In order to generalize our prior calculation to include the energetic disorder in a physical way, we shall consider electronic-energy differences between adjacent sites which are multiples of the phonon energy. In this way we avoid an unphysical situation which arises in the limit of narrow vibrational dispersion for the Molecular-Crystal Model; namely, with phonons restricted in energy to essentially $\hbar\omega_0$ an arbitrarily large number of phonons is required to conserve energy when the local-site energy difference is not a multiple of $\hbar\omega_0$. The jump rate can be readily calculated for such a situation in a way which is analogous to that described for the ordered situation, yielding[(19)]

$$R = \left(\frac{2\pi J^2}{\hbar^2\omega_0}\right) e^{-2S} e^{\mp\beta\Delta/2} \sum_{n=-\infty}^{\infty} \left\{ I_{\Delta/\hbar\omega_0}\left[\left(\frac{2\mathcal{E}_b}{\hbar\omega_0}\right) A_n \operatorname{csch}(\beta\hbar\omega_0/2)\right] - \delta_{\Delta,0} \right\}, \tag{7.61}$$

where Δ is the magnitude of the difference between the electronic energies of the initially occupied and unoccupied sites, $\Delta/\hbar\omega_0$ is an integer, and the upper sign in the term $\exp(\mp\beta\Delta/2)$ refers to a jump upward in energy while the lower sign refers to a jump downward in energy. Figures 7.9 and 7.10 show the respective results of calculating the upward and downward transition rates (measured in units of $2\pi J^2/\hbar^2\omega_0$) as a function of $\hbar\omega_0/kT \equiv \theta/T$ for $\mathcal{E}_b/\hbar\omega_0 = 10$ and various values of the ratio $\Delta/\hbar\omega_0$.

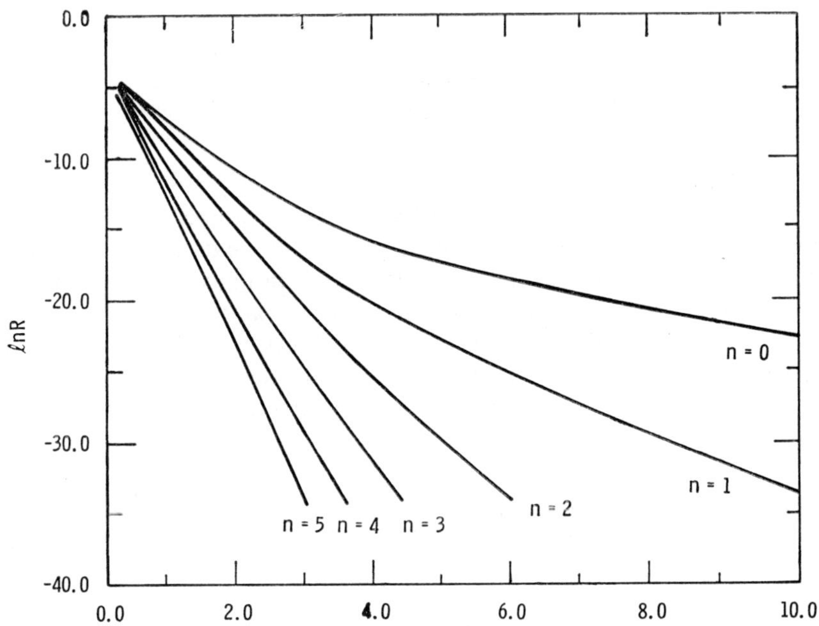

FIG. 7.9. The natural logarithm of the jump rate, in units of $2\pi J^2/\hbar^2\omega_0$, versus θ/T is plotted for the situation in which the carrier hops upward in energy by the amount $\Delta = 2n\hbar\omega_0$. Here $\mathcal{E}_b/\hbar\omega_0 = 10$ and $\omega_b/\omega_0 = 0{\cdot}1$.

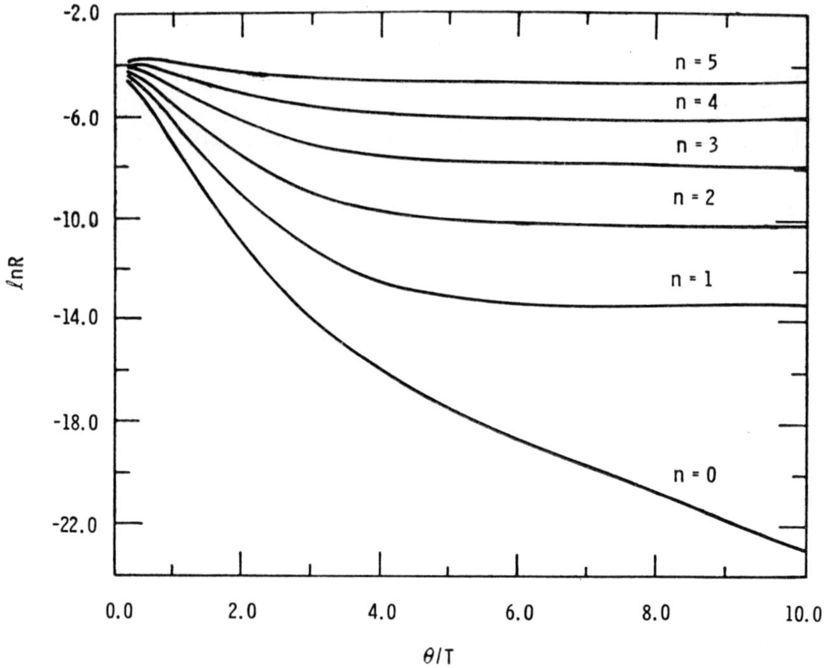

FIG. 7.10. A graph, analogous to Fig. 7.9, depicting the situation in which the carrier hops downward in energy. The magnitude of the electronic energy change in units of $\hbar\omega_0$ is $2n$.

As one might expect, a hop upward at low temperatures becomes increasingly difficult as one lowers the temperature, due to the decreased probability of there being phonons available for absorption in the jumping-up process. Also it is not surprising that as the energy discrepancy increases the jump rate decreases, indicating the decreased probability of multiphonon absorption with an increase in the number of phonons needed to satisfy the conservation of energy requirements. The rate characterizing an upward jump at low temperatures is given by

$$R = \left(\frac{2\pi J^2 \mathrm{e}^{-2S}}{\hbar^2\omega_0}\right) \mathrm{e}^{-\beta\Delta} \left(\frac{2\mathcal{E}_b}{\hbar\omega_0}\right)^{\Delta/\hbar\omega_0} \frac{1}{\Gamma\left[\left(\frac{\Delta}{\hbar\omega_0}\right)+1\right]} \sum_{n=-\infty}^{\infty} (A_n)^{\Delta/\hbar\omega_0} \qquad (7.62)$$

for $\Delta/\hbar\omega_0 \neq 0$. It should be noted that the n-summation converges in a three-dimensional model since, $A_n \propto n^{-3/2}$, even for one-phonon absorption [see eqn (83) of reference 11]. In a model of lesser dimensionality problems will be encountered with the convergence for one-phonon absorption.

The rate characterizing a downward hop becomes temperature-independent at low temperatures due to the fact that the required energy loss can be accomplished by the spontaneous emission of phonons. Because the parameter which governs the emission of phonons, namely $(2\mathcal{E}_b/\hbar\omega_0)$, is taken to be much larger than unity, the rate characterizing the spontaneous emission of phonons increases with the energy differential between the initial and final sites. An explicit expression for the rate of a downward hop at low temperatures is

$$R = \left(\frac{2\pi J^2\,\mathrm{e}^{-2S}}{\hbar^2\omega_0}\right)\left(\frac{2\mathcal{E}_b}{\hbar\omega_0}\right)^{\Delta/\hbar\omega_0}\frac{1}{\Gamma\left[\left(\dfrac{\Delta}{\hbar\omega_0}\right)+1\right]}\sum_{n=-\infty}^{\infty}(A_n)^{\Delta/\hbar\omega_0}, \tag{7.63}$$

where again $\Delta/\hbar\omega_0 \neq 0$. The previously made comments about the n-summation clearly are also applicable here.

In considering the high-temperature approximation for the jump rate we cannot assume that the argument of the Bessel function is large for all values of n, since when n is sufficiently large A_n, and hence the argument of the Bessel function, approaches zero. However, at high enough temperatures so that the arguments of the terms which give the principal contributions to the summation are all large compared with unity one finds that the jump rate in a disordered material is simply related to the jump rate in an ordered material:

$$R(\text{disordered}) = \mathrm{e}^{\mp\beta\Delta/2}\,R(\text{ordered}), \tag{7.64}$$

where the upper sign in the exponent refers to an upward jump and the lower sign refers to a downward hop.

7.9. The Semiclassical Theory

We have seen (in Section 7.8) that at low temperatures the small-polaron jump rate may be thought of as arising essentially from the lowest order (non-vanishing) phonon-assisted process by which a carrier may tunnel between adjacent sites. At high temperatures we have seen that the character of the jump rate is quite different from its low-temperature behaviour; here the activation energy which is associated with a hop, essentially $[(A^2/4M\omega_0{}^2) \pm \Delta/2]$, is independent of the energy of a phonon, $\hbar\omega_0$. Thus, one is led to suspect that in this high-temperature regime (in which the probability of multiphonon processes is not negligible, as it is in the low-temperature regime) one may treat the vibratory motion of the configurational coordinates, the x_g's, classically. Indeed, some time ago Holstein developed a semiclassical approach to calculating the jump rate in the high-temperature regime which is based on just such a notion.[16] It is from this semiclassical theory that one may first obtain a physical feeling for the nature of the jump process and then glean an understanding of the limitations of the perturbation approach to calculating the jump rate.

Let us now proceed to summarize the essential features of Holstein's semiclassical theory. The observation that is the central ingredient of this approach is that (in the classical limit) an electron can only be transferred between adjacent sites when the two local electronic energy levels associated with the carrier residing on either one of the two neighbouring sites, **g** and **g** + **h**, are equal:

$$-Ax_{\mathbf{g}}(t) = -Ax_{\mathbf{g}+\mathbf{h}}(t); \tag{7.65}$$

here the classical nature of this approach is manifested by $x_{\mathbf{g}}(t)$ and $x_{\mathbf{g}+\mathbf{h}}(t)$ being explicit functions of time. Thus in the classical limit the carrier only has an opportunity to hop when, amid the myriad of distortional configurations that are assumed by the oscillating $x_{\mathbf{g}}$'s, such a "coincidence event" occurs. While in the classical limit a coincidence event may be thought of as instantaneous, in fact the "duration" of a coincidence event is governed by the Heisenberg uncertainty relation. Namely, if the carrier "sees" (measures) the energy "equality" for a time interval Δt_c about a coincidence event, the minimum uncertainty in the energy measurement is $\sim\hbar/\Delta t_c$. However, during a time Δt_c the difference in the local electronic energies, $-A(x_{\mathbf{g}} - x_{\mathbf{g}+\mathbf{h}})$, generally changes by an amount $\sim Av\Delta t_c$, where

$$v = \frac{\mathrm{d}}{\mathrm{d}t}\left[(x_{\mathbf{g}}(t) - x_{\mathbf{g}+\mathbf{h}}(t)]\right|_{t=t_c}. \tag{7.66}$$

v is essentially a typical vibrational velocity. Equating the minimum uncertainty in the energy determination which is associated with a measurement of duration Δt_c to the spread in electronic energies which the carrier witnesses in a time Δt_c, one obtains

$$\frac{\hbar}{\Delta t_c} = Av\Delta t_c. \tag{7.67}$$

Utilizing this expression, one obtains Holstein's expression for the "duration" of a coincidence event:[16]

$$\Delta t_c \approx (\hbar/Av)^{\frac{1}{2}}. \tag{7.68}$$

The magnitude of the quantum-mechanical transition amplitude associated with a carrier moving between two sites as a result of a coincidence event between the sites is (in the perturbation scheme in which J is taken to be small) essentially $(J/\hbar)\,\Delta t_c$. That is, the magnitude of the transition amplitude is the product of the rate which governs the transfer of a carrier from one *equivalent* site to another, $J/\hbar$, and the duration of a coincidence. It is assumed in this perturbative treatment that $J\Delta t_c/\hbar \ll 1$.

Since the local atomic vibrations associated with the distortion to which the electron is coupled are nearly periodic, a coincidence event at time zero may often be followed by another coincidence event at a time $\sim 2\pi/\omega_0$ later. Thus the total amplitude for a hop will generally be a sum of amplitudes, each of which is associated with one of the coincidence events that occur between the time a particle is prepared (at time zero) as being on a site and the time at which one is observing the system. Therefore the transition probability that may be associated with the most recent coincidence event includes not only the square of the amplitude associated with that particular coincidence event, but also cross terms which are essentially products of the amplitudes associated with different coincidence events. If there is sufficient vibrational dispersion each such cross term may be shown to yield a contribution to the jump rate which is small compared with that of the squared term. The condition for the cross terms associated with successive coincidence events to be small compared with the squared term may be shown to be the previously obtained condition [see eqn (7.60)]

$$(2\pi)\left(\frac{\omega_b}{\omega_0}\right)\frac{(\mathcal{E}_2\,kT)^{\frac{1}{2}}}{(\hbar\omega_0)} \gg 1. \tag{7.69}$$

It may therefore be stated that the role played by vibrational dispersion is that of providing a mechanism which tends to destroy the correlation between an electron's responses to different coincidence events.

Assuming that the vibrational dispersion (and dimensionality) are sufficiently great to justify treating the carrier's response to a given coincidence event as independent of prior coincidence events, one may write the jump rate in the high-temperature regime as

$$R = P\left[\left(\frac{\omega_0}{2\pi}\right)e^{-\mathcal{E}_A/kT}\right], \tag{7.70}$$

where the square-bracketed term is the rate at which a coincidence event involving the initially occupied site and an unoccupied site occurs, $\mathcal{E}_A$ is the minimum energy required to produce the distortion associated with the coincidence event, and P is the probability that the electron will hop, given the coincidence event:

$$P = \frac{J^2}{\hbar^2}\left(\frac{2\pi\,\hbar\sqrt{\pi}}{A\sqrt{kT/M}}\right) \approx \left(\frac{J}{\hbar}\,\Delta t_c\right)^2. \tag{7.71}$$

A more complete discussion of this approach is found in Holstein's original calculation.[16]

It is clear from the interpretation of P as the probability of a carrier hop when a coincidence event occurs that P can never be allowed to be greater than unity. This observation enables us to obtain a measure of the maximum value of J for which the perturbation approach may be applied (in the high-temperature regime); namely, the perturbation approach is only valid for $P < 1$. Thus, for $J^2 < \hbar A\sqrt{k/TM}$ the carrier moves so slowly that it cannot always follow the atomic motion. In this situation the carrier may continue to reside on a site even if the occupancy of a neighbouring site becomes energetically more favourable; it simply responds too slowly to avail itself of the opportunity to hop. Outside of this regime, for $J^2 > [\hbar A\sqrt{k/TM}]/(2\pi^{3/2})$, the carrier can move with sufficient speed so as to permit it to adjust in general to the motion of the lattice; this is the *adiabatic* regime, for which we set $P = 1$. This situation is discussed in considerable detail in Appendix III of the Holstein paper[16] and in a paper by Emin and Holstein.[23] It can be seen from the discussion of Appendix I of the latter paper that the time which characterizes the destruction of a potential well that will trap the electron at the initial site and the creation of a comparable potential well at an adjacent site is $t_{\text{atomic}} \sim (J/A)/v \sim J/A\sqrt{kT/M}$. Thus, since the time required for the transfer of an electron between two adjacent equivalent sites is $t_{\text{transfer}} \sim \hbar/J$, one may rewrite the condition for the carrier to follow the lattice, $P > 1$, as $t_{\text{atomic}} > t_{\text{transfer}}$. The condition $P > 1$ defines the regime of validity of the adiabatic approximation while the condition $P < 1$ defines the non-adiabatic regime. In general, the utilization of the non-adiabatic theory (the perturbation approach) is valid only for very small values of the electronic bandwidth parameter, $J \lesssim 10^{-2}$ eV. It should be cautioned that although we have employed the simplification of treating J as independent of the x_g's it is not generally constant. Thus the value of J which will occur in the (high-temperature) rate formula and in the adiabaticity condition more closely corresponds to a coincidence situation than the value calculated from considering the lattice to be rigid.

So far our discussion of the semiclassical approach to the small-polaron jump rate in the high-temperature regime has not explicitly dealt with the effects of disorder. Proceeding as before to consider the effects of disorder of the local electronic energies, we focus our attention upon how such disorder affects the calculation of the activation energy which occurs in the expression for the jump rate. It is recalled that in the previously discussed quantum-mechanical calculation (Section 7.8) the sole modification of the jump rate arising from the inclusion of disorder was the presence of the additional activation factor $\exp(\mp\beta\Delta/2)$ in eqn (7.61) provided that (a) the dominant contribution to the n-summation came from the $n = 0$ term [or at least the small-n terms, $(n \lesssim \omega_0/\omega_b)$] and that (b) $\mathcal{E}_2 kT \gg \Delta^2$. To begin our semiclassical calculation of the activation energy associated with a hop in our model of a disordered system, the physical significance of the activation energy must be borne in mind. Namely, the activation energy associated with a hop is the minimum distortion-related energy that is required in order to produce a coincidence event. The distortion-related energy associated with a carrier occupying a particular site is[(24)]

$$\mathcal{E}_{\text{distortion}} = \frac{M\omega_0^2}{2}\left(x_0 - \frac{A}{M\omega_0^2}\right)^2 + \sum_{g\neq 0} \frac{M\omega_0^2 x_g^2}{2}, \qquad (7.72)$$

where we have for simplicity neglected the vibrational dispersion. The configuration coordinate associated with the occupied site (taken to be the zeroth site) oscillates about a displaced origin that reflects the distortion of that site due to the presence of the carrier. The condition which characterizes a coincidence between the occupied site and a particular neighbouring site (site 1 say) is

$$\mathcal{E}_0 - Ax_0 = \mathcal{E}_1 - Ax_1, \qquad (7.73)$$

i.e., the electronic energy levels of the two sites are equal. Redefining Δ as $\mathcal{E}_1 - \mathcal{E}_0$ (rather than the magnitude of this energy difference) and eliminating x_1 in eqn (7.72) by utilizing the coincidence condition, eqn (7.73), one has

$$\mathcal{E}_{\text{distortion}} = \frac{M\omega_0^2}{2}\left(x_0 - \frac{A}{M\omega_0^2}\right)^2 + \frac{M\omega_0^2}{2}\left(x_0 + \frac{\Delta}{A}\right) + \sum_{g\neq 0,1} \frac{M\omega_0^2 x_g^2}{2}, \qquad (7.74)$$

where the first two terms are the distortion-related energies of sites zero and one, respectively. Minimizing $\mathcal{E}_{\text{distortion}}$ with respect to x_0 and the other x_g's, one obtains the result that the distortional energies (which occur in the minimum energy-coincidence configuration) of the two involved sites are equal to each other, each being equal to $\frac{1}{4}[\mathcal{E}_2 + (\Delta/2) + (\Delta^2/16\mathcal{E}_2)]$; the activation energy associated with a hop between sites 0 and 1 is therefore $[\mathcal{E}_2 + (\Delta/2) + (\Delta^2/16\mathcal{E}_2)]$, where $\mathcal{E}_2 = A^2/2M\omega_0^2$. The jump rate between two sites whose local electronic levels are separated by Δ differs from the jump rate between equivalent sites by the factor

$$\exp[\beta\Delta/2] \exp\left[\frac{\beta\Delta^2}{16\mathcal{E}_2}\right],$$

which in the appropriate limit, $\beta\Delta^2/\mathcal{E}_2 \ll 1$, agrees with the result obtained by the quantum-mechanical calculation.

If one relaxes the restriction that the local parameters A and ω_0 are unaltered by either the disorder or the presence of the carrier, the distortional energies associated with the two sites are no longer equal to one another. Rather they are in the ratio

$$\frac{\mathcal{E}_{\text{coincidence}}(0)}{\mathcal{E}_{\text{coincidence}}(1)} = \frac{(A_0/\omega_{00})^2}{(A_1/\omega_{01})^2}, \tag{7.75}$$

where the indices of A and ω_0 refer to the involved sites. The hopping activation energy is then

$$\left[\mathcal{E}_2{}^0 + \frac{\Delta}{2} + \frac{\Delta^2}{16\mathcal{E}_2{}^0}\right] \Big/ \left[1 + \frac{\mathcal{E}_2{}^1 - \mathcal{E}_2{}^0}{2\mathcal{E}_2{}^0}\right], \tag{7.76}$$

where $\mathcal{E}_2{}^0 \equiv A_0{}^2/4M\omega_{00}{}^2$ and $\mathcal{E}_2{}^1 \equiv A_1{}^2/4M\omega_{01}{}^2$. It is not an unreasonable first approximation to neglect the effects associated with this more elaborate model of the disorder, since the alterations of the theory resulting from its inclusion do not appear to be large.

Finally, it should be added that in our discussion of the calculation of the jump rate we have taken the energy associated with a carrier being on a site to be that appropriate to an isolated molecular site, namely $\mathcal{E}(x_1) = -Ax_1$. Actually there are corrections to this expression which result from the fact that the molecules are not isolated, but such additional terms are only of significance in the adiabatic regime (sufficiently large J). For instance, for a two-site model within the adiabatic approximation the electronic energy associated with a particle that is allowed to occupy site 1 or 2 is given by [see eqn (1.5) of reference 23]

$$\mathcal{E}(x_1, x_2) = -\frac{A(x_1 + x_2)}{2} - \left[\frac{A^2(x_1 - x_2)^2}{4} + J^2\right]^{\frac{1}{2}}. \tag{7.77}$$

If a particle resides on site 1 ($x_1 \approx A/M\omega_0{}^2, x_2 \approx 0$) then $\mathcal{E}(x_1, x_2)|_{\text{site}\,1} \approx -Ax_1 \approx -A^2/M\omega_0{}^2\,[1 + J^2/(A^2\,M\omega_0{}^2)^2]^{\frac{1}{2}} \approx -A^2/M\omega_0{}^2$. However, if one considers the minimum energy-coincidence configuration, $x_1 = x_2 = A/2M\omega_0{}^2$, then $\mathcal{E}(x_1 = x_2) = -A^2/2M\omega_0{}^2 - J$. Thus the activation energy calculated for the adiabatic regime is reduced from the previously obtained value (for the nonadiabatic regime), $\mathcal{E}_A = A^2/2M\omega_0{}^2$, by the amount J:

$$\mathcal{E}(x_1 = x_2) - \mathcal{E}(x_1, x_2)|_{\text{site}\,1} \approx A^2/2M\omega_0{}^2 - J, \tag{7.78}$$

where $A^2/2M\omega_0{}^2 \gg J$. A discussion of the calculation of the jump rate in the adiabatic regime is found in Appendix I of reference 23.

7.10. LATTICE RELAXATION EFFECTS: CORRELATED SMALL-POLARON MOTION

A basic assumption in the herein discussed theories of small-polaron hopping motion, as well as in analogous treatments of atomic diffusion and excitonic motion, is that each hop of a carrier may be treated as independent of its prior hops. Preliminary considerations (outlined on page 3648 of reference 11) based on the semiclassical occurrence probability approach indicate that it is a well-founded procedure to assume that the electron's *responses* to successive coincidence events involving different pairs of sites are uncorrelated:

$$P(1 \to 2, 0;\ 2 \to 3, t) = P(1 \to 2)\, P(2 \to 3). \tag{7.79}$$

The probability that an electron will hop from site 1 to site 2 at time zero given a coincidence between these two sites at $t = 0$, and then hop from site 2 to site 3 at a later time t provided there is a coincidence involving sites 2 and 3 at time t, is simply the product of the jump probabilities associated with the two coincidence events. However, the occurrence of successive coincidence events between an occupied site and any neighbouring site will in general be correlated. In other words, a coincidence event between a pair of sites at a given time will have a greater than random probability of being followed by another coincidence event (involving at least the occupied site) at a time $\sim 2\pi/\omega_0$ later.

To understand why this is so let us refer to Fig. 7.11, in which the three stages of a hop in a periodic system are illustrated. Consider the situation depicted in Fig. 7.11(a), in which the equilibrium atomic position for an excess charge in an array of diatomic molecules is illustrated; the length of the vertical line at each site is a schematic way of representing the displacements of the atoms of each molecule. The excess charge is denoted by the black dot. In this figure the "irrelevant" thermal vibrations of the molecules have been suppressed for simplicity.

As we have previously discussed (in Section 7.9), the hop of a carrier to an adjacent site is associated with an atomic distortion which brings the electronic energy level of the occupied site and that of a neighbouring site into momentary coincidence. As illustrated in Fig. 7.11(b), such a coincidence event involves a substantial (thermal) distortion of the relevant diatomic molecules. The minimum energy required to produce such a situation is the activation energy associated with a hop.

Having achieved the coincidence situation illustrated in Fig. 7.11(b), it is observed that, in the usual picture of small-polaron hopping, the jump is completed when the distorted molecules relax, dissipating an energy that equals the activation energy to the remainder of the material and leaving the carrier once again trapped at a single site. Figure 7.11(c) depicts the situation where such a hop to a "new" site has occurred.

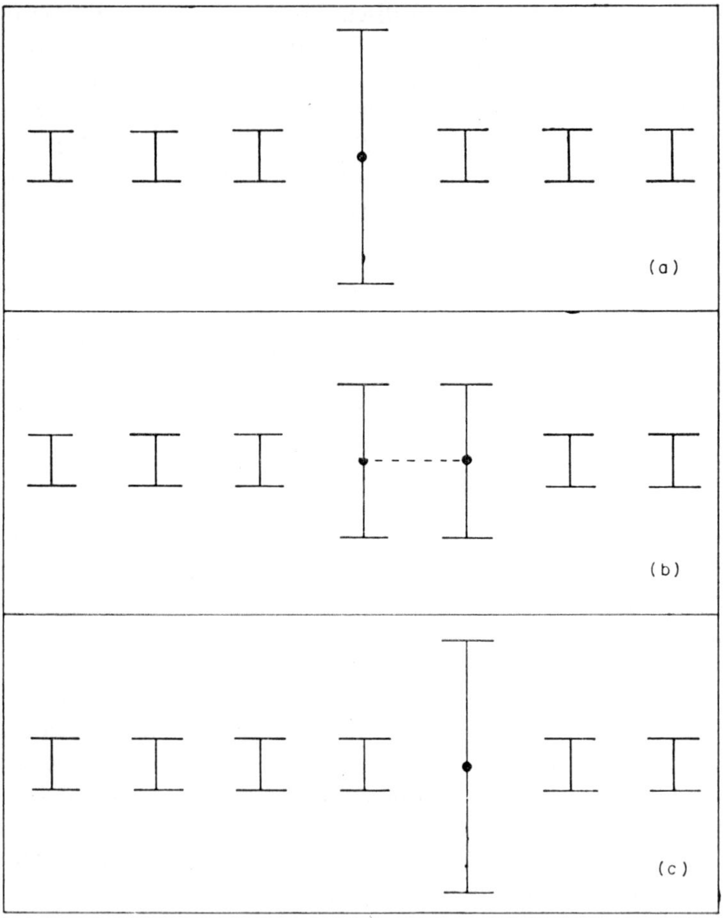

FIG. 7.11. The small-polaron jump process. (After Ref. 25.)

It is clear that such a picture is only reasonable if the time characterizing the relaxation of the lattice is short compared with the mean time between hops.[21] If one considers only the effect of one hop on its successor, it can be shown that the activation energy characterizing the second hop is reduced if the distortion associated with the first hop has not fully dissipated before the second hop occurs.[20,25] Thus it is useful to consider the activation-like energy which is the additional energy required to produce a coincidence event occurring at a time t after an initial coincidence event at time zero. Specifically, one finds this energy to be[11]

$$\mathcal{E}_A(t) = \mathcal{E}_2[1 - G_{\mathbf{h}}(t)]/[1 + G_{\mathbf{h}}(t)], \quad (7.80)$$

where $\mathcal{E}_2$ is the activation energy associated with a two-site coincidence event (and hence an elementary hop) in a relaxed lattice, and $G_{\mathbf{h}}(t)$ is the "relaxation function" pertaining to a coincidence between the occupied site and a neighbouring site at position **h**. The relaxation function $G_{\mathbf{h}}(t)$ is related to the previously introduced "relaxation amplitude" and "relaxation phase angle" [see eqn (7.54)] by the relation

$$G_h(t) = A_h(t) \cos[\omega_0 t - \phi(\omega_b t)], \quad (7.81)$$

where **h** refers to the neighbouring site which was coincident with the occupied site at $t = 0$; a detailed discussion of the relaxation function is found in References 11 and 20. To demonstrate the relationship between the relaxation of the relevant distortions and the relaxation function itself, the following result is observed [eqn (32) in reference 11]:

$$x_{\mathbf{h}}(t) - x(t) = [x_{\mathbf{h}}(\infty) - x(\infty)][1 - G_{\mathbf{h}}(t)], \quad (7.82)$$

where $x_{\mathbf{h}}(t)$ and $x(t)$ are the molecular distortion coordinates referring to site **h** and the occupied site, respectively. While the relaxation function has been studied in detail for a periodic system, the concern here will only be with those general features that may be relevant to a system which is not highly ordered. To begin, it is noted that for $t = 0$ we have $G_{\mathbf{h}}(0) = 1/2$ and $\mathcal{E}_A(0) = \mathcal{E}_2/3$ when **h** refers to a noncoincident site.[11] If **h** refers to the unoccupied coincident site, $G_{\mathbf{h}}(0) = 1$ and $\mathcal{E}_A(0) = 0$, thereby manifesting the feature that an immediate return hop requires no additional energy since the coincidence configuration already exists.[11] As the initial distortion dissipates ($t \to \infty$), $G_{\mathbf{h}}(t)$ approaches zero and at the same time $\mathcal{E}_A(t)$ approaches the value appropriate to a relaxed lattice, $\mathcal{E}_2$. Furthermore, the time characterizing the decline of the amplitude of $G_{\mathbf{h}}(t)$ depends very strongly on the dimensionality of the system. For instance, in a linear chain model the lattice relaxation involves many transfers of vibrational energy back and forth between the initially distorted sites and their neighbours.[11] However, in a three-dimensional system it is found that the relaxation proceeds much more efficiently due to the increased number of neighbours which are available to receive the vibrational energy. In fact, the demise of the initial coincidence distortion is essentially completed when energy is transferred to only the immediately adjacent sites. The time characterizing this energy transfer is the reciprocal of the optical phonon bandwidth parameter ω_b.[11] This feature is apparent from a comparison of Figs. 7.12 and 7.13, in which a relaxation amplitude for a linear-chain molecular crystal and for a three-dimensional crystal respectively are plotted.

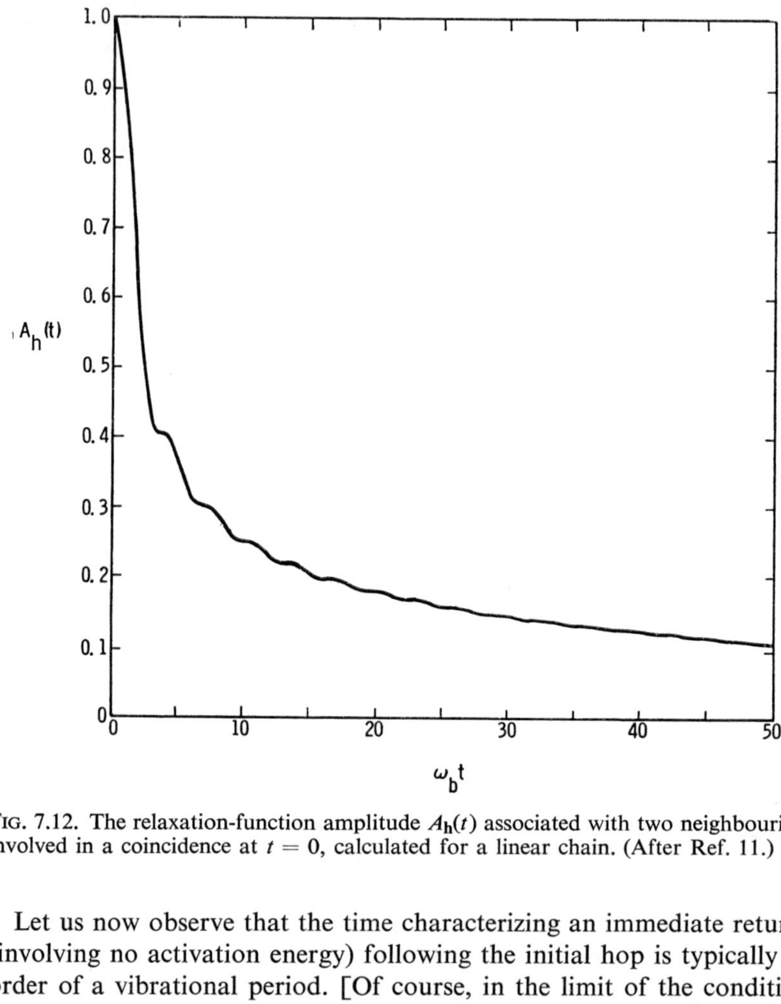

FIG. 7.12. The relaxation-function amplitude $A_h(t)$ associated with two neighbouring sites involved in a coincidence at $t = 0$, calculated for a linear chain. (After Ref. 11.)

Let us now observe that the time characterizing an immediate return hop (involving no activation energy) following the initial hop is typically of the order of a vibrational period. [Of course, in the limit of the condition for the validity of the nonadiabatic theory being very well fulfilled, the return-jump time will be longer than $2\pi/\omega_0$.] In that this time is generally considerably shorter than the time required for the coincidence configuration to relax substantially, the carrier should be viewed, within this two-hop model, as moving back and forth between the two coincident sites. In this situation a contribution to the net diffusion of the particle occurs when the carrier alters its back-and-forth jumping motion to hop to another (third) site. The rate characterizing such a jump is that appropriate to an unrelaxed lattice, that is, the diffusive activation energy is roughly a third of that appropriate to a jump in a relaxed lattice. From detailed consideration of the type outlined here it is concluded that both the drift and Hall[20,21] mobility activation energies associated with the hopping motion of small polarons may (but need not) be sufficiently small so as to be effectively counterbalanced by the pre-exponential temperature dependence of the mobility expressions. Such a situation is illustrated in Fig. 7.14, in which the drift mobility is plotted for two choices of the activation energy.

Some additional comments are now in order prior to applying this theory to disordered materials. In that is has been demonstrated that the process by which a small polaron may move between adjacent sites essentially involves only the vibrational and electronic coupling between neighbour

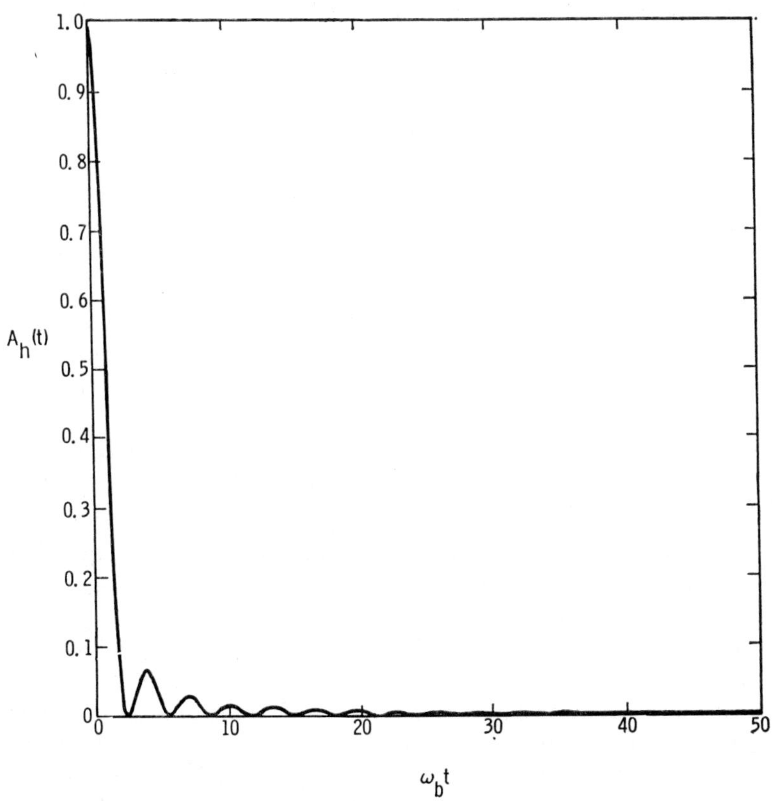

FIG. 7.13. The relaxation-function amplitude $A_h(t)$ associated with the two neighbouring sites involved in a coincidence at $t = 0$, calculated for a three-dimensional simple cubic array of sites. (After Ref. 11.)

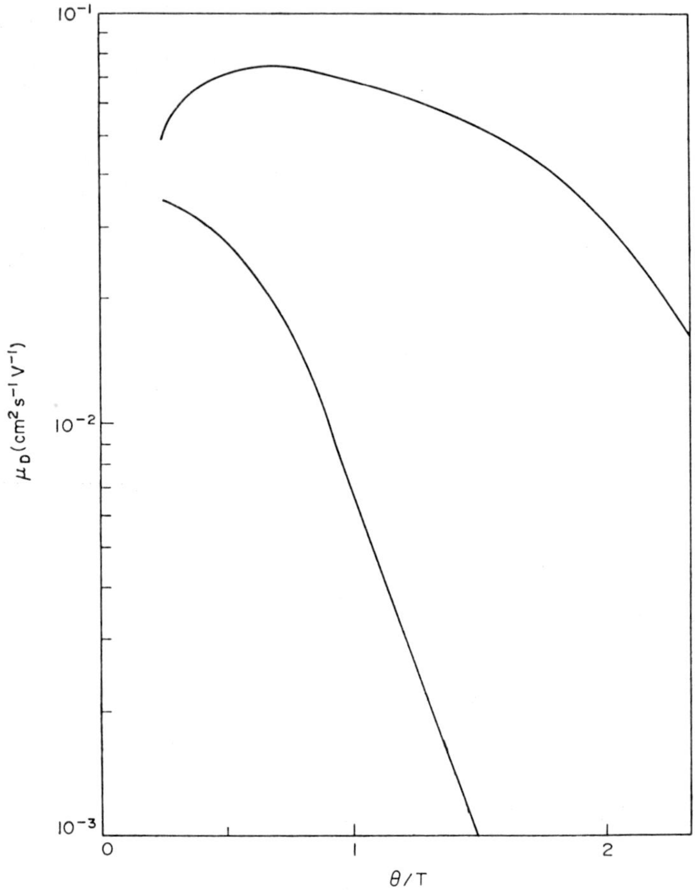

FIG. 7.14. Typical drift mobility curves: $\mathcal{E}_2 \equiv \mathcal{E}_b/2 = 5k\theta$ in the upper curve and $\mathcal{E}_2 = 15k\theta$ in the lower curve. The electronic transfer integral J equals $k\theta$ in these plots; $k\theta$ is the optical-vibration energy. (After Ref. 20.)

sites, it is seen that our simplification of assuming the exact equality of *all* local vibrational frequencies and coupling constants plays no essential role in our results. However, it should be noted that a difference in the local vibrational frequencies of the sites involves in the hopping and concomitant relaxation process does affect our considerations. In such a situation not only is the phase relationship between the oscillations of adjacent molecules altered but the vibrational-energy transfer process is shifted "off-resonance," thereby slowing down the lattice relaxation process.[25] Studies of systems possessing disorder of the vibrational parameters indicate that lattice relaxation effects play a comparable or larger role in such systems than in the strictly periodic systems. Thus while the mobilities associated with the hopping of small polarons are generally less than $ea^2 \nu_{\text{vib}}/kT \sim 1\text{cm}^2\,\text{V}^{-1}\,\text{sec}^{-1}$, a clearly activated temperature dependence *need not* be manifested unless the hopping time is considerably longer than the lattice relaxation time; such a condition is not expected to be fulfilled in a wide assortment of low-mobility materials. Finally, it should be observed that these considerations demonstrate the interrelationship of charge transport and the transport of vibrational energy. In particular, to understand the small-polaron hopping mechanism in disordered materials it is generally necessary to consider in detail the transport of vibrational energy in such substances. Finally, it should be mentioned that the method which has been employed in references 11, 20, 21 and 25 to incorporate lattice relaxation effects into the small-polaron hopping theory has been "bootstrap" in nature. It remains a difficult problem to study such effects via a general linear response formalism. In addition, one must bear in mind the fact that the theory has been developed for a carrier interacting with optical phonons in the weak dispersion limit; the theory will be somewhat altered if one considers more elaborate models in which the carrier interacts strongly with a number of different vibrational modes.

7.11. The Hall Mobility of the Small-Polaron

The measurement of the Hall mobility is a technique which may be of considerable value in elucidating the mechanism that characterizes the motion of a charge carrier. In particular, since it is quite clear that one is generally unable to infer much about the magnitude or temperature dependence of the carrier mobility from a conductivity measurement alone, one must study those phenomena which yield more direct information about the carrier's transport mechanism. The measurement of the Hall mobility is one such technique for doing this.

In that the theory of the Hall mobility of the small polaron is a topic which by itself could easily provide sufficient material for a lecture series, we shall restrict ourselves here to a presentation of the principal results and conclusions of the theory, and refer the reader to the original literature for more detailed information. To begin, let us review what is meant by the Hall mobility. The microscopic mobility (referred to as the drift mobility in the small-polaron literature) provides a measure of the response of a carrier to the application of a small electric field, E; $\mu_D \equiv v_D/E$, where μ_D and v_D are respectively the drift mobility and drift velocity associated with a single excess carrier. The Hall mobility, on the other hand, provides a measure of the response of a charge carrier to an applied magnetic field. More precisely, if one creates a net current flow via the application of an electric field and in addition applies a small magnetic field perpendicular to the current direction, the particles will experience a net deflection which is perpendicular to the direction of motion in the absence of the magnetic field. The Hall mobility is then defined in Gaussian units as

$$\mu_H \equiv v_\perp / \left(\frac{v_D H}{c}\right), \tag{7.83}$$

where $v_\perp$ is the component of carrier velocity which is perpendicular to the electric field direction and $(v_D H/c)$ may be thought of as an effective electric field. In general one should not expect the carrier's response to an electric field to be the same as that to a magnetic field; hence usually $\mu_H \neq \mu_D$. Furthermore, in the simplest way of thinking about the Hall effect, the Hall mobility will be unaffected by the presence of traps since the Lorentz force only operates on a moving carrier. Hence, in some sense, the Hall mobility is an intrinsic quantity.

In the present discussion we shall restrict our considerations to the high-temperature (semiclassical) regime, in which the atomic motion may be treated classically. As before, let us first review the results obtained for an ordered array and then comment on the inclusion of the effects of disorder.

In calculating the Hall mobility one must proceed beyond the study of charge-transfer processes which only involve two sites (the initial and final site) and consider those processes which involve at least one additional site. Actually, the simplest system wherein one can study the Hall effect is one in which the basic unit consists of three sites located at the vertices of an equilaterial triangle. Friedman and Holstein[24] have utilized the occurrence probability approach in calculating the high-temperature small-polaron Hall mobility (arising from an excess electron) in the hexagonal lattice structure, which is composed of such three-site units. They find that the resulting Hall mobility is thermally activated, with an activation energy one-third of that which is associated with the drift mobility, namely $\mathcal{E}_2/3$. Furthermore, the Hall mobility is found to be typically much larger than the drift mobility. Specifically, the formulae which they obtain are:

$$\begin{aligned} \mu_H &= \frac{ea^2}{\hbar} J \left[\frac{\pi}{12kT\,\mathcal{E}_2}\right]^{\frac{1}{2}} e^{-\mathcal{E}_2/3kT}, \\ \frac{\mu_H}{\mu_D} &= \frac{2}{3\sqrt{3}} \left(\frac{kT}{J}\right) e^{2\mathcal{E}_2/3kT}. \end{aligned} \tag{7.84}$$

A plot of μ_H versus $\hbar\omega_0/kT = \theta/T$ is presented in Fig. 7.15. A number of calculations based on an evaluation of the Kubo formula,[18] which yield contradictory results, have been published.[26,27] More recently Holstein and Friedman have reobtained their results from the Kubo formula;[12] a subsequent publication by Firsov now also reports agreement with the initial Friedman–Holstein result.[28] Thus the confusion associated with this calculation now appears to be resolved.

The calculation of the small-polaron Hall mobility associated with an excess electron in a simple cubic lattice has also been carried out by the occurrence probability approach.[29] What is interesting about this calcula-

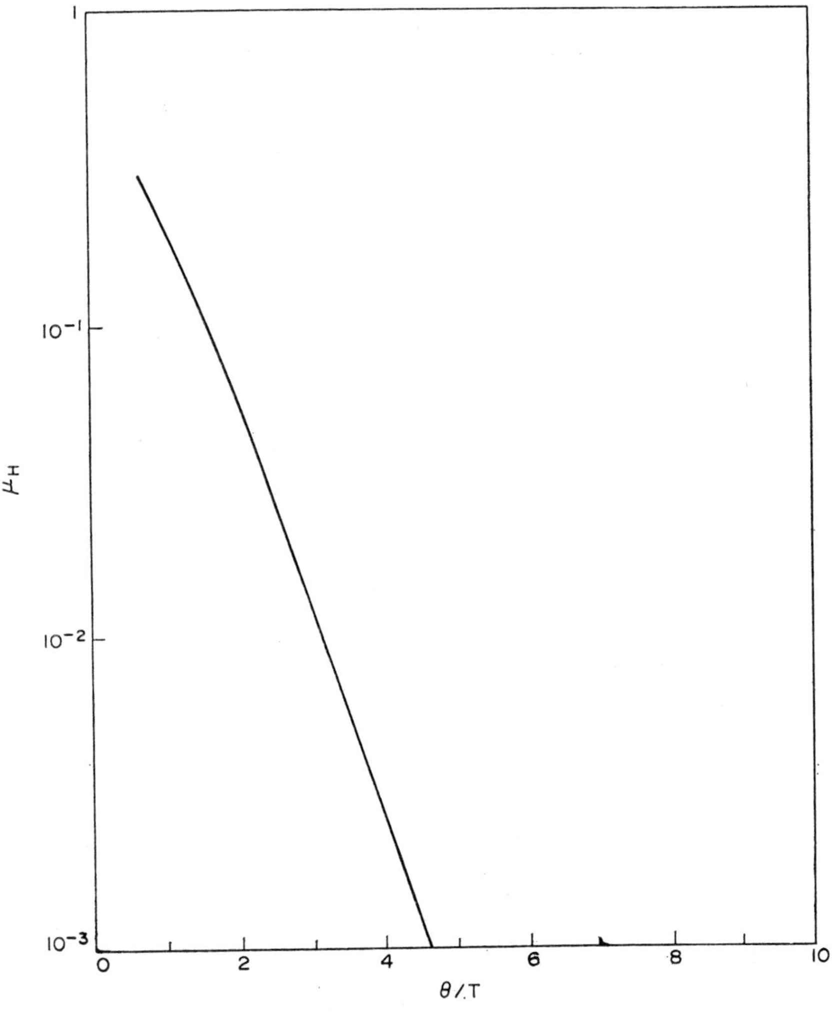

FIG. 7.15. The Hall mobility in an hexagonal lattice as calculated in the nonadiabatic theory of Friedman and Holstein [in units of $(ea^2/2J)(\omega_0/2\pi)$] is plotted against θ/T for the choice of parameters $\mathcal{E}_2 = 5\hbar\omega_0$, $J = \hbar\omega_0$. (After Ref. 24.)

tion is that it illustrates the fact that the local geometry may have a profound effect on the temperature dependence manifested by the Hall mobility. For the cubic geometry in the limit of high temperature[29]

$$\mu_H \propto \left(\frac{\theta}{T}\right)^{3/2} e^{-\mathcal{E}_2/3kT}. \tag{7.85}$$

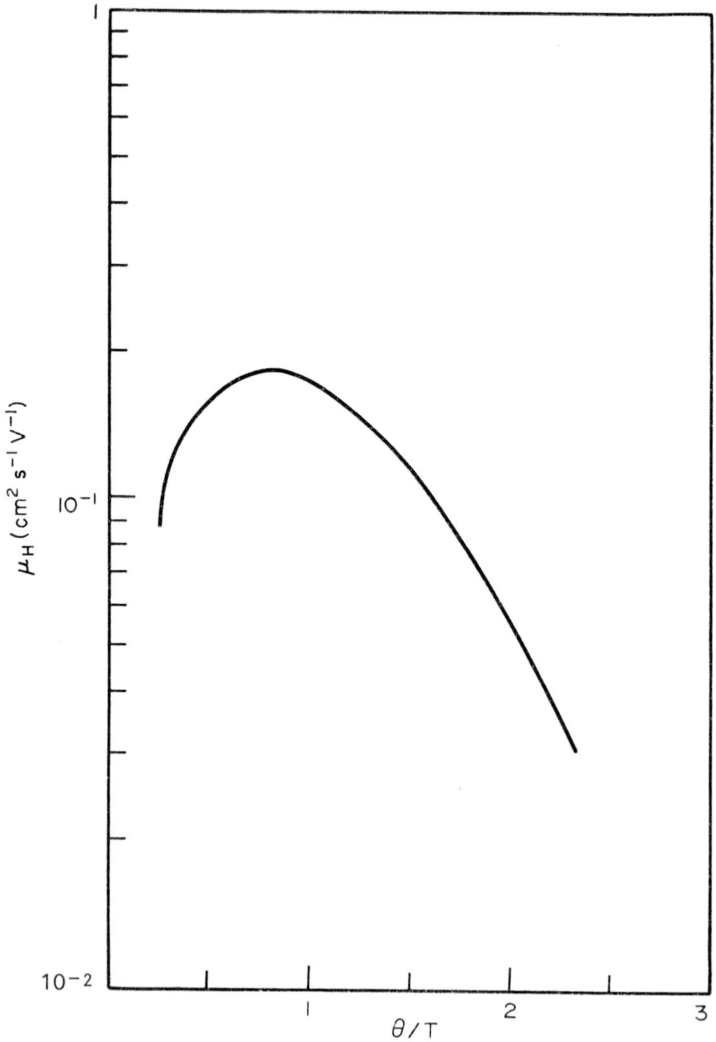

FIG. 7.16. The Hall mobility in a simple cubic lattice as calculated via the nonadiabatic theory is plotted versus θ/T for the parameters $J = \hbar\omega_0 = 0{\cdot}04$ eV, $\mathcal{E}_2 = 5\hbar\omega_0$, and $a = 4$ Å. (After Ref. 29.)

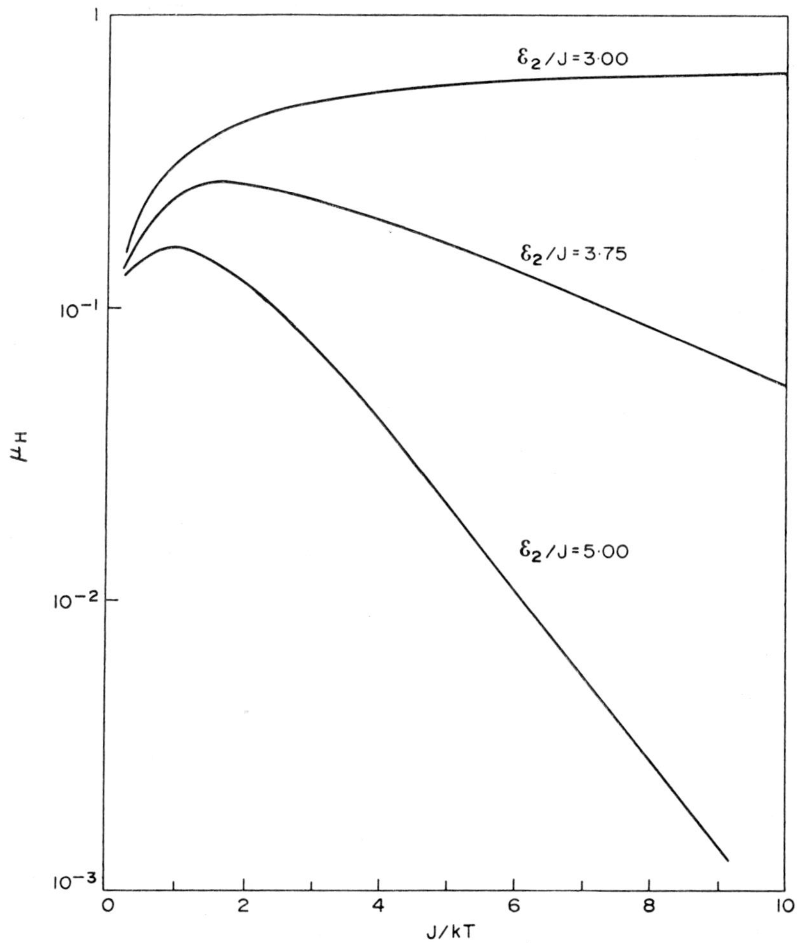

FIG. 7.17. The Hall mobility of an excess electron in the triangular lattice structure [in units of $(ea^2/2J)(\omega_0/2\pi)$], calculated in the adiabatic approximation, versus J/kT for three values of the parameter $\mathcal{E}_2/J$. (After Ref. 23.)

A graph of the complete expression (Fig. 7.16), plotted for the same values of the physical parameters that were used in the graph of the Friedman–Holstein result, illustrates this point; the Hall mobility of the small polaron for this geometry may vary slowly with temperature over a considerable temperature range and will decrease with temperature above $T = 2\mathcal{E}_2/9k$. In a calculation of the Hall mobility in a cubic lattice reported by Bryksin and Firsov[30] they obtain a temperature dependence that is identical to the Friedman–Holstein result rather than to the result illustrated in Fig. 7.16.

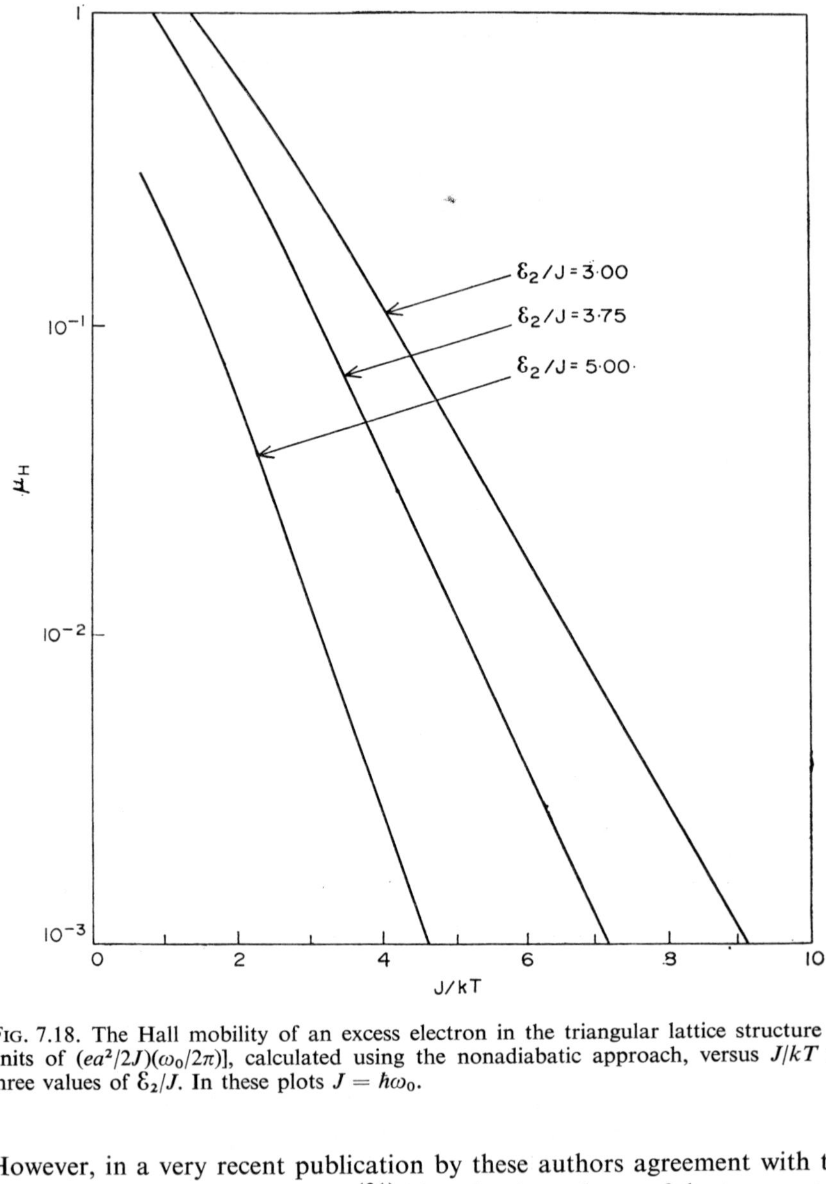

FIG. 7.18. The Hall mobility of an excess electron in the triangular lattice structure [in units of $(ea^2/2J)(\omega_0/2\pi)$], calculated using the nonadiabatic approach, versus J/kT for three values of $\mathcal{E}_2/J$. In these plots $J = \hbar\omega_0$.

However, in a very recent publication by these authors agreement with the result presented here is obtained.(31) Thus the dependence of the temperature dependence of the Hall mobility on the local geometry has now also been established.

As has been pointed out previously (Section 7.9), the application of the perturbation treatment which has been utilized in obtaining these results

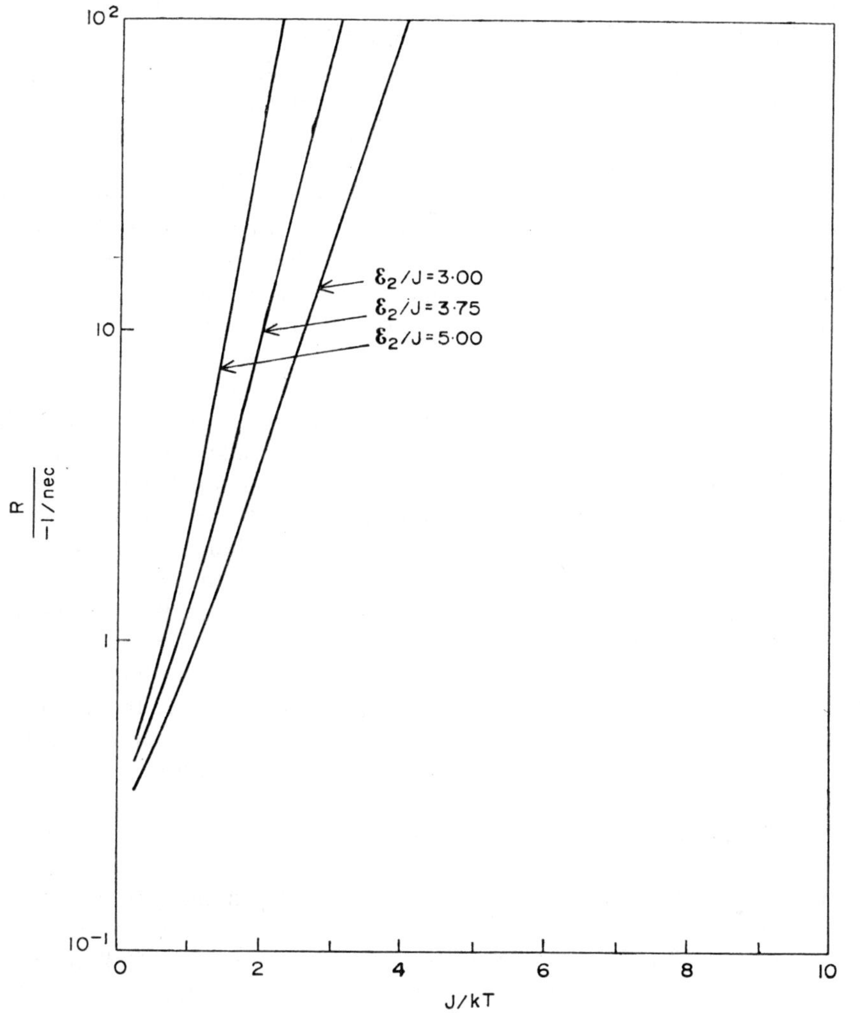

FIG. 7.19. The temperature dependence of the reduced three-site Hall coefficient, $R/(-1/n_e ec)$, appropriate to an excess *electron* in the triangular array plotted against J/kT. The reduced Hall coefficient is the ratio of the Hall mobility to the drift mobility. (After Ref. 23.)

may well be restricted to rather small values of J. Thus it is of interest to calculate the high-temperature Hall mobility within the adiabatic approximation. This has been done for the case of an excess electron moving in a triangular array.[23] A principal feature of these results is that for reasonable choices of the physical parameters the activated behaviour of the Hall mobility is much weaker than in the Friedman–Holstein perturbation calculation. Specifically, the Hall mobility activation energy is less than that found in the perturbation limit by an amount $\approx J$, i.e. $E_{\text{Hall}}(\text{adiabatic}) = \mathcal{E}_2/3 - J$. As a result, the Hall mobility associated with the hopping motion of an (electron) small polaron can easily be a decreasing function of temperature over a considerable range of temperature. Such a situation is illustrated in Fig. 7.17, in which Hall mobility versus reciprocal temperature curves obtained from the adiabatic theory are plotted. For comparison the curves obtained from inserting identical choices of the physical parameters into the Friedman–Holstein result are shown in Fig. 7.18. In addition, the adiabatic theory demonstrates that the small-polaron Hall mobility need not always be larger than the drift mobility; this is manifested at the high-temperature end of the plot of the reduced Hall constant $R/(-1/n_c ec) = \mu_H/\mu_D$ (Fig. 7.19), where R is the Hall constant, n_c is the density of electrons, and c is the speed of light. It should be borne in mind that while the Hall and drift mobilities calculated via the perturbation method are strongly dependent on J, the results of the adiabatic theory (which may be relevant to a large number of experimental situations) do not necessarily indicate such a strong J-dependence. Indeed, the form of the J-dependence in the adiabatic theory is quite different from that which arises in the perturbation theory. Finally, it should be mentioned that while there are well developed criteria for the applicability of the perturbation and adiabatic approaches as far as the calculation of the two-site jump rate is concerned, a quantitative theory of the transition region between the regimes of validity of the two approaches has not been carried out. A qualitative discussion of this transition regime is found in Section 9 of reference 23.

The above discussion of the Hall effect has been restricted to the situation in which the carrier is an excess electron. We must now approach the question of the applicability of the Hall effect calculation to the circumstance in which the charge carrier is associated with an electron deficiency. It has been demonstrated[(32)] that for an approach to the calculation of the Hall mobility which is perturbative in powers of J the sign of the Hall mobility for holes is opposite to that of electrons for the four-site situation (which is encountered in studying cubic materials), but is the *same* as that for electrons when the predominant local geometry is of the three-site variety. Thus from an application of the Friedman–Holstein calculations to the case of an electron deficiency one finds the Hall effect to be that which would correspond to the carriers being electrons, while a determination of the charge of the carriers from the sign of the thermoelectric power would indicate the carriers to be "holes."[(32)] Therefore the theory predicts a sign "anomaly" associated with the hopping motion of "holes" in materials characterized by a local three-site geometry: p-type thermoelectric power but n-type Hall mobility.

For excess electrons both the thermoelectric power and Hall mobility are n-type. These results hinge on the hopping nature of the conduction and not specifically on whether the carriers are small polarons. Thus such an anomaly also arises in a study of impurity conduction.[32] Finally, it should be remarked that although the adiabatic theory for the Hall effect in a three-site configuration[23] is only applicable to the case of excess electrons, preliminary studies by this author indicate that the above-mentioned sign anomaly associated with the motion of "holes" also is a feature of this adiabatic theory. Furthermore, the reduction of the Hall mobility activation energy for electrons in the three-site adiabatic theory by the amount J from the Friedman–Holstein result is a feature which only arises in the treatment of the carriers as being excess electrons. For a deficiency of electrons in the three-site configurations preliminary considerations indicate that the Hall mobility activation energy is essentially $\mathcal{E}_2/3$ rather than $\mathcal{E}_2/3 - J$ (the result for excess electrons). Up to this point our discussion has neglected effects associated with the dispersion of vibrational frequencies. At sufficiently high temperatures the quantum-mechanical corrections to the Hall mobility will be small, just as they are in the calculation of the simple two-site jump rate. However the classical lattice relaxation phenomena may, as before, play a role when the jump time, $>(2\pi/\omega_0)$, is less than or comparable to the time which characterizes the relaxation of the coincidence distortion in our model $(\sim\omega_b{}^{-1})$. In this situation the Hall and drift mobility activation energy will always be less than the activation energy which characterizes a two-site hop [see page 1333 of reference 20]. Hence the magnitude of the activation energies associated with small-polaron hopping motion may be rather small ($\approx 10^{-2}$ eV). If this situation prevails, the temperature dependence of the preexponential factor will generally manifest itself in the rate expression, thereby yielding mobilities which vary slowly with temperature over several-hundred-degree temperature variations and which may even decrease with increasing temperature.

The introduction of disorder into the calculation of the Hall mobility requires solving an even more complicated percolation problem than that involved in calculating the dc conductivity, since the processes which contribute to the Hall mobility involve more than pairs of sites. However, in a number of transport experiments performed on materials at and around room temperature the presence of sizable regions which are characterized as possessing short-range order may yield a situation in which carriers by and large move in regions for which $\mathcal{E}_2 \gg \Delta$ and $kT \gtrsim \Delta$. Since, as in the case of the two-site jump rate, the presence of disorder produces alterations of the order of Δ in the activation energies associated with local contributions to the Hall mobility, the above-mentioned situation may be treatable as "nearly ordered" and the formulae which have been presented here may be qualitatively correct (at sufficiently high temperatures). Thus in light of the great inherent difficulty of solving the general percolation problem, especially if one tries to take account of the presence of short-range order, extremely little has been done in realistically calculating the Hall mobility or even the dc conductivity from first principles, with or without the presence of small polarons. At the present stage of our study of disordered materials we are reduced to comparing our speculations (rather than well-developed theories) with experiment.

7.12. Thermoelectric Power

A technique which can be used together with a dc conductivity measurement to elicit information about the conduction mechanism is that of measuring the thermoelectric power. In a thermoelectric power experiment one measures the e.m.f. that is developed across a material when a thermal gradient is maintained across the sample. One approach to the determination of the thermoelectric power associated with the (high-temperature) multiphonon small-polaron hopping regime is to calculate the voltage difference that is developed between two adjacent sites which are characterized as being separated by an infinitesimal temperature difference, ΔT.[33] Following such a procedure, one can show that the thermoelectric power associated with a charge q moving in the molecular-crystal-model lattice is simply

$$S = \frac{-k}{q}\left(\frac{\mathcal{E}_0}{kT}\right), \tag{7.86}$$

where $\mathcal{E}_0$ is the energy which characterizes the thermal generation of carriers. In performing this calculation it is noted that there is no transfer of vibrational energy associated with a small-polaron hop. This is because the contribution (see Section 7.9) to the activation energy arising from the distortion of the initial site remains equal to the energy attributable to the distortion of the second site even for non-zero values of the local-site energy difference, Δ.

However, there are a number of more complicated models for which a non-zero transfer of vibrational energy is associated with a small-polaron hop. The first such situation can occur in a periodic system and is associated with permitting the presence of the carrier to alter the stiffness, $M\omega_0{}^2$, of the occupied site from the value that it has when the site is unoccupied. Then, depending on whether the carrier increases or decreases the local stiffness, the energy $\mathcal{E}_0$ in eqn (7.86) is diminished or augmented by an amount that is some fraction of the activation energy for the hop. Another situation in which energy is added to or subtracted from $\mathcal{E}_0$ occurs when there is disorder in the local electron–lattice coupling constants and/or in the local vibrational frequencies. Finally, it should be mentioned that in the regime in which hops are correlated due to the previously discussed lattice-relaxation phenomena the bulk of the energy required for a hop is supplied by the "final" site, and hence a net transfer of vibrational energy ($\sim\mathcal{E}_2/3$) in the direction opposite to the direction of the carrier's jump is associated with each hop. While variations of the local stiffness and of the local electron–lattice coupling constant as well as effects associated with lattice-relaxation phenomena do in principle occur to some extent, it is not at all clear at the present time whether they are of much significance.

It should be noted that eqn (7.86) is based on the assumptions that the width of the band of small-polaron levels is very small compared with kT and that the energy of each carrier level is independent of temperature. More generally the method utilized to derive eqn (7.86) yields $S = (-k/q)\,T[\partial(\ln n_c)/\partial T]_{\mathcal{E}\,(\mathrm{Fermi})}$. Evaluating this formula only yields a vanishing "heat of transport" constant in the limit of the width of the band of small-polaron states being very small compared with kT. However, since in a noncrystalline solid the width of the energy distribution of the small-polaron states (essentially the disorder energy) is not necessarily small compared with kT there will be an additional non-negligible contribution to the thermoelectric power which will depend on the details of the energy distribution. For a band which is wide compared with kT but very dense compared with the typical conduction bands of solids this will generally yield a sizeable, almost constant, contribution; for example, consideration of some simple models of energy distribution functions suggests estimates of ~ 1 to 10 for the "heat of transport" term. In addition, evaluating the above general formula indicates that a temperature variation of the small-polaron energy levels cannot give rise to a strictly constant contribution to the thermoelectric power.

Finally, one should recognize the fact that the question of a heat of transport contribution arising from a carrier-induced thermal current (analogous to the phonon-drag phenomenon in crystals) has not as yet been treated for small-polarons in either crystalline or noncrystalline substances. The low mobility, strong electron–lattice interaction, and relatively high density of charge carriers associated with the possibility of small-polaron motion in some noncrystalline materials suggests that some consideration should be devoted to this issue.

7.13. AC Conductivity

It is the intent of this Section to present the results of studying the high-frequency ($\omega \gg \omega_0$) conductivity attributable to the motion of small polarons. To begin, let us employ the same approach as has been used in performing the calculations that are described in Section 7.8. Namely, we shall consider the perturbation approach in the limit of the band of vibrational frequencies becoming arbitrarily small, and concomitantly restrict our attention to those external frequencies which are an integral multiple of the basic vibrational frequencies, i.e. let $\omega = m\omega_0$, where m is an integer. This restriction insures energy conservation in a process in which a photon and a finite number of phonons are involved. One then readily obtains from a direct evaluation of the Kubo formula[18] the result,[19]

$$\sigma_{xx}(\omega) = \left(\frac{n_c e^2 a^2}{kT}\right)\left(\frac{2\pi J^2}{\hbar^2 \omega_0}\right)\left[\frac{\sinh(\beta\hbar\omega/2)}{(\beta\hbar\omega/2)}\right] e^{-2S}$$

$$\times \sum_{n=-\infty}^{\infty} \left\{ I_m\left[\left(\frac{2\mathcal{E}_b}{\hbar\omega_0}\right) A(n)\,\mathrm{csch}\left(\frac{\beta\hbar\omega_0}{2}\right)\right] - \delta_{m,0}\right\}, \qquad (7.87)$$

where n_c is the density of carriers, and the delta function in the curly brackets

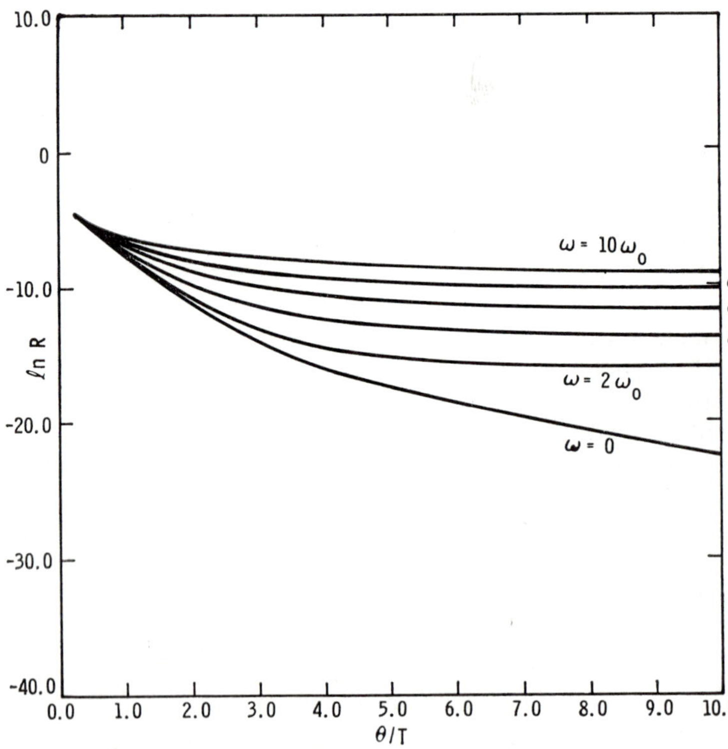

FIG. 7.20. The natural logarithm of the two-site jump rate plotted versus θ/T for various frequencies of the external field. Here $\mathcal{E}_b = 10\hbar\omega_0$ and $\omega_b/\omega_0 = 0{\cdot}1$, and R is expressed in units of $2\pi J^2/\hbar\omega_0$.

insures that in the dc limit ($m = 0$) only "non-diagonal" transitions are considered in calculating the conductivity. In the high temperature limit the major contribution to the ac conductivity is, as in the dc case, essentially that of the multiphonon processes, once again suggesting that the vibrational motion may be treated classically. Thus the restriction of ω/ω_0 to an integer value is only a computational convenience in this regime, and one may connect the discrete set of points $\omega = m\omega_0$ to obtain a smooth curve of $\sigma_{xx}(\omega)$ versus ω which rises with frequency to a maximum at $\omega = 2\mathcal{E}_b/\hbar$; this peak was first described in the work of Reik and Heese.[34]

At low temperatures the processes which contribute to the ac conductivity generally involve the spontaneous emission of enough phonons to make up for the energy absorbed by the carrier from a photon. Thus in the low-temperature regime (and small vibrational-dispersion limit) the $\sigma_{xx}(\omega)$ versus ω curve is a series of peaks centred at $\omega = m\omega_0$. The envelope of these peaks is itself peaked at $\omega \approx 2\mathcal{E}_b/\hbar$. It should be stressed that the details of the low-temperature structure are strongly dependent on the phonon frequencies which are coupled to the carrier and hence are very much model-dependent. In Fig. 7.20 curves of the frequency-dependent jump rate

$$\sigma_{xx}(\omega)/(n_c e^2 a^2/kT),$$

in units of $2\pi J^2/\hbar\omega_0$, are plotted against $\hbar\omega_0/kT \equiv \theta/T$ for various frequencies. It should be noted that as the temperature is lowered thermally activated processes freeze out and the ac conductivity becomes independent of temperature. Both the peaking of the ac conductivity about high-frequencies $\sim 2\mathcal{E}_b/\hbar$ and the loss of the temperature dependence of the high-frequency conductivity at low temperatures are features which are expected to be manifested by small polarons in disordered materials.

7.14. Discussion and Summary

Based on studies of small-polaron formation in model crystalline systems, it has been suggested that the presence of disorder in noncrystalline systems tends to slow down and localize a carrier, thereby aiding small-polaron formation. Such a general argument is at most a statement of the plausibility of finding small polarons in some disordered materials. Clearly, the occurrence of such a situation is dependent on the precise system studied. Indeed, the complexities of and differences between various disordered inhomogeneous materials render it prohibitively difficult to argue more strongly than this without recourse to specific experimental data. Thus, having established the possibility that the carriers in some chalcogenide glasses are hole-like small polarons,[35] some of the ramifications of such a situation will be discussed and ultimately compared with some electrical-transport measurements.

In trying to ascertain whether small polarons are formed in a particular noncrystalline solid it is desirable to investigate the transport properties of the material at high temperatures. One reason for this is that the effects of the disorder (as manifested by fluctuations of the local-site energies) on the percolation problem play a much less significant role at such temperatures than at lower temperatures. Another very important reason for focusing attention on high-temperature measurements is that the occurrence of distortion-related activation energies, $\sim\mathcal{E}_b/2$, is a feature of only the high-temperature regime (see Section 7.8). At lower temperatures, where the multiphonon processes are frozen out, the phonon-assisted hopping of small polarons does not show a temperature dependence which would distinguish the hopping from that which characterizes jumps between non-polaronic localized states. Thus the high-temperature regime provides a much better opportunity to compare experiment with the small-polaron theory than is afforded at low temperature.

The electrical conductivity associated with high-temperature small-polaron hopping motion within the disordered material (or dominant constituent of a glass—see Section 7.1) will essentially be thermally activated ($\sigma \simeq \sigma_0 \, e^{-\mathcal{E}/kT}$), the activation energy $\mathcal{E}$ being the sum of the energy associated with the production of an equilibrium number of carriers, $\mathcal{E}_0$, and the activation energy associated with the intrinsic hopping of small polarons, $W_{\mathbf{hop}}$. The activation energy contribution $W_{\mathbf{hop}}$ will in general vary little between glasses of the same dominant material, while $\mathcal{E}_0$ may change considerably from glass to glass. The value of σ_0 will generally be given by $(Ne^2 a^2 \nu/kT)\, P$, where N is the number density of equivalent carrier sites in the dominant material ($\sim 10^{22}\,\text{cm}^{-3}$), ν is a vibrational frequency, a is an intersite spacing, and P is a factor which for the adiabatic theory of small-polaron motion[11,12] is unity [in the nonadiabatic regime it is less than unity (see Section 7.9)]. Since the constants involved in σ_0 change little from material to material, σ_0 will vary little between glasses; its value, extrapolated from measurements near 300 K, will be within an order of magnitude of 10^2–10^3 mho/cm. It is important to note that this feature is in agreement with numerous experimental observations in chalcogenide glass systems. Furthermore, concomitant with the low mobilities that are associated with small-polaron motion, the theory predicts a much higher carrier density than that which characterizes non-polaronic theories.

By studying the thermoelectric power, S, (see Section 7.12) as a function of temperature in the intrinsic-semiconductor regime, one can gather information about the carrier-generation energy $\mathcal{E}_0$. Specifically, in the simplest model of small-polaron hopping motion, where the alteration of the local vibrational frequencies associated with the presence of the charge carrier and the disorder inherent in the noncrystalline material are neglected, one may directly relate the thermoelectric power to $\mathcal{E}_0$ using eqn (7.86), [see Section 7.12]. Furthermore, when $\mathcal{E}_0$ is essentially independent of temperature the thermoelectric power should vary linearly with reciprocal temperature. The value of the slope yields $\mathcal{E}_0$, and this energy is less than the conductivity-activation energy by the amount W_{hop}. It should be remarked that this feature has also often been observed experimentally.

Digressing from discussing electrical transport phenomena for a moment, it should be pointed out that for the case of intrinsic carriers $\mathcal{E}_0$ will be less than half of the optical gap. This is because the gap which arises in calculating the probability that a given carrier level is occupied is related to the energy difference between the electron and hole *small-polaron* levels (assuming that both electrons and holes form small polarons), while the optical gap is associated with electronic transitions in which the atoms of the material are unable to adjust to the presence of the carriers (the Franck–Condon principle). This feature of the theory is illustrated in Fig. 7.21, in

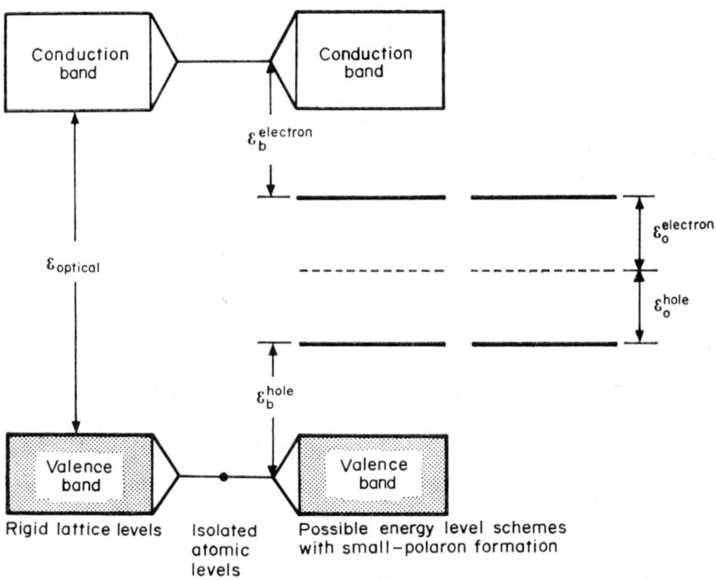

FIG. 7.21. In the forming of a regular lattice the isolated atomic energy levels are broadened out into bands (left side of diagram). The excitation of an electron from the valence to the conduction band by incident radiation is associated with the energy $\mathcal{E}_{optical}$; in the excitation process the atoms are essentially held fixed in their rigid-lattice positions. However, once the atoms have an opportunity to adjust to the presence of the carrier the possible energy schemes for a coupled carrier-lattice system in which small polarons are formed (and play an essential role in the transport phenomena) are illustrated on the right side of the figure. The dashed line indicates the Fermi level which, since $kT \gg$ small-polaron bandwidth, generally lies midway between the electron and "hole" small-polaron bands.

which the energy levels associated with the rigid-lattice situation as well as the two possible energy schemes which correspond to observable small-polaron motion are shown.† Due to the high density of states associated with the small-polaron levels, we have $\mathcal{E}_0^{\text{hole}} \approx \mathcal{E}_0^{\text{electron}}$. Thus, as is generally observed, the optical gap is not simply related to either the carrier-generation energy or the conductivity-activation energy. Finally, I might comment that in this model the luminescence spectrum is expected to be peaked about an energy which is below the excitation energy.

Proceeding to review very briefly some of the salient features of the theory of the small-polaron Hall mobility, it is first noted that the Hall mobility in general will differ substantially from the (trap-free) mobility which is involved in the conductivity expression. In particular, unlike the "drift" (trap-free conductivity) mobility, the magnitude, temperature dependence, and even the sign of the Hall mobility are dependent on the local (or short-range order) of the material. Thus, let us focus our attention on a hole-like small polaron moving within the triangular-type geometry which characterizes a number of disordered materials including many chalcogenide glasses. For such a situation the simplest theory (see Section 7.11) predicts that the motion of hole-like small polarons will yield an *n*-type Hall effect. Thus, the frequently observed anomaly of *p*-type thermoelectric power and *n*-type Hall mobility is a natural (but not uniquely a) consequence of hole-like small-polaron hopping motion. Finally, the theory predicts that the Hall mobility should be low, below 0·1 to 1·0 $\text{cm}^2\ \text{V}^{-1}\ \text{sec}^{-1}$, and characterized by a small activation energy; in the simplest model it is roughly one-third the hopping energy W_{hop}. It should be stressed that the Hall mobility is not affected by simple charge traps, as the influence of the magnetic field is felt only when the carrier moves.

In summary, the model advanced here suggests that electrical transport in some noncrystalline materials may occur via small polarons hopping within a "dominant constituent" which is characterized by short-range order and relatively high carrier conductivity and mobilities. When such a situation occurs, the predictions of the theory are (1) that the magnitude of the pre-exponential factor of the thermally-activated conductivity will be 10^2–10^3 mho/cm, (2) that there will be a difference between the characteristic conductivity activation energy ($\mathcal{E}$) and thermoelectric-power energy ($\mathcal{E}_0$) which is equal to the hopping activation energy (W_{hop}), (3) that this deduced hopping energy will be essentially a characteristic of the dominant constituent

† A conceivable situation (illustrated in Fig. 7.21) is that in which carriers can exist in either (low mobility) small-polaron states or in (higher mobility) rigid-lattice states. If the energy gap between the two groups of states is small enough compared with W_{hop}, charge transport may be dominated by charge carriers in the "conduction" levels. While such a situation has not been considered in the present work, it may ultimately be relevant to some noncrystalline system.

and thus vary little between materials of the same dominant constituent, and finally that (4) the Hall mobility associated with hole-like motion will, for the appropriate geometry, be n-type and will possess an activation energy which is a fraction of the hopping energy (essentially $1/3\ W_{\text{hop}}$ in the simple model studied here).

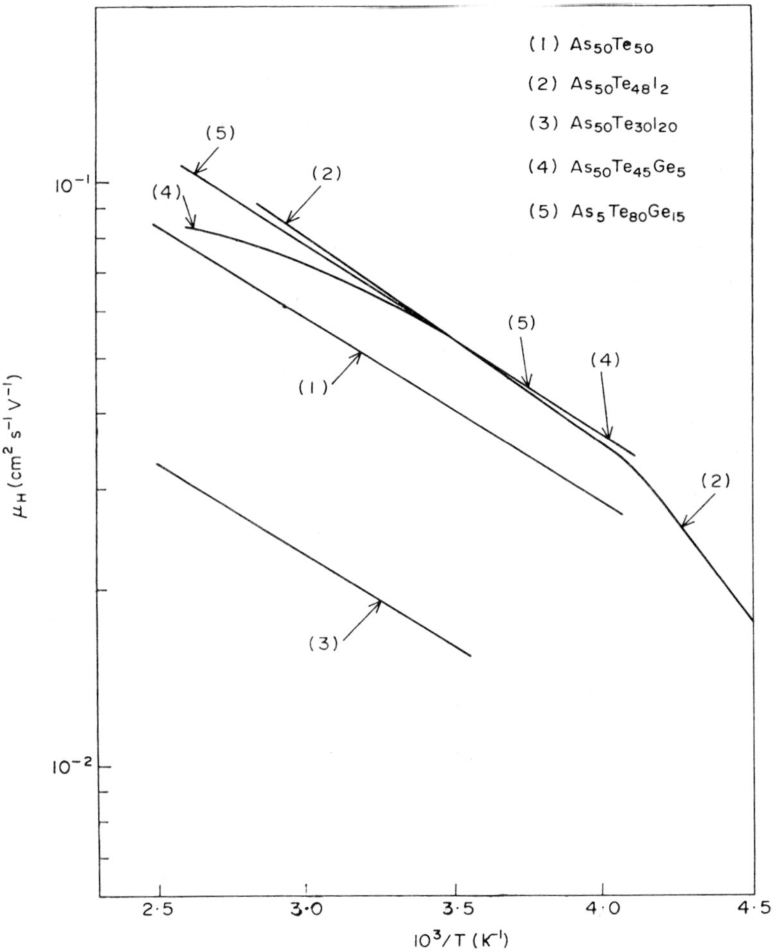

FIG. 7.22. The Hall mobility of some chalcogenide glasses as reported in Ref. 35.

Having discussed the most likely observable predictions of the theory, a few of the experimental results which bear on the theory will now be briefly reviewed. Specifically, I want to consider experiments carried out on a series of glasses all of which contain substantial amounts of arsenic and tellurium; namely, the AsTeI, AsTeGe and AsTe glasses.[35,36,37] The principal conclusion drawn from the structural experiments is that a common structural entity associated with arsenic and tellurium is present in all these glasses. The electrical conductivity of all of these glasses is found to be thermally activated. The thermoelectric power is p-type and has the expected $1/kT$-type temperature dependence at high temperatures. An illustrative comparison of the measured energies is shown in Table 7.1 for a few of the glasses studied. It is noted that there is little variation in the deduced value of W_{hop}. Measurements of the Hall mobility, shown in Fig. 7.22, once again yield a very similar thermally-activated temperature dependence for the wide variety of glasses that have been studied. The sign of the Hall mobility is anomalous (n-type) and its activation energy is roughly one-third of W_{hop}. Similar Hall mobility results have been reported by Roilos[38] for a series of AsTeSe glasses; his Hall mobility curves would lie parallel to those shown here. Furthermore, as may be expected from the annealing out of some trap-type defects, the electrical conductivity is somewhat altered by slight annealing of these glasses while the Hall mobility is essentially unaffected by this annealing.[39] Finally, the fact that recent electrical transport measurements performed on bulk samples of noncrystalline As_2Te_3 yield very similar results[37] to those obtained on the AsTeI, AsTeGe and AsTeSe systems while measurements on the non-arsenic containing chalcogenide glass,[35] $Ge_{17}Te_{83}$, yield very different results provides support for the notion that some As–Te structural unit is the dominant constituent of these chalcogenide glasses.

TABLE 7.1. The conductivity activation energy $\mathcal{E}_0 + W_{hop}$, the thermoelectric power energy $\mathcal{E}_0$, both accurate to $\pm 0{\cdot}01$ eV, and the deduced hopping energy W_{hop} all in eV.

	$\mathcal{E}_0 + W_{hop}$	$\mathcal{E}_0$	W_{hop}
$As_{50}Te_{50}$	0·44	0·25	0·19
$As_{50}Te_{48}I_2$	0·36	0·16	0·20
$As_{50}Te_{30}I_{20}$	0·45	0·27	0·18
$As_{50}Te_{45}Ge_5$	0·46	0·29	0·17
$As_5Te_{80}Ge_{15}$	0·43	0·24	0·19

Thus, measurements on the AsTe, AsTeI, and AsTeSe systems support the notion that, in these materials, the carriers are hole-like small polarons. In particular, the magnitude of the pre-exponential factor of the thermally-activated electrical conductivity, the composition-independent difference between the characteristic conductivity activation energy and the thermo-electric-power energy, and the small activation energy and anomalous sign of the Hall mobility provide experimental arguments for the validity of our model of these glasses. Indeed, the occurrence of many of these features in other disordered systems leads one to suspect that small-polaron formation may occur in many more noncrystalline materials than just those mentioned here.

Finally, it should be borne in mind that obtaining an understanding of the physical processes occurring in noncrystalline materials such as glasses is an extremely ambitious task. Only the plausibility of small-polaron formation in such systems has been argued. While our initial results are encouraging, much more work is required both to further investigate the questions of small-polaron formation and motion in noncrystalline materials and to ascertain the applicability of these ideas to specific systems.

Acknowledgement

The support of the U.S. Atomic Energy Commission in the work described in this article is gratefully acknowledged.

References

1. Lipskis, K., Sakalas, A., and Viscakus, J., *Phys. Status Solidi* (*a*) **4,** K217 (1971).
2. Juretschke, H. J., Landauer, R. and Swanson, J. A., *J. Appl. Phys.* **27,** 838 (1956).
3. Mott, N. F., and Gurney R. W., "Electronic Processes in Ionic Crystals." Dover, New York, 1964, p. 86.
4. Gibbons, D. J. and Spear, W. E., *J. Phys. Chem. Solids* **27,** 1917 (1966).
5. Street, G. B. and Gill, W. D., *Phys. Status Solidi* **18,** 601 (1966).
6. Crevecoeur C. and de Wit, H. J., *J. Phys. Chem. Solids* **31,** 783 (1970).
7. Castner, T. G. and Kanzig, W., *Phys. Chem. Solids* **3,** 178 (1957).
8. Holstein, T., *Ann. Phys.* (*N.Y.*) **8,** 325 (1959).
9. Toyozawa, Y., *Progr. Theor. Phys.* (*Kyoto*) **26,** 29 (1961).
10. Holstein, T., Private Communication.
11. Emin, D., *Phys. Rev.* **B4,** 3639 (1971).
12. Holstein, T and Friedman, L., *Phys. Rev.* **165,** 1910 (1968).
13. Emin, D., *Phys. Rev. Lett.* **28,** 604 (1972).
14. Kittel, C., "Quantum Theory of Solids." Wiley, New York, 1963, Ch. 7, eqn. (16).
15. Emin, D., *Advan. Phys.*, to be published.
16. Holstein, T., *Ann. Phys.* (*N.Y.*) **8,** 343 (1959).
17. Anderson, P. W., *Nature* **235,** 163 (1972).
18. Kubo, R., *J. Phys. Soc. Japan* **12,** 570 (1957), eqn. (5.11).
19. Emin, D., to be published.
20. Emin, D., *Phys. Rev.* **B3,** 1321 (1971).
21. Emin, D. *Phys., Rev. Lett.* **25,** 1751 (1970).
22. de Wit, H. J., *Philips Res. Repts.* **23,** 449 (1968).
23. Emin, D. and Holstein, T., *Ann. Phys.* (*N.Y.*) **53,** 439 (1969).
24. Friedman, L. and Holstein, T., *Ann. Phys.* (*N.Y.*) **21,** 494 (1963).
25. Emin, D., *J. Non-Cryst. Solids* **8-10,** 511 (1972).

26. Firsov, Yu. A., *Fiz. Tverd. Tela* **5,** 2199 (1963), English translation in *Sov. Phys. Solid State* **5,** 1566 (1964).
27. Schnakenberg, J., *Z. Physik* 185, **123** (1965).
28. Firsov, Yu. A., *Fiz. Tverd. Tela.* **10,** 3027 (1968), English translation in *Sov. Phys. Solid State* **10,** 2387 (1969).
29. Emin, D., *Ann. Phys.* (*N.Y.*) **64,** 336 (1971).
30. Bryksin, V. V. and Firsov, Yu. A., *Fiz. Tverd. Tela.* **12,** 627 (1970). English translation *Sov. Phys. Solid State* **12,** 480 (1970).
31. Bryksin, V. V. and Firsov, Yu. A., *Fiz. Tverd. Tela.* **14,** 463 (1972), English translation *Sov. Phys. Solid State* **14,** 384 (1972).
32. Holstein, T., *Phil. Mag.,* **27,** 225, (1973).
33. Austin, I. G. and Mott, N. F., *Advan. Phys.* **18,** 41 (1969).
34. Reik, H. G. and Heese, D., *J. Phys. Chem. Solids* **28,** 581 (1967).
35. Emin, D., Seager, C. H. and Quinn, R. K., *Phys. Rev. Lett.* **28,** 813 (1972).
36. Seager, C. H., Emin, D. and Quinn, R. K., *J. Non-Cryst. Solids* **8-10,** 341 (1972).
37. Seager, C. H., Emin, D. and Quinn, R. K., to be published.
38. Roilos, M., *J. Non-Cryst. Solids* **6,** 5 (1971).
39. Roilos, M., Meimaris, D. and Zigiris, K., *J. Non-Cryst. Solids* **2,** 29 (1972).

8. AC CONDUCTIVITY AND HIGH FIELD EFFECTS

A. K. JONSCHER

Chelsea College, University of London,
Pulton Place, London, SW6 5PR

8.1. FREQUENCY EFFECTS IN HOPPING TRANSPORT

8.1.1. INTRODUCTION

The first observations of a frequency dependent conductivity, $\sigma(\omega)$ in a hopping system were reported by Pollak and Geballe[1] for compensated crystalline silicon in the impurity conduction range. The dependence could be expressed by the empirical relation

$$\sigma(\omega) \propto \omega^n \tag{8.1}$$

where ω is the radian frequency and the exponent n was found to have a value close to 0·8. The theory of this phenomenon was treated by Pollak in several papers[2–4] which analyse the hopping motion of electrons between pairs of localized sites and also extended hopping over many sites. More recent developments in the field of amorphous semiconductors have focussed attention on the frequency dependence of conductivity in this much wider class of materials, many of which show a behaviour similar to that given by eqn (8.1).

This subject was reviewed recently by the present author[5] and it was pointed out that this type of frequency dependence is found not only in disordered, glassy and amorphous solids, but also in ordered molecular solids. The carriers involved are not necessarily electrons, but may be polarons, protons or ions. Attention was drawn to the fact that the exponent n is not a constant for all systems, as is assumed by many people, but is a function of temperature, approaching unity at low temperatures and decreasing to 0·5 or less at higher temperatures. The point was also made that eqn (8.1) implies the dielectric loss dependence $\kappa'' \propto \omega^{n-1}$ which is almost independent of frequency if $n \approx 1$. This type of loss is found experimentally in a wide range of dielectric materials of technological importance.

A recent paper by Pollak and Pike[6] raises similar points, arguing that the conductivity of many glasses is due to ionic hopping and that significant information is contained in the departures of the exponent n from unity.

Other theoretical treatments of the frequency dependent conductivity in hopping systems were given by Mott[7], Austin and Mott,[8] Davis and Mott,[9] and also more recently by Scher and Lax[10,11] who have extended to ac conditions the dc theory of Miller and Abrahams[12] and by Butcher.[13] The principal drawback of all these theories, in the opinion of the present author, is the difficulty of obtaining an adequate physical insight into the mechanisms involved, especially in view of the extremely wide range of physical systems in which they are found.

The present treatment, which follows the argument given in the earlier paper[5] and develops it further, approaches the problem in the time-domain as distinct from the frequency-domain used by other authors. It is argued that the time-domain gives a more tangible physical picture of the phenomena concerned and thus provides a welcome alternative approach which is fully equivalent to the frequency-domain approach in the linear regime, i.e. at relatively low electric fields. The situation becomes more complicated in the non-linear regime and this topic is discussed elsewhere.[14]

The present paper begins with a review of a wide range of experimental data relating to frequency-dependent hopping systems and proceeds then to develop a qualitative picture of the underlying physical mechanisms. The discussion is based on a very general concept of hopping of any kind of charge carrier and without specific reference to any form of host matrix. This approach is believed to embrace all possible types of frequency dependent mechanisms which may be expected in poorly conducting media. The primary objective is to develop a basic understanding of the frequency dependence rather than to attempt a quantitative analysis.

8.1.2. REVIEW OF EXPERIMENTAL EVIDENCE

We begin our survey with measurements on a crystalline impurity semiconductor NiO doped with Li which introduces holes jumping between Ni^{++} ions surrounding the Li impurities.(15) Figure 8.1(a) gives the frequency dependence for three different Li concentrations, together with the fitted curve of the form

$$\sigma_1(\omega) = \sigma(\omega) - \sigma_0 \propto \frac{\omega^2}{\omega^2 + 1/\tau^2}. \tag{8.2}$$

Here $\sigma_1(\omega)$ is the ac component of the total conductivity, σ_0 is the dc value, τ is a relaxation time which in this instance is $\tau = 2{\cdot}2 \times 10^{-10}$ s. This type of frequency dependence is exactly that expected of a Debye dipole and the corresponding dependence of the real part of the dielectric permittivity is

$$\kappa'(\omega) - \kappa'(\infty) \propto \frac{1}{\omega^2 + 1/\tau^2} \tag{8.3}$$

where $\kappa'(\infty)$ is the value at frequencies $\omega \gg 1/\tau$. Figure 8.1(b) gives the corresponding temperature dependence and shows clearly that the ac component $\sigma_1(\omega)$ becomes almost independent of temperature while the dc component has a high activation energy.

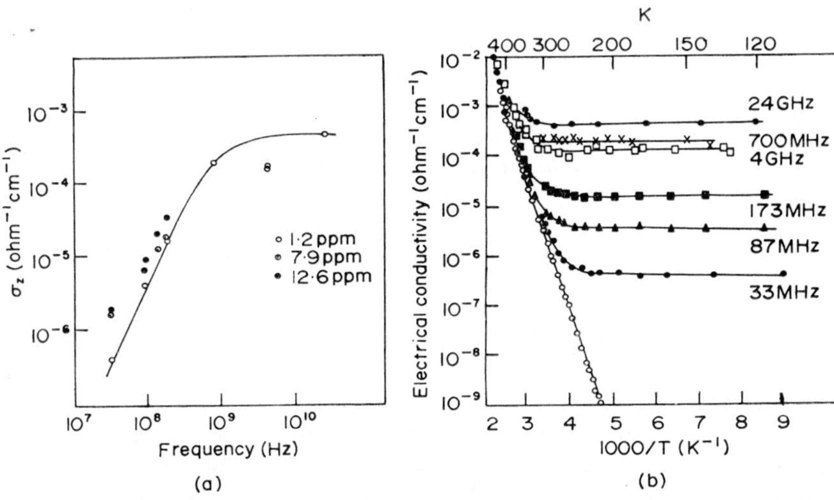

FIG. 8.1 (a) Frequency dependence of conductivity of Li-doped NiO in the low temperature range where $\sigma \approx \sigma_1$. The solid curve is calculated from eqn (8.2) with $\tau = 2{\cdot}2 \times 10^{-10}$ sec. (b) The dc and high frequency conductivities of 1·2 ppm Li-doped NiO. (From Ref. 15.)

Our next example refers to measurements on solid polymeric CS_2[16] and shows three frequency ranges, Fig. 8.2. At relatively low frequencies the dependence follows eqn (8.1) with $n = 0{\cdot}65$ at 394 K to $n = 0{\cdot}75$ at 300 K. The dc level is indicated by an arrow. Above 1MHz a dependence of the type given by eqn (8.2) applies with a saturation frequency similar to that in Fig. 8.1. The high frequency data around the knee of the saturation region were taken by contactless methods in a slotted line and in *X*-band waveguide so that no series contact and lead resistances were involved. These data extrapolate very well the data taken with conventional bridge techniques. The variation of κ' is shown on a linear scale and indicates an almost constant value except for a small resonance around 10^8 Hz.

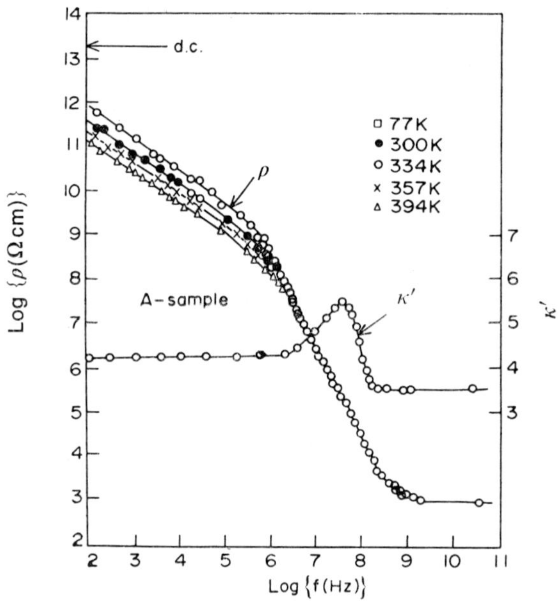

FIG. 8.2. The electrical resistivity of solid polymeric CS_2, with the dc level shown. The permittivity κ' is also shown on a *linear* scale. (From Ref. 16.)

The examples given in Figs. 8.1 and 8.2 represent the only well-documented cases of reliable ω^2 dependence followed by saturation that have come to our notice. There are many other examples in the literature where a ω^2 dependence has been reported but in most of them there is serious doubt that spurious series resistance is responsible for this result.[5,17]

Various chalcogenide glasses have been studied in considerable detail and, allowing for spurious contact effect, show invariably a dependence of the type given by eqn (8.1) with exponents in the range $0{\cdot}7 < n < 1{\cdot}08$[17–20]

(Fig. 8.3(a)). There is a suggestion that saturation may set in above 10^8 Hz, but this point is not clearly established as yet.[21] Figure 8.3(b) shows the corresponding temperature dependence, with σ_1 becoming progressively less temperature dependent with increasing frequency.

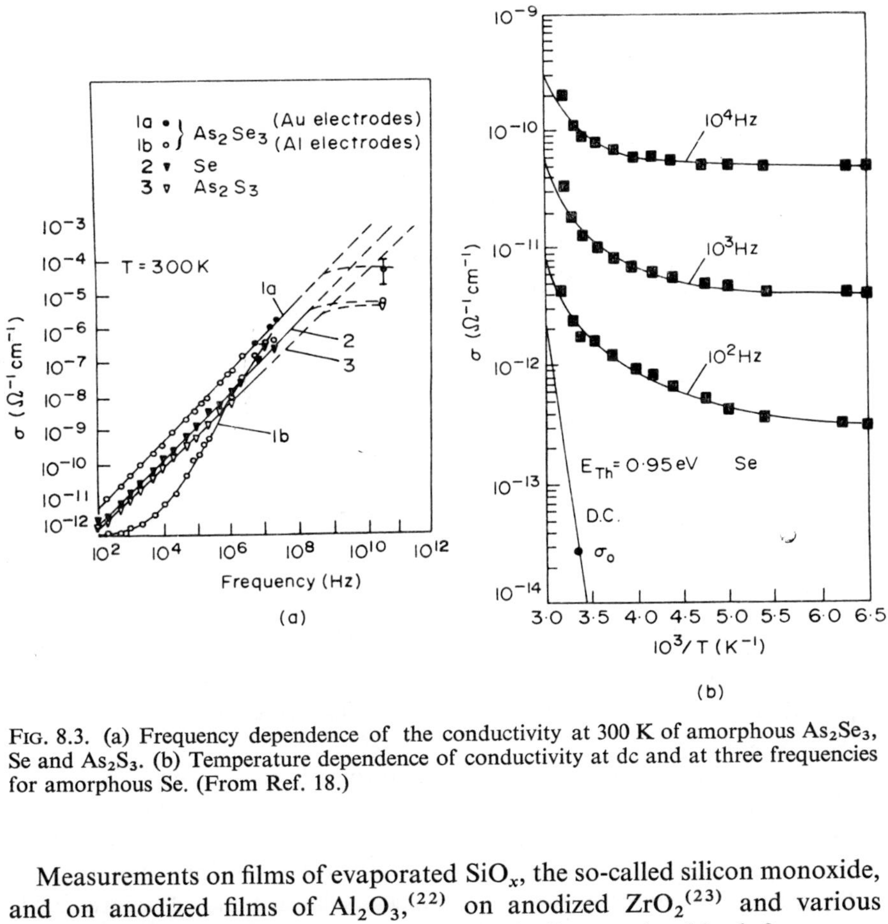

FIG. 8.3. (a) Frequency dependence of the conductivity at 300 K of amorphous As_2Se_3, Se and As_2S_3. (b) Temperature dependence of conductivity at dc and at three frequencies for amorphous Se. (From Ref. 18.)

Measurements on films of evaporated SiO_x, the so-called silicon monoxide, and on anodized films of Al_2O_3,[22] on anodized ZrO_2[23] and various other disordered materials show essentially similar trends of both frequency- and temperature-dependence.

Figure 8.4 shows an example of superposition of a power-law dependence of the type given in eqn (8.1) with $n \approx 1$ and with very little temperature dependence and of a typical Debye peak in loss with an activation energy of 1·1eV. The Debye peak which shows the characteristic relatively narrow frequency range is accompanied by a strong dispersion of permittivity, as

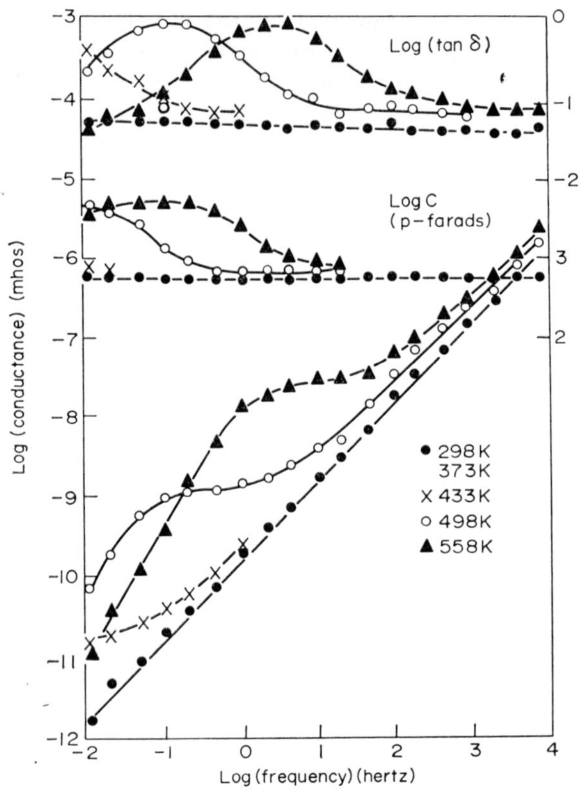

FIG. 8.4. Superposition of temperature-dependent loss peaks on a flat spectrum of almost temperature-independent loss in silicon dioxide thermally grown on silicon. Also plotted is the conductivity $\sigma(\omega)$ and capacitance $C(\omega)$. Measurements by F. M. Nazar. (Ref. 24.)

expected from eqn (8.3). The measurements refer to glow-discharge-deposited SiO_2.[(24)]

Mica which has a crystalline layer structure and may be considered to be an impurity conductor shows similar behaviour.[(25)] Measurements on the organic ferroelectric crystals of triglycene sulphate have revealed an exponent $n = 0{\cdot}8$ over 8 decades of frequency from 10^{-3} Hz to 10^5 Hz.[(26)]

Figure 8.5 shows data for multi-molecular-layer films of stearic acid, $CH_3\,(CH_2)_{16}\,COOH$, extending over a frequency range 10^{-4}–10^2 Hz for several temperatures.[(27)] The significant feature is the good power-law dependence, coupled with very weak temperature dependence, except for a slight variation of the exponent n.

A rather striking example of the same familiar power-law dependence of conductivity was observed in vitreous aqueous solutions conducting by protons.[(28)]

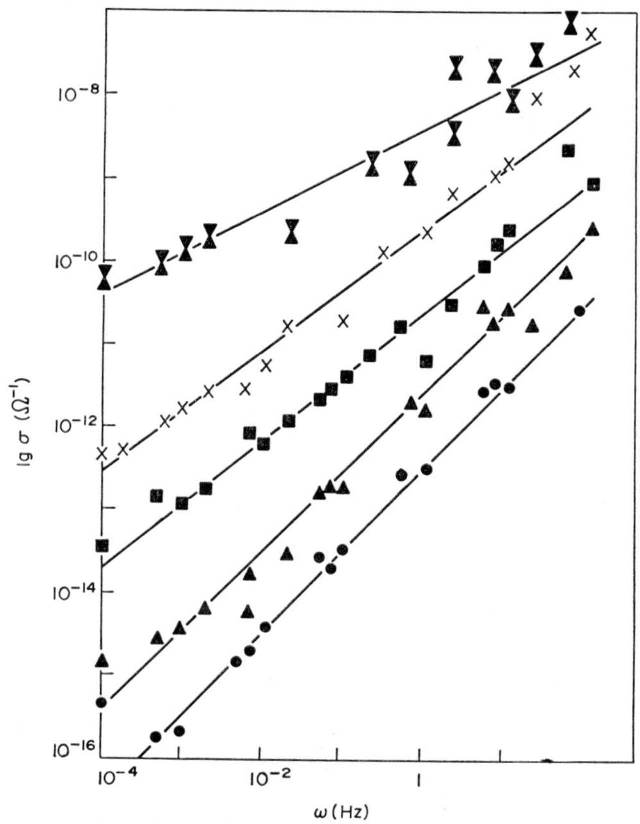

FIG. 8.5. Frequency dependence of conductance in multi-molecular layers of stearic acid at various temperatures. Film thickness 270 Å, area 1 mm². The conductance scale refers to the lowest temperature, higher temperature characteristics being successively displaced upwards by one decade each for clarity. ● 105 K, ▲ 295 K, ■ 338 K, × 362 K, ⧗ 373 K. (After Ref. 27.)

We conclude this review with some examples of the behaviour of dielectric loss in typical insulators. Figure 8.6 gives a collection of data for the frequency dependence of the loss tangent, $\tan\delta = \kappa''/\kappa'$ for a range of materials, some of which are ceramic, some organic and some are sputtered or evaporated inorganic dielectrics.[29–32] Their common feature is a very weak dependence of loss on frequency over a wide range of frequencies, indicating values of n in eqn (8.1) slightly less than unity.

The experimental results quoted above may be summarized in the following way:

(i) A very wide range of poorly conducting solids show in addition to dc conductivity σ_0 also a frequency dependent ac conductivity $\sigma_1(\omega)$ which

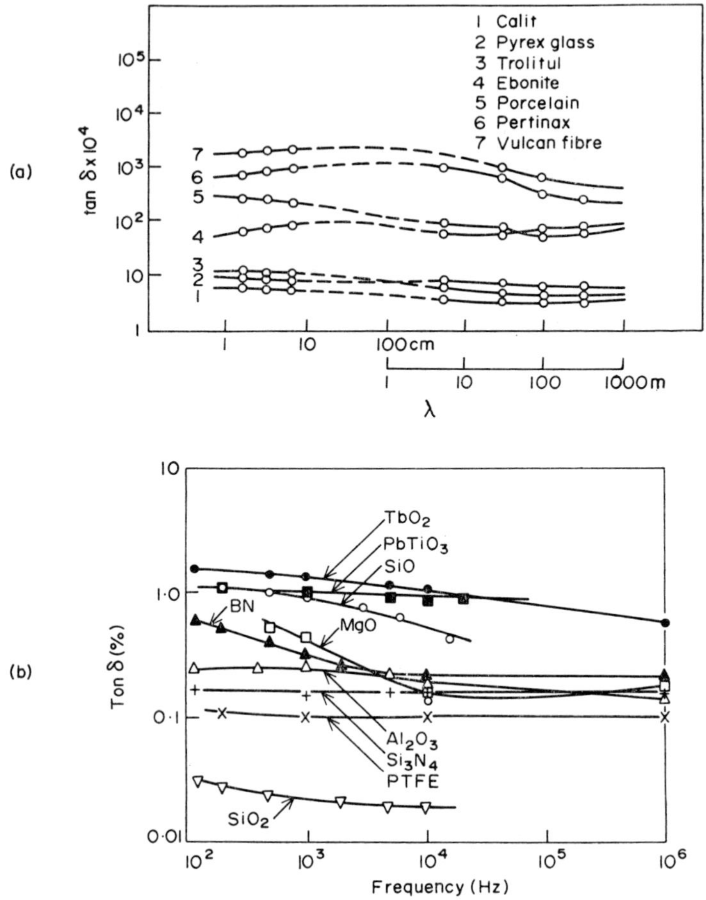

FIG. 8.6. Collected ac loss data for a range of dielectric materials, showing a relatively flat frequency dependence. (a) From Ref. 30. (b) From Ref. 29.

either follows a power-law relation of the type given by eqn (8.1) with the exponent $n < 1$, or a dependence as ω^2 followed by saturation, as given by eqn (8.2).

(ii) There is no evident correlation between the magnitudes of the dc and ac components of conductivity for any particular material. $\sigma_1(\omega)$ has a weaker temperature dependence than σ_0 and this dependence becomes progressively weaker as the frequency is increased. The main effect is to reduce the value of the exponent n from almost unity at low temperatures to smaller values at high temperatures. The ω^2 dependence is very little temperature dependent.

(iii) The ω^n frequency dependence is not necessarily a characteristic of any particular type of mechanism of low-mobility conduction. It is found in different host matrices and for different types of charge carrier. It does not appear to depend on the presence or absence of any particular type of disorder.

(iv) The two types of dependence given by eqn (8.1) and eqn (8.2) may be superimposed, the latter being characteristic of Debye-type loss mechanisms.

(v) The ω^n dependence of conductivity is usually associated with an almost frequency-independent dielectric permittivity κ'.

8.1.3. THE STOCHASTIC MODEL OF HOPPING

We propose to base our discussion on the analysis of hopping behaviour of materials as a series of stochastic events in time.[5] The time- and frequency-dependence of the electric current are intimately related in the linear regime through the Fourier transformation. Let $i(t)$ be the transient part of the time-dependent current due to a unit-step field excitation $E(t) = 1(t)$ turning on at time $t = 0$. The total current is

$$I(t) = i(t) + i_\infty \tag{8.4}$$

where i_∞ is the direct current flowing after infinite time. The frequency dependent conductivity is then given by the Fourier transform

$$\sigma_1(\omega) = \omega \int_0^\infty i(t) \sin \omega t \, \mathrm{d}t \tag{8.5}$$

Although most treatments of the frequency dependence of conductivity operate in the frequency domain, we prefer the time domain as having more direct physical significance.

We now wish to consider the effect on the external circuit of an "elementary" hopping event, which we represent as an instantaneous transition of a charge q from site i to site j in the host matrix, Fig. 8.7. The effect on the external circuit is a delta-function current followed by a time-dependent relaxation due to the response of the dielectric. We represent this response as

$$G(t) = a\delta(t) + g(t) \tag{8.6}$$

where the constant a is determined from the condition

$$\int_0^\infty G(t) \, \mathrm{d}t = 1. \tag{8.7}$$

The elementary current is then given by the relation

$$i(t) = (q\,r_{ij}/w)\,G(t) = AG(t) \tag{8.8}$$

where r_{ij} and w are defined in Fig. 8.7.

We are not concerned at this stage with the detailed form of the relaxation function $g(t)$ beyond stating that it must be a decaying function—which may be oscillatory—with a decay time τ. We would expect $1/\tau$ to lie in the short microwave or infra-red region of the spectrum, although in special circumstances much lower relaxation frequencies are possible.

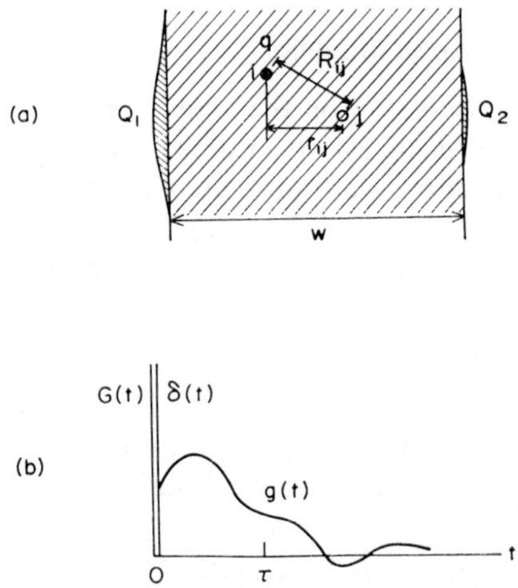

FIG. 8.7. (a) The model of a hopping medium. (b) The time response $G(t)$ to a hopping event. (From Ref. 5.)

The contribution of the elementary current, eqn (8.8), to the frequency response of the system involves the Fourier integrals

$$c(\omega) = \int_0^\infty G(t)\cos\omega t\,\mathrm{d}t, \tag{8.9}$$

$$s(\omega) = \int_0^\infty G(t)\sin\omega t\,\mathrm{d}t \tag{8.10}$$

multiplied by the appropriate amplitude A. We note that the transforms given by eqns (8.2) and (8.3) correspond to $G(t) = \mathrm{e}^{-t/\tau}$ but very similar forms would be obtained for a wide class of decaying response functions which all give

$$c(\omega) \approx \text{const}, \; s(\omega) \approx \omega\tau \quad \text{for} \quad \omega \ll 1/\tau. \tag{8.11}$$

Our analysis will not depend critically on the detailed form of $c(\omega)$ and $s(\omega)$.

The hopping system under consideration comprises a large number of localized sites and in the absence of an external field the available carriers execute random hops between neighbouring pairs of sites giving rise to a random signal which may be represented, with reference to Fig. (8.8a), by

$$I(t) = \sum_m A_m G(t - t_m) \tag{8.12}$$

where the index m counts the events as seen in the external circuit. With zero external field we must have

$$\sum_m A_m = 0$$

taken over any sufficiently large number of consecutive events.

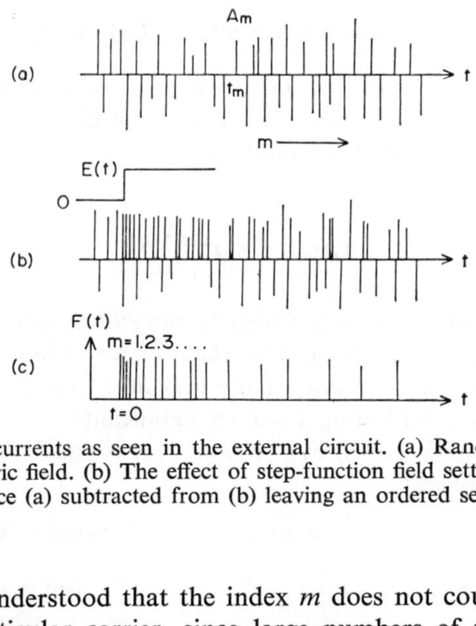

FIG. 8.8. Hopping currents as seen in the external circuit. (a) Random sequence in the absence of an electric field. (b) The effect of step-function field setting in at time $t = 0$. (c) Random sequence (a) subtracted from (b) leaving an ordered sequence.

It should be understood that the index m does not count the successive hops of any particular carrier, since large numbers of these carriers are hopping "in parallel." In our analysis we shall assume that the amplitudes A_m do not vary significantly between hops, since the distances r_{ij} for the most likely transitions do not vary by large amounts.

We now consider the effect of the application of a step function field at $t = 0$. We assume that the magnitude of this field is sufficiently small so that only linear effects need be considered, and this means:

$$qEr_{ij} \ll kT. \tag{8.13}$$

This implies that, to a first approximation, the random events are perturbed relatively little by the superposition of a para-random sequence of "ordered" amplitudes in the direction of the applied field, Fig. 8.8(c). The sequence of ordered amplitudes will be denoted by $F(t)$ and in the following analysis the index m will be used to count these amplitudes only.

Immediately after the application of the field there is a rapid rush of hopping events but their frequency decreases with time and eventually reaches a quasi-steady-state corresponding to dc conduction after infinite time. This quasi-steady-state contribution is subtracted in defining $F(t)$ so that $F(t)$ has no hopping events as $t \to \infty$.

We now note that a delayed event $G(t - t_m)$ has the Fourier transforms

$$c(\omega; t_m) = c(\omega) \cos \omega t_m - s(\omega) \sin \omega t_m \tag{8.14}$$

and

$$s(\omega; t_m) = s(\omega) \cos \omega t_m + c(\omega) \sin \omega t_m. \tag{8.15}$$

The detailed analysis of this problem which involves a summation of the contributions $A_m G(t - t_m)$ in eqn (8.12) has been given elsewhere.[5] At this point we shall merely quote the final results of this analysis which may be stated as follows:

$$\sigma_1(\omega) \approx A\omega^2 \sum_{m=1}^{M} t_m \tag{8.16}$$

where the last event M to be included in the summation is defined by the relation $1/t_M = \omega$. The implication is that for any given frequency ω, the conductivity is given by the sum of the first M events as defined, the subsequent events not contributing anything significant.

Three model infinite sequences of events were found[5] to have the following frequency responses:

linear	$t_m = am,$	$\sigma_1(\omega) = \text{const}$	(8.17)
quadratic	$t_m = am^2,$	$\sigma_1(\omega) \propto \omega^{0 \cdot 5}$	(8.18)
exponential	$t_m = a \exp(bm),$	$\sigma_1(\omega) \propto \omega$	(8.19)

The first of these is trivial and corresponds to dc, the last gives the limiting form of the exponent $n = 1$, the quadratic form gives the value of n found at higher temperatures. In general, any power law of the form given by eqn (8.1) can be generated by the sequence

$$t_m = am^{1/(1-n)} \tag{8.20}$$

provided $0 < n < 1$. We have asserted[5] that no values of the exponent n greater than unity can be generated by physically plausible sequences of events t_m, with the exception of $n = 2$. We suggest that the available experimental evidence supports the view that $n > 1$ is not a physically admissible range of exponents. This is also true of loss tangent data, with the exception of frequency ranges where loss peaks intervene.

With regard to $n = 2$, it is easy to show that any sequence of a finite number of hopping events terminating at time t_{lim} gives $\sigma_1(\omega) \propto \omega^2$ for $\omega < 1/t_{\text{lim}}$, (Fig. 8.9). For frequencies $\omega > 1/t_{\text{lim}}$ the frequency dependence is determined by the detailed nature of the series and is the same as given by eqns (8.17)–(8.20).

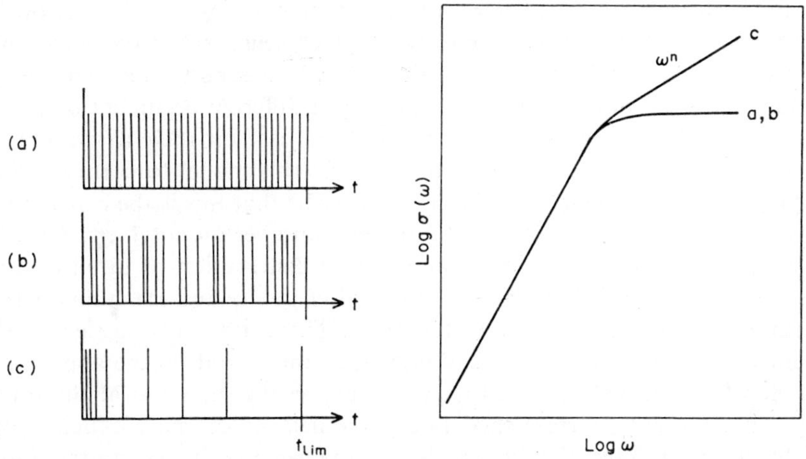

FIG. 8.9. Typical hopping sequences in response to a step-function excitation with a limited number of hops finishing at time t_{lim}, and the corresponding frequency dependences of conductivity, $\sigma(\omega)$. (a) Regular limited sequence. (b) Pseudo-regular limited sequence. (c) Limited sequence with increasing intervals.

8.1.4. RESPONSE OF SPECIFIC SYSTEMS

We are now in a position to examine in detail the expected frequency response of specific hopping models. We shall start with the simplest system consisting of an assembly of non-interacting identical pairs of centres, with a common relaxation time, τ. The application of a step-function field will cause those pairs that have a carrier "up-stream" to hop after random delays t_m which will be centred at τ, and after all hops have been completed there will be no further events. The situation will be as shown in Fig. 8.9(a) or (b) and the frequency response will be that of a system of Debye dipoles, from which our system is indistinguishable both physically and mathematically.

Physical examples of such systems of identical pairs may be found in the case of Li doped NiO,[15] where holes jump between Ni ions surrounding a Li impurity, so that distances of all hops are identical and no dc conduction can arise from this mechanism.

As the next example consider an infinite chain of identical steps with a single electron on it. Since the chain is infinite, the sequence of hops represents a pseudo-regular series which is equivalent to eqn (8.17) and has a flat frequency response from dc to the relaxation frequency $1/\tau$ of the dielectric medium. This fact is well known from the theory of the random walk on an infinite chain with an alternating force field. Physically this case might be approximated by the hopping of a charge carrier on a very long polymer chain.

A finite chain of identical steps, on the other hand, with one electron on it gives rise to a limited sequence of hops which ceases when the carrier has reached the end of the chain. We would therefore expect a ω^2 dependence for this model, as was found theoretically by a different argument by Beeby and Hayes,[33] although some doubt has been cast on their derivation by Butcher.[13] Experimentally, one would expect this model to correspond to hopping on finite polymeric chains and it may be that this is the explanation for the ω^2 region in the results of Chan and Jonscher on polymeric CS_2.[16]

Next consider a random three-dimensional distribution of localized trap sites forming a network with many branching points and extending across the sample from one electrode to the other. There are many carriers in the system which came from suitable donor-like centres and on the application of a step function field they all begin to hop in the direction of the force. Very soon, however, certain carriers become delayed on more difficult hops and block the passage of other carriers in the same path, forcing them into the next more difficult branch which again in turn becomes blocked, and so on. In this way our infinite branched network becomes progressively less conducting and the sequence of hopping events $G(t - t_m)$ becomes more extended in time without actually ever finishing. The steady current after infinite time, which is not included in this series, consists of the passage of carriers across the relatively few easy "percolation paths" extending from one electrode to the other, after all the other paths became blocked.

This infinite random branched network would therefore lead to the ω^n dependence with the exponent n depending upon temperature but not exceeding unity.

A set of non-interacting pairs with a very wide distribution of pair spacings and relaxation times would give rise to a similar behaviour as the random network—the firing times would go on indefinitely, provided the relaxation times were distributed on a sufficiently wide scale. This corresponds directly to the Pollak–Geballe model and by allowing interaction between pairs we obtain the random network model.

We now return to a finite regular chain but with several carriers on it coming from suitable charge-compensating donor-like centres, but still filling only a small proportion of the available sites. If there were no correlation effects the response of the system would consist essentially of a superposition of the individual ω^2 dependences, i.e. it would go as ω^2. However, the repulsive interaction between carriers may provide a significant slowing down of the process when the leading carrier takes longer over one hop, the others may not spontaneously follow at their own pace but must await the clearing of the path by the first. This effect is especially pronounced in random chain hopping, as pointed out earlier,[5] but it may also provide an explanation of ω^n dependence on a regular finite chain with more than one carrier. There exists to our knowledge no theory of this type of transport.

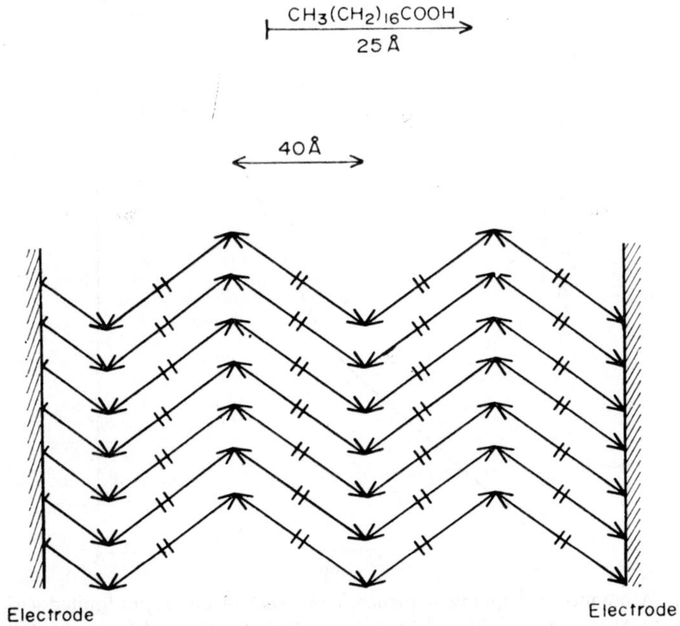

FIG. 8.10. Schematic arrangement of molecules in multi-layer films of stearic acid, $CH_3(CH_2)_{16}COOH$, in which the molecules are represented as arrows. The length of each molecule is 25 Å.

Turning next to stearic acid films,[24] we are dealing with relatively short chains of molecules, arranged as shown schematically in Fig. 8.10. High-field evidence suggests the donor-like character of some centres, presumably the COOH groups, spaced by approximately 40 Å, and this spacing is confirmed by X-ray evidence. At low fields, $E < 10^4$ V/cm, the energy difference across the "sublattice" spacing of 40 Å corresponds to a few meV, hence less than kT down to the lowest experimentally investigated temperature. The possibility of branching is very evident if one path becomes blocked by a slow carrier, and this may explain the observed frequency dependence.

This explanation, if confirmed, would establish the important concept of ω^n dependence on a *regular* network. It would also imply that variable range hopping is not a necessary condition for the observation of the Pollak–Geballe type of frequency dependence as ω^n.

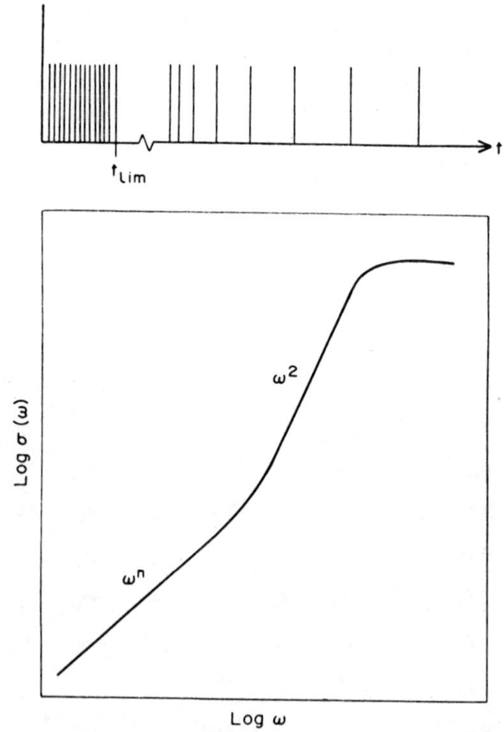

FIG. 8.11. A composite hopping sequence, consisting of one rapid limited series finishing in a very short time t_{lim}, followed by a much slower unlimited sequence with increasing time intervals. The time scale to the right of the break in the axis is understood to be much slower than to the left. The corresponding frequency dependence of conductivity is also shown.

Our last model represents a system in which a very rapid limited sequence is followed by a much slower infinite sequence with increasing period, (Fig. 8.11). The expected response now is a superposition of ω^n dependence appropriate to the slow sequence at low frequencies, and an ω^2 dependence corresponding to the limited sequence at high frequencies. Physically this may be represented by a branched network of hopping sites and a separate set of dipole-like pairs with a much shorter relaxation time.

8.1.5. HOPPING IN MICRO-INHOMOGENEOUS SYSTEMS

Fritzsche has suggested[34] that the band edges of strongly disordered and amorphous materials may be severely distorted in the sense of producing a random array of potential valleys which are only partially interconnected. These valleys may be well conducting, but they are separated by potential barriers of random height and width. At the same time the band gap itself may not necessarily contain any significant density of localized levels.

Our model of ac conduction may be applied to this situation by considering that each of the valleys is a conducting path of finite extent which can only accept another carrier from outside by emitting one carrier elsewhere. In that sense the finite spatial extent of the conducting islands does not change their basic similarity with point centres such as traps, which cannot take on more than one carrier. Just as the randomly spaced traps give rise to an ac conductivity of the ω^n type, so the finite islands behave in exactly the same way. This explains why materials such as selenium, which are regarded by Fritzsche as falling into this category, show the same type of frequency dependence as other materials in which deep levels are believed to exist.

It is worth pointing out the similarity between this model and the Maxwell–Wagner system of conducting particles dispersed in a non-conducting matrix. A Maxwell–Wagner system with a very wide distribution of resistive and capacitive parameters could show a similar frequency dependence to that given by eqn (8.1). The difference lies in the scale of dimensions of the conducting regions, which may become of atomic dimensions in our model, and also in the distribution of the height and width of the barrier parameters.

8.1.6. TEMPERATURE DEPENDENCE OF AC CONDUCTIVITY

We consider next the interpretation of the experimentally observed weak temperature dependence of ac conductivity, at least in the case of electronic conduction, since data for ionic systems are much less complete. In the limit of the high frequency ω^2 dependence there is no variation with temperature and this may be understood in terms of tunnelling transitions between closely regularly spaced sites. With the ω^n relation the temperature dependence increases towards lower frequencies and is weaker than dc. This may be understood in terms of the effect of increasing thermal activation of the rate-limiting hops—the most difficult ones are going to be influenced most strongly. This will result in a reduction of the exponent in the power series determining the t_m sequence, eqn (8.20), thus reducing the exponent n. The most difficult hops involved in dc conduction require the highest thermal activation, with ac this becomes progressively less critical.

8.1.7. CONCLUSIONS

Our survey of the experimental data relating to the frequency dependence of ac conductivity has revealed a considerable similarity of behaviour for a very wide range of materials, ordered and disordered, conducting by electrons, ions, polarons, etc., involving various types of chemical bonds and various electronic energy level structures. We conclude that the frequency dependence $\sigma_1 \propto \omega^n$ must represent a very general law, not confined to any particular host matrix, type of carrier, or type of disorder. It is proposed that a qualitative explanation of the observed behaviour and a better physical understanding of the mechanisms involved can be obtained by considering a stochastic model of hopping in time and space. Frequency transforms are obtained for various specific models of hopping sequences and these are related to experimentally known systems.

It is concluded that the requirement for the appearance of the ω^n dependence of conductivity is the existence of a wide distribution of transition probabilities for carriers. This may typically but not exclusively be obtained in a three-dimensional irregular distribution of localized levels for electronic hopping. It may arise in an ordered system with a distribution of probabilities and with correlation effects forbidding too close an approach of two carriers. Finally, it may arise as a result of the presence of incompletely interconnected microscopic conducting regions separated by randomly distributed barriers. The ω^n region is not necessarily characteristic of localized level hopping or of any other particular model of conduction. It is predicted that the exponent n may not exceed unity in this process.

The alternative dispersion law giving $\sigma_1 \propto \omega^2$ followed by saturation is characteristic of any limited sequence of hopping events under the application of a step-function field and as such is evidence of the presence of molecular dipolar loss mechanisms or two-centre hopping.

The dc conductivity σ_0 represents the limiting form of transport where carriers move continuously from one electrode to the other and are replenished and removed as required. There is no evident correlation between the magnitudes of the dc conductivity and of the ac conductivity and their temperature dependences are different.†

† See Note Added in Proof on p. 359.

8.2. High Field Effects

8.2.1. INTRODUCTION

The study of the electrical conductivity of any medium gives us an insight into the transport mechanisms for the prevailing carrier species, be they electrons, holes, polarons or ions. The starting point for experimental work and for theoretical analysis is usually the low-field condition, defined by the requirements that the applied field should not significantly disturb the thermal equilibrium of the system. In this situation one may treat transport as a small perturbation of the equilibrium system,[36] usually through Boltzmann's equation, with the distribution of electrons in energy remaining to first order unchanged in comparison with the equilibrium situation. When dealing with a free-carrier system, such as a semiconductor or a metal, one derives an expression for the mobility μ of the carriers, in terms of their effective mass m^*, charge q and mean collision time τ_p: $\mu = q\tau_p/m^*$. The density of free carriers is assumed to be the equilibrium density n_0, so that the resulting low-field conductivity is given by:

$$\sigma_0 = q\mu n_0. \tag{8.21}$$

The application of a high field to a free carrier system may influence both the mobility and the number of available carriers. The mobility may be influenced by the change in the effective distribution of carriers in energy, which could be described in terms of a rise of carrier temperature T_c above the lattice temperature T_0. This may be seen most readily by considering that an electric field E acting on a carrier system moving with an average velocity $v_d = \mu E$ will give rise to a power input $q\mu E^2$. Under steady state conditions this power input has to be dissipated to the lattice through collisions which take the energy away from the carriers.

An energy relaxation time τ_ε may be defined by the relation

$$q\mu E^2 = \frac{k(T_c - T_0)}{\tau_\varepsilon} \tag{8.22}$$

equating the power output from the field to the power loss to the lattice. Since the conditions for energy relaxation are quite different from those for momentum randomisation τ_ε is not necessarily equal to τ_p, the time τ_ε being usually much longer.[33,38] Depending upon the prevailing scattering mechanism, the mobility may either increase or decrease with increasing electric field due to this carrier heating effect. It appears, however, that at fields in excess of some 10^3–10^4 V/cm, the proportionality between drift velocity and field is largely lost and v_d increases much more slowly, saturating in most instances at values of the order of 10^7 cm/s.[39]

At much higher electric fields, of the order of 10^6 V/cm, the heating of the carrier population by the electric field becomes sufficiently effective to impart to a finite number of carriers energies comparable with the width of the forbidden gap, thus leading to rapid multiplication of carrier numbers by the avalanche process.[40] This causes a corresponding rise of current with field.

In the case of free-carrier systems high field effects are most readily observable in materials with a small number of equilibrium carriers since heating effects are kept to reasonable proportions. Highly conducting solids do not readily lend themselves to this type of study. For the same reason, the study of high field effects is particularly favoured in low-conductivity solids, many of which are such because of their low mobility. It is here that the amorphous semiconductors come into consideration, since the mobility of localized carriers in them is very low. Some of these materials may have so low an equilibrium conductivity that they may be classified as insulators and the only observable transport occurs under high field conditions. The present paper will be concerned with a review of the theory and experimental results relating to high field conduction in a wide range of low mobility solids, most of which are amorphous or heavily disordered.

8.2.2. CLASSIFICATION OF HIGH FIELD PHENOMENA

In the context of low mobility solids it is possible to distinguish four broad classes of phenomena which could arise in the presence of high electric fields to the system.

(i) Contact-induced effects, e.g. injection of non-equilibrium carriers, space charge limited flow, blocking effects, tunnelling through the barriers, etc.

(ii) Purely thermal effects arising from Joule heating of the sample.

(iii) Bulk effects due to the enhancement of the conduction by the high electric field, without external excitation by light, etc., or with such excitation.

(iv) Catastrophic bulk effects leading to strongly localized filamentary conduction with or without resulting thermal effects and structural changes—breakdown, switching, etc.

The problem of contacts will be discussed very briefly in the following Section, but the remainder of the first category of effects will not be dealt with here since they form a vast subject in their own right for which there is ample literature.[41–45] The second category of purely thermal effects will not be considered either, except for mentioning the obvious precautions which have to be taken avoid overheating of samples.

The last category of catastrophic effects, whether they are reversible or irreversible, forms the subject of two well-developed directions of study: dielectric breakdown effects[(46)] and switching phenomena in amorphous semiconductors which are being dealt with in the context of the present Summer School.[(47)] Accordingly, we do not propose to discuss them at present.

This only leaves the third category of bulk effects due to high electric fields. It is inevitable that in most cases the experiments have to be performed on relatively thin films of the materials under study, to avoid excessive dissipation of heat and to keep the required potential difference across the sample down to manageable values. The resulting "sandwich" geometry, metal—semiconductor—metal, makes it impracticable to apply separate potential probes to eliminate any contact effects and considerable caution is necessary in the interpretation of experimental results on that account.[(48)]

8.2.3. CONTACTS TO LOW MOBILITY SOLIDS

An electrical contact represents an interface between two different materials, one of which in our case is non-metallic and the other metallic. In general, a potential difference exists between the two materials even in the absence of an external bias. This may be due to a difference between the contact potentials of the two materials, or to the formation of a surface charge on the semiconductor due to the peculiar charge conditions of the electronic surface states.[(46)] In either case, the resulting potential difference is screened within a distance related to the "screening length" λ in the semiconductor, which depends on the number of ionizable centres in the vicinity of the interface. This number is in turn dependent on the density of available electronic states in the forbidden gap of the material in question, especially in the range of energy close to the Fermi level.

If the density of ionizable centres is high, in excess of $10^{19}\,\text{cm}^{-3}$ say, a barrier of 1eV height would have a width of 100 Å, the resulting energy barrier being as shown in Fig. 8.12(a). In this situation it is very likely that the dominant process of carrier injection into the semiconductor is by direct tunnelling near the Fermi level into the localized levels in the forbidden gap of the material. In this situation the observed behaviour would not depend strongly on contact phenomena. If the semiconductor has very few localized levels in the forbidden gap, then the screening length is very large and may exceed the total thickness of a relatively thin sample, (Fig. 8.12(b)). The potential of the bulk material is now determined entirely by the contact potential difference and the transport properties depend on the motion in the conduction or valence bands of carriers injected over the contact barrier.

The presence of contact barriers leads to two kinds of effects—asymmetry with respect to the polarity of the applied bias and the failure of the voltage—

current characteristics to scale according to the thickness of the sample, since barrier phenomena depend on voltage while bulk effects depend on field. A test for symmetry, for correct thickness dependence and for independence of the type of metal electrode used should eliminate any doubts as to the significance of contact effects in any particular experiment.

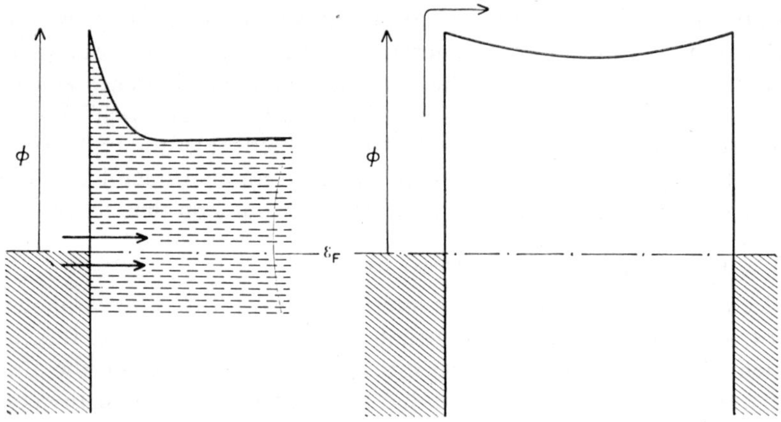

FIG. 8.12. The contact between a metal and a semiconductor, with a potential barrier ϕ for emission of electrons from the Fermi level $\mathcal{E}_F$ in the metal into the conduction band of the semiconductor. (a) Semiconductor with a high density of ionizable centres, the screening length is short and the width of any barrier present is insignificant. Tunnelling directly into the localized levels may occur. (b) Semiconductor with very few ionizable centres. the screening length is much larger than the sample thickness.

8.2.4. BULK HIGH FIELD EFFECTS

We begin our discussion of bulk high field effects with a brief review of the types of localized levels which are likely to be found in the forbidden gap of a disordered solid. It is possible to distinguish two basic types, according to the state of charge of the centre which is assumed to be capable of existing in two states, one of which contains one electron more than the other. The electron-rich state will be referred to as "occupied" or "full" and the other as "unoccupied" or "empty". A centre is called a donor if it is neutral when full and positively charged when empty; it is called a trap if it is neutral when empty and negatively charged when full. The basic difference for our present purpose consists in the nature of the forces binding the electron to the centre—in a donor the force is Coulombic and therefore of relatively long range, in a trap it is non-Coulombic and of much shorter range.

A disordered solid may contain a wide distribution of both donors and traps in energy and in space and Fig. 8.13 shows a schematic representation of such a distribution, donors being represented by squares, traps by circles and shading denoting full centres. All transitions giving rise to transport must be between full and empty centres and the most energetically favourable are those involving the smallest change of energy. In low field transport there is no significant difference between donors and traps and all transitions marked in Fig. 8.13 contribute to the flow of current. At high fields, however, an important distinction arises between traps and donors and we shall distinguish the following cases:

(i) donor–donor hopping;

(ii) trap–trap hopping;

(iii) donor–trap hopping.

These various cases will be discussed in turn.

(i) *Donor–donor hopping*

At low electric fields this mechanism is responsible for conduction in compensated crystalline silicon and germanium in the impurity conduction range, i.e. at very low temperatures.[50,51]

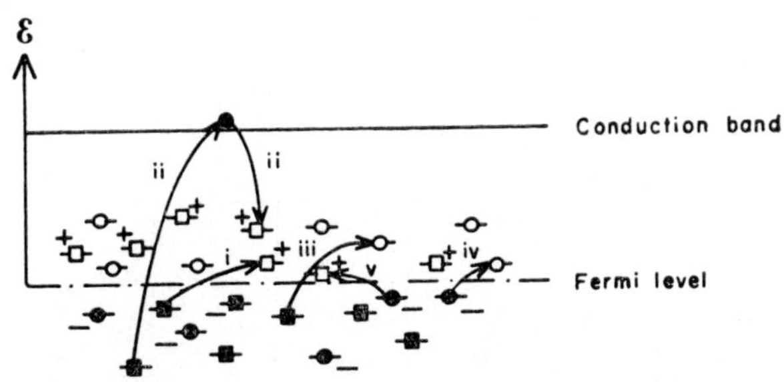

FIG. 8.13. Various electronic transitions occurring in a system of donors (square symbols) and traps (round symbols) distributed in space and in energy. Filled contours denote occupation by electrons. (i) Donor–donor hopping. (ii) Excitation between donor and conduction band. (iii) Donor–trap hopping. (iv) Trap–trap hopping. (v) Trap–donor hopping (From Ref. 49.)

The presence of a high electric field E will affect the potential barrier of a Coulombic centre situated at the point $\mathbf{r} = 0$ in the following manner:

$$\phi(\mathbf{r}) = -\frac{e^2}{4\pi\kappa' r} - eE\cdot\mathbf{r} \tag{8.23}$$

as shown in Fig. 8.14. The maximum of the potential barrier in the direction of the electric field occurs at a point $x_m = \beta/2E^{\frac{1}{2}}$ and the lowering with respect to the zero level in the absence of field is

$$\phi_m = -e\beta E^{\frac{1}{2}} \quad \text{where} \quad \beta = (e/\pi\kappa')^{\frac{1}{2}}. \tag{8.24}$$

If the ground state of the donor is at ϕ_i in the absence of the field, the probability of emission from this ground state due to thermal excitation is

$$p \propto \exp(-\phi_i/kT)\exp\left(\frac{e\beta E^{\frac{1}{2}}}{kT}\right). \tag{8.25}$$

The electric current arising from the transport of the emitted carriers in the conduction band will depend not only on P but also on the distance l moved by the carriers before recapture in donor centres further downstream. If l is the average path before recapture, we may postulate that $l = \text{const}$, if the distance travelled is independent of the field, or $l \propto E^a$, where a is an exponent of the order of 1 or 2. The current will therefore be of the form:

$$I = A(T)\,E^a \exp\left(\frac{e\beta E^{\frac{1}{2}}}{kT}\right) \tag{8.26}$$

where $A(T)$ is a function of temperature incorporating, among other factors, the first exponential term in eqn (8.25).

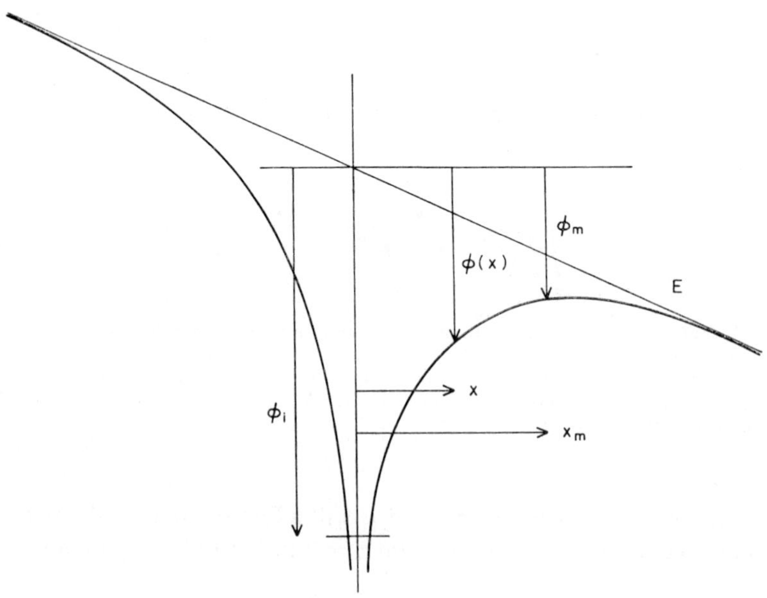

FIG. 8.14. The potential of a Coulombic centre in the presence of an electric field E. x is the distance measured in the direction of E. ϕ_i is the ground state.

Since the dominant field dependence in eqn (8.26) is through the exponential term, the E^a factor giving a relatively weak contribution, it is possible to write eqn (8.26) in the following form:

$$\log I = \text{const} + \frac{e\beta}{kT} E^{\frac{1}{2}}. \tag{8.27}$$

This shows that a graph of $\log I$ versus $E^{\frac{1}{2}}$ ought to give a straight line with a slope $e\beta/kT$ which should be easily checked experimentally, with the value of β taken from eqn (8.24). This mechanism is known as Poole–Frenkel conduction(52) and has been observed in a range of amorphous materials, such as SiO,(53) Ge,(54) C,(55) to name only a few.

A different situation arises if the Coulombic centres are situated so close together that their fields overlap. In this case the lowering of the barrier is given by eEs, where $2s$ is the distance between centres. One would therefore expect this model to be characterized by a current-field dependence of the type:

$$\log I = \text{const} + \frac{es}{kT} E \tag{8.28}$$

and this is known as the Poole effect.(56) Again, this type of behaviour has been found in a number of materials.(56,57)

The simple theory discussed so far provided for emission over the lowest point of the barrier, directly in line with the electric field. A more realistic assumption is that the emission attempts are distributed isotropically in all directions and that they succeed according to the barrier lowering in any particular direction. This treatment has been applied(52,58) and has been further extended by Hill(56) to include emission "upstream", i.e. against the field. The final result is that the exponential term in eqn (8.26) is replaced by a function

$$f(\alpha) = \alpha \cosh \alpha - \sinh \alpha \tag{8.29}$$

where $\alpha = e\beta E^{\frac{1}{2}}/kT$ for the Poole–Frenkel mechanism and $\alpha = esE/kT$ for the Poole mechanism. Hill expresses the current in a reduced form

$$J = I\,T^{-b} \exp\left(\frac{e\phi_i}{kT}\right) \tag{8.30}$$

where the T^{-b} term corresponds to $1/A(T)$ in eqn (8.26), ϕ_i is the donor activation energy in the absence of field. With these notations it is possible to write a general equation

$$J = f(\alpha) \tag{8.31}$$

which contains both field and temperature dependence implicitly through the variable α. In this way it is possible to reduce a number of $I - E$ characteristics corresponding to a range of temperatures to one single curve. The parameters b and ϕ_i are determined by curve fitting and the final result for data on SiO[53] is shown in Fig. 8.15. The solid line represents the graph of $f(\alpha)$ expected theoretically and it is clear that the agreement with this generalized theory is excellent over many decades of the reduced parameters. This type of reduction represents an excellent and rigorous check on the applic-

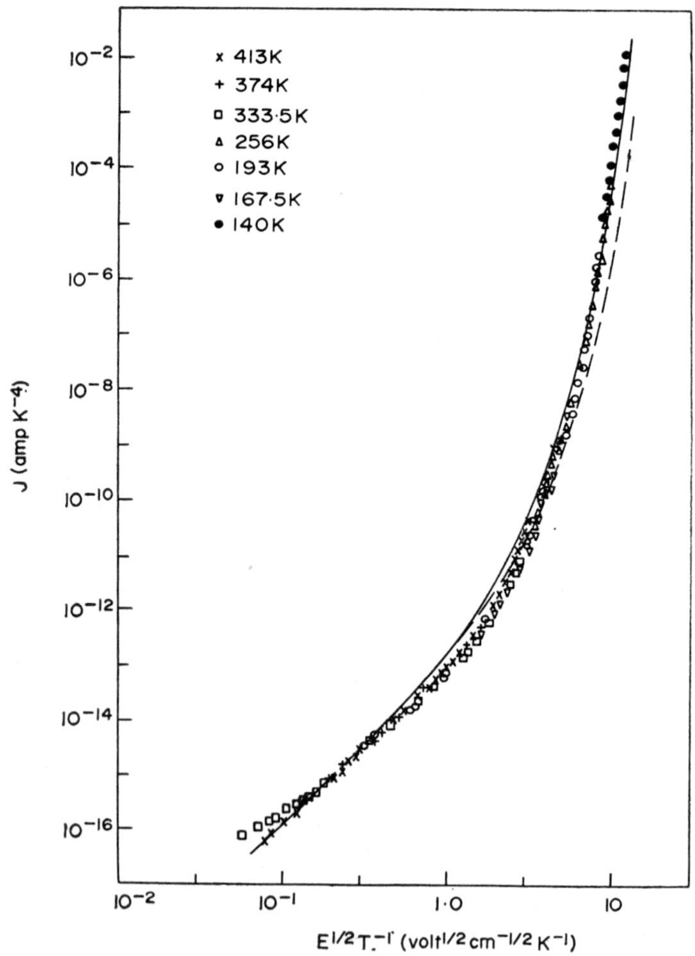

FIG. 8.15. The dependence of the reduced current $J = IT^{-4} \exp(-0{\cdot}35/kT)$ upon $E^{1/2}T^{-1}$ The data for SiO from reference 18, given by Hill (Ref. 56.) The curves are $\alpha^2 \sin \alpha$ and $\alpha \cosh \alpha - \sinh \alpha$. (From Ref. 56.)

ability of the Poole–Frenkel or Poole theory to any given set of experimental data and Hill was able to show that the published interpretation of some data in terms of Poole–Frenkel mechanism was in fact in error.

Exactly similar procedures may be applied to the case of Poole mechanism, by normalizing to E/T, instead of $E^{\frac{1}{2}}/T$ and Fig. 8.16 shows the results for published data on Cr_2O_3.[59] The details of the curve-fitting method are given by Hill.[60]

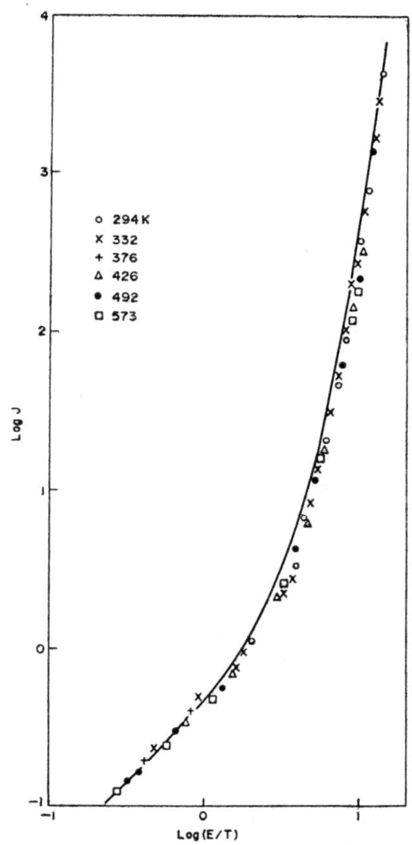

FIG. 8.16. Normalized plot according to eqn (8.31) of experimental data for amorphous Cr_2O_3 from Barbe and Herman (Ref. 59), showing Poole effect. (Courtesy R. M. Hill.)

The very good agreement between theory and experiment shown especially in the Hill procedure is rather surprising in view of the relatively crude model which is used throughout. Not only is the procedure purely classical but the assumption is made that the system is homogeneous, which is most unlikely to be the case in real amorphous materials in which strong internal fields must be presumed to exist. No specific assumptions are made regarding the field dependence of the capture process, although this could be a significant factor.[61] Also some objections were raised to the form of the $f(\alpha)$ function used by Hill.[61,62] One point which has attracted considerable attention in the past is the disagreement between the experimentally determined value of the co-efficient β and the value expected from theory. The experimental value falls between 0·5 and 1 times the theoretical. A number of possible explanations have been suggested, including Schottky emission either from the electrode or from some internal islands, some special distribution of trapping centres which would necessitate the use of $2kT$ instead of kT in the expression for the slope in eqn (8.27), the concept of effective temperature for the carriers under the action of a high electric field,[63,64] $T_c > T_0$, which would have the same effect on the slope as lowering of β, and finally the possibility that the ionized donor may be recharged very soon after emitting the electron.[65] The only common statement that can be made about these various theories is that, for all their ingenuity, there is at present no means of testing them experimentally. The systems on which measurements can be made are so poorly characterized that it does not appear profitable to refine the theory much beyond its present state to explain the experimental results.

(ii) *Trap–trap hopping*

By contrast with the relatively well-developed understanding of the Poole–Frenkel and Poole mechanisms in donor-dominated hopping at high electric fields, the behaviour of hopping between traps is much less well understood. The high field has a much smaller direct effect on the strongly localized barrier of a trapping centre than on the comparatively much more extended Coulombic potential of a donor. The most pronounced effect that one would expect in this case is that the distribution of carriers in energy among the traps will be affected if the energy gained per hop is appreciably larger than the thermal energy:

$$eEr_{ij} > kT. \tag{8.32}$$

Figure 8.17 shows an assumed distribution of trapping levels in energy in the presence of a high electric field with three types of transitions indicated by arrows. On average, the carriers are unlikely to lose as much energy to the lattice as they gain from the field and this leads to an increase of their "effective temperature" in the traps. This situation is analogous to hot electrons in a free band which are described by eqn (8.22) but the energy relaxation time τ_ε is much longer in localized levels. It was suggested[64] that the further apart in energy the localized levels are the longer the relaxation time, since it is more difficult to dissipate this energy to the lattice. As the density of levels in energy increases towards the band edges, so the energy relaxation time decreases towards the free band value which may be 10^{-14}–10^{-13} sec.

This heating of the carrier population in localized levels by a high electric field would bring about an increase of mobility with field, since the more energetic electrons are moving in regions of higher mobility. In those cases where a field-dependence of mobility has been postulated in amorphous materials it is very likely that the mechanism responsible is of this type.

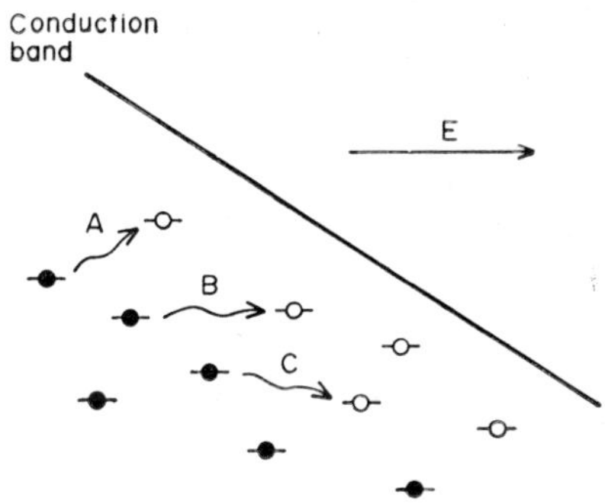

FIG. 8.17. Electronic transitions in a high electric field between full and empty traps, involving a slight gain of energy from phonons (A), conservation of energy (B), and a slight loss of energy to the lattice (C). (From Ref. 64.)

(iii) *Donor–trap hopping*

This mechanism has been introduced to explain some experimental observations of high field photo-conduction in SiO,[63] although its relevance to dark conduction is fairly clear. It applies in situations where there exist a large density of trapping levels distributed in energy together with some donor levels which effectively supply carriers to these traps. Figure 8.18 shows a donor surrounded by trapping levels in the presence of a high field. The electron does not have to be excited to the "band edge" for emission to take place. It is sufficient that it should acquire enough energy to move in to the upper range of localized traps in which the mobility is relatively high. It then moves by the trap-hopping process described above and may move even closer to the "band edge", without necessarily ever reaching it, because of the rapidly decreasing energy relaxation time. In the final process of re-capture in an empty donor site, the thermalization of the carrier has to be taken into account,[64] and it has been postulated that the carrier may be re-emitted before reaching the donor ground state. This would account for the effective temperature in the Poole–Frenkel mechanism which may be

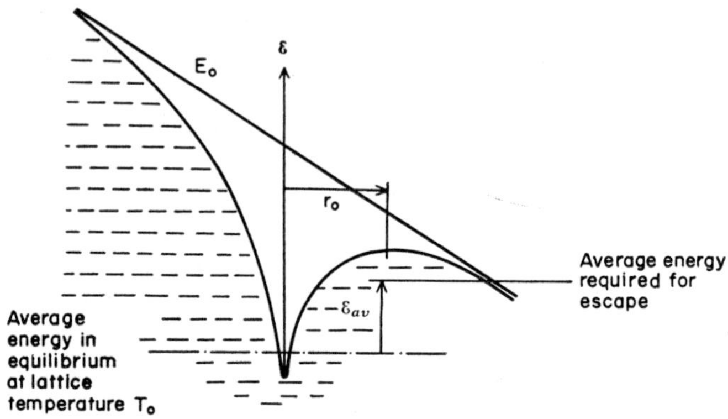

FIG. 8.18. The modified model of Poole–Frenkel conduction involving emission of an electron from a Coulombic centre via localized traps. $\mathcal{E}_{av}$ is the average energy required for escape when the electron has reached the critical radius r_0. (From Ref. 63.)

responsible for the abnormally low values of the slope of the $\log I$ versus $E^{\frac{1}{2}}$ graphs mentioned earlier. A different but related approach to the interpretation of high field photoconductivity in amorphous Se and Te postulates a field-dependent quantum efficiency.[66,67]

Most of the known examples of Poole–Frenkel or Poole mechanism of high field conduction occur in amorphous semiconductors and it is interesting to consider the reasons for the absence of this effect in crystalline materials. Simple consideration of energy balance in the presence of a high field shows[68] that the field-lowering mechanism for carrier emission is in competition with the carrier heating mechanism which leads to the avalanche ionization process either from impurity levels or across the energy gap. It is easily shown that in high mobility semiconductors and for low ionization energies the avalanche process has a much higher probability than the field-lowering mechanism. Only in very low mobility solids is the probability of avalanche ionization significantly lower than the Poole–Frenkel mechanism. Experiments on layered crystalline structures show that in the direction normal to the layers, i.e. in the direction of low mobility, there appears to be good evidence of the latter mechanisms.[69,70]

8.2.5. HIGH FIELD EFFECTS IN POLARONIC CONDUCTION

It is interesting to consider the possible effect of high electric fields on polaronic conduction in low mobility solids.[71,72] A small polaron may be considered to form on a neutral centre, i.e. trap in our nomenclature, since an electron on a donor site is dominated by the strong Coulombic field and any polarization effects of the electron on the medium would be negligible. The effect of a high electric field may therefore be considered to be two-fold.

On the one hand, the field may enhance the supply of electrons from donors to the medium, and their further transport between donor centres may then proceed by polaronic transport or otherwise, but the end result would be indistinguishable from ordinary Poole–Frenkel conduction by "bare" electrons not accompanied by their polarization clouds. Secondly, however, the field may affect the hopping probability of the polarons themselves, by causing a lowering of the potential barrier of the short-range field between nearest neighbours. This is a Poole-type effect and would be expected to follow eqn (8.28). Effects of this type have been reported[(71)] but it is not possible to determine with complete assurance the extent of polaron formation in these situations.

8.2.6. CONCLUSIONS

This brief review of high field effects in disordered and amorphous semiconductors shows that donor–donor and donor–trap hopping become dominated by field-assisted thermal emission of carriers from donor-like centres and that our quantitative understanding of these processes is relatively good, especially in view of the very uncertain knowledge of the detailed microscopic conditions in these materials. As with many other measurements on amorphous semiconductors, the problem of characterizing the materials under study makes more refined theoretical understanding impossible at the present stage. It appears, in general, that the field-assisted emission from donors and the establishment of an effective temperature in the localized trapping states are the main consequences of high fields in bulk conduction.

The point should be stressed, however, that these bulk phenomena are only one special aspect of high-field phenomena and that other effects arise as a result of contact-induced processes which are not reviewed in the present context. Nor have we touched upon the potentially important and interesting subject of high-field transient and non-linear processes in high alternating fields.[(14)]

Note added in proof

An alternative explanation of the ω^n dependence has recently been put forward by the present author[(35)] in terms of carrier transfer from interfaces either between the sample and electrodes or between grains"" in the bulk. This model can be analysed on similar lines to those outlined in Section 8.1.3 for the bulk, and may explain the otherwise remarkable insensitivity of ac properties to the structure and dc properties of low mobility solids.

REFERENCES

1. Pollak, M. and Geballe, T. H., *Phys. Rev.* **122,** 1745 (1961).
2. Pollak, M., *Phys. Rev.* **133,** A564 (1964).

3. Pollak, M., *Phys. Rev.* **138,** A1822 (1965).
4. Pollak, M., *Phil. Mag.* **23,** 519 (1971).
5. Jonscher, A. K., *J. Non-Cryst. Solids* **8-10,** 293 (1972).
6. Pollak, M. and Pike, G. E., *Phys. Rev. Letters* **28,** 1449 (1972).
7. Mott, N. F., *Phil. Mag.* **19,** 835 (1969).
8. Austin, I. G. and Mott, N. F., *Advan. Phys.* **18,** 41 (1969).
9. Davis, E. A. and Mott, N. F., *Phil. Mag.* **19,** 903 (1971).
10. Scher, H. and Lax, M., *J. Non-Cryst. Solids* **8-10,** 497 (1972).
11. Scher, H. and Lax, M., *Phys. Rev.* Vol. **7B,** (1973).
12. Miller, A. and Abrahams, E. *Phys. Rev.* **120,** 745 (1960).
13. Butcher, P. N., *J. Phys. C: Solid State Phys.* **5,** 1817 (1972).
14. Jonscher, A. K., in "Electrets, Charge Storage and Transport in Dielectrics", ed. M. M. Perlman, El. Chem. Soc., (1973).
15. Kabashima, S. and Kawakubo, T., *J. Phys. Soc. Japan* **24,** 493 (1968).
16. Chan, W. S. and Jonscher, A. K., *Phys. Stat. Solidi* **32,** 749 (1969).
17. Fritzsche, H., *J. Non-Cryst. Solids* **6,** 49 (1971).
18. Lakatos, A. I. and Abkowitz, M., *Phys. Rev. B* **3,** 1791 (1971).
19. Polanco, J. I., Roberts, G. G. and Myers, M. B., *Phil. Mag.* **25,** 117 (1972).
20. Roberts, G. G. and Polanco, J. I., *Solid State Comm.* **10,** 709 (1972).
21. Owen, A. E., private communication.
22. Argall, F. and Jonscher, A. K., *Thin Solid Films* **2,** 185 (1968).
23. Chan, W. S. and Loh, C-K., *Thin Solid Films* **6,** 91 (1970).
24. Nazar, F. M., PhD. Thesis, University of London, 1973.
25. Morse, C. T., private communication.
26. Whall, T. E., private communication.
27. Nathoo, M. H. and Jonscher, A. K., *J. Phys. C: Solid State Phys.* **4,** L301 (1971).
28. Angell, C. A., Bressell, R. D. and Gammell, P., *J. Non-Cryst. Solids* **7,** 295 (1972).
29. Harrop, P. J., Wood, G. C. and Pearson, C., *Thin Solid Films* **2,** 457 (1968).
30. Gevers, M., *Philips Res. Rept.* **1,** 279 (1945/46).
31. Gevers, M., *Philips Res. Rept.* **1,** 447 (1945/46).
32. Gevers, M., *Philips Res. Rept.* **1,** 298 (1945/46).
33. Beeby J. L. and Hayes, T. M., *J. Phys. C: Solid State Phys.* **4,** 1759 (1971).
34. Fritzsche, H., this Volume. Chapter 2.
35. Jonscher, A. K., *J. Phys. C: Solid State Phys.*, **6,** L235 (1973).
36. Ziman, J. M., "Electrons and Phonons." Oxford, 1960.
37. Smith, R. A., "Semiconductors", Cambridge, 1959.
38. Jonscher, A. K. and Rodger, K. S., *J. Phys. C: Solid State Phys.* **8,** 4, 194 (1971).
39. Papers on Carrier Drift Velocities in Silicon at High Electric Fields, *I.E.E.E. Trans. ED*-14 **1,** 37-49 (1967).
40. Sze, S. M., "Physics of Semiconductor Devices." Wiley, 1969.
41. Lampert, M. A. and Mark, P., "Current Injection in Solids." Academic Press.
42. Tredgold, R. H., "Space Charge Conduction in Solids." Elsevier, Amsterdam, 1966.
43. Polanco, J. I. and Roberts, G. G., *Phil. Mag.* **25,** 17 (1972).
44. Roberts, G. G., this Volume, Chapter 12.
45. Roberts, G. G. and Polanco, J. I., *Phys. Stat. Sol.* (*a*) **1,** 409 (1970).
46. Klein, N., *Advan. Phys..* **21,** 605 (1972).
47. Fritzsche, H., this volume, Chapter 18.
48. Jonscher, A. K., *J. El. Chem. Soc.* **116,** 217C (1969).

49. Jonscher, A. K., *J. Vac. Sci. Techn.* **8,** 135 (1971).
50. Fritzsche, H. and Cuevas, M., *Phys. Rev.* **119,** 1238 (1960).
51. Miller, A. and Abrahams, M., *Phys. Rev.* **120,** 745 (1960).
52. Jonscher, A. K. *Thin Solid Films* **1,** 213 (1967).
53. Servini, A. and Jonscher, A. K., *Thin Solid Films* **3,** 341 (1969).
54. Morgan, M. and Walley, P. A., *Phil. Mag.* **23,** 661 (1971).
55. Morgan, M., *Thin Solid Films* **7,** 313 (1971).
56. Hill, R. M., *Phil. Mag.* **23,** 59 (1971).
57. Nathoo, M. H. and Jonscher, A. K., *J. Phys. C: Solid St. Phys.* **4,** L301 (1971).
58. Hartke, J. L., *J. Appl. Phys.* **39,** 4871 (1968).
59. Barbe, D. F. and Herman, D. S., *J. Appl. Phys.* **41,** 3116 (1970).
60. Hill, R. M., *Thin Solid Films* **7,** R57 (1971).
61. Murgatroyd, P. N., *Thin Solid Films* **11,** 125 (1972).
62. Camphausen, D. L., Connell, G. A. N. and Paul, W., *J. Non-Cryst. Solids* **8-10,** 223 (1972).
63. Jonscher, A. K. and Ansari, A. A. *Phil. Mag.* **23,** 181 (1971).
64. Jonscher, A. K., *J. Phys. C: Solid St. Phys.* **4,** 1331 (1971).
65. Hill, R. M., *Thin Solid Films* **8,** R21 (1971).
66. Davis, E. A., *J. Non-Cryst. Solids* **4,** 1969) 107 (1970).
67. Pai, D. and Ing. S. W., *Phys. Rev.* **173,** 729 (1968).
68. Jonscher, A. K., *J. Phys. C: Solid St. Phys.* **3,** L159 (1970).
69. Abdulaev, G. B. *et al.*, *Phys. Stat. Sol.* **20,** 421 (1967).
70. Abdulaev, G. B. *et al.*, Proc. 9th Intern. Conf., Phys. Semiconductors, Moscow, 1968 (Vol. 2, 1237) (1968).
71. Austin, I. G. and Mott, N. F., *Advances in Physics* **18,** 41 (1969).
72. Emin, D., this Volume Chapter 7.

9. THEORY OF THE HALL EFFECT OF AMORPHOUS SEMICONDUCTORS IN THE RANDOM PHASE MODEL

LIONEL FRIEDMAN

Cavendish Laboratory, University of Cambridge, Cambridge

9.1. INTRODUCTION

The difficulty in characterizing the structure of amorphous semiconductors is well known. Hopefully, one feature characteristic of the amorphous state is the absence of long range crystalline order but the presence of short range order, although there is some evidence that even this may not always be the case.[1] Therefore, in trying to understand the electronic states and conduction mechanisms of amorphous semiconductors at the present time, it is best to focus on basic features common to a wide variety of materials and to try to explain these by simple, even phenomenological, models which hopefully take into account the basic physics. Such is the case for the present work.

It is widely believed that the loss of long range order results in two distinct classes of states. First, there are localized states which contribute to the electrical conductivity only at finite temperatures and/or finite frequencies of the driving electric field, and are characterized by a wave function which has a finite spatial extent. Secondly, there are extended states which contribute to the electrical conductivity for d.c. electric fields even at the absolute zero of temperature if there are carriers present in these states; they are described by a wave function which in some manner extends through the material as a whole and gives no preferred location to the charge carrier. Recent studies support Mott's hypothesis[2] that there is a critical energy ($\mathcal{E}_c$) separating the two classes of states, though there remains some disagreement about whether the mobility as a function of energy is discontinuous at $\mathcal{E}_c$ or tends to zero continuously.[3] In the present work, it is simply assumed that as long as the temperature is not too low, the extended states above $\mathcal{E}_c$ dominate the conductivity because of their larger mobility; however, even here, experimental results and theoretical considerations will later be cited that this may not always be the case.

9.2. Calculation of the Conductivity Tensor

Even though the non-localized states are extended, they are not of course Bloch states because of the absence of long range order. There is likely to be cellular disorder, in which the local energies of the individual atoms are spread over a wide range, topological disorder characterized by variations in bond lengths and orientations, and compositional disorder as well. The solution of Schrödinger's equation for this case is of course prohibitive. The present work, like most analyses on this topic, is based on the so-called Anderson model[4] in which only cellular disorder is taken into account; the overlap integrals between neighbouring atoms are taken to be equal to a single constant.[5] Electronic motion in this regime is assumed to be diffusive[6] in which the carriers perform a random walk from site to site. Phase information is lost over one lattice spacing, a, which is therefore of the order of the mean free path, L.

Mott's procedure[7] for taking this feature into account is basically the following. He starts from the so-called Kubo–Greenwood formula

$$\sigma = \int d\mathcal{E}\, \sigma(\mathcal{E}) \left(- \frac{\partial f_0}{\partial \mathcal{E}} \right), \tag{9.1}$$

where f_0 is the equilibrium electron distribution function and

$$\sigma(\mathcal{E}) \propto \sum_{\kappa,\kappa'} |\langle \kappa' | p | \kappa' \rangle|^2 \, \delta(\mathcal{E}_\kappa - \mathcal{E})\, \delta(\mathcal{E}_{\kappa'} - \mathcal{E}), \tag{9.2}$$

where κ, κ' label the one electron states (as yet unspecified) and p is the electron momentum operator in the direction of the applied electric field. The sample is then divided into $\mathcal{N} = (V/a^3)$ volume elements, where V is the total sample volume. The loss of phase information over $L \sim a$ is then taken into account by writing the above momentum matrix element

$$D = (V/a^3)^{\frac{1}{2}}\, \delta, \tag{9.3}$$

where δ is taken over the volume a^3. The absolute square of D is then the sum of $\mathcal{N}$ non-interfering contribution from each volume element. Making an estimate for δ, Mott obtains[8]

$$\sigma = \frac{2\pi}{3} \frac{e^2 \hbar^3 a}{m^2} \{g(\mathcal{E}_c)\}^2 \exp\left\{-\frac{\mathcal{E}_c - \mathcal{E}_F}{kT}\right\}, \tag{9.4}$$

where $\mathcal{E}_c$ has been previously defined, $g(\mathcal{E}_c)$ is the density of electron states at this energy, and the Fermi energy $\mathcal{E}_F$ is many kT below $\mathcal{E}_c$, corresponding to the amorphous semiconductor situation. The rest of the symbols have their conventional meanings. It is to be emphasized that the density of states concept remains valid in spite of lack of specification of the one electron states.

The identical result was obtained by Hindley,[9] who was the first to introduce the random phase model (R.P.M.). Here, the one electron states $|\kappa\rangle$ are written as a linear superposition of local atomic or Wannier functions $|n\rangle$ as follows:

$$|\kappa\rangle = \sum_n a_{\kappa n} |n\rangle. \tag{9.5}$$

If the system were periodic, the coefficients $a_{\kappa n}$ would be equal to

$$a_{\kappa n} = a_{kn} = \mathcal{N}^{-\frac{1}{2}} e^{i\mathbf{k}\cdot\mathbf{n}},$$

there would be perfect phase coherence and the electron mean free path would be infinite in the absence of scattering mechanism. In the present case, however, Hindley writes

$$a_{\kappa n} = a_{\kappa n} e^{i\phi_{\kappa n}}, \tag{9.6}$$

where the phase angles $\phi_{\kappa n}$ occur at random in the interval $(0, 2\pi)$ while

$$\langle |a_{\kappa n}|^2 \rangle = 1/\mathcal{N} \tag{9.7}$$

since the states are extended, where $\langle \ldots \rangle$ denotes an ensemble average. Taking matrix elements in the Kubo–Greenwood formula and averaging over phases and magnitudes, Hindley obtains Mott's result, eqn (9.4) as previously mentioned. In essence, the scattering over an interatomic distance is built into the wave function from the start, and the agreement with Mott's procedure is not surprising. Hindley has also used the R.P.M. to compute electron energy loss by hot electrons in amorphous solids,[10] though the parameters of the theory are not sufficiently known to predict the critical breakdown field within an order of magnitude.

In view of the success of this simple model in predicting the electrical conductivity, it appeared of interest to calculate the off-diagonal elements of the conductivity tensor from a suitable version of Kubo's formula in order to predict the Hall effect.[11] This is the subject of the remainder of this paper. However, before so doing, the essential experimental results for the Hall effect in amorphous semiconductors are reviewed, in particular those features which are common to a wide variety of materials.

9.3. The Hall Effect

Typical Hall effect data[12] is shown in Fig. 9.1. It is noted that the Hall mobility μ_H is only weakly dependent on temperature. Over this same temperature range the electrical conductivity is thermally activated and shows substantial variations by several orders of magnitude. The magnitude of the Hall mobility is small, of the order of a few tenths $cm^2\ s^{-1}\ V^{-1}$ or less, and is less than the corresponding estimated conductivity mobility.[13] Finally, with few exceptions, the sign of the Hall coefficient is negative, while the sign of the thermopower is positive (the so-called "*n–p* anomaly"). Thus, if the sign of the thermopower is taken to yield the sign of the charge carrier, it follows that the sign of the Hall effect does not. It will be seen how this is possible within the framework of the present theory.[14]

Since the off-diagonal elements of the conductivity tensor are needed to calculate the Hall effect, the "Kubo–Greenwood" formula does not suffice.

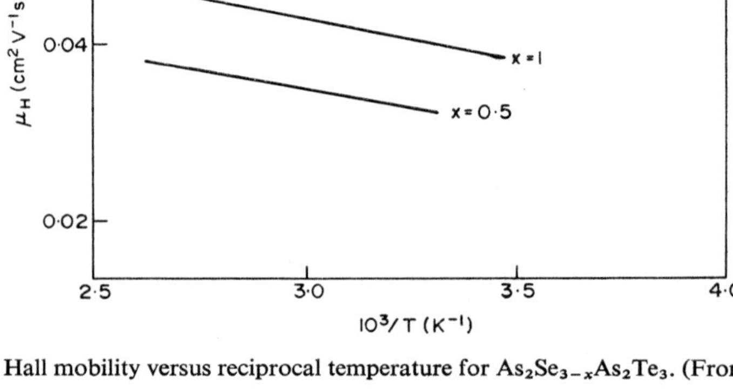

FIG. 9.1. Hall mobility versus reciprocal temperature for $As_2Se_{3-x}As_2Te_3$. (From Ref. 12.)

Rather, use is made of a version of the Kubo formula used to study the Hall effect in a localized (small polaron) representation.[15] It reads

$$\sigma_{ij}(\omega) = \frac{e^2 \hbar}{i\Omega V} [\mathscr{F}_{ij}(\Omega) - \mathscr{F}_{ij}(0)]. \tag{9.8}$$

Here, $\Omega = \hbar(\omega + is)$, $s = 0^+$, is a complex frequency and $\mathscr{F}_{ij}(\Omega)$ is the analytic continuation into the complex Ω-plane of the thermodynamic velocity correlation function

$$\mathscr{F}_{ij}(\hbar\omega_r) = \int_0^\beta du\, e^{\hbar\omega_r u} \frac{1}{Z} \operatorname{Tr} \{e^{-H(\beta-u)} v_i\, e^{-Hu} v_j\}, \tag{9.9}$$

where $\omega_r = 2\pi ir/\beta\hbar$, $r =$ an integer. Also, $\beta \equiv 1/kT$, Tr denotes the trace over one electron state, and Z is the partition function.

The Anderson model is explicit in the Hamiltonian in the absence of applied fields whose matrix elements in the local basis are

$$\langle n' |H_0| n\rangle = U_n \delta_{n'n} - J(1 - \delta_{n'n}). \tag{9.10}$$

Here, U_n is the electronic energy at site n and has some statistical spread. On the other hand, the transfer integral $(-J)$ has one and the same value between all pairs of nearest neighbour sites n and n'. First calculating the diagonal elements of eqn (9.8), the Mott result eqn (9.4) is obtained.[16] Dividing by the charge carrier density above $\mathcal{E}_c$,

$$n = kT\, g(\mathcal{E}_c)\, e^{-(\mathcal{E}_c - \mathcal{E}_F)/kT}, \tag{9.11}$$

one obtains the conductivity mobility

$$\mu_c = \left(\frac{2\pi}{3}\right)\left(\frac{ea^2}{\hbar}\right) z \left(\frac{J}{kT}\right) [a^3 J g(\mathcal{E}_c)], \tag{9.12}$$

where z is the coordination number. This quantity is of importance in order to compare it with the Hall mobility which is calculated next.

The transverse conductivity $(i = x \neq j = y)$ is calculated from eqn (9.8) in the presence of both applied electric and magnetic d.c. fields. The matrix elements of the (field dependent) perturbing Hamiltonian corresponding to eqn (9.10) are

$$\langle n' |H'| n\rangle = (J_{n'n}^{(H)} - J_{n'n}^{(0)})(1 - \delta_{n'n}) + e\,\mathbf{E} \cdot \mathbf{n}\, \delta_{n'n}. \tag{9.13}$$

The second term is just the electric field energy at site $\mathbf{n}$. The first (magnetic field dependent) term follows from the fact that, as has been shown,[17] the sole effect of an applied magnetic field, to first order, is to modify the phases of the transfer integrals according to the formula

$$J^{(H)}_{n'n} = J^{(0)}_{n'n}\, e^{i\alpha_{n'n}} \qquad (J^{(0)}_{n'n} = -J) \tag{9.14}$$

where

$$\alpha_{n'n} = \frac{e}{\hbar c}\mathbf{H}\,.\,\tfrac{1}{2}[\mathbf{n}\times\mathbf{n}'] \tag{9.15}$$

is the magnetic flux (in units of $-\hbar c/e$) enclosed by the triangle consisting of the origin[18] and the sites **n** and **n**′. These phase factors give rise to the Hall effect in hopping-type conduction due to small polarons[15,17] and impurity conduction,[19] and yield the conventional Lorentz force term in the classical limit of propagating wave packets.[20] Expanding the exponentials to first order in H' and taking the average over the random phases, eqn (9.6), σ_{xy} is obtained. From this and $\sigma_{xx} = \sigma$ found previously, one finds the Hall angle

$$\theta_H = \sigma_{xy}/\sigma_{xx}$$

and the Hall mobility

$$\mu_H = \frac{c}{H}\theta_H.$$

The result is

$$\mu_H = \left(4\pi\eta\frac{\bar{z}}{z}\right)\left(\frac{e^2 a}{\hbar}\right)[a^3 Jg(\mathcal{E}_c)], \tag{9.16}$$

and, comparing with eqn (9.12), one gets

$$\frac{\mu_H}{\mu_c} = \frac{6kT}{J}\eta\frac{\bar{z}}{z^2}\,. \tag{9.17}$$

The additional symbols introduced are $\bar{z}$, which is the average number of closed three site paths about an arbitrarily chosen site, and η, which is a constant less than but of order unity, and is defined by

$$\langle (A_{kji}{}^{(z)})^2\rangle = \eta a^4,$$

where $A^{(z)}_{kji}$ is the area of the triangle (ijk) projected on to the plane normal to the applied magnetic field and $\langle\ldots\rangle$ denotes an average over all three site configurations. Equations (9.12), (9.16) and (9.17) are the essential results of the present investigation. Their consequences will now be investigated.

9.4. DISCUSSION

First, for representative values of the parameters ($kT = 1/40$ eV, $\eta = 1/3$, $2zJ = 1$eV, $z = \bar{z} = 6$), the ratio given by eqn (9.17) is

$$\frac{\mu_H}{\mu_c} \approx \frac{1}{10},$$

which is in accord with measured values of μ_H and estimates of μ_c and the fact that the Hall coefficient is too small.[(21)] Second, it is noted that eqn (9.16) is explicitly independent of temperature, although the parameters z, $\bar{z}$, and $g(\mathcal{E}_c)$ may be (weakly) temperature dependent. This is not the case for the conductivity mobility as seen from eqn (9.12). Third, the expected sign of the Hall effect will be discussed. As is evident from the earlier discussion, the random phase model is equivalent to an incoherent superposition of jump processes. Therefore, the sign of the Hall effect is that of an elemental (three site) jump process. It would be too much of a digression to discuss this question in detail. Suffice it to say that it has been shown[(22)] that the *relative* sign of the Hall effect for holes as compared to the sign for electrons is given by the factor

$$-(-1)^n, \tag{9.18}$$

where n is the minimum number of sites in a closed path. Taking triangular closed paths ($n = 3$) as in the present work, it follows that the sign of the Hall effect for holes should be the *same* as that for electrons, namely negative. Thus, if conduction is predominantly hole-like, the sign of the Hall effect is negative. This would explain the "*n–p* anomaly".

Next, the results of the present investigation are compared with other theoretical and experimental investigations on this subject. The first is a calculation of the Hall effect of an excess hole or electron in a half filled band in the atomic limit of the Hubbard model. This has been treated recently by Brinkman and Rice[(23)] for a square lattice. The first point is that these authors find that

$$\frac{\mu_H}{\mu_c} \propto \frac{kT}{zJ},$$

in agreement with our result shown in eqn (9.17). Since the Hubbard Hamiltonian in the atomic limit is essentially a strong scattering limit where $L \sim a$, this agreement is not surprising, but is encouraging in view of the different formal approaches. The second point is that these authors find that the sign of the Hall effect for holes is opposite that for electrons. According to the authors, the basic contribution is associated with a four site walk around an elementary square of their lattice. Hence, $n = 4$ and, according to eqn (9.18), the sign for holes should be opposite that for electrons. Presumably, for a triangular lattice ($n = 3$), the sign of the Hall effect for holes would be the same as that for electrons, as in the present work.

Finally, there are experimental results[12,24,25] which show that, in some of the chalcogenide glasses, the Hall mobility is not strictly temperature independent, but exhibits a weakly activated temperature dependence with $\mathcal{E} \approx 0{\cdot}03$–$0{\cdot}06$ eV. It has been suggested[25] on the basis of conductivity and thermopower results as well as Hall effect in certain Te-based chalcogenide glasses, that the hole-like charge carriers are in fact self-trapped small polarons, and physical arguments are given for the small value of the activation energy. The explanation of the "n–p" anomaly is the same as that of the present theory. A number of amorphous materials should be studied to see if this small activation energy occurs only in certain classes of materials or is a common feature which has been overlooked. In such a case, it must be understood why the extended states play no role, in which case the familiar Mott–C.F.O. model would have to be revised.[26] The classification of the dominant mechanism awaits further experimental and theoretical work.

Footnotes and References

1. Rudee, M. L. and Howie, A., *Phil. Mag.* **25,** 1001 (1972).
2. Mott, N. F., *Adv. Phys.* **16,** 49 (1967).
3. Mott, N. F., *Phil. Mag.* **22,** 7 (1970). See also Fritzsche, H. this volume Chapter 2.1.
4. Anderson, P. W., *Phys. Rev.* **109,** 1492 (1958).
5. The case of purely topological disorder has been considered only for the purpose of establishing lower bounds on the band gap of an amorphous solid characterized by short range order. See Weaire, D. and Thorpe, M. F. *Phys. Rev.* **B4,** 2508 (1971).
6. Cohen, M. H., *J. Non-Cryst. Solids* **2,** 432 (1970).
7. Mott, N. F. and Davis, E. A., *Electronic Processes in Non-Crystalline Materials* (Clarendon Press, Oxford, 1971), Chapter 2.
8. The proportionality between the electrical conductivity and the square of the density of states is best verified for the case of liquid metals where degenerate statistics are obeyed (not treated in the present review). For the case of liquid Te, for example, the electrical conductivity is proportional to the square of the Knight shift. See Cabane, B. and Froidevaux, C. *Phys. Letters* **29A,** 512 (1969).
9. Hindley, N. K., *J. Non-Cryst. Solids* **5,** 17 (1970).
10. Hindley, N. K., *J. Non-Cryst. Solids* **5,** 31 (1970).
11. Friedman, L., *J. Non-Cryst. Solids* **7,** 329 (1971).
12. Roilos, M., *J. Non-Cryst. Solids* **6,** 5 (1971).
13. The conductivity mobility refers to the intrinsic drift mobility in the absence of trapping effects. For an order of magnitude estimate of this quantity, see reference 7.

14. It is also possible in hopping type conduction of localized charge carriers to be discussed at the end of this paper.
15. Holstein, T. and Friedman, L., *Phys. Rev.* **165,** 1019 (1968).
The present investigation makes direct use of the results of this paper.
16. With the replacement $\hbar^2/ma^2 \rightarrow z^{\frac{1}{2}}J$ in eqn (9.4) where z is the coordination number, the result of the random phase calculation is obtained.
17. Friedman, L. and Holstein, T., *Ann. Phys. (N.Y.)* **21,** 494 (1963.
18. Although the individual $\alpha_{n'n}$ are dependent on the gauge of the vector potential, their sum around a closed path is proportional to the magnetic flux through the area enclosed by the path, a gauge invarient quantity. Quantities of physical interest are proportional to the sum around the closed path.
19. Holstein, T., *Phys. Rev.* **124,** 1329 (1961).
20. Friedman, L., *Phys. Rev.* **131,** 2445 (1963).
21. This applies best to the (degenerate) liquid metal situation. The Hall coefficient is too small compared with the value expected from the number of valence electrons. According to the present theory, it must be reduced by the factor $(\mu_H/\mu_c) < 1$.
22. Holstein, T., *Phil. Mag.* **27,** 225 (1973). Also see Austin, I. G. and Mott, N. F., *Adv. Phys.* **18,** 41 (1969) and Friedman, L. in *Proc. International Conference on Conduction in Low Mobility Materials,* Eilat, Israel (Taylor & Francis, London, 1971).
23. Brinkman, W. F. and Rice, T. M., *Phys. Rev.* **B4,** 1566 (1971).
24. Nagels, P., Callaerts, R., Denayer, M., de Coninck, R., *J. Non-Cryst. Solids* **4,** 295 (1970).
25. Emin, D., Seager, C. H. and Quinn, R. K., *Phys. Rev. Letters* **28,** 813 (1972).
26. In particular, polaronic transport within the tail of the distribution of localized states would yield too small a value of the pre-exponential σ_0 of the electrical conductivity. The correct order of magnitude would require that all the atoms of the solid contribute to the narrow small polaron band.

10. SOME ELECTRONIC PROPERTIES OF AMORPHOUS Si FILMS*

P. G. LECOMBER, A. MADAN AND W. E. SPEAR,

Carnegie Laboratory of Physics,
University of Dundee,
Dundee, Scotland

10.1. INTRODUCTION

The electrical properties of amorphous silicon (*a*–Si) have in recent years been investigated in a number of laboratories. The general interpretation of many features of this work is complicated by the fact that the results depend significantly both on the method of preparation and on the preparation conditions. For example vacuum evaporated(1,2,3) and sputtered(3) layers possess relatively high conductivities (of the order of 10^{-3} $(\Omega\,\text{cm})^{-1}$), whereas specimens prepared by the r.f. glow discharge decomposition of silane(4,5,6) are many orders of magnitude less conducting. Their properties also depend

*Some of the work included in this paper has recently been published in *J. Non-Cryst. Solids* **11**, 219 (1972)

markedly on the substrate temperature during deposition as is clearly shown by the work of Chittick.[5]

In this paper we should like to present the results of a detailed investigation of the electronic transport properties of *a*–Si specimens prepared by the glow discharge technique, using a wide range of deposition temperatures. Some of the preliminary results have already been briefly reported.[6] An important feature of the present work lies in the correlation of the results from drift mobility and both dark- and photo-conductivity measurements on the same specimens. With this combined approach it has been possible to make meaningful deductions about electronic states and the transport mechanisms. In Section 10.4.1 a model of the localized state distribution is put forward, based on the transport and photoconductivity results and on some of the information from recent field effect experiments.[7] It is particularly encouraging that the proposed model appears to be applicable not only to the glow discharge specimens, but also to films prepared by evaporation or sputtering. As shown in Section 10.4.4, thermo-electric results[2] and the generally observed annealing behaviour[3] of all types of *a*–Si specimens can be interpreted satisfactorily on this basis.

10.2. Experimental Methods

The specimens used in the present work were prepared by the decomposition of silane, a method largely developed by the S.T.L. Group.[4] Pure silane gas flowed at about 0·1 $cm^3\ s^{-1}$ into the top of a vertical quartz tube, 5 cm in diameter and 40 cm in length, via a needle valve and a flowmeter. The gas was continuously pumped away at the lower end, resulting in a residual pressure of typically 0·5 torr. An r.f. coil, powered from a 5 MHz generator, surrounded the middle portion of the quartz tube. Just below the coil a molybdenum substrate holder was supported inside the tube. Its temperature could be raised by an electrical heater. The quantity T_d used in this paper, refers to the temperature of the substrate holder and is, therefore, likely to be only an approximate measure of the true substrate temperature during deposition.

Amorphous Si layers, about 1 μm thick, were deposited on glass or quartz substrates at rates of a few Å s^{-1}. The film thickness was obtained from the difference in weight of the substrate before and after deposition, using a density of 2·0 g cm^{-3} (Ref. 5). We estimated from optical measurements that the films were uniform in thickness to within 10 per cent.

Samples were prepared in both the sandwich and gap cell configurations. In the latter Al–Al electrodes were evaporated onto the Si film and provided contacts that resulted in linear current–voltage characteristics over the range of steady fields, from 10^2 to 10^4 V cm^{-1}, used in these experiments. In the sandwich cell configuration Al provided a good top contact, but did not lead to ohmic behaviour when used as a bottom electrode. Au was more satisfactory in this respect for samples deposited at temperatures below about 370 K. However, above this temperature gold tends to aid recrystallisation of the amorphous film[(8)] and for this reason we used Mo bottom electrodes for our higher temperature samples. The Mo–Al combination was not perfectly ohmic, but by restricting measurements to a field range above 10^3 V cm^{-1}, the values of conductivity and activation energy obtained were essentially the same as those from a gap cell on the same specimen.

The drift mobility technique and its application to crystalline and non-crystalline solids has been discussed in some detail by Spear[(9)] and only a brief outline of this experimental method will be given here. 5 keV electron pulses, 10 to 30 ns in duration, generated electron–hole pairs near the top electrode. Excess electrons were extracted by a synchronised field pulse and their transit time t_t across the specimen measured directly on a fast oscilloscope after suitable amplification. The interpretation of the observed current signal has been discussed by Spear and LeComber.[(7)] In most specimens the pulse showed a reasonably well-defined discontinuity associated with the extraction of a coherent group of injected carriers. The transit time determined by this feature is inversely proportional to the applied field up to the highest values used and the measurements lead, therefore, to a meaningful electron drift mobility. The discontinuity in the current signal is followed by a long tail which appears to be caused by the thermal release of electrons trapped during transit in deeper lying localized states.

It is worth emphasising that the two types of measurements, conductivity and drift mobility, used in the present investigation, refer to different electronic distributions. In the mobility experiments, one deals with a group of injected excess carriers giving rise to a local density, orders of magnitude higher than that of the thermally generated carriers. In an amorphous solid the mobility of these carriers may depend on the energetic position of the states associated with the transport. Marked differences in the distribution of carriers amongst the states can, therefore, result in substantially different mobilities. In such a case it will not be possible to correlate the results of conductivity and drift mobility measurements. This problem arises in the present work with specimens prepared at low T_d and will be referred to again in Section 10.4.3.

In the steady state photoconductivity measurements, carrier densities are also increased by external excitation above the thermal equilibrium values. However, with the comparatively low light intensities used in these experiments, it is unlikely that the results refer to electron distributions substantially different from thermal equilibrium.

10.3. Experimental Results

10.3.1. Electrical Conductivity

The temperature dependence of the electrical conductivity is shown in Fig. 10.1 for two samples deposited in the sandwich configuration at temperatures of $T_d = 500$ K (sample (b)) and $T_d = 460$ K (sample (c)). For the purpose of discussion it will be useful to divide the curves into three regions. As indicated in the insert, region 1 extends from about room temperature to the temperature T_1; the conductivity is an activated process with an activation energy $\mathcal{E}_1$ of 0·62 eV for these particular samples. Between T_1 and T_2 (region 2), the conductivity is still activated, but with a reduced energy which we shall denote by $\mathcal{E}_2$. The change in $\mathcal{E}$ near T_1 is more clearly brought out with the extended scales used in the insert. Below T_2, in region 3, the curves gradually level out with decreasing T and can no longer be described by a simple activated mechanism.

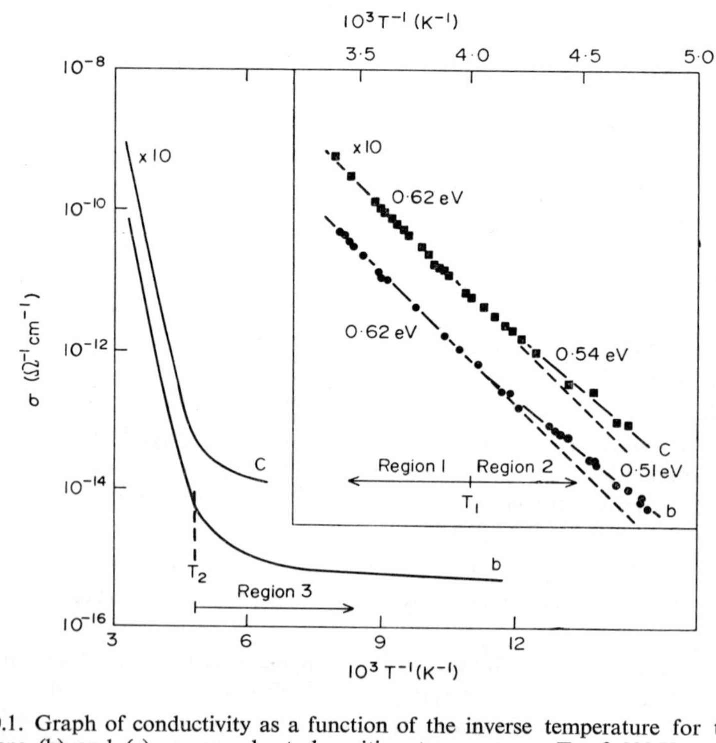

FIG. 10.1. Graph of conductivity as a function of the inverse temperature for two *a*-Si specimens (b) and (c), prepared at deposition temperatures T_d of 500 K and 460 K respectively. The insert gives a more detailed plot of the results near the temperature T_1 at which the activation energy changes from $\mathcal{E}_1$ to $\mathcal{E}_2$. (After Ref. 16.)

This behaviour pattern is typical of all specimens deposited above 350 K. With samples prepared below this temperature, regions 1 and 2 tend to coalesce and the change in gradient at T_1 is no longer observable. In these specimens the curves will be characterised by the single activation energy $\mathcal{E}_1'$.

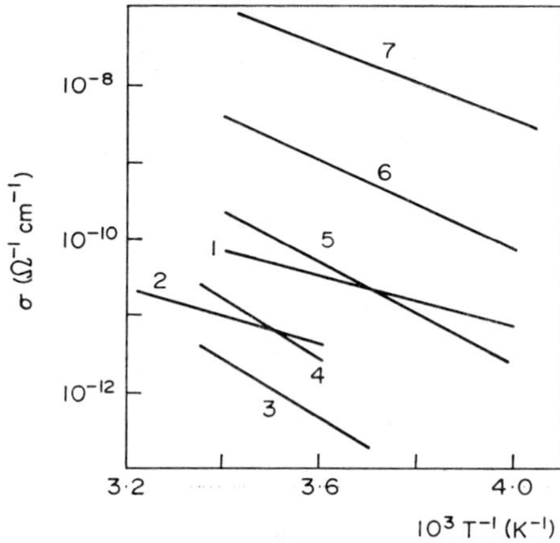

	T_d (K)	$\mathcal{E}$ (eV)	σ_0 $(\Omega^{-1}cm^{-1})$
1	310	0·33	3×10^{-5}
2	350	0·36	3×10^{-5}
3	350	0·75	19
4	400	0·77	270
5	470	0·65	34
6	600	0·57	18
7	660	0·48	13

FIG. 10.2. Temperature dependence of the conductivity in region 1 for seven specimens prepared at progressively higher deposition temperature, T_d. Values of T_d, activation energy and pre-exponential factor (eqn (10.1)) are given. (After Ref. 16.)

Figure 10.2 shows the temperature dependence of conductivity in region 1 for seven specimens prepared at progressively higher substrate temperatures, which are listed together with the measured activation energies. The interesting result is that with increasing T_d, the conductivity goes through a minimum; near this substrate temperature ($\simeq$ 350 K) there occurs a remarkable jump in the activation energy from $\mathcal{E}_1'$ to $\mathcal{E}_1$. The same applies to the pre-exponential factor σ_0, defined by

$$\sigma = \sigma_0 \exp(-\mathcal{E}/kT). \tag{10.1}$$

This behaviour is clearly brought out in Fig. 10.3 which summarises the data for more than forty specimens. The room temperature conductivity σ_{RT}, the activation energies $\mathcal{E}_1$ and $\mathcal{E}_1'$, and σ_0 are all shown as a function of T_d. Evidently, there is a fundamental difference in the transport properties of specimens prepared at substrate temperatures above and below about 350 K. This will be discussed in detail in Section 10.4.3.

In the course of the experiments it was found that for specimens with $T_d > 350$ K the value of T_1 (Fig. 10.1) remained essentially independent of T_d. This was not the case for the temperature T_2 at which the conductivity changed from an activation energy $\mathcal{E}_2$ to the essentially unactivated process at the lowest temperatures used. This transition was not very clearly defined and it also depended to some extent on the thermal history of the sample.

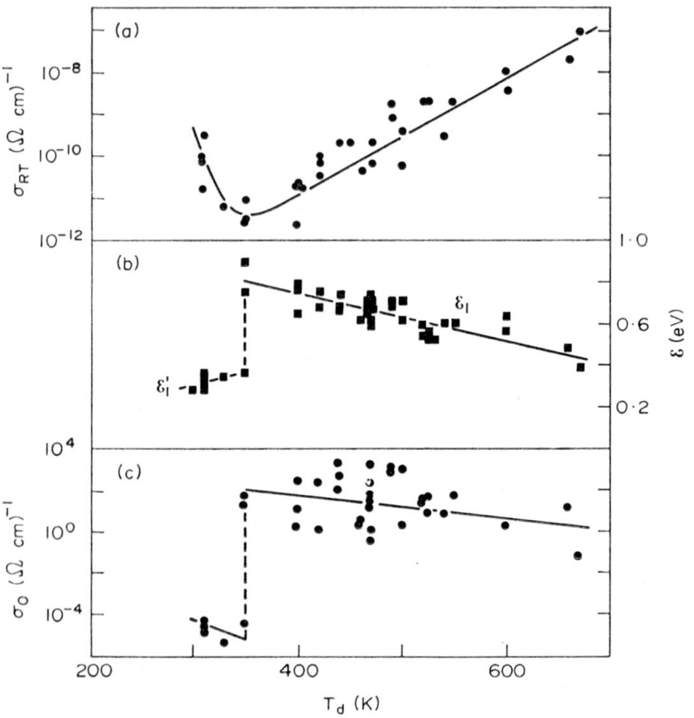

FIG. 10.3. (a) Room temperature conductivity σ_{RT}, (b) activation energy in region 1 and (c) pre-exponential factor σ_0 as a function of deposition temperature T_d for over 40 *a*-Si specimens. (After Ref. 16.)

10.3.2. DRIFT MOBILITY

Drift mobility measurements were carried out on a number of the samples used in the above conductivity experiments. Figure 10.4 shows the electron drift mobility μ for the two samples (b) and (c) referred to already in Fig. 10.1 and for two other specimens, (a) and (d), both deposited at 440 K. Clearly the curves can be divided into two regions by the well-defined change in gradient. The significant result is that this occurs between 243 and 252 K and corresponds, therefore, closely to the temperature T_1 defined by the conductivity curves in Fig. 10.1. It can be seen that between room temperature and T_1 the electron mobility is controlled by an activation energy of about 0·18 eV. Below T_1, the slope is reduced to about 0·1 eV. It was not possible to carry out drift mobility measurements below 180 K because the diffusive tail on the current pulse obscured the coherent transit at these temperatures.

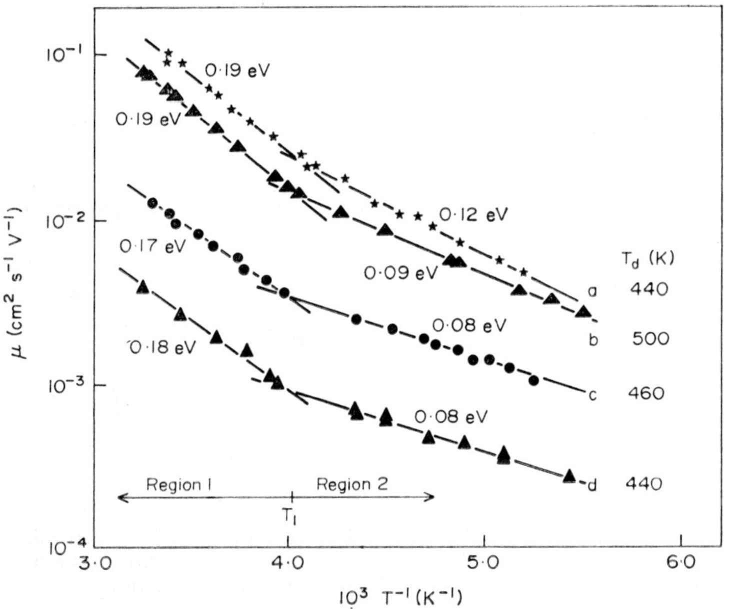

FIG. 10.4. Temperature dependence of the electron drift mobility for four *a*-Si specimens prepared at the values of T_d indicated. (b) and (c) refer to the same specimens as in Fig.10.1. (After Ref. 16.)

The data for a number of specimens prepared at different substrate temperatures are summarised in Fig. 10.5. The room temperature electron mobility lies between 10^{-2} and 10^{-1} cm^2 s^{-1} V^{-1} with no obvious systematic dependence on T_d. Figure 10.5(b) correlates the values of the mobility activation energy in region 1 ($T > T_1$), and Fig. 10.5(c) those in region 2 ($T < T_1$). These activation energies are denoted by $\mathcal{E}_c - \mathcal{E}_A$ and W_e respectively, for reasons discussed in the next section. Both are constant within the experimental error of $\pm 0{\cdot}01$ eV, with average values of $\mathcal{E}_c - \mathcal{E}_A = 0{\cdot}18$ eV and $W_e = 0{\cdot}10$ eV. It can be concluded that the electron transport at room temperature is essentially independent of the deposition temperature and certainly cannot account for the six orders of magnitude change in the pre-exponential term σ_0 shown by Fig. 10.3(c).

When the applied electric field was reversed, the movement of positive charge carriers could be observed, particularly in specimens prepared at the lowest substrate temperatures. Unfortunately, transits were not sufficiently well defined for meaningful hole mobility measurements.

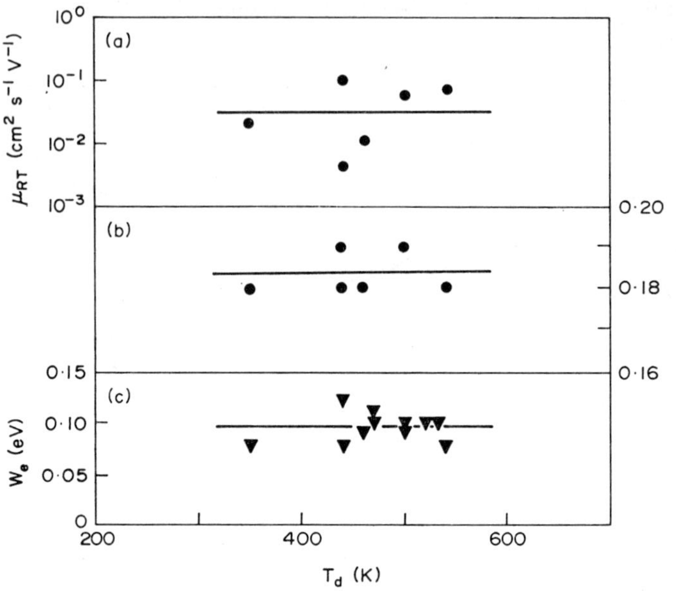

FIG. 10.5. (a) Room temperature electron mobility μ_{RT}, (b) the mobility activation energy $\mathcal{E}_c - \mathcal{E}_A$ in region 1, and (c) the hopping activation energy W_e in region 2 as a function of deposition temperature T_d for a number of specimens. (After Ref. 16.)

10.3.3. PHOTOCONDUCTIVITY †

Steady photoconductivity measurements were carried out at room temperature between photon energies of 0·5 and 3·0 eV on a series of specimens prepared at different T_d. In most of these experiments gap cells with an electrode spacing of 1·5 mm were used. The photocurrent i_p, defined as the difference in the current reading with and without illumination, was ohmic for most specimens.

Figure 10.6 shows the results with the corresponding T_d values given on the right. $\mathcal{N}_0(1-R)$ denotes the number of incident photons per second, corrected for surface reflection. The ordinate represents therefore the number of charge carriers flowing around the circuit for each photon entering the specimen.

The following features are of particular interest:

(i) The photoresponse of the group of high temperature specimens approaches unity at photon energies between 1·8 and 2·0 eV, for the values

† We are most grateful to Dr. R. Loveland and Mr. A. Al-Sharbati for allowing us to include some of their recent photconductivity results in this paper. A more detailed account of the optical work will be published in the *Journal of Non-Crystalline Solids.*

of the applied field used. With higher fields, this value of unity could be exceeded.

(ii) Specimens prepared below about 500 K exhibit a drastic decrease in photoresponse at all wavelengths. A limiting case is an *evaporated* specimen, 14E, which was deposited on a substrate at 550 K.

(iii) Practically all the curves exhibit a fairly rapid rise above about 1·5 eV.

(iv) Most of the curves show a shoulder between 1·1 and 1·3 eV.

(v) The photoresponse disappears below about 0·6 eV.

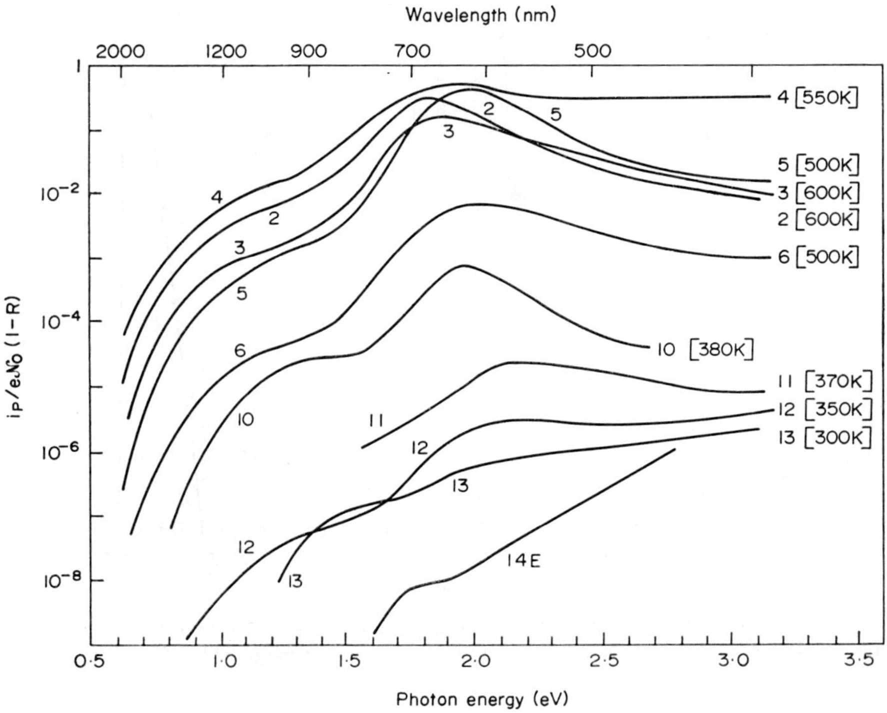

FIG. 10.6. The number of photogenerated carriers flowing in the circuit per incident photon plotted against photon energy for a series of specimens prepared at the values of T_d given. Applied fields between 3 and 10 kV cm^{-1} were used. 14E refers to an evaporated specimen.

It can be shown that in the weakly absorbing region below about 1·7 eV, the photoresponse is an approximate, though very sensitive, measure of the absorption coefficient α. In analysing absorption data from amorphous solids, $(\alpha h\nu)^{\frac{1}{2}}$ is generally regarded as a theoretically meaningful variable.[10]

The corresponding quantity, using the photoconductive data of Fig. 10.6, is plotted in Fig. 10.7 for two typical specimens in the weakly absorbing region. Extrapolating the well-defined straight portion of the graph leads to values of the energy $\mathcal{E}_0$ between 1·5 and 1·7 eV for all specimens, irrespective of T_d. The "tail" of the curves, also shown multiplied by a factor of five in Fig. 10.7, extrapolates to values of $\mathcal{E}_1^{opt}$ between 0·6 and 0·8 eV. It is significant that $\mathcal{E}_1^{opt}$ and its dependence on T_d correlates well with that of $\mathcal{E}_1$ in Fig. 10.3(b) obtained from the conductivity measurements.

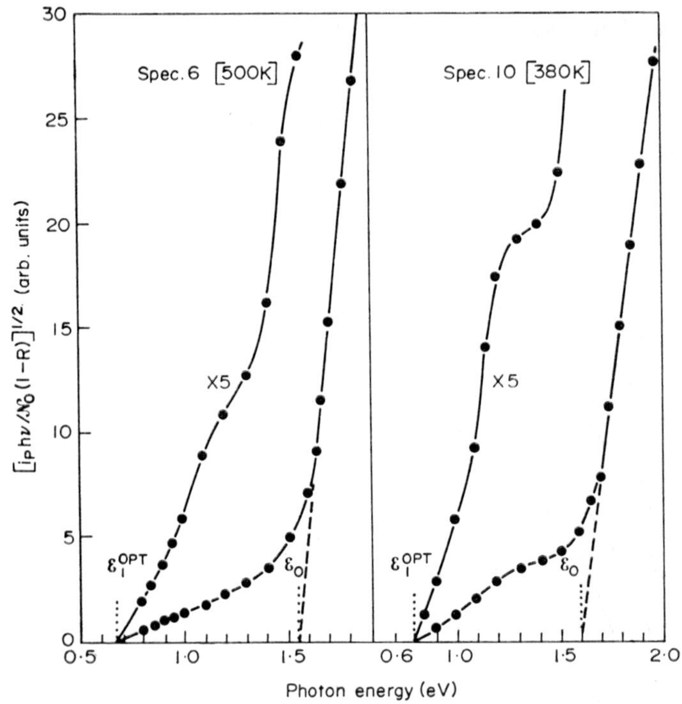

FIG. 10.7. Plot of $[i_p h\nu/\mathcal{N}_0(1 - R)]^{1/2}$ versus photon energy for two specimens from Fig. 10.6. The "tails" of the photoresponse are also shown magnified $\times 5$.

10.4. DISCUSSION

We shall begin with a description of the suggested model and then apply it to a detailed interpretation of the experimental results in the temperature regions defined above. The dependence of the transport properties on the deposition temperatures will be discussed and finally an attempt will be made to relate the results obtained from evaporated and sputtered Si films to the present interpretation.

10.4.1. MODEL USED IN THE INTERPRETATION

This is shown in Fig. 10.8(a) where the density of state function $g(\mathcal{E})$ (i.e. the number of electron states per unit volume per unit energy range) is plotted against the electron energy. The essential features are those of the general model of an amorphous solid which has evolved in recent years through the work of Mott, Cohen and others.[10] However, the figure also incorporates some of the details deduced from recently published field effect measurements on silane deposited specimens.[7]

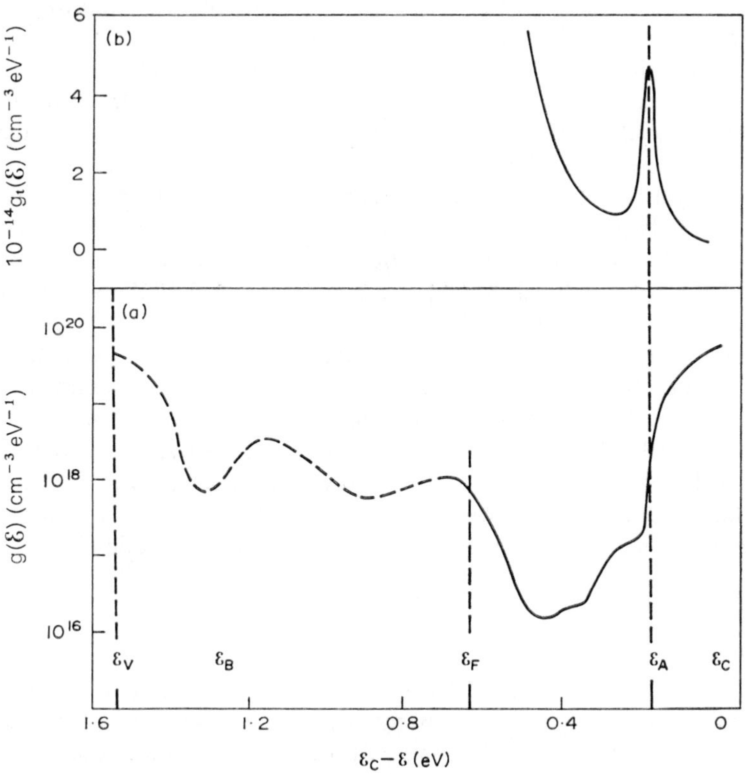

FIG. 10.8. (a) Model used in the interpretation. The density of state function $g(\mathcal{E})$ is plotted against the electron energy referred to $\mathcal{E}_c$. (b) Shows the distribution $g_t(\mathcal{E})$ of electrons trapped in localized states, calculated according to eqn (10.5) for a quasi-Fermi energy $\mathcal{E}_F' = 0{\cdot}4$ eV.

In the extended states just above $\mathcal{E}_c$ (or below $\mathcal{E}_v$), the carrier mean free path is of the order of the interatomic spacing, a. Transport through these states may be likened to Brownian motion and has been treated by Cohen[11] and by Mott.[12] An approximate upper limit to the mobility is given by

$$\mu_0 = \frac{ea^2}{6kT}\nu_{el} \tag{10.2}$$

where ν_{el} is a typical electronic frequency ($\sim 10^{15}$ Hz). Equation (10.2) predicts a mobility $\mu_0 \simeq 5\,\text{cm}^2\,\text{s}^{-1}\,\text{V}^{-1}$ at room temperature for transport in states just above $\mathcal{E}_c$ (or below $\mathcal{E}_v$).

Between $\mathcal{E}_c$ and $\mathcal{E}_v$ the states are localized. As shown in Section 10.4.2, conductivity and mobility measurements lead to $g(\mathcal{E}_c) \simeq 5 \times 10^{19}\,\text{cm}^{-3}\,\text{eV}^{-1}$. Below $\mathcal{E}_c$ there occurs a rapid decrease in $g(\mathcal{E})$, which according to the field effect measurements[7] approaches a value of about $10^{18}\,\text{cm}^{-3}\,\text{eV}^{-1}$ about 0·18 eV below $\mathcal{E}_c$. This energy will be denoted by $\mathcal{E}_A$ in the following. The above measurements also indicate the presence of some structure in $g(\mathcal{E})$ between $\mathcal{E}_A$ and the Fermi energy $\mathcal{E}_F$, which appears to depend on the deposition temperature T_d. In Fig. 10.8(a) $\mathcal{E}_F$ is shown at 0·65 eV below $\mathcal{E}_c$, which would correspond to a specimen prepared at $T_d \simeq 500$ K. The photoconductive data give some information on $g(\mathcal{E})$ between $\mathcal{E}_F$ and $\mathcal{E}_v$ which will be discussed briefly at the end of Section 10.4.2. Apart from this, it has been assumed that $g(\mathcal{E})$ between $\mathcal{E}_v$ and the energy $\mathcal{E}_B$ is essentially similar to the distribution in the range $\mathcal{E}_c$ to $\mathcal{E}_A$.

We should like to suggest that there are two distinct groups of localized states in our model. First, the distribution adjoining the extended states, between $\mathcal{E}_c$ and $\mathcal{E}_A$ and between $\mathcal{E}_v$ and $\mathcal{E}_B$, which originates from the lack of long range order in the solid; secondly, the localized states in the central region, between $\mathcal{E}_A$ and $\mathcal{E}_B$, which are likely to be associated in an elemental amorphous semiconductor with reasonably well defined structural defects. Further reference will be made to these two groups in the following sections.

10.4.2. INTERPRETATION OF RESULTS

Region 1

Let us first consider the results for specimens deposited at relatively high temperatures ($T_d \simeq 500$ K) shown by the conductivity and drift mobility curves of Figs. 10.1 and 10.4. There is no doubt from the mobility experiments that any injected current will be predominantly carried by electrons. In the presence of excess electrons the distribution can be described by a quasi-Fermi level $\mathcal{E}_F'$ which, according to the injection levels used in our experiment, lies between 0·5 and 0·4 eV below $\mathcal{E}_c$. The conductivity measurements refer to thermal equilibrium conditions, but the evidence from field effect experiments and the negative sign of the thermo-electric power[13] clearly indicates that in these specimens electrons make the major contribution to the conductivity.

We therefore suggest that σ in region 1 (Fig. 10.1) is associated with the transport of electrons through the extended states just above $\mathcal{E}_c$. The conductivity is given by:

$$\sigma_1 = e\,\mu_0\,N_c \exp(\delta/k) \exp\left[-\frac{(\mathcal{E}_c - \mathcal{E}_F)_0}{kT}\right] \quad (10.3)$$

where it is assumed that $\mathcal{E}_c - \mathcal{E}_F$ depends linearly on temperature[10] such that

$$\mathcal{E}_c - \mathcal{E}_F = (\mathcal{E}_c - \mathcal{E}_F)_0 - \delta T. \quad (10.4)$$

We therefore identify $\mathcal{E}_1$ with $(\mathcal{E}_c - \mathcal{E}_F)_0$ and σ_0 with $e\mu_0 N_c \exp \delta/k$.

The magnitude and temperature dependence of μ in region 1 is determined by the interaction of the excess electrons during transit with the localized states below $\mathcal{E}_c$. The fact that an electron transit can be observed means that a fairly coherent group of carriers drifting in the extended states is being extracted. In order to maintain a sufficient degree of coherence, the carriers can interact, through trapping and thermal release, only with states in a narrow energy range. In an amorphous solid it is an essential condition that this energy range is associated with a maximum in the distribution of trapped carriers. This is given by

$$g_t(\mathcal{E}) = g(\mathcal{E})\,\{1 + \exp[(\mathcal{E} - \mathcal{E}_F')/kT]\}^{-1} \quad (10.5)$$

if a local steady-state distribution at any time during transit is assumed. Figure 10.8(b) shows $g_t(\mathcal{E})$ calculated according to eqn (10.5) from the distribution of states. A well-defined maximum in $g_t(\mathcal{E})$ is found within about kT of the energy $\mathcal{E}_A$. Its occurrence is linked with the rapid drop in $g(\mathcal{E})$, which in an elemental amorphous material is likely to mark the transition between the two types of localized states referred to in Section 10.4.1.

The temperature dependence of the measured drift mobility is essentially determined by the thermal release time of trapped electrons from the maximum of the distribution close to $\mathcal{E}_A$; we thus identify the observed mobility activation energy in region 1 with $\mathcal{E}_c - \mathcal{E}_A$.

The drift mobility, controlled by multiple trapping, can be expressed by the following equation, which applies to a number of simple distributions of localized states:[10]

$$\mu = \mu_0\,\beta \exp\left[-\frac{\mathcal{E}_c - \mathcal{E}_A}{kT}\right] \quad (10.6)$$

μ is the observed drift mobility and β a constant depending on the distribution between $\mathcal{E}_c$ and $\mathcal{E}_A$. In the case of a single level of centres at $\mathcal{E}_A$, $\beta = N_c/N_t$, where N_t is their density. In this form eqn (10.6) has been widely used in the interpretation of drift mobility data on crystalline materials.[9] In the present case, we can set $N_t \simeq g(\mathcal{E}_A)\,kT$ as a first approximation, so that

$$\beta \simeq g(\mathcal{E}_c)/g(\mathcal{E}_A);$$

from the data given in Fig. 10.8(a) β is about 50, and with eqn (10.6) leads to $\mu_0 \simeq 3\,\text{cm}^2\,\text{s}^{-1}\,\text{V}^{-1}$. Evidently the detailed shape of the distribution between $\mathcal{E}_c$ and $\mathcal{E}_A$ is of no particular importance in determining μ, because only states within one or two kT of $\mathcal{E}_A$ are involved in the trap-controlled transport of the coherent group of carriers. A linear distribution of states, with its origin at $g(\mathcal{E}_A) = 0$ and extending beyond $\mathcal{E}_c$, has been considered by Mott and Davis.[10] Here $\beta = (\mathcal{E}_c - \mathcal{E}_A)/kT$ and eqn (10.6) leads to $\mu_0 \simeq 16\,\text{cm}^2\,\text{s}^{-1}\,\text{V}^{-1}$. Similar results are obtained when other simple distributions are assumed. These values are in reasonable agreement with the μ_0 predicted from eqn (10.2) so that the present results support the theoretical model[11,12] referred to above for transport through the extended states near the localization limit.

From the expression for σ_0, a value for $g(\mathcal{E}_c)$ can now be estimated. Using $\sigma_0 = 10\,\Omega^{-1}\,\text{cm}^{-1}$, $\mu_0 = 5\,\text{cm}^2\,\text{s}^{-1}\,\text{V}^{-1}$ and $e^{\delta/k} \simeq 10$ (Ref. 10), we obtain $g(\mathcal{E}_c) = 5 \times 10^{19}\,\text{cm}^{-3}\,\text{eV}^{-1}$ (see Ref. 14).

Regions 2 *and* 3

With decreasing temperature, the probability of thermal release from localized states near $\mathcal{E}_A$ becomes rapidly smaller, and eventually it will be more likely that an electron hops to a neighbouring site in the distribution. One would therefore expect a change in transport mechanism from conduction through the extended states to transport by phonon assisted hopping through regions of the localized state spectrum. We would like to suggest that this transition occurs around the temperature T_1 characterised by a decrease in slope of both the conductivity and mobility curves, (Figs. 10.1 and 10.4). The fact that the change in slope occurs simultaneously in both these measurements implies that one is dealing here with a transition from one conduction mechanism to another. For example, any shift with temperature in the position of the maximum near $\mathcal{E}_A$, would modify μ, but should not be reflected in the measured conductivity.

The phonon assisted hopping mobility is given by an expression of the form

$$\mu_H = (\mu_H)_0 \exp(-W_e/kT) \tag{10.7}$$

$(\mu_H)_0$ is only weakly temperature dependent as compared to the exponential term and contains a phonon frequency and a term expressing the wave function overlap between hopping sites; W_e is the energy difference between the sites and is given by the gradient of the drift mobility curves in region 2 (Fig. 10.4). With reasonable values for the parameters, eqn (10.7) does lead to hopping mobilities between 10^{-3} and 10^{-2} $cm^2 s^{-1} V^{-1}$ in satisfactory agreement with the measured values. Suppose that the hopping transport in region 2 proceeds through states around an energy $\mathcal{E}_x$. Thus the activation energy of the conductivity changes from

$$|\mathcal{E}_1| = (\mathcal{E}_c - \mathcal{E}_F)_0 \qquad \text{to} \qquad |\mathcal{E}_2| = (\mathcal{E}_x - \mathcal{E}_F)_0 + W_e$$

in the neighbourhood of T_1. Taking an average over all our results,

$$|\mathcal{E}_1| - |\mathcal{E}_2| = \mathcal{E}_c - \mathcal{E}_x - W_e$$

is found to be $(0{\cdot}09 \pm 0{\cdot}02)$ eV. The corresponding change in the mobility activation energy is from $\mathcal{E}_c - \mathcal{E}_A$ to W_e. The drift mobility experiments show that $\mathcal{E}_c - \mathcal{E}_A - W_e$ has an average value of $(0{\cdot}08 \pm 0{\cdot}02)$ eV. These results lead, therefore, to the conclusion that $\mathcal{E}_x \simeq \mathcal{E}_A$, which is not unexpected in view of the maximum in $g_t(\mathcal{E})$ at $\mathcal{E}_A$ (Fig. 10.8(b)).

The above numerical agreement lends support to our interpretation of the data in terms of a transition from a conduction process in which the electrons drift in extended states to a transport that is controlled by the hopping of electrons in localized states near $\mathcal{E}_A$.

The conductivity data in region 3 ($T < T_2$) show a gradual decrease of $\mathcal{E}_x - \mathcal{E}_F + W_e$ towards the lower temperatures. It implies that the predominant hopping path in the localized state distribution (at $\mathcal{E}_x$) moves progressively closer towards the Fermi energy. At the same time W_e decreases. For hopping close to $\mathcal{E}_F$ one would expect that $\sigma \propto \exp[-B/T^{\frac{1}{4}}]$ (see Ref. 12). However, our low temperature data extend over too limited a range for a meaningful test of this relation.

Photoconductivity

The interpretation of the room temperature photoconductivity data in Figs. 10.6 and 10.7 rests on the conclusion drawn from the electrical measurements, that the predominant current path at this temperature is through the extended electron states near $\mathcal{E}_c$. This is supported by the fact that the photoresponse sets in at $\mathcal{E}_1^{opt} \simeq \mathcal{E}_1$, which must therefore be associated with transitions from filled states just below $\mathcal{E}_F$ to $\mathcal{E}_c$. The shoulder in Fig. 10.6, also brought out clearly in Fig. 10.7, suggests the presence of a local maximum in $g(\mathcal{E})$ around 1·2 eV. The consistent values of $\mathcal{E}_0$ obtained from plots such as those in Fig. 10.7, place the onset of transitions from $\mathcal{E}_v$ to $\mathcal{E}_c$ at about 1·6 eV.

10.4.3. DEPENDENCE ON DEPOSITION TEMPERATURE

Having interpreted in some detail the results for specimens deposited at $T_d \simeq 500$ K, the discussion will now be extended to the data summarised in Figs 10.3 and 10.5, showing the dependence of room temperature conductivity and mobility on deposition temperature.

An important conclusion, which simplifies the following discussion, can be drawn from Fig. 10.5, namely that T_d has little effect upon μ_{RT}, the position of $\mathcal{E}_A$ or the value of W_e. This means that the localized state distribution between $\mathcal{E}_c$ and $\mathcal{E}_A$, and particularly around $\mathcal{E}_A$, cannot be greatly affected by the substrate temperature, which would be reasonable if these states are primarily associated with the lack of long range order as suggested in Section 10.4.1.

The conductivity data, on the other hand, is strongly influenced by T_d, as is evident from Fig. 10.3(a). The increase in σ_{RT} for $T_d > 350$ K is essentially caused by the movement of the Fermi level from 0·85 eV to about 0·5 eV below $\mathcal{E}_c$ (Fig. 10.3(b)). The pre-exponential factor σ_0 varies comparatively slowly with T_d in this range. We should like to suggest that the movement of $\mathcal{E}_F$ is caused by changes in the localized state distribution between $\mathcal{E}_A$ and $\mathcal{E}_B$, which is mainly determined by structural defects. The field effect experiments referred to above have provided some independent evidence for this suggestion. They show a substantial increase in $g(\mathcal{E})$ around the Fermi level in specimens deposited at progressively lower T_d.

A very similar deduction can be made from the photoconductivity results of Fig. 10.6. The response at a given photon energy, say 2·0 eV, is mainly determined by the electron lifetime with respect to localized states, presumably near $\mathcal{E}_F$, through which the recombination takes place. The rapidly decreasing photoconductivity observed on lowering T_d, represents conclusive evidence for a marked increase in localized state density. This applies particularly to the evaporated specimen (14E), included in Fig. 10.6.

The data of Fig. 10.3 shows an abrupt change in the general behaviour pattern at $T_d \simeq 350$ K. The most reasonable and consistent explanation appears to be that at $(\mathcal{E}_c - \mathcal{E}_F)_0 \gtrsim 0{\cdot}8$ eV, holes, rather than electrons, begin to make the predominant contribution to the conductivity. The remarkable decrease in σ_0, by about six orders of magnitude, makes it unlikely that the hole transport at room temperature proceeds through the extended hole states. We postulate that holes move by hopping via localized states around an energy $\mathcal{E}_y$. The observed activation energy of the conductivity for such specimens would then be $|\mathcal{E}_1'| = \mathcal{E}_F - \mathcal{E}_y + W_h$, where W_h represents the hopping activation energy for holes. The drop in σ_0 implies that the ratio

$$\mu_0\, g(\mathcal{E}_c)/\mu_h\, g(\mathcal{E}_y) \sim 10^6,$$

giving

$$\mu_h \sim 10^{-3}\, \text{cm}^2\, \text{V}^{-1}\, \text{s}^{-1},$$

a magnitude well within the hopping regime. This estimate is based on the assumption that $g(\mathcal{E}_y) \sim g(\mathcal{E}_A)$.

Although hole transits could not be resolved, we did observe the movement of positive charge carriers in specimens prepared at low T_d. Further evidence for the onset of appreciable hole transport in such samples comes from the field effect experiments, which showed a response for negative potentials applied to the field electrode. Because of the low conductivities involved, thermoelectric power measurements are extremely difficult. Some preliminary measurements in our laboratory[13] seem to indicate a positive α, for the lowest temperature samples, whereas α is undoubtedly negative for those prepared at high T_d.

An estimate of $\mathcal{E}_c - \mathcal{E}_v$ and the position of $\mathcal{E}_y$ can be made as follows. For samples deposited just above 350 K, $(\mathcal{E}_c - \mathcal{E}_F)_0 \simeq 0{\cdot}85$ eV and this should not change appreciably for a T_d somewhat below 350 K. In the latter case we have

$$|\mathcal{E}_1'| = \mathcal{E}_F - \mathcal{E}_y + W_h \simeq 0{\cdot}35\, \text{eV},$$

so that

$$\mathcal{E}_c - \mathcal{E}_y + W_h \simeq 1{\cdot}2\, \text{eV}.$$

If, as is likely, $W_h \simeq W_e$, $\mathcal{E}_c - \mathcal{E}_y \simeq 1{\cdot}1$ eV. Also, since we observe a hopping transport of holes, the conductivity associated with that process must dominate over the conductivity of holes in the extended states. This leads to the inequality

$$\sigma_{RT} \gtrsim \sigma_0 \exp\left[-\frac{(\mathcal{E}_F - \mathcal{E}_v)_0}{kT}\right]$$

with $\sigma_{RT} \simeq 5 \times 10^{-12}\, (\Omega\, \text{cm})^{-1}$, and a σ_0 similar to that for electrons in extended states, $(\mathcal{E}_F - \mathcal{E}_v)_0 \gtrsim 0{\cdot}7$ eV. One then obtains $\mathcal{E}_c - \mathcal{E}_v \gtrsim 1{\cdot}55$ eV.

This estimate is consistent with the value of $\mathcal{E}_0 \simeq 1{\cdot}6$ eV deduced from the photoconductive results. It is also interesting to note that the latter suggest a local maximum in $g(\mathcal{E})$ close to $\mathcal{E}_y$, at which the predominant hopping transport of holes occurs.

Our conductivity correlation graph (Fig. 10.3) is in good general agreement with the previously published results of Chittick.[5] However, two significant points in the interpretation given by Chittick and subsequently by Mott and Davis,[10] require revision in the light of our transport measurements. The first concerns the gradual increase in $|\mathcal{E}_1|$ as T_d is decreased, which the above authors associate with a change in $\mathcal{E}_c - \mathcal{E}_A$. The second point relates to the abrupt change in $\mathcal{E}_1$ at $T_d \simeq 350$ K; it was suggested that this marked a change from electron conduction at $\mathcal{E}_c$ to electron hopping conduction at $\mathcal{E}_A$.

Both points disagree with the electron drift mobility results summarised in Fig. 10.5. $\mathcal{E}_c - \mathcal{E}_A$ is essentially independent of T_d over the complete experimental range. Also, there is no evidence for an abrupt change in μ at

$$T_d \simeq 350 \text{ K}.$$

In fact, a specimen deposited somewhat below 350 K, which gave

$$|\mathcal{E}_1'| \simeq 0{\cdot}35 \text{ eV} \quad \text{and} \quad \sigma_0 \simeq 3 \times 10^{-5}\,(\Omega\,\text{cm})^{-1},$$

possessed the same room temperature drift mobility as other specimens prepared at T_d up to 550 K. Thus the abrupt changes in conductivity activation energy and σ_0 cannot be associated with a change in the electron transport. Experimental results below $T_d = 350$ K illustrate the point made in Section 10.2, that conductivity and drift mobility measurements can refer to different transport situations. Electron drift mobility results are in this instance governed by a quasi–Fermi level $\mathcal{E}_F'$, lying well above the $\mathcal{E}_F$ at which the transition to predominant hole conduction occurs.

10.4.4. APPLICATION TO SPECIMENS PREPARED BY OTHER METHODS

We shall now attempt to generalize our interpretation by including *a*-Si specimens prepared by vacuum evaporation and by r.f. sputtering. Such specimens possess very much higher conductivities than those investigated in this paper and it is suggested that within the general pattern of Fig. 10.3(a) the room temperature conductivities would lie along the rapidly rising curve to the left of the minimum. Thus evaporated or sputtered specimens would correspond to low "effective T_d" values, probably between 200 and 300 K. According to the discussion in the previous section, this means that such specimens should possess a high density of localized states fairly close to $\mathcal{E}_v$, which holds $\mathcal{E}_F$ in this region. Their electrical properties should thus be determined by the transport of holes in localized or extended states. The results of Grigorovici *et al.*[2] on evaporated *a*-Si lead to an interesting correlation. At room temperature the measurements gave a conductivity activation energy of 0·3 eV and a positive thermoelectric power, α. Above 500 K the same samples showed an increased activation energy of 0·8 eV and a negative α. On the basis of Fig. 10·3 we would interpret the room temperature results as a hopping transport of holes around an energy $\mathcal{E}_y$, where

$$\mathcal{E}_F - \mathcal{E}_y + W_h = 0{\cdot}3 \text{ eV}.$$

The suggestions put forward in this Section also resolve the apparent anomaly in the annealing behaviour of *a*-Si prepared by different techniques. Glow-discharge silicon usually increases in conductivity upon annealing. The conductivity of evaporated or sputtered films, on the other hand, generally decreases.(3) The obvious explanation is provided by Fig. 10.3(a). In both types of specimen, annealing causes a reduction in the density of the structural defects and allows $\mathcal{E}_F$ to move to a higher energy. The effect of annealing is thus in the same direction as an increase in T_d. Evaporated or sputtered films, corresponding to points on the left of the conductivity minimum, will therefore experience a decrease in σ on annealing. A glow-discharge specimen, normally prepared above 350 K, will lie to the right of the minimum so that σ will increase.

Evidently the electrical properties of *a*-Si are critically dependent on the density of a group of deep lying states, which we have tentatively associated with reasonably well defined structural defects. It has been suggested that such states are formed by dangling bonds and Brodsky(15) pointed out that their relatively low density in glow discharge specimens may have been caused by the presence of nascent hydrogen during deposition, tending to saturate the dangling bonds. We tried to test this suggestion in the following experiment. An *a*-Si specimen, prepared by vacuum evaporation and having a conductivity of $10^{-2}\,(\Omega\,\text{cm})^{-1}$ was placed in a hydrogen glow discharge in the apparatus normally used for the silane decomposition. After nine hours of discharge treatment, both the conductivity and activation energy were found to be the same within experimental error. This result throws considerable doubt on Brodsky's suggestion and the nature of these states remains an unsolved problem at present.

10.5. Conclusions

(i) The transport properties of *a*-Si films prepared by glow discharge decomposition of silane, depend critically on the density of deep lying localized states, probably associated with structural defects. Their density, and with it the position of the Fermi energy, is determined by the deposition temperature T_d.

(ii) In specimens prepared above about 350 K, the current is carried by electrons. For $T_d \lesssim 350$ K, where $\mathcal{E}_c - \mathcal{E}_F \gtrsim 0{\cdot}8$ eV, a transition to predominant hole conduction takes place.

(iii) From the temperature dependence of the drift mobility and conductivity it is concluded that at room temperature electron transport proceeds through the extended states with a mobility of between 3 and 16 $cm^2\ s^{-1}\ V^{-1}$. Below about 250 K, electrons hop through localized states situated around 0·18 eV below $\mathcal{E}_c$.

(iv) The suggested model appears to be applicable to the interpretation of transport results in *a*-Si films prepared by other methods.

ACKNOWLEDGEMENTS

The authors are greatly indebted to Stewart Kinmond and Michael Ormestad for their help with the specimen preparation and the conductivity measurements. We should also like to thank the Science Research Council for support of this project.

REFERENCES

1. Walley, P. A., *Thin Solid Films* **2,** 327 (1968).
2. Grigorovici, R., *Mater. Res. Bull.* **3,** 13 (1968).
3. Brodsky, M. H., Title, R. S., Weiser, K. and Petit, G. D., *Phys. Rev.* **B1,** 2632 (1970).
4. Chittick, R. C., Alexander, J. H. and Sterling, H. F. *J., Electrochem. Soc.* **116,** 77 (1969).
5. Chittick, R. C., *J. Non-Cryst. Solids* **3,** 255 (1970).
6. Le Comber, P. G. and Spear, W. E., *Phys. Rev. Letters* **25,** 509 (1970).
7. Spear, W. E. and Le Comber, P. G., *J. Non-Cryst. Solids* **8-10,** 727 (1972).
8. Bosnell, J. R. and Voisey, V. C., *Thin Solid Films* **6,** 161 (1970).
9. Spear, W. E., *J. Non-Cryst. Solids* **1,** 197 (1969).
10. See, for example, Mott, N. F. and Davis E. A., "Electronic Processes in Non-Crystalline Materials." Clarendon Press, Oxford, 1971.
11. Cohen, M. H., *J. Non-Cryst. Solids* **2,** 432 (1970).
12. Mott, N. F., *Phil. Mag.* **19,** 835 (1969).
13. Jones, D., to be published.
14. The value of $g(\mathcal{E}_c)$ obtained here is lower than that given in Ref. 7 where the temperature dependence of the Fermi energy had not been allowed for.
15. Brodsky, M. H., private communication; see also Ref. 3.
16. LeComber, P. G., Madan, A. and Spear, W. E., *J. Non-Cryst Solids* **11,** 219 (1972).

11. AMORPHOUS TRANSITION METAL OXIDES

I. G. AUSTIN AND E. S. GARBETT

Department of Physics,
The University, Sheffield, S3 7RH

11.1. INTRODUCTION

Many transition metal oxides form glasses with P_2O_5, in which the transition metal ion is a major constituent ($\gtrsim$ 50 mol per cent). Comprehensive studies[1–5] have been made on the vanadates (V_2O_5–P_2O_5), and to a lesser extent on the iron and copper phosphates. Glasses based on Ti, Cr, Mn, Co, Ni, Mo and W have also been examined[1,5] briefly, as well as binary phases like V_2O_5–TeO_2 and ternary structures.

These materials show semiconducting behaviour when the transition metal ion is present in more than one valence state; for example, as V^{4+} and V^{5+} ions in the vanadates. A similar "mixed valence" principle applies in the crystalline 3*d* oxides, and it is now well established[6,7] that in many of these oxides (e.g. MnO, ferrites, garnets, etc), transport in the 3*d* levels is by a hopping mechanism. The *d*-overlap is small, and there is good evidence that the carriers form dielectric or Jahn–Teller small polarons. Trapping may also occur due to the spin state of a carrier.

In the 3*d* glasses, we expect the 3*d* bands to form localised states in the Anderson sense, and any polaron hopping energy (W_H) will be increased by a disorder term (W_D). There is good evidence for this type of model in 3*d* glasses, but it is difficult to determine the size of the W_D term with certainty.

In this paper, we base our discussion on the vanadate system. First, we summarise the essential experimental features, and the theory of polaron hopping in a glass—including variable range arguments and the problem of averaging when W_D is large. The electrical and optical evidence is then discussed and new data are presented on the optical and high-field behaviour.

11.2. General Features of the Experimental Data

(a) The dc conductivity of the $3d$ glasses varies as $\exp - \mathcal{E}/kT$ over the region 100–700 K. At $T > 300$ K, $\mathcal{E}$ ranges from $\sim 0{\cdot}4$ eV in the vanadates to $\sim 1{\cdot}2$ eV in glasses with Ni, Co or Cu ions.[5]

(b) $\mathcal{E}$ falls off with decreasing temperature—in the vanadates from $\sim 0{\cdot}4$ eV to $\sim 0{\cdot}05$ eV. In contrast, $\mathcal{E}$ in the crystals is constant (Fig. 11.1).

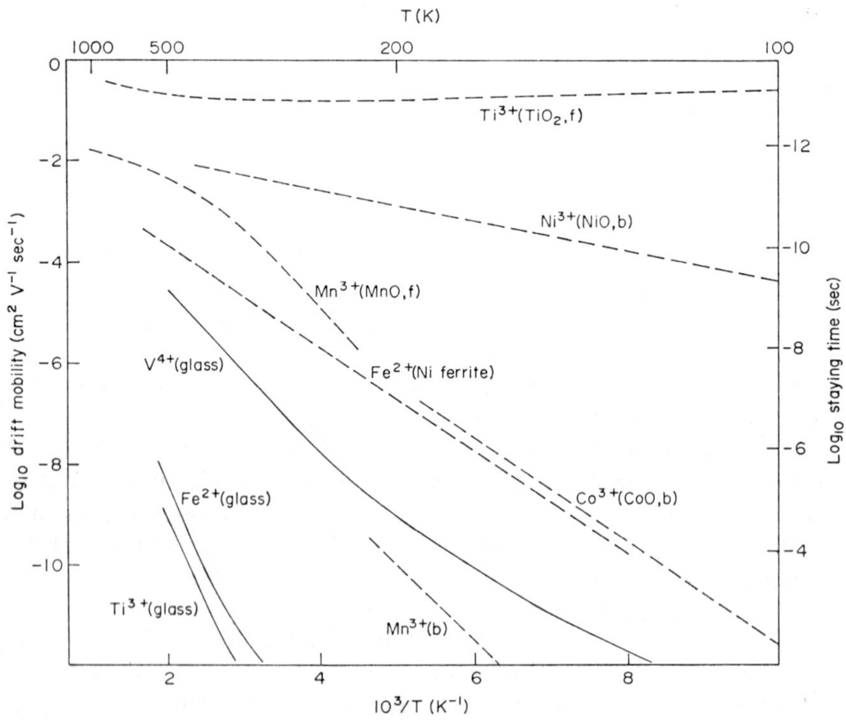

FIG. 11.1. Drift mobilities of $3d$ carriers in various crystalline (– – –) and amorphous (———) transition metal oxides, calculated from conductivity data (Ref. 6). f denotes free carriers; b, bound carriers hopping around impurity centres.

(c) The conductivity mobility (μ) is smaller, and the activation energy is larger than in the corresponding crystal (Fig. 11.1).

(d) The thermopower[2] (α) [Fig. 11.2(a)] and esr signals[8] are independent of temperature for $T \gtrsim 150$ K, indicating a constant carrier concentration. The increase in α for $T \lesssim 150$ K is not reflected in the conductivity, and this resembles the behaviour of certain ferrites [Fig. 11.2(b), $T < 300$ K].

(e) The ac conductivity[9,10] is of the form $\sigma \propto \omega^n$ (where $n = 0{\cdot}85$,) and is almost independent of temperature at frequencies $> 10^8$ Hz. The static dielectric constant is large (20–40) in the vanadates and increases with vanadium content.[5]

(f) Infrared,[11,20] nmr[12] and esr[8] studies show that the vanadium ion has a 5-fold oxygen coordination, as in V_2O_5. In some of the VO_5 pyramids, a PO_4 unit probably replaces an apex oxygen, but is not clear how the VO_5 units are linked. Phase separation[13] has been observed at high vanadium concentrations and low annealing temperatures.

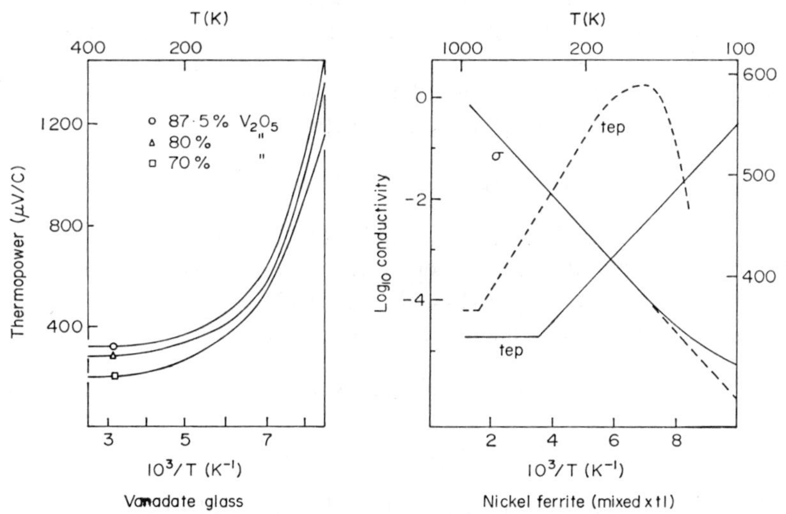

FIG. 11.2. (a) Thermopower in various vanadate glasses (Ref. 2). (b) Thermopower (α) and conductivity in nickel ferrite (Ref. 14) ($Ni_{1-x}Fe_{2+x}O_4$, with $x = 0{\cdot}06$). Also shown (----) is α for $x = 0{\cdot}046$, showing turn-over, probably due to impurity conduction.

11.3. POLARON THEORY

(i) *High and low temperatures*

At $T \geqslant \theta/4$, the intersite hopping probability for a small polaron is[3]

$$P = c(1-c)\,\nu \exp - (W_H + \tfrac{1}{2}W_D)/kT \qquad (11.1)$$

where

$$W_H = W' (\tanh \tfrac{1}{4}\hbar\omega_0 \beta / \tfrac{1}{4}\hbar\omega_0 \beta). \tag{11.2}$$

Here c is the occupation probability, W_H the polaron hopping energy, W_D the disorder potential between sites, ω_0 ($\equiv k\theta/\hbar$) the optical phonon frequency and $\beta = 1/kT$. For adiabatic transfer $\nu = \omega_0/2\pi$, and in the non-adiabatic case

$$\nu = \pi^{\frac{1}{2}}(J_0{}^2/2\hbar)(W_H kT)^{-\frac{1}{2}} \exp - 2\alpha R \tag{11.3}$$

$J_0 \exp - \alpha R$ is the overlap integral between sites, separated by R, and α the fall-off with distance. Thus in the non-adiabatic case

$$P \propto \exp - (2\alpha R + W_H/kT + W_D/2kT). \tag{11.4}$$

W_H in eqn (11.2) falls with temperature because the hopping is a multi-phonon process; at $T \lesssim \theta/4$ single optical phonon processes dominate and $W_H \rightarrow h\omega_0$. However, recent calculations by Emin[15] show that the decrease predicted by eqn (11.2) is too rapid, and W_H is essentially constant down to $\theta/3$, especially if there is appreciable phonon dispersion. At lower temperatures, $W_H \rightarrow \hbar\omega_0$ rapidly.

At the lowest temperatures, the polaron hops with one or more acoustic phonons making up the energy difference between sites, and[16]

$$P \propto \exp - (2\alpha R + W_D/kT). \tag{11.5}$$

The carrier may also hop preferentially beyond the nearest-neighbour distance (R_0) to a site at a lower energy; W_D in eqn (11.5) is then a function of R. Following Mott,[2] we put $R = pR_0$ in eqn (11.5), where p is dimensionless and $W_D \rightarrow W_D/p^3$. Maximising the exponential gives

$$P \propto \exp - 2{\cdot}4 \, [W_D (\alpha R_0)^3/kT]^{1/4} \tag{11.6}$$

with

$$p^4 = (\alpha R_0)^{-1} (3 W_D/2kT). \tag{11.7}$$

Thus if $p > 1$, the "activation energy" in a $1/T$ plot falls off as $T^{3/4}$ [Fig. 11.3(a)].

(ii) *Averaging*

Formulae like eqn (11.5) must be averaged to obtain the bulk mobility. This difficult problem was first treated by Miller–Abrahams (MA) in the limits $W_D < kT$ and $\alpha R < 1$, which imply nearest neighbour hops. They assumed that conduction occurs down parallel (non-intersecting) chains. At each link the electron chooses the jump of least impedance and the net resistivity is given by the average value of the smallest impedance link. Their calculation

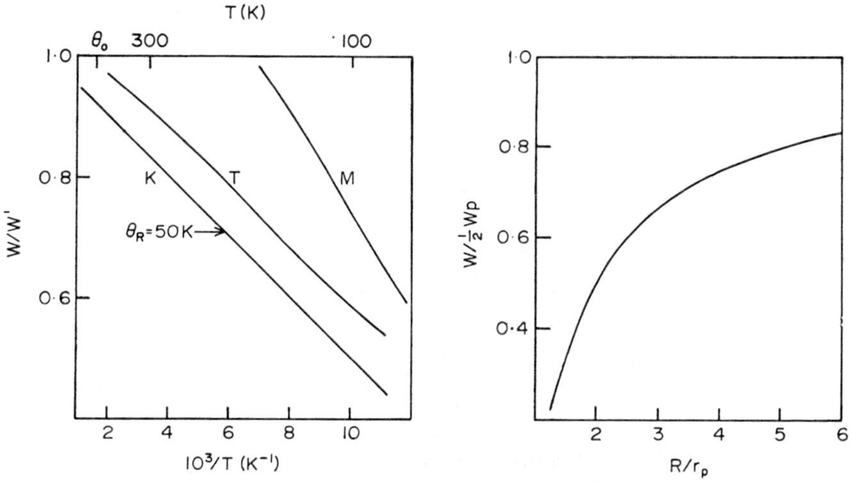

FIG. 11.3. (a) Temperature dependence of the hopping energy for various models. T, small polarons (eqn (11.2) with $\theta_0 = 600$ K); K, Killias model (eqn (11.11) with $\theta_R = 50$ K); M, variable range hopping (Mott, eqn (11.6)). (b) Variation of polaron hopping energy with distance, eqn (11.9).

shows that the activation energy is independent of temperature, but the average of the tunnel term is complicated and of the form $\exp - (\alpha R)^{3/2}$.

In the limit $W_D > kT$, the MA theory does not predict a constant activation energy,[17] and the conductivity varies roughly as $\exp - BT^{\frac{1}{4}}$. But the MA method of averaging has various shortcomings in this region.[17] Thus the restriction that at *each* hop the electron selects the easiest jump leads ultimately to some difficult hops. These would be avoided in a percolation model, or by tunnelling to a more distant site, as in eqn (11.6). Several authors[17] have tried to adapt percolation theory to this situation and their results indicate a conductivity approaching the form $A \exp - BT^{1/4}$, but at present it is difficult to calculate the constants in a rigorous manner.

The averaging of eqn (11.4) for small polarons will also depend on the size of the W_D/kT term. But a further complication arises because W_H is also a function of R—as first noted Killias.[18]

(iii) *Random jump distances—Killias' model*

At the instant of hopping, polarisation clouds are created at the two sites, and these overlap giving a partial reduction in W_H. Killias[18] proved formally that W_H increases linearly with site separation, thus

$$W_H = W(R_0) + a\Delta R \qquad (11.8)$$

Mott,[2,3] using simple electrostatics, showed that

$$W_H = \frac{e^2}{4\kappa_p}\left\{\frac{1}{r_p} - \frac{1}{R}\right\} \tag{11.9}$$

where r_p is the polaron radius, $\kappa_p^{-1} = (\kappa_\infty^{-1} - \kappa_0^{-1})$, and κ_∞ and κ_0 are the high and low frequency dielectric constants.

Killias[18] applied eqn (11.8) to the high temperature formula eqn (11.4). He assumed nearest-neighbour hopping, but allowed for fluctuations in W_H due to random variations in ΔR. The net mobility was calculated by taking a simple average of the form

$$\left[\exp - \frac{W(R_0)}{kT}\right]\left\langle \exp - \frac{a\,\Delta R}{kT}\right\rangle_{\text{av}} \tag{11.10}$$

assuming a Gaussian distribution in the ΔR. Since the ΔR are symmetric, we can simplify his argument and write

$$\langle \exp - x\rangle = \tfrac{1}{2}\langle \exp x + \exp - x\rangle$$

$$\simeq \exp \bar{x}^2/2 \qquad (\text{for } \bar{x} \lesssim 1\cdot 5)$$

where $\bar{x} = a\,\Delta\bar{R}/kT$, and $\Delta\bar{R}$ is an rms value. Thus the mobility varies as $\exp - W/kT$, where

$$W = W(R_0)\,[1 - \theta_R/T] \tag{11.11}$$

and

$$k\theta_R = \frac{a^2(\Delta\bar{R})^2}{2W(R_0)} \tag{11.12}$$

eqn (11.11) again predicts an activation energy which falls with temperature [Fig. 11.3(a)], and was applied by Killias to ferrous ion glasses.[18]

It is not easy to evaluate the physical significance of the simple averaging procedure in eqn (11.10), but it is clear that it favours short (easy) hops, and in this sense it resembles the MA theory for $W_D < kT$. However, the fluctuations ΔR generate random potentials of magnitude $(a\Delta R)$, and if these are large compared with kT, we expect a breakdown of the Killias' formula.

It is useful to estimate the parameter "a" in eqn (11.8). Using eqn (11.9) we find $a = dW/dR \simeq W_H/2R_0$, assuming[3] $R_0/r_p \sim 3$. Thus for eqn (11.12) we obtain

$$k\theta_R \sim \tfrac{1}{8}W_H\,(\Delta\bar{R}/R_0)^2. \tag{11.13}$$

Finally, we note that the linear Killias term in eqn (11.8) can be combined with the tunnel term in eqn (11.4). Thus α is effectively increased to

$$(\alpha + a/2kT) \tag{11.14}$$

where $a \sim W_H/2R_0$, and W_H falls with T. Typically, for a vanadate glass at 500 K, $W_H \sim 0{\cdot}4$ eV, $R_0 = 4$ Å, giving $a/2kT \sim 0{\cdot}5$ Å^{-1}. At 100 K, $W_H \to \hbar\omega_0$ ($\sim$0·05 eV), and $a/2kT \sim 0{\cdot}3$ Å^{-1}.

11.4. DC Conductivity

Typical conductivity data for the vanadates are shown in Fig. 11.4.

(i) *High temperatures*

The general trend in Fig. (11.4) is consistent with eqns (11.1)–(11.4) and a number of workers[9,10] have fitted their data in this way, obtaining $\theta \simeq 600$ K, and $W_D \sim 0{\cdot}1$eV. However, Emin's recent theory[15] shows that the fall-off in W_H due to phonon freeze-out is much slower than predicted by eqn (11.2). No experimental test of this point has been made yet, but is interesting to note that in crystalline materials like nickel ferrite[14] and YIG, the hopping energy is remarkably constant from 900 K down to $\sim$ 150 K [Figs. 11.1 and 11.2(b)], although $\theta \sim 800$ K. Other $3d$ carriers in crystals show similar behaviour (Fig. 11.1). This provides some support for Emin's theory, since there is good evidence that the carriers are small polarons. It seems likely then, that in eqns (11.2 and 11.4), W_H is constant down to $\sim \theta/3$, or 150–300 K assuming $\theta = 500$–1000 K.

We turn now to the Killias model. Figure 11.3(a) shows that eqn (11.11) predicts a similar temperature-dependence to eqn (11.2), for $\theta_R \sim 40$ K, and thus fits the conductivity data equally well[10,19] for $T > 300$ K. This gives $W(R_0) = 0{\cdot}45$ eV in eqn (11.11). From eqn (11.12) we find

$$a\Delta\bar{R} = \sqrt{2W_H.k\theta_R} \tag{11.15}$$

giving $a\Delta\bar{R} \sim 0{\cdot}06$ eV. Also, from eqn (11.13), we calculate $\Delta\bar{R}/R_0 \sim 0{\cdot}1$. Little is known about the detailed structure of these glasses, but a 10 per cent fluctuation in the vanadium spacing is feasible. From this, we conclude that the Killias model offers a plausible explanation of the fall-off in hopping energy, but the fluctuations ($a\Delta\bar{R}$) are larger than kT. Thus the elementary averaging procedure in eqn (11.10) should be replaced by a percolation argument.

From the high temperature conductivity and eqn (11.1) (with a geometrical factor of 1/6), we estimate $\nu = 5 \times 10^{12}$ sec^{-1}, and from eqn (11.3), $J = J_0 \exp - \alpha R_0 \sim 0{\cdot}03$ eV. The optical phonon frequencies[20] are $\sim 10^{13}$ sec^{-1}, so that we are rather close to the adiabatic limit. When the V_2O_6 content

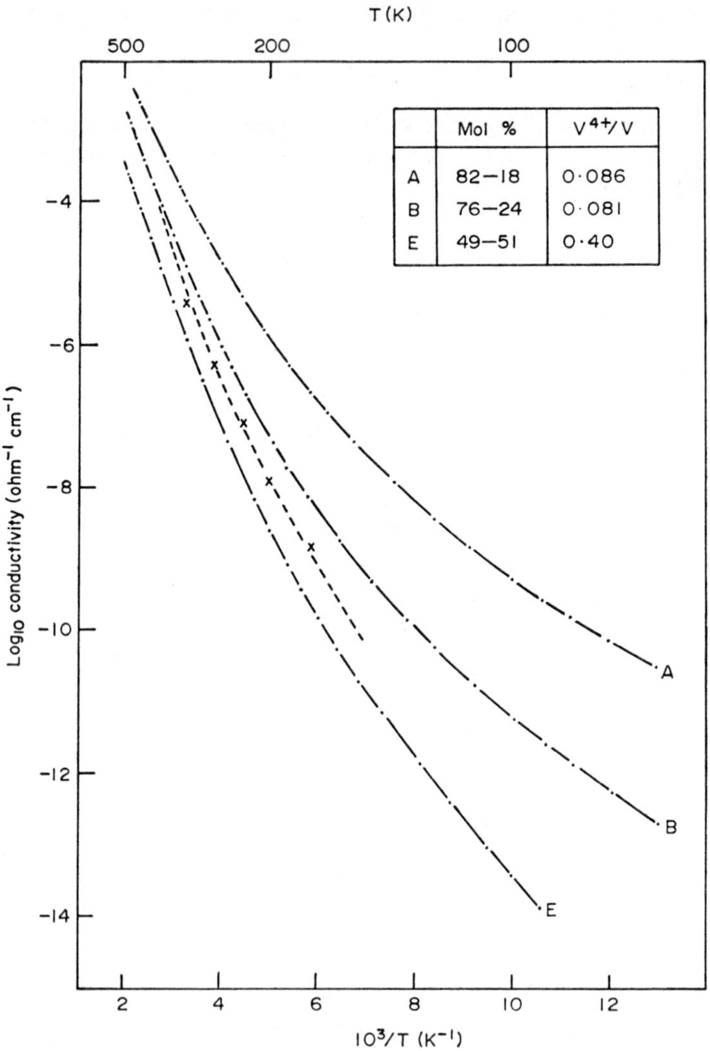

FIG. 11.4. Log conductivity versus $1/T$ for various vanadate glasses (Ref. 11). – – –, plot of eqn (11.1) assuming $W_H = 0{\cdot}35$ eV, $W_D = 0{\cdot}08$ eV, $\theta_0 = 620$ K.

changes from 50 to 90 mol per cent, ν increases by a factor of about 3,[(5)] and the mean V–V spacing is reduced by $\Delta R_0 = 0{\cdot}7$ Å. Assuming $\exp - 2\alpha\Delta R_0 = 0{\cdot}3$, we estimate $\alpha \sim 1\text{Å}^{-1}$. This argument is rough, since ν may be modified by averaging or correlation [Section 11.4], but we can at least set an *upper* limit to α of 2–3 Å^{-1}.

(ii) *Low temperatures*

At the lowest temperatures, the plots in Fig. (11.4) show a curvature with slopes (W) of 0.05–0.1 eV. We consider two possibilities—(a) hopping to distant sites, in which case $W < W_D$, and (b) hopping to nearest neighbours, with $W \simeq W_D$. In both cases, an exp $- \hbar\omega_0/kT$ factor may be present as well. However, there is no evidence of such a term in the ac conductivity at low temperatures. Also, the dc conductivity at low temperatures is sensitive to preparation methods, annealing etc., suggesting that a W_D term is dominant.

To test assumption (a), we show in Fig. 11.5 some $T^{1/4}$ plots down to 50 K. These are linear, except below 70 K, indicating a fit with eqn (11.6) with

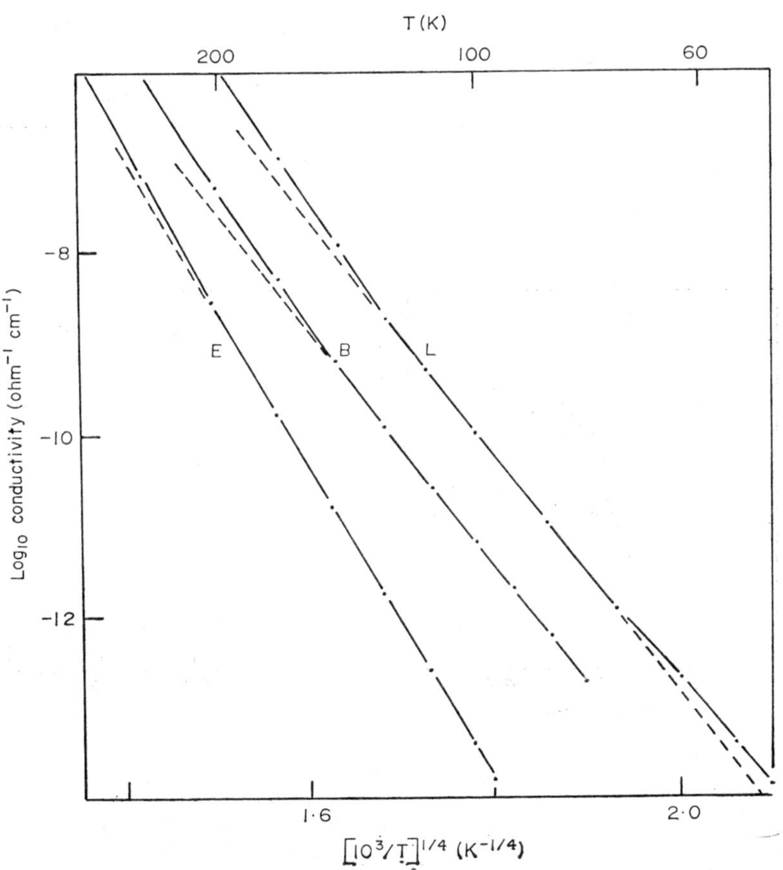

FIG. 11.5. Log conductivity versus $T^{1/4}$ for various vanadate glasses. Samples E and B from Fig. 11.4; L from Linsley *et al.* (Ref. 10).

$W_D(\alpha R_0)^3 = 960$, where $R_0 = 4$ Å. However, in Section 11.4 we inferred that $\alpha < 2$–3 Å^{-1}, and the Killias contribution to α [eqn (11.14)] is < 1 Å^{-1}; thus the $T^{1/4}$ analysis implies $W_D > 0{\cdot}6$ eV. A disorder term of this size would imply drastic fluctuations in the cation environment, and is much larger than the values ($W_D \lesssim 0{\cdot}1$ eV) estimated from thermopower[2,3] and dielectric loss.[9] From this we conclude that the $T^{1/4}$ analysis is *not* valid for $T > 50$ K.

Assumption (b) is therefore preferable, but if $W_D \lesssim 0{\cdot}1$ eV as the data in Fig. (11.4) imply, then W_D may still be large compared with kT. In this case, a percolation model might account for the curvature in the $1/T$ plot, but we have no good theory.† Alternatively, the data might be fitted by an expression of the form $T^n \exp - W_D/kT$, if the hopping is acoustic and multiphonon.

Some 3*d* glasses (Fe, Cu etc.) show evidence of much larger W_D values,[2,3,21] and in one model of switching W_D can be altered by field-induced changes in structure.[21] Probably there are basic structural differences[5] between glasses based on 3*d* ions like Fe, Ni, Cu which form cubic oxides, and those like V, Mo which readily form covalent complexes and chains.

(iii) *Concentration dependence*

Figure 11.6(a) shows the increase in W with average V–V spacing in a series of glasses where the V_2O_5 content decreases from 90 to 50 mole per

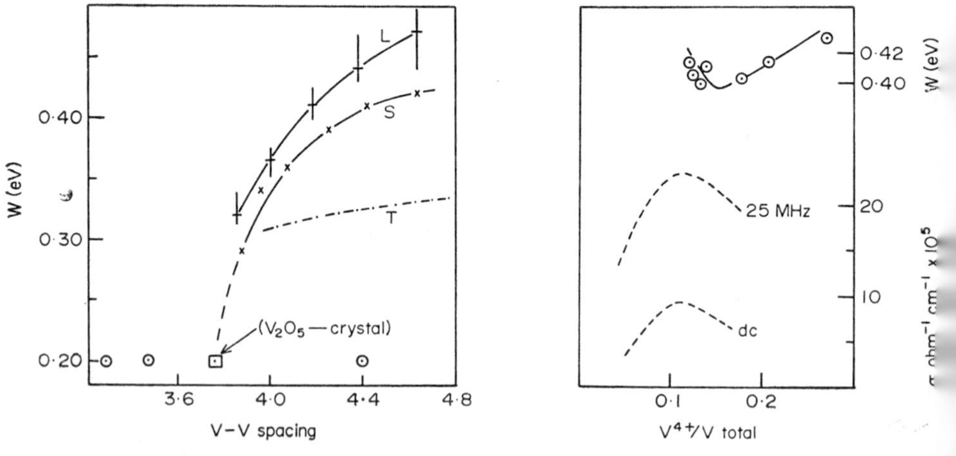

FIG. 11.6. (a) High temperature activation energy in the vanadate glasses as a function of mean spacing between vanadium ions. *S*, Sayer *et al.* (Ref. 5); *L*, Linsley *et al.* (Ref. 10). Also shown, following Sayer (Ref. 5), are the V–V spacings along the main axes in crystalline V_2O_5, and the mean value (□). *T* is calculated from eqn (11.9), assuming $r_p = 1{\cdot}5$ Å. (b) High temperature activation energy W and conductivity, as a function of V^{4+} fraction (Ref. 10). ⊙, 70 mol %, ---, 90 mol %.

† For a recent theory, see Ref. 25.

cent. The variation predicted by eqn (11.9), assuming $r_p \sim 1.5$ Å,[5] is much too small (line T in Fig. 11.6(a)). However, the high frequency dielectric constant (κ_∞), decreases by 20 per cent over this composition range,[5] and this may account [in eqn (11.9)] for much of the change in W_H.[2,3]

Equation (11.1) has a maximum, at $c = 1/2$, where the V^{4+} and V^{5+} concentrations are equal. This is not consistent with the conductivity data which show a maximum at $c = 0{\cdot}15$—which correlates with a weak minimum in the activation energy[10] [Fig. 11.6(b)]. Various explanations have been proposed for this;[4] some based on the idea that a fraction of the V^{4+} ions are firmly trapped in complexes.[10] But this only alters c in the pre-exponential in eqn (11.1). Others have stressed the importance of polaron–polaron interactions at high carrier concentrations.[4,5] Thus when each V^{4+} carrier has at least one V^{4+} ion as a nearest neighbour, the polarisation clouds will partially cancel, leading to a decrease in W_H with increasing concentration. Also, the random fields set up by the V^{4+} carriers should increase with c. This might explain the shallow minimum ($\sim 0{\cdot}02$ eV) in W, but no quantitative theory has been attempted. Sayer *et al.*,[5] have proposed that correlation effects due to short-range Coulomb repulsion will modify c in eqn (11.1) to $c(1-c)^q$, and there is some evidence for such a variation with $q \sim 6$.

11.5. Optical Properties

We turn now to the optical behaviour. For a small polaron, theory predicts[3,6] a Gaussian-like absorption at $h\nu = 2W_p$. Here W_p is the polaron binding energy, and the energy for thermal transfer (W_H) is $\leqslant \frac{1}{2}W_p$. Thus $W_{\text{opt}} \leqslant 4W_H$. The oscillator strength is approximately[6]

$$f \simeq z\left(\frac{m_0 R^2}{\hbar^2}\right)\frac{(2J)^2}{h\nu} \tag{11.16}$$

where z is the number of nearest neighbours. In crystals like TiO_2, $R \sim 5$ Å, $h\nu \sim 1$ eV, $J \sim 0{\cdot}1$ eV, giving $f \sim 0{\cdot}05$. This absorption corresponds to a Franck–Condon excitation of the electron out of its polarisation well. In a system like the vanadate glass, it is equivalent to an optical charge transfer between V^{4+} and V^{5+} ions, with broadening due to disorder.[22]

Recently, this type of absorption has been studied in a number of 3d crystals and ferrites [Fig. 11.7], and the correlation between optical and thermal energies examined.[6] Thus in materials like nickel ferrite and YIG, the Fe^{2+} carriers show a peak at ~ 1 eV [Fig. 11.7(b)], and the dc mobility shows a hopping energy of 0·2 eV [Fig. 11.1]. This implies $W_p = 0{\cdot}5$ eV and $W_{\text{opt}} \simeq 4W_H$.

One problem in interpreting spectra like those in Fig. 11.7 is that in some

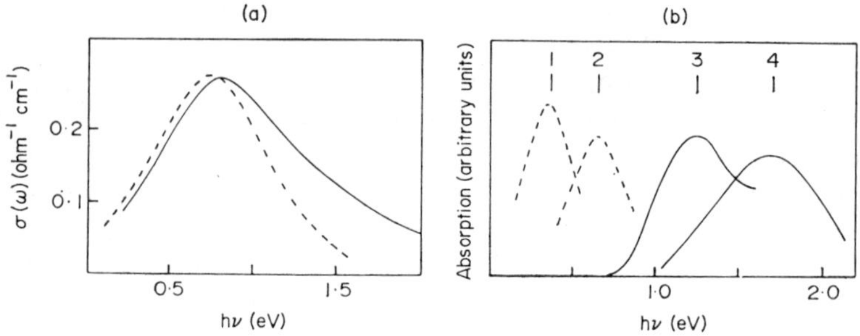

FIG. 11.7. Optical absorption associated with $3d$ carriers in various crystalline semiconductors. (a) Ti^{3+} carriers in TiO_2; ——, experiment; – – –, theory. (b) Fe^{2+}–Fe^{3+} transfer in (1) Fe_3O_4, (2) ferrous phosphate (vivianite), (3) yttrium iron garnet (YIG), and (4) Prussian blue. Full references are given in Austin and Gamble, Ref. 6.

ions the peak could arise through an internal (d–d) excitation of the $3d$ carrier. Recent work[22,24] shows that electro-optical studies can help to distinguish between an internal optical excitation at a single site, and an excitation between two *different* sites, since the latter is more sensitive to an external electric field.

No optical absorption studies of charge carriers have been reported for the $3d$ semiconducting glasses, but recently we have observed[23] spectra of this type in the vanadates and a glass containing manganese. Thin films of the vanadates, about 5 microns thick, were prepared by melting the glass and squeezing it between polished quartz flats. Absorption data for a sample with different V^{4+} concentrations are shown in Fig. 11.8(a). At $h\nu > 2{\cdot}5$ eV, we observe a steep edge, with a long tail extending to lower energies; this agrees with earlier work by Anderson and Compton.[20] In addition, our spectra show a broad peak at 1·8 eV. The integrated intensity of the peak is proportional to the V^{4+} (d^1) concentration, giving $f \sim 4\times 10^{-4}$, and $J \sim 0{\cdot}01$ eV [from eqn (11.16)]. The V^{5+} (d°) ion has no d-electron absorption.

The dc hopping energy for this glass is 0·4 eV. Thus, assuming W_D is small and $W_H = 0{\cdot}4$ eV, the small polaron model predicts a peak at 1·6 eV, in fair agreement with Fig. 11.8(a). The theory[3] also predicts that the conductivity at the maximum of the peak (σ_{max}) is roughly

$$\sigma_{max}/\sigma_{dc} = (kT/4W_H)\exp W_H/kT. \tag{11.17}$$

For this glass, $\sigma_{dc} \simeq 10^{-5}$ ohm^{-1} cm^{-1} at 300 K, and $W_H = 0{\cdot}4$ eV. Thus the peak absorption coefficient ($375\,\sigma_{max}/n$) is ~ 300 cm^{-1}, which is close to the measured value of 200 cm^{-1} in Fig. 11.8(a). The calculation is sensitive however to errors in W_H in eqn (11.17).

It seems plausible therefore, to interpret the 1·8 eV peak as evidence for small polaron formation in this material, with $W_p = 0{\cdot}9$ eV. We note however, that the oscillator strength is rather low in comparison with 3*d* crystals (Table 11.1), and at this stage we have certainly not ruled out the possibility of a crystal field transition semiconducting within the V^{4+} ion.

Similar data for a semiconducting manganese glass is shown in Fig. 11.8(b). In this system, only Mn^{2+} ions are present when the glass is fully reduced, but in the oxidised state Mn^{3+} are also present. The main part of the broad peak at 2·8 eV ($f \sim 10^{-4}$) is probably a crystal field transition of the Mn^{3+} ion.[28] Thus any optical charge transfer of the Mn^{2+}—Mn^{3+} type is too weak to observe, implying $f \lesssim 10^{-5}$ and $J \lesssim 3 \times 10^{-3}$ eV; alternatively it occurs at $h\nu > 4$ eV.

TABLE 11.1. Optical and thermal activation energies of 3*d* carriers

	Carrier	f	J_{opt}	W_{opt}	W_H
Vanadate glass	V^{4+}	$\sim 4 \times 10^{-4}$	~0·01	1·8	0·4
Mn glass	Mn^{3+}		~0·01	~2·8	1
Garnet[6]	Fe^{2+}	$< 10^{-2}$	~0·05	1·1	0·2
(J_{opt} and all W in eV)					

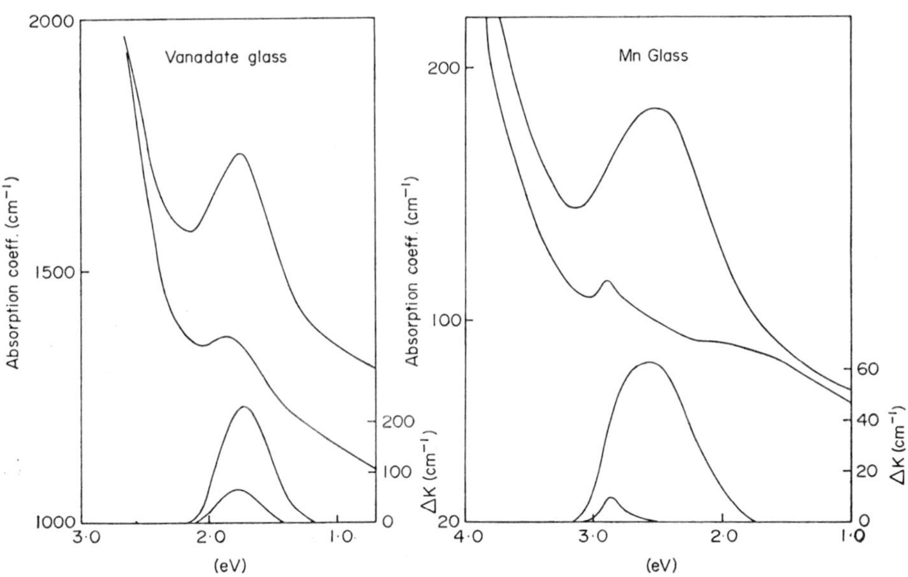

FIG. 11.8. Optical absorption in 3*d* glasses (Austin and Garbett, Ref. 22). (a) Vanadate, 70 mol % V_2O_5. V^{4+} fractions (c) are ~0·1 and 0·3. (b) Manganese glass; mol %—SiO_2 (70), CaO + Mgo (5), MnO (~12), + Na_2O. Upper sample oxidised; ~3 mol % Mn^{3+}.

11.6. High Electric Field Studies

In amorphous semiconductors like As_2Se_3, the high-field conductivity often shows non-ohmic behaviour of the Poole–Frenkel type at $E \gtrsim 10^4$ V/cm, due to the field emission of carriers from localised states into extended states. In contrast, in the small polaron model where *all* the states are localised, the high field simply modifies the hopping rate eqn (11.1). The mobility in a field E is then given by[(6)]

$$\mu(E) = \mu(0) \exp \left\{ \frac{-e^2 R^2 E^2}{16 W_H kT} \right\} \frac{\sinh eRE/2kT}{eRE/2kT} \tag{11.18}$$

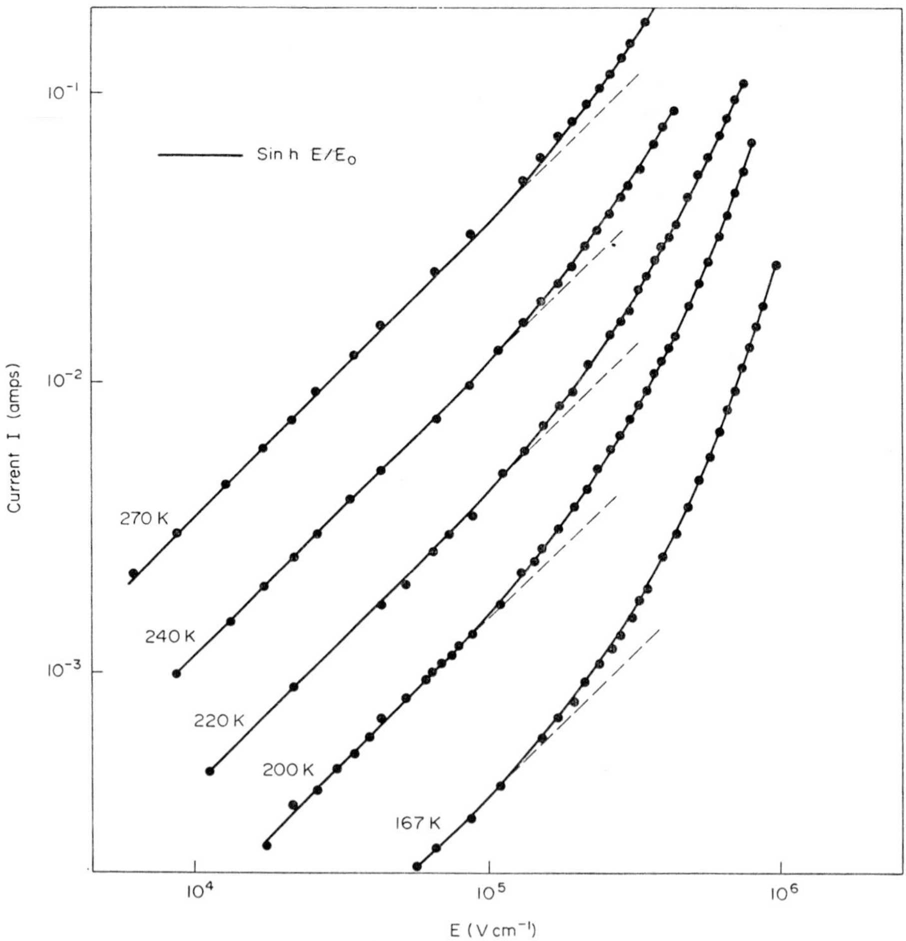

Fig. 11.9. Log I versus log E in 80 mol % vanadate glass, $c \sim 0{\cdot}15$. Gold contacts. Pulsed measurements.

The sinh term implies non-ohmic behaviour when $eRE \sim kT$, and for $R \sim 5$ Å, $T \sim 300$ K, this requires $E \sim 5 \times 10^5$ V/cm. We have observed sinh behaviour of this type in several crystalline ferrites,[(6)] although the calculated jump parameter R is often larger than anticipated, perhaps due to a local field correction, or a field-dependence in J [eqn (11.4)]. A similar problem arises in high-field ionic conduction in glasses.

Figure 11.9 shows some data for a vandadate glass.[(26)] The curves give a good fit with a sinh expression, but again R is large ($\sim$ 18 Å). However, from eqn (11.7), we obtain $p^4 \simeq W_D/0{\cdot}1\alpha$ at $\sim$ 200 K. Also, in Section 11.4 we inferred $W_6 < 0{\cdot}1$ eV and $\alpha < 2$ Å^{-1}—the Killias term in eqn (11.14) is probably negligible ($< 0{\cdot}5$ Å^{-1}). This gives $p^4 < 0{\cdot}5$, and strongly implies nearest neighbour hopping at $T > 200$ K. Thus we conclude, as for the ferrites, that eqn (11.81) needs modification.[(26)]

11.7. CONCLUSIONS

The electrical and optical data on the vanadate glasses are consistent with a small polaron model of nearest-neighbour hopping, from 50 to 700 K. W_D may be as low as $\sim 0{\cdot}05$ eV. The fall-off in W_H is probably a Killias effect above 200–300 K, and phonon freeze-out at lower temperatures. The disorder potentials ($a\Delta\bar{R}$, W_D) are large compared with kT at all temperatures, suggesting a model in which the carriers percolate down easy chains. The comparatively low J values suggest that the d-overlap is sensitive to disorder.

ACKNOWLEDGEMENTS

We are indebted to Mr. C. Drake for supplying the manganese glass, to Drs. D. Emin, M. Sayer and C. E. Turner for valuable discussions, and to Dr. Sayer for access to unpublished data.

REFERENCES

1. Mackenzie, J. D., "Modern Aspects of the Vitreous State." Vol. 3, p. 126 (ed. J. D. Mackenzie) London, Butterworths, 1965.
2. Mott, N. F., *J. Non-Cryst. Solids* **1,** 1 (1968).
3. Austin, I. G. and Mott, N. F., *Adv. in Physics* **18,** 41 (1969).
4. Adler, D., *Critical Reviews in Solid State Sciences* **2,** 317 (1971).
5. Sayer, M. and Mansingh, A., *Phys. Rev.* **6B,** 4629 (1972).
6. Austin, I. G. and Gamble R., Eilat Conference on Conduction in Low Mobility Materials, page 1. Taylor & Francis, London, 1971.
7. Bosman, A. J. and van Daal, H. J., *Adv. in Phys.* **19,** 1 (1970).

8. Lynch, G. F., Sayer, M., Segel, S. L. and Rosenblatt, G., *J. App. Phys.* **42,** 2587 (1971).
9. Sayer, M., Mansingh, A., Reyes, J. M. and Rosenblatt, G. *J. App. Phys.* **42,** 2857 (1971).
10. Linsley, G. S., Owen, A. E. and Hayatee, F. M., *J. Non-Cryst. Solids* **4,** 208 (1970).
11. Sayer, M., Mansingh, A., Reyes, J. M. and Lynch, G. F., Eilat Conference on Conduction in Low Mobility Materials, page 115. Taylor & Francis, London, (1971.
12. Landsberger, F. R. and Bray, P. J., *J. Chem. Phys.* **53,** 2757 (1970).
13. Anderson, G. W. and Luehrs, F. U., *J. App. Phys.* **39,** 1634 (1968).
14. Griffiths, B. A., Elwell, D. and Parker R., *Phil. Mag.* **22,** 163 (1970).
15. Emin, D., this volume, Chapter 7.
16. Schnakenberg, J., *Phys. Stat. Solidi* **28,** 623 (1968).
17. Jones, R. and Schaich, W., *J. Phys. C: Solid State* **5,** 43 (1972).
18. Killias, H. R., *Physics Letters* **20,** 5 (1966).
19. Owen, A. E., (1972) private communication.
20. Anderson, G. W. and Compton, W. D., *J. Chem. Phys.* **52,** 6166 (1970).
21. Drake, C. F. and Scanlan, I. F., *J. Non-Cryst. Solids* **4,** 234 (1970).
22. Austin, I. G., *J. Phys. C: Solid State* **5,** 1687 (1972).
23. Austin, I. G. and Garbett, E. S., (unpublished).
24. Reik, H. G., *Solid State Communications* **8,** 1737 (1970).
25. Adler, D., Flora, L. P. and Santuria, S. D., *Solid State Communications* **12,** 9 (1973).
26. Sayer, M., Private communication. See also Ref. 27.
27. In a recent study of high-field conduction in the *d*-glasses, Austin and Sayer propose that the effective field in eqn (11.18) is enhanced due to the existence of 'hard hops.' (*J. Phys. C*: *Solid State* (1973), to be published.)
28. McClure, D. S., *Solid State Physics* **9,** 399 (1959).

12. NONEXTRINSIC AND SPACE CHARGE LIMITED CURRENTS IN SEMI-INSULATORS

G. G. ROBERTS

School of Physical Sciences, New University of Ulster, Coleraine, Northern Ireland

12.1. INTRODUCTION

Certain requirements must be met for space charge limited (SCL) current flow in a solid to take place and be detected. Understandably, the first stringent condition is that the electrodes make ohmic contact to the material. That is, rather than producing a barrier at the surface, the electrode must provide a plentiful supply of carriers which can participate in the conduction process as and when they are needed. In addition the contribution of the thermally generated carriers to the total current must be small.

The complete mathematical analysis of time independent SCL conduction in solids is so complex that no explicit expressions have yet been obtained relating the current and voltage. Mott and Gurney[1] were the first to emphasize the importance of an injecting contact between a metal and an insulator and they provided an approximate expression relating the current density J, voltage V and thickness L in a trap-free insulator. A slight modification of their theory results in the following equation

$$J = \frac{9}{8}\theta_n \kappa \mu_0 \frac{V^2}{L^3}. \tag{12.1}$$

In this equation κ is the permittivity, μ_0 is the microscopic mobility and θ_n

is the fraction of total carriers (all electrons above the Fermi level) which are free. An equivalent equation could be written for holes.

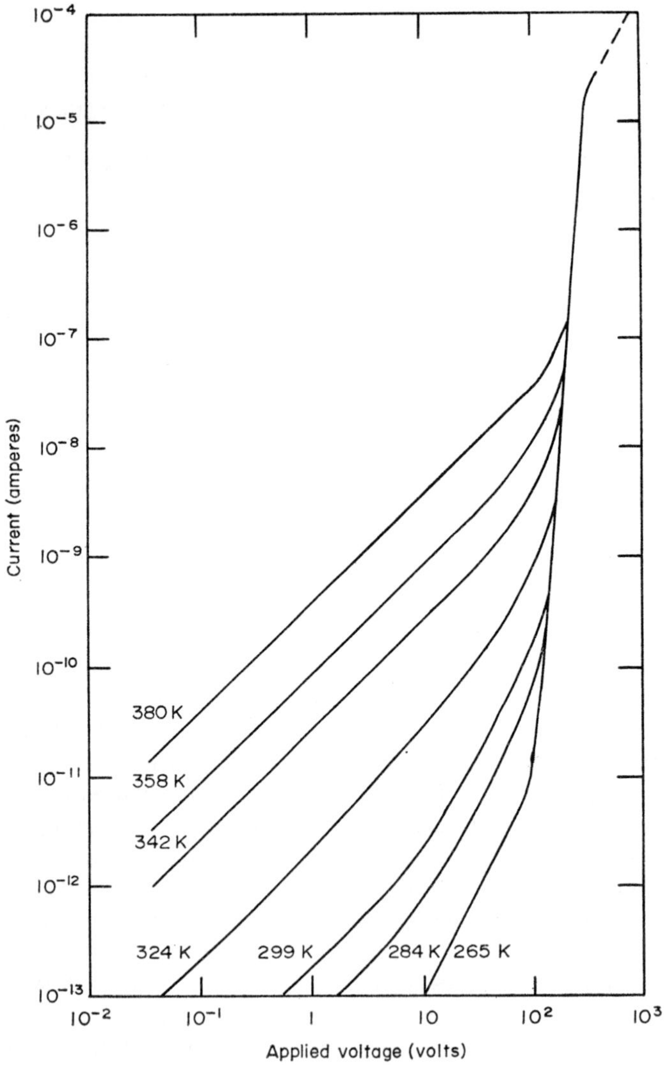

FIG. 12.1. Current–voltage curves at several different temperatures for single crystal mercury sulphide. (After Roberts and Schmidlin, Ref. 2.)

The introduction of deep trapping centres in the crystal can lead to a higher power dependence of current on voltage than the square law relationship shown in eqn (12.1). Figure 12.1 shows a set of current–voltage curves at several different temperatures for single crystal mercury sulphide.[2] The shape of these curves conforms to the general picture envisaged by Lampert[3] for a model containing a discrete level of trapping states which is gradually filled by injected charge. Two threshold voltages of interest may be defined. The voltage $V_{\Omega c}$ where the current converts from ohmic to SCL behaviour can be found by equating the two currents.

$$V_{\Omega c} = \frac{8}{9}\frac{enL^2}{\kappa\theta_n} \tag{12.2}$$

where n is the density of free carriers. The trap-filled limit voltage V_{TFL} may be related to the trap density N_m responsible for the steep rise in the current–voltage characteristic.

$$V_{\mathrm{TFL}} = \frac{2N_m eL^2}{3\kappa}\,. \tag{12.3}$$

Several authors[4–7] have discussed the effect on the SCL current characteristic of taking into account the decrease in the effective trap depth by an electric field. It may be shown that close compensation and high field values, two conditions likely to be satisfied in experiments on amorphous semiconducting films, can modify quite appreciably the transition voltages defined above. It should be emphasized however that in such materials the presence of a high density of states could effectively prevent the build up of an accumulation layer over the several atomic spacings necessary to fulfil the normal requirements of SCL conduction.

Equation (12.1) is a special case of the scaling law for bulk space charge currents in a homogeneous medium[8–10]

$$J = AL\,(V/L^2)^m \quad \text{or more generally} \quad J/L = \text{function}\,(V/L^2) \tag{12.4}$$

where m is a constant which clearly need not necessarily be an integer. For example in the case of minority carrier injection $m = \frac{1}{2}$.[8] The law is valid provided diffusion currents are neglected; high field effects such as Poole–Frenkel and impact ionization are ignored; there is no contact limitation to the current and the mobility is independent of electric field.

It is of some importance to be able to distinguish between non-ohmic behaviour due to space charge injection and that arising from other physical processes. The obvious method is to investigate the dependence on thickness of a set of current–voltage curves and see if eqn (12.4) is satisfied. That is, plots of J/L versus V/L^2 should superimpose. To eliminate any possible influence of the contact in establishing this scaling law it is preferable to change the thickness by successively etching away the non-injecting electrode only.

It is instructive to consider SCL conduction in a solid containing a level whose density of states falls off exponentially with energy. This may be expressed mathematically by the equation

$$g(\mathcal{E}) = \frac{N}{kT_c} \exp{(\mathcal{E}/kT_c)} \tag{12.5}$$

where $g(\mathcal{E})$ is the density of states per unit energy range, N is the total density and T_c characterizes the way in which the concentration falls off with energy. The result, first obtained by Mark and Helfrich,[11] $J = BV^{l+1}/L^{2l+1}$, where B is a constant at a particular temperature and $l = T_c/T$, may be seen to be a special case of the general scaling law. Other types of trap distribution may be analyzed in a similar fashion.[12,13]

12.2. Activation Energies for Ohmic and SCL Conduction in Semi-insulators

A recent theory discussing all possible interpretations of a simple activation energy for electrical conduction has been given by Roberts and Schmidlin.[2] In physical terms, one of the requirements for a simple activation energy to be manifest in electrical conduction is that specific energy levels must contribute overwhelmingly to the partition functions for electrons and holes. At each temperature there must be only one dominant level for each carrier and it must be remembered that such states can arise from entities other than donors and acceptors. For an n-type material, if the excess donor concentration is small in comparison with the concentration of electrons excited from the dominant hole level to the dominant electron level, then the Fermi energy is located between the two dominant levels in such a way as to make the concentration of electrons in the dominant electron level equal to the concentration of holes in the dominant hole level. This is what is termed the nonextrinsic situation and it arises when

$$(N_d - N_a)^2 \ll 4Q_e Q_p \tag{12.6}$$

where N_d and N_a are the concentrations of donors and acceptors in the system and Q_e and Q_p are the partition functions for electrons and holes respectively.

In general both ohmic and SCL currents are thermally activated, their corresponding activation energies being contained in the quantities n and θ_n respectively. Roberts and Schmidlin show that the density of free carriers n

controlling the ohmic activation energy in an n-type material when the criterion expressed in eqn (12.6) is satisfied, is given by the equation

$$n = N_c \left(\frac{N_q}{N_m}\right)^{\frac{1}{2}} \exp\left[-(\mathcal{E}_c - \mathcal{E}_m)/2kT\right] \exp\left[-(\mathcal{E}_c - \mathcal{E}_q)/2kT\right] \qquad (12.7)$$

whereas the activation energy of a current in the SCL region is controlled by

$$\theta_n = \frac{N_c}{N_m} \exp\left[-(\mathcal{E}_c - \mathcal{E}_m)/kT\right] \qquad (12.8)$$

In the above equations the subscripts c, m and q correspond to conduction band, dominant electron and dominant hole states respectively. The two distinct solutions arise because the activation energy for electrical conduction generally involves some combination of the energies required to create carriers in the dominant levels and to raise carriers from the dominant levels to their corresponding transport bands. The difference in activation energies can be understood by the fact that the activation energy for creation can be suppressed by injecting enough electrons to overwhelm the nonextrinsic concentration. Identity of the above two activation energies occurs only when the inequality in eqn (12.6) is not satisfied and the material is extrinsic.

The direct comparison between the two cases, depending whether or not eqn (12.6) is satisfied, leads to the following theorem: different (or identical) activation energies for ohmic and SCL conduction is both necessary and sufficient for ohmic conduction to be nonextrinsic (or extrinsic). The importance of this theorem is that it provides one with an experimental method of uniquely determining the nature of ohmic conduction and establishing unequivocally the location of the dominant energy levels in the solid.

From eqn (12.2) it may be seen that the voltage $V_{\Omega c}$ has a thermal activation energy equal to the difference between the activation energies for the ohmic and SCL regions. Accordingly it is equivalent to say that a temperature dependent $V_{\Omega c}$ implies a sample that is nonextrinsic. Inspection of Fig. 12.1 shows that this is clearly the situation in mercury sulphide, the onset of the quadratic voltage dependence being shifted to lower fields at low temperatures. The activation energies for $V = 2$ volts where the current is ohmic and $V = 100$ volts where the current is SCL are equal to 0·85 eV and 0·62 eV respectively. The different activation energies confirm that the material is nonextrinsic. Equation (12.3) for V_{TFL} gives a trap density $N_m = 10^{13}\ \mathrm{cm}^{-3}$ while the high field square law region may be analyzed to give $\mu_0 = 7\ \mathrm{cm}^2\ \mathrm{V}^{-1}\ \mathrm{s}^{-1}$. Using the procedure implied by eqns (12.7) and (12.8) we obtain $(\mathcal{E}_c - \mathcal{E}_m) = 0{\cdot}62$ eV and $(\mathcal{E}_c - \mathcal{E}_q) = 1{\cdot}08$ eV. Further information that can be extracted from the experimental data and eqns (12.1) and (12.2) is $N_q = 10^{16}\ \mathrm{cm}^{-3}$ and $N_c = 10^{19}\ \mathrm{cm}^{-3}$. It is interesting to note that if the region following a trap-filled limit voltage is accessible experimentally, then the effective mass of the majority carrier can be determined.

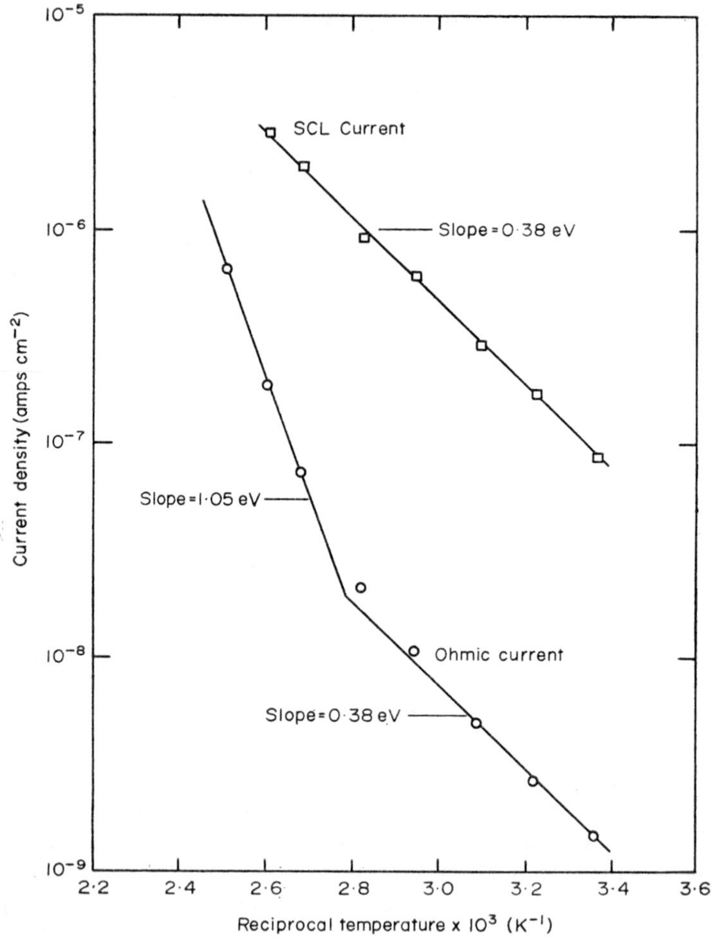

FIG. 12.2. Variation of current density with temperature for $V = 10$ volts (Ohmic region) and $V = 400$ volts (square law region) for β phthalocyanine single crystal. (After Barbe and Westgate[14]).

In some solids, well defined activation energies can be identified in different temperature regions. Of course, each temperature interval in which a simple activation energy is observable can be analyzed using the scheme presented above for mercury sulphide. However, the transition temperature itself, defined by the intersection of neighbouring extrapolated linear segments in the activation energy plot, is an additional bit of information which can be useful to establish additional relations between dominant state densities. There are three prime causes for a transition (i) when the two largest terms in a partition function become equal (ii) when the majority carrier changes sign (iii) when the inequality between the contributed and non-extrinsic carrier concentrations reverses. Only the last of the three is exemplified here as it is the new feature resulting from the Roberts–Schmidlin analysis. From eqn (12.6) it is seen to occur when $(N_d - N_a)^2 = 4Q_e Q_p$. Some recent experimental data on metal-free phthalocyanine single crystals provide an excellent illustration of this situation.(14) Figure 12.2 shows Barbe and Westgate's activation energy curves for samples where the correct thickness and voltage dependences for both ohmic and SCL currents have been established. The activation energy of the ohmic current through the crystal is seen to change from 0·38 eV below 363K to 1·05 eV above this temperature. By contrast the activation energy of the SCL current is 0·38 eV throughout the temperature range studied. The temperature induced change in the ohmic activation energy must be interpreted in terms of an extrinsic–nonextrinsic transition as the temperature increases through 363K. As in the case of mercury sulphide it is possible to determine the important state densities and their location in the energy gap. The two dominant statistical levels are calculated to be at 0·38 eV and 1·72 eV from the conduction band edge and to have densities 4×10^{19} cm^{-3} and $1{\cdot}5 \times 10^{21}$ cm^{-3} respectively. The well established value of the band gap of this material is 1·72 eV. This, together with the fact that the density of the level at this energy is approximately equal to the molecular density of metal-free phthalocyanine, suggests that the dominant hole level is the appropriate transport "band".

12.3. Activation Energies for Electrical Conduction in Amorphous Semiconductors

(i) *General discussion*

Most amorphous semiconductors exhibit an exponential temperature dependence of the electrical conductivity. That is,

$$\sigma = \sigma_0 \exp(-\Delta\mathcal{E}/kT). \tag{12.9}$$

In the majority of cases $\Delta\mathcal{E}$ has been reported to be approximately equal to half the optical gap. In addition, such materials appear to remain "intrinsic" even with the incorporation of a few per cent of foreign atoms. At least two models to explain the positioning of the Fermi level near the centre of the gap have been proposed. Davis and Mott(15) have assumed a peak in the density of localized states near mid band gap while Fritzsche(16) has suggested that potential fluctuations which spread the range of localized states from the band edge to the centre of the gap are responsible for the observed behaviour. However, it should be stressed that optical absorption and photoemission experiments have failed to reveal the presence of large densities of localized states within the gap.

When interpreting conductivity data on amorphous semiconductors one should always consider whether the observed activation energy is due to the mobility of the majority charge carriers and/or to their generation mechanism. As far as the temperature dependence of the carrier densities are concerned the theoretical arguments presented by Roberts and Schmidlin[2] are equally applicable to ordered and disordered systems. The following statement would, for example, be expected to hold in both cases. "In narrow band gap materials the dominant levels tend to be the transport bands (with or without tails) whereas in wider band gap materials some other localized levels within the band gap tend to become dominant instead." It has been pointed out that in amorphous solids, foreign atoms of different valency tend to form their specific coordination environment and hence satisfy their valence requirements. However, in these materials it is easy to visualize the possibility of $GeTe_2$, for example, behaving as an entity with its own characteristic energy level when a glass is formed containing the elements germanium and tellurium. This level would probably not be discrete but be confined to a narrow band of energies and have approximately a Gaussian state distribution. Two levels of this type orginating due to particular structural groupings, one controlling the electron and the other the hole partition function, may be responsible for determining the position of the Fermi level in many amorphous solids. Their densities could be quite low, for it is easy to compute that 10^{10} levels per cm^3 located $20\,kT$ (0·5 eV at room temperature) from a transport band or 10^2 levels per cm^3 located $40\,kT$ (1 eV at room temperature) from a transport band are sufficient to dominate the statistics. Impurities or structural defects such as dangling bonds which can act as donors, acceptors or neutral traps, might also contribute to the type of energy level scheme envisaged here.

Emin *et al.*[17] have presented fairly convincing arguments that polaron effects can play a vital role in determining charge transport in some classes of amorphous semiconductors. Polaron bands with energies lying within the energy gap could therefore control the position of the Fermi level in such cases. The high densities of states involved ($10^{22}\,cm^{-3}$) together with the fact that typically the electron and hole polaron bands might be situated at most about 0·4 eV apart,[18] would certainly ensure that the criterion for nonextrinsic conduction defined in eqn (12.6) be met. Of course, it is possible that these polaron bands could dominate the carrier statistics but have too low a mobility associated with them to completely control the conductivity.

In view of the high state densities associated with the polaron bands it seems unlikely that sufficient charge could be injected in these cases to raise the Fermi level. In any event this would change the observed activation energy by less than 0·1eV.

The feature of the Roberts–Schmidlin analysis which is implicit in the two possible models suggested here is that two levels lying within the band gap can very easily dominate the carrier statistics and lead to nonextrinsic conduction. The densities of these levels may be large enough to control the position of the Fermi level and yet at the same time be too low to be detected optically. By contrast, a single distribution of states as suggested by Davis and Mott(15) would have to contain an unrealistically high concentration to pin the Fermi level. It is interesting to note that Main and Owen(19) have succeeded in interpreting their photoconductivity data on chalcogenide compounds in terms of the energy level picture proposed here. Such a model could also account for the recent data obtained by Egerton.(20)

(ii) *The Meyer–Neldel Rule or Compensation Law*

For many types of solid, experimental evidence suggests that a correlation exists between the activation energies and pre-exponential factors for electrical conduction σ_0, of the following form

$$\log \sigma_0 = \alpha \Delta \mathcal{E} + \beta \tag{12.10}$$

where α and β are constants.(21) Rosenberg *et al* (22) have demonstrated that this formula can also be valid for a single material. For this case they have shown that eqn (12.10) is incomplete and have suggested instead a three constant equation to explain their results.

$$\sigma(T) = \sigma_0' \exp(\Delta \mathcal{E}/kT_0) \exp(-\Delta \mathcal{E}/kT). \tag{12.11}$$

This was necessary for although their activation energy curves were obtained over only a limited range of temperature, some, when extrapolated to higher temperatures, intercepted at a focal point T_0 and not at an infinite temperature. Many amorphous chalcogenide semiconductors have a common σ_0 and therefore a focal point which occurs at $T_0 = \infty$.

A model(21) that can explain this result is one in which the conduction band tails a finite energy distance towards the valence band but where the Fermi level is also controlled by a fixed dominant hole level deeper in the gap. For a continuous band whose density of states varies exponentially with energy [see eqn (12.5)] and extends between the two energies $\mathcal{E}_m$ and $\mathcal{E}_c$, the number of free carriers is given by

$$n = \int_{\mathcal{E}_m}^{\mathcal{E}_c} \frac{N}{kT_c} \exp(\mathcal{E}/kT_c) \exp[(\mathcal{E}_f - \mathcal{E})/kT]\, d\mathcal{E} \qquad (12.12)$$

The upper limit of the integral may be extended to infinity provided the band is sufficiently wide. With this approximation eqn (12.12) becomes

$$n = n_0' \exp[(\mathcal{E}_m - \mathcal{E}_f)/kT_c] \exp[-(\mathcal{E}_m - \mathcal{E}_f)/kT]. \qquad (12.13)$$

A direct comparison of eqns (12.11) and (12.13) is now possible. Two main points of interest are that $T_0 \equiv T_c$ and $\Delta\mathcal{E} \equiv (\mathcal{E}_m - \mathcal{E}_f)$. For those amorphous chalcogenides where $T_0 \simeq \infty$ the above result would suggest that T_c is also large and therefore that band states down to $\mathcal{E}_m$ tail off uniformly in energy. It is a reasonable assumption that $\mathcal{E}_m$ in an amorphous semiconductor defines the energy at which the conduction band states become localized. Therefore it would appear that the activation energy one observes, if the above model is valid, is exactly equal to the energy $(\mathcal{E}_m - \mathcal{E}_f)$. This result could explain both the observation of the Meyer–Neldel Rule in a single material or a group of materials. Consider, for example:

(a) A mixed chalcogenide semiconductor system $A_2X_3 : B_2X_3$.

(b) A range of different chalcogenides containing the same group VI element.

(c) A single amorphous film prepared or annealed under different conditions.

(d) A single amorphous film whose activation energy is measured at various high electric field strengths.

In all four cases the basic requirement to observe the rule is that the nature of the band tailing is similar and T_c remains unaltered. In cases (a) through (d) the variation in activation energy can result simply from the Fermi level being a different distance from the transport band for each different composition or material respectively. In cases (c) and (d), however, the Meyer–Neldel Rule could also occur due to a change in the position of $\mathcal{E}_m$. That is, annealing or the application of a large electric field could alter the energy at which the states become localized, while at the same time unaffecting the position of the Fermi level. Careful activation energy measurements on the four types of systems indicated above could yield information concerning the density of states in the conduction band tails.

Thermally assisted tunneling (TAT) can also provide an explanation for the Meyer–Neldel Rule.[21] In this case the focal point is obtained by plotting activation energy curves for TAT through a potential barrier across which various voltages are applied. TAT may be an important mechanism in amorphous semiconductors since it could control the number of electrons in localized states below the mobility edge that may reach higher energy states where they can contribute to the conduction. Roberts and Polanco[23] have shown that for a nearly uniform distribution of band tail states the resulting current–voltage characteristic is of the Poole–Frenkel or Schottky type. However, if the mobility of the charge carrier is also a function of energy then other power laws result. Thermally assisted tunnelling in band tail states with the additional feature of a field or energy dependent mobility seems in fact the most likely explanation for the linear log J versus E characteristics observed in some chalcogenide glasses at low electric fields.[24] Marshall and Owen[25] have also concluded from drift mobility studies that their observation of a field dependent charge carrier mobility in some disordered materials may be associated in some way with a field dependent mobility edge.

12.4. Sub-linear Current–Voltage Characteristics

In a recent publication Van Roosbroeck and Casey[26] have discussed carrier transport in a semiconductor where the dielectric relaxation time τ_d exceeds the diffusion length lifetime τ_0. These authors have suggested that the principal result of their theory, "recombinative space charge injection", should be commonly observed in amorphous semiconductors in view of the long relaxation times involved. (The small signal solution had actually been derived in an earlier paper by Van Roosbroeck,[27] by considering an injected pulse of electron-hole pairs in an infinite semiconductor.)

The basic idea is that if $\tau_d \gg \tau_0$ the transport is characterized by a relatively rapid recombination rate which is determined by τ_0 and not by dielectric relaxation. The recombining excess current carriers of one kind cause local depletion of the carriers of the opposite sign. The resulting concentration distributions persist for a time long compared with the lifetime and drift under the action of the applied electric field until ultimately decaying through dielectric relaxation. The depletion drift region, which extends the distance the majority carriers drift in time τ_d and in some cases can reach the anode, ends in a recombination front. The proposed model leads to a sublinear, approximately half power law dependence of current on voltage. To date there are two reports[28,29] of experimental current–voltage curves in chalcogenide glasses showing this particular relationship.

Roberts[8] has provided both computational and analytical solutions to the problem of minority carrier injection into a semiconductor. If the relatively unimportant contribution of the recombination front is ignored then this problem is essentially the same as that considered by Van Roosbroeck and Casey. The simplest starting equations that yield the required J proportional $V^{\frac{1}{2}}$ relation are the following:

$$J = e\mu E\,(n + p); \frac{dE}{dx} = \frac{-(n - p)}{\kappa}\,; np = C. \quad (12.14)$$

Reducing the temperature should commonly convert many materials from being controlled by lifetime considerations, where local space charge neutrality is a good approximation, to relaxation type semiconductors. The resulting change of shape inferred by the Van Roosbroeck analysis is not, however, observed experimentally. It is certainly true that in most high resistivity materials one would anticipate τ_d to be much larger than τ_0. Therefore the reason why the sublinear dependence is so rarely observed is probably associated with the crucial $np = C$ relation assumed in both the theoretical treatments by Roberts and Van Roosbroeck and Casey. As has already been argued, in higher resistivity materials which usually have wide band gaps, levels other than the transport bands usually control the carrier statistics. The assumption of equality of capture and release rates in traps and consequently $np = C$ in such materials therefore seems improbable.

Resistance limited currents[(30)] have been proposed as an alternative explanation for sub-linear curves in solids which also show low field ohmic behaviour. The model proposed is that of a material fitted with two blocking contacts, whose effect is masked at low voltages by the series resistance of the sample. That is, at low field values, the current is limited by the carrier concentration at the contact but the current–voltage curve is linear because most of the voltage is dropped across the bulk series resistance. As the voltage is increased the depletion layer at the reverse biased end of the sample extends further into the bulk and eventually a stage is reached where the current converts from being resistance limited to contact limited. When this occurs the ideal reverse bias diode characteristic (J proportional $V^{\frac{1}{2}}$) is followed. Beyond a certain value of field the Schottky barrier breaks down and the current increases more rapidly than linearly with applied voltage. However, it is important to recognize that when this occurs, the series resistance still limits the current to values below the extrapolated ohmic region. When the current reaches a value equal to this limit, oscillations might be expected to occur.

Resistance limited currents have been observed in single crystal trigonal selenium[(30)] and the existence of two back to back Schottky diodes separated by a series resistance has been confirmed by capacitance measurements on the same sample.[(31)] It is possible that the gallium arsenide data put forward by Van Roosbroeck and Casey to illustrate their theory might also be interpreted in this manner. It is perhaps significant that oscillations were observed in their sample at a current value corresponding to the extrapolated ohmic region.

References

1. Mott, N. F. and Gurney, R. W., "Electronic Processes in Ionic Crystals." Oxford University Press, London, 1940. p. 172.
2. Roberts, G. G. and Schmidlin, F. W., *Phys. Rev.* **180,** 785 (1969).
3. Lampert, M. A., *Phys. Rev.* **103,** 1648 (1956).
4. Murgatroyd, P. N., *J. Phys. D: Appl. Phys.* **3,** 151 (1970).
5. Raykerus, P. A., *Radio Engrs. and Electron Physics (U.S.A.)* **16,** 675 (1971).
6. Boon, M. R., *Thin Solid Films* **9,** 457 (1972).
7. Barbe, D. F. and McRae, H. F., *J. Vac. Sci. and Tech.* **9,** 70 (1971).
8. Roberts, G. G., *Phys. Stat. Sol.* **27,** 209 (1968).
9. Murgatroyd, P. N., *J. Phys. D. Appl. Phys.* **3,** 1488 (1970).
10. Lampert, M. A. and Mark, P., "Current Injection in Solids." Academic Press, New York, 1970.
11. Mark, P. and Helfrich, W., *J. Appl. Phys.* **33,** 205 (1962).
12. Hwang, W. and Kao, K. C., *Solis St. Electronics* **15,** 523 (1972).
13. Nespurek, S. and Smejtek, P., *Czech. J. Phys.* **B22,** 160 (1972).
14. Barbe, D. F. and Westgate, C. R., *Solid St. Comm.* **7,** 563 (1969).
15. Davis, E. A. and Mott, N. F., *Phil. Mag.* **22,** 903 (1970).
16. Fritzsche, H., *J. Non-Cryst. Solids* **6,** 49 (1971).
17. Emin, D., Seager, C. H. and Quinn, R. K., *Phys. Rev. Lett.* **28,** 813 (1972).
18. Emin, D., private communication.
19. Main, C. and Owen, A. E., this volume, Chapter 16.
20. Egerton, R. F., *Appl. Phys. Lett.* **19,** 203 (1971).
21. Roberts, G. G., *J. Phys. C: Solid St. Physics* **4,** 3167 (1971).
22. Rosenberg, B., Bhowmik, B. B., Harder, H. C. and Postow, E., *J. Chem. Physics* **49,** 4108 (1968).
23. Roberts, G. G. and Polanco, J. I., *Phys. Stat. Sol.*. **1a,** 409 (1970) and **13a,** 603 (1972).
24. de Wit, H. J. and Crevecoeur, C., *Solid St. Electronics* **15,** 729 (1972).
25. Marshall, J. M. and Owen, A. E., *Phys. Stat. Sol.* **12a,** 181 (1972).
26. van Roosbroeck, W. and Casey, H. C., *Phys. Rev.* **B5,** 2154 (1972).
27. van Roosbroeck, W., *Phys. Rev.* **123,** 474 (1961).
28. Polanco, J. I., Roberts, G. G. and Myers, M. B., *Phil. Mag.* **25,** 117 (1972).
29. Fritzsche, H., private communication.
30. Schmidlin, F. W., Roberts, G. G. and Lakatos, A. I., *Appl. Phys. Lett.* **13,** 353 (1968).
31. Lakatos, A. I. and Roberts, G. G., *J. Appl. Phys.* **39,** 5308 (1968).

IV

OPTICAL AND RELATED PROPERTIES

13. OPTICAL PROPERTIES OF AMORPHOUS SEMICONDUCTORS

E. A. DAVIS

Cavendish Laboratory, University of Cambridge, Cambridge, England

13.1. INTRODUCTION

Optical absorption in solids or liquids can occur by several different mechanisms, all of which involve coupling of the electric vector of the incident radiation to dipole moments in the material and a consequent transfer of energy. Analysis of optical absorption spectra, with the primary variable being the incident photon energy and secondary variables including temperature, pressure, and electric or magnetic fields, provides essential information on phonon and electron states in the material whether this is crystalline, amorphous or liquid.

13.2. ABSORPTION PROCESSES

Figure 13.1 illustrates schematically most of the absorption processes which will be discussed later with reference to specific materials. Two ordinates are shown: the optical absorption coefficient α and the a.c. electrical conductivity σ related to each other by:

$$\alpha = 4\pi\sigma / n_0 c \qquad \text{(c.g.s.)}$$

where n_0 is the real refractive index and c the velocity of light. In Fig. 13.1 a value of $n_0 \sim 4$ has been used. The angular frequency of the incident radiation is denoted by ω.

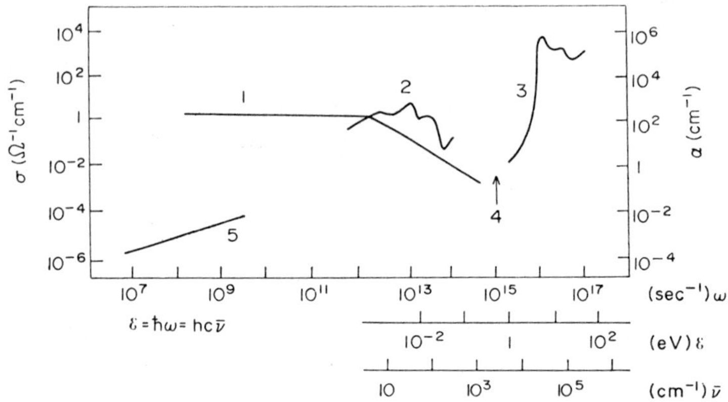

FIG. 13.1. Schematic representation of optical absorption processes. (See text.)

Process 1: *Free carrier absorption*

This is observed in metals and semiconductors whenever there is a sufficient density of free carriers n in a single band. Its frequency dependence generally follows the Drude–Lorentz relation

$$\sigma(\omega) = \sigma(0)/(1 + \omega^2 \tau^2)$$

where $\sigma(0)$ is the dc conductivity, and τ the relaxation time, or time between scattering events. If the effective mass of the carriers is m^* then

$$\sigma(0) = ne^2\tau/m^*.$$

When $\omega\tau \ll 1$, $\sigma(\omega) \to \sigma(0)$ and when $\omega\tau \gg 1$, $\sigma(\omega) \propto 1/\omega^2$ i.e. proportional to the square of the wavelength of the incident radiation.

This *intraband absorption* would be forbidden if it were not for a relaxation of selection rules which occurs in the presence of impurities or phonons. The scattering processes that determine $\sigma(0)$ are in fact the same as those that allow free carrier absorption to occur.

In amorphous materials free carrier absorption *has* been observed but *not* the fall-off as $1/\omega^2$, probably because τ is very short $\sim 10^{-15}$–10^{-16} sec.

Process 2: *Lattice* (*or phonon*) *absorption*

Vibrational modes of the atoms in solids and liquids give rise to absorption of radiation normally in the infra-red region of the spectrum. Three processes can be distinguished.

(i) "Reststrahl". This occurs in ionic materials by creation of single phonons of near zero wave vector due to coupling of oscillating dipole moments (e.g. $Na^+ Cl^-$) to the electric field of the radiation.

(ii) "Multiphonon" processes occur in ionic and covalent crystals. One vibrational mode induces instantaneous charges on the atoms and a second mode simultaneously causes a vibration of the induced charges. This produces an electric moment of second order which couples to the radiation field. Absorption occurs at sum and difference frequencies of the two phonons. Anharmonic forces are necessary to allow coupling between the modes.

(iii) "One-phonon" processes can occur in covalent materials. The creation of a single phonon will be accompanied by a change in dipole moment if there are three or more atoms per unit cell. In addition, by destroying periodicity, defects or impurities allow one-phonon processes to occur.

In amorphous materials these lattice absorption processes are retained to a degree which depends on the material. Generally speaking, fine structure present in the spectra of corresponding crystalline materials is lost. In addition one-phonon processes may possibly be allowed at very low frequencies due to the disorder present.

Process 3: *Electronic interband transitions*

The excitation of electrons from filled to empty states by photon absorption is of primary importance in semiconductors. The onset of these optical absorption processes is marked by a sharp rise in α over many decades—the fundamental absorption edge—to a level $\sim 10^6 \text{ cm}^{-1}$ at a photon energy corresponding to the energy gap. In crystals the transitions can be "direct" or "indirect", the latter being phonon-assisted. The form of the edge depends on the symmetry of the wave functions and the effective masses at the edges of the valence and conduction bands. In addition, excitonic effects may be important and can lead, in some cases, to discrete line spectra at energies just below the fundamental edge. Beyond the edge, the absorption coefficient remains high, the photons causing excitations from states deeper into the valence band to higher lying states in the conduction bands. Eventually in the ultra-violet and X-ray regions of the spectrum, transitions from atomic core states to the conduction bands occur.

In amorphous semiconductors a fundamental edge is observed although its shape is often quite different from that occurring in crystals. The k-conservation selection rule appears to be relaxed and no discrete exciton states have been observed, although the Coulomb interaction between electron and hole must still exist and is probably of importance. At higher photon energies much of the fine structure present in the crystalline spectrum is lost.

In addition to the above primary absorption processes there may be absorption occurring on the low energy side of the fundamental edge (in the region marked as *Process* 4 in Fig. 13.1) due to transitions from or to impurity or defect levels in the gap. An absorption process in amorphous materials is sometimes observed at very low frequencies as an ac loss in the electrical conductivity. This process, with a frequency dependence close to $\omega^{0 \cdot 8}$ is sketched as *Process* 5 in Fig. 13.1. In many cases it is probably due to electrons hopping between localized states at the Fermi level although other mechanisms with which it might be associated exist.

Measurements other than optical absorption, but related to it, which also convey information on electronic states and transitions between them are photoemission, electron energy loss, electromodulation spectroscopy and photoconductivity. Results using some of these techniques will be called upon when appropriate in the remainder of this article, which will review and attempt to interpret experimental observations on amorphous and liquid semiconductors.

13.3. AC Conductivity

An ac conductivity which increases approximately linearly with frequency has been observed in several amorphous semiconductors. Results obtained by Rockstad[1] are shown in Fig. 13.2. Similar behaviour has also been observed in other chalcogenide glasses[2] and, by one group, in amorphous Ge.[3] There are two quite distinct models that have been proposed to explain this effect. One is that it is a Debye-type loss associated with electrons hopping back and forth between localized states at the Fermi level. This process is known to occur in the case of impurity conduction in heavily doped semiconductors and the theory of Pollak and Geballe[4] was originally developed to explain results for these materials. The theory was adapted to the situation

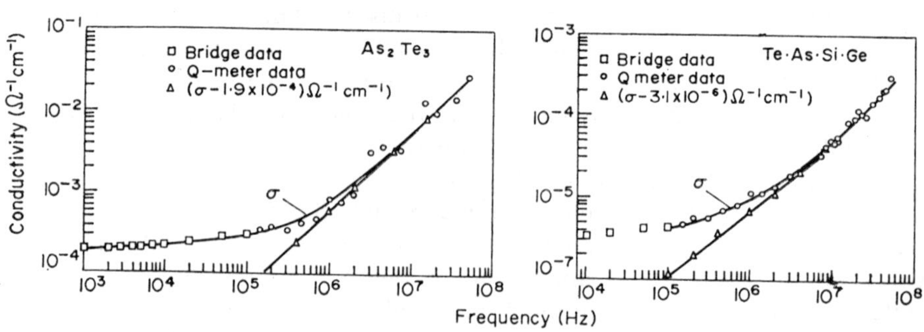

Fig. 13.2. Frequency dependence of electrical conductivity in amorphous As_2Te_3 and $Te_{48}As_{30}Si_{12}Ge_{10}$. (From Ref. 1.)

expected in amorphous semiconductors by Austin and Mott[5] and subsequent treatments[6-8] have followed. The formula given by Davis and Mott[2] is:

$$\sigma(\omega) = \tfrac{1}{3}\,\pi e^2\,kT\,\{g(\mathcal{E}_F)\}^2\,\alpha^{-5}\,\omega\,\{\ln(\nu_{\rm ph}/\omega)\}^4$$

where $g(\mathcal{E}_F)$ is the density of states ($\mathrm{cm^{-3}\,eV^{-1}}$) at the Fermi level, α^{-1} is the radial extent of the localized wave function and $\nu_{\rm ph}$ is the appropriate phonon frequency. The factor $\omega\{\ln(\nu_{\rm ph}/\omega)\}^4$ behaves like $\omega^{0\cdot8}$ from very low values of ω up to $\sim\nu_{\rm ph}/10^4$ and this is approximately the frequency dependence found. Applying the above formula (with $\alpha^{-1} = 8$ Å) to the results shown in Fig. 13.2 leads to estimates of $g(\mathcal{E}_F)$ between 10^{19} and 10^{21} $\mathrm{cm^{-3}\,eV^{-1}}$. In amorphous As_2S_3[9] a similar interpretation yields approximately $10^{18}\,\mathrm{cm^{-3}\,eV^{-1}}$ and in amorphous Ge, $\sim 5\times10^{21}\,\mathrm{cm^{-3}\,eV^{-1}}$. These estimates are rather higher than values obtained by other methods (see Chapters 2 and 15) and corrections to the theory may be necessary. Effects which were neglected in the derivation of the above formula include multiple hopping, correlation and polaron effects, all of which may be important.

An alternative model which could account for the $\omega^{0\cdot8}$ behaviour involves tunnelling of *atoms*. The central hypothesis of the model (which was originally devised to explain the linear variation with temperature of the specific heat observed in many glasses at low temperatures[10]) is that there should be a certain number of atoms, or groups of atoms, which can sit more or less equally well in two equilibrium positions. The barrier separating the two minima in energy should be sufficiently great to prohibit resonant tunnelling, but small enough so that thermal equilibrium can occur during the time span of the experiment (say $10^{-10}\,s < t < 10^3\,s$). It is also required that the pairs of minima under question are accidentally degenerate to within a few kT. The probability distribution of the energy difference between *all* pairs will of course cover a much wider range; what is necessary is that this distribution be *smooth* on the scale of kT. Pollak and Pike[11] have recently discussed the possibility that this tunnelling of atoms can account for the ac loss observed in many glasses, assuming of course that the transition involves a net motion of charge. They also stress the similarity of the analysis with the electron tunnelling model, particularly if lattice relaxation effects are important in the electronic mechanism so that transitions between two states are activated in *both* directions. It appears easier to explain on a hopping *over* a barrier model the observation in several materials that the power of ω increases from $\sim 0\cdot8$ towards unity as the temperature is lowered.

The suggestion that the ac conductivity behaviour in certain glasses may be an atomic rather than an electronic process was, I believe, first made by Austin and Garbett.[12] These authors found, in the far infra-red vibrational spectrum of amorphous As_2Se_3, a long wavelength tail extending to about 1000 μm which was not present in the corresponding spectrum of the crystal (see Fig. 13.5). By plotting their results along with ac conductivity data of Ivkin and Kolomiets[(13)] it can be seen in Fig. 13.3 that the magnitudes of the absorption processes allow for the possibility of the a.c. results being associated with the continuation of a very low energy tail to the phonon absorption processes. Of course, there is a large (and rather inaccessible) frequency range between the two sets of measurements and further work is required to see if the a.c. process saturates in the region of, about 10^{11} s^{-1}, as it should if the mechanism is electronic.

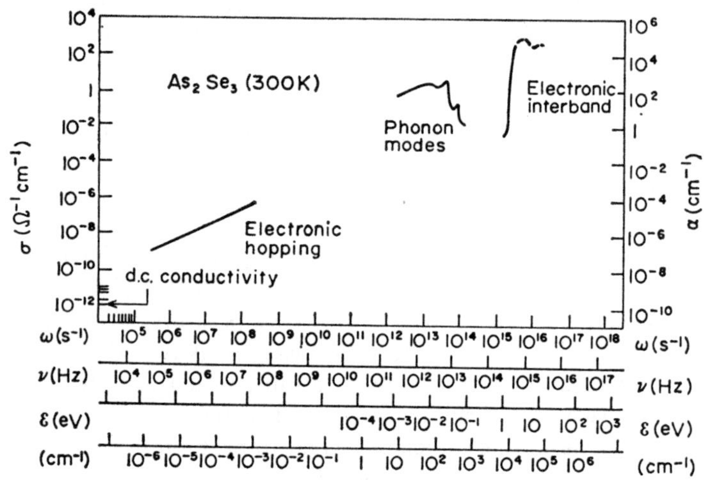

FIG. 13.3. AC conductivity and optical absorption over a broad frequency range in amorphous As_2Se_3. (From Ref. 2.)

Although it would appear that ac loss processes observed in the more insulating glasses could well have their origin in atomic processes, a rather surprising correlation between the ac and dc conductivity exists[(14)] in several chalcogenide glasses which suggests that, in the more conducting materials at least, an electronic mechanism is to be preferred. This correlation is shown in Fig. 13.4 where the results of several authors have been used.[(15)] In this figure the right hand ordinate is the approximate observed activation energy for dc electrical conduction and the lower abscissa gives the estimated density of states at the Fermi level using the formula given above. At the present time it is not clear why the materials with smaller energy gaps (given approximately by $2\mathcal{E}$) should have a larger density of states at the Fermi level. One may perhaps suggest that if these states arise from unsatisfied bonds then a larger number of these may be expected in materials which have smaller gaps and hence probably contain weaker bonds.

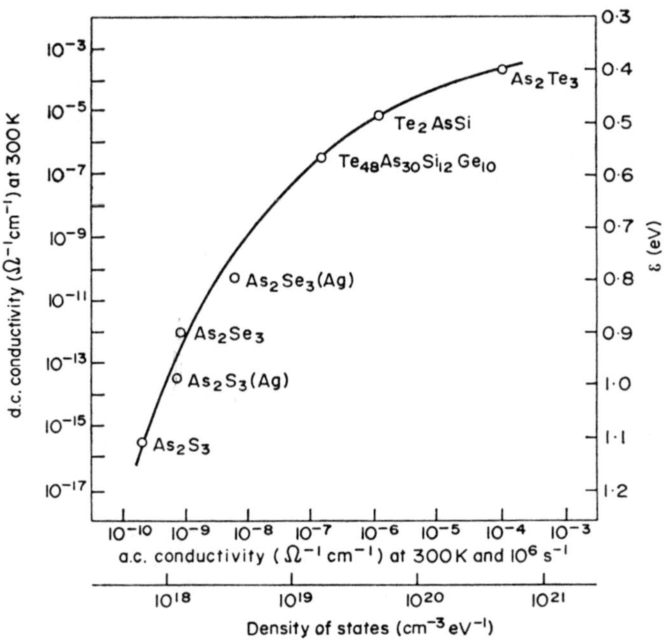

FIG. 13.4. Plot of dc against ac conductivity at $\omega = 10^6\ s^{-1}$ in various chalcogenide glasses at room temperatures. Also shown are $g(\mathcal{E}_F)$, the estimated density of states at $\mathcal{E}_F$, and approximate values of $\mathcal{E}$, the activation energy for d.c. conduction. Note that $\mathcal{E}$ is approximately half the mobility or optical gap in these materials. (From Ref. 14.)

13.4. ABSORPTION BY VIBRATIONAL MODES

A large spectrum of behaviour for the effects of disorder on absorption or scattering (Raman activity) due to atomic vibrations is encountered. For those materials which can be classed as "molecular" there is little difference between the gross features of the absorption in the amorphous and crystalline phases of the same material. This is attributed to the preservation in the amorphous state of molecular units, the vibrational modes of which can account for dominant features in the spectrum for the crystal. Good examples of this class of material appear to be As_2S_3 and As_2Se_3. At the other extreme are materials like Ge, Si and $CdGeAs_2$: in the crystalline state the spectra of these substances are dominated by long range atomic vibrations which are lost on disordering. This is not to say that the phonon density of states is *necessarily* different, in fact for Ge and Si a close similarity between the phonon density of states in the crystalline and amorphous states has been deduced.[27] It is rather a breakdown of selection rules for optical transitions, similar to that encountered for electronic interband transitions (see following

Sections), which occurs. This may also be the case in As_2Se_3, for example, but localized molecular-type vibrations weakly coupled to the network presumably give rise to strong peaks in the phonon density of states and hence in the i.r. spectrum. Amorphous Se and Te appear to show intermediate behaviour between the two extremes, although in the case of Se there are complications due to the possible occurrence in the amorphous state of molecular units not present in a particular crystalline form.

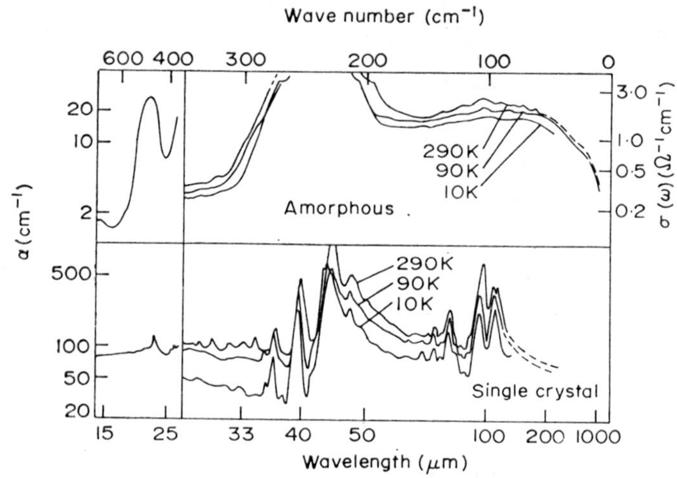

FIG. 13.5. Vibrational spectra of glassy and crystalline As_2Se_3. (From Ref. 12.)

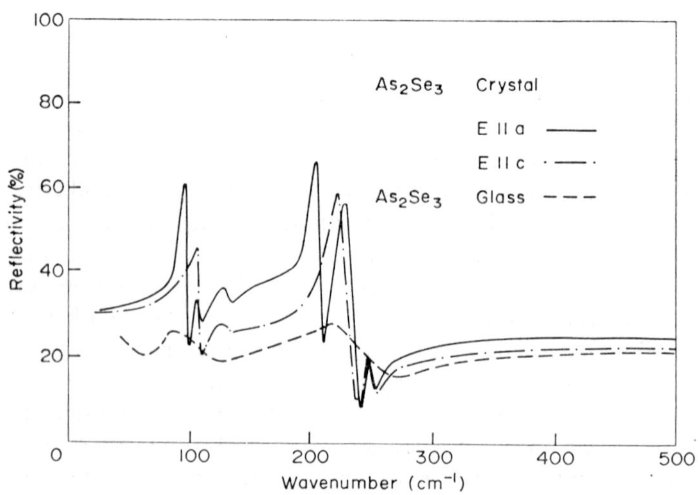

FIG. 13.6. Reflectivity of crystalline and vitreous As_2Se_3 from 40 to 500 cm^{-1}. (From Ref. 16.)

The infra-red transmission spectrum of amorphous and crystalline As_2Se_3, as determined by Austin and Garbett,[12] is shown in Fig. 13.5. Although much fine structure is lost, gross features near 100 cm^{-1} and 220 cm^{-1} are retained. Mitchell, Bishop and Taylor[16] among others,[17–19] have measured the reflectivity spectra. Their results, shown in Fig. 13.6, confirm the retention of these dominant reststrahlen bands. Furthermore, these peaks persist in the liquid state up to at least 673K, indicating that molecular units remain intact and that intermolecular coupling, which presumably breaks down above the glass transition temperature, is characterised by a weak force constant and does not contribute significantly, at least in this region of the i.r. spectrum.

Onomichi *et al.*,[19] who also measured the transmission and reflectivity spectra of amorphous As_2Se_3, fit their reststrahlen bands to a Gaussian line shape and suggest that this, rather than the normal Lorentzian fit, indicates a Gaussian distribution of Se–Se bond lengths and hence force constants. More recently Lucovsky[20] has shown that the 220 cm^{-1} band has a weaker line superimposed on its high energy side at approximately 250 cm^{-1} and that a Lorentzian line shape for each is quite satisfactory. Lucovsky has identified all the phonon modes as local modes of a molecular like unit consisting of $AsSe_3$ tetrahedra cross linked to each other via a single Se atom. Furthermore, Lucovsky has deduced that the bonding is mainly covalent, the infra-red activity arising from a dynamic rather than a static charge, that is to say an electric charge flows out of the bond when under tension and into the bond under compression. Similar spectra have been observed for amorphous As_2S_3.[20]

The i.r. transmission spectrum of amorphous Se is compared with that of α-monoclinic Se in Fig. 13.7.[21] The correspondence of several features in these spectra has been taken as evidence for the existence of Se_8 ring molecules in amorphous Se. Other structure is correlated with absorption bands in the trigonal form of Se and hence with vibrational modes of chain-like species. Although studies of Raman spectra[22] have been considered to support these ideas, recent Raman work in Te,[23] for which no crystalline form containing rings exists, suggests that Raman lines do shift on disordering and casts some doubt on direct comparison of spectra in this manner. A more detailed discussion of this is given by Dr. Mort in Chapter 14 and by Lucovsky in Ref. 89.

As for As_2Se_3, the dominant absorption bands observed in amorphous Se persist in the liquid state[24] indicating again the preservation of molecular like units. For the glass $Tl_2SeAs_2Te_3$, however, the dominant peaks disappear[25] abruptly at a temperature of 480K, approximately 120K above the glass transition temperature: glasses containing Tl and Te apparently form less tightly bound molecular units. The infra-red absorption spectrum

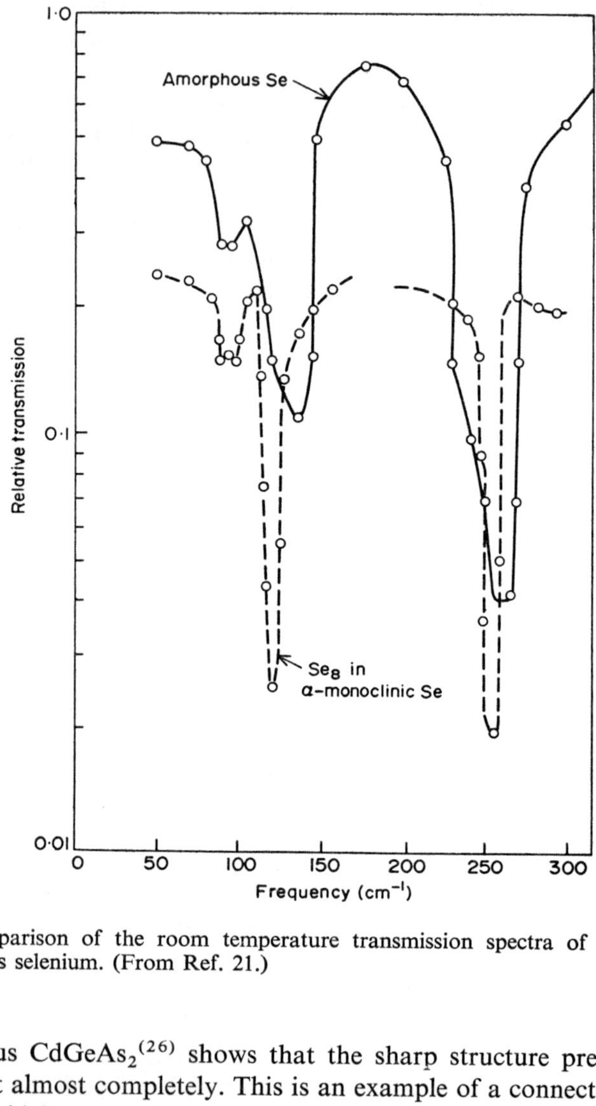

FIG. 13.7. Comparison of the room temperature transmission spectra of α-monoclinic and amorphous selenium. (From Ref. 21.)

of amorphous $CdGeAs_2$[26] shows that the sharp structure present in the crystal is lost almost completely. This is an example of a connected network material in which no molecular-like units exist. The same is true for Ge, Si and III–V compounds. Figure 13.8 shows the Raman spectra[27] of amorphous Si; the first order Raman spectrum of crystalline Si exhibits one sharp line at the zone centre TO–LO phonon energy (vertical dashed line). In Fig. 13.8(b) the phonon density of states for crystalline Si (shown dotted) is convoluted with a Gaussian of half-width 30 cm^{-1} to approximate the situation in amorphous Si in which there are minor variations in bond lengths and angles. The similarity between the solid curves in Fig. 13.8(a) and (b) suggests that for the amorphous material all vibrational modes are equally likely to take part in Raman scattering. However, the shift in peak positions can be taken to indicate some difference between the phonon density of states in the crystalline and amorphous forms.

The tetrahedrally coordinated glass SiO_2 may be expected to behave in a similar way to Ge and Si. However, the sharp absorption lines observed for this material[28] suggest dominance of vibrational modes associated with SiO_4 tetrahedra. The fundamental difference between the structure of amorphous Ge and fused silica lies in the bridging oxygen atoms which link tetrahedra in the latter. In this connection SiO_2 is similar to As_2Se_3.[20]

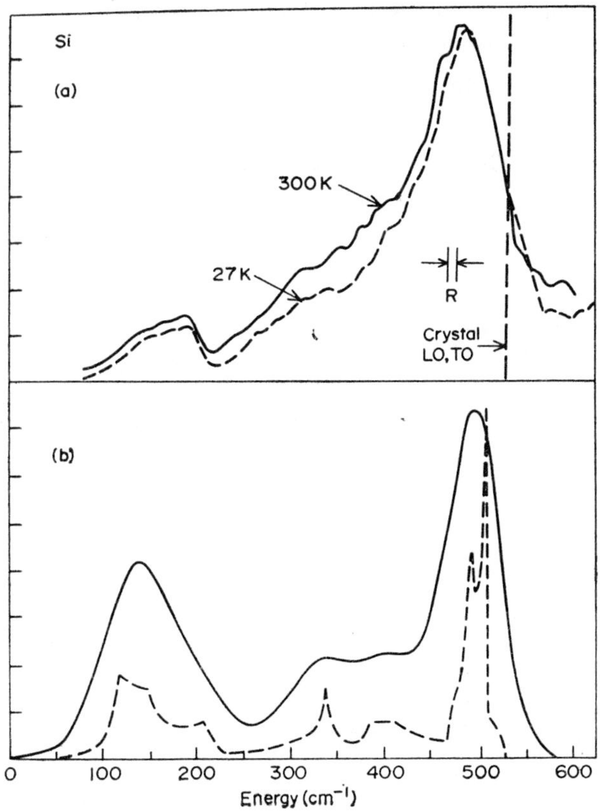

FIG. 13.8. (a) Reduced Raman spectra of amorphous Si at two temperatures. The first order Raman spectrum of crystalline Si has only a single sharp line at the zone centre TO–LO phonon energy, shown by the vertical dashed line. (b) Dashed line, one phonon density of states of crystalline Si; solid line, density of vibrational states for amorphous Si; calculated by convoluting the dashed curve with a Gaussian of half-width 30 cm^{-1}. (From Ref. 27.)

13.5. Free Carrier Absorption

Figure 13.9 shows the optical absorption near the fundamental edge in amorphous As_2Se_3.[29] In the solid glassy state the absorption coefficient obeys the spectral Urbach rule (see Section 13.7) and there is a parallel shift to lower photon energies as the temperature is raised. Curves A and B refer to two different specimens. In the liquid state (curves 4–11) strong absorption occurs at much lower energies and it may be considered that the edge is broadening and shifting, due either to an increase in random electric field intensity (see Section 13.7) or to the creation of additional electron states in the forbidden gap i.e. the filling in of a pseudogap in the density of states. However, the magnitude of α at a fixed photon energy (0·5 eV ~ 4000 cm^{-1}) is found to be proportional to the dc electrical conductivity, at least for

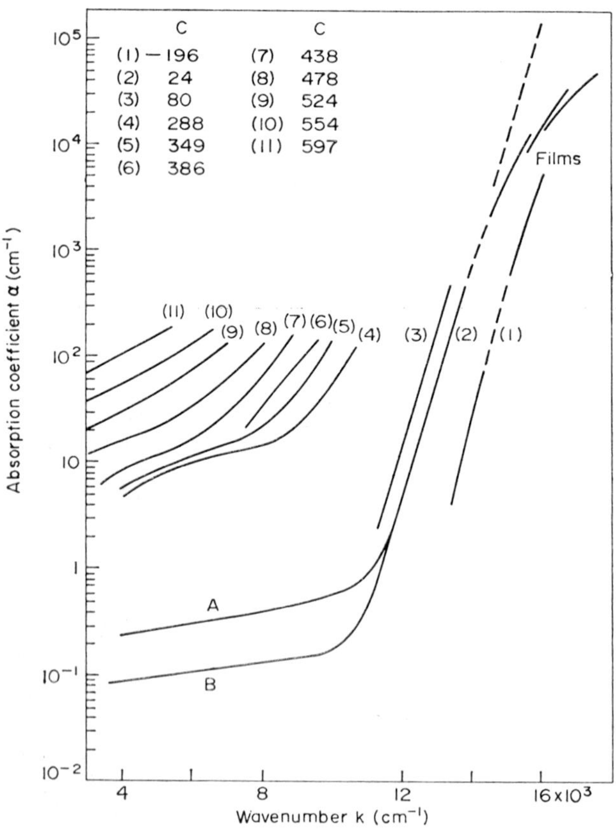

FIG. 13.9. Optical absorption in liquid and glassy As_2Se_3 as a function of temperature. (From Ref. 29.)

temperatures above about 450°C. This is shown in Fig. 13.10. There is thus a strong possibility that free-carrier absorption is being observed.

As mentioned in Section 13.2, free carrier absorption in amorphous semiconductors is not, on account of the short relaxation time, expected to follow the normal λ^2 behaviour, and one might expect to find $\sigma(\omega)$ in this range to be independent of energy and equal to $\sigma(0)$. Mott and Davis[2] have suggested that the increase in $\sigma(\omega)$ with photon energy (shown in Fig. 13.9) arises from an increasing density of available final states as carriers are excited deeper into their respective band (it is not known whether the intraband absorption is due to holes or electrons). A simple analysis assuming a band edge in which the density of states varies linearly with energy predicted a slope proportional to T^{-1}. However, this does not agree with the data of Fig. 13.9 and further work on this problem is necessary.

Edmond[30] has also observed a similar behaviour in solid glasses of the system $As_2(SeTe)_3$. In all cases the high frequency conductivity was found to be about twenty times the dc values, a result which may suggest the importance of polaronic effects.

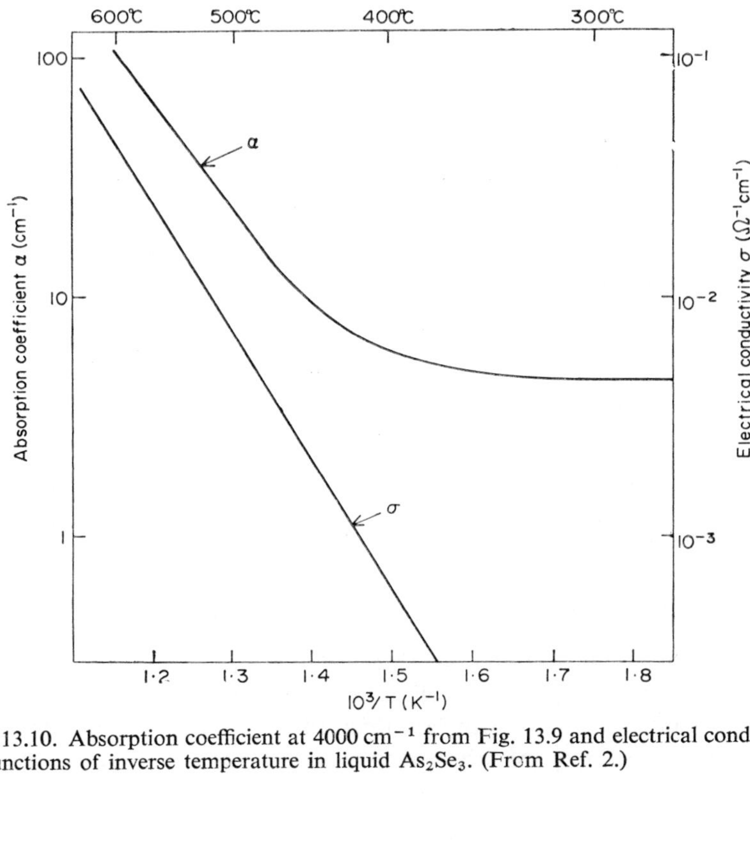

FIG. 13.10. Absorption coefficient at 4000 cm^{-1} from Fig. 13.9 and electrical conductivity as functions of inverse temperature in liquid As_2Se_3. (From Ref. 2.)

Free carrier absorption has also been observed[31] in $Tl_2SeAs_2Te_3$. The temperature dependence of the absorption coefficient and the dc electrical conductivity for this material are shown in Fig. 13.11. Identical activation energies of 0·35 eV are found. However, in this glass no dependence of α on wavelength was discovered. In view of the smaller photon energies involved compared to As_2Se_3 and the limited wavelength range studied, a large variation is not to be expected. Again $\sigma(\omega)/\sigma(0)$ was found to be greater than unity, in this case the ratio is about 8.

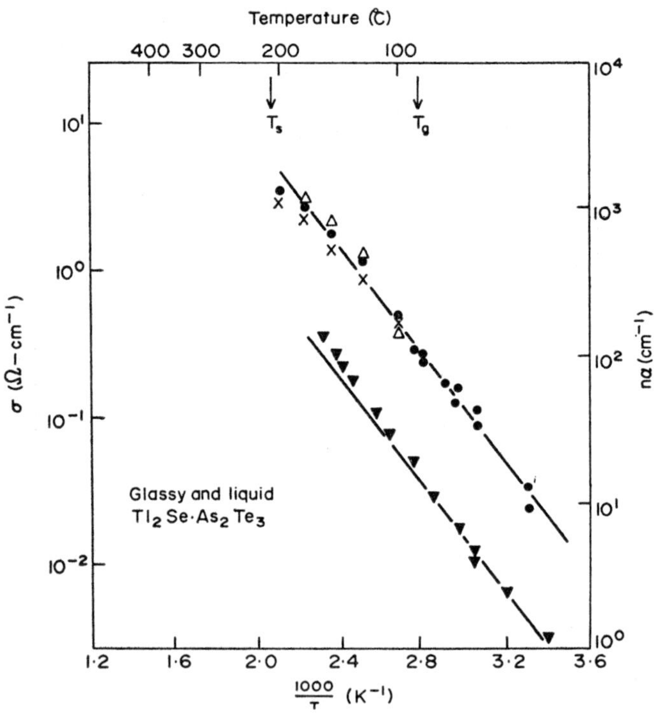

FIG. 13.11. Conductivity at dc and optical (infrared) frequencies versus reciprocal temperature in glassy and liquid $Tl_2SeAs_2Te_3$. The open triangles, circles and crosses represent data taken at 5, 10 and 15 μm (2000, 1000 and 666 cm^{-1}) respectively. The solid triangles represent dc measurements. (From Ref. 16.)

13.6. Absorption Involving States in the Gap

The sharp optical absorption edge observed in carefully prepared evaporated films of amorphous Ge (see Section 13.7) has been used[32] to suggest a density of states in the gap of less than 10^{17} cm^{-3} eV^{-1}. This figure was arrived at by assuming that the matrix elements for transitions from or to these states are the same as that for band states. The figure is contradictory to that deduced from electrical conductivity measurements by Knotek[33] on similar samples: assuming that conduction occurs by variable range hopping of carriers at the Fermi level, as suggested by the plots of ln σ versus $T^{-1/4}$, one obtains a density of localized states $\sim 10^{19}\ \mathrm{cm^{-3}\ eV^{-1}}$. A similar, or higher value, is also implied by other experiments (see Chapters 2 and 15), although none of these can be interpreted without ambiguity and some were

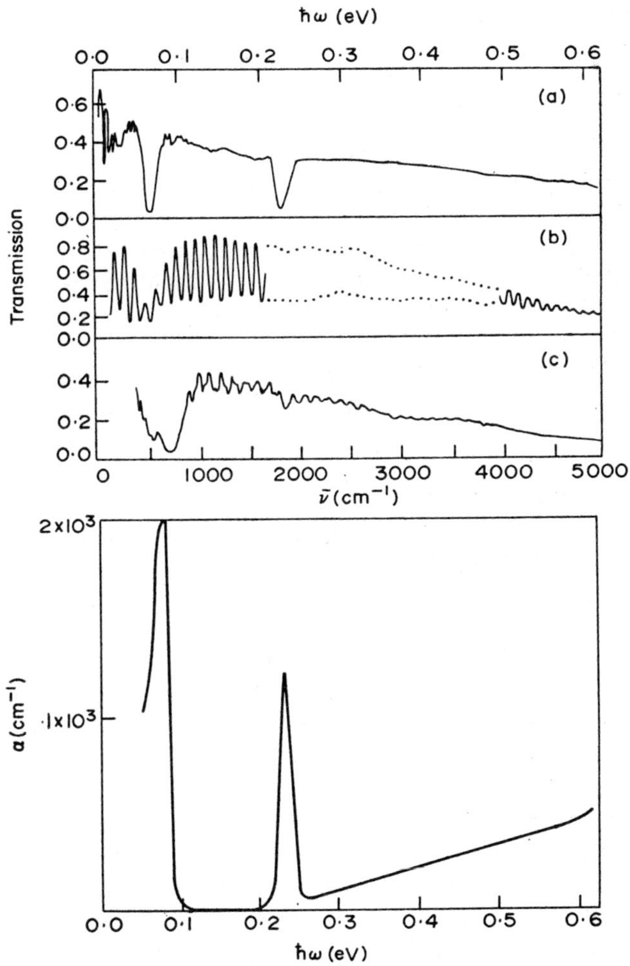

FIG. 13.12. Upper: Infrared transmission spectra of amorphous germanium prepared by three different methods (a) Electrolytic amorphous Ge, thickness = 27 μm; (b) evaporated amorphous Ge, thickness = 12 μm; (c) sputtered amorphous Ge, thickness = 35 μm. Lower: Infrared absorption spectrum of amorphous germanium prepared by glow-discharge deposition. Film thickness = 4·1 μm. (From Refs. 36 and 37.)

made on samples which were not prepared under the conditions necessary to obtain the sharp absorption edge.

A simple, although perhaps not altogether satisfactory, explanation of the anomaly is that a density of states of say 10^{19} cm^{-3} eV^{-1} does indeed occur in these samples of amorphous Ge and that the states are concentrated into a narrow band of width ~0·01–0·1eV around the Fermi level $\mathcal{E}_F$. (Observation of the $T^{-1/4}$ behaviour requires localized states in a range of energy $\sim 3kT$ around $\mathcal{E}_F$). An integrated density of 10^{17}–10^{18} cm^{-3} may thus exist near $\mathcal{E}_F$ and be optically undetectable, particularly if there is a suppression of the matrix elements connecting such states to the bands. A narrow band of states was originally proposed by Davis and Mott(14) as a general feature of amorphous semiconductors. The states are assumed to be an "extrinsic" feature of the films in that they are associated with defects, in particular unsatisfied or dangling bonds. Thus their density will vary from sample to sample and in any one sample as a function of annealing. The model is not unique and other explanations may suggest themselves. Fritzsche(34) has proposed that electrostatic potential fluctuations could be invoked to reconcile a high density of states at the Fermi level with good transparency on the low energy side of the absorption edge. It is difficult to design experiments which would test this idea decisively. Anderson(35) has suggested that relaxation effects may be important; the model here being that to make a transition into a localized gap state requires an additional energy associated with the lattice distortion which accompanies occupancy i.e. a Franck–Condon shift.

Optical absorption which is possibly associated with gap states in amorphous Ge has been observed in samples prepared by glow discharge decomposition of germane(36) as well as by electrolytic deposition and sputtering(37) (Fig. 13.12). The recurrence of the absorption peak at 0·23 eV observed in different samples by different workers is striking (the peak at 0·07 eV is probably a vibrational mode) and it appears likely that this is associated with a native defect present in high densities in some specimens at least. However, the appearance of an absorption *peak* cannot be taken as evidence for a *single* set of levels in the gap which should give rise to a threshold followed by a continuum.

Brodsky(38) has recently provided evidence for "extrinsic" absorption processes near the edge of evaporated amorphous Si. Figure 13.13 shows the correlation found between the shape of the absorption edge and the concentration of spins observed by e.s.r. The spins are believed to arise from unsatisfied bonds residing on internal surfaces of voids, and their concentration can be varied by choosing different rates of evaporation and substrate temperatures. For samples with a spin concentration $\sim 10^{20}$ cm^{-3} the edge is found to be relatively broad; for lower spin concentrations the edge is sharper and displays features near 0·6 eV and 1·2 eV. The solid curve is not an observed edge, but one obtained by extrapolating data from eleven samples to that expected for zero spin concentration. The implication here is that a defect-free film of amorphous Si would have a sharp edge near 1·8 eV.

Optical absorption edge data on chalcogenide glasses are complicated by the presence of exponential behaviour which probably arises from exciton effects. Although such edges *could* be interpreted in terms of exponential band tailing, the evidence against this is quite strong[(39)] (see references in Section 13.7). However, the existence of weak band tails giving rise to optical absorption lying beneath the exponential edge cannot be ruled out. Measurements by Perron[(40)] on liquid Se–Te alloys, for which there is independent evidence for the formation of a pseudogap at high temperatures, show a monotonic variation of the slope of the optical edge as a function of either Te content or temperature. Perron interprets these results in terms of band tailing.

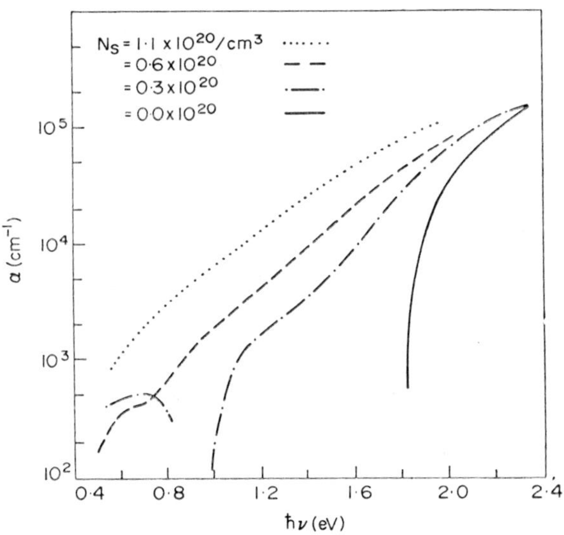

FIG. 13.13. Absorption coefficient versus photon energy for three representative films and the extrapolated spectrum of spin-free amorphous Si. (From Ref. 38.)

Tauc has reported[(38)] that at absorption coefficients below about 1cm^{-1}, several chalcogenide glasses exhibit a "foot" to the normal exponential edge, [Fig. 13.14]. The temperature dependence of this residual absorption does not, at least in amorphous As_2S_3, behave like that expected for free carrier absorption, and in principle could arise from weak band tailing. The total number of states involved here is only $\sim 10^{16}$–10^{17} cm^{-3}, under the assumption again of normal matrix elements.

There are many other experimental results concerned with the question of band tailing in amorphous semiconductors. However, it seems fair to conclude that at the present time there is little decisive evidence from optical experiments for the large density of states in the gap deduced from other measurements.

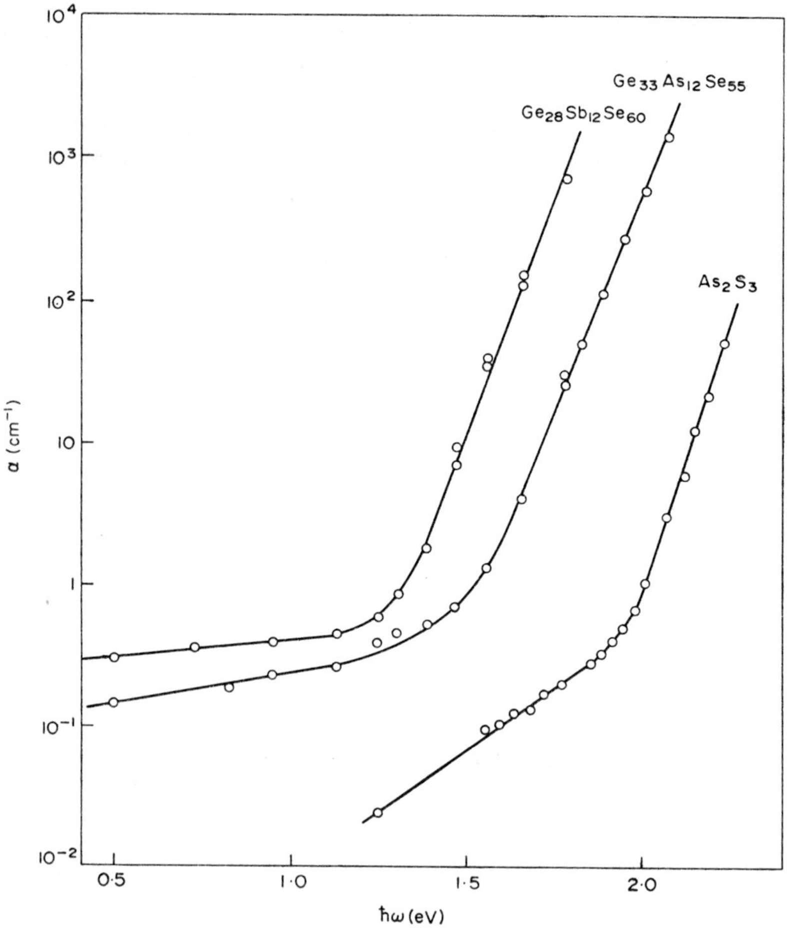

FIG. 13.14. Optical absorption versus photon energy for three glasses at room temperature. (From Ref. 39.)

13.7. The Absorption Edge

As mentioned in the last Section, the optical absorption edges in most chalcogenide glasses are characterised by an absorption coefficient which increases exponentially with photon energy and this feature will be the main subject of this Section. Amorphous Ge, Si and III–V compounds exhibit absorption edges the form of which is very sample (and apparently laboratory!) dependent, ranging from very sharp to very broad. A selection of edges observed for amorphous Ge is displayed in Fig. 13.15 and recent measurements not shown here appear in Refs 41, 42 and 43. Correlations between the shape and position of the edge in amorphous Ge and such variables as substrate temperature, annealing, oxygen content, etc. have been investigated, but the overall picture is still far from clear. The edges in Ge and similar materials will be referred to again only in connection with the *absence* of exponential edges in certain carefully prepared samples.

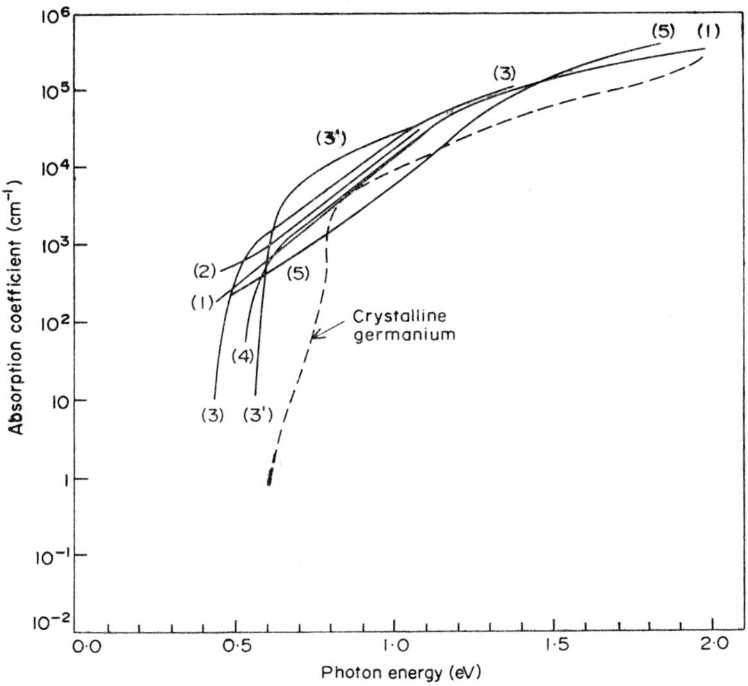

FIG. 13.15. Room temperature optical absorption edges in amorphous germanium as reported by various authors. (1) Chittick, Ref. 36; (2) Clark, Ref. 84; (3) and (3′) Spicer and Donovan, Ref. 60; (4) Chopra and Bahl, Ref. 3; (5) Tauc, Ref. 85. The dashed curve for the crystal is from Dash and Newman, Ref. 86. (From Ref. 2.)

Although the excellent transparency of many glasses has been known for centuries, the implication of this, namely that there exist in these materials electron energy gaps containing very few states, poses the question that Ziman raised at the Cambridge Conference a few years ago.[44] How, in a material which has no long range order, can the electrons be scattered in a sufficiently coherent manner to give rise to energy gaps in the electron density of states spectrum? Ziman was, of course, thinking in terms of energy gaps as they arise in the band structure of crystalline materials, namely by Bragg reflection at periodic lattice planes.

Since this question was raised, we have had the theorems of Weaire.[(45)] and Heine[(46)] which showed (Chapter 3 and Section 13.9) that, *using a tight-binding formulation*, energy gaps in fully connected tetrahedrally connected networks are to be expected. Ziman's own group has shown[(47)] (Chapter 3), *using a multiple scattering formalism*, that short range order is sufficient to give, if not true gaps, at least pseudogaps, again in systems containing tetrahedrally connected units, but this time arranged in clusters. These are theoretical methods which can be used in systems without translational periodicity and are already enjoying a great deal of success in describing the electronic density of states in amorphous semiconductors (Section 13.9).

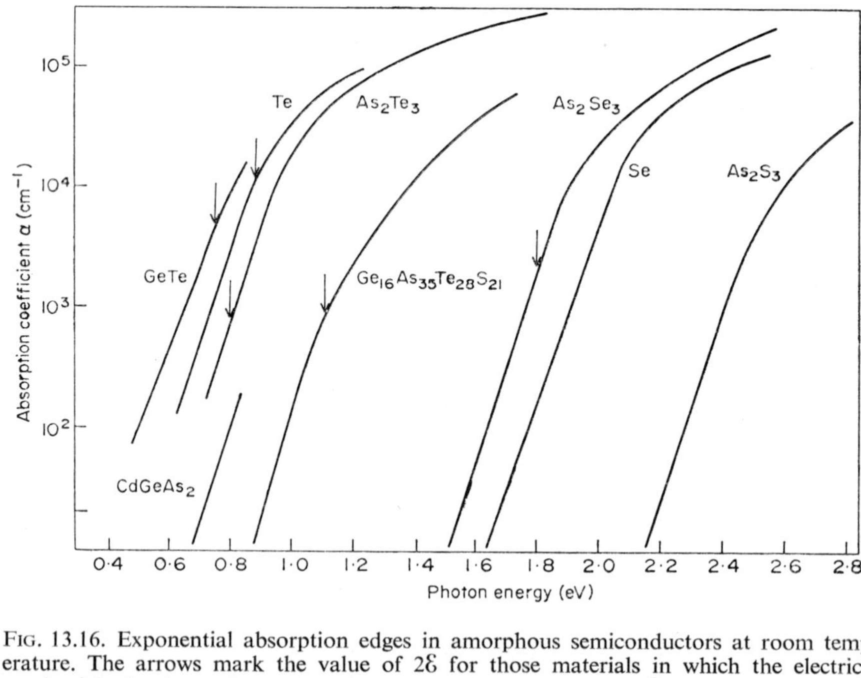

FIG. 13.16. Exponential absorption edges in amorphous semiconductors at room temperature. The arrows mark the value of $2\mathcal{E}$ for those materials in which the electrical conductivity has been observed to obey the relation $\sigma = C \exp(-\mathcal{E}/kT)$. (From Ref. 2.)

The most generally observed form of the optical absorption edge in amorphous chalcogenide semiconductors is of the type shown in Fig. 13.16. For all these materials the edge is exponential over several decades in optical absorption coefficient α, i.e.

$$\alpha = \text{const} \exp(\Gamma\omega).$$

This is the so-called "Spectral Urbach rule". It is not unknown in crystalline materials and indeed Urbach first proposed[48] his empirical rule to describe the edge in crystals of alkali halides. The full Urbach rule includes a temperature dependence of the slope

$$\alpha = \text{const} \exp\left(\frac{\gamma' (\hbar\omega - \mathcal{E}_0)}{kT}\right)$$

where T is the absolute temperature down to a critical value T_0 and equal to T_0 for lower temperatures. In this form $\mathcal{E}_0$ is the focal point of a family of isotherms each having a slope γ'/kT. More generalised forms of Urbach's law have appeared in the literature.[49,2] One notes that if we associate $\mathcal{E}_0$ with an energy gap and this varies linearly with temperature as $\mathcal{E}_0 = \mathcal{E}_0(0) - \beta T$, a focal point $\mathcal{E}_0(0)$ still results. Although the full Urbach rule is obeyed in many crystalline materials (e.g. alkali halides, CdS and trigonal Se) there is considerable uncertainty as to its physical origin. This situation exists in spite of numerous theoretical attempts to account for it (see references in 49).

The edges shown in Fig. 13.16 do not follow the temperature dependence given by Urbach's expression. The general rule is an almost parallel shift of the edge with increasing temperature, although in the liquid state of Se, As_2Se_3 and Se–Te alloys the slope does decrease, but with a weaker temperature dependence than T^{-1}. It is therefore not normally possible to locate an energy $\mathcal{E}_0$ from the exponential part of the edge in amorphous semiconductors.

The most satisfactory explanation to date of the Urbach rule is that proposed by Dow and Redfield.[50,51] These authors have solved the problem of optical absorption associated with excitons in the presence of a uniform electric field. Their results are shown in Fig. 13.17. The left-hand side of this figure shows a quantity proportional to the absorption coefficient plotted versus the photon energy (expressed in units of the exciton Rydberg) for various values of electric field characterised by the parameter f. For zero field the exciton absorption is represented by a delta function at $\mathcal{E} = -1$. As the field increases the exciton line undergoes a shift first to lower energies (Stark effect) followed by one to higher energies (ionization). Simultaneously the line is broadened and, as can be seen from the plots, the low energy side is accurately described by an exponential behaviour.

The parameter f is the electric field expressed as the potential energy difference across the exciton radius divided by the exciton Rydberg i.e.

$$f = E\,|e|\,a/R, \qquad a = \frac{\hbar^2 \kappa^2}{m^* e^2}, \qquad R = \frac{e^2}{2\kappa a} = \frac{m^* e^4}{2\kappa^2 \hbar^2}$$

where m^* is the reduced mass of the electron–hole pair and κ is the dielectric constant. Thus for a field E in V cm^{-1}

$$f = 3{\cdot}89 \times 10^{-10} (m/m^*)^2 \kappa^3 E$$

and for $m^* = m$, $\kappa = 6$ the curve marked $f = 0{\cdot}6$ corresponds to a real field $\sim 7 \times 10^6$ V cm^{-1}.

It is worth noting that the classical criterion for direct field ionization of the exciton at $E_c = \kappa R^2/4e^3$ corresponds to an f value $\sim 0{\cdot}125$. However, a discernible exciton peak remains for fields almost an order of magnitude

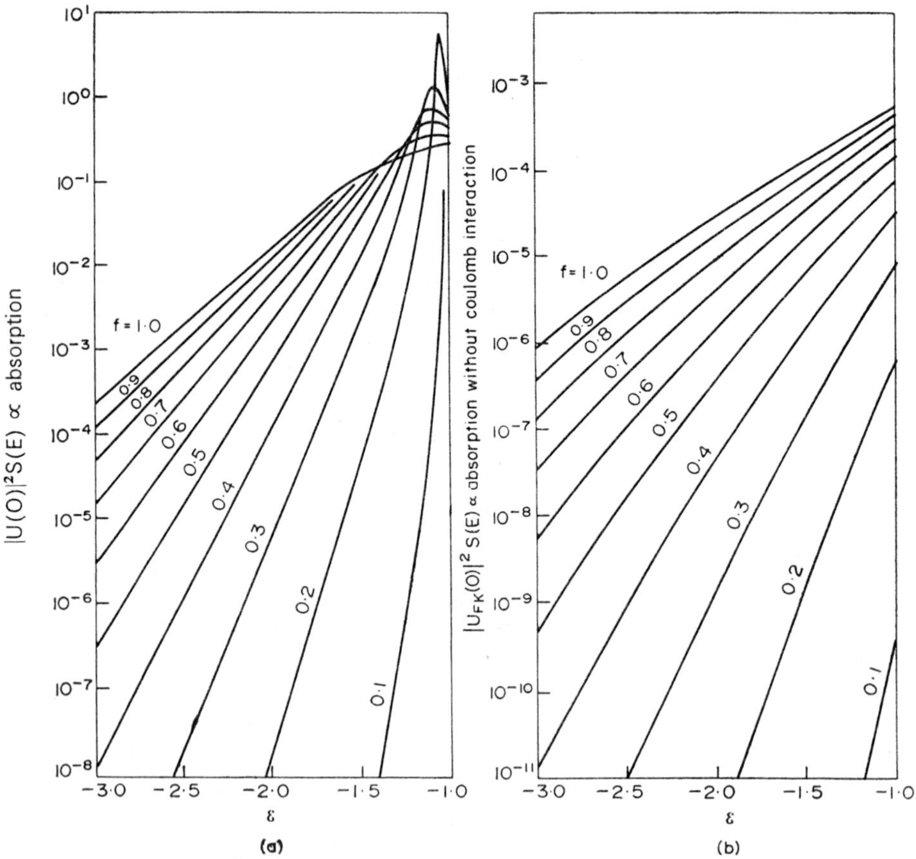

FIG. 13.17. Dow and Redfield's theory for optical absorption in the presence of an electric field: (a) exciton effects included; (b) exciton effects not included (Franz–Keldysh result). The parameter f is a measure of the strength of the electric field. The energy $\mathcal{E}$ is measured from the energy gap in units of the binding energy of the unperturbed exciton. (From Ref. 50.)

larger than this. The influence of the electron-hole interaction on the shape of the absorption curves can be seen by comparing the left hand side of Fig. 13.17 with the curves on the right for which the Coulomb interaction has been "turned off", giving the normal Franz–Keldysh result:

$$\alpha = \text{const} \exp \{C(\hbar\omega - \mathcal{E}_0)^{3/2}\}.$$

If Dow and Redfield's theory is accepted as an explanation of the spectral Urbach behaviour found in many amorphous semiconductors then it is

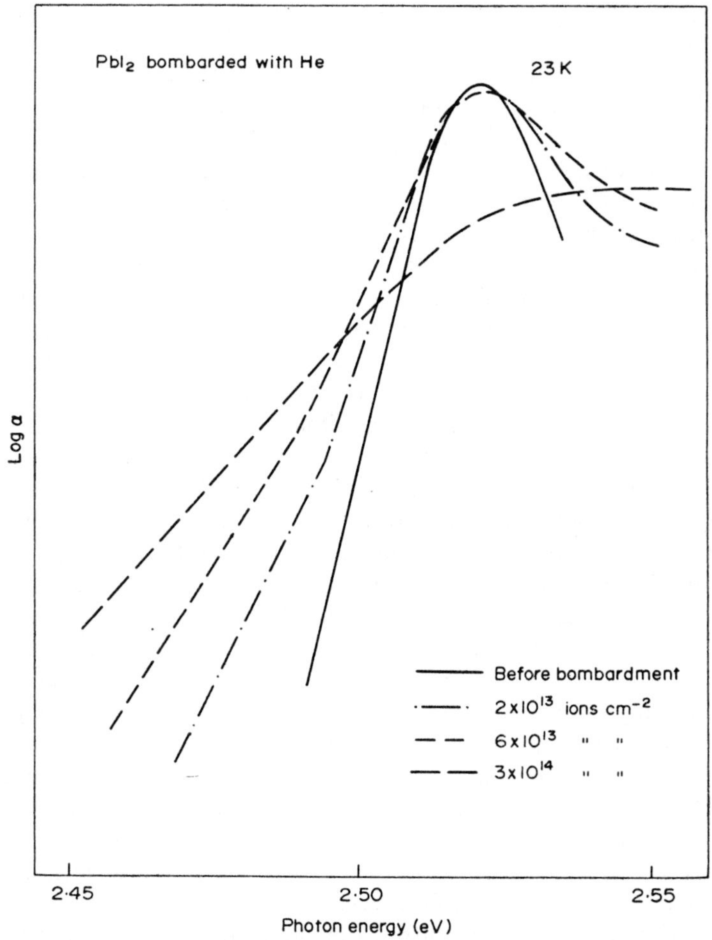

FIG. 13.18. Optical absorption edge of crystalline PbI_2 measured at various stages of bombardment with He ions. The excitonic feature broadens into an exponential tail. (From Ref. 52.)

necessary to consider the origin of the electric field. Presumably one is dealing with *random* fields and the first question to ask is if the exponential behaviour emerging out of the above theory survives the necessary averaging process. In a recent publication[(51)] the answer appears to be in the affirmative.

An experiment pertinent to the effect of disorder on exciton absorption has been made by Olley[(52)] [Fig. 13.18]. A single crystal of PbI_2, exhibiting a sharp exciton absorption line, was bombarded with helium ions and the absorption spectrum measured as a function of dose. With increasing bombardment the exciton is quenched and broadened and its leading edge becomes exponential as the material becomes more disordered. Similar behaviour has been observed in other materials. Presumably one is introducing random electric field associated with the creation of defects.

The slopes of the absorption edges shown in Fig. 13.16 lie between $\Gamma \sim 15 - 22\,\mathrm{eV}^{-1}$ and there do not appear to be many materials exhibiting values outside of this range. For $\Gamma = 17\,\mathrm{eV}^{-1}$, Dow and Redfield's theory yields, for materials with $\kappa = 6$ and $m^* = m$, an f value $\sim 0{\cdot}5$. Thus we are considering electric fields $\sim 10^6$–$10^7\,\mathrm{V\,cm}^{-1}$. These are high fields and it is still an open question whether they can arise from dangling bonds, voids, bent bonds or other imperfections in amorphous materials. Other problems which remain are (a) why the fields required are so similar in such a variety of materials, (b) what is the spatial scale of the potential fluctuations; are they of the kind suggested by Fritzsche[(34)] for example, (c) what is the effect of temperature on these fields, (d) why, if such fields exist, does one observe non-ohmic effects in many amorphous semiconductors (and also drastically alter the quantum efficiency in amorphous Se[(53)]) at relatively low fields ($\sim 10^3\,\mathrm{V\,cm}^{-1}$)? Clearly much more work is needed on these matters.

The existence of sharp absorption edges in carefully prepared samples of amorphous Ge demands special attention. There seems to be no reason why random electric fields should not occur in this material; the density of defect centres is probably larger than in the chalcogenides and small voids may be present even in samples deposited onto high temperature substrates. The exciton binding energy in crystalline Ge is ~ 2 meV, an exceedingly small value arising partly because of the small effective mass of the electron–hole pair ($m^* \sim 0{\cdot}033$ m) but partly because of a large dielectric constant (15.8). Although it is not clear what value to take for the mass in amorphous Ge, it might be considered that the absence of an exponential edge similar to that observed in the chalcogenides arises from the higher dielectric constant in this material making exciton effects negligible. However, there is some difficulty with this suggestion. From the curves of Fig. 13.7(a), the slope of the optical absorption edge can be found to be given roughly by the relation

$$\Gamma = \mathrm{const}\,\{(m^*)^{(2b-1)}/\kappa^{(3-2b)}\}\,E^{-b}$$

where b lies between about 0·78 and unity. Thus, for the same magnitude of electric field E, the slope decreases with decreasing effective mass and increasing dielectric constant. To obtain a sharp edge necessitates the assumption of random electric fields considerably lower in magnitude than estimated for the chalcogenides. Alternatively it could be that high fields do exist in the vicinity of defects, but their spatial extent is small (on account of the high dielectric constant); their effect on optical properties may thus be reduced in importance relative to the chalcogenides.

13.8. Interband Absorption

At photon energies somewhat higher than that at which the exponential behaviour is observed in chalcogenide glasses, and in other materials which do not exhibit an exponential edge, absorption presumably occurs by the excitation of electrons across the gap into band states and the possibility exists of extracting information about the form of the density of states at the band edges.

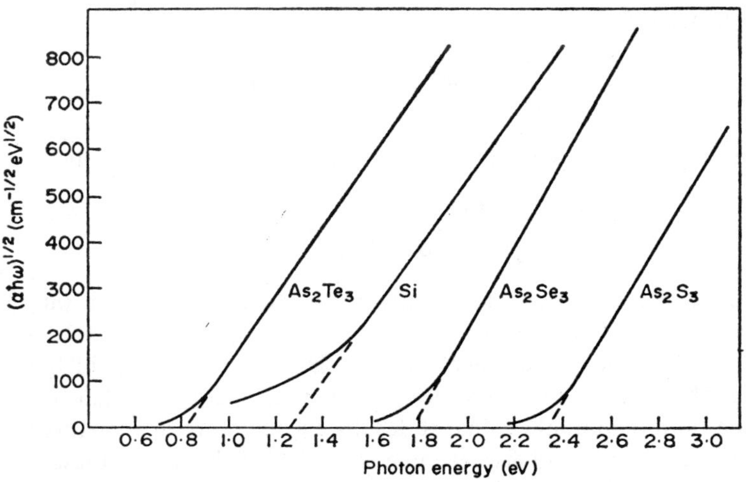

FIG. 13.19. Examples of absorption edges whose functional dependence on photon energy is given by $\alpha\hbar\omega = B(\hbar\omega - \mathcal{E}_0)^2$. Data for As_2Te_3 from Rockstad, Ref. 1; Si from Brodsky, Title, Weiser, and Pettit, Ref. 87; values of the slope and intercept for these materials (From Ref. 2.)

A common, although not universal, behaviour of the absorption coefficient in this energy range is

$$\alpha = \text{const}\ (\hbar\omega - \mathcal{E}_0)^2/\hbar\omega.$$

A few examples are shown in Fig. 13.19. Using this relationship one can define a gap $\mathcal{E}_0$, but whether this represents a true gap in the density of states, a mobility gap or something related to these energies, remains to be investigated.

Under the assumption that k is not a meaningful quantum number for amorphous semiconductors and that the k-conservation selection rule which operates in crystals is relaxed, the absorption coefficient is expected to be proportional to the convolution of state densities integrated over all energies

$$\alpha = C \int g_v(\mathcal{E})\, g_c\,(\mathcal{E} + \hbar\omega)\, |D|^2\, d\mathcal{E}/\hbar\omega.$$

If we assume that the matrix elements D associated with the transitions are independent of energy, then if the density of states at the bottom of the conduction band is represented by $g_c(\mathcal{E}) = \text{const}\ (\mathcal{E} - \mathcal{E}_A)^S$ and at the top of the valence band by $g_v(\mathcal{E}) = \text{const}\ (\mathcal{E}_B - \mathcal{E})^P$, we have

$$\alpha = C' \int_{\mathcal{E}_A - \hbar\omega}^{\mathcal{E}_B} (\mathcal{E}_B - \mathcal{E})^P\, (\mathcal{E} + \hbar\omega - \mathcal{E}_A)^S\, d\mathcal{E}/\hbar\omega.$$

Making the substitution $y = (\mathcal{E}_A - \hbar\omega - \mathcal{E})/(\mathcal{E}_A - \hbar\omega - \mathcal{E}_B)$ and denoting the gap by $\mathcal{E}_0 = \mathcal{E}_A - \mathcal{E}_B$:

$$\alpha = C'' \left[\int_0^1 (1 - y)^P\, y^S\, dy \right] (\hbar\omega - \mathcal{E}_0)^{P+S+1}/\hbar\omega.$$

The quantity in the square brackets is a known integral, namely

$$\frac{\Gamma(S+1)\,\Gamma(P+1)}{\Gamma(S+P+2)}.$$

The quadratic variation of $\alpha\hbar\omega$ against $\hbar\omega$ can therefore be interpreted in terms of a parabolic nature for the band edges i.e. $S = P = \frac{1}{2}$. Alternatively one can assume that there are localized states existing in band tails for which the density of states varies linearly with energy and that transitions for which both the initial or final states are localized are improbable because of the spatial separation between such states.(14) If the last of these assumptions does *not* hold, a variation of α as

$$\alpha = \text{const}\ (\hbar\omega - \mathcal{E}_0)^3/\hbar\omega$$

is obtained and this may be the explanation for the absorption edges observed in certain multicomponent glasses by Fagen(54) [Fig. 13.20].

The magnitude of the absorption coefficient expected on these models has been calculated by Davis and Mott(14) using a value for the matrix elements derived for electrical conductivity (see also Tauc and Menthe(39)). Fair agreement with experiment was obtained.

Although the procedures outlined above for extracting an $\mathcal{E}_0$ from the form of the optical absorption edge at high energies are somewhat arbitrary, they can be useful experimentally in that they provide *some* measure of the gap and can be used at the very least as a marker in studies of the effect of temperature and pressure on absorption edges.

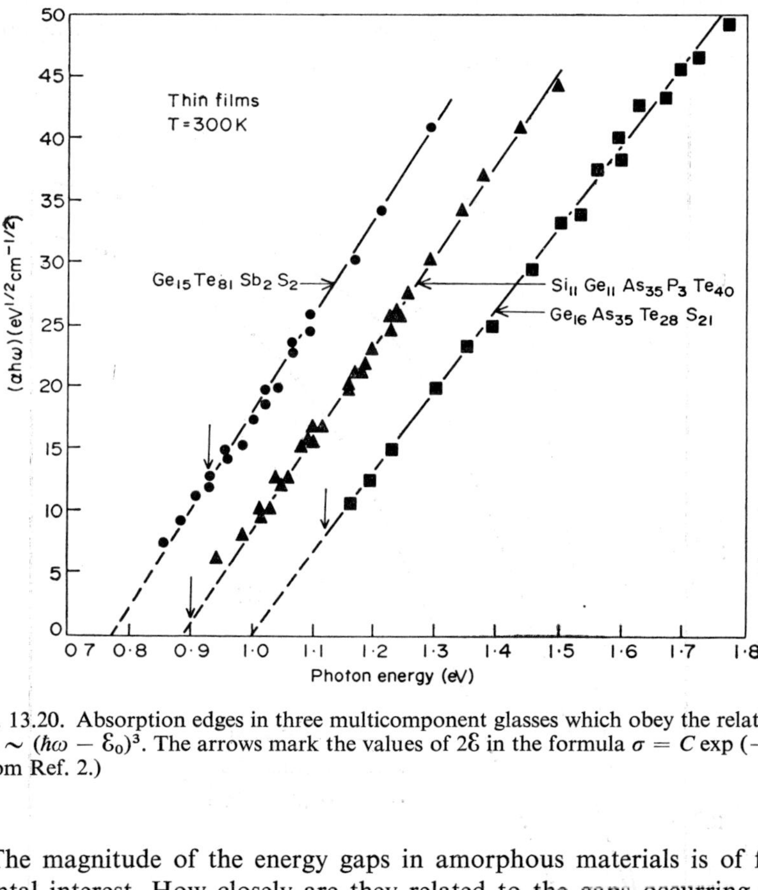

FIG. 13.20. Absorption edges in three multicomponent glasses which obey the relationship $\alpha\hbar\omega \sim (\hbar\omega - \mathcal{E}_0)^3$. The arrows mark the values of $2\mathcal{E}$ in the formula $\sigma = C \exp(-\mathcal{E}/kT)$. (From Ref. 2.)

The magnitude of the energy gaps in amorphous materials is of fundamental interest. How closely are they related to the gaps occurring in the corresponding crystalline state (when this exists)? A general rule appears to be that, if the local atomic order is not appreciably altered in the amorphous phase, then the gaps of the two states are not appreciably different. Although this may well be expected in "tight binding materials" for which one can associate the gap with the energy to break a local bond, it is a little surprising to find it working well in materials for which the gap in the crystalline state is known to occur as an "accident" of the band structure.

Connell and Paul[42] find little correspondence between the energy gaps or their pressure coefficients in crystalline and amorphous forms of Si, Ge and III–V compounds. For GaP in particular they find a large discrepancy (0·42 eV amorphous, 2·22 eV crystalline). They suggest that this may be explained by the occurrence of "wrong bonds" due to the presence of 5-fold rings in the amorphous structure. For example, in GaP one cannot avoid the occurrence of Ga—Ga or P—P bonds if 5-atom rings exist. Connell and Paul regard the average or Penn gap as holding more significance as far as optical properties are concerned.

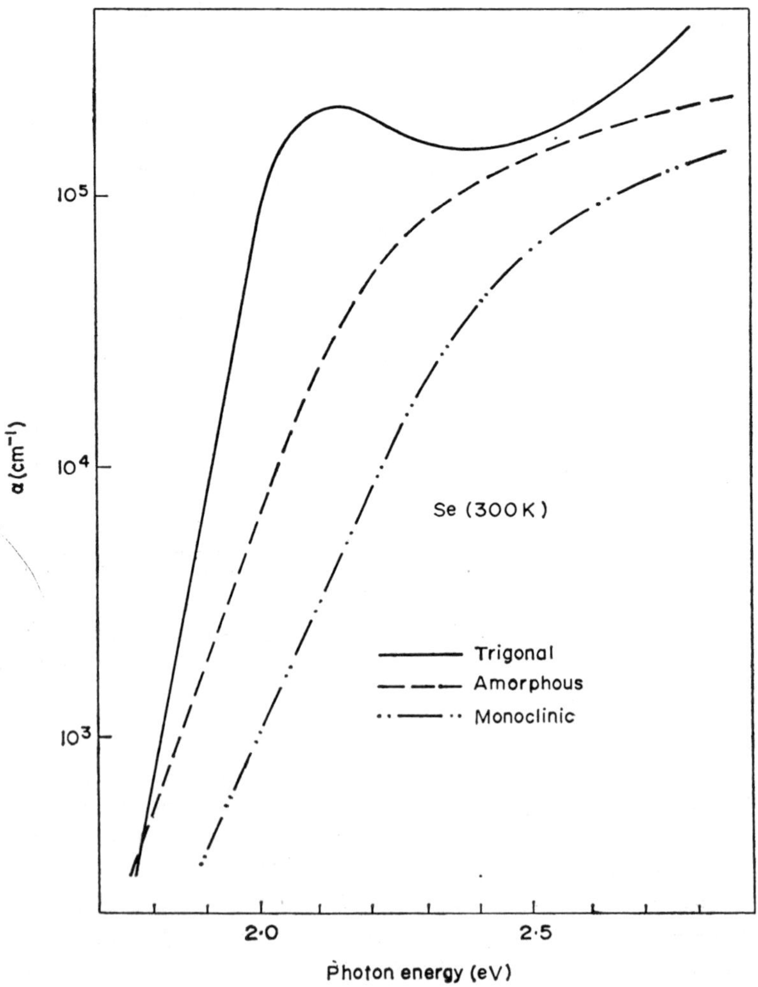

FIG. 13.21. Optical absorption edges of amorphous, trigonal and α-monoclinic selenium at 300 K. (From Ref. 55.)

In crystalline Te the smallest gap occurs in a direction in the Brillouin zone which does not correspond to that along the chain structure. This suggests that interaction between chains is important and it is not surprising to find a large change in the gap ($\sim$0·75 eV amorphous, $\sim$0·25 eV crystalline) in the amorphous state: even if the chain structure is preserved to some degree, the interaction between chains must change. In Se the chain interaction is less and the gaps in the trigonal crystal and amorphous states are similar ($\sim$2 eV) (Fig. 13.21). Crystalline As is a semimetal, amorphous films of As are quite transparent below about 1·5 eV;[55] clearly a change of local structure suggests itself here. In crystals for which the smallest gap is indirect, the breakdown in the k-selection rule for the amorphous state leads to high absorption coefficient at photon energies corresponding to this gap. Early results on Si seemed to confirm this idea (Fig. 13.22); however the recent results of Brodsky[38] on this material have now to be taken into consideration (Fig. 13.13).

13.9. Density of Electron States in Amorphous Semiconductors

For the purposes of understanding electrical conductivity and other transport properties of semiconductors we are of course primarily interested in the nature and density of *electron states at the band edges* because it is in these that the current is normally carried. From the same point of view, we are also interested in *gap states* because, except in very pure "intrinsic" material, these control the position of the Fermi level and hence the statistics as well as the trapping and recombination kinetics. At low temperatures electrons in these states may also carry the majority of the current by the process of thermally assisted tunnelling between centres (as in impurity conduction), particularly under ac conditions.

However, from a more general and fundamental point of view we are also interested in electronic states throughout the whole of the valence and conduction bands. In crystalline materials we are concerned not only with the density of electron states, but also how these are distributed in k-space. The $\mathcal{E}$–k diagrams of many crystalline semiconductors have been calculated from a knowledge of the wave functions of the constituent atoms and the structure. In an amorphous semiconductor, for which k is unlikely to be a good quantum number, we are interested in comparing simply the density of states with that of the material in its crystalline forms, and in finding out how to calculate it without the aid of translational periodicity to help us with the mathematics.

In this Section we are going to be concerned almost entirely with germanium and silicon. It is for these elements (particularly Ge) that we have most information, experimentally and theoretically, for both the crystalline and

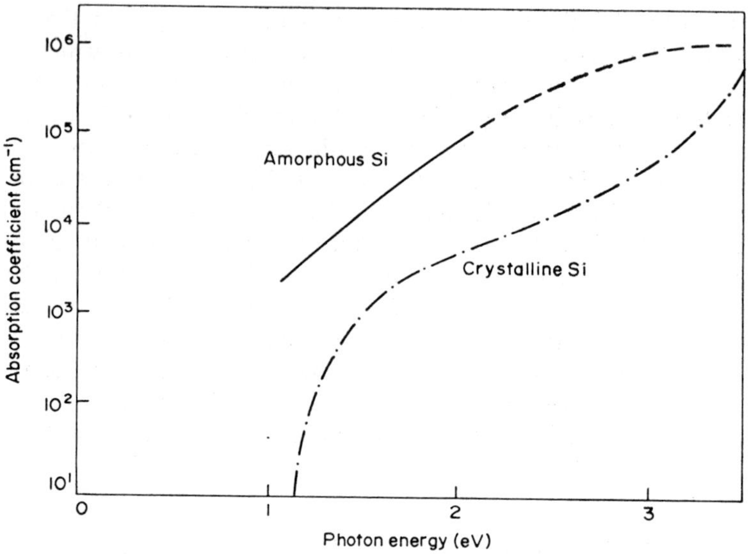

FIG. 13.22. Comparison of room temperature optical absorption edges in amorphous (Beaglehole and Zavetova, Ref. 88) and crystalline (Dash and Newman, Ref. 86) silicon. (From Ref. 56.)

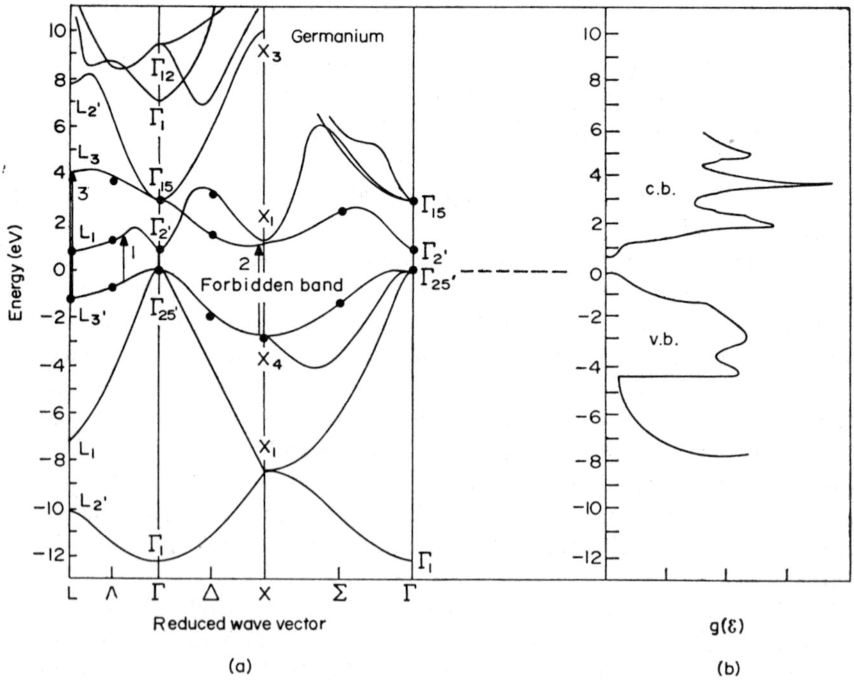

FIG. 13.23. Electronic band structure of crystalline Ge (from Herman *et al.*, Ref. 61), showing the principal interband transitions contributing to the κ'' spectrum of Fig. 13.24 and the density of states in the valence and conduction bands. (From Ref. 2.)

amorphous states. Much of what is said may also apply to III–V compounds, and some of the principles to amorphous materials in general. Considerable progress has been made in understanding the electronic structure of amorphous selenium, but lack of space forbids a detailed description of this work (see references 57–59 and 90).

Figure 13.23 shows the calculated band structure and density of states for crystalline germanium with the normal diamond structure. The smallest gap ($\sim$0·66 eV) is indirect $\Gamma_{25'}$–L_1; the smallest direct gap ($\sim$0·8 eV) is at Γ. Electronic transitions across these gaps are observed by transmission measurements at the absorption edge [see Fig. 13.15]. At higher photon energies the transitions which contribution most strongly to the absorption spectrum are (as in most crystals) vertical ones [1, 2 and 3 in Fig. 13.23] at critical points in the zone. Critical points occur where the valence and conduction

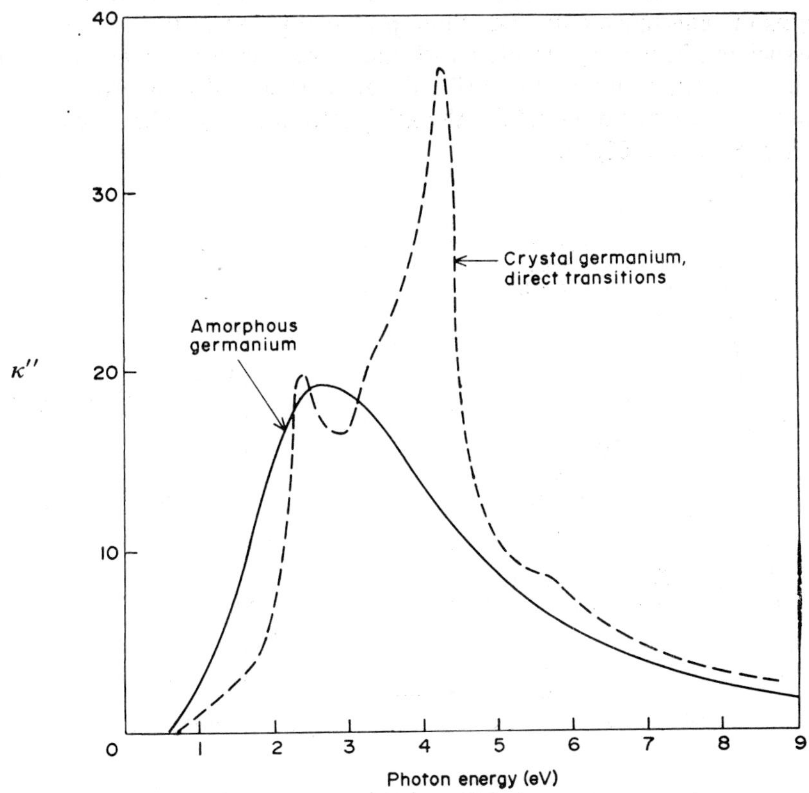

FIG. 13.24. Spectral dependence of the imaginary part of the dielectric constant κ'' as determined by the Kramers–Kronig analysis of reflectance data for amorphous Ge (solid line) and crystalline Ge (dashed line). (From Ref. 60.)

bands are parallel in k space i.e. where the joint or combined densities of states has maximum values. However, the absorption spectrum does not simply reproduce the joint density of states because the transition probability is not the same for all transitions. In particular, the matrix elements for transitions at the X point in Ge are considerably enhanced by an Umklapp effect.

The observed absorption spectrum for crystalline Ge from the fundamental edge up to 9 eV is shown by the dashed line in Fig. 13.24.[60] The ordinate here is the imaginary part of the dielectric constant κ'' related to the absorption coefficient α and the conductivity σ by:

$$\kappa'' = n_0\,\alpha/\omega = 4\pi\sigma/\omega.$$

Because of the difficulty in preparing samples of crystalline Ge sufficiently thin for transmission spectroscopy, κ'' is obtained by a Kramers–Kronig analysis of reflectance data. The three principal peaks in this spectrum can be associated in a rough fashion with the three transitions marked in Fig. 13.23, however, more of the Brillouin zone is actually involved. Recent attempts to account in detail for the κ'' spectrum of crystalline Ge are described in references 61, 62.

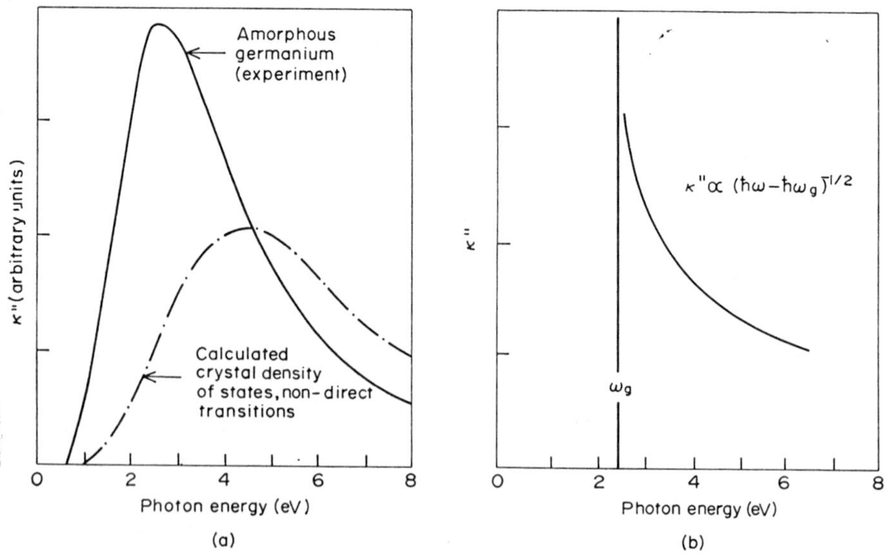

FIG. 13.25. (a) The imaginary part of the dielectric constant κ'' in amorphous germanium as determined experimentally (solid line) and as calculated from the crystalline density of states, assuming non-direct transitions and constant matrix elements (dashed and dotted line). (b) Expected form of κ'' on the Penn model. (From Ref. 2.)

In contrast to the spectrum associated with crystalline Ge, that for amorphous Ge is relatively featureless, consisting of a single broad peak with a maximum occurring at about 2·65 eV [solid line, Fig. 13.24]. Similar results have also been found for amorphous Si.

As a first attempt to account for this spectrum one might use a non-direct constant matrix element model. That is to say, one relaxes the *k*-conservation selection rule and allows all transitions, vertical and non-vertical, to contribute with equal probability. This is equivalent simply to convoluting the densities of states in the conduction and valence bands as shown in Fig. 13.23(b). The result[63] of this procedure for Ge is shown in Fig. 13.25(a). A featureless broad peak is indeed obtained, but the position of the maximum (~4·5 eV) clearly occurs at too high a photon energy to account for the data.

Unlike optical absorption which always involves the *product* of the initial and final state densities, photoemission experiments are able, in principle, to separate out the conduction and valence band densities of states. This arises because in photoemission one measures the energy distribution of photoemitted electrons for each photon energy, thereby providing additional information [see, for example, reference 2 and Chapter 15 by Dr. Mort].

Figure 13.26 shows the density of states in amorphous Ge as determined by photoemission compared with the calculated curve for the crystal.[64]

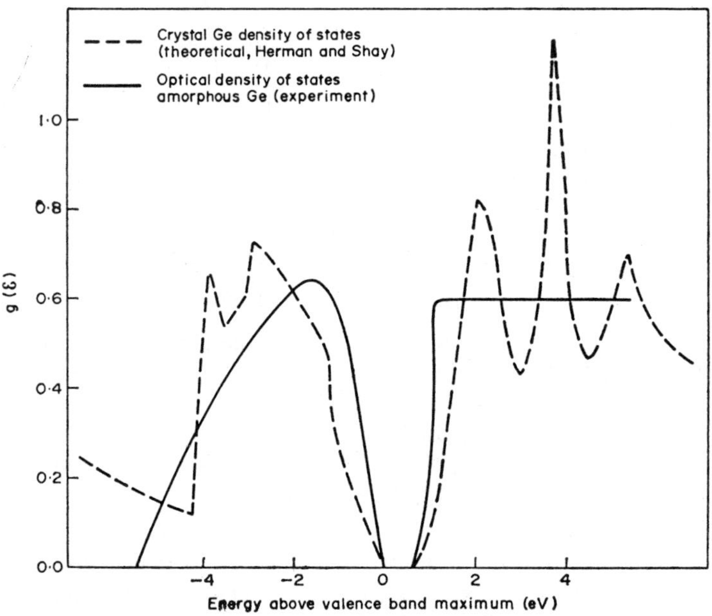

FIG. 13.26. Optical density of states for amorphous Ge as determined by photo-emission, and calculated density of states for crystalline Ge. (From Ref. 64.)

One notices that for the valence band, down to 4 eV from the band edge, the density of states in amorphous Ge is roughly similar to that in the crystal, but fine structure is lost and the maximum density occurs 1–2 eV closer to the band edge. For the conduction band all structure observed for the crystal is apparently lost. It is difficult to be certain about the density of states near the bottom of the conduction band because states below the vacuum level at the surface are inaccessible to photoemission techniques, and according to Spicer a weak maximum in the density of states at the band edge cannot be ruled out.

Recent measurements,(65) using X-ray emission spectroscopy, have probed the density of states to lower values of energy in the valence band. They have revealed a second broad peak in place of two broad peaks present in the crystal. This is shown in Fig. 13.37 and will be discussed later.

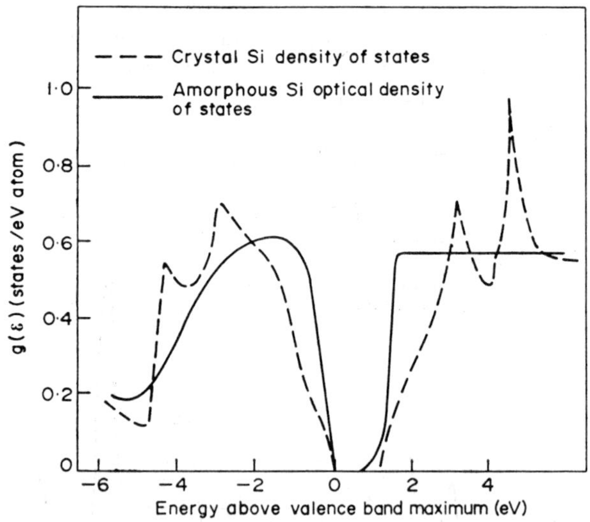

FIG. 13.27. Optical density of states (solid curve) of amorphous Si, calculated by using photo-emission results compared to the density of states (dashed curve) calculated by Herman *et al.*, Ref. 61, for crystalline Si. (From Ref. 66.)

It is perhaps not surprising that very similar results have been found for amorphous silicon.(66) Figure 13.27 compares the density of states with that calculated for crystalline Si having the diamond structure. The features discussed above for Ge i.e. a smooth peak in the valence band shifted to higher energies relative to the structure for the crystal and a featureless conduction band, are again observed. The loss of structure in the conduction band density of states for amorphous Si has also been confirmed by measurement of the optical absorption spectra at high photon energies using synchrotron radiation. These results by Brown and Rustgi,[67] are shown in Fig. 13.28. The spectrum for both crystalline (a) and amorphous (b) silicon has been decomposed into two components separated by 0·60 eV, corresponding to a spin–orbit split 2p-core level which is the initial state for the transition. For the crystal the three features in the conduction band density of states are observed and, apart from considerable enhancement due to the exciton effects at the edge, there is good agreement with the theoretical curve (shown dotted and positioned to line up with the L_{III} threshold). For the amorphous film all the structure beyond the edge is lost.

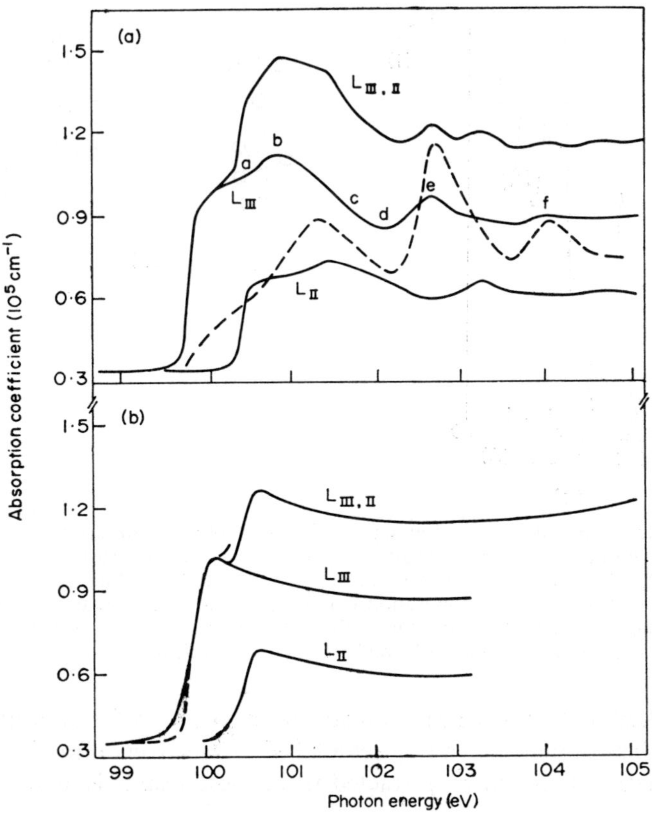

FIG. 13.28. (a) High-resolution soft X-ray absorption spectrum of crystalline silicon shown in the uppermost curve. The spectrum is resolved into L_{III} and L_{II} components just below this uppermost curve. The dashed line shows the theoretical conduction band state density according to Kane. (b) High resolution absorption spectrum of amorphous silicon. The dashed line is the crystalline spectrum very near threshold as given in (a) above. (From Ref. 67.)

What theoretical approaches have been used to calculate the density of states in amorphous semiconductors and to explain in particular the experimental observations of Figs. 13.25, 13.26 and 13.27? Before we answer this question let us consider what is known about the "structure" of amorphous Ge and Si.

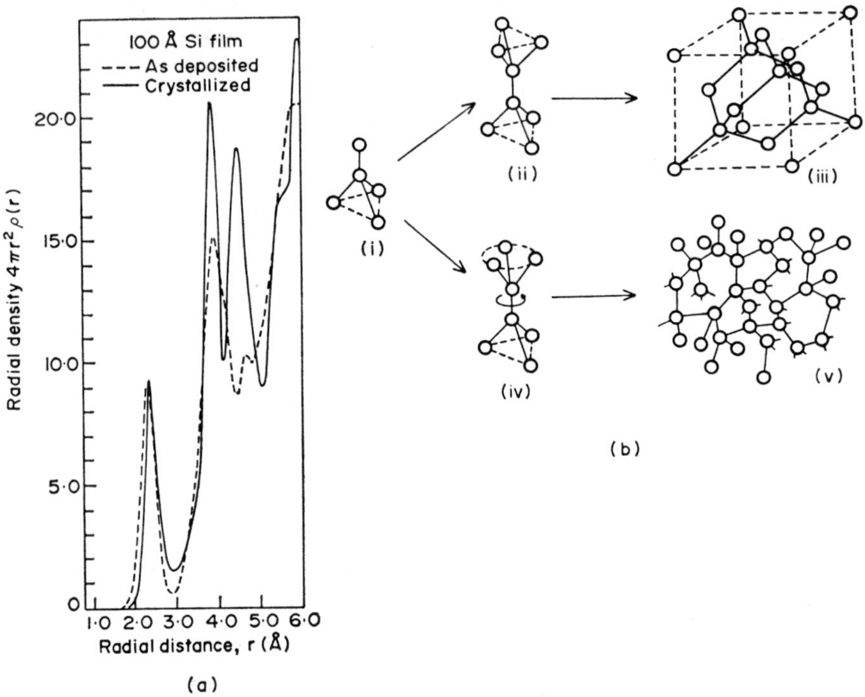

FIG. 13.29. Left: Radial distribution curves for amorphous and crystalline silicon as determined from analysis of electron diffraction data. (From Ref. 68.)
Right: Structural units in crystalline silicon (or germanium). Stacking of the tetrahedral unit (i) in a staggered bond configuration (ii), leads to the crystal (diamond lattice) (iii): stacking of neighbouring tetrahedral units allowing arbitrary rotations about the common bond (iv) leads to a random network model (v). (From Ref. 69.)

Figure 13.29(a) shows the radial distribution function for amorphous and crystalline Si as obtained by a Fourier analysis of electron diffraction data.[68] The information that can be extracted from these results is as follows.

(a) The coincidence of the first peak in the curves shows that the nearest neighbour separation (2·35Å) in the amorphous phase is unchanged (to within 1 or 2 per cent) from that in the crystal. Furthermore, the area under the peaks gives us the number of nearest neighbours which is 4 in both cases; i.e. the tetrahedral coordination is also retained.

(b) The second peak occurs at 3·86Å and its area gives 12 for the number of second nearest neighbours. However, for this shell of atoms, the breadth of the peak is wider in the amorphous film and one concludes that, after allowing for the effect of phonon broadening, there is a static mean square displacement $\sim 0{\cdot}031\text{Å}^2$. Moss and Graczyk assign this to bond-angle distortions which can be estimated to be $\sim 20°$ off the normal tetrahedral angle of 109° 28′.

(c) The third peak present in the curve for the crystal is characteristic of the diamond structure, which is obtained by stacking neighbouring tetrahedral units in the so-called staggered bond configuration [Fig. 13.29(b)]. Its almost complete absence in the curve for the film signifies that this stacking is not retained to any large degree in the amorphous state.

Unfortunately there is insufficient information available from electron or X-ray diffraction data to say much more with any certainty about the structure of amorphous Si or Ge. It is possible to fit the radial distribution function using a structure built up from an assembly of "amorphons" i.e. 20 atom units within which all neighbouring triads of bonds are in the eclipsed configuration.(70) Another and more plausible structure is that of a fully-connected network in which all possible angles of rotation about the common bond between neighbouring triads are allowed [Fig. 13.29(b)]. A model of this kind has been constructed by Polk.(71) A feature of both models is the presence of a large number of rings containing 5 atoms which do not occur in the diamond structure.

Electron microscopy studies of amorphous Ge by Rudee (see Ref. 72) have revealed bright spots of diameters >10Å in dark field and lattice fringes in interference micrographs. These features are not easily accounted for on a random network model,(73) although attempts have been made to do so,(74) and a structure composed of microcrystallites clearly remains a possibility.

Rather than attempting to fit the radial distribution function (r.d.f.), it is perhaps even more important to try to fit the experimentally determined diffraction curves from which the r.d.f. is derived. Calculations by Weinstein and Davis(75) have shown that there is little correspondence between experiment and theoretical curves based on microcrystallite models using the diamond, wurtzite, Ge III and several other structures. All models yield curves similar to those observed when the crystallites are made sufficiently small, but not when they are as large as 16Å—the size of the ordered domains suggested by Rudee and Howie. One can conclude, therefore, that if there is more order in amorphous Ge than is inherent in the random network model, its nature remains yet to be discovered.

Real films of amorphous Ge and Si are known to contain broken bonds, voids and other imperfections. These defects significantly affect the density, electrical properties and the form of the optical absorption edge (see Section 13.7). However, they do not appear to alter to any great extent the density of states in the valence and conduction bands, anymore, in fact, than they do in crystalline samples.

Theoretical attempts to calculate the optical absorption and density of states appropriate to amorphous Ge and Si can be divided into three fundamentally different approaches.

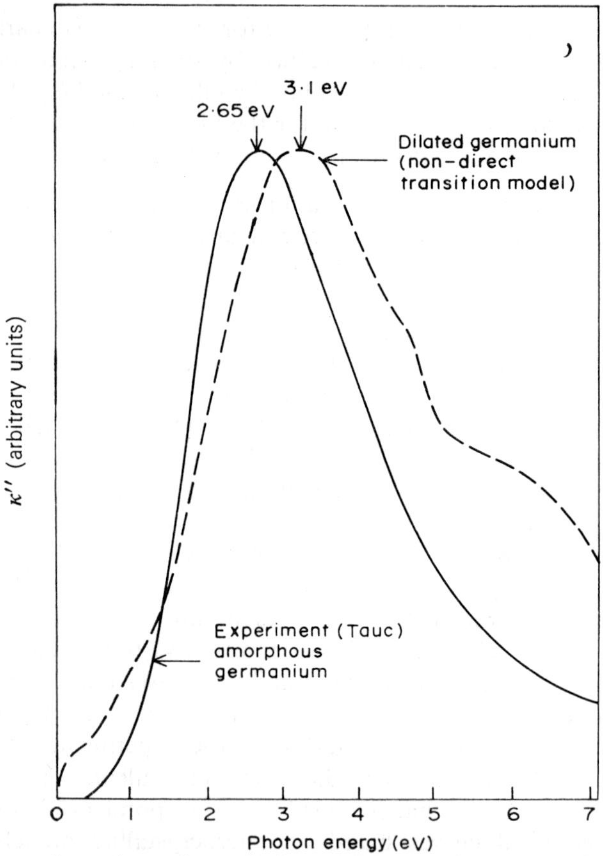

FIG.13.30. Comparison of experimental κ'' curve for amorphous germanium and theoretical κ'' curve based on the band structure of a dilated Ge lattice and the model of non-direct transitions and constant matrix elements. (From Ref. 63.)

1. Those based on various modifications to the band structure of the crystal.

2. Application of the theory of multiple scattering within ordered clusters of atoms.

3. Tight binding or LCAO methods.

The second of these methods has been fully described by Greenwood (Chapter 3) who also outlined some results using the third approach. It was however, methods based on modification to the band structure of the diamond lattice, which first attracted the attention of theoreticians and these will be described first here.

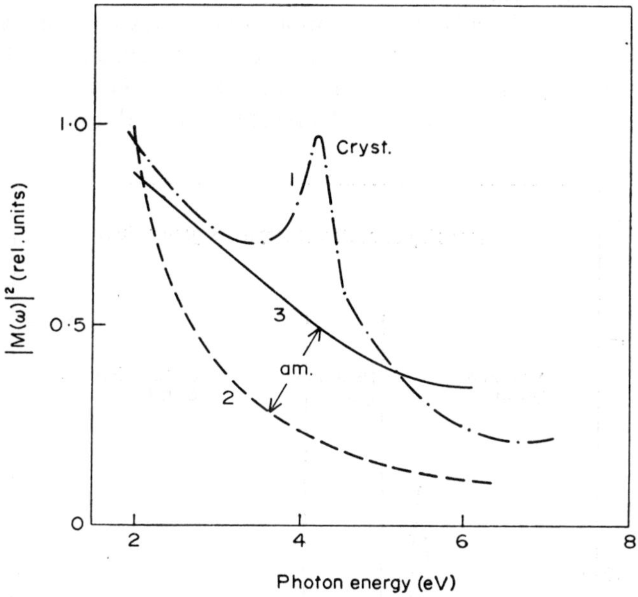

FIG. 13.31. Energy dependence of matrix elements (1) crystal, (2) amorphous with convolution of crystalline density of states, (3) amorphous with convolution of amorphous density of states. (From Ref. 56.)

In 1968 Herman and Van Dyke suggested[63] that a fair description of the absorption spectrum characteristic of amorphous Ge could be obtained by using the band structure of a dilated diamond structure and relaxing the k-conservation selection rule. Their result, using a lattice constant increased by 11 per cent, is shown in Fig. 13.30. A featureless spectrum is obtained, but the peak in κ'' occurs at 3·1 eV, i.e. at a lower energy than obtained without dilation, but still too high to account for the experimental observations. A 33 per cent reduction in density is certainly an overestimate for carefully prepared films, and in fact, the nearest neighbour separation is known to change by at most a few per cent. An additional indication that this approach is not valid is that, for the dilated lattice, Ge becomes a semimetal.

A second attempt to account for the κ'' spectrum of amorphous Ge was made by Brust. Brust[76] suggested a *partial* loss of the uniqueness of the k

vector in the band structure of the diamond lattice. By introducing a somewhat arbitrary parameter to characterise this and also energy-dependent matrix elements, good agreement with experiment could be obtained.

Kramer[77] and Maschke and Thomas,[57] have proposed a similar approach. These authors pointed out that the large peak near 4·5 eV in the spectrum associated with crystalline Ge [Fig. 13.24] is considerably enhanced by Umklapp effects. Calculated pseudopotential matrix elements as a function of energy for crystalline Ge are as shown by the dotted curve in Fig. 13.31. For amorphous Ge, on the other hand, Maschke and Thomas suggest that the energy dependence of the matrix elements is a smooth function as shown

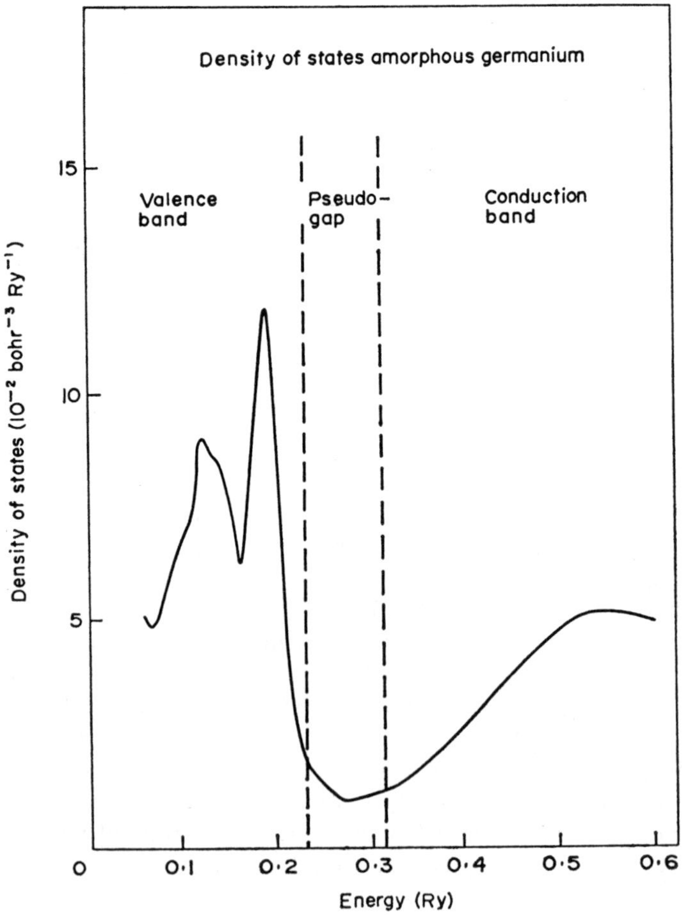

FIG. 13.32. Hypothetical density of states for amorphous Ge obtained by averaging with equal weights the results for the staggered and eclipsed configurations. (From Ref. 79.)

by the solid curve in Fig. 13.31. Folding this variation into the κ'' spectrum, obtained on a non-direct transition model using the crystalline band structure, leads to a spectrum in fair agreement with experiment. However, Maschke and Thomas conclude that it is still necessary to take into account a change in the density of states in going from the crystalline to amorphous form (see also reference 56).

A rather different approach to the problem has been proposed by Phillips,[(78)] who suggested that only the average or Penn gap appropriate to crystalline Ge is retained in the amorphous form. This isotropic model leads to a form for the optical absorption shown in Fig. 13.25(b). By introducing a broadening parameter Phillips showed how the basic form of the κ'' spectrum could be approximated. However, the density of states expected on this model is unlike that determined by photoemission (Fig. 13.26) as will be shown later.

Application of the theory of multiple scattering within ordered clusters of atoms to a calculation of the density of states in amorphous semiconductors has been fully described in Chapter 3 and will not be discussed further here. The principle result for amorphous Ge as far as we are concerned here, is shown in Fig. 13.32.[(79)] This is an average of the density of states calculated for an assembly of clusters each containing eight atoms; half of the clusters being in the staggered configuration and half in the eclipsed. The achievement of a pseudogap (not a true zero in the density of states) of approximately the correct magnitude using this method is very impressive. However, the details of the curve do not agree with experimental observations and further refinements in the theory are clearly necessary before a detailed comparison is attempted.

Weaire and Thorpe's proof,[(80)] using an LCAO method, of the existence of true energy gaps in tetrahedrally connected networks of Ge or Si atoms, attracted a lot of attention. However, it is known that tight binding methods are not particularly successful in *predicting* band structures of materials like Ge, in which the overlap of wave functions between neighbouring atoms is large, and nearly free electron methods have been shown to work much better. Nevertheless, the application of tight binding methods to various real and hypothetical tetrahedrally coordinated structures has given real insight into which features in the density of states arise from nearest and next-nearest neighbour interactions, into the role of boundary conditions and many other questions of interest.

Weaire and Thorpe[(81)] have calculated the electronic density of states for the following tetrahedrally connected lattices.

(i) *Diamond*. The result is shown in Fig. 13.33. The delta function singularities are a consequence of considering only nearest neighbour bond overlap which leads to the uppermost state in the valence band having a flat $\mathcal{E}$–k

spectrum instead of one which curves down from the Γ point. This is clearly the state which, when properly treated, becomes the principle hump at the top of the valence band between 0 and 4·5 eV in Fig. 13.23(b). The other two peaks in the band are in qualitative agreement with those found in more realistic band structure calculation of Ge with the diamond structure. They are not shown in Fig. 13.23(b) but appear later in Fig. 13.37(b). It is to be emphasised that the method employed here is not expected to give a realistic

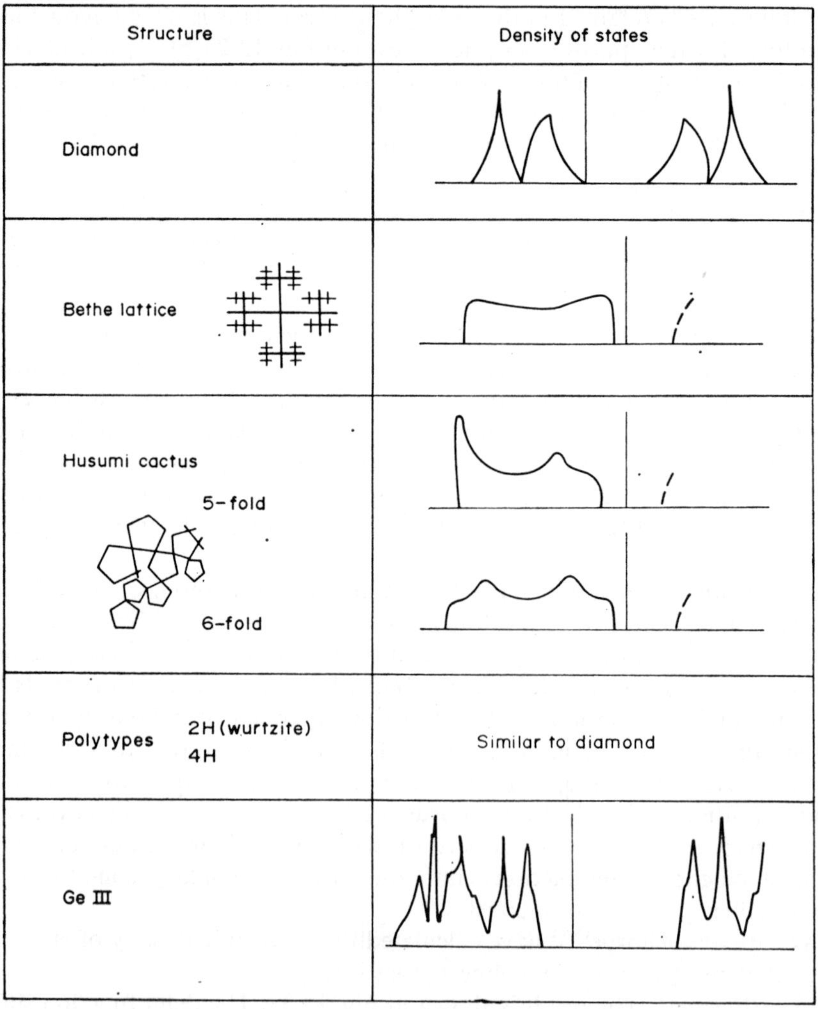

FIG. 13.33. Schematic representation of the density of states in Ge for various structures as calculated by Weaire and Thorpe and, for Ge III, by Joannopoulos and Cohen, using a tight-binding formulation.

description of the conduction band.

(ii) *Bethe lattice.* This is a hypothetical, infinitely branched, tree-like structure with no closed rings, illustrated in Fig. 13.33. Calculations using this structure lead to a flat featureless spectrum for the density of states.

(iii) *Husumi cacti.* These are also hypothetical lattices built up of connected *n*-fold rings. The results for 5- and 6-fold rings are shown in Fig. 13.33. These calculations suggest that the two broad peaks found in the valence band for the diamond lattice are attributable to the sixfold rings and that a random structure with 5-, 6- and 7-fold rings would yield a washed out spectrum, since the different rings tend to enhance the spectrum at different energies.

Let us break off for the moment to consider how various structures can be characterized in terms of the ratio of 5- and 6-fold rings.[81] In Fig. 13.34(a) the size of each square represents the proportion of atoms having a particular

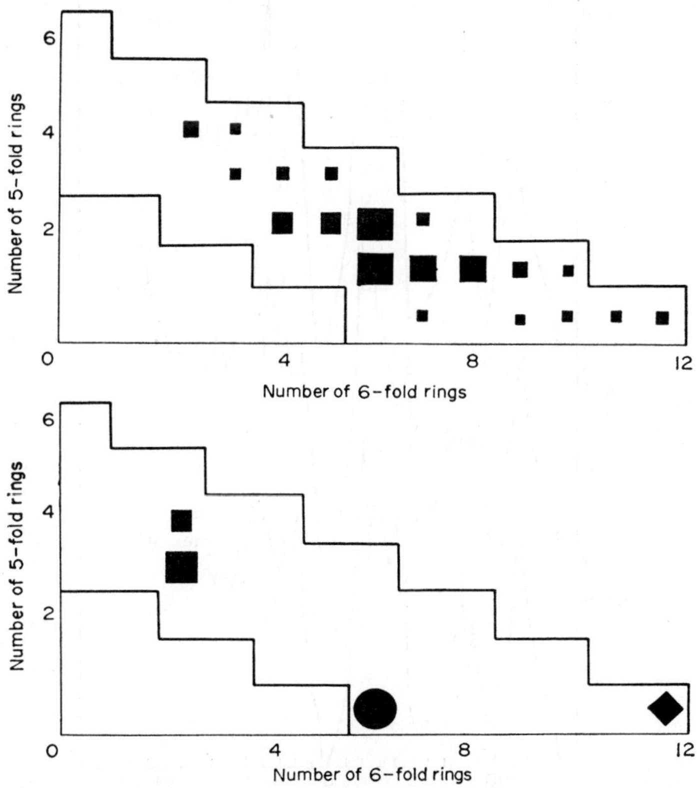

FIG. 13.34. (a) The size of each square represents the proportion of atoms having a particular number of 5- and 6-fold rings passing through it, in the model constructed by D. Polk. (b) A similar plot for the diamond (◆) Si III (●) and Ge III (■) structures. (From Ref. 81.)

number of 5- and 6-fold rings passing through them in the random network model constructed by Polk. This is not an exact representation of the structure because rings containing a larger number of atoms do in fact occur. Furthermore, there is no reason to suppose that Polk's model is unique, and it seems quite plausible to suggest that models using the same rules of construction could be built which might show a rather different distribution. The stepped lines are reasonable bounds on the number of 5- and 6-fold rings which can pass through a given site without unreasonable bond angle distortions. Figure 13.34(b) is a similar plot showing the distributions in the diamond,

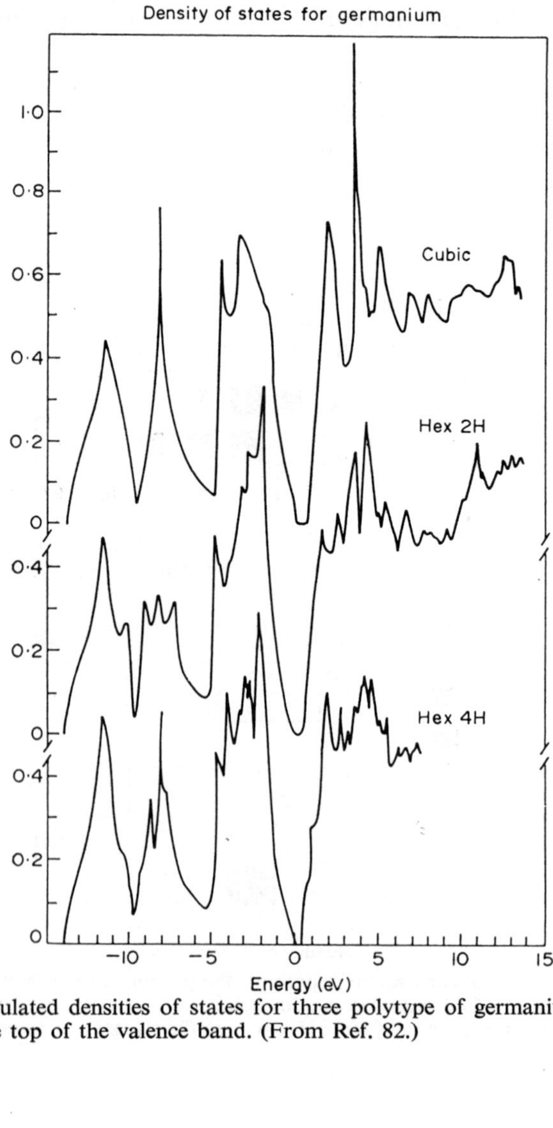

FIG. 13.35. Calculated densities of states for three polytype of germanium. The zero of energy is at the top of the valence band. (From Ref. 82.)

Si III and Ge III structures. The latter two are phases of Ge which can be obtained at high pressures and will be referred to again later.

The polytype structures, $2H$ (wurtzite) and $4H$, have not been treated by Weaire and Thorpe, but they suggest that the results should not differ significantly from those obtained for the diamond–cubic structure. This, in fact, has been confirmed by the pseudopotential calculations of Ortenburger, Rudge and Herman[(82)] shown in Fig. 13.35. In view of the presence of a proportion of bonds being in the eclipsed configuration (but not 5-fold rings) in these two polytypes, it was originally thought that their band structures may be a better starting point for understanding the density of states in amorphous Ge.[(56, 2)] Indeed one does notice a slight shift in the uppermost peak of the valence band towards the gap and a considerable loss of structure in the conduction band. However, calculations of κ'' on a non-direct constant matrix elements model using these band structures still produces a spectrum which peaks at too high an energy.

(iv) *Ge III*. This polytype of Ge has a tetragonal structure and contains 12 atoms per primitive cell. It consists of a mixture of 5-, 6- and 7-fold rings, and there are distortions of the tetrahedral bond angles by up to 20°. As shown in Fig. 13.34(b) there are two different environments for the atoms. The density of states for this structure using a tight binding model is included in Fig. 13.36 which also shows recent empirical pseudopotential calculations by

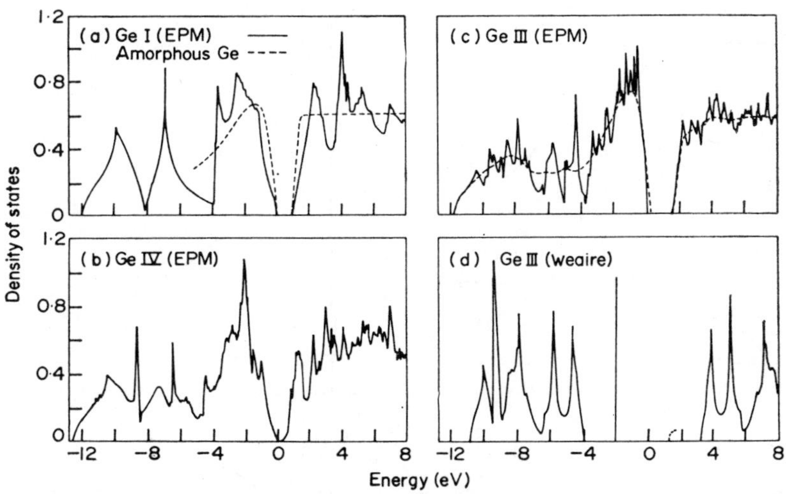

FIG. 13.36. Density of states per eV per atom for different polytypes of Ge: (a) Ge I (diamond cubic) using the Empirical Pseudopotential Method (EPM) and amorphous Ge (from Ref. 64); (b) Ge IV (Si III) using the EPM; (c) Ge III using the EPM (solid line) and an averaged density of states (dashed line); (d) Ge III using the tight binding model of Weaire. (From Ref. 83.)

Joannopoulos and Cohen[83] of the band structure of Ge I (diamond), Ge IV (also known as Si III) and Ge III. The most striking feature [Fig. 13.36(c)] is the mixing in Ge III of the two lowermost peaks in the valence band

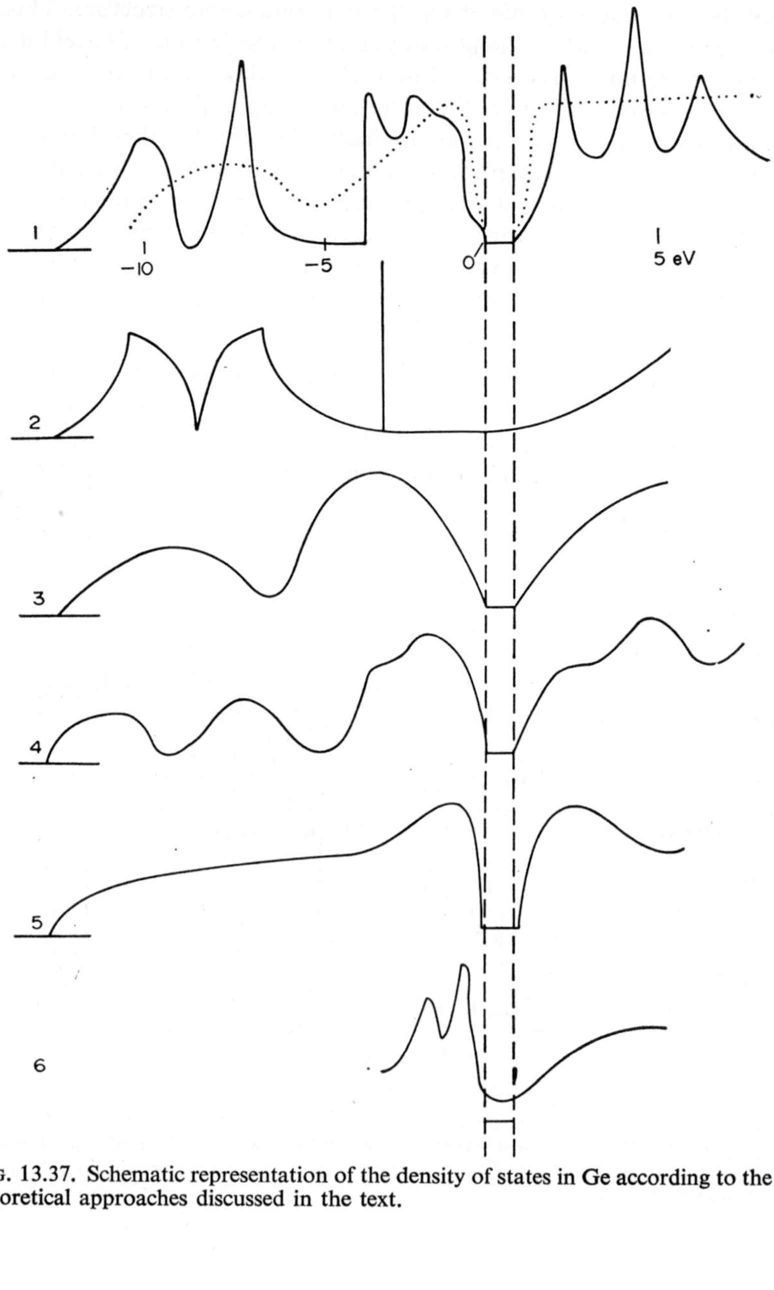

FIG. 13.37. Schematic representation of the density of states in Ge according to the various theoretical approaches discussed in the text.

found in the diamond structure. A single broad feature embracing these two peaks has recently been found using soft X-ray emission spectroscopy.[65] Other notable features are the shift of the uppermost peak in the valence band towards the gap and the relatively smooth conduction band, trends which were also evident in calculations on the 2H and 4H polytypes (Fig. 13.35). It should be emphasised that the sharp peaks in the calculated spectrum are primarily due to Bragg gaps and would be smoothed out in a structure with no peridocity.

What are we to conclude from these calculations? It would be rash to suggest that amorphous Ge consists of Ge III crystallites. It is true that the density of states in Ge III approximates closely to that observed in amorphous Ge, and it follows, therefore, that one can account for the κ'' spectrum using the same model. However, calculations of the diffraction curve from Ge III by Weinstein[75] do not lend support to this idea. It would seem preferable to conclude that the similarity of the density of states in Ge III to that in amorphous Ge arises because of similar features in their structures, namely the presence of 5-, 6- and 7-fold rings and considerable bond angle distortions. It would therefore seem instructive to calculate and measure further properties (e.g. vibrational spectra) of this polytype to test for further correspondences with those of amorphous Ge. It would also be interesting to see if a random network of the kind constructed by Polk could be built in which the distribution of 5- and 6-fold rings came closer to that present in Ge III.

In summary, Fig. 13.37 illustrates schematically the density of states in amorphous Ge as calculated or suggested by the variety of approaches discussed in this Section. Curve 1 shows that calculated for the diamond-cubic structure (solid) and that determined experimentally for amorphous Ge (dotted). Curve 2 shows the results of Weaire's tight binding calculations for the diamond cubic structure and curve 3 that suggested by Weaire and Thorpe for amorphous Ge on the basis of their calculations for a variety of lattices. Curve 4 is the density of states expected using a modified crystal approach and curve 5 that for the isotropic Penn model. Finally curve 6 shows the density of states calculated using the theory of scattering from ordered clusters. Although the latter method appears in principle to be the most promising fundamental approach for the future, it is the tight binding methods which have given us most insight into the gross features of the density of states in amorphous Ge at the present time.

13.10. Concluding Remarks

In a review of this nature, which attempts to cover a very wide spectrum (literally) of observations, it is inevitable that one cannot do justice either in description or discussion to individual worker's contributions to the subject. The author trusts that critical readers will consult the original references for further details and in some instances for different explanations.

There can be little doubt that our understanding of the non-crystalline state has advanced considerably over the last five years; the recentness of most of the references is an indication of this. As for crystalline materials, optical measurements have played a major role in the development of this understanding. It seems particularly appropriate that this should be the case here, for what property of glass is more evident that its transparency?

Gratitude is due to many colleagues who by discussion have helped in the preparation of this article.

References

1. Rockstad, H. K., *J. Non.-Cryst. Solids* **2,** 192 (1970).
2. Mott, N. F. and Davis, E. A., "Electronic Processes in Non-crystalline Materials." Oxford, 1971.
3. Chopra, K. L. and Bahl, S. K., *Phys. Rev.* **B1,** 2545 (1970).
4. Pollak, M. and Geballe, T. H., *Phys. Rev.* **122,** 1742 (1961).
5. Austin, I. G. and Mott, N. F., *Adv. Phys.* **18,** 41 (1969).
6. Scher, H. and Lax, M. *J., Non-Cryst. Solids* **8-10,** 497 (1972). *Phys. Rev.* to be published.
7. Pollak, M., *Phil. Mag.* **23,** 519 (1971).
8. Butcher, P. N., to be published.
9. Street, R. A. and Yoffe, A. D., *J. Non-Cryst. Solids* **8-10,** 745 (1972).
10. Anderson, P. W., Halperin, B. I. and Varma, C. M., *Phil. Mag.* **25,** 1 (1972).
11. Pollak, M. and Pike, G. E., *Phys. Rev. Lett.* **28,** 1449 (1972).
12. Austin, I. G. and Garbett, E. S. *Phil. Mag.* **23,** 17 (1971).
13. Ivkin, E. V. and Kolomiets, B. T., *J. Non-Cryst. Solids* **3,** 41 (1970).
14. Davis, E. A. and Mott, N. F. ,*Phil Mag.* **22,** 903 (1970).
15. See Rockstad, H. K., *Solid State Commun.* **9,** 2233 (1971). *J. Non-Cryst. Solids* **8-10,** 621 (1972).
16. Mitchell, D. L., Bishop, S. G. and Taylor, P. C., *J. Non-Cryst. Solids* **8-10,** 231 (1972).
17. Felty, E. J., Lucovsky, G. and Myers, M. B., *Solid State Commun.* **5,** 555 (1967).
18. Zltakin, L. B. and Markov, Y. F., *Phys. Stat. Sol.* **4(a),** 391 (1971).
19. Onomichi, M., Arai, T. and Kudo, K., *J. Non-Cryst. Solids* **6,** 362 (1971).
20. Lucovsky, G., *Phys. Rev.* **B6,** 1480 (1972).
21. Lucovsky, G., *Mater. Res. Bull.* **4,** 505 (1968).
22. Ward, A. T., *Adv. in Chem.* **110,** 163 (1972).
23. Brodsky, M. H., Gambino, R. J., Smith, J. E. and Yacoby, Y., *Phys. Stat. Sol.* **B52,** 609 (1972).
24. Caldwell, R. S. and Fan, H. Y., *Phys. Rev.* **114,** 664 (1959).
25. Taylor, P. C., Bishop, S. G. and Mitchell, D. L., *Phys. Rev. Lett.* **27,** 414 (1971).
26. Goryunova, N. A., Cross, E. F., Zlatkin, L. B. and Ivanov, E. K., *J. Non-Cryst. Solids* **4,** 57 (1970).
27. Smith, J. E., Brodsky, M. H., Crowder, B. L. and Nathan, M. I., *J. Non-Cryst. Solids* **8-10,** 179 (1972).
28. Bell, R. J. and Dean, P., *Discuss. Faraday Soc.* **50,** 55 (1971).

29. Edmond, J. T., *Br. J. Appl. Phys.* **17,** 979 (1966).
30. Edmond, J. T., private communication.
31. Bishop, S. G., Taylor, P. C., Mitchell, D. L. and Slack, L. H., *J. Non-Cryst. Solids* **5,** 351 (1971).
32. Spicer, W. E. and Donovan, T. M., *J. Non-Cryst. Solids* **2,** 66 (1970).
33. Knotek, M., private communication.
34. Fritzsche, H., *J. Non-Cryst. Solids* **6,** 49 (1971).
35. Anderson, P. W., *Nature Phys. Sci.* **235,** 163 (1972).
36. Chittick, R. C., *J. Non-Cryst. Solids* **3,** 255 (1970).
37. Tauc, J., Zallen, R. and Slade, M., *J. Non-Cryst. Solids* **4,** 279 (1970).
38. Brodsky, M. H., Kaplan, D. and Ziegler, J. F., Proc. XI Int. Conf. Semicond. Warsaw (1972), PWN–Polish Scientific Publishers.
39. Tauc, J. and Menthe, A., *J. Non-Cryst. Solids* **8-10,** 569 (1972).
40. Perron, J. C. private communication.
41. Theye, M. L., *Mater. Res. Bull.* **6,** 103 (1971).
42. Connell, G. A. N. and Paul, W., *J. Non-Cryst. Solids* **8-10,** 215 (1972).
43. Koc. S., Renner, O., Zavetova, M. and Zemek, J., *Czech. J. Phys.* **B22,** (1972).
44. Ziman, J. M., *J. Non-Cryst. Solids* **4,** 426 (1970).
45. Weaire, D., *Phys. Rev. Lett.* **26,** 1541 (1971).
46. Heine, V.,, *J. Phys. C.* **4,** L221 (1971).
47. See Keller, J. and Ziman, J. M., *J. Non-Cryst. Solids* **8-10,** 111 (1972).
48. Urbach, F., *Phys. Rev.* **92,** 1324 (1953).
49. See Hopfield, J. J., *Comments in Solids St. Phys.* **1,** 16 (1968).
50. Dow, J. D. and Redfield, D., *Phys. Rev.* **B1,** 3358 (1970).
51. Dow, J. D. and Redfield, D., *Phys. Rev.* **B5,** 594 (1972).
52. Olley, J., Ph.D. Thesis, University of Cambridge (1972), unpublished.
53. Tabak, M. D. and Warter, P. J., *Phys. Rev.* **173,** 899 (1968).
54. Fagen, E. A., private communication.
55. Knights, J. C., private communication.
56. Stuke, J., Proceedings of the 10th International Conference on the Physics of Semiconductors, Cambridge, Massachusetts (ed. Keller, Hensel and Stern). United States Energy Commission (1970).
57. Maschke, K. and Thomas, P., *Phys. Stat. Sol.* **41,** 743 (1970).
58. Kramer, B., *Phys. Stat. Sol.* **41,** 725 (1970).
59. Laude, L. D. and Fitton, B., *J. Non-Cryst. Solids* **8-10,** 971 (1972).
60. Spicer, W. E. and Donovan, T. M., *J. Non-Cryst. Solids* **2,** 66 (1970).
61. Herman, F., Kortum, R. L., Kuglin, C. D. and Shay, J. L., p 503 in "II-V Semiconducting Compounds, 1967 International Conference", (ed. Thomas, Benjamin), N.Y., 1967.
62. Van Dyke, J. P., *Phys. Rev.* **B5,** 1489 (1972).
63. Herman, F. and Van Dyke, J. P., *Phys. Rev. Lett.* **21,** 1575 (1968).
64. Donovan, T. M. and Spicer, W. E., *Phys. Rev. Lett.* **21,** 1572 (1968).
65. Ley, L., Kowalczyk, S., Pollak, R. and Shirley, D. A., *Phys. Rev. Lett.* **29,** 1088 (1972): also Grobman, W. D. and Eastman, D. E., *Phys. Rev. Lett.* **29,** 1508 (1972).
66. Pierce, D. T. and Spicer, W. E., *Phys. Rev.* **B5,** 3017 (1972).
67. Brown, F. C. and Rustgi, O. P., *Phys. Rev. Lett.* **28,** 497 (1972).
68. Moss, S. C. and Graczyk, J. F., Proceedings of the 10th International Conference on the Physics of Semiconductors, Cambridge, Massachusetts (ed. Keller, Hensel and Stern). United States Energy Commission (1970).

69. Davis, E. A., *Endeavour* **30,** 55 (1971).
70. Grigorovici, R., *J. Non-Cryst. Solids* **1,** 303 (1969).
71. Polk, D. E., *J. Non-Cryst. Solids* **5,** 365 (1971).
72. Rudee, M. L. and Howie, A., *Phil. Mag.* **25,** 1001 (1972).
73. Howie, A., Krivanek, O. and Rudee, M. L., to be published.
74. Chaudhari, P., Graczyk, J. F. and Charbrau, H. P., *Phys. Rev. Lett.* **29,** 425 (1972).
75. Weinstein, F. C. and Davis, E. A., to be published.
76. Brust, D., *Phys. Rev. Lett.* **23,** 1232 (1969).
77. Kramer, B., *J. Non-Cryst. Solids* **8-10,** 659 (1972).
78. Phillips, J. C., *Phys. Stat. Sol.* **B44,** K1 (1971).
79. McGill, T. C. and Klima, J., *Phys. Rev.* **B5,** 1517 (1972).
80. Weaire, D. and Thorpe, M. F., *Phys. Rev.* **B4,** 2508 (1971).
81. Weaire, D. and Thorpe, M. F., "Computational Methods for Large Molecules and Localised States in Solids". (ed. Herman, McLean and Nesbet), Plenum Press, N. Y., 1972.
82. Ortenburger, I. B., Rudge, W. E. and Herman, F., *J. Non-Cryst. Solids* **8-10,** 653 (1972).
83. Joannopoulos, J. D. and Cohen, M. L., *Solid State Commun.* **11,** 549 (1972).
84. Clark, A. H., *Phys. Rev.* **154,** 750 (1967).
85. Tauc, J., "Optical Properties of Solids". (ed. F. Abeles), North-Holland, Amsterdam, 1970.
86. Dash, W. C. and Newman, R., *Phys. Rev.* **99,** 1151 (1955).
87. Brodsky, M. H., Title, R. S., Weiser, K. and Pettit, G. D., *Phys. Rev.* **B1,** 2632 (1970).
88. Beaglehole, D. and Zavetova, M., *J. Non-Cryst. Solids* **4,** 272 (1970).
89. Lucovsky, G., *Phys. Stat. Sol.* **B49,** 633 (1972).
90. Chen, I., *Phys. Rev.* **B7,** 3672 (1973).

14. RAMAN SPECTROSCOPY

J. MORT

Xerox Corporation, Rochester Corporate Research Centre, Rochester, N.Y., U.S.A.

14.1. INTRODUCTION

Much of our concern has been devoted to the effect that disorder in solids has on the electronic energy level structure of solids. These effects can be probed by studying by a wide variety of techniques phenomena such as dc and ac electrical conductivity, charge transport, photo-conductivity, internal and external photoemission, and optical properties in the energy range covering electronic transitions. These studies can be, and have been, very profitable even without asking the natural question as to the specific nature of the atomic disorder which is present in a particular amorphous solid. This more intimate information about the relative arrangement of the atoms within a solid can be revealed by a study of its vibrational energy level structure. This information can be gleaned in some instances from infra-red absorption measurements as discussed by Dr. Davis in Chapter 13. In lattices with inversion symmetry it is fortunate that those vibrational modes which are not infra-red are Raman active. In this sense infra-red and Raman spectroscopy are complementary. The phenomenon of Raman scattering has thus proven to be a very powerful tool in investigating the vibrational spectrum of crystals and has now been applied to the same end for the amorphous state. It is the purpose of this lecture to discuss only those features of the Raman Effect which are necessary to appreciate the important role that this technique can play in any comprehensive study of the amorphous state. The discussion will be confined to first-order Raman scattering[1,2] from optical modes, thus excluding totally acoustic mode or Brillouin scattering.[3] This general introductory discussion will form Section 14.2 of the lecture, followed in Section 14.3 by some discussion of experimental techniques. Section 14.4 will be a presentation and discussion of results of Raman studies on Se, chalcogenide alloys, Si and Te in both the crystalline and amorphous forms.

14.2. General Concepts of Raman Scattering

In order for light of a given frequency to be absorbed at either a resonant electronic or vibrational frequency there must be a change in dipole moment between the initial and final states. In the case of electronic transitions this is assured in the transition between stationary states. In the case of vibrational transitions, which because the motion of atoms is involved occurs in the infra-red part of the spectrum, no absorption will take place unless a change in dipole moment of the molecule occurs during the vibration. Provided this condition is met the mode of vibration is said to be infra-red active and an absorption band will be observable. Consider the case of the hydrogen chloride molecule. The hydrogen is positively charged and at a distance d the chlorine is negatively charged. This molecule, therefore, has a permanent dipole moment, even in the absence of any vibration, $\mu = ed$. During a vibrational mode the distance d and hence the dipole moment μ is modulated. Such a mode is infra-red active. For the case of the hydrogen molecule, no permanent dipole moment exists so that no change in μ occurs during

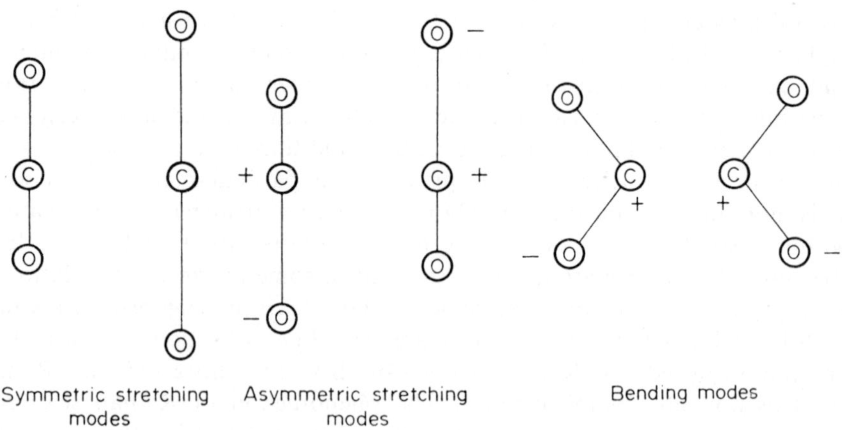

Fig. 14.1. Fundamental vibrational modes of the CO_2 molecule which possesses no permanent dipole moment. During the bending and asymmetric stretching vibrations a net dipole moment exists. The fourth fundamental vibration is a bending mode where each atom moves in a direction perpendicular to that illustrated.

vibration and the vibration is infra-red inactive. As a third case we consider the CO_2 molecule, Fig. 14.1, where the positive charge is centred at the carbon atom and the centre of negative charge can be visualized as lying between the two oxygen atoms. As is the case for the H_2 molecule, CO_2 has no permanent dipole moment. The symmetric stretching mode which produces no net charge displacement is infra-red inactive. On the other hand, the asymetric stretching and bending modes do change the dipole moment and are infra-red active.

Thus far we have considered only absorption of light which is resonant in frequency with a vibrational mode of a molecule. More generally, however, when an electromagnetic wave with a photon energy of $h\nu$ interacts with a molecule, the phenomenon of light scattering must be considered.

For an electromagnetic wave of frequency ν_1, the associated varying electric field E is given by

$$E = E_0 \sin 2\pi\nu_1 t. \tag{14.1}$$

For the case of a spherically symmetric atom, the induced dipole moment $\mu = \alpha E$, (α is the polarizability) will fluctuate with the same frequency as the driving electromagnetic wave

$$\mu = \alpha E_0 \sin 2\pi\nu_1 t. \tag{14.2}$$

Classically, as a vibrating dipole, the atom can now radiate an e.m. wave with the frequency of the exciting radiation. This is called Rayleigh scattering. The blueness of the sky which arises from Rayleigh scattering of sunlight from air molecules is not due to a change in individual photon energies but arises from the greater scattering of shorter wavelengths.

In general most molecules do not have a spherically symmetric structure, so that the polarizability can be different in the x, y, z directions and is a tensor

$$\begin{aligned}
\mu_x &= \alpha_{xx} E_x + \alpha_{xy} E_y + \alpha_{xz} E_z \\
\mu_y &= \alpha_{yx} E_x + \alpha_{yy} E_y + \alpha_{yz} E_z \\
\mu_z &= \alpha_{zx} E_x + \alpha_{zy} E_y + \alpha_{zz} E_z.
\end{aligned} \tag{14.3}$$

Since the polarizability tensor is symmetric, i.e. $\alpha_{xy} = \alpha_{yx}, \alpha_{yz} = \alpha_{zy}$ and $\alpha_{zx} = \alpha_{xz}$, one can choose three coordinates x', y', and z' such that only $\alpha_{x'x'}$, $\alpha_{y'y'}$ and $\alpha_{z'z'}$ are non-zero. The coordinates define three orthogonal directions in the molecule for which the induced dipole moments are parallel to the electric field. The locus of points formed by plotting $1/\sqrt{\alpha}$ in any direction from the origin defines the surface of a polarizability ellipsoid with axes, x', y' and z'. For a totally unsymmetric molecule, $\alpha_{x'x'} \neq \alpha_{y'y'} \neq \alpha_{z'z'}$, the ellipsoid has 3 axes of unequal length. A rotational ellipsoid results if two of the axes are equal. If all the three axes are equal then a polarizability sphere exists. If this polarizability ellipsoid changes as a result of molecular vibration or rotation a Raman spectrum will occur.

As an example, consider a diatomic molecule which is successively compressed and extended. For small displacements the polarizability is given by

$$\alpha = \alpha_0 + \frac{\partial \alpha}{\partial(\Delta r)} \Delta r \tag{14.4}$$

where α_0 is the equilibrium polarizability, Δr is the change in internuclear distance at any time. Considering the vibration to be a simple harmonic motion of frequency ν_v

$$\Delta r = a \sin 2\pi \nu_v t. \tag{14.5}$$

Thus

$$\alpha = \alpha_0 + \frac{\partial \alpha}{\partial(\Delta r)} a \sin 2\pi \nu_v t \tag{14.6}$$

substituting for α in eqn (14.2)

$$\mu = \alpha_0 E_0 \sin 2\pi \nu_1 t + a E_0 \frac{\partial \alpha}{\partial(\Delta r)} (\sin 2\pi \nu_v t) \sin (2\pi \nu_1 t)$$

and

$$\mu = \alpha_0 E_0 \sin 2\pi \nu_1 t + \frac{aE_0}{2} \frac{\partial \alpha}{\partial(\Delta r)}$$

$$\cos 2\pi (\nu_1 - \nu_v) t - \frac{aE_0}{2} \frac{\partial \alpha}{\partial(\Delta r)} \cos 2\pi (\nu_1 + \nu_v) t \tag{14.7}$$

From this equation we see that the induced dipole moment in the molecule varies with three frequencies and classically can emit radiation at these frequencies. The first is the unshifted frequency ν_1 and this is the Rayleigh scattered wave whose intensity depends on α_0. The two displaced frequencies give rise to the first order (single phonon) Raman scattering. The decreased frequency is the Stokes line and the increased frequency is the anti-Stokes line. The intensities of the Raman lines depend on the magnitude of the change of polarizability associated with particular vibrational modes. Since at room temperature most molecules are in the vibrational ground state, the anti-Stokes lines, which originate from higher energy states, are typically much lower in intensity.

The conservation of momentum plays an important role in the understanding of Raman scattering in crystals as in all interaction of light with solids. Figure 14.2(a) shows the dispersion curves for the optical and acoustic lattice modes in a solid. In an ideal, i.e. perfectly periodic, solid, lattice dynamics tells us that because of the periodicity of the lattice the normal-mode vibrations are waves of infinite extent. The frequency of the phonon

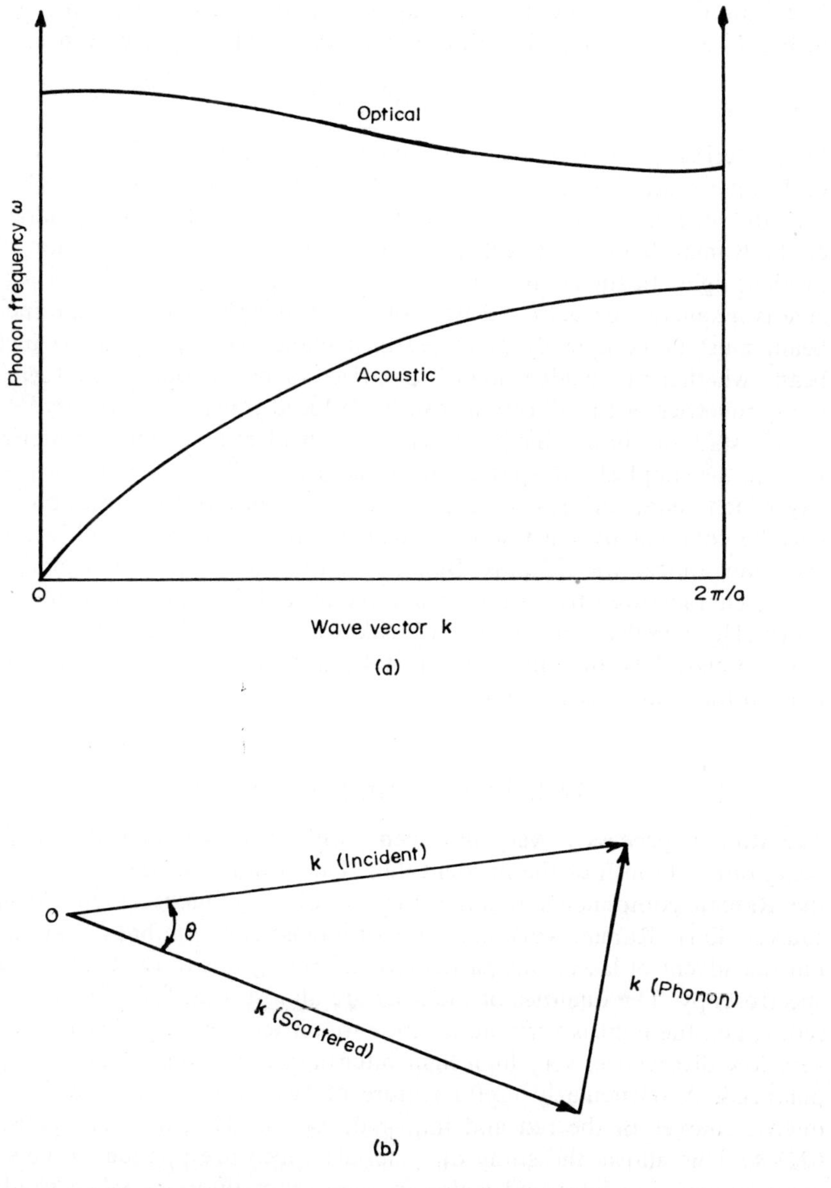

FIG. 14.2. (a) Phonon dispersion curves indicating the optical and acoustic branches. (b) Conservation of momentum diagram.

(quantum of vibrational energy) varies with momentum throughout the Brillouin zone out to the zone edge determined by a wave vector $\mathbf{k}_{max} = 2\pi/a$, where a is the crystal lattice constant. For typical values of a, $\mathbf{k}_{max} \sim 10^8\ cm^{-1}$. The conservation of momentum diagram for the scattering event is shown in Fig. 14.2(b). For a visible photon with an associated wave vector

$$\nu/c \sim 10^5\ cm^{-1},$$

the scattering can only create or destroy phonons with a $\mathbf{k} \sim 0$. Thus Rayleigh or Raman scattering in a crystal will be confined to only certain phonons near the zone centre. Information can also be deduced from the polarization of the Raman lines for different crystal orientations and polarization of the incident light. In the simplest case, if the polarizability ellipsoid for a molecule is a sphere, the scattered light observed at right angles to the incident beam must be completely polarized in a plane orthogonal to the incident beam, whether the incident light is polarized or not. From such information the symmetries of the vibrations can be deduced using group theory.[(1)]

This state of affairs changes when one considers amorphous materials.[(4)] Even in the simplest amorphous materials, the translational order or periodicity which characterizes crystals is lost. Thus typically the vibration modes may be coherent over a few tens of angstroms, distances which are much less than optical wavelengths. Because of this such vibrational modes will not be characterized by a simple wave vector and the $\mathbf{k}$ selection rule breaks down. The ramifications of such a breakdown in the selection rule is that in amorphous solids, in contrast to crystals, all vibrational modes may contribute to the Raman scattering.

14.3. Experimental Technique

The Raman process is very inefficient, with scattered intensities typically being one millionth of the incident flux. This means that the observation of the Raman components requires the use of very intense monochromatic sources. Early Raman work made use of intense emission lines in arc lamps, but the advent of lasers was particularly advantageous to the field of Raman spectroscopy. The qualities of the laser are almost ideal for Raman spectroscopy, i.e., the light is very monochromatic, it has very small beam size and very low divergence, very high light intensities are available and it is plane polarized. A particularly useful feature of laser sources is that they offer intense sources in the red and infra-red, e.g. the He–Ne laser operates at 6328 Å. This allows the study of materials which are opaque to the strong mercury emission lines and which in some cases fluoresce when irradiated with short wavelengths.

A representative experimental arrangement for measuring Raman scattering is shown in Fig. 14.3. Although many variations occur, this represents the essential experimental requirements. The laser beam, chopped at some suitably chosen frequency to allow the use of phase sensitive detection, is focussed on to the sample. The component of the light scattered at 90° to

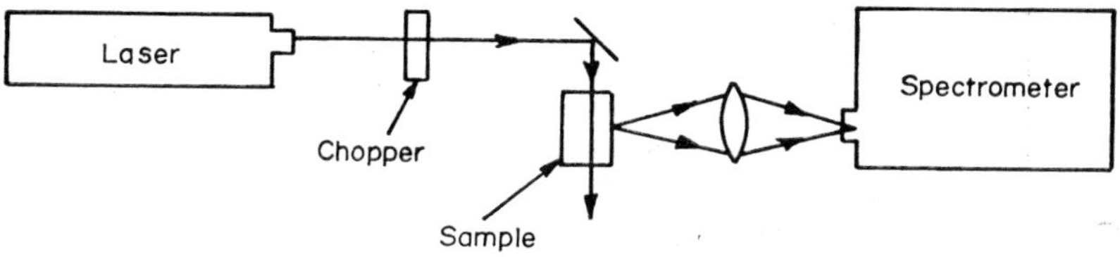

FIG. 14.3. Basic experimental arrangement for observing the Raman effect.

the direction of light propagation is collected by a lens system and spectrally analyzed by a spectrometer. Certain materials which absorb even the red and infra-red laser lines and associated Raman lines, such as Si, must be studied in reflection or using a back-scattering technique. Here again the features of a laser are particularly accommodating in this respect. Since the laser may be focussed to a very small spot, the power densities at the focus can be extremely high and the absolute amount of back-scattered light is detectable. Such high power dissipation in some instances can be extremely troublesome. In the case of amorphous Se, for example, the accompanying rise in temperature can lead to conversion into polycrystalline trigonal selenium at best and at worst to complete destruction of the sample.

14.4. EXPERIMENTAL RESULTS

(i) *Selenium and Arsenic–Selenium Alloys*

The first measurements of Raman scattering for amorphous Se and the crystalline allotropes, viz. monoclinic and trigonal selenium, were reported by Lucovsky *et al.*[5] Later measurements were reported by Ward[6] and included some polarization studies. Figure 14.4 shows the Raman spectra for amorphous Se and the two crystalline allotropes measured by Ward.[6] The Raman spectra for the amorphous phase is dominated by a peak at $250\,cm^{-1}$. This peak is also characteristic of the spectra for monoclinic selenium. The basic structural unit of monoclinic selenium determined from X-ray analysis is an eight membered puckered ring, Se_8. The Raman peak at $250\,cm^{-1}$ is therefore assigned to totally symmetric vibrational modes of the Se_8 ring on the basis of this evidence[5] and the fact that this line is polarization dependent.[6] A weak shoulder at $239\,cm^{-1}$ is observed in

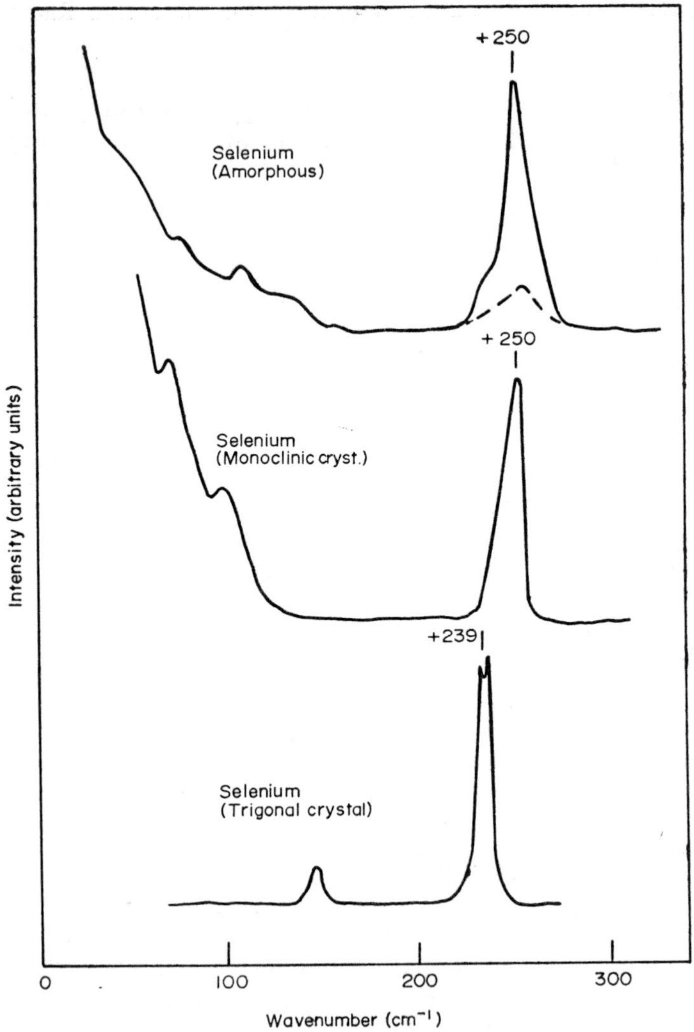

FIG. 14.4. Raman spectra of crystalline and amorphous forms of selenium. The dashed curve for amorphous Se corresponds to a different polarization of the incident light in which the *E* vector is parallel to the direction of observation. (After Ward, Ref. 6.)

amorphous Se and this coincides with the dominant peak for crystalline trigonal selenium. The structural unit known from X-ray analysis to be characteristic of trigonal selenium is that of infinite helical chains of selenium atoms. The greatly reduced strength of this peak in amorphous selenium is assumed to originate from the loss of long range order in the chains. It can be concluded therefore from Raman scattering studies of amorphous Se that this elemental amorphous insulator is unusual in containing both structural units, i.e. Se_8 rings and chains, which characterize the two crystal allotropes of Se.

Figure 14.5 presents experimental results of the Raman spectra of two compound chalcogenides, As_2S_3 and As_2Se_3, for both the amorphous[6] and crystalline[7] forms. The laser wavelength λ_0 is different for the two materials because the regions of optical transparency differ. As for the case

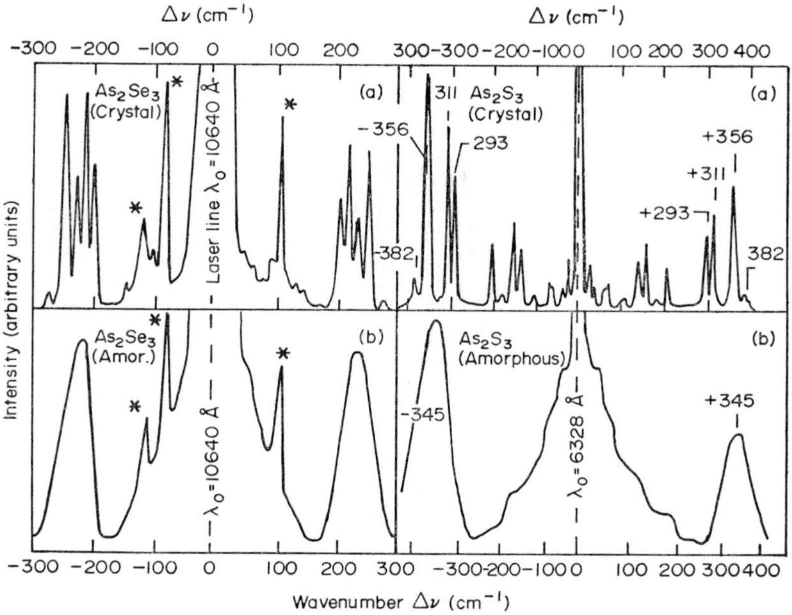

FIG. 14.5. Raman spectra of (a) crystalline As_2S_3 and As_2Se_3 (after Zallen *et al.*, Ref. 7), (b) amorphous As_2S_3 and As_2Se_3 (after Ward, Ref. 6). The laser wavelengths λ_0 is different for the sulphide and selenide because they are transparent in different regions of the spectrum

of Se, the striking fact is that the dominant peaks in the amorphous spectra are spectrally close to the positions they possess in the crystalline form. Characteristically, however, they are much broader. In the case of As_2S_3, for example, the four strong lines at 293, 311, 356 and 382 cm^{-1} observed in the crystal must presumably be correlated with the single broad peak at 345 cm^{-1}. Similar observations may be made for As_2Se_3. In fact, as pointed out by Zallen *et al.*,[7] there is a rigid shift of the spectra by a scaling factor of 0·7. This is a manifestation of the close similarity in structure between As_2S_3 and As_2Se_3. The close relationship between the amorphous and crystalline Raman spectra identifies the important vibrational modes as being associated with the fundamental structural unit of the crystal, i.e.

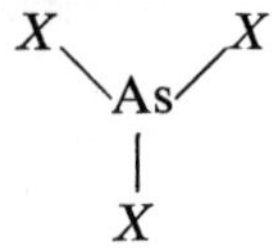

where X = S or Se. In the crystal form these units are connected with each

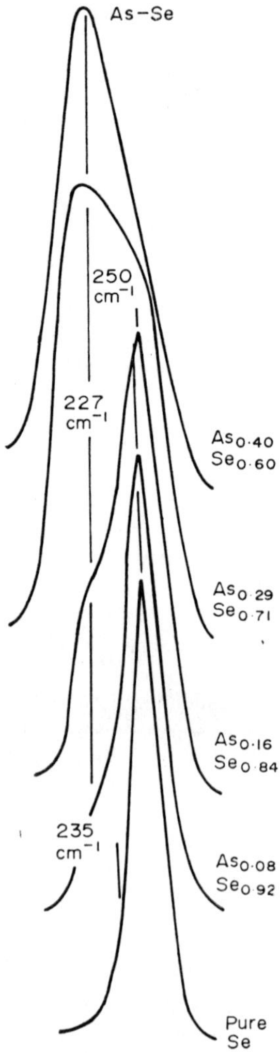

FIG. 14.6. Composition dependence of the Raman spectra of amorphous As–Se alloys. (After Ward, Ref. 6.)

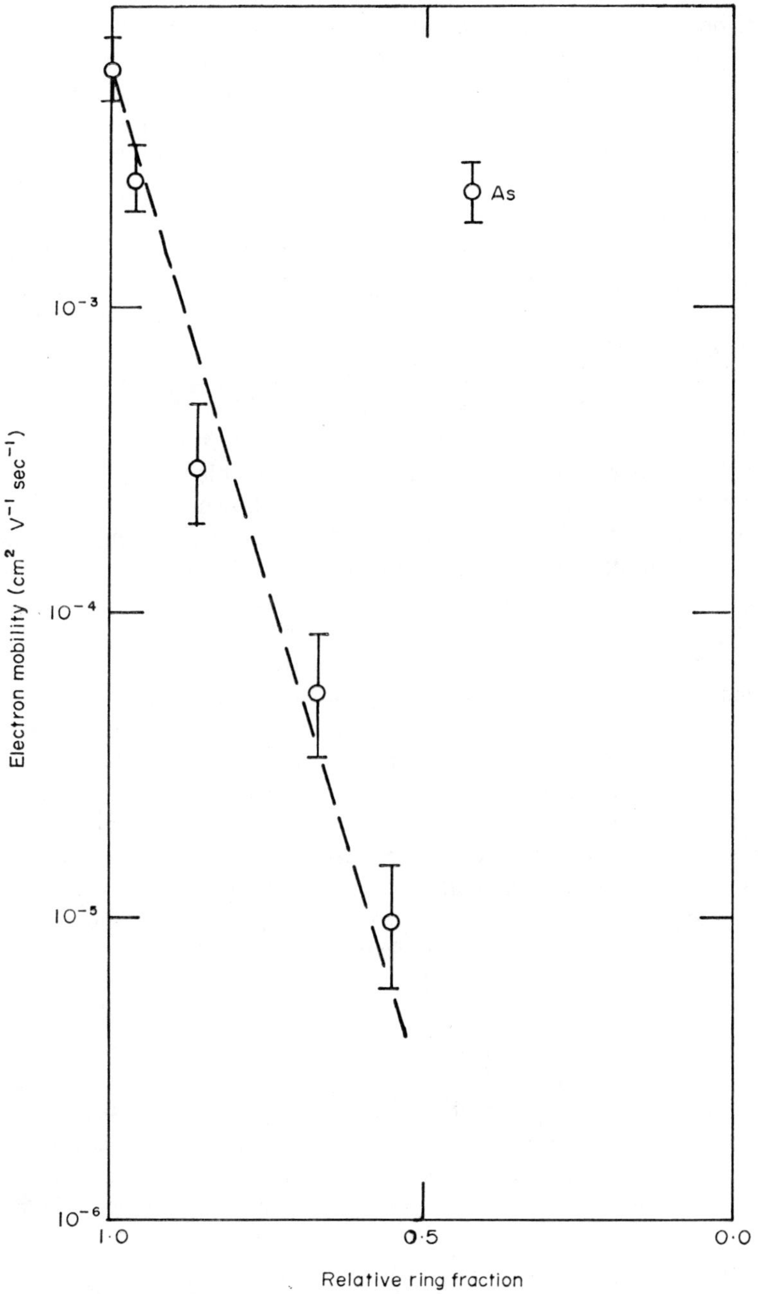

FIG. 14.7. Correlation between electron drift mobility and relative Se_8 ring concentration in amorphous As–Se alloys. (After Schottmiller *et al.*, Ref. 9.)

As atom bonded to three S or Se atoms and each S or Se atom is bonded to two As atoms and so on. In this way two dimensional layers are formed. In the amorphous state the Raman studies establish that the fundamental As–S or As–Se pyramid structure is preserved but does not allow any conclusion to be drawn as to whether any of the characteristic layer structure of the crystal persists. However, complementary optical absorption studies suggest that this is the case.(8)

Figure 14.6 shows the Raman spectra for As–Se alloys.(6) Comparison with Figs 14.4 and 14.5 shows that as the As content of the alloy is increased, the Raman line at 250 cm^{-1} (due to Se_8 rings) decreases in intensity while the 227 cm^{-1} line (due to As–Se network) grows. At 40 atomic per cent As (i.e. As_2Se_3) essentially no Se_8 rings remain and all the Se atoms form part of the branched As–Se network. Ward showed that the intensity of the 250 cm^{-1} Raman line is directly related to the concentration of Se_8 molecules.(6) With this relationship established it is possible to relate the density of Se_8 rings to electron drift mobility.(9) This is shown in Fig. 14.7 where it can be seen that the electron drift mobility depends exponentially on the density of Se_8 molecules. This is somewhat difficult to interpret simply in terms of an increasing density of traps, produced by the addition of As, decreasing a trap-controlled drift mobility, since the mobility would then be a linear inverse function of the density of traps. On the basis of a trap-controlled mobility, therefore, the only possibilities are that a geometric increase in trap density is produced by the incorporation of each As atom or that the As results in the reduction of the microscopic band mobility through a reduction in the density of states within the conduction band. Alternatively, the thermally activated drift mobility could be associated with a hopping between Se_8 molecules. In this model the introduction of As results in a reduction of the density of Se_8 molecules and hence an exponentially decreasing overlap of Se_8 wavefunctions. The independence of the electron drift mobility in amorphous Se on pressure(10) is not consistent with the association of this mobility with a hopping mode of transport however. It is interesting that the activation energy of the electron drift mobility in monoclinic selenium crystals, which is associated with a trap-controlled mobility, is very close to that of the electron drift mobility in amorphous Se.(11) In addition, no appreciable electron transport has ever been observed in As_2Se_3.(12) These measurements have brought out how it is possible to relate the results of Raman spectroscopy to transport processes in amorphous Se, although as yet no unambiguous interpretation is possible.

(ii) *Amorphous Si*

Smith *et al.*(13) have measured the Raman spectra of amorphous Si and related tetrahedrally bonded semiconductors. The measurements were made at several temperatures but very little change was observed in going from room temperature to 27K. The spectrum consists of two broad peaks at 480 cm^{-1} and 170 cm^{-1}. Crystallization of the amorphous films by heating resulted in the observation of the normal crystalline Raman spectrum with a sharp line at 520 cm^{-1}. An approximate form for the intensity $I(\omega)$ of a Raman line at wave number ω for a disordered solid has been derived by Shuker and Gammon[4] who showed that

$$I(\omega) = \sum_b C_b\,[n(\omega, T) + 1]\,\rho_b(\omega)/\omega \tag{14.8}$$

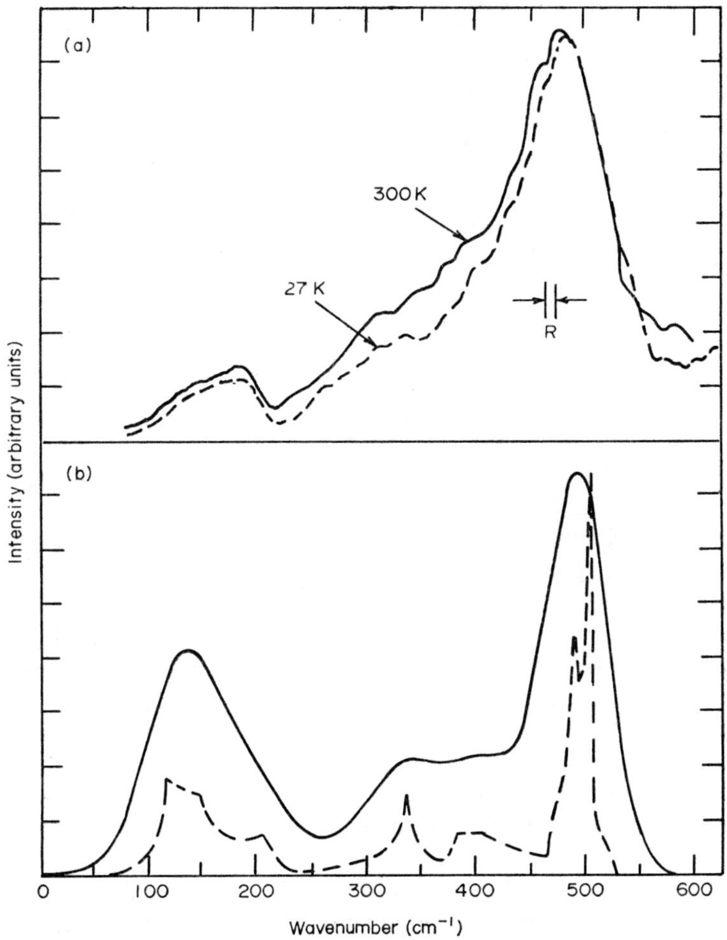

FIG. 14.8. (a) Raman spectra of amorphous Si at 27 K and 300 K reduced by $[n(\omega, T) + 1]/\omega$. Resolution is indicated by R. (b) Single phonon density of states of crystalline Si; dashed line is $\rho_c(\omega)$ from results of Dolling and Cowley (Ref. 15) and solid line is $\rho_a(\omega)$, obtained by broadening $\rho_c(\omega)$ as described in the text. (After Smith *et al.*, Ref. 13.)

where $n(\omega, T)$ is the Bose–Einstein distribution and $\rho_b(\omega)$ is the density of vibrational states in band b. C_b allows for different vibrational bands to couple with varying strengths to the radiation. This equation was derived on the basis that all vibrational modes in a band couple to the light; this being a direct consequence of the non-conservation of **k**. By reducing the raw Raman spectra by $[n(\omega, T) + 1]/\omega$, an experimental measure of the density of vibrational states in amorphous Si was obtained. This is shown in Fig. 14.8(a) for room temperature and 27K. The consistency of these curves support the interpretation of the spectra in terms of first order scattering. The dashed curve in Fig. 14.8(b) shows for comparison the vibrational density of states for crystalline silicon ρ_c, obtained from the results of Dolling and Cowley.(14) However, no calculated density of vibrational states of amorphous Si was available for comparison with the experimentally determined spectra. Smith *et al.*(13) therefore calculated the one-phonon density of states employing the phonon density of states for crystalline Si but incorporating a smeared distribution or disorder parameter δ. The smeared out function is given by

$$\rho_a(\omega) = \int_{-\infty}^{\infty} G(\delta, \omega - \omega')\, \rho_c(\omega')\, d\omega \tag{14.9}$$

where

$$G(\delta, \omega - \omega') = (2\pi)^{-\frac{1}{2}}\, \delta^{-1} \exp[-(\omega - \omega')^2/2\delta^2]. \tag{14.10}$$

An estimate for the value of δ was obtained by a comparison of the reduced Raman spectra [Fig. 14.8(a)] with ρ_c. From the dashed curve in Fig. 14.8(b) the phonon distribution for the crystal is seen to end abruptly at 520 cm^{-1}. In the case of the reduced experimental Raman spectra, however, considerable scattering intensity still exists 25 cm^{-1} above this cut-off value. On this basis Smith *et al.*(13,14) chose this value as an upper bound for δ. With this value and eqns (14.9) and (14.10) the solid curve in Fig. 14.8(b) was calculated and represents the one-phonon density of states of crystalline Si slightly smeared out, $\rho_a(\omega)$. The reduced experimental Raman spectra and $\rho_a(\omega)$ are generally similar in peak position and shape. The relative values of the major peaks, however, are not satisfactorily reproduced. This was attributed to assuming the constants C_b were the same for the entire spectra.(13) Nevertheless, the general agreement was taken to be suggestive that the phonon spectrum of the crystal is largely determined by near-neighbour interactions which are preserved in the disordered state. No gross changes were observed in the Raman spectra after annealing at temperatures of 400°C.

(iii) *Amorphous Te*

An investigation of this amorphous element is of interest in relation to the studies of amorphous Se. Te and Se are isomorphous but unlike Se no evidence has been found for the existence of a ring component in any form of Te. The potential complication of having a co-mixture of rings and chains

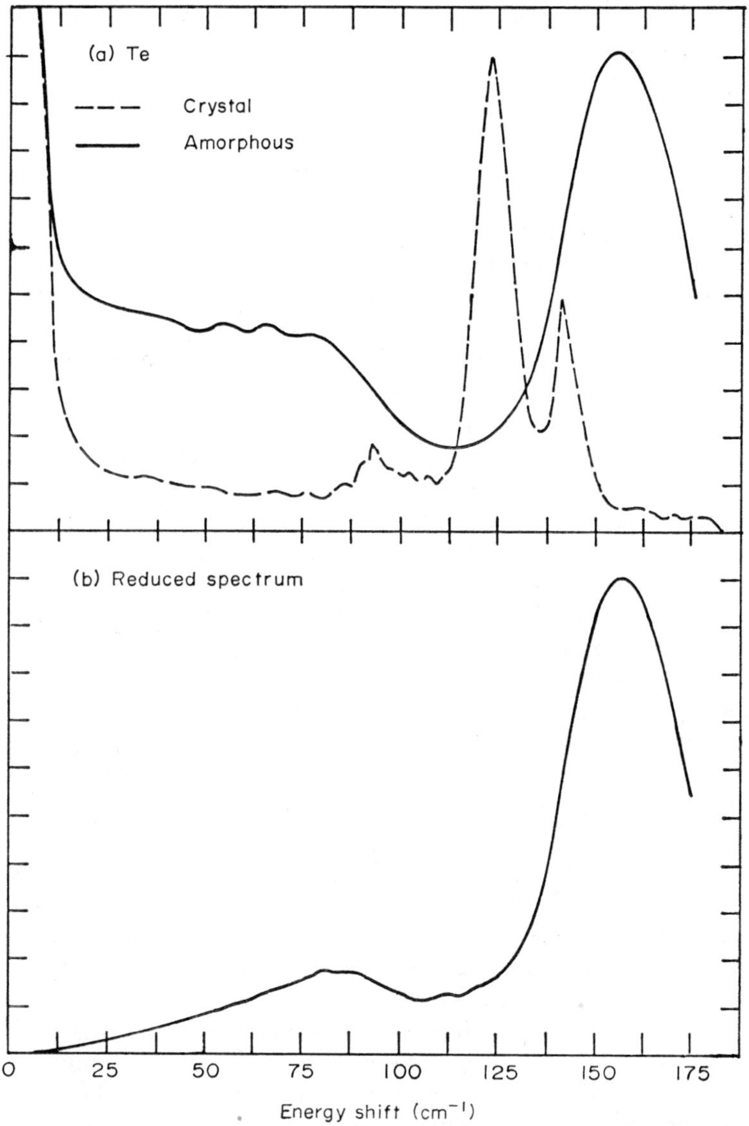

FIG. 14.9. (a) Raman spectra of amprphous Te (solid line) and polycrystalline Te (dashed line) at 77 K. (b) The reduced Raman spectrum of amorphous Te. (After Brodsky *et al.*, Ref. 15.)

is therefore avoided. Figure 14.9(a) shows results by Brodsky *et al.*[15] for the Raman spectrum of amorphous and crystalline tellurium measured at 77K. The latter data was obtained by warming the amorphous sample to room temperature, (amorphous Te converts to the crystalline form above 10°C), and then recooling. Figure 14.9(b) gives the reduced Raman spectrum obtained in an analogous manner to that for amorphous Si. It is clear that considerable differences exist between the spectra for amorphous and crystalline tellurium. In particular, the dominant Raman line for the amorphous phase lies at a higher energy than those in the crystalline sample. Brodsky *et al.*[15] calculated the one phonon density of states of crystalline Te and introduced a broadening parameter as was done for Si. The calculated spectrum has barely resolved maxima at 30 and 80 cm^{-1} and a well defined peak at 130 cm^{-1}. The 30 and 80 cm^{-1} peaks were associated with the asymmetrical peak at 90 cm^{-1} in Fig. 14.9(b) and the 130 cm^{-1} was associated with the 157 cm^{-1} peak.

The difference observed for the Raman lines between the ordered and disordered phases of Te was interpreted by Brodsky *et al.*[15] in terms of the shell model of vibrating atoms. Cochran's[16] homopolar shell model is comprised of two interactions. The first involves the Coulombic interaction between the dipoles created by the relative motion of the electronic shells to their associated cores. The second, while being determined by short range interactions, involves the coupling of a core vibration to that of another far away via the shells. According to Brodsky *et al.*[15] these long range interactions, which depress the phonon frequency of a single Te chain, are stronger in the crystalline form than in the amorphous phase due to the loss of long range order in the latter. A similar displacement, i.e. with the dominant amorphous peak lying at higher energy than that for the crystal, occurs if one compares amorphous Se and trigonal selenium (see Fig. 14.4). Brodsky *et al.*[15] raised the possibility that the dominant peak in the Raman spectrum of amorphous Se might be interpretable solely in terms of vibrations within a chain and that the corresponding peak appears at lower energy in the crystalline trigonal form because of the long range interaction between chains. The shift is smaller than in the case of Te because of the weaker inter-chain coupling in trigonal Se. With this interpretation the coincidence in energy of the dominant peak in amorphous Se and monoclinic Se is presumably co-incidental. The situation is complicated by the fact that no polarization effects were observed by Smith *et al.*[17] for amorphous Se which is inconsistent with the work described earlier by Ward.[6] The technique used by Smith *et al.*[17] was different in that it involved the back-scattering of strongly absorbed light in thin film samples; but the origin of the disagreement has not been identified. This is clearly an important question to resolve since the observation of a strong polarization effect in the 250 cm^{-1} line of amorphous

Se is critical in identifying the presence of a symmetric molecular species. In the view of Lucovsky,[18] for materials such as Si which form a continuous tetrahedrally bonded, but disordered network, an identifiable molecular species does not exist so that the Raman effect does not involve the normal polarization selection rules. Where a molecular species does occur then even in the disordered state polarization effects will be observable.

ACKNOWLEDGEMENTS

The author is indebted to Dr. A. T. Ward for helpful discussions. Thanks are extended to Dr. J. E. Smith, Jr., for his comments and providing a preprint of the work on amorphous Te.

REFERENCES

1. Herzberg, G., Infrared and Raman Spectra, Von Nostrand Co. Inc., New York, 1945.
2. Colthup, N. B., Daly, L. M. and Wiberley, S. E., Introduction to Infrared and Raman Spectroscopy, Academic Press, New York and London, 1964.
3. Proc. Int. Conf. on Light Scattering Spectra of Solids, (ed. Wright, B. G.), Springer-Verlag New York Inc., New York, 1969.
4. Shuker, R. and Gammon, R. W., *Phys. Rev. Letters* **25,** 222 (1970).
5. Lucovsky, G., Mooradian, A., Taylor, W., Wright, G. B. and Keezer, R. C. *Solid State Commun.* **5,** 113 (1967).
6. Ward, A. T., Proc. IIIrd Mardi Gras Symposium on Theoretical Chemistry, New Orleans (1971). *Adv. in Chem.* No. 110. 163 (1972).
7. Zallen, R., Slade, M. L. and Ward, A. T., *Phys. Rev.* **B3,** 4257 (1971).
8. Zallen, R., Drews, R., Emerald, R., and Slade, M. L., *Phys. Rev. Letters* **26,** 2564 (1971).
9. Schottmiller, J., Tabak, M. D., Lucovsky, G. and Ward, A. T., *J. Non-Cryst. Solids* **4,** 80 (1970).
10. Dolezalek, F. K. and Spear, W. E., *J. Non-Cryst. Solids* **4,** 97 (1970).
11. Spear, W. E., *J. Phys. Chem. Solids* **21,** 110 (1961).
12. Scharfe, M. E., *Phys. Rev.* **B2,** 5025 (1970).
13. Smith, J. E., Jr., Brodsky, M. H., Crowder, B. L., Nathan, M. I. and Pinczuk, A., *Phys. Rev. Letters* **26,** 642 (1971).
14. Dolling, G. and Cowley, R. A., *Proc. Phys. Soc.* (London) **88,** 463 (1966).
15. Brodsky, M. H., Gambino, R. J., Smith, J. E., Jr. and Yacoby, Y., *Physica Status Solidi* **52,** 609 (1972).
16. Cochran, W., *Proc. Roy. Soc.* **A253,** 260 (1959).
17. Smith, J. E., Jr., private communication.
18. Lucovsky, G., Optic Modes in As_2S_3 and As_2Se_3, to be published in *Phys. Rev.*

15. PHOTOELECTRONIC PROPERTIES OF AMORPHOUS MATERIALS

J. MORT

Xerox Corporation, Rochester Corporate Research Centre, Rochester, N.Y., U.S.A.

15.1. PHOTOCONDUCTIVITY: AN INTRODUCTION

The phenomenon of photoconductivity has played a major role in the development of our understanding of the photogeneration and transport of free carriers in crystalline solids. Under conditions of thermal equilibrium, the thermally generated carriers in a solid are distributed between available energy states according to Fermi statistics. A certain fraction of these electrons and holes are in conduction states and determine the dark electrical conductivity of the solid. This dark conductivity can only be measured if the measurement technique does not itself perturb the thermal equilibrium, e.g. by injection of charge from applied electrodes. The absorption of light which involves the photogeneration of free carriers disturbs the thermal equilibrium and leads to an enhanced number of electrons and holes in conduction states. Under constant illumination a steady-state photocurrent will be reached in which the rate of photogeneration is equalled by the various recombination processes through which the carriers are restored to the normal equilibrium condition. This disturbance of the equilibrium distribution of carriers by photoexcitation is the phenomenon of photoconductivity. It is this intrinsic feature of photoconductivity which makes it such a powerful technique in probing the range of energy states which exist in a solid and with which the carriers can interact.

Two very fundamental and distinct photoconductive processes can occur in solids. The simplest to conceive is one in which free electrons and holes are drawn by an impressed electric field to their appropriate electrode. Assuming unit quantum efficiency for photogeneration, the maximum photocurrent which can be drawn is equal to the photon flux. Although conceptually easy to understand, this type of so-called primary photocurrent can be observed only under certain experimental conditions. Examples of this type of photocurrent and the information it can give will be discussed later. The other class of photocurrents, and by far the more common, occurs when the extraction of a carrier at one electrode is accompanied by replenishment at the other. In this situation, as opposed to a primary photocurrent, the lifetime of the excess carrier does not expire when it reaches its respective electrode, since it is replaced by a carrier of like-kind. The carrier lifetime in this process is only terminated by the recombination of a free electron and hole. The actual recombination process can, and usually is, preceded by the trapping of one or both of the carriers. One central consequence of this characteristic of secondary photocurrents is that the maximum photocurrent density that can be drawn can exceed the photon flux. Much of the pioneering work in this area is due to Rose and for a much more detailed discussion of these processes and their ramifications the reader is directed to the excellent book by Rose.[1]

Before discussing experimental results on specific materials a brief review will be given of the various experimental techniques that are used for studying photoconductivity in solids. The techniques will be discussed within the framework of two types of excitation viz., non-uniform and uniform photoexcitation.

15.2. Techniques of Photoconductivity Measurements

(A) Non-Uniform Excitation

In this type of excitation the sample is usually sandwiched between two electrodes and the illumination is perpendicular to the plane of the electrode and consists of highly absorbed wavelengths. The arrangement is indicated in Fig. 15.1(a). As discussed elsewhere (see e.g. Chapter 10), excitation of free carriers can be accomplished by other than light, e.g. electron beam or X-rays. Since a large part of our discussion will be concerned with the mechanism of photogeneration in amorphous solids, we will not consider such alternative excitations, as the photogeneration mechanisms in these cases are quite complex and involve many secondary processes.

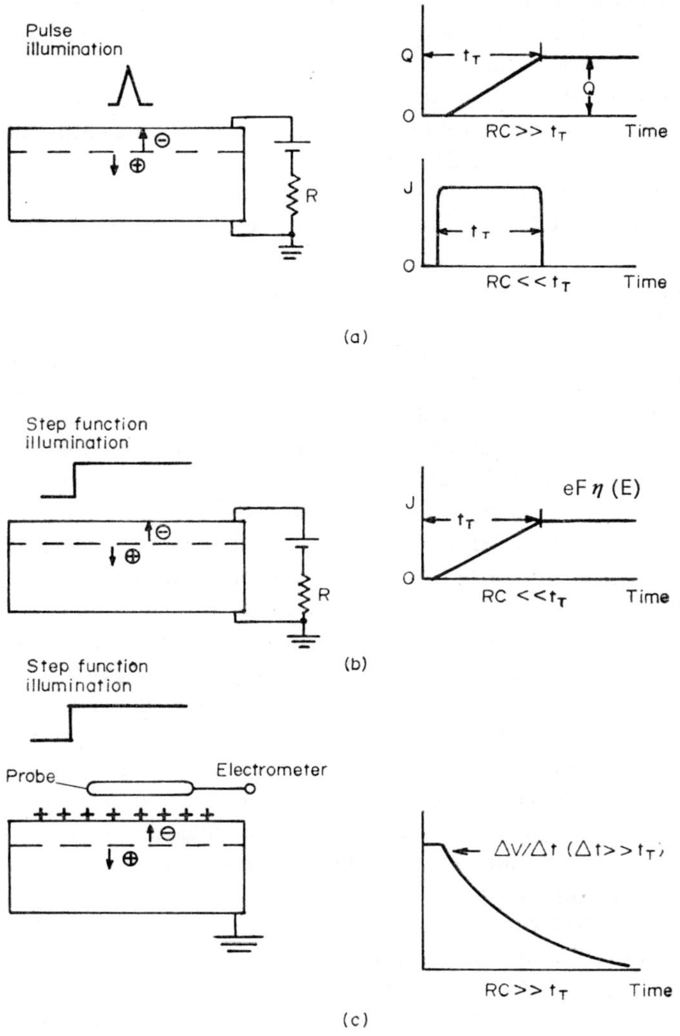

FIG. 15.1. Schematic representation of three different experimental techniques and observed responses under non-uniform illumination conditions: (a) Flash illumination with either $RC \gg t_T$ or $RC \ll t_T$ (R is the load resistor and C the stray and sample capacitance). (b) Step-function illumination where the rise of the step is much less than t_T and $RC \ll t_T$. (c) Xerographic discharge where the experimentally observed response is the photo-discharge of the surface potential.

(i) *Time of Flight Technique*

This is the most important technique which employs non-uniform excitation and since being developed by Spear[(2)] has been widely and successfully applied to the study of amorphous materials. As earlier lectures (see Chapter 10) have dealt with the technique, particularly as it applies to the measurement of drift mobilities, the discussion here will be brief and in the form of a recapitulation. The technique consists of using a short duration flash of light which produces free electron-hole pairs in an excitation depth small compared with the total sample thickness [see Fig. 15.1(a)]. Provided the flash duration is less than the transit time t_T of the carriers, which in turn is less than any bulk deep trapping lifetime, a time resolved transit of the carriers should be observable if the apparatus has sufficient time resolution. By choosing appropriate polarity of the illuminated electrode, and if the dielectric relaxation time t_D of the material is larger than t_T, the drift of either carrier without the complication of bulk recombination can, in principle, be studied independently of the other. This is a primary advantage of this technique together with the fact that it allows a separate measurement of the velocity of the carriers and their number. This is illustrated in Fig. 15.1(a) where the integrated transit signal i.e. accumulation of charge at the bottom electrode versus time, is shown. By choosing the RC time constant of the circuit, (C is stray and sample capacitance), to be much larger than the transit time t_T, the transit is observed in the open-circuit or V mode as a linear rising signal which saturates at t_T. From the viewpoint of studying photogeneration processes the important fact here is that the height of the integrated signal is proportional to the number of charges which left the excitation region, Q. This number is in general determined by the photogeneration step itself, and by trapping and recombination processes. The observed voltage signal V equals Q/C. Alternatively, in the current or I-mode ($RC \ll t_T$), it is the integrated area under the current pulse which equals Q. This technique therefore allows a direct measure of Q as a function of many important parameters, e.g. light intensity, excitation wavelength, electric field and temperature.

The transition from the transient to the steady-state current can be followed if a step function illumination of flux F is used where the risetime of the step is much less than t_T. For the small signal case it takes one transit time to reach the equilibrium current. This is sketched in Fig. 15.1(b). Here in the *current* mode the signal is a linearly rising ramp function with the constant current after one transit equal to the emission-limited current $J_e = eF\eta(E)$.[3]

These type of measurements can be made with the impressed electric field applied with two electrodes (usually metals) which are blocking, i.e. they do not inject significant charge, and an external power supply as shown in Fig. 15.1(a). Alternatively, the field can be derived by laying down a corona charge (blocking electrode) on the free surface, as in the xerographic process. This is shown in Fig. 15.1(c). This arrangement ensures that $RC \gg t_T$ and that the measurement will be made in the V-mode. In the limit of small signals, i.e. $Q \ll CV_0$, this measurement can give information about the *supply* of carriers.

This can be done by measuring the initial slope $(dV/dt)_{t=0}$ using a capacitive probe as the sample is discharged by continuous illumination [see Fig. 15.1(c)]. Provided the initial dV/dt is measured in a time larger than the first transit and tangentially so the sample voltage at that time is constant, then $J_e = CdV/dt = eF\eta(E)$ and it is possible to determine η.[4] The validity of this technique is checked by measuring CdV/dt as a function of thickness. In the absence of any bulk limitation i.e., the discharge is emission limited and controlled only by the injection rate of carriers into the bulk from the excitation region, CdV/dt at a given field (V/L) will be independent of thickness. This technique has the very useful advantage that the illuminated electrode (corona charge) is non-absorbing so that secondary photoinjection currents will not be a complication.

However, if the injected amount of charge $Q \gtrsim 0{\cdot}1\, CV_0$, then the free charge in transit in the sample, which in the case of non-uniform excitation and $t_D \gg t_T$ is not accompanied by neutralizing charge of the opposite sign, begins to distort the internal field and the space-change-perturbed region is reached. Ultimately when $Q = CV_0$, the field at the illuminated surface is driven to zero and the number of charges injected into the sample bulk becomes independent of light intensity. This limitation due to the transit of CV_0 of free charge is called the trap-free space-charge-limited region. At this point no information can be obtained concerning the mechanisms operative in the excitation region since the signal is only determined by the sample dimensions, applied field and the carrier mobility. A detailed discussion of the small and large signal cases in the I and V modes can be found in the literature.[3]

(ii) *Steady-State Measurements*

As demonstrated in the previous Section an important consequence of non-uniform illumination in insulating photoconductors is that the space charge limitations may be encountered. In the case of trap-free space-charge-limited currents (TFSCLC) this limitation can often be difficult to achieve. The total requirement is that CV_0 of charge be generated in a time less than t_T to ensure that the total CV_0 moves in one transit. Given a small enough t_T, which for a given sample thickness and applied field requires a large mobility, available light sources may not be sufficiently intense.

This whole picture changes with steady-state photocurrents produced by non-uniform illumination. So far it has been assumed that bulk deep trapping is unimportant. The primary advantage of the time of flight technique is that it is possible in a number of amorphous materials, where bulk deep trapping is relatively unimportant, to make the transit time much less than the bulk deep trapping lifetime, τ. In each transit, therefore, only a very few carriers are lost to deep traps. Even in these cases, however, the situation can change radically if one attempts to draw steady-state currents. At light intensities such that each transit is a small signal case, it is possible over many transits, i.e. the steady-state, to accumulate a CV_0 of trapped charge in deep traps by the steady integration of a very small number of charges trapped per transit. This effect, i.e. a transition from an emission-limited to a trap-limited space-charge-limited current (TLSCLC) has been measured in amorphous Se.[5] As in the TFSCLC case, the measurement of these photocurrents can give no information about the photogeneration process, although in principle, of course, such interaction with deep traps permits a probing of these states. The important point to be stressed here is that to derive the maximum information from a photoconductivity measurement, the experimenter must satisfy himself in what regime his measurements are being made in order to draw meaningful conclusions from the observations. A variety of experimental techniques exist to probe such deep states. These include observation of electrical glow curves[6] and measurements of the photo-induced noise spectrum.[7]

(*B*) *Uniform Excitation*

Uniform excitation refers to the case where electron-hole pairs are photoexcited essentially uniformly between the electrodes. This is a question of degree in the case of the sandwich configuration since totally uniform excitation precludes the possibility of any absorption at all. In a surface cell configuration, however, where the electric field is perpendicular to the direction of propagation of the light even strongly absorbed light results in uniform photoexcitation in the inter-electrode space. Contrary to the case of non-uniform excitation, electrical neutrality, except under special circumstances, is assured and the space-charge limitations discussed earlier will not be encountered. In addition, it is characteristic of this type of excitation that recombination processes can occur throughout the inter-electrode distance and can be studied up to very high light intensities, often beyond those which in the non-uniform excitation case would lead to space-charge-effects. Volume-excited photocurrents in sandwich cells are often very small because of the correlation between uniformity of excitation and lack of absorption. For this reason the surface cell configuration is a popular choice to study uniformly excited photocurrents. Despite the relative simplicity of making this type of sample these measurements suffer from two deficiencies—some uncertainty about the uniformity of the electric field and the fact that the surface of the sample is a significant part of the photo-excited region.

As before in the case of non-uniform excitation, if replenishment of the carriers at the electrodes does not occur, the maximum photocurrent that can be drawn is equal to the photon flux. If replenishment does occur then photoconductive gains in excess of unity may be observed. The phenomena are usually further complicated by the occurrence of trapping and a whole spectrum of effects may be observed contingent on a wide range of parameters—which carriers are replenished, which are mobile, the precise distribution of trapping states, etc. For a comprehensive treatment of the range of phenomena which are possible the reader is directed to the numerous texts dealing with these problems.[1,6]

15.3. Photoconductivity Results for Amorphous Materials

The literature on photoconductivity in amorphous solids is already quite extensive and is growing rapidly. It would be quite impossible to discuss more than a few representative pieces of work in this area. As illustrations of the application of some of the techniques outlined above, the discussion will centre on three amorphous solids, amorphous Se, amorphous As_2Se_3 and an amorphous alloy As_2Te_2Se. We shall first discuss photoconductivity in these solids and then turn to a brief discussion of photoluminescence.

(A) *Photoconductivity and Photogeneration*

(i) *Amorphous* Se *and* As_2Se_3

Amorphous Se is probably the best characterized amorphous photoconductor in terms of its optical, charge transport and photoconductivity properties. Since the optical and charge transport measurements are discussed elsewhere, they will be referred to here as they bear on the question of the photogeneration process.

A striking characteristic of amorphous Se is that a spectral gap exists between the optical absorption and photogeneration efficiency edges.[8] Initial measurements of the voltage dependence of the photodischarge rate of amorphous Se, with the illuminated surface positively charged, were interpreted in terms of bulk trapping of the holes. It was presumed, therefore, that the observed photocurrent was limited by how *far* the holes could move. Tabak and Warter[9] using the time of flight technique were able to measure directly the Schubweg $\mu\tau E$ where τ is the lifetime with respect to deep traps. The results of these measurements showed unequivocally that this was not the parameter limiting the photodischarge, since for fields much smaller than those for which the photodischarge was field dependent, the Schubweg for holes still exceeded the sample thickness. These results therefore led to the conclusion that the field dependence of the discharge was due to the supply of carriers out of the excitation region being field dependent. These conclusions were verified by the measurements made by Tabak and Warter shown in Fig. 15.2.[9] The upper curve in this figure shows the height of the integrated transit signal photogenerated by 4300Å light as a function of field. The lower curve is the peak photocurrent induced by a millisecond light pulse for two different thicknesses. This measurement corresponds to the type of measurement described in Fig. 15.1(b), i.e., the millisecond pulse is greater than t_T so that the emission-limited current is established but sufficiently short that a transition into TLSCL current does not occur. For excitation wavelengths larger than 4500Å the square root dependence was followed by a very rapid change in apparent quantum efficiency with field. As discussed by

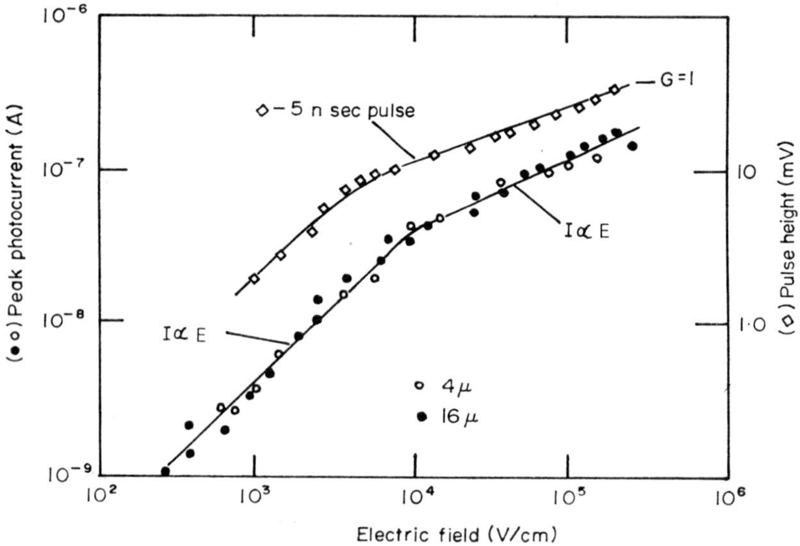

FIG. 15.2. Experimental results for amorphous Se. Upper curve is a plot of the height of the integrated hole transit signal produced by a 5 nsec pulse versus electric field. The lower curve is a plot of the hole photocurrent produced by a 1 msec pulse versus field for two thicknesses. (After Tabak and Warter, Ref. 9.)

Tabak and Warter[9] essentially two possibilities exist for the field dependent carrier supply. Either free carriers are generated and then lost via some recombination process or the photogeneration act itself is field dependent. These authors[9] presented facts to support the conclusion that the latter model was correct, i.e., the photogeneration process in amorphous Se is field-controlled. It was noted that the number of photogenerated carriers is linear in light intensity so that any conventional recombination process must be monomolecular. The characteristic shape of the number of collected carriers as a function of field, linear up to fields $\gtrsim 10^4$ V/cm and then approximately a square root dependence, is not significantly affected by changes in the depth of the excitation region. The very effective recombination centres required by assumption of a monomolecular recombination process would be very effective deep traps for one or other of the carriers in the bulk. Since direct measurements of the bulk deep trapping lifetime for both carriers[9] led to values of several tens of microseconds, this is not the case unless the density of recombination centres (or their type) in the surface region is different from that in the bulk. This possibility was ruled out because of the independence of the carrier supply-field plot on exciting wavelength and the inability to change it by mechanical abrasion.

Pai and Ing[10] also reported the observation of field dependent photogeneration in amorphous Se. In addition they reported that the photogeneration process was thermally activated with an activation energy which decreased with increasing excitation photon energy.

Scharfe,[4] again by use of the time of flight technique and xerographic discharge method, has reported that the supply of carriers from the excitation region in amorphous As_2Se_3 is field dependent. The field dependence of carrier supply is very similar to that for amorphous Se. The subsidiary experiments performed on amorphous Se, to determine that the field dependent supply is actually due to a field controlled photogeneration step, have not been carried out for As_2Se_3. Indeed it is known that the electron range in As_2Se_3 is extremely small[4] so that, unlike the case of amorphous Se, no special density of surface states need be invoked to account for substantial electron trapping which presumably occurs in the excitation region. For As_2Se_3, therefore, monomolecular recombination processes cannot be so easily discounted. Nevertheless, in the case of As_2Se_3 the quantum efficiency, even in the region of strong absorption, still changes with photon energy suggesting the occurrence of non band-to-band transitions.[7]

We now turn to possible interpretations of these measured photoconductive properties of amorphous Se and As_2Se_3 and we first consider the former material. Broadly speaking, there are two distinct phenomena which have to be accounted for—the large non-photoconducting absorption and the field dependence of the photogeneration process. In any attempt to understand the non-photoconducting absorption one must be aware that amorphous selenium is an atypical amorphous material in that it is basically a two-phase disordered molecular solid. This fact was first established by Lucovsky[11] who showed that vitreous Se is a mixture of polymeric chains and Se_8 ring molecules. Some experimental and theoretical evidence associates the non-photoconducting absorption with localized molecular excitations within the Se_8 rings, although this has recently been questioned by Adler[12] and Davis.[13] Nevertheless, this non-photoconducting absorption does not appear characteristic of the amorphous state, i.e. due to excitations between localized tail states, since a very similar, in fact larger, non-photoconducting absorption occurs in orthorhombic sulphur crystals consisting of S_8 rings.[14]

The relation of these localized, excitonic transitions to the photogeneration process in amorphous Se centres around three alternative models. The first proposed by Pai and Ing[10] assumes that the so-called non-photoconducting transitions can lead to free carriers provided sufficient energy to dissociate the localized excitation is available from a combination of exciting photon energy, temperature and electric field. This model leads to the conclusion that with increasing excitation energy a decreasing thermal activation energy will be required and this energy can be modified by a Poole–Frenkel lowering of the residual binding energy. This model could account for most of Pai and Ing's observations.[10] The field dependence of the quantum efficiency which enters through the Poole–Frenkel effect could not, however, explain the observed field dependence at low fields. The second model, originally proposed by Hartke and Regensburger[8] and subsequently extended by

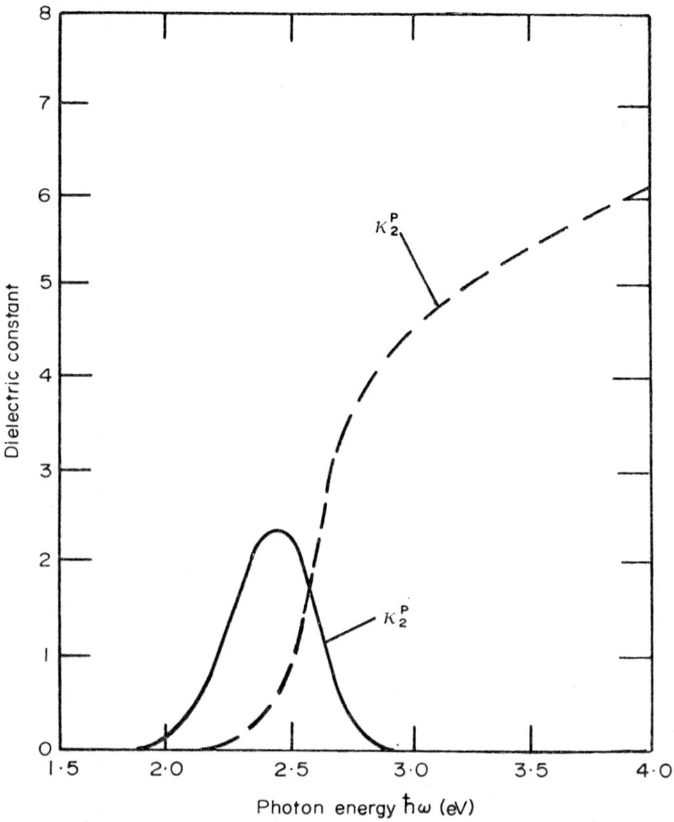

FIG. 15.3. The photon energy dependence of the imaginary parts of the complex dielectric constant for amorphous Se. $\kappa_2{}^N$ characterizes the non-photoconducting process and $\kappa_2{}^P$ the photoconductive process. (After Lucovsky, Ref. 15.)

Lucovsky,[15] proposed that two distinct, competing absorption processes occur in amorphous Se. One involves localized-excitations which never lead to free carriers and the other involves transitions between extended or band-like states and hence unit quantum efficiency for photogeneration (see Fig. 15.3).

In this model the shift of the absorption edge with temperature is essentially due to the shift of the extended states or quantum efficiency edge. The functional form of the imaginary part of the dielectric constant associated with the photoconductive transitions is such that the change in quantum efficiency with temperature increases with decreasing photon energy. In this way the dependence of the observed activation energy for photogeneration with wavelength observed by Pai and Ing[10] can be accounted for. The sharp upturn in the field dependence of the quantum efficiency observed only at larger wavelengths by Tabak and Warter[9] is interpreted in terms of a Franz–Keldysh shift of the photoconductive edge. Finally, a third model has been proposed by Davis and Mott[16] which is based on the reduced probability in the amorphous phase for the separation of an electron–hole pair. When an electron is excited from a localized state in the valence band to a non-localized state in the conduction band, it can only be considered free when it has diffused a distance R out of the attractive potential of the trapped hole, such that its binding energy is only a few kT. The thermalization time for this diffusion is expected to be very short because of the rapid loss of energy of a hot carrier in an amorphous solid due to non-conservation of $\mathbf{k}$. Thus a minimum energy in excess of the optical gap energy must be initially given to the electron to assure that it becomes free. Davis and Mott[16] estimated this excess energy using appropriate constants for Se to be $\sim 0{\cdot}5$ eV. Thus a gap between the optical absorption edge and photoconductivity edge of $\sim 0{\cdot}5$ eV was predicted and which is indeed observed. For this model to account for the observed field dependence, even for photon energies greater than 3·0 eV, it would have to be assumed that initial excitations between extended states always relax to bound states. In any case, this model like all the others, cannot by itself account for the field dependent photogeneration at fields less than 10^4 V cm^{-1}, where Poole–Frenkel effects should be negligible.

Of the models described, the one by Lucovsky[15] specifically excludes the non-photoconducting absorption as arising from the amorphous nature of selenium e.g., tail states, but attributes it to the molecular nature of this elemental material. In the case of As_2Se_3, on the other hand, Lucovsky[15] suggests that it may be more appropriate to describe the photoconductivity in this material with a model involving one electron states extending into the forbidden gap. Certainly in an alloy such as this, compositional as well as translational disorder may be important.

The only serious attempt to explain the low field behaviour of the photogeneration process, such as that observed in amorphous Se, is that by Warter.[17] The core of his argument is that following the absorption process an intermediate excitation must occur which can result in a free carrier or decay (initial or geminate recombination). Fields less than a critical value $E_c = (kT/e)/r_c$ (r_c is the Coulomb capture radius) would contribute less than thermal energy and therefore the photogeneration would be field independent and show a low field saturation. As pointed out by Warter,[17] at fields smaller than $E_c/10$ the photogeneration in Se shows no sign of saturation. This apparent paradox was resolved by invoking the concept of dimensionality. In three dimensions, even in the limit of zero applied field, some carriers will escape and become free. In the lower two or one dimensional space, however, even these carriers which escaped in three dimensions at zero applied field never became free. Thus contrary to the case of three dimensions, in a random walk constrained to one or two dimensions the smallest external field will affect the escape rate of a carrier. Warter[17] therefore proposed that the field dependence of photogeneration at low fields is related to local asymmetry in the carrier transport properties such that the path available to a carrier is more nearly one dimensional than three dimensional. As mentioned by Warter,[17] some low field limit to the photogeneration efficiency may exist because of the observation of fatigue effects in Se produced by large exposures at zero applied field.

(ii) As_2Te_2Se

We now turn to a discussion of photoconductivity in amorphous chalcogenides induced by uniform excitation. Weiser *et al.*[18] have made photoconductivity measurements on the alloy As_2Te_2Se using a surface cell geometry for the sample. Two pertinent measurements they made are shown in Fig. 15.4. Figure 15.4(a) shows the photoconductance induced by a He–Ne laser (6328Å) as a function of photon flux, F, over seven orders of magnitude at two different temperatures. In both cases a linear dependence on flux is followed by a well-defined square root dependence. The transition between the two regions occurs at a lower flux at the lower temperature. Figure 15.4(b) shows the temperature dependence of the photoconductance at two light fluxes, one which lies in the linear regime and the other in the square root regime over the temperature range covered. The temperature dependence is approximately exponential in each case but of opposite sign. Returning to Fig. 15.4(a), if a light flux is chosen so that as a function of temperature it shifts from the linear to square root region, the photoconductance as a function of temperature will exhibit a maximum. Similar behaviour has been reported in As_2Se_3 by Shaw,[19] Marshall and Owen[20] and in a quaternary alloy by Fagen and Fritzsche.[21]

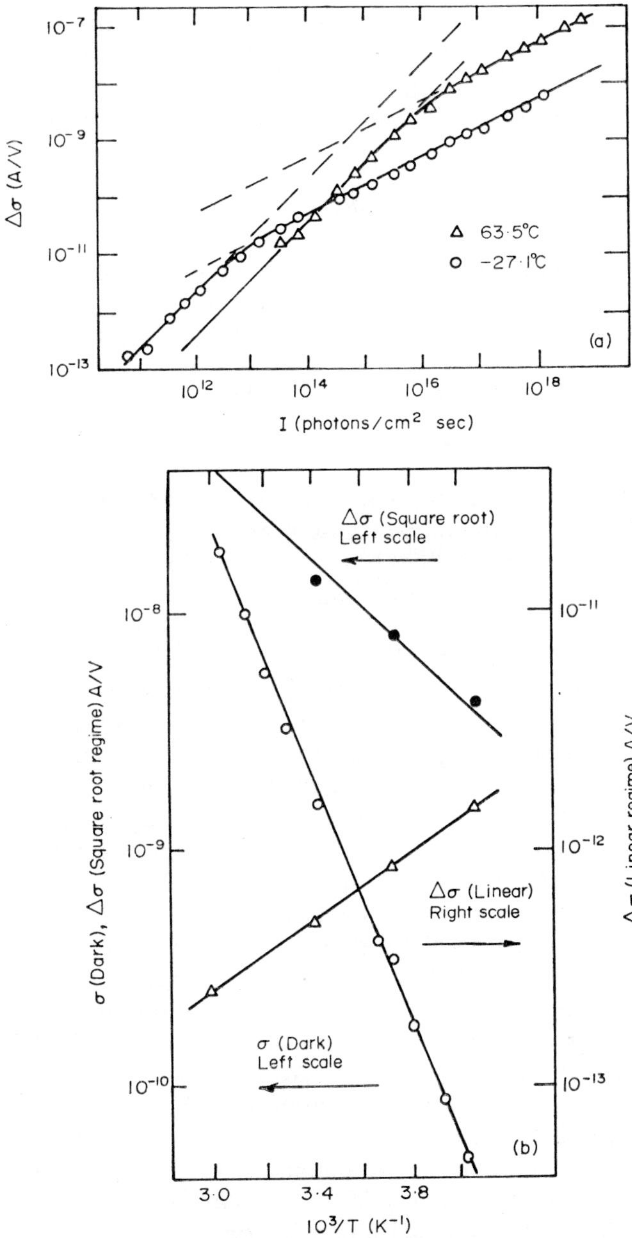

(a)

(b)

FIG. 15.4. (a) Photoconductance versus photon flux at two temperatures for As_2Te_2Se, using He–Ne laser (6328 Å) excitation. (b) Photoconductance (linear and square root) regime) and dark conductance versus inverse temperature for As_2Te_2Se. (After Weiser *et al.*, Ref. 18.)

As discussed by Weiser *et al.*,[18] the light flux dependence suggests a transition from a monomolecular to bimolecular recombination process as the light flux increases. Assuming the material to be intrinsic ($n_0 = p_0$) and that charge neutrality exists ($\Delta n = \Delta p$), the rate of change of the excess carrier density Δn is

$$\mathrm{d}(\Delta n)/\mathrm{d}t = -b(\Delta n^2 + 2n_0 \Delta n) + G \tag{15.1}$$

where b is the recombination constant, G is the uniform photoexcitation rate and n_0 is the density of thermal carriers available to recombine with the excess carrier density. In the steady state $\mathrm{d}(\Delta n)/\mathrm{d}t = 0$ so that

$$\Delta n_{ss} = (G/b)^{\frac{1}{2}} \quad \text{if} \quad \Delta n \gg n_0 \tag{15.2}$$

$$\Delta n_{ss} = G/2bn_0 \quad \text{if} \quad \Delta n \ll n_0 \tag{15.3}$$

Δn is the total excess carrier density, some of which may occupy trapping states so that only a fraction of Δn (those in the transport states) may contribute to the photoconductivity. The assignment of the non-linear flux dependence to a bimolecular process was verified by studying the decay of the photocurrent induced by a very fast GaAs laser pulse of sufficient intensity so that $\Delta n \propto F^{\frac{1}{2}}$. An excellent fit was obtained with the theoretical decay rate expected for a bimolecular recombination process. In order to explain the range of behaviours observed for the photoconductivity as a function of light flux and temperature, Weiser *et al.*[18] proposed a recombination model which they feel has a general validity for amorphous semiconductors. In addition to the concept of a mobility edge, which has been discussed in early lectures, they introduced the concept of a recombination edge. The exciting light induces transitions between states above the optical gap. The photocarriers thermalize by losing kinetic energy via phonon emission and populate the localized states. That fraction of carriers, which in the steady state exist in states above the mobility edge, are the determining ones for the observed photocurrents. The thermalization rate becomes progressively slower as the spatial separation of the localized states increases and the wave-function overlap decreases. At a characteristic energy $\mathcal{E}_R$, this thermalization rate equals the recombination rate. $\mathcal{E}_R$ is taken to be symmetric about the midgap for electrons and holes and defines the recombination edge. Excess carriers can only recombine with carriers of opposite sign if those carriers occupy states above $\mathcal{E}_R$. Carriers in states below $\mathcal{E}_R$ are considered too localized to communicate with oppositely charged carriers. Assuming an exponential decay of tail states with energy below the mobility edge, Weiser *et al.*[18] derived the following expression for the density of excess carriers in mobile states $\Delta n_c(x)$.

For high light intensities, with $\mathcal{E}_0$ as a band tailing constant

$$\Delta n_c(x) = (G/b)^{\frac{1}{2}} \exp [(\mathcal{E}_c - \mathcal{E}_R)/\mathcal{E}_0] \exp - [(\mathcal{E}_c - \mathcal{E}_R)/kT]. \qquad (15.4)$$

For low light intensities

$$\Delta n_c(x) = (G/2b) \exp (\Delta\mathcal{E}_R - \tfrac{1}{2}\Delta\mathcal{E}_c)/kT \qquad (15.5)$$

where $\Delta\mathcal{E}_R$ and $\Delta\mathcal{E}_c$ are the recombination and mobility gaps respectively. $\Delta n_c(x)$ is the local excess mobile carrier concentration but integration over the sample thickness does not affect the conclusions to be drawn.

From eqn (15.4) it is seen that at high light levels the photocurrent will be proportional to $F^{\frac{1}{2}}$. Assuming the mobility is weakly temperature dependent, the photocurrent will increase with increasing temperature because the fraction of excess carriers above the mobility edge increases. From Fig. 15.4(b) $\mathcal{E}_c - \mathcal{E}_R = 0{\cdot}18$ eV, so that this is the energy separation between the mobility and recombination edges. For low light levels [eqn (15.5)] an additional temperature dependence is involved because the recombination lifetime, determined by the value of n_0 [eqn (15.3)], is now temperature dependent. From Fig. 15.4(b), the activation energy $\Delta\mathcal{E}_R - \frac{1}{2}\Delta\mathcal{E}_c = 0{\cdot}13$ eV. It is thus easy to find that $\Delta\mathcal{E}_c/2 = 0{\cdot}49$ eV which is in good agreement with the value of 0·47 eV determined from the dark conductivity measurements and the assumption that the material is intrinsic. For As_2Se_3, measurements by Marshall and Owen[(20)] can be interpreted in terms of the recombination level lying 0·37 eV below the corresponding band edge.

(B) *Photoluminescence*

No *ad hoc* assumption can be made as to whether photoluminescence will be observed in amorphous solids and, if so, at what energy the emission should peak. In fact, in the case of amorphous Se no luminescence has been observed. This is not the case for As_2Se_3[(22)] or As_2Te_2Se[(23)] and other chalcogenide alloys.[(24)] On the basis of the Weiser, Fischer and Brodsky recombination edge model, however, if radiative recombination did occur it would be expected to peak with a photon energy equal to $\Delta\mathcal{E}_c - 2(\mathcal{E}_c - \mathcal{E}_R)$. Since these values are determined from the temperature dependence of the dark conductivity and photoconductivity respectively, a direct test of the model requires this data together with photoluminescence studies on the same material. Only in the cases of As_2Te_2Se[(23)] and As_2Se_3[(20,22,25)] has this been done. On the basis of dark and photoconductivity measurements the calculated emission peak for As_2Te_2Se is 0·58 eV[(18)] and for As_2Se_3 is 1·1eV.[(20)] These compare with experimentally observed values of 0·60 eV and 0·90 eV for As_2Te_2Se[(23)] and As_2Se_3[(22,25)] respectively. The shape of the photoluminescent spectra for As_2Te_2Se is shown in Fig. 15.5.[(23)] This

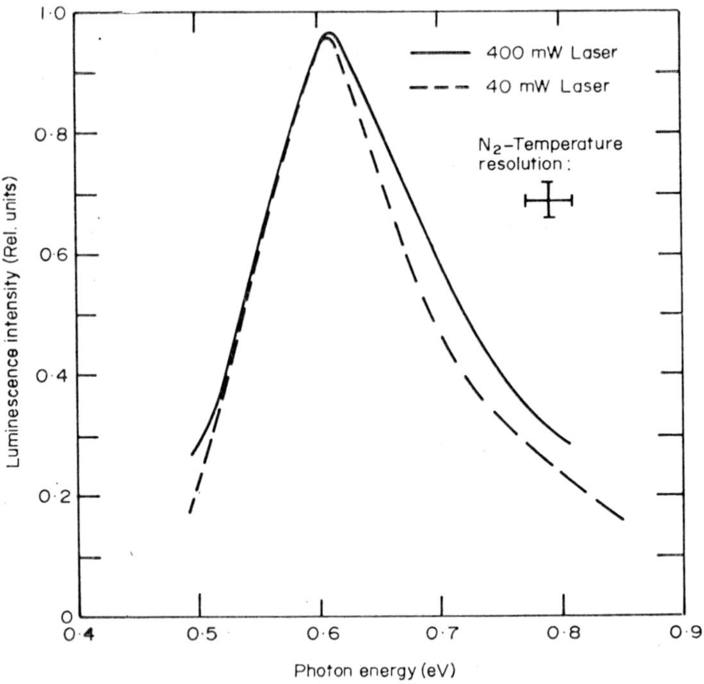

FIG. 15.5. Photoluminescence of As_2Te_2Se film, using a Nd doped YAG laser as the light source. (After Fischer *et al.*, Ref. 23.)

width is about 0·17 eV, i.e. approximately 20 per cent of the conductivity gap. This ratio is approximately the same for As_2Se_3. Coulombic interactions could account for this broadening since a Coulomb energy of $e^2/\kappa r$ would be lost during the annihilation.[26] A range of distances between the carriers prior to recombination, which would be expected in a disordered material, could thus lead to a broadened emission. Additional observations, such as the almost total quenching of the luminescence at room temperature, are not yet well understood.[26]

15.4. PHOTOCONDUCTIVITY: A SUMMARY

In this lecture an attempt has been made to give an overall view of the types of photoconductive behaviour that can be encountered in solids in general. It has been the writer's experience that this overall view of the general phenomena of photoconductivity is a helpful prerequisite as an aid in understanding, in any comparative framework, the many diverse experimental conditions and results encountered in the literature. No attempt was made, nor was it possible within the timescale, to cover every possible contingency. Sufficient groundwork was laid, however, to then allow some discussion of specific experimental observations made in amorphous solids. Rather than attempt a review of the numerous studies made on many amorphous solids, it was decided that a more helpful approach would be to review, in some detail, a few representative amorphous materials. A very detailed review of the many experimental studies on amorphous materials has been made by Adler.[12]

15.5. Internal and External Photoemission—Introduction

As with most experimental techniques that are being applied to amorphous materials, those of internal and external photoemission were initially developed in the study of crystalline solids. The aim was to learn something about the electronic energy level structure in such solids beyond that which could be learned by optical spectroscopy. The power of these new methods is that, in principle, they are capable of determining the absolute energies of initial and final states. This is not a capability that optical spectroscopy possesses, since in this case one is limited to determining only energy differences between the initial and final states of an optically induced transition. Internal photoemission is the phenomenon of injection of photo-excited carriers from one solid into another with which it is in contiguous contact. External photoemission is the ejection of photo-excited carriers from a solid into vacuum. A characteristic feature of internal photoemission is that one can study the injection of both electrons and holes. This is not the case for external photoemission where one is limited to the study of electron ejection. This lecture will have the following format. In Section 15.6 the general principles and techniques for studying internal photoemission will be presented. Experimental results of these types of measurements on amorphous Se will be discussed in Section 15.7. In Section 15.8 attention will turn to the technique of external photoemission, followed in Section 15.9 by a discussion of results and interpretation of external photoemission measurements on amorphous Se, amorphous As_2Se_3, amorphous Ge and amorphous Si.

15.6. Principles and Techniques for Measuring Internal Photoemission

(i) *General Principles*

Figure 15.6 is a schematic energy level diagram which illustrates the process of internal photoemission of either electrons or holes from a metal into an

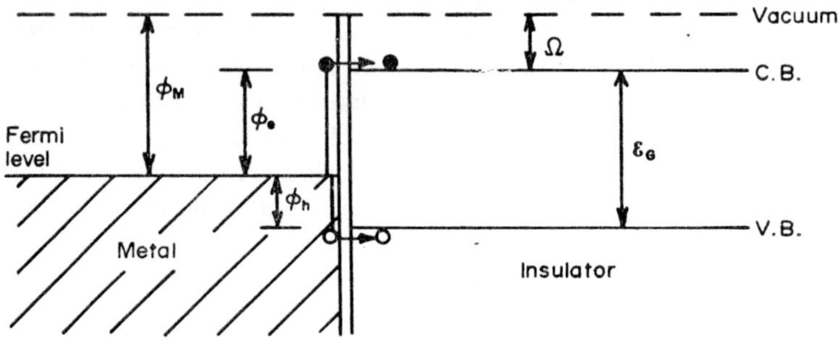

FIG. 15.6. Energy-band diagram for the metal–insulator interface. ϕ_M is the metal work function, Ω is the electron affinity of the insulator. ϕ_h and ϕ_e are the barriers for emission of holes and electrons, respectively, from the metal into the insulator. (After Mort and Lakatos, Ref. 33.)

insulator. Normally, in the absence of impurity photoconduction, the onset of photoconductivity would occur with photons of energy equal to $\mathcal{E}_g$. After contact is made between the metal and insulator, barriers ϕ_e and ϕ_h exist for the injection of electrons and holes respectively which are less than $\mathcal{E}_g$. By absorbing photons with energy equal to or greater than the appropriate barrier height, and with sufficient momentum normal to the interface, carriers have the possibility of entering the conduction states of the insulator. If both of these barriers heights can be measured with the same metal, their sum defines the minimum energy separation between electron and hole conduction states. One identifying feature of the occurrence of internal photoemission is the observation of photocurrents for photon energies less than $\mathcal{E}_g$. The competing possibility for this observation—impurity photoconductivity—can be distinguished from true internal photoemission on the basis of simple experimental tests. The recognition of the phenomenon of internal photoemission was first made by Mott and Gurney[(27)] in interpreting photoconductivity studies on NaCl crystals containing sodium particles. Following the pioneering work of Gilleo,[(28)] and Williams and Bube,[(29)] the technique has been applied to a wide range of solids. A review article by Williams[(30)] considers the results of many of these studies. As discussed by Williams,[(30)] when small barrier heights are involved it can be difficult to distinguish between photoemission and enhanced thermionic emission from the metal. Thus, if the barrier height is approximately 0·4 eV, a temperature change as small as 10^{-4} °C in the metal will produce a change in the thermionic current of $\sim 10^{-5}$ A cm^{-2}. Such a small temperature rise could occur just from the absorption of light by the metal. With an optimistic estimate for the quantum yield of photoemission of 0·1 and with a light flux of 10^{15} photons cm^{-2} sec^{-1}, the maximum photoemission current would also be $\sim 10^{-5}$ A cm^{-2}. Generally speaking, when the barrier height is

$\gtrsim 40\,kT$ it is easy to distinguish photoemission currents from those due to enhanced thermionic emission; when the barrier height is $\lesssim 20\,kT$ an experimental distinction becomes very difficult.

For most systems that have been studied the spectral dependence of the quantum yield Y is proportional to $(h\nu - h\nu_0)^2$ where $h\nu_0$ is the threshold energy and is equal to the barrier height. The quantum yield is defined from the observed current per absorbed photon in the metal. It is implicitly assumed in this definition that the measured current is emission-limited so that the bulk of the insulator plays no determining role in the magnitude of the current and serves as a collector. For this to be so, the carriers should have a range appreciably bigger than the sample thickness so that no bulk deep trapping occurs, and the currents should be linear in light intensity indicating the absence of space-charge-limitations. With these conditions fulfilled the currents will be independent of time (after one transit time) and equal the supply rate of photoemitted carriers into the insulator. A plot of $\sqrt{Y}$ versus photon energy gives an intercept equal to the barrier height. This type of spectral dependence should only occur for the case where both the emitter (metal) and collector (insulator) have bands wider than about 0·5 eV. The derivation of the functional form of the spectral dependence for this case and the case where either or both bands are narrow can be found in the review article by Williams.[30]

Even under conditions where the photoemission current is emission-limited, it is generally found that the current is field dependent. In some cases it is found that the current shows a linear increase with field followed by a saturation.[31,32] In addition the quantum efficiency, even under saturation conditions, is usually substantially less than unity. Both these observations indicate some limitations to the internal photoemission process. The low quantum efficiency is due to the range of a hot carrier in a metal. The internal photoemission of a carrier involves the excitation of a carrier to states above the Fermi surface. This non-equilibrium condition will relax and the hot carrier has a characteristic range before it thermalizes and returns to the Fermi surface. If the film thickness exceeds the range then most carriers will be scattered into energy states below the threshold energy before they reach the metal-insulator interface. Typical values for the electron range, which is a strong function of the excess electron energy, in gold are ~700Å for 1eV of excess energy and $\lesssim$ 70Å for a 5 eV electron. Clearly the thickness of the deposited metal film in general will control photoinjection, and if sufficiently thick prevent it from taking place.

The origin of the field dependence of the internal photoemission current can be understood by considering the potential profile at the metal-insulator interface. A potential barrier exists due to the mismatch of the metal Fermi level and, in this case, the conduction states of the insulator. A superposition of this barrier and the image potential of the electrons leads to a potential maximum. A carrier leaving the metal with the appropriate energy and momentum normal to the interface will cross the potential barrier. If, before the carrier falls several kT/e below the maximum, it undergoes a scattering event with the lattice vibrations of the insulator, the carrier will have a high probability of returning to the metal. A recent paper by Mort *et al.*[32] shows that the functional dependence of the number of carriers photo-emitted into an insulator, Q, as a function of field is of the form

$$Q = \frac{\mu E}{v_R + \mu E} Q_0 \tag{15.6}$$

Q_0 is the number of carriers starting with sufficient x-directed momentum to reach X_m, μ is the microscopic mobility, v_R is the effective recombination velocity. Note that Q increases linearly with E at low fields, but saturates at high fields (where $\mu E > v_R$). Extrapolation of the low-field region to intersect the saturation flux defines a saturation field E_s where

$$E_s = v_R/\mu \simeq v/4\mu \tag{15.7}$$

where v is the thermal velocity of a carrier. Such a behaviour was observed in CdS single crystals and gave very reasonable estimates of the microscopic mobility.[32] This is also the explanation of previously observed saturation in internal photoemission current with field in SiO_2.[31]

Having briefly reviewed the major feature of internal photoemission, we now turn to the two basic techniques employed to study this phenomenon, namely steady-state and transient methods.

(ii) *Steady-State Technique*

To simplify the analysis of the observations one of the electrodes between which the insulator is sandwiched should be neutral, i.e. does not photo-inject carriers. A non-absorbing electrode is thus required and NESA conducting glass or water fulfill this requirement well. The other electrode is an evaporated metal film of appropriate thickness. The sample is irradiated with light of variable wavelength and with a bias voltage applied the current is measured by an electrometer and displayed on the Y axis of the recorder. On the X axis the wavelength of the light from the monochromator is displayed as the spectrum is scanned. By choosing the appropriate polarity of bias either electron or hole photoinjection can be sought. By applying corrections for the absorption and reflectance of the metal film and spectral variation of the light intensity, the quantum yield for internal photoemission can be determined. The data is thus plotted to give a linear curve versus photon energy and the extrapolation of this linear curve yields the threshold energy or barrier height.

(iii) *Transient Technique*

The transient technique differs from the steady-state approach in both the excitation and detection. A 10μsec light pulse from a xenon flash tube is appropriately filtered using narrow band pass filters. With a sufficiently long RC time constant the charge displacement is integrated to yield directly the transit time of the photoemitted carriers but more importantly, in this context, the number of photoemitted carriers which transit the sample. The spatially narrow sheet of charge required in this type of measurement is well defined by the absorption depth of light in the metal film. This technique is a very close simulation of a transient thermionic experiment.

15.7. Internal Photoemission from Metals into Amorphous Selenium

The barrier heights for photoemission of electrons and holes from various metals into amorphous Se were measured by Mort and Lakatos[33] using the technique outlined in Section 15.6(ii). The results are summarized in Table 15.1. The sum of the hole threshold and the metal work function should equal, referring to Fig. 15.6, the photoemission threshold into vacuum. From the results for the three metals an average value of 5·7 eV is found for this

Table 15.1. Thresholds for photoemission of electrons and holes from various metals into amorphous Se[33]

Metal	Threshold for holes ϕ_h (eV)	Threshold for electrons ϕ_e (eV)	Thresholds for vacuum photoemission $\phi_M + \phi_h$ (eV)	"Photoinjection" band gap $\phi_h + \phi_e$ (eV)
Au	0·85 ± 0·15	1·30 ± 0·10	6·07 ± 0·15	2·15 ± 0·25
Cu	1·10 ± 0·10	1·00 ± 0·10	5·55 ± 0·10	2·10 ± 0·20
Al	1·40 ± 0·15	0·95 ± 0·10	5·60 ± 0·15	2·35 ± 0·25
			Average	Average
			5·64 ± 0·15	2·20 ± 0·25

parameter. As will be discussed in Section 15.9, this is in fair accord with external photoemission measurements. It was found that for gold the barrier height for holes is less than that for electrons whereas for aluminium the opposite is true. This is strong evidence against the pinning of the Fermi level due to a large density of localized states in the forbidden gap of amorphous Se. The shift in thresholds between aluminium and gold (~0·5 eV) is significantly smaller than would be expected from the work function difference of these two metals (~1·0 eV). The reason for this discrepancy has been explained by external photoemission measurements as will be discussed later. The sum of the electron and hole barrier heights for a given metal has a mean value of 2·2 + 0·2 eV. This is somewhat less than the band gap of 2·47 eV determined from quantum efficiency measurements.(8)

Transient internal photoemission of holes from metals into amorphous Se has also been reported.(33) The number of transported holes, Q_h, per pulse was always much less than CV, so that small-signal conditions prevailed. For photon energies greater than about 2·2 eV, the photoemission of holes was obscured by the much more efficient internal photogeneration process in selenium. Below ~2·1 eV, the field dependence of Q_h is essentially linear in field and does not show the characteristic field dependence of internal photogeneration observed by Tabak and Warter.(9) The conclusion is that two different mechanisms control the internal photogeneration and internal photoemission of carriers and this was borne out by temperature dependence studies. For photon energies less than about 2·1 eV, where photoemission of holes predominates, the activation energy is very small. For internal photogeneration, on the other hand, for these photon energies there is an appreciable activation energy.(10) Bulk limitations or interfacial trapping effects as a cause of the field dependent number of photoemitted carriers were discounted on the basis of their inconsistency with a number of experimental observations.(33)

One process which had to be considered because of the amorphous nature of the selenium was the photoemission of carriers into localized states lying below the conducting states of amorphous selenium. A field dependent capture or release rate could thereby account for the field dependent supply. Two qualitatively different types of states have to be considered and these are illustrated in Figs. 15.7(a) and (b). The one-electron tail states, Fig. 15.7(a), could arise from the disordered state of an amorphous material, as discussed by others, while the second model Fig. 15.7(b), involved states arising from localized correlated transitions. These transitions could be associated with molecular transitions as discussed in the lecture on photoconductivity and photoluminescence. It can be seen that in the one-electron states model, carriers can be photoemitted into those states which are unoccupied. In the case of the correlated states, however, no photoemission is possible, since without prior excitation these states do not exist and the act of absorption only creates occupied states. It is thus concluded that correlated states cannot play a limiting role in determining the supply of photoemitted carriers, but that one electron tail states could. This conclusion, coupled with the observation that different mechanisms control the supply of photoemitted and internally photogenerated holes, suggests that it is localized correlated states rather than one-electron tail states which determine the non-photoconducting

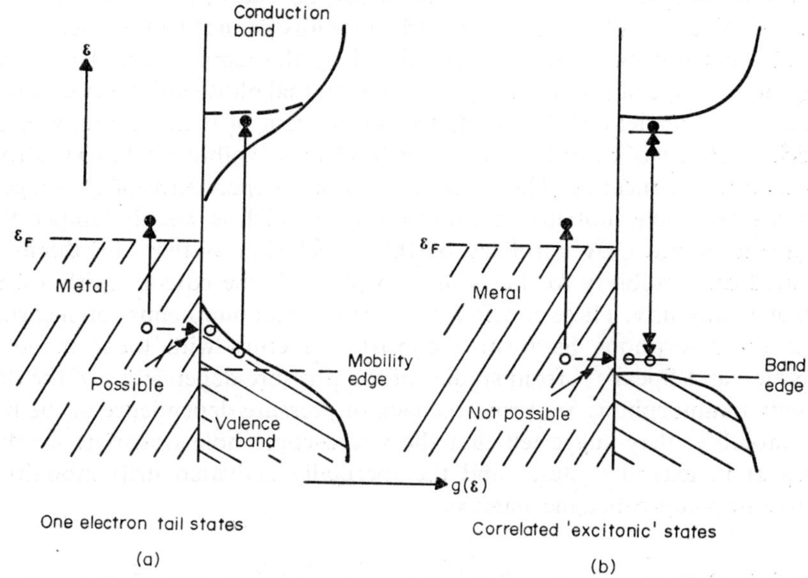

FIG. 15.7. The two alternative band models proposed to account for the absorption and photoconductive properties of amorphous Se. (a) One electron tail states originating from the disorder of the amorphous state. These states are extensions of the normal band states. (b) Correlated "excitonic" states deriving from localized optically induced transitions (After Mort and Lakatos, Ref. 33.)

absorption in amorphous Se. If photoemission into localized tail states does occur, then one must account for the very weak temperature dependence of Q_h for photon energies less than 2·1 eV. This must mean that the amount of tailing is negligible or that the tail states control the drift mobility via a trap-controlled process. In this mechanism, since free and trapped carriers come into equilibrium in a time much less than a transit time, no carriers could be lost and the number of collected carriers would be independent of both temperature and field. However, in this model photoemission into such states assures communication with states above the mobility edge so that in general one would expect a photoinjection band gap that is less than that deduced from internal photogeneration efficiency measurements.

The origin of the field dependent supply of photoemitted carriers must lie elsewhere. Mort *et al.*[32] attributed the field dependent supply to the back scattering of carriers from the amorphous Se into the metal as described in Section 15.6. The absence of a saturation of Q_h with field must be explained on the basis of eqn (15.7) in terms of the magnitude of the microscopic mobility in amorphous Se. If it is assumed that the measured drift mobility in amorphous Se[34] at room temperature, which equals 0·14 $cm^2\ V^{-1}\ sec^{-1}$, is the microscopic mobility, then the predicted saturation field would be about 10^7 V cm^{-1}. The fact that the drift mobility cannot be the microscopic mobility in amorphous Se is substantiated by the temperature dependence of Q_h. In the range of photon energies where internal photoemission dominates the supply of carriers, Q_h is essentially independent of temperature, whereas the drift mobility of holes in amorphous Se always exhibits a relatively strong temperature dependence. The weak temperature dependence of Q_h suggests a microscopic hole mobility in amorphous Se which is weakly temperature dependent at most. A corollary of this conclusion is that the thermally activated drift mobility for holes in amorphous Se be equated with a trap-controlled mobility. These conclusions from internal photoemission measurements on amorphous Se form a consistent picture with those made by Dolezalek and Spear[(35)] from studies of the pressure dependence of the drift mobility in amorphous Se. From the lack of pressure dependence of the hole drift mobility, they suggested that the microscopic hole transport involves transport in extended states and the thermally activated drift mobility is due to a trap-controlled mechanism.

15.8. Principles and Techniques for Measuring External Photoemission

A familiar example of external photoemission is the photoelectric effect, discovered in 1887 by Hertz, which led Einstein in 1905 to propose the quantum theory of light. The relevance and applicability of this phenomena to the elucidation of the electronic structure of solids was not recognized until much later, and it is only in the last decade, notably through the efforts of Spicer and his co-workers,[(36)] that substantial progress has been made in understanding the fundamental processes involved. There is no attempt here to give any rigorous discussion of either the experimental techniques or theoretical ramifications of external photoemission. The aim is rather to give a simple introduction to the phenomenon and its applicability to the determination of electronic energy level structures in a few representative amorphous solids.

In Fig. 15.8(a), the essentials of the experimental technique are shown. Photons pass through a small hole in a spherical collector and strike the sample surface. The electrons which are photoemitted above a threshold photon energy flow towards the collector and their energy is analyzed by the application of a retarding potential to the collector. By the superposition of a modulation voltage on the collector the electron energy distribution $dn/d\mathcal{E}$ can be determined. The thresholds for most solids lie at photon energies in excess of about 5 eV. In addition, interesting information about the electronic state densities deeper in the first filled and lower valence states are only

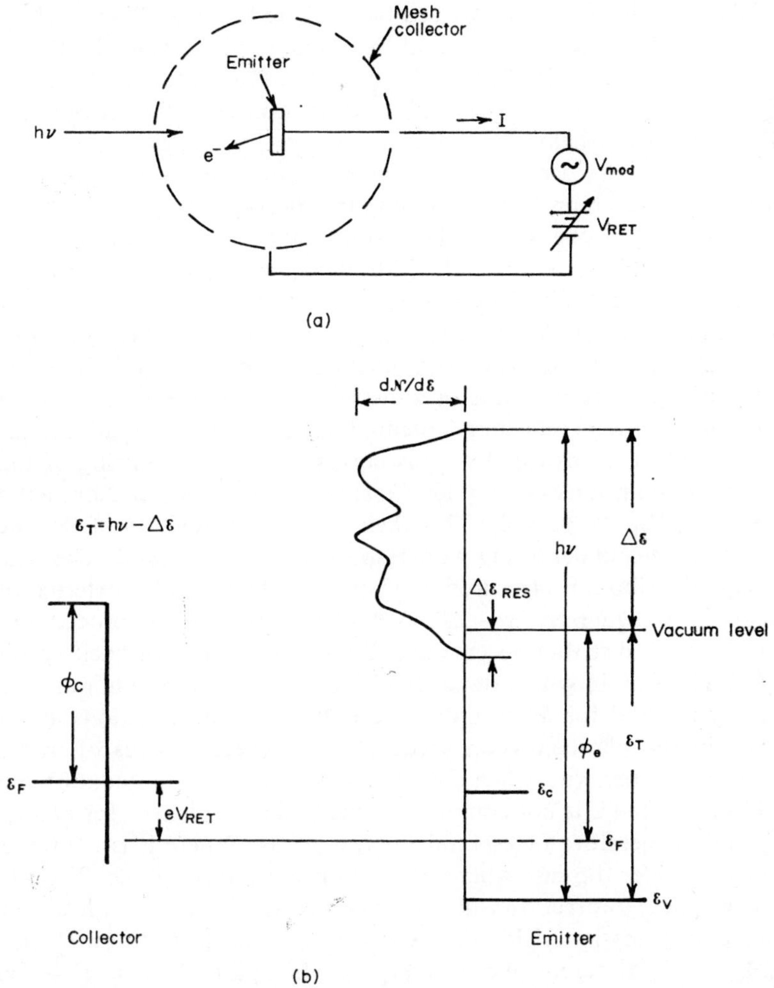

FIG. 15.8. Schematic representation of the technique of external photoemission. (a) Experimental arrangement. (b) Relative energy levels of emitter and collector.

accessible with photons energies much greater than 5 eV. For this reason, light sources such as hydrogen and helium glow discharges are necessary to obtain sufficient intensities of these high energy photons. To avoid the absorption bands of air, a vacuum monochromator must be used and the sample and collector must also be in an evacuated environment to allow the collection of the photoemitted electrons. Surface contamination can lead to results which are in no way characteristic of the solid under investigation. For this reason, the sample must be freshly made in the case of amorphous materials, or a clean surface exposed by cleaving for crystalline solids. These steps must occur in an ultra-high, oil free vacuum of typically 2×10^{-10} torr. The relative energy levels of the collector and emitter (the sample) are shown in Fig. 15.8(b). For the emitter the work function ϕ_e is the energy at which the Fermi level lies below the vacuum energy. The external photoemission threshold, $\mathcal{E}_T$, is the energy of the valence band edge with respect to the vacuum level. This represents the minimum photon energy required to excite an electron out of the solid. The resolution broadening of the analyzer, $\mathcal{E}_{\mathrm{RES}}$, which is defined as the full width at half maximum is typically about 0·15 eV. This is determined by a measurement of the rise of the photoemission yield in the threshold edge as a function of photon energy using an ultra-clean metal as the emitter. The work function of the collector ϕ_c is determined by again using either the same or a dissimilar metal as emitter. For a given photon energy $h\nu$, a retarding potential V_{RET} is applied until the photon emission current drops rapidly. This corresponds to the raising of the collector potential until its vacuum level just begins to lie higher than that of the emitter, i.e. $eV_{\mathrm{RET}} + \phi_c = \phi_e$. This then defines the position of the vacuum level of the solid under study with respect to its Fermi level. The width of the energy distribution measured is $\Delta\mathcal{E}$ and the threshold for external photoemission $\mathcal{E}_T$ is equal to $h\nu - \Delta\mathcal{E}$. A typical experimental procedure is to measure energy distribution curves (EDCs) for a range of exciting photon energies. The way in which these EDCs vary with exciting photon energy is a primary method for determining the nature of the transitions and states involved in the photoemission process. Two general classes of transitions must be considered, viz., when the wave vector **k** is a good quantum number and the case when **k** is not conserved. This latter case is true for amorphous solids and is therefore of especial interest to us. Where **k** is conserved we will only consider the case where $\Delta\mathbf{k} = 0$ for the transition, i.e. direct transitions in **k** space; indirect transitions for which $\Delta\mathbf{k} \neq 0$ typically lead to much smaller photoemission yields than competing direct transitions. If the measured EDCs for different photon energies, when plotted versus $(\mathcal{E} - h\nu)$, are displaced from one another this identifies that **k** is a good selection rule and that the transitions are direct. If the EDCs superimpose when plotted versus $(\mathcal{E} - h\nu)$ then either direct transitions occur from a flat valence band or **k** is not a good quantum number and non-direct transitions are involved. In either case the initial states involved in the transitions are the same and the emitted carriers have an increased kinetic energy equal to the increase in photon energy. Another characteristic feature of the dependence of EDCs on photon energy is whether already revealed structure undergoes change with increasing energy of the exciting photon energy. If this is not observed it means either that transitions from valence states occur directly into vacuum or that transitions occur into conduction bands within the solid which possess little structure in their distribution. Direct excitation into vacuum from amorphous solids is very unlikely in view of the very small scattering lengths for electrons in a disordered material. In this case an excited carrier will pass out into vacuum via some conduction state in the solid. In the event that revealed structure does change with photon energy, then the electrons must be making transitions into conduction states with significant structure in their energy distribution. Unlike the other examples, in this case the external photoemission does not only reflect the density of states in the valence band but rather the joint density of conduction and balence bands. One final phenomenon observed in external photoemission that must be discussed is the contribution to the EDCs of inelastically scattered electrons. Such electrons may still be sufficiently energetic to escape from the solid and contribute to the EDCs. The number of these electrons increases towards the low energy side of the distribution and becomes more pronounced as the exciting photon energy increases. This is a complicating factor in attempting to probe even deeper into the solid with higher photon energies. Depending on the scattering mechanism the number of scattered electrons increases steadily as the low energy edge of the EDC is approached. The combination of this rise and the cut-off at the threshold gives rise to a peak in the EDC which is an artifact and reflects nothing about the density distribution of states in the solid.

15.9. External Photoemission Results for Amorphous Se, As_2Se_3, Ge and Si

Studies of external photoemission on amorphous films of Se and As_2Se_3, evaporated in high vacuum immediately prior to measurement, have been reported by Nielsen.[37] Figure 15.9 shows the EDCs of Se, As_2Se_3 and Au. The energy scale is plotted with zero energy being the Fermi energy. In these plots the cut-off at high negative energy represents carriers with the lowest kinetic energy ($\sim -\Delta\mathcal{E}_{RES}$) while the cut-off at low negative energy represents carriers coming from the top of the valence band and having maximum kinetic energy. The films here are sufficiently thick (>150Å) that the EDCs are independent of any further increase in thickness. Nielsen[37] established, by measuring the EDCs as a function of exciting photon energy, that the transitions are non-direct because the EDCs did not shift with photon energy. The upper limit of these curves therefore represents the energy distribution of states at the top of the valence band. Despite the fact that only a fraction of the valence states are revealed with the photon energy used (7·7 eV), it is clear that the energy distribution of valence states in these solids are quite broad. This fact can also be inferred from optical studies, particularly vacuum uv reflectivity.[38,39] As was discussed in earlier lectures the main effect of a

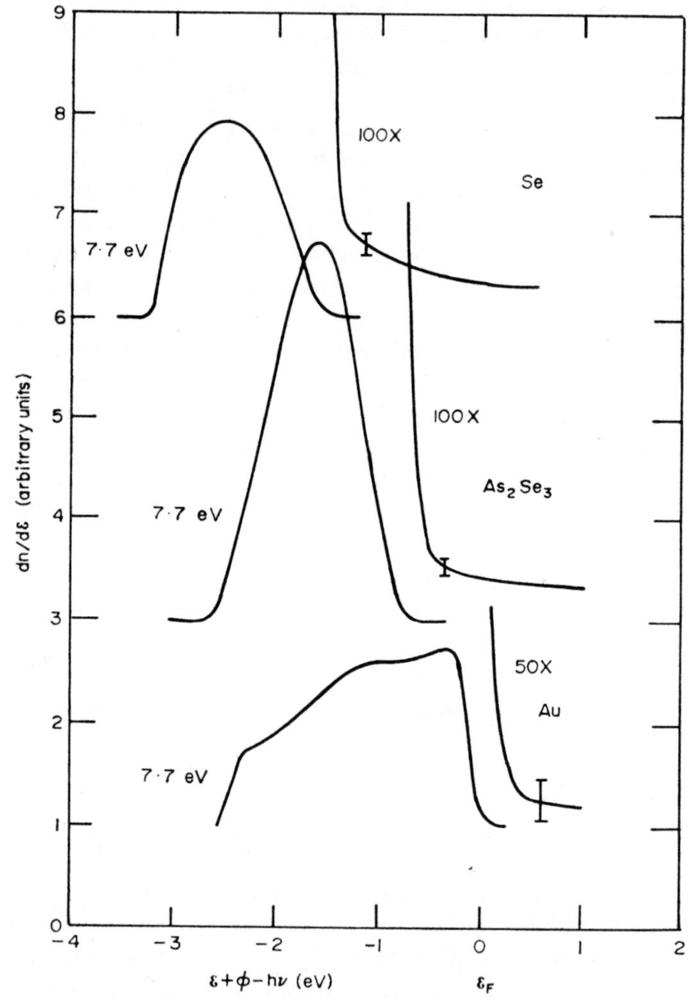

FIG. 15.9. Energy distribution curves for amorphous Se, As_2Se_3 and Au obtained for a photon energy of 7·7 eV. The error bars on the 50 and 100× expansions correspond to the *p–p* noise present on these curves. The energy scale gives the initial energy of the photoelectrons relative to the Fermi level. (After Nielsen, Ref. 37.)

disordered lattice is to broaden any structure in the energy band structure without significantly affecting the grosser features. Considerable discussion in earlier lectures has dealt with the question of band-tailing, i.e. the tailing of localized states from the band edges into the forbidden gap. Within the limits to be prescribed, such tailing should reflect itself in the EDCs of amorphous materials. Returning to Fig. 15.9, the EDCs at the upper limit are shown on a magnified scale for the two amorphous materials and Au. The finite $dn/d\mathcal{E}$ observed above the Fermi level must arise from analyzer characteristics (such as reverse currents). With corrections for this made, the optical density of states near the Fermi level is found to be less than 2×10^{-3} of the peak valence band density in Se and 4×10^{-4} of the peak valence band density of As_2Se_3. The small amount of tailing which is observed is probably instrumental since it also occurs in the EDC for Au. In absolute terms Nielsen[(37)] put an upper limit on the density of states in the gap of 10^{20} eV^{-1} cm^{-3} for Se and 2×10^{19} eV^{-1} cm^{-3} for As_2Se_3. As previously discussed, external photoemission studies are optical measurements for which the energy of the initial state may be determined. As such they should be sensitive to the local potential fluctuations which exist in the disordered state. An expectation of such potential fluctuations would be the observation of a tail in the energy distribution of the photoemitted carriers. Such tailing was not observed in amorphous Se or As_2Se_3 by Nielsen.[(37)] This is not too surprising in the case of an elemental solid since only translational disorder is to be expected. However, the infra-red studies of Lucovsky, *et al.*[(11)] imply the existence of two molecular species in amorphous Se—Se_8 rings and chains. In this sense amorphous Se may be an atypical amorphous elemental solid and therefore the absence of tailing would be more difficult to understand. However, no tailing is observed in the case of As_2Se_3 in which in addition to translational disorder one expects the occurrence of compositional disorder. In all candour it should be pointed out that the sensitivity of external photoemission measurements in general can only rule out densities of states in excess of 10^{19} eV^{-1} cm^{-3}.

For the case of Se and As_2Se_3, comparative studies of the energy distribution of states in single crystals have not been carried out to date. This is not the case for Si and Ge where very extensive measurements of energy distribution of states have been made for both the crystalline and amorphous state by Spicer and his co-workers. The complications of compositional disordered and phase separation are absent in these elemental amorphous materials. In addition, since the short range order should be comparable to that in the crystal, these elements are model systems to study the effect of loss of long range order on the electronic energy level structure. The room temperature curves in Figs. 15.10(a) and (b) show the EDCs for amorphous Ge and Si at a fixed photon energy of 8·6 eV as measured by Pierce *et al.*[(40)] In this instance the energy scale gives directly the electron energy above the valence band maximum, i.e. in this plot the high energy cut-off equals the photon energy. As was the case for Se and As_2Se_3, it was established that no shift in the EDCs occurs with change in photon energy indicating that non-direct transitions are involved. The EDCs of both elements are characterized by a single broad peak. In the case of amorphous Ge this peak lies 1·25 eV

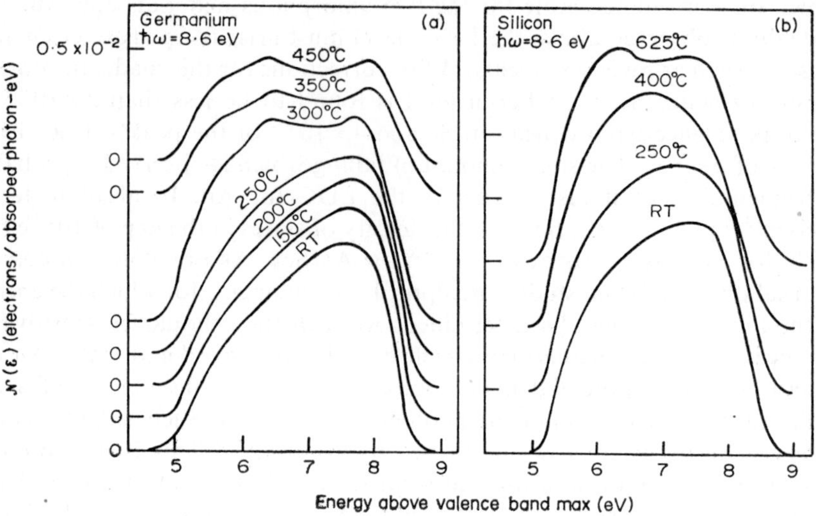

FIG. 15.10. Energy distribution curves obtained with a fixed photon energy of 8·6 eV for a range of temperatures showing the transition from EDCs characteristic of the amorphous phase to those of polycrystalline films. (a) Germanium. (b) Silicon. (After Pierce *et al.*, Ref. 40.)

below the valence band maximum. By measuring a clean Au sample, Pierce *et al.* showed that amorphous Ge and Si show no more tailing than does Au. They placed an upper limit on the density of states in the gap at $\gtrsim 3 \times 10^{19}$ eV^{-1} cm^{-3}. This result is in conflict with measurements of Peterson *et al.*[(41)] who reported the observation of a large degree of tailing in amorphous Si. Pierce *et al.*[(40)] suggest that this is not an intrinsic property of amorphous Si but that it may be due to impurities or defects such as voids. The other curves in Figs. 15.10(a) and (b) show the effect of successive annealing steps on the EDCs of amorphous Ge and Si. In the case of amorphous Ge no significant change occurs up to 250°C, but an abrupt change occurs at 300°C and EDCs characteristic of the crystalline phase emerge. A more gradual change occurs in amorphous Si until the crystalline EDCs are obtained at 625°C. The measurements show directly the effect of loss of long range order on the electronic energy level structure of these two amorphous elemental solids.

In addition to the use of external photoemission to study the intrinsic energy distribution of states within a solid, the technique can be used to study *in situ* the relative electronic energy level structure of two solids in contiguous contact. The measurement of Nielsen[(37)] on amorphous Se and As_2Se_3 on gold and aluminium substrates illustrate this technique, which consists of successive depositions of the materials and the determination of EDCs as a function of thickness. Figure 15.11 shows the position of the vacuum level and valence band edge of Se and As_2Se_3 respectively, deposited on aluminium (oxidised by room air) and gold. In the upper part of Fig. 15.11 it can be seen that the valence band edge for Se is, within experimental error, at a fixed energy for all thicknesses. The slight curvature in the case of the

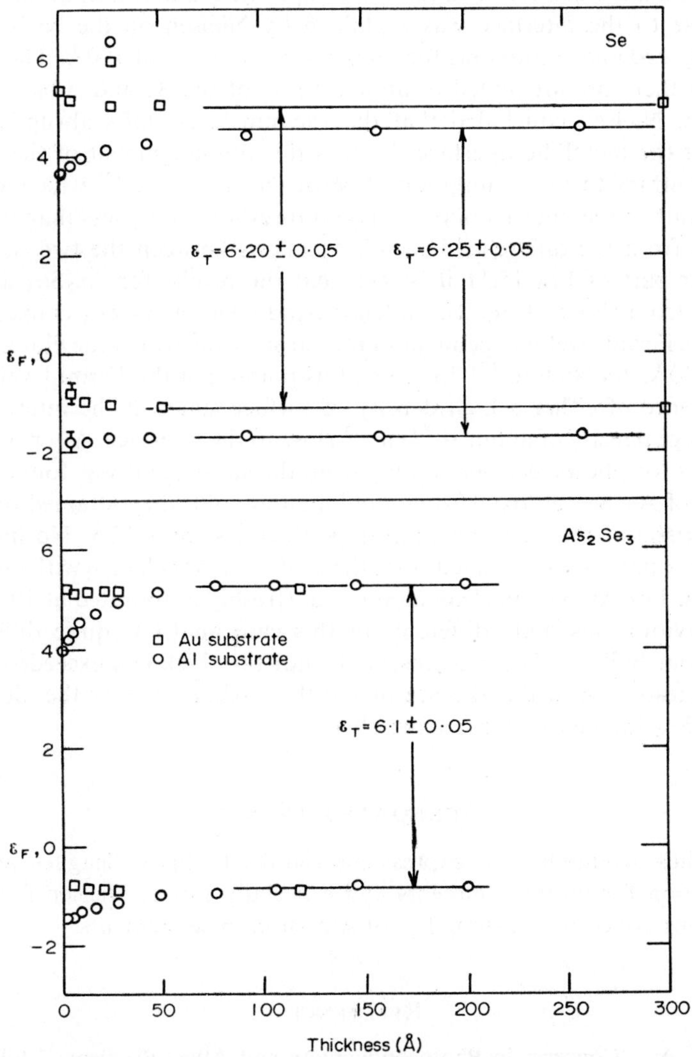

FIG. 15.11. Vacuum level and valence band energies for Se (top) and As_2Se_3 (bottom) deposited on Au and Al substrates, as a function of the average thickness of the films. (After Nielsen, Ref. 37.)

gold substrate at very small Se thicknesses is due to some electrons being photoemitted from the gold above the Se band edge and escaping. The difference of 0·55 eV in the valence band edge position for Se on Au and Al is in good accord with the internal photoemission measurements of Mort and Lakatos.[33] The presence of this energy difference and the constancy of the edge positions with increasing thickness indicates that the Fermi level is not pinned by a large density of states in the gap. The gradual shift in the vacuum level close to the interface was explained by Nielsen on the basis that Se films only become continuous for thicknesses in excess of 100Å. The gradual shift was therefore attributed to an averaging of the Se and substrate work functions. Nielsen concluded that the vacuum level shifts abruptly within ~30Å of the metal–Se interface.[37] It is this non-alignment of the vacuum levels, contrary to the assumption of Mort and Lakatos,[33] which accounts for the shift of the internal photoemission threshold being less than would be expected from the difference in work functions between the two metals. In the lower part of Fig. 15.11 it is seen that the results for As_2Se_3 are quite different from those of Se. The valence band edge of As_2Se_3 is fixed 0·9 eV below the Fermi level independent of the substrate provided the film thickness exceeds 50Å. As Nielsen[37] discusses, this pinning of the Fermi level implies the existence of either a high density of surface states at the interface or a bulk charge density of about 10^{20} cm^{-3} close to the interface. After correcting the EDCs for photoelectrons coming from the substrate it was found that, in the case of As_2Se_3 on oxidised aluminium and gold, they attained the shape characteristic of the bulk for an average thickness of ~12Å. No indication of alloy formation was detected since the EDCs did not change with increasing thickness. For As_2Se_3 on clean aluminium (freshly evaporated at 10^{-10} torr) the behaviour was quite different. In this case an EDC quite different in shape from bulk As_2Se_3 was observed until the thickness exceeded ~30Å. This suggests that in the absence of the thin oxidizing layer the aluminium and As_2Se_3 interact to form some alloy.

Acknowledgements

The author would like to express his thanks to his colleagues at Xerox Corporation for useful comments and especially Dr. P. Nielsen for helpful discussions concerning external photoemission measurements.

References

1. Rose, A., "Concepts in Photoconductivity and Allied Problems." John Wiley & Sons, New York, 1963.
2. Spear, W. E., *J. Non-Cryst. Solids* **1**, 197 (1970).

3. Mort, J. and Scher, H., *J. Appl. Phys.* **42,** 3939 (1971).
4. Scharfe, M. E., *Phys. Rev.* **B2,** 5025 (1970).
5. Tabak, M. D. and Scharfe, M. E., *J. Appl. Phys.* **41,** 2114 (1970).
6. Bube, R. H., "Photoconductivity of Solids." John Wiley and Sons, New York, 1960.
7. Marshall, J. M., Main, C. and Owen, A. E., *J. Non-Cryst. Solids* **8-10,** 760 (1972).
   See also Chapter 16 of these proceedings.
8. Hartke, J. L. and Regensburger, P. J., *Phys. Rev.* **139,** A970 (1965).
9. Tabak, M. D. and Warter, Jr. P. J., *Phys. Rev.* **173,** 899 (1968).
10. Pai, D., and Ing, S. W., *Phys. Rev.* **173,** 729 (1968).
11. Lucovsky, G., Mooradian, A., Taylor, W., Wright, G. B. and Keezer, R. C., *Solid State Comm.* **5,** 113 (1967).
12. Adler, D., "Critical Reviews in Solid State Sciences." Vol. 2, Chemical Rubber Co., Cleveland, Ohio, 1971.
13. Davis, E. A., private communication.
14. Spear, W. E. and Adams, A. R., *J. Phys. Chem. Solids* **27,** 281 (1966).
15. Lucovsky, G., Proc. Xth Int. Conf. Phys. of Semicond., Cambridge, Mass. (1970, U.S. AEC Div. of Tech Inf. 1970, p. 799.
16. Davis, E. A. and Mott, N. F., *Phil. Mag.* **22,** 903 (1970).
17. Warter, Jr., P. J., Proc. IIIrd Int. Conf. Photocond., Palo Alto, California, 1969, Pergamon Press, Oxford, 1971, p. 311.
18. Weiser, K., Fischer, R. and Brodsky, M. H., Proc. Xth Int. Conf. Phys. of Semicond., Cambridge, Mass., (1970) U.S. AEC Div. of Tech. Inf. 1970, p. 667.
19. Shaw, R. F., *Ph.D Thesis*, Cambridge University, 1969 (unpublished).
20. Marshall, J. M. and Owen, A. E., (to be published).
21. Fagen, E. A. and Fritzsche, H., *J. Non-Cryst. Solids* **4,** 480 (1970).
22. Kolomiets, B. T., Mamontova, T. N. and Babaev, A. A., *J. Non-Cryst. Solids* **4,** 289 (1970).
23. Fischer, R., Heim, U., Stern, F. and Weiser, K., *Phys. Rev. Letters* **26,** 1182 (1971).
24. Kolomiets, B. T., Mamontova, T. N. and Babaev, A. A., *Phys. Stat. Sol.* **27,** K15 (1968).
25. Fischer, R., to be published.
26. Weiser, K., *J. non-Cryst. Solids* **8-10,** 922 (1972).
27. Mott, N. F. and Gurney, R. W., "Electronic Processes in Ionic Crystals." 2nd ed., p. 73, Oxford Univ. Press, London and New York, 1950.
28. Gilleo, M. A., *Phys. Rev.* **91,** 534 (1953).
29. Williams, R. and Bube, R. H., *J. Appl. Phys.* **31,** 968 (1960).
30. Williams, R., "Semiconductors and Semi-Metals." (ed. by Willardson, R. K. and Beer, A. C.), Vol. 6, Academic Press, New York and London, 1970.
31. Goodman, A. M., *Phys. Rev.* **144,** 588 (1966).
32. Mort, J., Schmidlin, F. and Lakatos, A. I., *J. Appl. Phys.* **42,** 5761 (1971).
33. Mort, J. and Lakatos, A. I., *J. Non-Cryst. Solids* **4,** 97 (1970).
34. Spear, W. E., *Proc. Phys. Soc.* (London) **B76,** 876 (1960).
35. Dolezalek, F. K. and Spear, W. E., *J. Non-Cryst. Solids* **4,** 97 (1970).
36. Spicer, W. E. and Eden, R. C., Proc. IXth Int. Conf. on Semicond., Moscow, 1968 ("Nauka", Leningrad, 1968) p. 65.
37. Nielsen, P., *Solid State Comminucation* **9,** 1745 (1971).

38. Leiga, A. G., *J. Opt. Soc. America* **58,** 1441 (1968).
39. Zallen, R., Drews, R., Emerald, R. and Slade, M. L., *Phys. Rev. Letters* **26,** 2564 (1971).
40. Pierce, D. T., Ribbing, C. G. and Spicer, W. E., *J. Non-Cryst. Solids* **8-10,** 959 (1972).
41. Peterson, C. W., Dinan, J. M. and Fischer, T. E., *Phys. Rev. Letters* **25,** 861 (1970).

16. PHOTOCONDUCTIVITY AND NOISE IN CHALCOGENIDE GLASSES

C. MAIN AND A. E. OWEN

School of Engineering Science,
University of Edinburgh, Edinburgh,
Scotland.

16.1. INTRODUCTION

Several variants of the Cohen–Fritzsche–Ovshinsky (C.F.O.)[1] model of band structure—that is, a continuous distribution of localised states in the mobility gap—have been used to interpret photoconductivity data in chalcogenide glasses[2–7]. (See, for example, the discussion by Dr. Mort in Chapter 15.) The treatments differ in the distributions of states and electronic transitions involved—thus references 2–5 and 7 *assume* tails extending through the middle of the mobility gap while Mott and Davis[6] propose more limited "linear" tails. In all cases, however, the assumption at the outset of tails-of-states has been an essential feature in the interpretation and in some cases it has also been necessary to invoke a change in the nature of the localised states at some specific energy.

In the present paper the authors take a contrasting approach. Photoconductivity data on two simple amorphous chalcogenide compounds is interpreted by a band model with more or less discrete sets of localised states. *The minimum number of localised levels required to interpret the experimental data is introduced.* The data require two such levels making up, with the conduction and valence bands, a basic four-level model. The purpose of this paper is to see how far such a model is consistent with experiment. In the following illustrations the four levels are represented in the conventional way by lines at unique energies. This is, of course, a simplification. The localised states will certainly be spread around a mean energy and the band edges will probably be "smeared out" by at least a small degree of tailing. Moreover, there may be other localised states not revealed by the present experiments.

It is relevant to note that Marshall and Owen adopted a similar approach in their paper on As_2Se_3[8]. Also, Mott and Davis[9,10] estimate that the extent of band tailing in simple chalcogenide glasses is less than 0·2 eV. while mobility measurements on some of these materials reveal much deeper states[8,11,12] presumably *not* associated with such tails.

16.2. Rate Analysis

(i) The free and trapped carrier densities under optical excitation are obtained from the rate of change of occupation in terms of the transition rates (of electrons and holes) into and out of a particular level. It is assumed that the

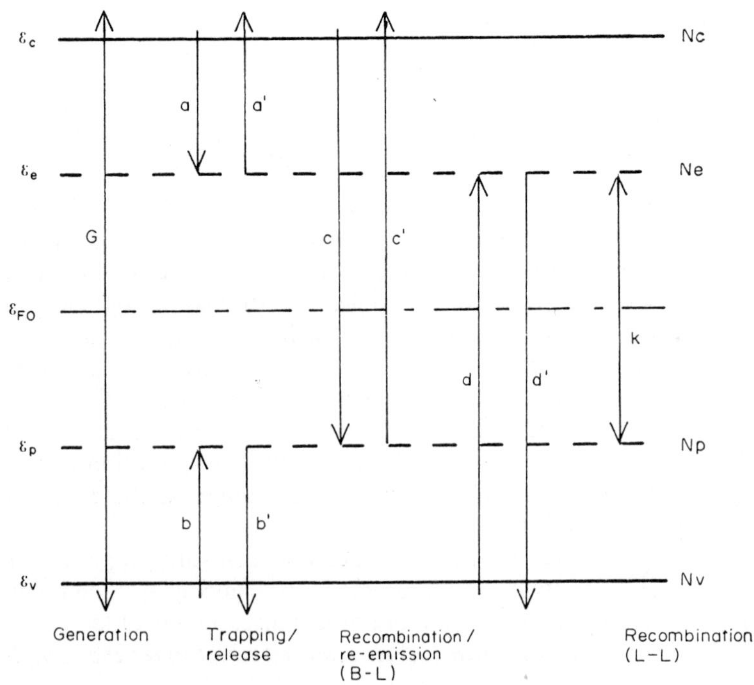

FIG. 16.1. Basic 4-level semiconductor model showing electron and hole transitions.

localised states are monovalent, and that the following discussion can be applied to centres which are charged when "occupied" (by the relevant carrier—hole or electron) or neutral, as in the C.F.O. model. It is also assumed that dc conduction is dominated by holes, as the great mass of evidence suggests.[13,14]

All the relevant parameters for the rate analysis are shown in Fig. 16.1, e.g., hole traps at energy $\mathcal{E}_p$ have density $N_p(\text{cm}^{-3})$ and fractional hole occupation $\bar{f}_p$. The capture coefficient of an unoccupied hole trap for a free hole is C_p, and C_e' is the capture coefficient of an occupied hole trap for a free electron. Thermally produced densities and occupancies are denoted by the subscript "0", while photo-produced densities and occupancies are denoted by the prefix "Δ". The figure is symmetrical so that transitions a, a', c and c' involve electrons while transitions b, b', d and d' involve holes. Similarly, f_{e_0} is the equilibrium Fermi occupation function for electron traps, while $\bar{f}_{p_0}$ is the complement of the Fermi function for hole traps. Processes a and a' are trapping and release rates for electrons, and b and b' are trapping and release rates for holes. For example, the hole trapping and release rates can be written as

$$\begin{aligned} b &= pC_pN_p(1-\bar{f}_p) = p/\tau_t\ (\text{cm}^{-3}\text{sec}^{-1}) \\ b' &= p_1C_pN_p\bar{f}_p = N_pf_p/\tau_{rt}\ (\text{cm}^{-3}\text{sec}^{-1}) \end{aligned} \tag{16.1}$$

where τ_t, τ_{rt}, are trapping and release times for holes, and p_1 obtained from detailed balance considerations is

$$p_1 = N_v \exp\left(-\frac{\mathcal{E}_p - \mathcal{E}_v}{kT}\right) \tag{16.2}$$

All the other transition rates can be expressed in a similar way.

(ii) *Recombination*

Recombination in the 4-level semiconductor of Fig. 16.1 can occur in a number of ways, one of which usually predominates. Three "paths" have been considered. The first two, paths 'd' and 'c', are two-step processes, in which rapid trapping of a free carrier in a localized state is followed by the capture of a carrier of opposite sign, usually a slower process. Thus, path 'd' involves processes 'a' and 'd', while path 'c' involves processes 'b' and 'c'. This is referred to as Band-Localised (B–L) recombination, and it is a process familiar in ordinary semiconductor physics.[15–17]

The possibility of a third recombination path involving localised–localised (L–L) transitions, when ,as in amorphous semiconductors, the densities of localised states are high has been pointed out previously.[3,4,6,18,19] The rate for this process is $KN_p\bar{f}_pNef_e$ where K is a rate constant. Whether B–L or L–L recombination predominates in a given case is determined by the relative rates.

(iii) *Rate Equations*

The set of differential equations (D.E.) for the energy level occupancies together with the charge neutrality condition.

$$N + N_e f_e - p - N_p \bar{f}_p = 0 \quad \text{(neutral traps)} \tag{16.3}$$

give all the required information. For example, the valence band D.E. is

$$\frac{\mathrm{d}p}{\mathrm{d}t} = G - b + b' - d + d' = 0 \quad \text{in steady state.} \tag{16.4}$$

All terms are written as sums of thermal ('$_0$') photo-excited terms ('Δ') and solved for the latter. $G = g_0 + g$ where g is the photo-excitation rate.

A great simplification can be effected if the trapping interchange is much greater than the recombination rate. In particular, if $b'(a') \gg c(d)$, or k, then the localised levels lie well outwith the so-called demarcation levels,[20,21] and are in "quasi-thermal equilibrium" with the nearest band. Then,

$$\frac{\text{free holes}}{\text{trapped holes}} = \frac{p}{N_p f_p} = \frac{p_0}{N_p f_{p0}} = \frac{p_1}{N_p} \quad \text{(constant).} \tag{16.5}$$

In these circumstances, the free and trapped carrier densities are "synchronised" except for times less than the relevant *trapping* times, and the situation reduces to a "two-reservoir" problem. The overall rate equation for holes becomes

$$\frac{\mathrm{d}(p + N_p \bar{f}_p)}{\mathrm{d}t} = G - c - d - k \tag{16.6}$$

Finally, in the present situation, the trap densities are high, and the traps are not fully occupied under most experimental conditions, hence the following inequalities hold:

$$n, p \ll N_e f_e, N_p \bar{f}_p \ll N_e, N_p \tag{16.7}$$

Results of analysis: At low excitation levels ($\Delta p \ll p_0$ etc.) excess carriers recombine into a pool of thermally produced recombination centres (monomolecular kinetics), while at high excitation levels ($\Delta p \gg p_0$ etc.) the recombining species are proportional (bimolecular kinetics).

Table 16.1 shows the results of the kinetic analysis for Band–Localised, and Localised–Localised recombination.

(iv) *Band–Localised Recombination*:

The following discussion assumes path 'd' is dominant but similar answers are obtained for path 'c'. At low intensities, the rate equation is monomolecular, and the steady state solution for excess hole density Δp_m has the same features as observed experimentally i.e. a linear intensity relationship and exponential *increase* with $1/T$. The transient decay is exponential, and the decay time τ_{md} (where subscript d refers to the recombination path) has a marked temperature dependence. This is because the excess free holes actually involved in recombination constitute a small *temperature dependent* fraction of the total photoexcited hole density (free + trapped), and the recombination centre density (at $\mathcal{E}_e$) is *also* temperature dependent. As Rose pointed out, the photo decay time is much longer than the free recombination time τ_r.
i.e.

$$\tau_{md} = \tau_r \times \frac{\text{trapped holes}}{\text{free holes}} = \tau_r \times \left(\frac{N_p f_p}{p}\right) \tag{16.8}$$

where $\tau_r = (C_p' N_e f_e)^{-1}$, and both factors in eqn (16.8) are temperature dependent. At high excitation levels, the rate equation is bimolecular and the steady state solution for excess holes is of the form observed experimentally i.e. the square root intensity dependence, and an exponential *decrease* with $1/T$. Note that the "activation energy" predicted here is *half* the trap depth.

The decay is now non-exponential and has a long hyperbolic tail. In addition, the decay time τ_{bd}, defined as the time to decay to half value, is intensity dependent and much *less* temperature dependent than the monomolecular decay time. This behaviour results from the swamping of thermally produced carriers and occupied recombination centres by the photo-generated densities.

(v) *Localised–Localised Recombination*

Table 16.1 also lists the predicted photoconductivity features when L–L recombination prevails ($c, d \ll k$). Again, regions of monomolecular and bimolecular recombination are predicted, but quite different temperature dependences (for steady state and transient solutions) are obtained. For example, the steady state bimolecular solution for excess holes predicts an "activation energy" *equal* to the hole trap depth (c.f. $\frac{1}{2}$ trap depth for B–L

TABLE 16.1 Recombination Models

	Rate equation	Steady state	Decay
		1 Band–Localised	
Low intensity	$\frac{d\Delta p}{dt} = g\left(\frac{p_1}{N_p}\right) - \left(\frac{1}{\tau_{md}} + \frac{1}{\tau_{mc}}\right)\Delta p$	$\Delta p_m \propto g \exp\left(\frac{\mathcal{E}_e - \mathcal{E}_{F_0}}{kT}\right)$	$\Delta p(t) = \Delta p_m \exp(-t/\tau_{md})$
	$\tau_{md} = (2C_p' f_{p0} p_1)^{-1} : \tau_{mc} = \left(2C_e' n_1 \frac{N_p}{N_e} f_{p0}\right)^{-1}$		$\tau_{md} \propto \exp\left(\frac{\mathcal{E}_p - \mathcal{E}_v + \mathcal{E}_e - \mathcal{E}_{F_0}}{kT}\right)$
High intensity	$\frac{d\Delta p}{dt} = g\left(\frac{p_1}{N_p}\right) - (\gamma_d + \gamma_c)\Delta p^2$	$\Delta p_b \propto g^{\frac{1}{2}} \exp\left(-\frac{\frac{1}{2}(\mathcal{E}_p - \mathcal{E}_v)}{kT}\right)$	$\Delta p(t) = \Delta p_b(1 + \Delta p_b \gamma_d t)^{-1}$
	$\gamma_d = C_p' \quad \gamma_c = C_e' \frac{n_1}{p_1} \frac{N_p}{N_e}$	(Solutions for path d)	$\tau_{bd} \propto g^{-\frac{1}{2}} \exp\left(\frac{\frac{1}{2}(\mathcal{E}_p - \mathcal{E}_v)}{kT}\right)$
		2 Localised–Localised	
Low intensity	$\frac{d\Delta p}{dt} = g\left(\frac{p_1}{N_p}\right) - \frac{1}{\tau_{mL}}\Delta p$	$\Delta p_m \propto g \exp\left(\frac{\mathcal{E}_e - \mathcal{E}_{F_0} - (\mathcal{E}_p - \mathcal{E}_v)}{kT}\right)$	$\Delta p(t) = \Delta p_m \exp(-t/\tau_m)$
	$\tau_{mL} = (2KN_p f_{p0})^{-1}$		$\tau_{mL} \propto \exp\left(\frac{\mathcal{E}_e - \mathcal{E}_{F_0}}{kT}\right)$
High intensity	$\frac{d\Delta p}{dt} = g\left(\frac{p_1}{N_p}\right) - \gamma_L \Delta p^2$	$\Delta p_b \propto g^{\frac{1}{2}} \exp\left(-\frac{(\mathcal{E}_p - \mathcal{E}_v)}{kT}\right)$	$\Delta p(t) = \Delta p_b(1 + \Delta p_b \gamma_L t)^{-1}$
	$\gamma_L = \frac{KN_p}{p_1}$		$\tau_{bL} = (gK)^{-\frac{1}{2}}$

Note: The energy term ($\mathcal{E}$ [illegible] $-$ $\mathcal{E}$ [illegible]) is an approximation, wherever it appears, to $\frac{1}{2}(\mathcal{E}$ [illegible] $- \mathcal{E}$ [illegible]).

model), and the monomolecular decay time is much less temperature dependent as recombination does not involve a small temperature dependent fraction of the photoexcited carriers.

(vi) *Trap Limited Drift Mobility* μ_d

The trap limited hole drift mobility, given by the approximate relation

$$\mu_d \approx \mu_0 \times \frac{p}{N_p f_p} = \mu_0 \frac{p_1}{N_p} \tag{16.9}$$

(where μ_0 is the free mobility in the band) can be calculated from transient decay and growth measurements as follows.

The rate equations in Table 16.1 applied to a "step" illumination predict an initial region of linear growth of carrier density before recombination equilibrium is reached. As Ryvkin has shown,[22] there is also a region of rapid growth at very short times ($t \leqslant \tau_t$), but in the present case this cannot be resolved experimentally. The mobility is obtained from the photocurrent growth slope.

$$\left.\frac{\mathrm{d}\Delta I}{\mathrm{d}t}\right]_{t\sim 0} \propto \left.\frac{\mathrm{d}\Delta p}{\mathrm{d}t}\,\mu_0\right]_{t\sim 0} \propto g\left(\frac{p_1}{N_p}\,\mu_0\right) \propto g\,\mu_d. \tag{16.10}$$

Similarly, from the overall decay curve, the drift mobility can be obtained

$$\Delta I \text{ (steady state)} \propto g\,\frac{p_1}{N_p}\,\mu_0 \tau \propto \tau \mu_d \tag{16.11}$$

where τ is the relevant photo decay time.

(vii) *Distributed States*

The foregoing analysis can be applied to the C.F.O. model and distributed states. Simmons,[7] Weiser [3] and Arnoldussen *et al.*,[4] for example, all assume a model with more or less featureless tails of states and introduce transition mechanisms,[3] or discontinuities in the nature of the states[4] with energy, to produce a "peaking" in the excess carrier distribution at particular energies. The results are similar to those for a model with discrete levels.

16.3. Photoconductivity in Amorphous As_2Te_3

Samples were evaporated thin films (0·5–1·5 μm) with *coplanar* gold electrodes on 7059 borosilicate substrates (typical gap 25 μm).

The results obtained in Section 16.2 apply to uniform excitation throughout a homogeneous sample. Absorption of light, surface effects, and diffusion have been neglected. It is relatively easy to include absorption effects in both steady state and transient solutions, and this has been done. Surface effects and diffusion have been shown to have little influence.

(ii) *Amorphous Arsenic Telluride*

The optical absorption constant α, necessary for interpretation of photoconductivity data, was obtained from transmission measurements. The data were fitted to the relation[(23–25)]

$$\alpha(\nu) = \text{const } (h\nu - \mathcal{E}_g)^2/h\nu \qquad (16.12)$$

giving an optical gap of 1·0 eV at 0 K, and the temperature coefficient $-4{\cdot}7 \times 10^{-4}$ eV K^{-1}.

Dark conductivity measurements gave a conductivity activation energy of 0.42 eV, and room temperature conductivity of 10^{-4} ohm^{-1} cm^{-1}.

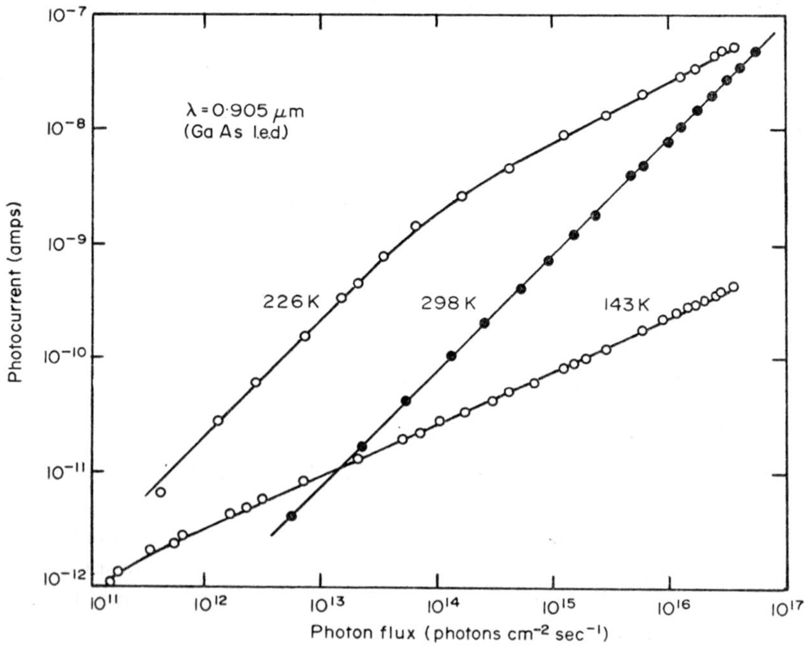

FIG. 16.2. Photocurrents versus incident photon flux for As_2Te_3.

The photocurrent versus intensity relation in Fig. 16.2 illustrates the transition from linear to square root behaviour, and the reversal of temperature dependence in the two regions. Figure 16.3 shows the temperature dependence of the steady state photocurrent and decay time. This demonstrates the reversal of temperature dependence predicted for the steady state, and the rapid fall of the decay time with temperature in the monomolecular region (0·55 eV slope). The low temperature behaviour of τ has not been fully analysed.

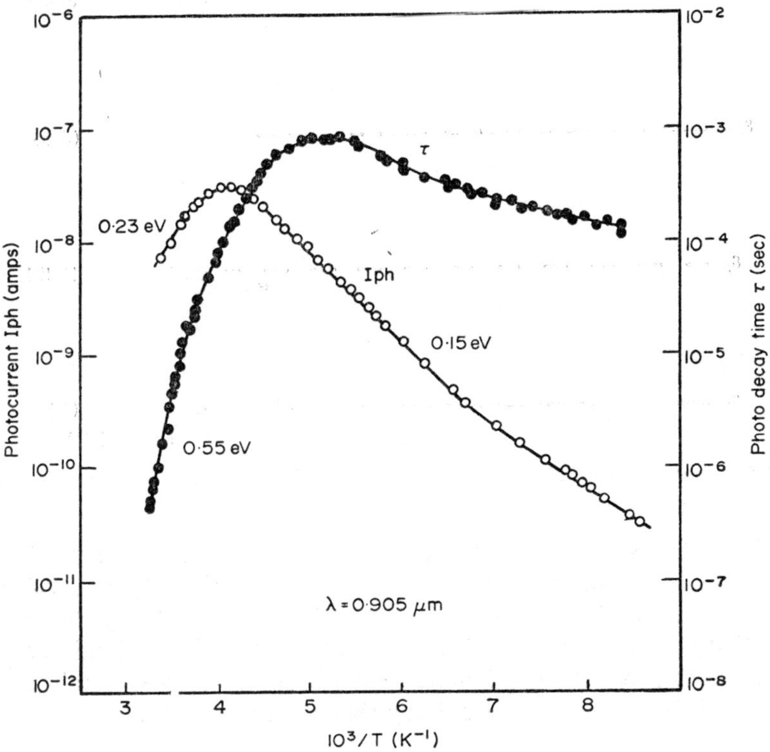

FIG 16.3 Photocurrent and decay time versus ($10^3/T$) for As_2Te_3. Photon flux $= 9 \times 10^{15}$ photons cm^{-2} sec^{-1}.

The hole trap limited mobility computed from eqn (16.11), and the monomolecular data, has a room temperature value of 10^{-2} cm^2 s^{-1} V^{-1}, and an activation energy of 0·33 eV, in good agreement with the data of Croitoru.[26]

Figure 16.4 illustrates schematically, the band diagram for 0 K, giving as close a fit as possible to experimental data—optical absorption, dark conductivity, and steady state photoconductivity—for B–L (path 'd') and L–L recombination. The gap is assumed to reduce uniformly with temperature. This diagram is drawn in the spirit of the simplifications mentioned in Section 16.1.

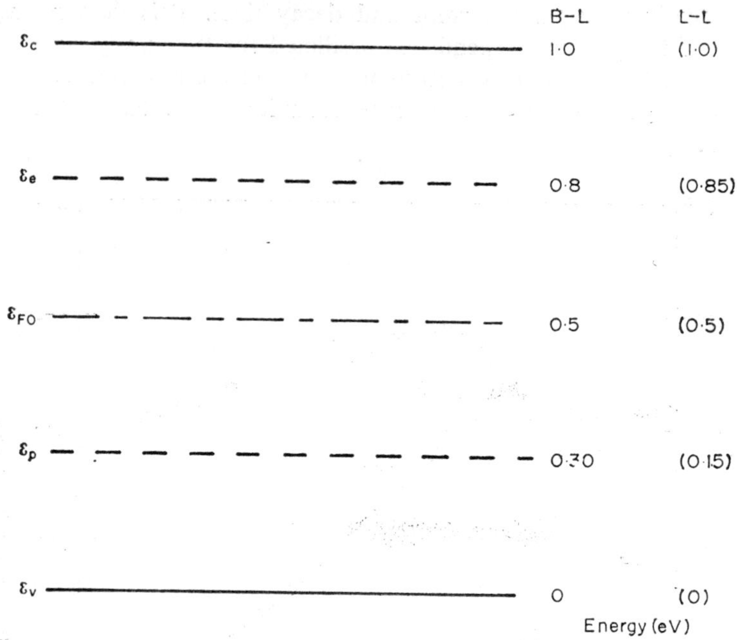

FIG. 16.4. Schematic band diagram for As_2Te_3. Energies shown for B–L and L–L recombination models.

Table 16.2 lists some of the "activation energies" predicted by the two models, and also the measured values. The quantities $\Delta\mathcal{E}_m$ and $\Delta\mathcal{E}_b$ were obtained from the slopes of the mono- and bimolecular steady state photoconductivity versus $(1/T)$, $\Delta\mathcal{E}_\mu$ is the trap-limited mobility activation energy, $\Delta\mathcal{E}_\sigma$ the dark conductivity activation energy, and $\Delta\mathcal{E}_{\tau m}$ was obtained from the slope of the monomolecular decay time τ_m versus $(1/T)$.

TABLE 16.2. Parameters for As_2Te_3 for B–L and L–L recombination

Activation energy (eV)	B–L	Measured	L–L
Monomolecular $\Delta\mathcal{E}_m$	0·20	0·23	0·20
Bimolecular $\Delta\mathcal{E}_b$	0·15	0·15	0·15
Mobility $\Delta\mathcal{E}_\mu$	0·30	0·32	0·15
Conductivity $\Delta\mathcal{E}_{F_0}$	0·5	0·42	0·5
Decay time $\Delta\mathcal{E}_{\tau_m}$	0·5	0·55	0·35

Both models require $\mathcal{E}_{F0}$ to be roughly in the middle of the gap, rather higher than its measured position. (Pinning of $\mathcal{E}_{F_0}$ by localised states is being investigated.) While the B–L model gives a good fit for the parameters (within 10%) the L–L model gives values for $\Delta\mathcal{E}_\mu$ and $\Delta\mathcal{E}_{\tau m}$ which are much smaller than the measured quantities.

It therefore seems probable that B–L recombination predominates. On this basis the trap density can be estimated from the mobility, conductivity, and temperature coefficient of the gap as $N_p \approx 3 \times 10^{18}\ \text{cm}^{-3}$, and the product $\mu_0 N_v \approx 1 \times 10^{21}\ \text{cm}^{-1}\text{s}^{-1}\text{V}^{-1}$. Thus if the free carrier mobility is $\mu_0 = 10\ \text{cm}^2\ \text{s}^{-1}\text{V}^{-1}$, $N_v \approx 10^{20}\ \text{cm}^{-3}$; both are reasonable figures.[25] The ratio of trapped to free holes at room temperature would be 10^3, the recombination time τ_r for a free hole about 10^{-9} sec at room temperature, and the recombination coefficient $C'_p \approx 10^{-8}\ \text{cm}^3\ \text{sec}^{-1}$ (rather high). Some of these results have been verified independently by noise measurements (see Section 16.4)

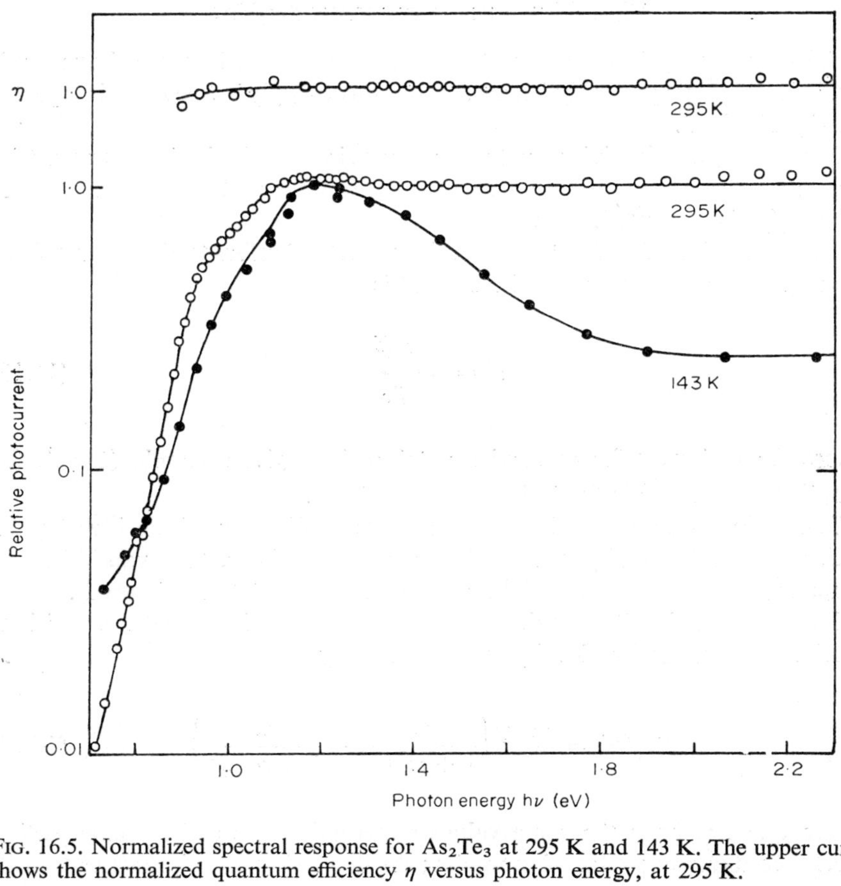

FIG. 16.5. Normalized spectral response for As_2Te_3 at 295 K and 143 K. The upper curve shows the normalized quantum efficiency η versus photon energy, at 295 K.

Spectral response and quantum efficiency

Figure 16.5 shows the normalised photocurrent per incident photon versus photon energy for amorphous As_2Te_3 at two temperatures. At 295 K, in the monomolecular region, the response rises to a constant level with photon energy, while at 143 K, in the bimolecular region, a peak occurs. The fall off at high photon energies is not due to surface effects, but results from the bimolecular bulk behaviour and the reduction of absorption depth with photon energy. Calculation of the photo-response per absorbed photon (relative quantum efficiency η) verifies this. At both temperatures this is constant for photon energies between 1·0 and 3·0 eV, implying an absolute value for η of unity.(2) At 143 K there is a slight fall in η at photon energies below 1·0 eV. In calculating the generation rate g, η was taken to be unity.

16.4. Noise in Amorphous As_2Te_3

(i) There are very few reports of noise measurements in amorphous semiconductors,(27–29) probably because of the surprisingly low level noise spectra usually obtained.

In practice, conductivity fluctuations give rise to noise in the dc current I_0, when an external source is connected to the sample, and the spectral distribution $S_i(f)$ is measured.

If the subscript "$_0$" refers to mean values and "Δ" refers to departures from the mean, the mean square noise current is

$$\overline{\Delta I^2} = I_0 \frac{\overline{\Delta p^2}}{p_0} \tag{16.13}$$

where Δp and p_0 refer to carrier *numbers* in a given sample (1 cm^3, for simplicity), and $\overline{\Delta p^2}$ is the hole number *variance*.

The main features of noise in amorphous As_2Te_3 have been investigated with the 4-level model proposed for photoconductivity. (See van Vliet and Fasset, (30) and van der Ziel(31) for the mathematical methods used).

The spectral density obtained for carrier fluctuations under conditions similar to those for low-level photoconductivity is

$$S_{\Delta p}(\omega) = \overline{4\Delta p^2} \left\{ \frac{\tau_r}{1+\omega^2\tau_m{}^2} + \frac{\tau_t}{1+\omega^2\tau_t{}^2} \right\} \tag{16.14}$$

where τ_m, τ_r, τ_t, have the same values and significance as before. Equation (16.14) contains two relaxation modes; the first refers to noise due to transitions from the reservoir of valence band and associated traps, to the reservoir of conduction band and associated traps(recombination component) while the second refers to noise due to trapping transitions *within* the first reservoir (trapping component).

The variance Δp^2 may be found for such a system by the generation-recombination theorem.[32,33] In this case, however, as $N_p f_p \gg p$, the individual carriers are generated independently, and poisson statistics apply.[34] The variance is then

$$\overline{\Delta p^2} = p_0 \qquad (16.15)$$

and this is the highest value which can arise in thermal equilibrium. As suspected, high trap densities result in a high *total* noise level. The trapping component accounts for most of the variance, but it is spread over a wide frequency range and thus has a very low *spectral density*. Only the recombination component could be resolved experimentally and it accounts for a small fraction of the total noise—hence the reports of very low noise levels in amorphous semiconductors.

(ii) *Experimental results*

Noise measurements were made on the same thin film samples used in the photoconductivity investigations. Standard techniques were used,[35] with

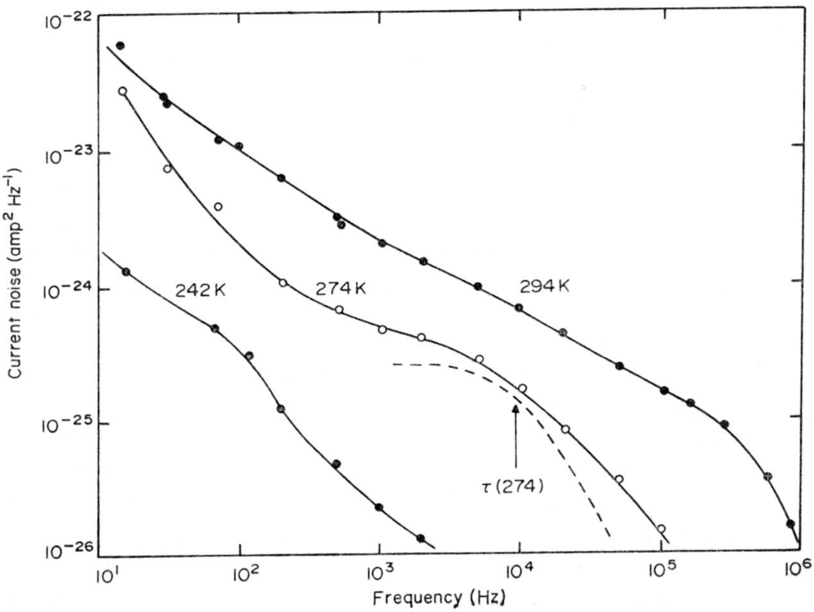

FIG. 16.6. Current noise spectra $S_i(f)$ for As_2Te_3.

low-noise wide-band and narrow-band circuitry (noise figure < 1 dB), and true r.m.s. detection.

Figure 16.6 shows the noise spectra obtained for As_2Te_3, normalised to constant applied voltage, for three temperatures. The relaxation mode (recombination component) appears as a "bump" on a featureless $1/f$ trend due probably to contacts. Good agreement was obtained between the magnitude *and* the temperature dependence of the noise "turnover" time constant, and the monomolecular photo decay time. The trapped hole *density* computed from these data (noise plateau and time constant) is, at room temperature,

$$N_p \bar{f}_{po} \approx 5{\cdot}5 \times 10^{16}\,\text{cm}^{-3}$$

resulting in a trap density

$$N_p \approx 3 \times 10^{18}\,\text{cm}^{-3}$$

if $(\mathcal{E}_p - \mathcal{E}_v) = 0{\cdot}3$ eV at 0 K. This is in good agreement with the photoconductivity results and is *independent* of estimates of μ_0 and N_v.

An upper limit to the trapping time τ_t of 5×10^{-11} sec can be estimated from the fact that the trapping noise plateau must be less than 10^{-26} $A^2\,Hz^{-1}$.

The capture cross section of the traps for holes σ_p, is given by $\tau_t = ({}_pNv\sigma_p)^{-1}$, where v is the thermal velocity of free holes ($\approx 10^7$ cm sec^{-1}). Thus $\sigma_p \approx 10^{-16}$ cm^2. This is large, considering the depth of the traps (0·3 eV, but according to Lax,[36] is possible with "cascade" capture, even for *neutral* centres.

16.5 Amorphous Arsenic Selenide

(i) A similar series of experiments was carried out on thin-film As_2Se_3 samples. The results are more difficult to interpret unambiguously, however, as several trapping levels may be effective.

Optical measurements reveal a 0 K optical gap of 2·0 eV and a temperature coefficient of $-5{\cdot}7 \times 10^{-4}$ eV K^{-1}. The dark conductivity is 6×10^{-13} $(\Omega\,\text{cm})^{-1}$ at room temperature, with activation energy 0·92 eV, in good agreement with other reports.[37]

Figure 16.7 shows the photocurrent as a function of photon flux for As_2Se_3, illustrating again the transition from linear to square-root behaviour. The reversal of temperature dependence also occurs, and is shown explicitly in Fig. 16.8, where the photocurrent at constant illumination is plotted versus $(10^3/T)$. The slopes on this plot give $-0{\cdot}37$ eV at low temperatures, and about $+0{\cdot}37$ eV at high temperatures but these values are sensitive to changes in the absorption constant with temperature at the photon energy used (1·9 eV). It is estimated that the value obtained from the low temperature slope could be as low as $-0{\cdot}32$ eV.

Time constant measurements show a rapid fall at high temperatures, corresponding to an activation energy of 0·8 eV, but relative insensitivity at low temperature. Note that the room temperature photo decay time, for moderate intensities (10^{15} photons cm^{-2} sec^{-1}), is about 10^{-3} sec compared with 10^{-6} sec for As_2Te_3.

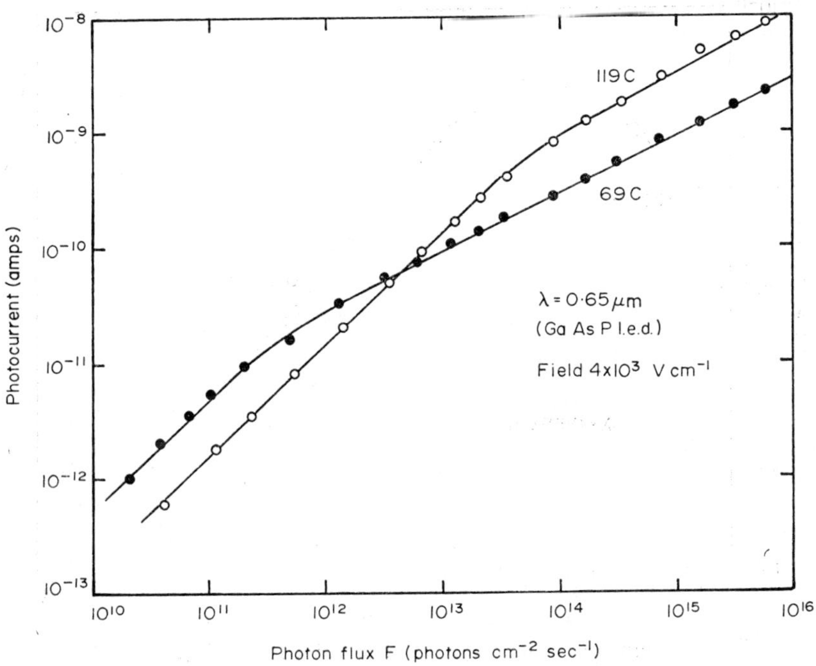

FIG. 16.7. Photocurrent versus photon flux for As_2Se_3. Excitation source: GaAsP light emitting diode (l.e.d.).

Trap limited drift mobilities, obtained from "initial slope" measurements [eqn (16.10)] give reasonable agreement with other workers(8,38) e.g. $\mu_d \approx 3 \times 10^{-5}$ cm^2 $sec^{-1}V^{-1}$ at room temperature and a field of 10^5 V cm^{-1}.

On the basis of these results, a schematic band diagram may be constructed as for As_2Te_3, but with L–L recombination predominating, as indicated by the relatively temperature independent bimolecular decay. Thus, the hole and electron traps would be 0·3 eV and 0·35 eV deep in a 2·0 eV gap. There is reason to suspect that the true picture for As_2Se_3 is not so simple, and that several trap species are present at different depths in the gap.(39)

(ii) *Spectral response*

Figure 16.9 shows the normalised photocurrent per *incident* photon for a thin-film (0·87 μm) As_2Se_3 sample, at room temperature. The peak observed is again due to bimolecular bulk conditions, now prevailing at *room* temperature for As_2Se_3. In the same figure is the computed photocurrent per absorbed photon (η). It can be seen that η levels off well above the optical gap, and is much less (0·2) at the optical absorption edge than its saturation value. This type of behaviour has been observed in Se.[40,41]

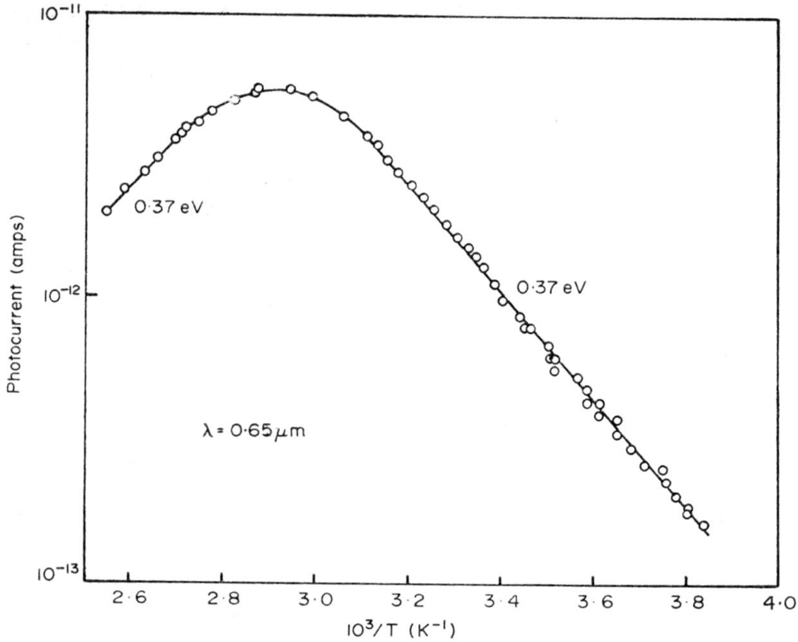

FIG. 16.8. Photocurrent versus $10^3/T$ for As_2Se_3.

Davis and Mott's interpretation involves separation of the photogenerated hole–electron pair, before geminate recombination occurs, in order that the photo-produced carriers contribute to the photocurrent. Other explanations for non-photoconducting absorption invoke the molecular nature of these materials, and excited molecular states.

(iii) *Noise in* As_2Se_3

Measurements have been made of noise in thin film sandwich samples of As_2Se_3 under optical excitation, and some of the results have already been reported. [42] Again, there is agreement between the relaxation modes in the noise spectra, and the photoconductivity time response, though detailed calculations have not yet been performed.

16.6 CONCLUSIONS AND SUMMARY

Photoconductivity data and current noise measurements in amorphous chalcogenide compounds can be adequately interpreted through a band model with discrete levels of localised states but it is not possible to distinguish on the basis of existing data between this model and the extensive tails-of-localised states used in previous interpretations. It should be noted, however, that in contrast to the "continuous tails-of-states" school of thought, the present approach does not require the additional arbitrary assumption of a change in the nature of the states at some specific energy.[4,5] We would suggest, therefore, that a model involving more or less well defined localised states presents the best approach to the problem.

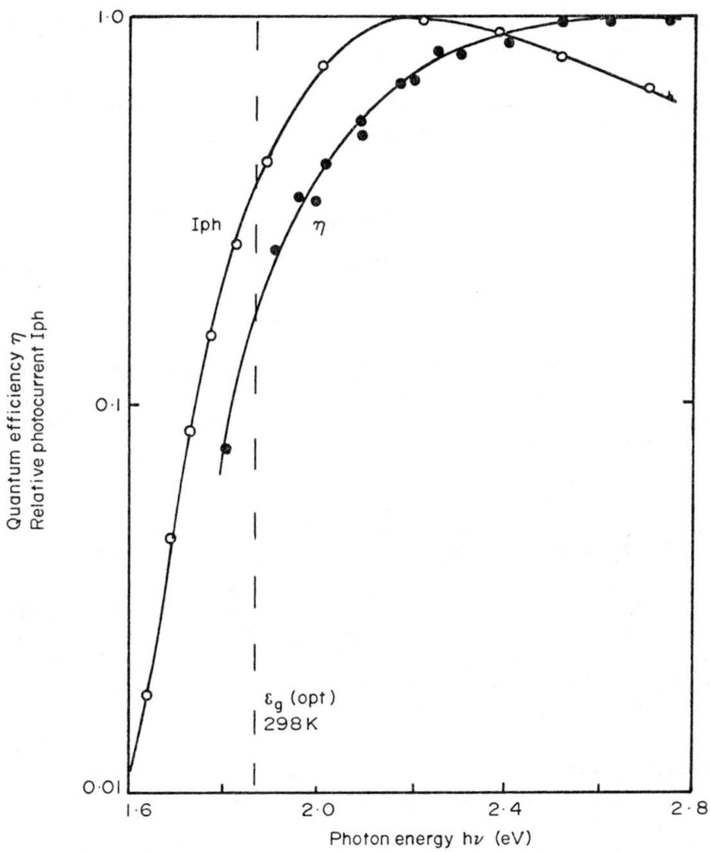

FIG. 16.9. Normalized spectral response and quantum efficiency for As_2Se_3 at room temperature.

These conclusions refer specifically to the relatively simple amorphous chalcogenide compounds such as As_2Se_3 and As_2Te_3. Structural studies have shown that these materials have a high degree of short range order and it would not be surprising if this is reflected in their electronic band structure. In complex alloy chalcogenide glasses, where there is compositional as well as structural disorder, band tailing may well be very extensive.

Although the same basic four-level model has been used for both compounds, there are significant differences in the photoconductive behaviour of As_2Se_3 and As_2Te_3 viz.

(a) In As_2Se_3 the decay times are much slower—milliseconds, compared with microseconds in As_2Te_3.

(b) At room temperature there seems to be a region of non-photoconductive absorption in As_2Se_3 but not in As_2Te_3.

(c) In As_2Te_3 the dominant recombination transitions are band–localised while in As_2Se_3 they *may be* localised–localised.

References

1. Cohen, M. H., Fritzsche, H. and Ovshinsky, S. R., *Phys. Rev. Letters* **22,** 1065 (1969).
2. Fagen, E. A. and Fritzsche, H., *J. Non-Cryst. Solids* **2,** 180 (1970). *ibid,* **4,** 480 (1970).
3. Weiser, K., Fischer, R. and Brodsky, M. H., Proceedings of the Tenth International Conference on the Physics of Semiconductors, Cambridge, Massachusetts, 1970, U.S. Atomic Energy Commission, Oak Ridge, Tennessee, 667 (1970).
4. Arnoldussen, T. C., Bube, R. H., Fagen, E. A. and Holmberg, S., *J. Appl. Phys.* **43,** 1798 (1972).
5. Howard, W. E. and Tsu, R., *Phys. Rev.* **B1,** 4709 (1970).
6. Mott, N. F. and Davis E. A., "Electronic processes in Non-crystalline Materials." p. 229 Oxford University Press, 1971.
7. Simmons, J. G. and Taylor, G. W., *J. Non-Cryst. Solids* **8-10,** 947 (1972).
8. Marshall, J. M. and Owen, A. E., *Phil. Mag.* **34,** 1281 (1971).
9. Davis, E. A. and Mott, N. F., *Phil. Mag.* **22,** 910 (1970).
10. Mott, N. F. and Davis, E. A.,' 'Electronic Processes in Non-crystalline Materials." p. 45 Oxford University Press, 1971.
11. Marshall, J. M. and Owen, A. E., *Phys. Stat. Sol. (a)* **12,** 181 (1972).
12. Grunwald, H. P. and Blakney, R. M. *Phys. Rev.* **165,** 1006 (1968).
13. Edmond, J. T., *Br. J. Appl. Phys.* **17,** 979 (1966).
14. Owen, A. E. and Robertson, J. M., *J. Non-Cryst. Solids* **2,** 40 (1970).
15. Shockley, W. and Read, W. T., *Phys. Rev.* **87,** 835 (1952).
16. Rose, A. ,*Phys. Rev.* **97,** 322 (1955).
17. Blakemore, J. S., "Semiconductor Statistics." Pergamon Press, Oxford, 1962.
18. Dobrego, V. P. and Ryvkin, S. M., *Sov. Phys. Solid State* **6,** 928 (1964).
19. Stern, F., *Phys. Rev.* **B3,** 2636 (1971).
20. Rose, A., "Concepts in Photoconductivity and Allied Problems." p. 26, Wiley, New York, 1963.

21. Bube, R. H., "Photoconductivity of Solids." p. 64, Wiley, New York, 1960.
22. Ryvkin, S. M., "Photoelectric Effects in Semiconductors." p. 129. Consultants Bureau, New York, 1964.
23. Davis, E. A. and Mott, N. F., *Phil. Mag.* **22,** 914 (1970).
24. Tauc, J., in "Optical Properties of Solids." (ed. F. Abeles), North-Holland, 1969.
25. Weiser, K. and Brodsky, M. H., *Phys. Rev.* **B1,** 791 (1970).
26. Croitoru, N., Vescan, L., Popescu, C. and Lazarescu, M., *J. Non-Cryst. Solids* **4,** 493 (1970).
27. Kornfel'd, M. I. and Sochova, L. S., *Fiz. Tver. Tela.* **1,** 1370 (1959).
28. Main, C. and Owen, A. E., *Phys. Stat. Sol.* (*a*) **1,** 297 (1970).
29 Kolomiets, B. T., Lebedev, E. A. and Tsinman, E. A., *Sov. Phys. Semiconductors* **5,** (1971).
30. Van Vliet, K. M. and Fasset, J. R., in "Fluctuation Phenomena in Solids." (ed. R. E. Burgess), Academic Press, 1965.
31. Van der Ziel, A., "Fluctuation Phenomena in Semiconductors." Butterworths, London, 1959.
32. Burgess, R. E., *Physica* **20,** 1007 (1954).
33. Van Vliet, K. M. and Blok, J., *Physica* **22,** 231 (1956).
34. Van der Ziel, A., "Fluctuation Phenomena in Semiconductors." p. 23, Butterworths, London, 1959.
35. Brophy, J. J., *J. Appl. Phys.* **25,** (1954).
36. Lax, M., *J. Phys. Chem. Solids* **8,** 66 (1959).
37. Edmond, J. T., *Br. J. Appl. Phys.* **17,** 979 (1966).
38. Scharfe, M. E., *Phys. Rev.* **B2,** 5025 (1970).
39. Kolomiets, B. T., Proc. 9th Internat. Conf. on the Physics of Semiconductors, Moscow, Vol. 2. Acad. Sci. USSR (Leningrad: "Nauka").
40. Davis, E. A. and Knights, J., to be published.
41. Hartke, J. L. and Regensburger, P. J., *Phys. Rev.* **139,** A970 (1965).
42. Marshall, J. M., Main, C. and Owen, A. E., *J. Non-Cryst. Solids* **8-10,** 760 (1972).

V

PHYSICAL BASIS OF SOME APPLICATIONS OF AMORPHOUS MATERIALS

17. THE TECHNOLOGICAL IMPORTANCE OF AMORPHOUS SOLIDS

C. H. L. GOODMAN

Standard Telecommunication Laboratories Limited,
Harlow, Essex, England

The present intense interest in amorphous solids, which, not without good reason, makes them a particularly active field in solid state physics, should not be allowed to obscure the fact that amorphous solids have been used in industry for many years, in some cases on quite a large scale. To this extent then the technological importance of amorphous solids is real, here and now, and so does not rest upon the promise of new phenomena and new applications. That promise is of course there, and some attempt will be made to discuss it later, but a brief survey of present applications is essential if the "applied" point of view is to be appreciated.

It must be stressed that if a material is in use it does not mean that it is understood. Technology has largely been empirical in its approach, and it is of the utmost importance that the scientific investigation of amorphous solids, as it creates new understanding, should be brought to bear on applied problems—the rewards could be great. Conversely, the knowledge of the existence of the numerous industrial applications of amorphous solids and of problems encountered with them could be stimulating to the scientist and suggest new lines of investigation. This kind of remark is all too often made, of course, but in the case of amorphous solids is particularly apt. It is not too much to say that it was academic interest in the dielectric properties of an amorphous solid in the late 1500's which led, by just such an interaction, to *all* our present electronics and electrical engineering. That solid was, of course, amber, and its Greek name "electron" has become almost a symbol of the age in which we live as a result. But, if amorphous solids were studied early on in the history of science, it was never with great understanding. It is only rather recently, with the crystalline solid state at least reasonably well understood, that it has become possible for physicists to recognise the outlines of the new research frontier of the amorphous solid state.

There is a variety of different sorts of amorphous solid, conveniently classified under three headings. There are glasses, usually well characterisable materials best considered as supercooled liquids; there are rapidly condensed vapours, which give a range of solid materials whose properties depend very much on the precise technique of formation; finally there are disordered solids, produced by reactions at temperatures far below those at which thermal diffusion and re-ordering can occur (anodic oxides are a good example). The common characteristic of these groups, apart from their lack of structure as shown by X-ray diffraction, is that all are metastable. Thermodynamically speaking, this means that any of these materials is less stable than a corresponding crystalline variant. Metastability, of course, does not mean instability, and amorphous solids for technological applications must obviously not transform under normal operating conditions. Even so this metastability can place definite limitations on the utilisation of a particular material and must always be borne in mind.

One of the widest present uses of amorphous solids is for their dielectric properties. They make particularly good insulators, probably because, for reasons that are dealt with elsewhere, the transport of charge through them occurs less readily than in crystalline materials. First let us consider disordered solids, if only because the anodic dielectrics, which are among the best insulators known, fall in this group. Nearly 100 years ago Wheatstone found that an oxide layer formed on aluminium when it was made the anode in an electrolytic bath, and that this oxide efficiently blocked current flow. A range of other metals were later found to behave similarly (Ta, Nb, Zr, Hf and the rare earth metals for example). The so-called electrolytic capacitors based on such anodic oxide dielectrics are mostly made from aluminium and tantalum, the former for cheapness and high capacitance, the latter where reliability and long term stability are important. The present sales of such capacitors in Britain amount to more than £14 million per annum which gives at least an indication of the importance of these components. Now while a great deal of research has been carried out on electrolytic capacitors over the years the physics underlying their characteristics seems rather poorly understood. Perhaps recent developments in the theory of the amorphous solid state may be able to throw some light here. Two examples must suffice:

(1) Such capacitors are "polar", behaving like rectifiers with good insulation only in the "reverse" direction. There seems to be no convincing explanation for this but it appears likely that the metal–oxide interface must be concerned, and that this will be "graded", so that as the composition changes from metal to oxide an energy gap will gradually open up giving a built in and directional band edge warping.

(2) Particularly with tantalum capacitors the maximum working voltage seems to be limited by "field-induced crystallisation"—the amorphous material breaks down and becomes crystalline. This behaviour is certainly

reminiscent of switching phenomena, and perhaps not the least valuable fruit to be gathered from the present great interest in switching could be an improvement in the performance of electrolytic capacitors.

Dielectrics can, of course, also be produced from the vapour, and in thin film form find application in integrated circuits and microelectronics. For example, silicon monoxide is used as a readily evaporatable dielectric for thin film capacitors, while silicon nitride plays a vital role in metal-insulator–semiconductor devices, where its virtual suppression of field induced diffusion of sodium ions is particularly important. A range of techniques is used for preparing such dielectrics: sputtering, evaporation, pyrolysis, glow discharge deposition, etc., and the properties of the resulting films can vary markedly with the technique used, the presence of residual traces of impurities in the deposition apparatus, the temperature of deposition, and even with the presence or absence of electric and magnetic fields. Such variations can cause considerable difficulty in reproducing in one establishment a result obtained in another, and has contributed not a little to some of the discordant results obtained in this field. This is not to decry vapour phase techniques for the preparation of amorphous materials, but rather, to emphasise that caution is essential in characterising the product, since even two nominally identical materials may well have significant differences.

Good conductors, as opposed to dielectrics, are rarely encountered in the amorphous state, at least in bulk form. "Electroless" nickel is perhaps the most commonly encountered example and is an amorphous, disordered, nickel–phosphorus alloy used in electronics as a contacting layer, e.g. with semiconductor devices. (Here, of course, it is not usually left in the amorphous state, but is heat treated to crystallise it.) Other transition element phosphides can be prepared similarly and are ferromagnetic, though with relatively low saturation magnetisation. Of course the "Beilby layer" on a polished metal is more or less amorphous, but seems to have been rather little studied recently.

A relatively little understood amorphous material which, again, conducts electricity and which has a long history of application is microphone carbon. Its importance is obvious, its mode of operation is not, and seems to be tied up not just with the pressure induced modulation of conduction at small areas of contact between the amorphous carbon granules in the microphone capsule but actually also with effects in amorphous surface films containing oxygen which could well be semiconducting.

Semiconducting amorphous solids produced from the vapour have just one or two rather specialised applications (e.g. Sb_2S_3 is used as a photoconducting thin film in vidicon television pickup tubes—here the main problem is "lag", or response time). There is, however, one application of overwhelming commercial importance—xerography. Amorphous photoconducting films of selenium or its alloys, deposited on the xerographic drums, are the basis of what has become an enormous business which, world-wide, amounts to about £600 million per annum. This, of course, entirely dwarfs all other applications of amorphous solids, and not surprisingly has led to a great deal of research on selenium and related materials. Despite this, even the staff at Xerox Research Laboratories, probably the main centre for such investigations, would not claim that selenium is well understood or well characterised. It seems in fact a singularly difficult solid to deal with. Perhaps the lapse of the basic patents held by Xerox will cause that organisation to search for other materials, presumably amorphous, which could perform even better than selenium alloys, while other competitors now entering the field, IBM for example, will be driven to investigate selenium and its alloys in yet greater detail. Problems yet to be solved include the ability to reproduce a grey scale without loss in speed of operation.

Finally we come to "glass". This is probably the group of amorphous materials with the longest history of continuous use. In the electronics industry it has largely been employed as a container material and here the main properties of importance have been ease of formability, control of thermal expansion to enable appropriate seals to be made, good electrical isolation properties and, possibly, high transparency. Empirical findings, such as the use of alkaline earth constituents to suppress transport of sodium ions for example, have given the glass technologist control of a vast range of materials which can be tailor made to suit particular applications. Even so, the physics of glass, why it forms, and what determines its properties, are still not too well understood. So long as no very sophisticated demands were made on glass this situation could be tolerated: glass offered practical solutions to problems at low cost. However, increasingly sophisticated demands are now being made. For example, every colour television set working on the PAL system requires a "delay line", a device for storing the video signal for a line scan interval. This delay must be extremely precise and constant at 63·943 μs, varying less than 5 nanoseconds whatever the temperature within the television set. This specification can be realised in terms of the velocity of sound in certain lead oxide based silicate glasses. Fitted with piezo-electric transducers to introduce the video signal at one end and pick it up at the other, a block of such glass is literally vital to the operation of the colour television sets one sees. The performance of this glass delay line was of course achieved empirically. There are, however, a number of important new applications possible for delay lines if the delay accuracy could be improved by at least an order of magnitude, if the sound velocity could be at least doubled, or if acoustic loss at gigahertz frequencies could be reduced. Although cut-and-try methods might eventually give these desired results it would be most helpful to have improved physical insight into the factors which determine, say, the Young's modulus, the density, the thermal expansion coefficient and the phonon spectra of glasses, which are basic parameters affecting the acoustic properties and which must be related to the structure of the glass and to its chemical bonding.

An even more remarkable electronic application of glass is in the channel plate amplifier. The active element in this vacuum device is a disc made up of a bundle of glass tubes all welded together, with their bores approximately perpendicular to the plane of the disc. The tube bore is small, of the order of 50 μm or less and the walls are thin enough for two thirds or more of the projected area to be "hole". The tubes are made of a glass which can be chemically treated to give a slightly conducting coating on each bore. When mounted in vacuum a potential is applied across the disc, the current flow along the conducting bores giving a potential gradient which accelerates any electron entering a tube. This tube then acts as a continuous dynode electron multiplier, the accelerated electron eventually hitting the tube wall and giving rise to secondaries which repeat the process to yield an avalanche. Gains of 10^5 or more are readily achieved which, in conjunction with a lens, a sensitive photoemissive cathode and a phosphor screen, enables one to see by starlight.

There is no other material than glass with which it could conceivably be possible to make this kind of structure, let alone achieve at the same time a reasonable match of secondary electron emission and electrical conductivity. However, it seems to be essentially a coincidence that such a combination of these parameters can be obtained. If it were properly understood what factors cause a high secondary emission coefficient for a glass, as well as the apparently somewhat simpler problem of what factors enable a suitable conductivity level to be achieved, then some of the defects of present channel plates might be eliminated and very much more efficient operation achieved.

In recent years attention has very much more been focussed on electrical conduction processes in glass. Kolomiets in Leningrad, Ovshinsky in the USA have been particularly active, and the phenomenon (or better, perhaps, phenomena) of switching has awakened enormous interest. Switching will of course be discussed in detail in other presentations; here we need only consider some of the technological implications.

Basically glass switches come in two forms—monostable (or "threshold") which requires a holding current ($\sim$ 100 μA) to maintain the "on" state, and bistable (or "memory") which is indefinitely stable in either condition. Both switch "on" when a threshold voltage is exceeded. Bistable switches are turned "off" when "on", by a fast rising current pulse ($>$ 100 mA, rise time $\leqslant$ 1 μs).

Now glass switches are essentially two-terminal devices. To the electronic engineer this is a disadvantage since it usually necessitates considerable complication of circuitry to separate the control signal from the information actually being switched. On this ground alone many circuit engineers are of the opinion that applications of glass switches are necessarily limited.

It has been found possible to obtain a three-terminal glass switch, with an intermediate electrode giving some degree of control over the switching action, but this approach, at least so far, does not appear to offer a controlling action comparable, say, to that obtained with a semiconductor controlled rectifier, nor is it clear if it can be incorporated in a structure capable of high speed switching.

One interesting possibility, however, arises from the fact that glass devices can be switched by a laser beam. If an optically switched device operating at low control power level is possible, excellent isolation could be achieved. (Laser beam switching incidentally has important implications in the field of xerography.)

The problem of isolation arises in another form in connection with the bistable device. This is sometimes talked of as it could be a replacement for the relay. While this may be true in some non-critical applications, such an argument overlooks just how effective a switch is the traditional electromechanical relay. In the "on" condition its resistance is about 10^{-2} ohms or less, while in the "off" condition it is 10^{12} ohms or more (ceramic or glass insulated relays show still higher insulance). Even more significant, a relay can "hold off" voltage spikes of 500 V or more, an important safety requirement with switches for use in telephone exchanges, for example.

By contrast, chalcogenide glass switches have an "on" resistance of about 1 ohm (considerably more in some cases) and an "off" resistance of 10^6 ohms or less. With transition metal oxide glass switches the latter rises to 10^8 ohms, but this is still three or four orders of magnitude below the insulation of even a cheap relay, while by definition high voltage spikes will switch on any glass switch so that protection against unwanted pick-up (for example, from lightening induced surges) would have to be incorporated.

The rapid turn-on time which can be obtained with glass switches is one of their most interesting properties—under high drive it can apparently be of the order of 1 nanosecond, and perhaps is limited more by the measurement apparatus than by the switch itself.

While this is true of monostable switches, rather more complicated behaviour is found with bistable devices in that a definite "lock-on time" is required (after the rapid fall in resistance which initiates the switching action) if the "on" condition is to be stabilised. This effect sets a minimum cycle time for such devices which with the earlier chalcogenide switches could be many milliseconds, but is now believed to have been reduced to somewhat nearer a microsecond. (With transition metal oxide glass switches somewhat faster cycle times can be achieved, but the basic behaviour is identical.)

There is still argument about the mechanism(s) of operation of glass switches, but whatever this turns out to be they can be remarkably reproducible in operation. Since these devices are not yet commercially available except as large read-mostly memory arrays, it is difficult to quote realistic figures. However, monostable switches can show a remarkably small variation from cycle to cycle in the critical voltage for switching "on" ($\sim \pm 1\%$, though with an appreciable temperature coefficient of about 1% per deg. C) and can operate for at least 10^9 cycles (10^{14} have been reported). Bistable switches show more variability in "on" voltage though still less than $\pm$ 10% and again can give 10^9 operations. Current handling ability is normally small—tens of milliamps—but a recent development is an oxide switch which can pass several amps in the "on" state (with, however, the consequence that a more than 10 amp pulse is required for turning "off").

Where then do glass switches have their application? There are some promising possibilities, e.g. for "read-only" or "read-seldom" memories for computers, for incorporation in the address circuitry for large electroluminescent arrays, and for measurement and control in automobile instrumentation, but much work remains to be done before such applications can be considered fully practical. This is not to say that switching in glass is of little practical interest; it is only to stress that new components and new technologies are only accepted if they offer significant technical *and* economic advantages over what is already in use, or if they enable some entirely new and highly desirable function to be achieved. For the moment the promise that the field of amorphous solids will produce new components and new technologies is readily believed in industry if only because it already has produced them in the past, but we cannot yet be certain that switching in glass is the next along the line. The general topic of the effects of high electric fields on amorphous solids is certainly intriguing and both interesting research findings and useful applications are very likely to be found. But a very wise man would be needed to predict just what these will be.

Another new field of application is glass for optical waveguides. Ordinary glass, of course, is not very transparent (just look at the green edge of a window pane), its optical absorption being due to transition element impurities, particularly iron. Such impurity absorption is broadened compared to that in a crystalline solid, and to that extent weakened at the peak wavelengths. However, to make glass with an absorption coefficient as low as the 10 km^{-1} specified for economic optical links (in order to maintain a reasonable distance between repeater stations) needs extremely low impurity levels, below 0·1 ppm and in the range previously required only for crystalline semiconducting materials. (Such emphasis on purity is a new departure for the field of amorphous solids where the tendency has been to marvel at how little effect there is on the physical properties of adding even gross amounts of impurities.) The fabrication in such high purity of the micrometre diameter fibres required for optical waveguides, which must also be free from inhomogeneities and clad with a matched glass of similar purity but lower refractive index, is stretching present techniques to the limit. It is only being undertaken because it offers the sole hope of reliable laser beam communication. The eventual goal, of ultra-wide-band information transmission, is far beyond any immediate requirements, but is almost certainly essential for the extrapolation of present trends to the end of the century. Perhaps the World then will be linked with a web of glass far more complex than the present mesh of our copper wire communication networks.

Finally, a plea to research workers to look at luminescence in amorphous solids. In particular, is it possible to obtain electroluminescence, the direct conversion of electrical energy into light, with appreciable efficiency and brightness? Surprisingly little seems to be known about anything of the sort, though what information is available suggests that luminescence in amorphous solids is weak. Is this necessarily so? What are the competing non-radiative processes? Can one use rare earth impurities to minimise the effect of the random variations in crystal field (as in the Nd laser glass)? Such questions, perhaps better framed, are important if only because were efficient, bright electroluminescence possible it could be of great practical interest, particularly as it should be relatively straightforward to produce large panels of the appropriate amorphous solid. Apart from the obvious application to "lighting" (high efficiency required to minimise power consumption, hence improbable, at least in the near future), there is the specialised but important requirement of the "large area display". Initially this could be an "information panel" for data presentation, showing reasonably large numbers of characters (several hundred); ultimately it might be the long awaited replacement for the cathode ray tube: the TV screen on the wall. But without more, actual information on luminescence in amorphous solids all this is speculation.

To end, let me quote from Francis Bacon's "Novum Organum", a document which is one of the great foundations of our present-day science and technology, and which, perhaps, is too little known to scientists today. Writing 400 years ago about the use of the scientific approach for making discoveries useful to mankind, he said, "It appears at first incredible that any such discovery should be made; and when it has been made it appears incredible that it should so long have escaped men's research. All of which afford good reason for the hope that the vast mass of inventions yet remains which may be deduced not only from the investigation of new modes of operation, but also from transferring, comparing and applying those already known". Certainly amorphous solids offer a new mode of operation to solid state physicists, equally certainly there is much to transfer, compare and apply; from which let us deduce a vast amount of discovery will result!

18. PHYSICS OF SWITCHING AND MEMORY DEVICES

H. FRITZSCHE

Department of Physics and James Franck Institute, The University of Chicago, Chicago, Illinois 60637.

18.1. INTRODUCTION

The discovery of nondestructive switching[1] in thin film amorphous semiconductors initiated a reexamination of the conditions for breakdown and switching in insulators and stimulated a surge of interest in the physics of noncrystalline materials.

During the past four years a variety of switching and memory effects have been observed in many different materials. I limit this discussion to chalcogenide alloy glasses which were studied most extensively and refer you to conference proceedings[2,3] and recent reviews[4–8] for more detailed information.

The earlier chapters of these Proceedings indicate that our understanding of even the simplest transport phenomena in amorphous semiconductors is quite incomplete. For example, we do not know the distribution, density, and capture cross-sections of the localized states in most materials, and many questions concerning the role of polaron effects, heterogenities, and of contacts are not settled. It is therefore not surprising that theories of electronic conduction at high electric fields and of the breakdown and switching process are rather speculative. These processes are very complicated because they involve non-uniform fields and large deviations from carrier equilibrium which in turn might modify the electron states. There are several recent reviews on electric breakdown in solids[(9–12)] which should be consulted to gain a larger perspective of this field.

It is worth noting that applications of amorphous semiconductors are not restricted to xerography and to the electrical switching and memory devices. Applications either realized or anticipated cover a wide spectrum; they include the continuous dynode electron multiplier (channeltron),[(13)] optical mass memories,[(14,15)] phase contrast holograms,[(16,17)] photographic imaging,[(14,18)] high energy particle detectors,[(19)] and ultrasonic delay lines.[(20)] Well known is the use of amorphous semiconductors as infrared windows and as photoconductors in vidicons, of Al_2O_3 and Ta_2O_5 in capacitors, and of SiO_2 as an insulator in the integrated circuit silicon technology. (See also Chapters 17, 19 and 20). Further development in all these fields depends on improved understanding and characterization of amorphous semiconductors.

18.2. Threshold Switch

Figures 18.1 (a) and (b) show the current–voltage characteristic and the time response to an applied voltage of the threshold switch.[(1–7)] The amorphous semiconductor is a sputtered film ($\sim 1\ \mu$m thick) of a stable chalcogenide alloy, such as $Te_{50}As_{30}Si_{10}Ge_{10}$, with a room temperature conductivity $10^{-5} > \sigma > 10^{-8}\ \Omega^{-1}cm^{-1}$. It is placed between carbon or metal electrodes of about $5 \times 10^{-6} cm^2$ cross-sectional area. When the voltage drop exceeds a threshold voltage V_t, or in case of an applied voltage pulse $V_p > V_t$ after a delay time t_D, switching occurs very rapidly within a switching time $t_s < 10^{-10}$ sec to the ON state of the characteristic. Provided the current exceeds a minimum, or holding, current I_h the device remains conducting with a dynamic resistance in the range 1–10 Ω and a voltage drop $V_h \sim 1$ V. However, if the current falls below I_h, the threshold switch returns to its original high resistance OFF state in about 1 μsec. In the following we summarize the experimental features of the device parameters. We restrict the discussion to devices which are less than a few microns thick. Thicker devices are more

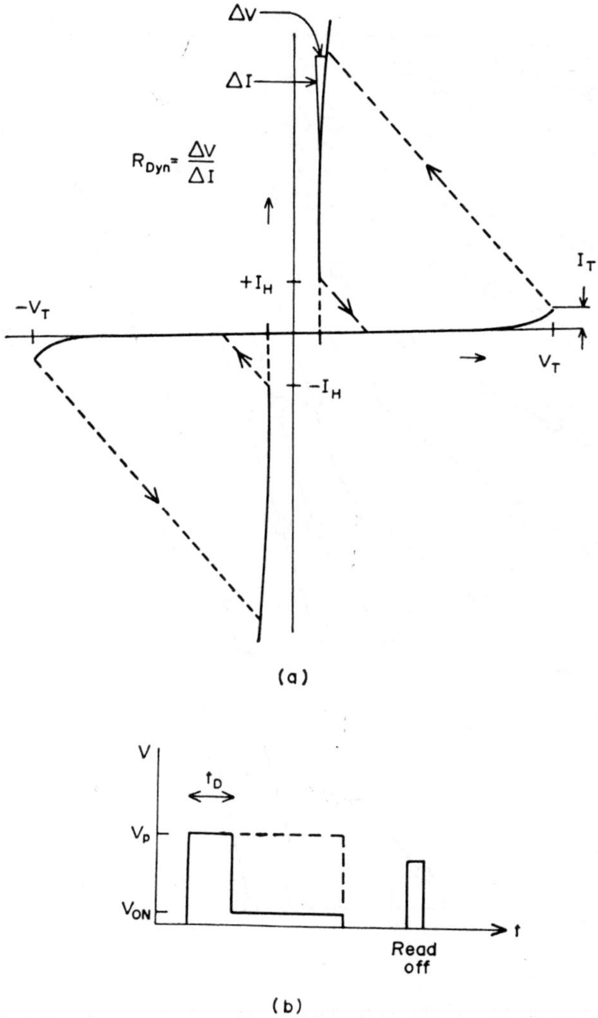

FIG. 18.1. (a) Current–voltage characteristic of a threshold switching device, (b) sketch of the time response to a voltage pulse $V_p > V_t$ and to an interrogating voltage pulse $V < V_t$.

difficult to characterize because they depend on the thermal boundary conditions on account of the low heat conductivity of the amorphous film.

18.2.1. THE OFF STATE

The conductance of a typical chalcogenide semiconductor increases with applied field E as shown in Fig. 18.2 for various temperatures.[21] These data

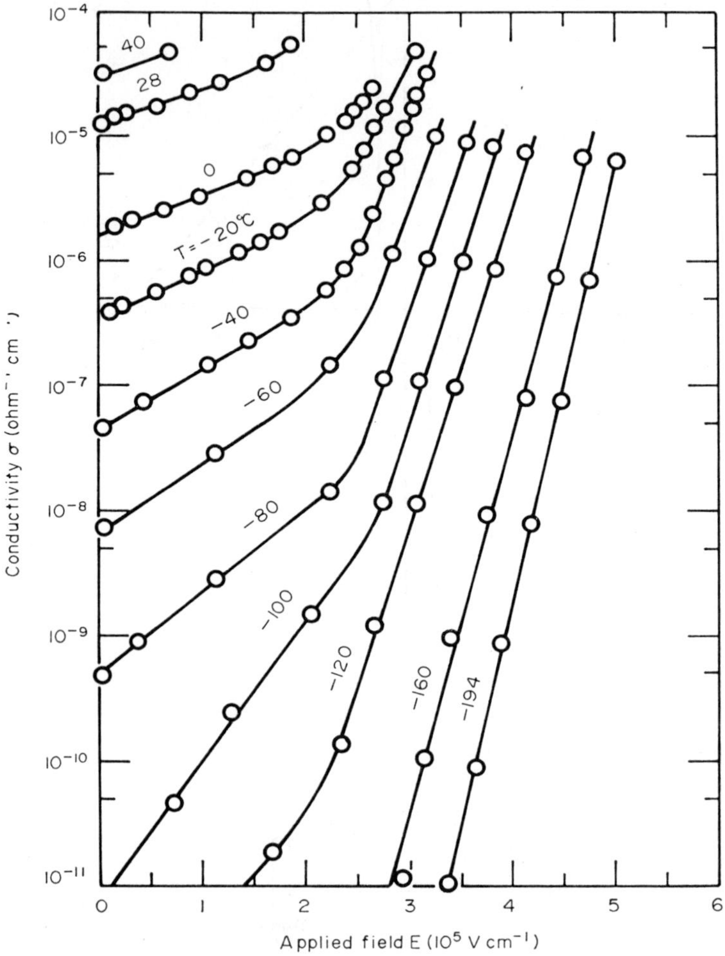

FIG. 18.2. Field dependence of the conductivity of a 1·5 μm thick sputtered layer of $Ta_{81}Ge_{15}Sb_2S_2$ at various T. Applied field was assumed to be uniform. (After Buckley ned Holmberg, Ref. 21.)

are not significantly influenced by Joule self-heating. High field conduction has been discussed earlier at this summer school, (see Chapter 8). The origin of the field effect is not yet established. One observes that the first nonohmic term is proportional to $|E|$. The conductance appears to increase as

$$\sigma = C \exp(-\Delta\varepsilon + \alpha_1 E)/kT \tag{18.1}$$

at low fields and above 2×10^5 V/cm as

$$\sigma = \text{const} \exp(\alpha_2 E). \tag{18.2}$$

The conductance is nearly the same for both polarities.

18.2.2. THRESHOLD VOLTAGE V_t

The temperature dependence of the voltage, current, and conductance at threshold is shown in Fig. 18.3.[21] At low T, these parameters are nearly temperature independent. According to Fig. 18.2, the conductance before switching is described by the steep exponential of eqn (18.2) in the low T region, whereas at higher temperatures eqn (18.1) describes the conductance up to the breakdown voltage.

The definition of V_t in the high T region is not unique; the value of V_t is first constant and then increases as the threshold point is approached with a faster rising voltage ramp. This suggests that some heating is taking place before switching and that the field distribution is nonuniform. This is not the

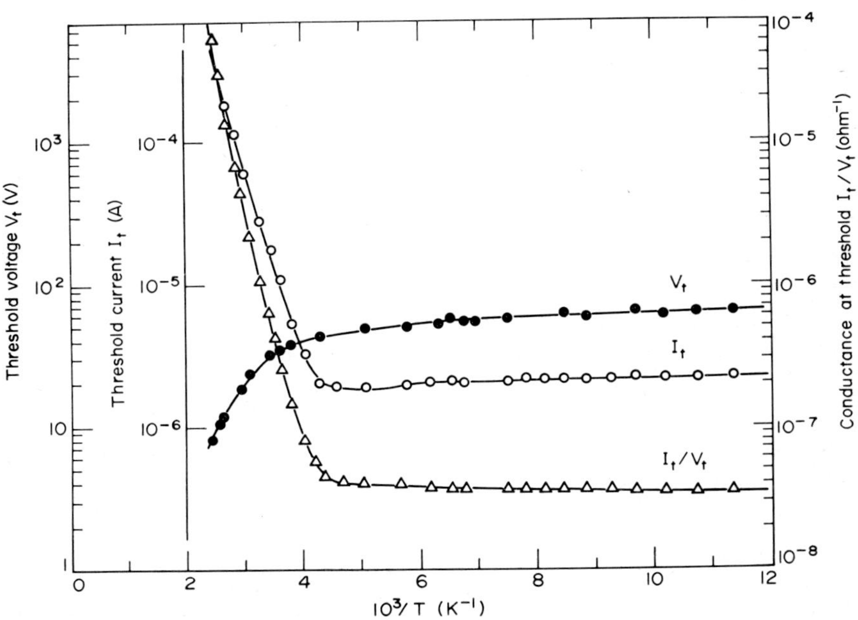

FIG. 18.3. Voltage, current and conductance at threshold as a function of reciprocal temperature for material shown in Fig. 18.2. (After Buckley and Holmberg, Ref. 21.)

case in the low T region. Here V_t is proportional to thickness for thicknesses less than a few micron. V_t values ranging between 3 and 400 V can be obtained with different materials.

18.2.3. DELAY TIME t_D

The delay time t_D decreases rapidly as the applied voltage pulse V_p is increased beyond V_t. For $V_p > 1.2\, V_t$ it decreases exponentially as

$$t_D = t_{D_0} \exp(-\alpha_3 V_p). \tag{18.3}$$

In the high T region the prefactor t_{D_0} decreases with increasing T. For $T = 300$ K and a film thickness of 1 μm, t_D decreases typically from 10^{-6} to 10^{-9} sec as V_p/V_t is increased from 1·2 to 2·0. Reversing the sign of V_p during the t_D interval has no noticeable effect on the value of t_D in the high T range.[22] This is as expected since polarity effects due to space charges can only be observed when the dielectric relaxation time is long compared to the time of observation. At low T the delay time is lengthened by reversal of V_p during t_D.[23]

One expects that the internal device temperature increases when the device is operated at large overvoltages V_p/V_t and higher pulse repetition rates. This should result in shorter delay times. One observes, on the contrary, an increase of t_D with increasing pulse frequency or if the pulse is preceded by a higher value switching pulse.[24–26]

18.2.4. THE ON STATE

In the conducting state the current flow is constricted to a filament, 1–3 μm in diameter, depending on the electrode separation and the current value. The voltage drop in the ON state is typically 1 V for Mo, W, and Ta electrodes and 1·5 V for carbon electrodes, independent of thickness. This shows that the potential drop occurs predominantly near one or both contacts. A small magnitude of V_h is desirable in order to minimize Joule heating for a given current value. Typical switches can tolerate short (~ 1 μsec) current pulses of several amperes. Steady currents should be kept below 50–100 mA depending on the cooling efficiency of the electrodes. The holding current is about $I_h \sim 0{\cdot}2$ mA.

The presence of the self-sustaining ON state is the important feature of switching in these amorphous semiconductors. It protects them from destructive burnout. The sustaining mechanism is therefore of particular interest. Henisch and his group have measured the transient response of the ON state.

If the ON state is completely interrupted for a short time t_i it can be reestablished without a new switching process. For longer interruptions (about 1–2 μsec) V_t approaches its normal value as illustrated in Fig. 18.4. Henisch and Pryor[23] find that this recovery curve remains essentially unchanged, except for a lengthening of t_i(max) as the ON current of the pulse before interruption is increased by a factor of 10. This indicates that processes

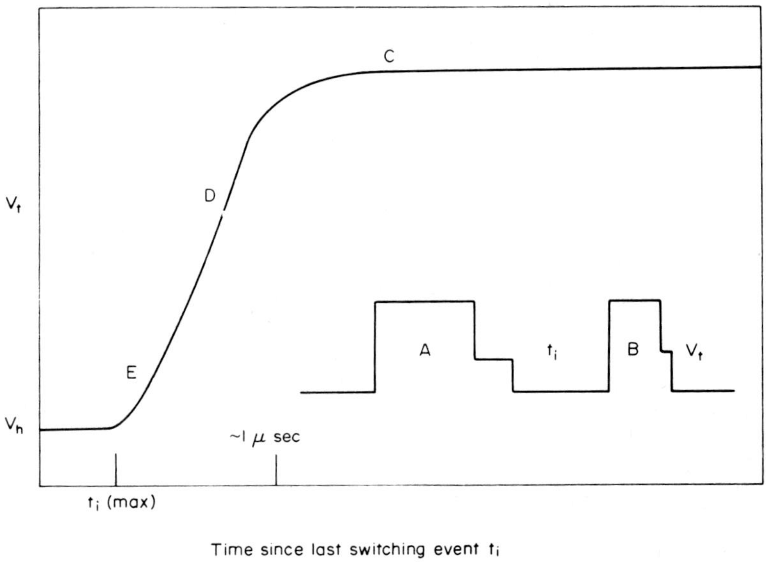

FIG. 18.4. Typical recovery curve of threshold voltage after switching process. The ON state can be reestablished after an interruption lasting less than t_i(max) without requiring a new switching process. (After Henisch and Pryor, Ref. 23.)

other than cooling influence the return to equilibrium. By keeping $t_i < t_i$(max), to avoid renewed switching, and applying a voltage $\pm V_i$ during t_i one can trace out a so-called transient ON characteristic[27] as shown in Figs. 18.5(a) and (b). For $-V_h < V < V_h$ the transient current is exceedingly small. The sharpness of the kinks in the transient ON characteristic depend more noticeably on the contact material than the gross features of the threshold switch.

If space charges near the contacts play an important role in sustaining the ON state one might expect that these space charges (if they reside in deeper traps) as well as the ON state could be frozen in at low temperatures (4·2 K). Such attempts have failed.[28]

18.2.5. DEVICE STRUCTURE AND PERFORMANCE

Attempts to develop threshold switches into commercial devices have not

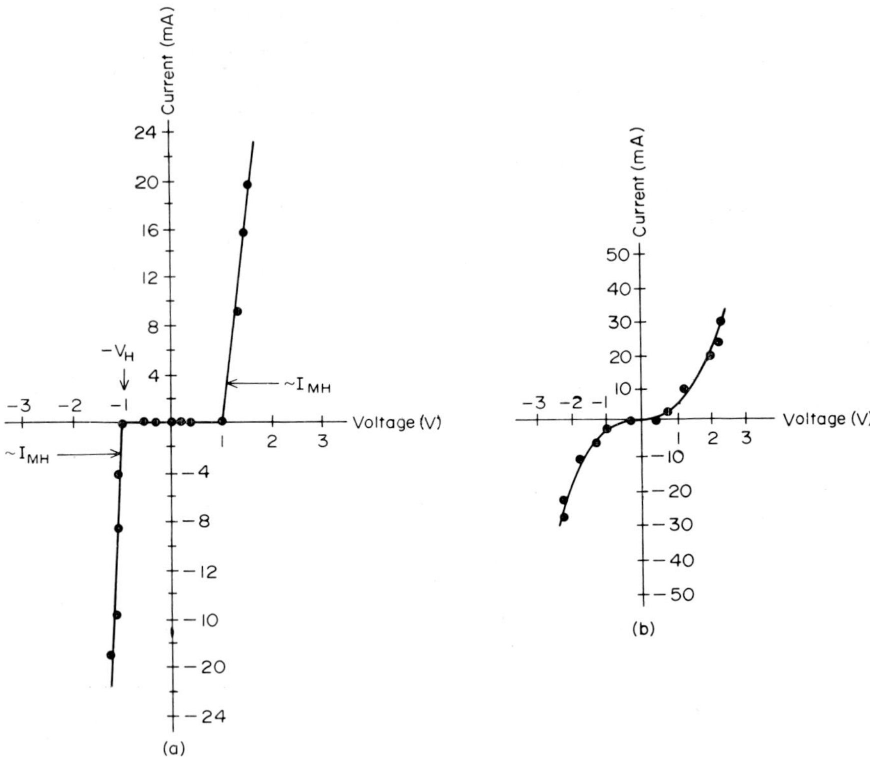

FIG. 18.5. Current–voltage characteristic during interruption period $t_i < t_i(\max)$ of the ON state of a threshold switch (a) with pyrolytic graphite electrodes and (b) with vitreous graphite electrodes. (After Pryor and Henisch, Ref. 27.)

yet been made. They appear well suited, for example, for switching electroluminescence and liquid crystal displays and for telephone cross point switching.[29] Threshold switches have been kept switching for more than four years using 60 Hz voltages, i.e. for more than 10^{10} switching operations. Poor reliability and life results when contacts such as Au, Ag, Cu, or other low melting metals are used which form alloys with the chalcogenide semiconductors.

A four terminal threshold switch was recently described by Van Landingham.[4] As shown in Fig. 18.6 a 0·15 μm thick carbon film is placed between a silicon substrate and a threshold gap device. Between the two is a 1 μm thick SiO_2 film which yields an isolation resistance of about 10^9 Ω and a coupling capacitance of the order of 0·01 pF. The carbon film is used as a heater. When the threshold switch is biased with an ac voltage below threshold, a pulse to the heater strip triggers switching. The device remains then in the switching

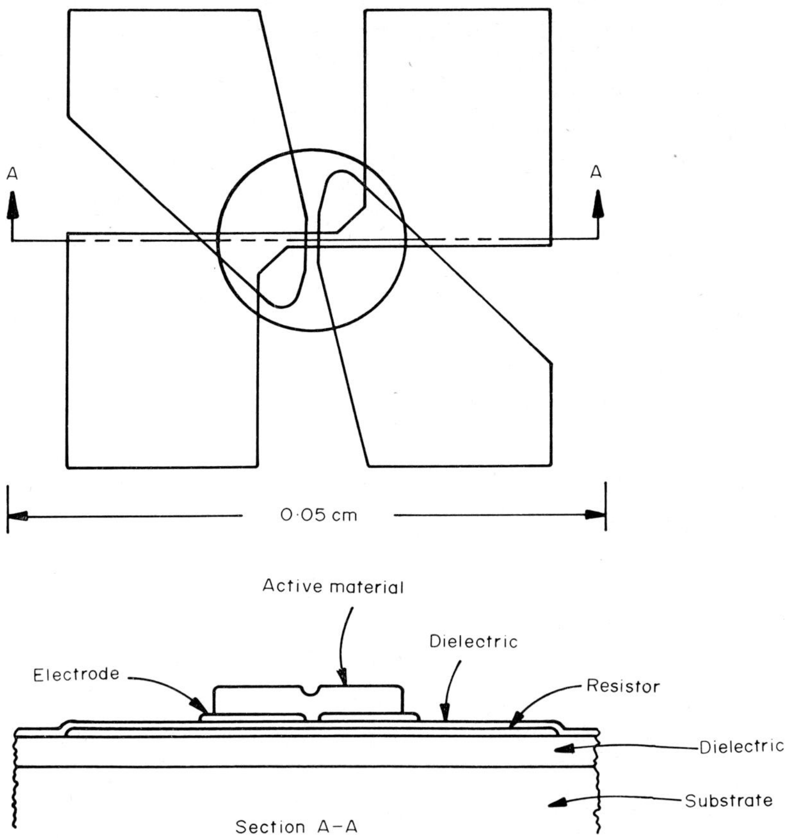

FIG. 18.6. Layout of four terminal threshold switch.

mode until interrupted for a period of order 5 μsec because V_t under load is less than under low duty cycle operation .The present design permits triggering with 10 μsec pulses of digital circuit voltage level.

18.3. MEMORY SWITCH

18.3.1. EXPERIMENTAL FEATURES

Figures 18.7(a) and (b) show the I–V characteristic of a memory switch[1–7] and the response to SET, RESET and interrogating READ pulses, respectively. For the memory switch an amorphous semiconductor is used whose structure can be changed reversibly between two structural states, the amorphous state which has a high resistance, and a microcrystalline state which

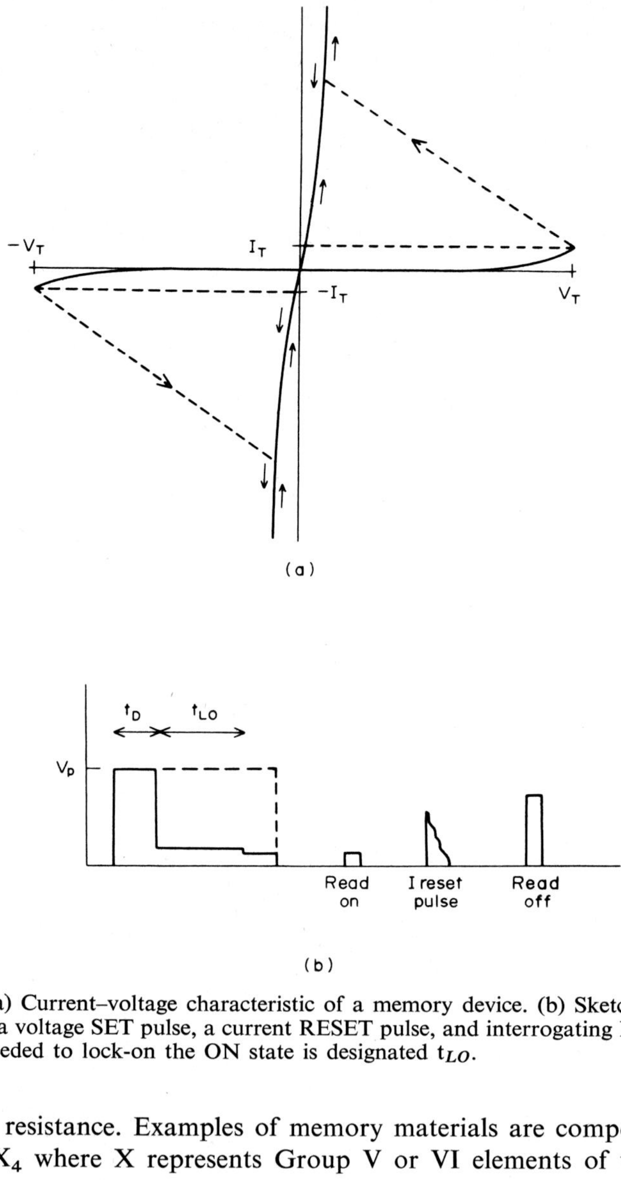

FIG. 18.7. (a) Current–voltage characteristic of a memory device. (b) Sketch of the time response to a voltage SET pulse, a current RESET pulse, and interrogating READ pulses. The time needed to lock-on the ON state is designated t_{LO}.

has a low resistance. Examples of memory materials are compositions like $Te_{81}Ge_{51}X_4$ where X represents Group V or VI elements of the periodic table.

The OFF state and the switching process of the memory device is the same as that of a threshold switch. However, once the memory switch is kept in the ON state (5–10 mA current level) for longer than about 10^{-3}–10^{-4} sec, it is SET and retains its conductive ON state even when the current level approa-

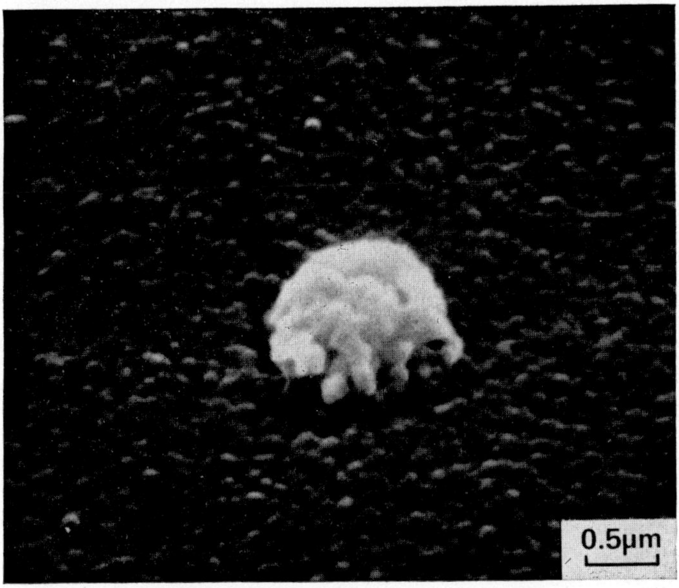

(a)

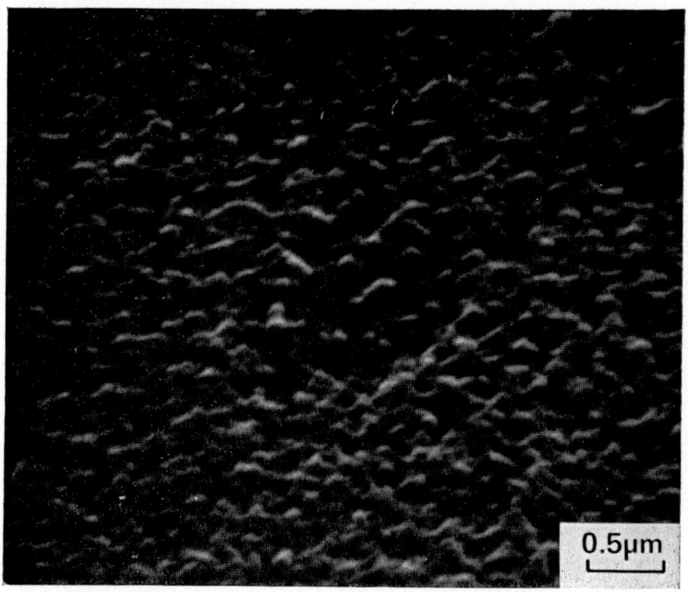

(b)

FIG. 18.8. Photomicrograph of etched memory material (a) of a device left in the SET state and (b) of a device left in the RESET state. (After Sie *et al.*, Ref. 30.)

ches zero or is reversed. A RESET pulse typically of 5 μsec duration and 120 mA current is required to return the memory switch to its high resistance OFF state.

The memory action in these devices is based on the reversible crystallization and revitrification of the material in the current filament. The filament region of the memory device can be made visible by etching off the upper electrode and some of the amorphous layer. A scanning electron microscope picture of such etched memory devices left in the SET and in the RESET states, respectively, are shown in Figs. 18.8(a) and (b).[30] The filament

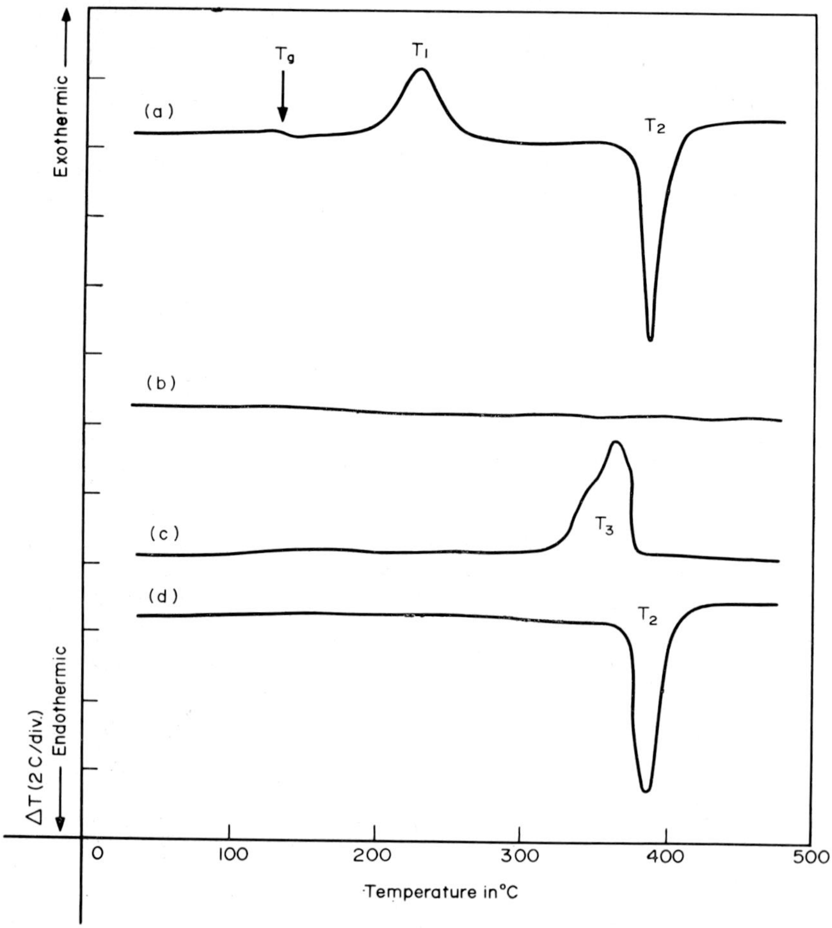

FIG. 18.9. Differential thermal analysis of a memory type chalcogenide alloy, $Te_{82}Ge_{16}Sb_2$: (a) heating from the amorphous state, (b) rapid cooling from the liquid, (c) slow cooling from the liquid, (d) heating from the crystallized phase. (After Fritzsche and Ovinsky, Ref. 31.)

diameter is about 2 μm. The microcrystalline material is phase separated into Te and GeTe rich regions. These are homogenized and revitrified by the short RESET pulse.

The characteristic difference between a structure stable material, used for threshold switches, and a structure reversible material used for memory switches can be seen most simply by differential thermal analysis (DTA).[31] By means of two thermocouples the DTA signal measures the temperature difference between the material and a standard of constant heat capacity as both are heated at a constant rate. The temperature difference reveals any exothermic reaction (crystallization), endothermic reaction (melting), or changes in heat capacity (glass transition) occuring in the sample. Figure 18.9 shows examples of DTA traces obtained with a memory type material at different heating and cooling rates. One observes the glass transition at T_g, crystallization at T_1, melting at T_2, and crystalline solidification at T_3. Trace (b) shows that fast cooling retains the "liquid" state down to room temperature, i.e. it causes vitrification. A structure stable glass, in contrast, can be heated far beyond the glass transition temperature T_g and slowly cooled without causing crystallization.

18.3.2. DEVICE STRUCTURE AND PERFORMANCE

Memory arrays[32] on the market are the 256 array (RM-256) and the 2048 bit hybrid package (HRM-2048) which consists of eight 256-bit arrays on a chip with internal "x"-line decoding. The RM-256 read mostly memory,[33,34] as shown diagrammatically in Fig. 18.10(a), is a 16 $\times$ 16 matrix with a diode-isolated memory switch at each crosspoint. Figure 18.10(b) is a close-up of the 3 $\times$ 3 mm Si chip showing some memory switches (round) and diodes (square). The individual memory switch measuring 30μm in diameter is a film of amorphous memory material sandwiched between two molybdenum electrodes and connected to the silicon diodes by aluminum metallization. These memory arrays are designated "Electrically Alterable Read Mostly Memory" (RMM) because the relatively long SET time ($\sim$ 1 msec) suggests applications which demand only intermittent change of the information content. However, a factor of 100 in alteration speed can be gained by SETTING the whole array at once and then using the fast RESET. The READ speed in contrast is very fast (50 nsec per bit) since one reads essentially a bistable resistor at digital circuit compatible voltage levels.

Among the advantages of the RMM are that (i) no power is required to maintain the information, (ii) information does not degrade during storage, (iii) information can be read without perturbing the state, (iv) it is directly compatible with DLT and TTL logic levels and (v) it is resistant to high energy particle radiation. Examples for applications include (a) telephone

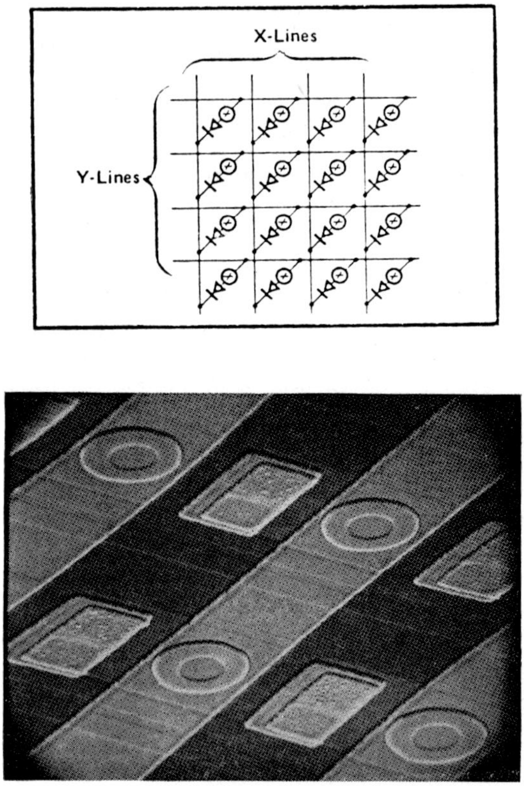

FIG. 18.10. (a) Partial schematic diagram of the RM-256 matrix (16×16). Symbol in circle represents Ovonic memory switch. (b) Scanning electron micrograph of part of the RM-256 memory array. Circles are memory switches, rectangles are *p–n* diodes. (After Neale *et al.*, Ref. 33.)

dial codes, (b) machine instructions, (b) microprogramming, (d) terminal encoding and decoding and (e) automatic polling.

Individual memory switches were subjected to more than 10^6 set–reset pulse trains as those shown in Fig. 18.7(b) without failure. Whole 16 × 16 memory arrays have presently a failure-free life of more than 10^6 set–reset cycles.

18.4. SWITCHING MECHANISMS

The switching process appears to be the same in switching and in memory devices. In order to be able to explain it in terms of the properties of the amorphous semiconductor it is essential that the material in the current filament has not undergone a change as the result of switching. As soon as a "forming" process occurs,[35] which causes a change in the OFF resistance and the threshold voltage, one suspects some alloying with the contacts, electromigration, or a change in structure of the amorphous film. Bosnell and Thomas[36] base their model of threshold switching exactly on these changes by assuming that phase separation and crystallization are essential for non-destructive and repetitive switching. In their model, a conductive crystalline matrix forms the ON state even in the threshold switch. When the device switches off as the current falls below I_h, these authors assume that the material changes to a phase separated material in which the crystallites are isolated from one another.

Although these ideas describe a failure mechanism rather than threshold switching, they emphasize the importance of ascertaining that the data to be explained are obtained on devices in which no material change occured.

It is also important to note that electronic and thermal phenomena do not scale in the same manner. Much data on filament formation was gathered by studying the switching process between two point contacts placed rather far apart on the surface of a semiconductor. Because of the low heat conductivity of amorphous materials, overheating, electromigration, and material changes are likely to occur in such a case and, in general, when the electrodes are more than a few microns apart and not well connected to a heat sink. These data are therefore not directly applicable to thin film devices.

The theories discussed below assume that these precautions were taken and that one deals with a homogeneous amorphous semiconductor. Two approaches are taken: switching is either explained as a Joule self-heating in which the high field effects on electronic conduction play an essential role or as an electronic process which has to take account of the fact that the temperature in the ON state filament rises above ambient. Different theories may be applicable in different regimes of temperature, electrode separation, and material properties.

18.4.1. ELECTROTHERMAL THEORY

The physical process is essentially the following. Joule heating raises the temperature inside the semiconductor. The resultant increase in conductance allows more current to flow through the heated regions with the consequence of enhanced Joule heating and increased concentration of current in the heated region. A new stationary state is established when the heat conducted away from the current filament equals the heat generated. Figure 18.2 makes it obvious that one should not use the low field ohmic conductivity but the conductance modified by high fields, space charges, field concentration near the electrodes, and, if present, current injection by the electrodes. At present, only a simplified form of the field dependence of the OFF state has been used in the calculations. The most comprehensive treatment of this problem is the recent work of Kroll[37] where references to the earlier literature can be found. Kroll finds that switching is initiated when a statistical fluctuation in either the current, thermal conductivity, temperature or conductivity reaches the size of the electrode separation and moves the system from the stable OFF branch of the characteristic to an adjacent unstable branch which corresponds to a current concentrated in a narrow filament. From this unstable branch the system switches to the ON state. For thin film devices Kroll finds that there is no stable negative resistance region. Switching off depends on nucleation and growth of a cold region. I_h may be larger or smaller than I_t.

The transient ON state (Fig. 18.5) and the recovery of V_t (Fig. 18.4) is explained by Kroll as follows. In the ON state the field is concentrated in the cold regions of widths d_c adjacent to the electrodes. For very short interruption intervals $t_i < t_i$ (max) the T-distribution does not change. The I–V characteristic is then determined by eqn (18.2) where $E = V/d_c$. Region D and E of Fig. 18.4 is interpreted to be determined by diffusion of the hot core, region C by cooling of the central region after the hot core has diffused away. The polarity effects,[23] the increase of t_D with increasing pulse repetition frequency,[24-26] and insensitivity of the recovery curve, Fig. 18.4, to increases in current[23] are not explained.

18.4.2. ELECTRONIC THEORIES

There are two very different kinds of electronic switching models. The models of the first category assume (i) that an excess electron and hole concentration is established by field dependent processes in the bulk or by injection from the electrodes and (ii) that this excess concentration in the bulk can be sustained after switching by high field regions near the contacts. The excess concentrations, which cause the bulk to be highly conducting (but neutral), can be described by quasi-Fermi levels for electrons and holes, split apart from the equilibrium Fermi energy. The time for equilibrium τ_0 is assumed to be longer than the dielectric relaxation time τ_{rel}.

The switching model of van Roosbroeck[38] belongs to the second category. It takes the opposite point of view and assumes $\tau_0 \ll \tau_{rel}$, which is the condition for a relaxation type semiconductor. As mentioned earlier (Chapter 2), minority carrier injection does not lead to a splitting into quasi-Fermi levels but to a shift of ε_F and thus to considerable space charges. In the ON state, the resistance and the voltage drop is limited to a narrow recombination front adjacent to the noninjecting electrode. Whether or not amorphous semiconductors are in the relaxation regime of conduction may be tested by exploring the dependence on electrode separation.

Space permits only a brief discussion of the electronic theories of the first category. Details are given in several reviews.[4,5,7,12]

For establishing the conductive plasma of excess carriers in the bulk one has considered avalanche excitation,[12,39] field enhanced hopping and excitation,[12,40] and double injection.[41] Any process which produces a high excess concentration of carriers will lead to space charges at the electrodes. The field at the electrodes increases until carriers can tunnel and replenish the charge carriers lost be recombination. In order to sustain the ON state by double injection or by tunneling it is necessary that the excess carrier lifetime is longer than the transit time.[42] Lucas' model of switching assumes that beyond a critical injection current both electron and hole traps are neutralized and recombination is thus sharply decreased. When the diffusion length grows to the size of the electrode separation, the conductance rises sharply.

Heywang and Haberland[43] use the percolation path model, discussed in Chapter 2, and point out that injected electrons fill predominantly positive traps along the electron percolation paths and injected holes fill their respective traps along the hole percolation paths. This causes an overall decrease of the potential fluctuations, a widening of the percolation paths and perhaps an increase in mobility. If this also leads to an increase in double injection, then it provides the feedback mechanism needed for the high conductance ON state.

ACKNOWLEDGEMENTS

It is a pleasure to acknowledge many clarifying discussions with S. R. Ovshinsky, N. F. Mott, E. A. Fagen, H. K. Henisch, J. de Neufville, R. G. Neale, and M. P. Shaw. The support of the Air Force Office of Scientific Research (AFSC), United States Air Force, under contract F44620–71–C–0025, and the general support of the Materials Research Laboratory by the National Science Foundation, is gratefully acknowledged.

REFERENCES

1. Ovshinsky, S. R., *Phys. Rev. Letters* **21,** 1450 (1968).
2. *J. Non-Cryst. Solids* Vol. **2, 4,** and **8-10.**
3. Proc. Intl. Congress Thin Films, Cannes, France, 1970.
4. I.E.E.E. Transactions on Electron Devices, Special issue on amorphous semiconductor devices, Feb. 1973.
5. Adler, D., *Critical Reviews in Solid State Sciences* **2,** 317 (1971). (Chem. Rubber Pub. Co., Cleveland, Ohio).

6. Fundamentals of Amorphous Semiconductors, Report of the National Academy of Sciences, 2101 Constitution Avenue, N.W., Washington, D.C. (1972).
7. "Amorphous and Liquid Semiconductors", (J. Tauc, ed.). Plenum Press (1973).
8. Dearnaley, G., Stoneham, A. M. and Morgan, D. V., *Rep. Progr. Phys.* **33,** 1129 (1970).
9. Stratton, R., *Progr. Dielectrics* **3,** 235 (1961).
10. Klein, N., *Thin Solid Films* **4,** 149 (1971).
11. Klein, N., to be published.
12. Mott, N. F., *Phil. Mag.* **24,** 911 (1971).
13. Adams, J. and Manley, B. W., *Electron. Eng.* **37,** 180 (1965).
14. Ovshinsky, S. R., Proc. Fifth Annual Natl. Conf. on Industrial Res., Sept. 18, p. 68 (1969).
15. Feinleib, J., de Neufville, J., Moss, S. C. and Ovshinsky, S. R., *Appl. Phys. Letters* **18,** 254 (1971).
16. Pearson, A. D. and Bagley, B. G., *Mat. Res. Bull.* **6,** 1041 (1971).
17. Keneman, S., *Appl. Phys. Letters* **19,** 205 (1971).
18. Ovshinsky, S. R. and Klose, P., *J. Non-Cryst. Solids* **8-10,** 892 (1972).
19. Srinivasan, G. R., Younger, S., Macedo, P. B. and Litovitz, T. A., *Phys. Rev. Letters* **27,** 1084 (1971).
20. Krause, J. T., Kurkjian, C. R., Pinnow, D. A. and Sigety, E. A., *Appl. Phys. Letters* **17,** 367 (1970).
21. Buckley, W. D. and Holmberg, S., to be published.
22. Shanks, R. R., *J. Non-Cryst. Solids* **2,** 504 (1970); Balberg, I., *Appl. Phys. Letters* **16,** 491 (1970).
23. Henisch, H. K. and Pryor, R. W., *Solid State Electronics* **14,** 765 (1971).
24. Haberland, D. R., *Solid State Electronics* **13,** 207, 451 (1970).
25. Csillag, A., in reference 4.
26. Lee, S. H. and Henisch, H. K., to be published.
27. Pryor, R. W. and Henisch, H. K., *J. Non-Cryst. Solids* **7,** 181 (1972) and *Phys. Stat. Sol.,* to be published.
28. Adler, D., 1972, private communication.
29. A number of circuit applications can be found in references 2 and 4.
30. Sie, S. H., Dugan, M. P. and Moss, S. C., *J. Non-Cryst. Solids* **8-10,** 877 (1972).
31. Fritzsche, H. and Ovshinsky, S. R., *J. Non-Cryst. Solids* **2,** 148 (1970).
32. Energy Conversion Devices, Inc., Troy, Michigan.
33. Neale, R. G., Nelson, D. L. and Moore, G. E., *Electronics* **43,** No. 20, 56 (1970).
34. Neale, R. G. and Aseltine, J. A., in reference 4.
35. Coward, L. A., *J. Non-Cryst. Solids* **6,** 107 (1971).
36. Bosnell, J. R. and Thomas, C. B., *Phil. Mag.* (in press).
37. Kroll, D. M., *Phys. Rev.* to be published.
38. van Roosbroeck, W., *Phys. Rev. Letters* **28,** 1120 (1972).
39. Hindley, H. K., *J. Non-Cryst. Solids* **5,** 17, 31 (1970).
40. Fritzsche, H. and Ovshinsky, S. R., *J. Non-Cryst. Solids* **2,** 393 (1970).
41. Henisch, H. K., Fagen, E. A. and Ovshinsky, S. R., *J. Non-Cryst. Solids* **4,** 538 (1970).
42. Lucas, I., *J. Non-Cryst. Solids* **6,** 136 (1971).
43. Heywang, W. and Haberland, D. R., *Solid State Electronics* **13,** 1077 (1970).

19. PHOTOSTRUCTURAL EFFECTS AND THE USE OF AMORPHOUS SEMICONDUCTORS FOR HOLOGRAPHY AND IMAGING

H. FRITZSCHE

Department of Physics and James Franck Institute,
The University of Chicago, Chicago, Illinois 60673.

19.1. INTRODUCTION

The differences, rather than the similarities, with crystals make the subject of amorphous semiconductors interesting. One fundamental difference is that the structure and the bonding configuration of an amorphous solid are not determined and fixed by thermodynamic equilibrium conditions but can be altered, sometimes reversibly, by some means which change the energy of the system. In this contribution we deal in particular with structural changes caused by light.

Many substances undergo physical changes under long exposure to sunlight. The degrading effects of light on lacquers and fabrics have been known for a long time. Bitumen of Judea was used by the early Sumerians, Assyrians and Babylonians as a bonding material in place of mortar, to caulk vessels and for setting jewels and mosaics. Bituminous and tar-like materials become non-tacky after exposure to sunlight as a result of photo-crosslinking of the polymers. Now we understand that Noah had good reason to pray for good weather when he built his ark. Also the Egyptians appreciated the effects of the sun when after wrapping their mummies in linen cloth which was dipped in a solution of bitumen of Judea and lavender oil, the material becomes non-tacky after lying in the sun, a certain advantage for handling mummies in hot climates.

J. N. Niepce, being an amateur Egyptologist, knew of bitumen of Judea. He dissolved this asphalt in lavender oil, coated a sheet of pewter with this mixture and let light shine through an ink drawing placed over it. The material hardened except for the black lines where the coating remained soft and soluble in lavender oil. In this manner he made the first lithographic plates (he called the process heliogravure). Around 1824 Niepce recorded the first permanent photograph with a pinhole camera. Later one used gelatine containing potassium bichromate for making lithographic plates. Here again photo-crosslinking renders the light exposed portions of the gelatin insoluble in hot water.

To a large extent photochemistry deals with polymeric materials. Best known in semiconductor technology are the photoresist materials which are essential for the manufacture of integrated circuits. But many inorganic materials are unstable on exposure to light, the silver halides used for photography, ferric salts used for blueprints, and, surprisingly, several crystalline and amorphous semiconductors, As_2S_3 and As_2Se_3 for example, whose properties are discussed in these Proceedings.

The fact that crystalline materials, such as As_2S_3 (orpiment) and As_4S_4 (realgar) are unstable on exposure to light has been known to minerologists since the end of last century. Solid state scientists have become aware of these effects only recently.

Although little is known in the literature about the physcial processes involved, I wish to give a brief introduction to this field and discuss some possible applications for holography, imaging, and optical memories. The field of local order changes in chalcogenide amorphous materials for imaging purposes has been under development for considerable time.(27)

If one wants to compare experimental results obtained in different laboratories, one has to keep in mind how much the preparation conditions, the thermal history, and also the exposure to radiation determine the local bond arrangement of an amorphous solid. On the other hand, the possibility of inducing reversibly some structural changes can be used to one's advantage when studying how the various material properties depend on such changes.

Depending on the stability of the material and the amount of light exposure, different types of structural changes are observed. These include (i) changes in local order or bonding and defect states which preserve the overall vitreous state, (ii) light induced phase separation, (iii) photonucleation, and (iv) photocrystallization. These will be discussed in the following sections.

19.2. Photo-Induced Structural Changes

Chang discovered[1] that the exponential absorption tail of vitreous selenium is shifted to lower photon energies and broadened after the selenium is exposed to white light of a tungsten lamp. Figure 19.1 shows the relative absorption

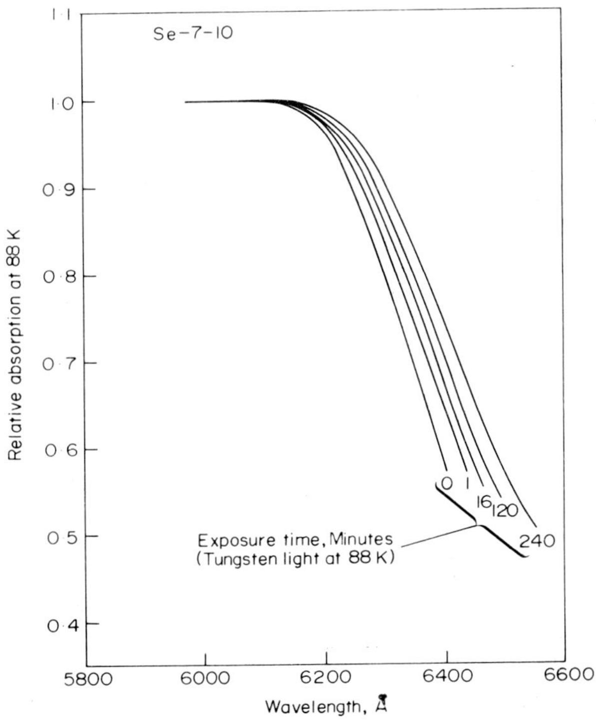

Fig. 19.1. Relative change of optical absorption of vitreous selenium after exposure to white tungsten light at 88 K. (After Chang, Ref. 1.)

measured at 88 K after exposing the sample for different durations. The sample temperature remained at 88 K. The spectral response of the irradiating light has a threshold which lies somewhat below the absorption edge $\mathscr{E}_0$ as seen in Fig. 19.2. The material irradiated at 88 K recovers its original state almost completely upon warming to room temperature. The changes induced by irradiation near the glass transition temperature T_g (303 K) fail to anneal out even after keeping the selenium at 320 K for several days. This indicates that the structural changes produced at low temperatures and near T_g are quite different.

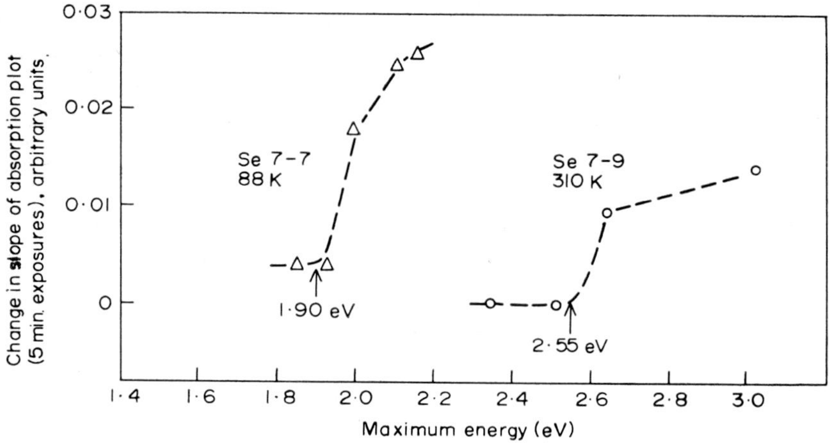

FIG. 19.2. Spectral sensitivity curve of irradiating light at 88 K and near the glass transition temperature. The temperature shift of the sensitivity threshold is opposite to that of the absorption edge. (After Chang, Ref. 1.)

The recovery curve of Fig. 19.3 was measured by first irradiating a sample at 88 K and then warming it to room temperature at a constant rate. Although the slope of the absorption curve plotted in Fig. 19.3 was measured at the respective annealing temperatures it appears that the annealing process occurs in two stages. The experiment would have been less ambiguous if the absorption curve had been measured at 88 K after bringing the sample to successively

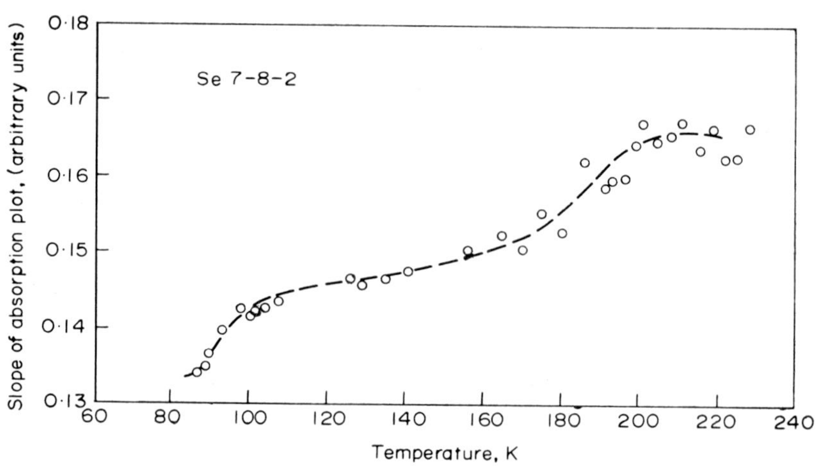

FIG. 19.3. Recovery of vitreous selenium after light exposure at 88 K. (After Chang, Ref. 1.)

higher temperatures. This procedure was followed by Zakis[2] who recently studied annealing of photostructural changes in evaporated As_2Se_3 films. Zakis finds two annealing steps between 80 and 400 K in his films.

A nearly parallel shift of the absorption edge toward lower photon energies occurs in a large number of arsenic–sulphur and arsenic–selenium glasses on exposure to photons of greater than band edge energy.[2–8,11] This shift is accompanied by a small increase in the refractive index and a decrease in electrical resistivity.[9] Some of these changes can be reversed by annealing the chalcogenide glasses near the glass transition temperature T_g.

In the following we present arguments and results from the work of Moss, de Neufville and Ovshinsky.[11] In order to understand the structural changes which give rise to the reversible and nonreversible effects, these authors point out that it is important to ask first how the absorption edge is affected by various preparation methods and annealing procedures. The chalcogenide glasses considered here can be produced (a) in bulk form by cooling from the melt, (b) as thin films by evaporation or (c) by sputtering. The absorption edge $\mathcal{E}_0$ of bulk glasses does not change noticeably by annealing near T_g unless they were prepared by rapid quenching. The absorption edge of sputtered or evaporated films of chalcogenide glasses lies usually at lower photon energies than $\mathcal{E}_0$ of the bulk glasses. Annealing near T_g shifts their absorption edge upwards toward $\mathcal{E}_0$.[10] As_2S_3 and As_2Se_3 and perhaps a few other glasses form interesting exceptions.[11] Although sputtered films of these materials follow the general rule mentioned above, evaporated As_2S_3 and As_2Se_3 have absorption edges which lie above $\mathcal{E}_0$ of the respective bulk materials. Annealing near T_g shifts these edges down toward $\mathcal{E}_0$. These annealing effects are obviously not reversible by simple means.

When freshly evaporated films of As_2S_3 or As_2Se_3 are exposed to light ($h\nu \gtrsim \mathcal{E}_0$), their absorption edge shifts down to $\mathcal{E}_0$ and even beyond $\mathcal{E}_0$ to lower energies. Freshly evaporated films can therefore be brought to their annealed structural state by exposure to light. This structural change is not reversible. At the same time some unspecified structural defects are produced by light, which can be removed by thermal annealing. These eventually cause the absorption edge to lie below $\mathcal{E}_0$. These defects might be the same as those which apparently exist in freshly sputtered As_2S_3 and As_2Se_3.

The structural change which occurs when freshly evaporated As_2S_3 and As_2Se_3 films are annealed can be observed by X-ray scattering.[11] Moss *et al.* find that the virgin films resemble a molecular glass consisting predominantly of As_4S_6 or As_4Se_6 molecules condensed from the vapour phase. This molecular glass has a higher absorption edge than the more cross-linked polymeric glass which is produced by annealing or illumination. In a sputtering process, atoms or smaller molecules seem to be deposited so that no molecular glass is formed. Also flash evaporation used for the deposition of ternary and quarternary alloy compositions tends to decrease the chance that larger molecules form.

We mentioned that As_2Se_3 and As_2S_3 are exceptional and that the absorption edge energy of many other virgin films of chalcogenide glasses increases with annealing. Light exposure of annealed material invariably produces an absorption edge shift to lower energies if heating effects are avoided. At high light intensities one not only has to worry about heating effects which cause annealing but also about thermal and/or light induced evaporation of the film.

The absorption edge shifts by about 0·1 eV in As_2S_3,(3) 0·04 eV in As_2Se_3, and 0·2 eV in $As_{40}Se_{50}Ge_{10}$(7) following exposures between 1–5 J/cm^2. Refractive index increases range between 0·01 and 0·05.(8)

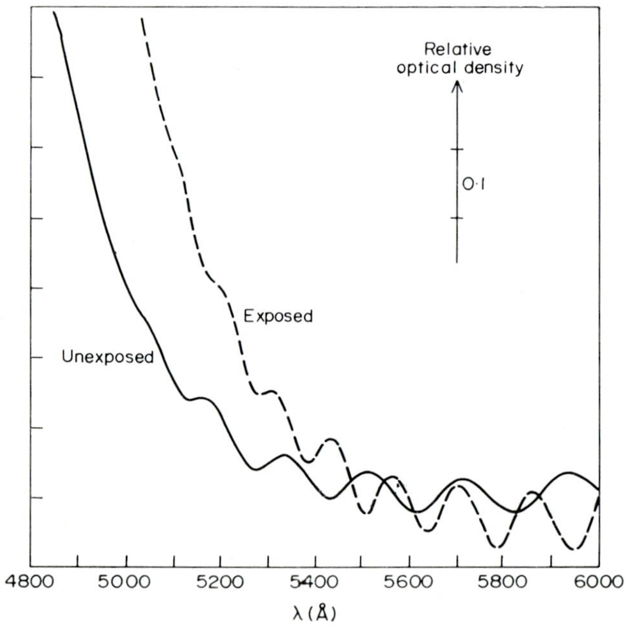

FIG. 19.4. Spectral dependence of the optical density of exposed and unexposed 2 μm thick evaporated As_2S_3 film. Exposure was by focussed microscopic lamp for 15 minutes. (After Keneman, Ref. 3.)

Figure 19.4 shows the optical density of a 2 μm thick evaporated As_2S_3 film before and after exposure for 15 minutes to a focussed microscope lamp. Keneman(3) reports storing holograms consisting of 2860lines/mm in a 2 μm thick As_2S_3 film with no sign of approaching the resolution limit. He obtained a first order diffraction efficiency of several percent with a 15 mW laser power in 10 sec. An 80 percent diffraction efficiency was obtained in a 10 μm thick film. The efficiency increases with the square of the film thickness.

High resolution holograms were also obtained by Ohmachi and Igo.[8] The diffraction efficiency reaches a maximum within the order of minutes.

The changes of the optical properties depend on the product of light intensity and exposure time. Deviations from this law of reciprocity occur at higher light intensity when thermal annealing and evaporation become noticeable.[5–8] The proportionality of the diffraction efficiency with the square of the film thickness shows that the changes occur in the film. The electrical transport properties, in particular the ac conductivity (see Chapter 8), are expected to be strongly affected by these structural changes. Street and Yoffe[12] found, for example, significant differences between the ac conductivity of evaporated films and that of bulk glasses of As_2S_3 and As_2Se_3. Evaporated films show a dipolar ac loss mechanism which may be caused by the motion of unlinked molecular chains according to these authors and which disappears after annealing.

19.3. Photodecomposition and Phase Separation

Berkes *et al.*[4] demonstrated that the chemical instability[13] of As_2Se_3 and As_2S_3 is associated with the photostructural changes. Figure 19.5 shows that

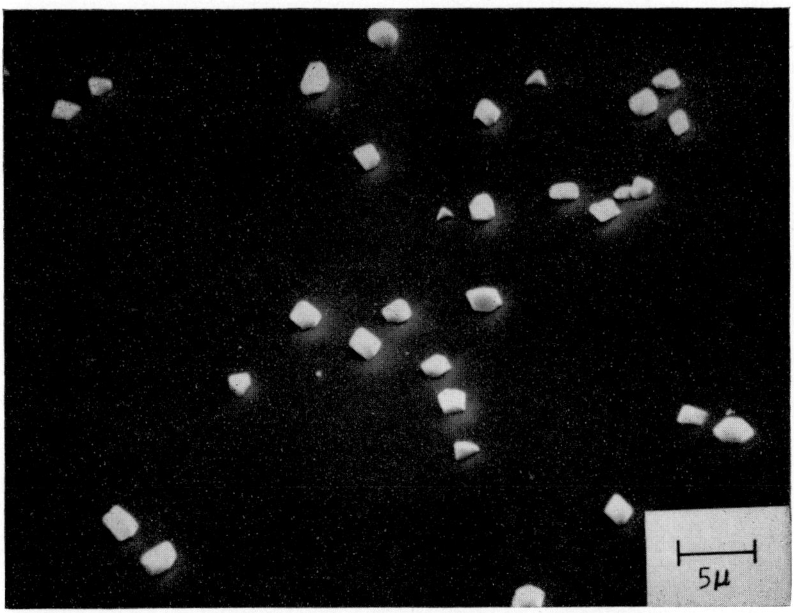

Fig. 19.5. Scanning electron micrograph of surface of evaporated As_2Se_3 film after it was exposed to laser light ($\lambda = 4579$ Å, $F = 10^{18}$ photons cm^{-2} sec^{-1}) for one hour at ambient conditions. The small crystals are cubic As_2O_3. (After Berkes *et al.*, Ref. 4.)

small cubic As_2O_3 crystals form on the surface of amorphous As_2Se_3 when it is illuminated in the presence of oxygen and water vapour. They suggest that the photons liberate some arsenic in the bulk according to

$$As_2Se_3 \xrightarrow{h\nu} x\,As + As_{2-x}Se_3 \tag{19.1}$$

with $0 < x < 2$. At the surface the arsenic readily oxidizes in the presence of water

$$4\,As + 3\,O_2 \xrightarrow{H_2O} 2\,As_2O_3 \tag{19.2}$$

Irradiation of As_2S_3 for several months produces at the surface sulphur crystals in addition to As_2O_3, which suggests the photoreaction[4]

$$As_2S_3 \xrightarrow{h\nu} 2\,As + 3S \tag{19.3}$$

Of all inorganic compounds the most photosensitive are the ones which are molecular and contain different bond strengths. These materials have usually an average valence number larger than five. Two requirements have to be fulfilled. First the recombination process must occur locally so that the energy is dissipated in a small molecular volume. Secondly the reaction products must have a chance to separate. This happens most easily in laminated structures which have weak bonds between layers or chains. It is difficult to achieve this in a three-dimensionally crosslinked network structure. The photosensitive materials combine a large energy gap with relatively low melting or glass transition temperatures. The recombination process can, for instance, be very local if the recombination occurs before the excited electron can separate from the hole and enter an extended or itinerant state of the material.

Brandes *et al.*[14] and Pearson and Bagley[15] studied hologram formation in arsenic–sulphur bulk glasses which they prepared by hot pressing small pieces at 150°C to a thickness of 0·045 cm. Exposures up to 10 J/cm^2 were used. They found an increase of diffraction efficiency as the sulphur content was increased to $As_{15}S_{85}$. Moreover, Pearson *et al.* observed three new peaks in the electron spin resonance spectra of irradiated samples. Two of these match quite well spin resonance peaks in irradiated pure, crystalline sulphur. They suggest that light enhances the depolymerization of the monomer. Glass irradiated for 2 minutes at 520 mW/cm^2 exhibited in the scanning calorimeter an endothermic peak at 113 ± 1°C which corresponds to the melting point of rhombic (α) sulphur. This peak occured in addition to the broad endotherm centered at 140°C, which is also observed in unirradiated samples, and is attributed to the polymerization transition of sulphur rings.

Pearson *et al.* conclude that precipitation of small rhombic sulphur microcrystals is produced by the illumination.

19.4. Photo-Induced Diffusion of Metals

Several metals, in particular Ag, readily diffuse into sulphur and selenium glasses at elevated temperatures (100–170°C). The diffusion of Ag and the formation of silver chalcogen compounds is greatly accelerated by illumination with light of greater than band edge energy.[16–19] It readily proceeds at room temperature at illumination levels as low as 10^{-4} J/cm^2.[17] This process is called photodoping. It appears to be intimately connected with the photostructural changes discussed in the previous sections.

The Ag–As_2S_3 system has been studied most extensively.[16–19] X-ray diffraction studies[19] reveal the formation of Ag_2S and As_2S_2 crystallites during the photodoping process. The resulting changes in refractive index and optical transmission can be used for hologram formation and photographic imaging. Many questions concerning the formation of a latent image and the role internal fields play in the kinetics of the photodoping process are presently being studied.

19.5. Light-Induced Flow and Topographic Changes

Changes in the structure and the degree of polymerization involve atomic motions and result in visco-elastic changes. Berkes *et al.*[4] reported, for instance, a smoothing of the surface of As_2Se_3 films in the illuminated areas. Feinleib *et al.*[20] observed reversible optical effects caused by scattering centres which were induced by flow in the amorphous phase of Se-rich Se–Te alloys. By focusing the light of a Krypton laser (6471 Å) on the interface between a glass substrate and 0·5–5 μm thick sputtered or evaporated Se alloy, these authors could alter the topography of the film at the interface in a reproducible and reversible manner. Light pulses from the laser produce spots of 1–5 μm diameter which appear dark in transmission, principally as a result of refraction and scattering loss of light. Etching with CS_2 also revealed an increase in polymerization in the exposed region. The optical effects are reversed by heating the spot region with the laser to somewhat above the glass transition temperature. This allows the material to relax and to become smooth.

Figure 19.6 shows that the spots produced by laser pulses (upper half of figure) can be erased (lower half) by less intense radiation from the same laser. The small spot size, large optical contrast, and the speed of spot writing (about 10^{-7} sec) suggest the use of these materials for optical mass memories.

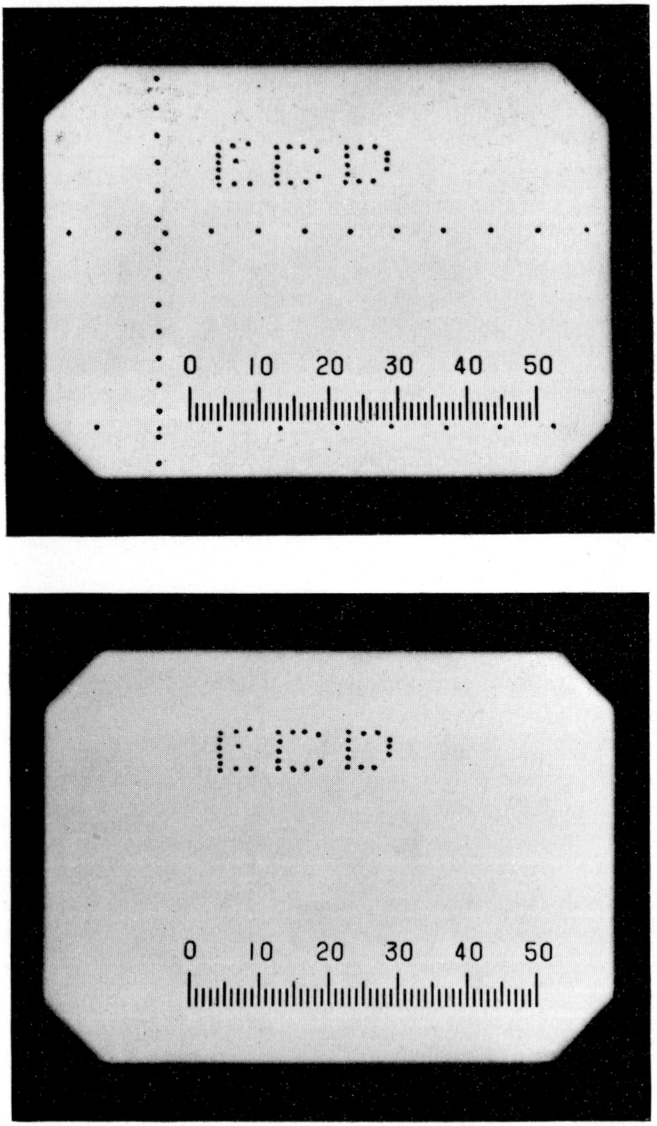

FIG. 19.6. Microscopic view in transmission of a glass substrate coated with an amorphous semiconductor film. The dark dots, about 2 μm in diameter, are written with a laser pulse within 10^{-7} sec. The same laser can erase dots as shown in the lower half of the picture. (Courtesy of S. R. Ovshinsky.)

19.6. Photocrystallization and Nucleation

The fact that the crystallization of Se is promoted and accelerated by light was discovered about 100 years ago.[21] Selenyi[22] observed that light enhances crystallization of amorphous Se at lower temperatures. He found that a positive crystallization image with good resolution but weak contrast could be obtained if the surface of the amorphous Se is first roughened with tissue paper. Dresner and Stringfellow[23] demonstrated that this increase of the crystal growth rate by light results from the production of electron–hole pairs in the vitreous Se and that the growth is controlled by the flow of holes in the interface field to the crystal boundary.

The experiments of Dresner and Stringfellow were carried out with rather low light intensities (about 3×10^{18} quanta $cm^{-2}sec^{-1}$) so that the growth rate (of order 50 μm/hour) could be visually measured. Feinleib *et al.*[24] showed that laser pulses of about 10^{-8} J can produce crystallization of a 1–2 μm diameter spot in less than 10^{-6} sec. These authors used 1–3 μm thick amorphous Ge–Te alloy alloy films supported by a glass plate. The structure change is confined to a small fraction of the film thickness at the glass–semiconductor interface where most of the light is absorbed.

In the amorphous–crystalline transformation the optical density is altered by several orders of magnitude for light near the absorption edge. The structural state of the spot can also be read out in reflection. The reflectivity changes by roughly a factor of 2 in Te-rich alloys.[25] One of the reasons why large changes in the optical properties occur is that the lack of long-range order in the vitreous state lifts the restriction, valid in crystals, that only optical transitions which conserve crystal momentum can occur. Moreover,

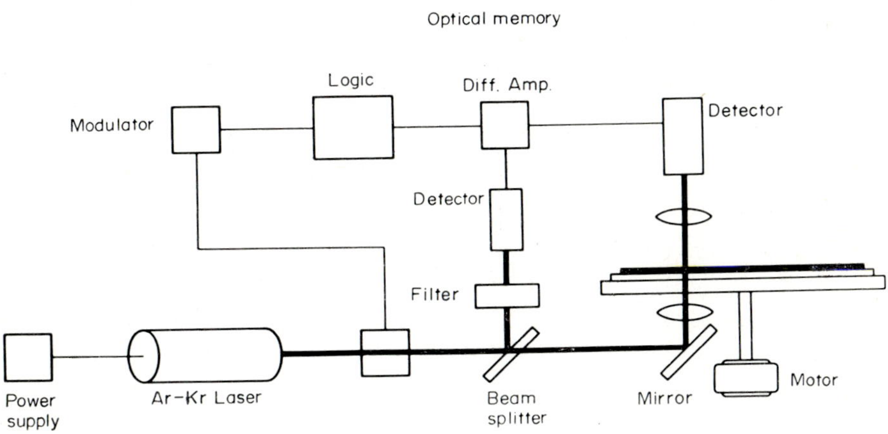

Fig. 19.7. Schematic layout of optical mass memory. The amorphous photostructural film is coated on the transparent disk rotated by the motor. (After Ovshinsky, Ref. 28.)

the polycrystalline structure of the transformed material scatters an appreciable amount of light out of the beam.

The stability and reversibility of the structure change, the small spot size, and the small inertia involved in moving the scanning laser beam recommends this system for high density optical memories. Figure 19.7 shows the layout of a laboratory prototype optical mass memory where the amorphous semiconductor film depositied on a glass disc is used in an arrangement resembling the familiar magnetic disk memory. The same laser beam is used for dot writing, erasing, and for reading the binary information. The low light intensity used for the read process does not affect the structural memory state at all.

A large area image can be produced by projecting an image onto the surface or by exposing the amorphous semiconductor layer through an appropriate mask. The sensitivity of an amorphous photostructural film can be enhanced by either first forming a latent image by photonucleation[26] at room temperature and then dry development for a few seconds by heat, or alternatively by exposing the film at elevated temperatures. A grey scale is provided by the variation of grain size with exposure. The resolution is determined by the density and size of crystallites in the exposed area. These depend on film thickness and composition. High resolution films (500 lines/mm) have a

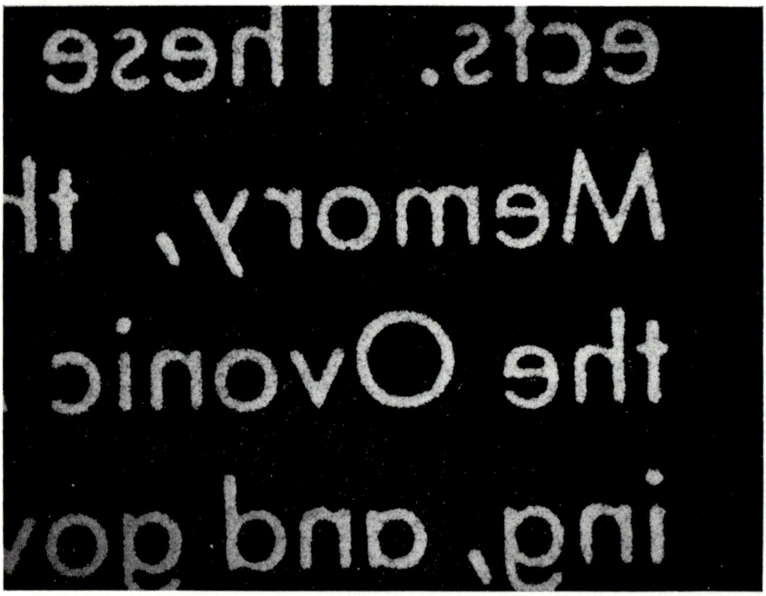

FIG. 19.8. Area of high contrast high resolution microfiche copy consisting of photostructural Te–alloy film on mylar. Resolution: 500 lines/mm. Optical contrast 30 to 1 after brief etch. (Courtesy of S. R. Ovshinsky.)

crystallite surface density of about $10^7 cm^{-2}$ with grains a few microns in diameter. The optical contrast in reflection is about 2:1. The optical contrast can be enhanced by a brief chemical etch.

Figure 19.8 shows a small area of a microfiche copy having a resolution of 500 lines/mm. The high contrast and resolution and the quick processing of amorphous photostructural films are great advantages for producing and copying projection material such as slides and microfiche.

19.7. CONCLUSIONS

The examples discussed here illustrate that a structural parameter affects the electrical, optical, mechanical and chemical properties of amorphous semiconductors and thus must be understood. At the same time this parameter opens up the possibility of using amorphous semiconductors for information storage, display, and imaging.

ACKNOWLEDGEMENTS

I wish to thank S. R. Ovshinsky, J. P. de Neufville and J. Zakis for many stimulating and illuminating discussions. The support of the Air Force Office of Scientific Research (AFSC), United States Air Force, under contract F44620–71–C–0025, and the general support of the Materials Research Laboratory by the National Science Foundation, is gratefully acknowledged.

REFERENCES

1. Chang, R., *Mat. Res. Bull.* **2,** 145 (1967).
2. Zakis, J., *Bull. Am. Phys. Soc.* March 1973, and to be published.
3. Keneman, S. A., *Appl. Phys. Letters* **19,** 205 (1971).
4. Berkes, J. S., Ing. Jr., S. W. and Hillegas, W. J., *J. Appl. Phys.* **42,** 4908 (1971).
5. Igo, T. and Toyoshima, Y., Proc. Third Conf. Solid State Devices, Tokyo 1971, p. 61.
6. Igo, T. and Toyoshima, Y., *Japan J. Appl. Phys.* **11,** 117 (1972).
7. Igo, T. and Toyoshima, Y., *J. Non-Cryst. Solids* **9,** 139 (1972).
8. Ohmachi, Y. and Igo, T., *Appl. Phys. Letters* **20,** 506 (1972).
9. Agarwal, S. C. and Fritzsche, H., *Bull. Am. Phys. Soc.* March 1973, and to be published.
10. de Neufville, J. P., *J. Non-Cryst. Solids* **8-10,** 85 (1972).
11. Moss, S. C., de Neufville, J. P. and Ovshinsky, S. R., *Bull. Am. Phys. Soc.* March 1973, and to be published.
12. Street, R. A. and Yoffe, A. D., *J. Non-Cryst. Solids* **8-10,** 745 (1972).
13. Trubisky, M. P. and Neyhart, J. H., *Appl. Opt. Suppl.* **3,** 59 (1969).
14. Brandes, R. G., Laming, F. P. and Pearson, A. D., *Appl. Optics* **9,** 1712 (1970).
15. Pearson, A. D. and Bagley, B. G., *Mat. Res. Bull.* **6,** 1041 (1971).

16. Kostyshin, M. T., Mikhaylovskaya, Ye. V. and Romanenko, P. F., *Soviet Physics Solid State* **8,** 451 (1966).
17. Kostyshin, M. T., Proc. Int. Congress on Photographic Science, Moscow, July 20, 1970.
18. Sakuma, H., Shimizu, I., Kokado, H. and Inoue, E., Proc. Third Conf. Solid State Devices, Tokyo 1971, p. 76.
19. Matsuda, A. and Kikuchi, M., Proc. Fourth Conf. Solid State Devices, Tokyo 1972.
20. Feinleib, J., Iwasa, S., Moss, S. C., de Neufville, J. P. and Ovshinsky, S. R., *J. Non-Cryst. Solids* **8-10,** 909 (1972).
21. Adams, W. G. and Day, R. E., *Proc. Roy. Soc.* p. 113 (1876).
22. Selenyi, P., *Nature* **161,** 522 (1948); *Acta Physica* **2,** 129 (1952).
23. Dresner, J. and Stringfellow, G. B., *J. Phys. Chem. Solids* **29,** 303 (1968).
24. Feinleib, J., de Neufville, J. P., Moss, S. C. and Ovshinsky, S. R., *Appl. Phys. Letters* **18,** 254 (1971).
25. Feinleib, J. and Ovshinsky, S. R., *J, Non-Cryst. Solids* **4,** 564 (1970).
26. Ovshinsky, S. R. and Klose, P., *J. Non-Cryst. Solids* **8-10,** 892 (1972).
27. Ovshinsky, S. R. *et al.*, at Energy Conversion Devices, Inc., (for example U.S. patent 3,530,441).
28. Ovshinsky, S. R., Proc. 5th Ann. Nat. Conf. Industrial Research, 1969, p. 86.

20. PHYSICS OF XEROGRAPHIC PHOTORECEPTORS

J. MORT

Xerox Corporation, Rochester Corporate Research Centre, Rochester, N.Y., U.S.A.

20.1. INTRODUCTION

For the most part our attention has so far focussed on the many fundamental questions which arise when trying to interpret and understand the basic properties of amorphous materials. It is now appropriate that the potential technological applications of these materials be considered. In the case of xerography that potential, in large measure, has already been realised. The rental and sales of copying and duplicating products that utilize amorphous chalcogenide photoreceptors constitute a billion-dollar industry. Beyond the financial aspects, the impact of amorphous materials on graphic communication at all levels of life, through their use in the xerographic process, has been pervasive.

The xerographic process uses a photoconductive insulator, such as amorphous Se, As_2Se_3 or a sensitized polymer. Some tens of microns of the insulator are deposited on an aluminum substrate. The free surface of the photoconductive insulator is corona charged to an initial potential $V_0 \sim 700$ Volts. Ideally Q_0, the initial surface charge density, is equal to CV_0 where C is the geometric capacitance. The charged photoreceptor is then exposed to a light and dark image formed by reflecting light from the document to be copied. Where light strikes the photoreceptor a reduction in the surface potential is caused by photo-current flow normal to the plate surface. In this way the surface potential distribution mirrors the light-dark content of the document. The resistivity of the photoreceptor must be high to minimize dark decay both normal and parallel to the surface. The latent electrostatic image on the photoreceptor is now rendered visible by use of a developer which consists of black toner particles $\sim 10\mu m$ in diameter and carrier beads $\sim 100\ \mu m$ in diameter. The toner particles, which are charged via tribo-electricity, are preferentially attracted by the surface image field at light-dark boundaries. The toner particles are next transferred from the photoreceptor to a sheet of paper by charging the latter with the opposite polarity charge to that of the toner and the toner image is made permanent by fusing the plastic based toner particles onto the paper. This can be done by either heat or solvent vapours. Since the transfer step is not 100% efficient, any remaining toner on the photoreceptor is next removed by mechanical wiping and electrostatic attraction. Finally, any remaining latent electrostatic image is removed by uniformly exposing the photoreceptor to a sufficiently intense erasure lamp to reduce the surface to an equipotential value of $\gtrsim 100$ Volts. After this step the photoreceptor is ready to repeat the cycle.

Having outlined the basic steps used in xerography to produce a copy of a document, our attention will be concentrated on the physical processes involved in the photoreceptor to obtain the latent electrostatic image. No further consideration will be given to the charging step, image transfer, image fusing, cleaning and erasure of which detailed discussions can be found elsewhere. [1,2] Parenthetically it may be noted that these crucial steps in the xerographic process impose very stringent criteria on an amorphous photoreceptor if it is to be acceptable for machine use. Amongst others, these include the ability to produce large area, defect free films of reproducible quality, with long term stability against crystalization under adverse conditions, mechanically strong against cracking and abrasion, and insensitivity to a range of humidities and temperatures. The fact that the rigid specifications in mechanical properties together with equally demanding electrical requirements are met is a testament to the unusual properties of amorphous materials.

For the purpose of laying sufficient groundwork for the more detailed discussion to come, we return briefly to the exposure and development steps. Figure 20.1(a) shows typical photodischarge curves for an amorphous selenium photoreceptor for two light intensities incident on it in the manner indicated in the insert. The potential difference or contrast potential between the area struck by the full light intensity (referred to as the "white background") and the attenuated light (corresponding to reflection from a dark image) as a function of time can be determined by the simple subtraction of the two discharge curves. Figure 20.1(b) is a plot of this contrast potential as a function of exposure (i.e. white background flux $F_0 \times t$). It can be seen that the contrast potential is small at low exposures (underexposed) goes through a maximum (optimum exposure) and decreases at large exposures (overexposed). Clearly a xerographic machine should be operated at the optimum exposure in order to achieve maximum contrast potential, because the density

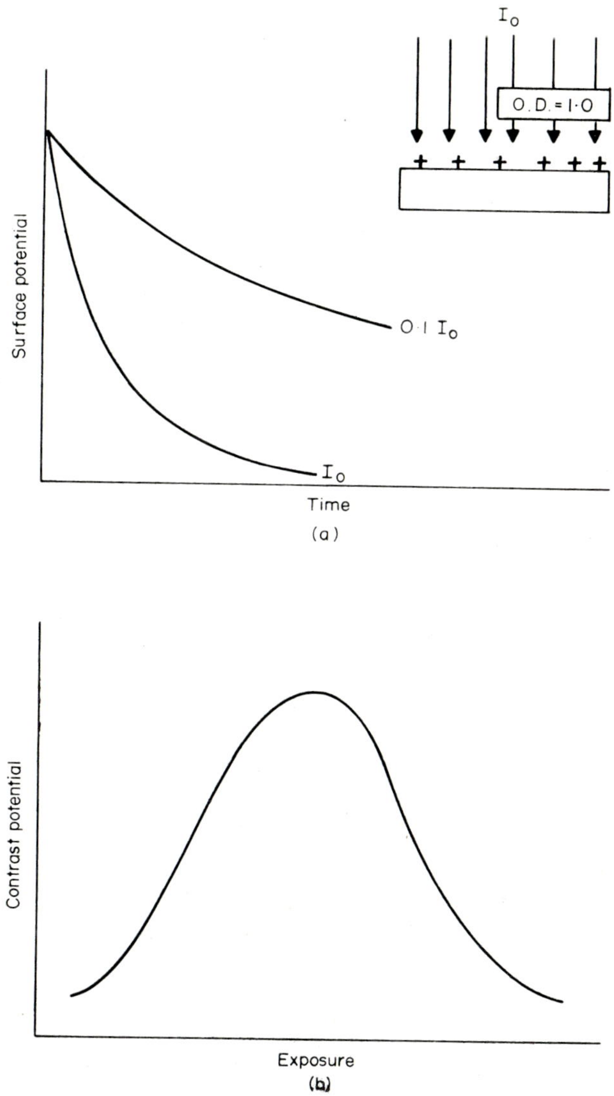

FIG. 20.1. (a) Schematic photodischarge curves versus time for an amorphous Se photoreceptor exposed to a light pattern of two intensities (insert). (b) Schematic contrast potential curve versus exposure.

of the developed image (attraction of toner particles by the fringe fields) below a certain contrast potential strongly depends on the absolute value of the contrast potential.[1] The sharpness of the contrast potential curve shown in Fig. 20.1(b) can be viewed as some measure of the degree of exposure latitude that is allowed for either under or overexposure before copy density is affected. This latitude for amorphous selenium is relatively narrow so that this photoreceptor, without more sophisticated exposure or development techniques, reproduces only a relatively narrow range of object densities.[1] As we shall see this property is directly due to the photogeneration process in amorphous Se.

20.2. Physics of Xerography

In the xerographic form of imaging, as mentioned already, a normally insulating photoconducting film is uniformly charged and then exposed to light to form a latent electrostatic image. The manner in which the surface potential drops from the initial potential V_0 due to exposure to light is a crucial step in the image forming process. This xerographic discharge characteristic, i.e. the time dependent surface potential during, or after, illumination depends intimately on the electronic properties of the photoconductor, e.g. carrier supply rate, mobility and lifetime. The mode of exposure, e.g. bulk or surface absorption of the light, flash or step function illumination, can also be determinants for the discharge characteristic.

For conceptual ease, we will consider the discharge of a photoreceptor exposed to uniform illumination in order to highlight the basic processes involved, whilst remembering that xerographically what is important is the contrast potential or potential difference that can be developed as a result of *non-uniform* illumination.

Assume that a photoreceptor with a geometric capacitance per unit area $C = \kappa/4\pi L$, where κ is the dielectric constant and L the sample thickness, has been positively charged to an initial potential V_0 and that a surface charge Q_0 resides on the surface. This surface charge can be fully discharged by displacing an amount of positive charge Q_0 from the top surface to the bottom electrode. This is accomplished by photogenerating free electron-hole pairs in the photoreceptor with light absorbed very close to the top surface. In this case holes then drift in the field across the photoreceptor thickness L. The electrons drift back through the excitation region and have little effect on the discharge characteristic.

If the schubweg of the carriers, which equals $\mu\tau E(t)$, where μ is the carrier mobility, τ is the lifetime with respect to deep traps and $E(t)$ is the field at time t, becomes less than L then the discharge will stop and a residual field or surface potential will remain. For practical reasons a material which exhibits significant bulk trapping is not a desirable xerographic photoreceptor because of limitations to the contrast potential which can be achieved. In addition, integration of charge in the deep traps over succeeding cycles can lead to residual buildup. For this reason we will consider only photoreceptors in which this is not a problem.

The important point about the xerographic discharge process is that it is a non-constant voltage process. Obviously as the discharge proceeds the field in which succeeding carriers drift becomes progressively smaller. Two extreme cases of the discharge process can occur. If the amount of charge which moves in one transit time $t_T = L/\mu E(t)$ is much less than $Q(t)$, the surface charge remaining at time t, then many transits must occur before the photoreceptor is fully discharged. This type of discharge is termed an *Emission-Limited* discharge since the discharge rate, i.e. dV/dt, is determined by the rate of supply or emission of carriers from the excitation region. Such a process will have the light intensity dependence of the photogeneration process in the photoreceptor. In the case of amorphous Se, for example, the xerographic discharge characteristic is linear over many decades of light intensity. If, however, an amount of charge Q_0 is generated, either by a light flash or step-function illumination, in a time less than the initial transit time of the carriers $t_T = L/\mu E_0$ then the total discharge will occur within one transit. As this sheet of charge, Q_0, moves away from the top surface, the electric field behind it in the excitation region falls to zero and additional carriers will be neither generated nor drift into the sample bulk. This extreme type of discharge is therefore light intensity independent and is the maximum rate of xerographic discharge for highly absorbed light. This is called a *Trap free space-charge-limited* (TFSCL) discharge.

Our discussion of photoinduced discharge characteristics therefore naturally separates into two categories, one defined as emission limited and the other space-charge limited. The distinction between the two cases is in the ratio of the amount of charge in transit to the charge remaining on the surface We will now outline the theory of the xerographic discharge process discussing in turn the case of emission-limited discharge and space-charge-limited discharge and comparing the predictions with experimental results.

20.2.1. EMISSION LIMITED DISCHARGE

(i) *Theory*

An expression[3,4] for the surface potential of a xerographic photoreceptor as a function of exposure (flux × time) can be derived with the following assumptions. First, the effects of bulk trapping will be neglected. For practical xerographic photoreceptors this approximation is valid down to voltages $\gtrsim$ 10 V for a 50μm film. Secondly, it is assumed that all the light is absorbed in a depth small compared with the sample thickness. Thirdly, the amount of charge in transit is small compared to the remaining surface charge at any time.

Under the assumptions listed above, the emission or injected current is related to the decay of the surface potential by

$$-\,dV/dt = e\mathcal{N}_e/C = J_e/C \tag{20.1}$$

where $\mathcal{N}_e$ is the numbers of carriers emitted per unit time per unit area. A V versus t relationship can be obtained by integrating (20.1) to give

$$t = -\,C\int_{V_0}^{V} 1/J_e\,dV \tag{20.2}$$

where V_0 is the initial ($t = 0$) surface potential.

For the integration of (20.2) we express J_e as a function of V. $\mathcal{N}_e$ can be written as

$$\mathcal{N}_e = F\,\eta\,(F, V) \tag{20.3}$$

where F is the light flux and η the quantum efficiency of carrier emission which is taken to include the photogeneration step together with any limitations due to recombination and injection barriers.

The quantum efficiency can be written as

$$\eta(V) = \left(\frac{V}{V_\eta}\right)^P \qquad V \leqslant V_\eta \tag{20.4}$$

$$\text{or} = 1 \qquad V \geqslant V_\eta \tag{20.5}$$

where V_η is the potential at and above which the quantum efficiency is unity. The xerographic process is one of unity gain since secondary photocurrents do not occur. The integration in eqn (20.2) can be carried out giving the following results for t:

for $V_0 \leqslant V_\eta$

$$t = \frac{C}{eF}\,\frac{V_\eta}{1-P}\,[(V_0/V_\eta)^{1-P} - (V/V_\eta)^{1-P}] \quad \text{if} \quad P \neq 1$$

or

$$t = \frac{C}{eF}\,V_\eta \ln (V_0/V) \quad \text{if} \quad P = 1 \tag{20.6}$$

for $V_0 > V_\eta$

$$t = \frac{-CV_\eta}{eF}\left[\left(\frac{V_0}{V_\eta}\right) + \frac{P}{1-P} - \frac{1}{1-P}\left(\frac{V}{V_\eta}\right)^{1-P}\right] \quad \text{if } P \neq 1$$

or

$$t = \frac{CV_\eta}{eF}\left[\left(\frac{V_0}{V_\eta}\right) - 1 + \ln\left(V_\eta/V\right)\right] \quad \text{if} \quad P = 1. \tag{20.7}$$

From the above equations we see that if the emission of carriers is independent of field (i.e. $P = 0$) the surface potential decays linearly with time

$$V = V_0 - \frac{eF\,t}{C}. \tag{20.8}$$

For $P > 0$, the V versus t curve deviates from the linear time decay and exhibits dispersion. The effect becomes more pronounced as the field dependence i.e. P increases, as illustrated in Fig. 20.2(a). Since, in general, photoreceptors with different P have different V_η, the discharge curves in Fig. 20.2(a) are plotted with the potential coordinate normalized to $V_\eta = 1$ and the abscissa expressed in photons. $V_0 = V_\eta = 1$ was chosen for the curves shown.

Figure 20.2(b) shows contrast potential curves versus exposure for a subject ot 1.0 density unit. In this case $V_0 = V_\eta = 800$ Volts. It can be seen as the field dependence of carrier emission increases from 0 to 2 that the exposure latitude increases. The optimum exposure shows a corresponding increase reflecting the fact that as the field dependence of the carrier supply increases, there is a decrease in the efficient utilization of the light. Sublinear intensity dependence of the carrier supply can also lead to dispersion of the photodischarge characteristics and hence increase latitude. For further details the reader is directed to the published work of Chen and Mort[3] and Chen, Mort and Tabak[4].

(ii) *Experimental Results*

The dispersion or curvature in the discharge characteristics has been observed for amorphous Se. Figure 20.3 shows the result of a photodischarge measurement on a 45 μm amorphous Se film exposed to 4300 Å light which is highly absorbed. The light intensity was 1×10^{12} photons cm^{-2}sec^{-1} The solid line is a plot of eqn (20.6) with $P = \frac{1}{2}$ and $E = V/L = 2{\cdot}8 \times 10^5$ V cm^{-1}. These values are the measured photogeneration data for amorphous Se[5] discussed in Chapter 15. It is clear, as first pointed out by Tabak and Warter[5], that the photodischarge curve of selenium can be fully explained in terms of its photogeneration properties and the dispersive character is

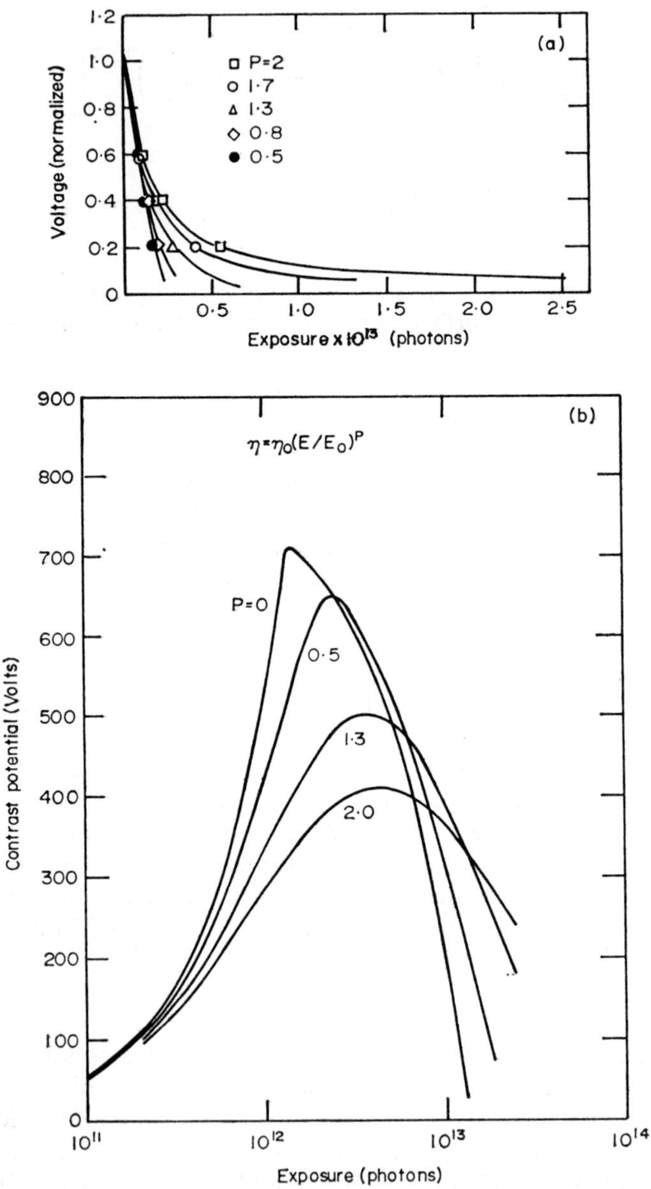

FIG. 20.2. (a) Theoretically calculated photodischarge curves versus time, eqn (20.6), for a series of model photoreceptors with different field dependences of the carrier supply *P*. (After Chen and Mort, Ref. 3.) (b) Contrast potential curves obtained with differing values of *P* considered in (a) for an input subject density of 1·0.

attributable to the field dependence of photogeneration of carriers in this amorphous material.

The measurements of Tabak and Warter[5] revealed that the bulk deep trapping lifetime for holes in xerographic grade amorphous Se is typically $\sim 50\ \mu$sec. Since the drift mobility is $\sim 0{\cdot}13\ cm^2\ sec^{-1}\ V^{-1}$, the schubweg $\mu\tau E$ approaches 50 μm, a usual photoreceptor thickness, only at fields $\sim 10^3\ V\ cm^{-1}$. This leads to a residual voltage due to bulk deep trapping $\sim$ 5 V which is no problem in the usual development systems.

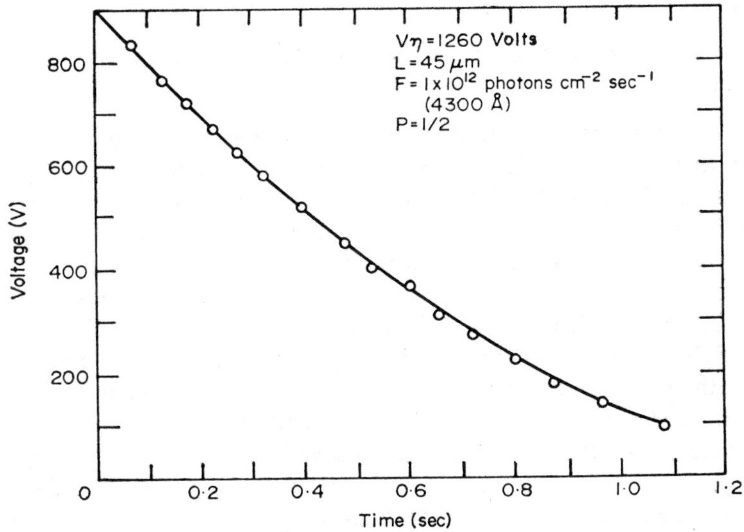

FIG. 20.3. Discharge curve (circles) of amorphous Se compared with that calculated from eqn (20.6) with $F = 1 \times 10^{12}$ photons $cm^{-2}\ sec^{-1}$, $\kappa = 6{\cdot}3$, $L = 45{\cdot}0\ \mu m$; $E_\eta = V_\eta/L = 2{\cdot}4 \times 10^5$ V/cm, $P = 0{\cdot}5$. (After Chen *et al.* Ref. 4).

Amorphous As_2Se_3 has some properties which are similar to amorphous Se but others which are quite different. It has a very similar field dependence in its supply of carriers out of the excitation region and has a comparable exposure latitude to amorphous Se. A comparison of the spectral dependence of the quantum efficiency for photogeneration for amorphous Se and amorphous As_2Se_3[5] shows that As_2Se_3 is considerably more panchromatic and would discharge faster than Se if both were exposed to the same white light source. Together with its increased panchromaticity As_2Se_3 has a relatively high dark discharge, but this is not a limitation and As_2Se_3 is successfully used in high speed copying machines. As_2Se_3 has significantly different transport properties from amorphous Se[7]. Like amorphous Se, it is operated with the exposed surface charged positively, i.e. hole motion produces the latent electrostatic image. The hole drift mobility is field dependent and at a field of 10^5 V cm^{-1} is $\sim 10^{-5}$ cm^2 sec^{-1} V^{-1}.[7] As discussed by Tabak *et al.*[6] this can lead to a transport limitation because at a sufficiently low surface potential the transit time of carriers can exceed the time for the exposed photoreceptor to enter the development zone in a machine. Although this does occur it is not a serious problem in machine usage because the transport limitation does not set in with a 60 μm sample until the surface potential has decayed to $\gtrsim$ 50 V. The fact that this effect does occur, however, is a direct consequence of the much lower hole mobility in As_2Se_3 compared with amorphous Se.

20.2.2. SPACE-CHARGE-PERTURBED AND SPACE-CHARGE-LIMITED DISCHARGE

(i) *Theory*

As we have seen, emission-limited discharges are characterized by the amount of charge in transit being less than $Q(t)$, the surface charge of the photoreceptor at any time. For total discharge, therefore, many transits must occur and the discharge can be visualized as occuring because of many small, equal, flash exposures. Since in the xerographic mode the surface potential is time dependent, each succeeding flash exposure occurs at a progressively lower initial surface potential. Ultimately, it is possible that during the discharge the amount of charge produced in a flash is a significant fraction of the remaining charge on the photoconductor. A discharge which initially starts as emission limited can therefore become space-charge-perturbed and ultimately space-charge-limited when the amount of free charge in the final transit is equal to $Q(t)$. In order for the total discharge to be space-charge-limited from the outset, a charge equal to the initial surface charge Q_0 must be generated in a time much less than the carrier transit time in the initial field. Apart from geometrical parameters, which will vary little from one photoreceptor to another, the determining factor as to whether a space-charge or emission-limited discharge will occur at a given light flux is determined by the drift mobility of the carriers. Since for holes in amorphous Se the drift mobility is 0·13 cm^2 s^{-1} V^{-1}, the space-charge-limited region cannot be reached with any reasonable steady state light source although it can be reached with very high intensity flash exposures. In other photoreceptors involving charge transport in polymers, where the mobilities are several orders of magnitude lower, this type of discharge can manifest itself for normally accessible light fluxes. Batra *et. al.* have analytically solved the space-charge-limited discharge under open-circuit conditions for strongly absorbed light and flash exposure.[5] Chen independently arrived at identical results but in the broader context of variable exposure modes, field dependent mobility and bulk photogeneration.[9] We will simply state some of his results, referring the reader to the paper of Chen[9] for details of the calculation.

The results can only be expressed analytically for the case of flash exposure, (Q_0 generated in times $\ll$ transit time t_T), and strongly absorbed light i.e. $\alpha L \to \infty$. For $t < t_T$, i.e. times less than the transit of the leading edge of the charge sheet where $t_T = L/\mu_0 E_0$ and μ_0 is the mobility at the initial field E_0,

$$\tilde{V}(\tilde{t}) = 1 - \frac{1}{\omega + 2}[1 - (1 - \rho_0)^{\omega+2}]\tilde{t} \tag{20.9}$$

$$\frac{d\tilde{V}}{d\tilde{t}}(\tilde{t}) = \frac{-1}{\omega + 2}[1 - (1 - \rho_0)^{\omega+2}] \tag{20.10}$$

where $\tilde{t} = t/t_T$, and ω is the field dependence of the mobility $\mu = \mu_0 (E/E_0)^{\omega}$. For $t > t_T$ but less than the time for total discharge,

$$\tilde{V}(\tilde{t}) = \frac{(1 - \rho_0)^{\omega+2}}{\omega + 2}\tilde{t} + \frac{\omega + 1}{\omega + 2}(1/\tilde{t})^{1/(\omega+1)} \tag{20.11}$$

$$\frac{d\tilde{V}(\tilde{t})}{d\tilde{t}} = -\frac{1}{\omega + 2}\left[\left(\frac{1}{\tilde{t}}\right)^{(\omega+2)/(\omega+1)} - (1 - \rho_0)^{\omega+2}\right] \tag{20.12}$$

where ρ_0 is the fraction of Q_0 emitted into the sample in the flash.

The crosses in Fig. 20.4(a) represent the values of $\tilde{V}(\tilde{t})$ calculated from the analytical formula, eqns (20.9) and (20.11) for a photoreceptor in which one CV_0 of charge is emitted in one-hundredth of a transit time. The upper curve is for field dependent ($\omega = 2$) mobility and the lower one for field independent ($\omega = 0$) mobility. In Fig. 20.4(b), $d\tilde{V}/d\tilde{t}$ versus $\tilde{t}$ is plotted in log scales. The crosses are calculated from the analytical formulae, eqns (20.10) and (20.12). A constant value exists for $d\tilde{V}/d\tilde{t}$ versus $\tilde{t}$ in log scales, up to $\tilde{t} = 1$ and then a straight line with a slope $-(\omega + 2)/(\omega + 1)$.

The surface potential after $\tilde{t} = 5$ does not change much with time. However, this quasi-residual potential depends strongly on ω, the field dependence of the mobility.(9) Thus, for a flash exposure, the field dependence of the mobility determines the residual potential and the mobility at E_0 determines how soon this value can be achieved. A fast discharge, therefore, requires not only high mobility but also low field dependence.(9)

A comparison will now be made between these theoretical results and experimentally observed space-charge-perturbed and space-charge-limited discharges of two different photoreceptors.

(ii) *Experimental Results*

(a) *Amorphous Se*

Amorphous Se has a well defined hole mobility $\sim 0{\cdot}13\ \mathrm{cm^2\ sec^{-1}\ V^{-1}}$ which is field independent ($\omega = 0$) at room temperature.[10] Equation (20.10) predicts and Fig. 20.4(b) shows that under conditions of flash exposure the discharge rate $d\tilde{V}/d\tilde{t}$ will be constant up to the transit of the leading edge of the charge sheet. The value of the discharge rate will be $J/C = \mu V_0^2/(2L^2)$ and for $\mu = 0{\cdot}13\,\text{cm}^2\,\text{sec}^{-1}\,\text{V}^{-1}$, $L = 50\,\mu\text{m}$ and $V_0 = 100$ V, dV/dt (for $t < t_T$) should be $\sim 2{\cdot}5 \times 10^7$ V/sec. For $t > t_T$, dV/dt will decrease $\alpha\, 1/t^2$. These predictions are in full accord with the experimental results of Batra *et al.*[8]

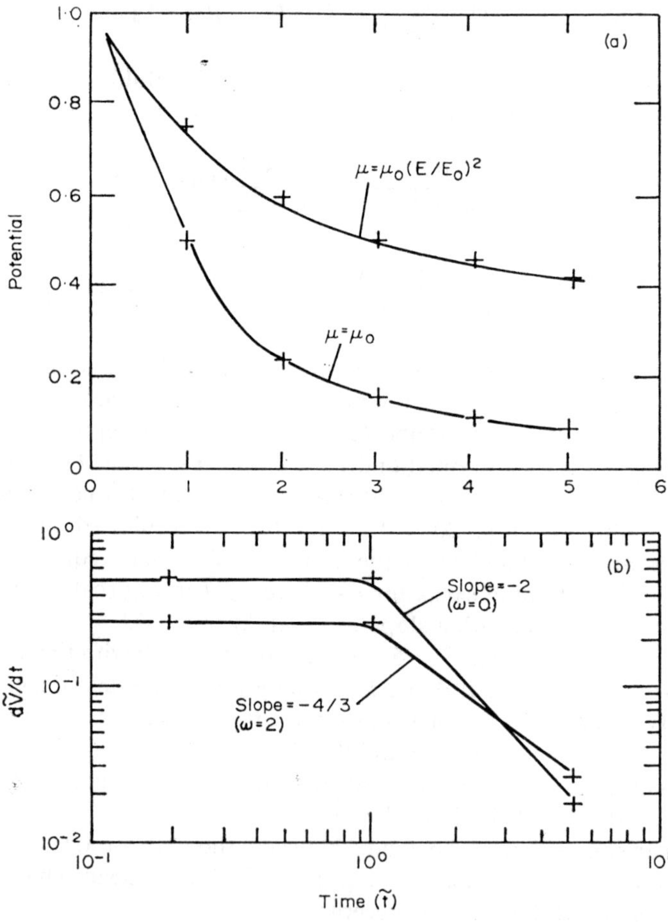

FIG. 20.4. (a) Discharge characteristics calculated (crosses) from eqns (20.9) and (20.11) for a photoreceptor with $\alpha L = 1000$, $P = 0{\cdot}5$, $\omega = 2$ (upper curve) and $\omega = 0$ (lower curve). One CV of charge is generated in $0{\cdot}01t_T$. (b) The dV/dt versus t curves for the discharges of (a). Crosses are calculated from eqns (20.10) and (20.12). (After Chen, Ref. 9.)

(b) *Se:Poly (N-vinylcarbazole) Composite Layers*

A heterogeneous photoreceptor employing these two materials consists of a thin ~ 2 μm amorphous Se evaporated over a thicker ~ 20 μm layer of the polymer coated from solution. The selenium makes this structure sensitive to visible light. The polymer has been shown to exhibit unusual hole transport characteristics and efficient hole injection can occur into it from amorphous Se.[11,12] The polymer is characterized by a very dispersive transport of holes, since in response to the injection of a sheet of holes a pronounced distribution of transit times is observed.[12] In addition the effective mobility is very low and field dependent.[12] The mobility of the fastest carriers is ~ 10^{-7}cm^2 sec^{-1}V^{-1} at a field of 10^5 V cm^{-1} and over a limited field range is proportional to V^2.

Figure 20.5 shows the dV/dt as a function of light intensity of a 22 μm PVK layer overlayed with 2 μm of amorphous Se. The data was taken with the illuminated selenium surface charged positively. The measurement was made at 4500 Å so that all the light is absorbed in the selenium layer.[13] At all light intensities and low fields a common value for dV/dt is found which is approximately proportional to V^4. This has been interpreted as the TFSCL discharge of PVK[12]. Since dV/dt equals J_s/C and from Childs law $J_s = 10^{-13}\,\kappa\,\mu E^2/L$, μ must be proportional to E^2. The mobility at any field can be calculated by using the appropriate value of J_s and Childs law, For the initial potential, corresponding to a field of $3{\cdot}3 \times 10^5$ V cm^{-1}, one finds that $\mu = 4 \times 10^{-7}$ cm^2 sec^{-1}V^{-1}. Figure 20.6 shows experimental plots of dV/dt versus time in log scales for the Se:PVK sample at four decreasing light fluxes.[13] In this case the analytical formulae eqns (20.10) and (20.12) are not generally applicable; instead an iterative calculation must be employed.[9] This is because in general a CV of free charge in the sample is reached only after a finite exposure. At very high light intensities, of course, Q_0 can be injected in times less than t_T and in this case the continued exposure in the step-function mode is not utilized. This simulates flash exposure and the analytical eqns (20.10) and (20.12) can be used. In this case dV/dt is constant up to time t_T, eqn (20.10), and then decreases such that dV/dt versus t in log scales is a straight line with slope $-(\omega + 2)/(\omega + 1)$. With decreasing values of flux, dV/dt starts with a smaller value but decreases less steeply even after $t = t_T$. The dV/dt values exceed those for the highest fluxes for a limited time but ultimately fall onto the straight line observed at the highest fluxes. This is simply a manifestation of the fact that at lower fluxes the sample becomes TFSCL at voltages much less than V_0 and together with the slow discharge times at lower fluxes this means that the transition occurs at longer times. These predictions are borne out by the data in Fig. 20.6.

The fact that the linear portions of the curves for the highest two fluxes coincide implys, that at the highest flux of $3{\cdot}1 \times 10^{14}$ photons cm^{-2}sec^{-1}, the sample becomes TFSCL in a time short compared with t_T. The slope of this line is $-4/3$ indicating that $\omega = 2$. Using the mobility value determined from Fig. 20.5 one can determine the constant dV/dt value $= \mu(E_0)E_0^2/4$ to be $1{\cdot}1 \times 10^4$ V sec^{-1} at the initial field. This is very close to the observed value. The extrapolation of the constant dV/dt line to intercept the extended slope $= -4/3$ line yields a value of $t_T = 2 \times 10^{-2}$ sec. Since $t_T = L/\mu E_0$, this gives $\mu = 3{\cdot}6 \times 10^{-7}$ cm^2 sec^{-1}V^{-1} at the initial field of $3{\cdot}3 \times 10^5$ V cm^{-1}.

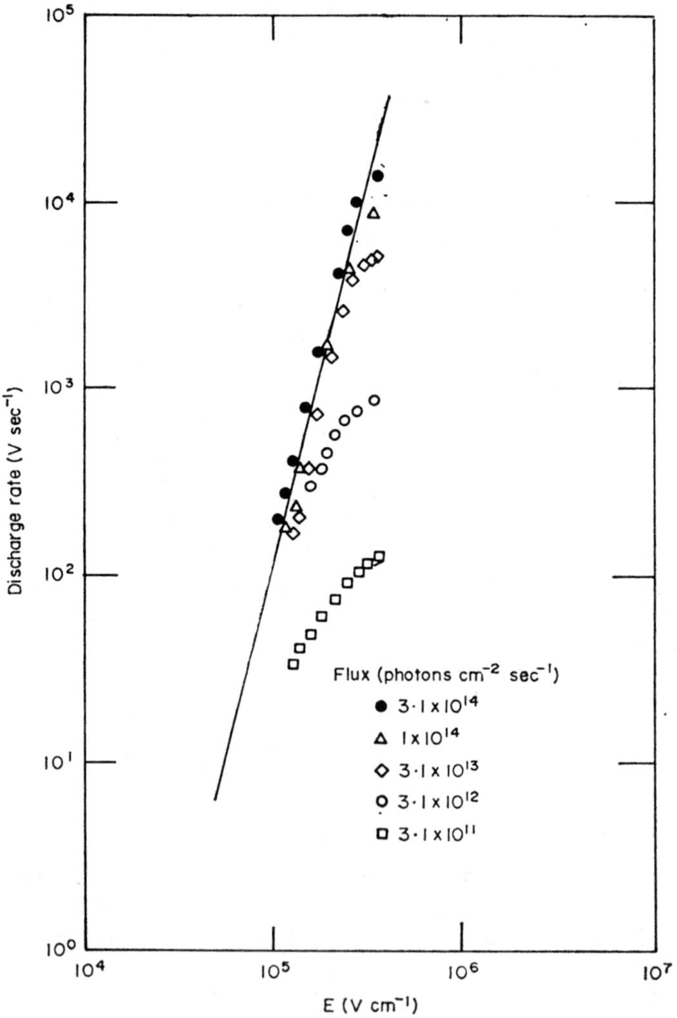

FIG. 20.5. dV/dt versus field for a 22 μm PVK film overcoated with 2 μm of amorphous Se at five different light intensities. $E = V(t)/L$. (After Mort *et al.*, Ref. 13.)

This value is in excellent accord with that calculated from Fig. 20.5 using the modified Childs law and both values are consistent with those obtained from time of flight measurements[12].

In an actual xerographic process, the photoreceptor is typically exposed for a finite time, which is chosen so that the white background light (corresponding to an object of zero optical density) generates approximately one CV_0 of charge. Light from darker objects is attenuated by $10^{-\mathrm{O.D.}}$ where O.D. is the optical density of the darker subjects. As was discussed earlier, it is the different discharge rates produced by differing input densities which lead to the contrast potentials so important in developing the latent image. The consequences for the contrast potential when a TFSCL xerographic discharge occurs is illustrated in Fig. 20.7.[13] Here the contrast potential for an input density of one has been measured at white background fluxes of $3{\cdot}1 \times 10^{14}$ photons $\mathrm{cm}^{-2}\,\mathrm{sec}^{-1}$, $3{\cdot}1 \times 10^{13}$ photons $\mathrm{cm}^{-2}\,\mathrm{sec}^{-1}$, and $3{\cdot}1 \times 10^{12}$ photons $\mathrm{cm}^{-2}\,\mathrm{sec}^{-1}$. At an input density of one the fluxes are one tenth of these values. For each intensity the contrast potential was measured in two

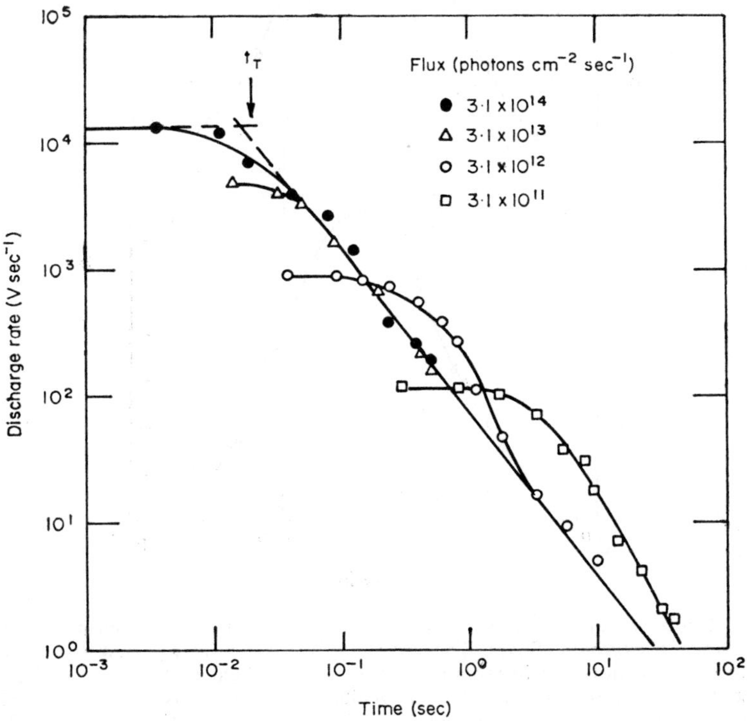

FIG. 20.6. Plot of dV/dt versus time for the Se:PVK sample measured at four light intensities. (After Mort *et al.*, Ref. 13.)

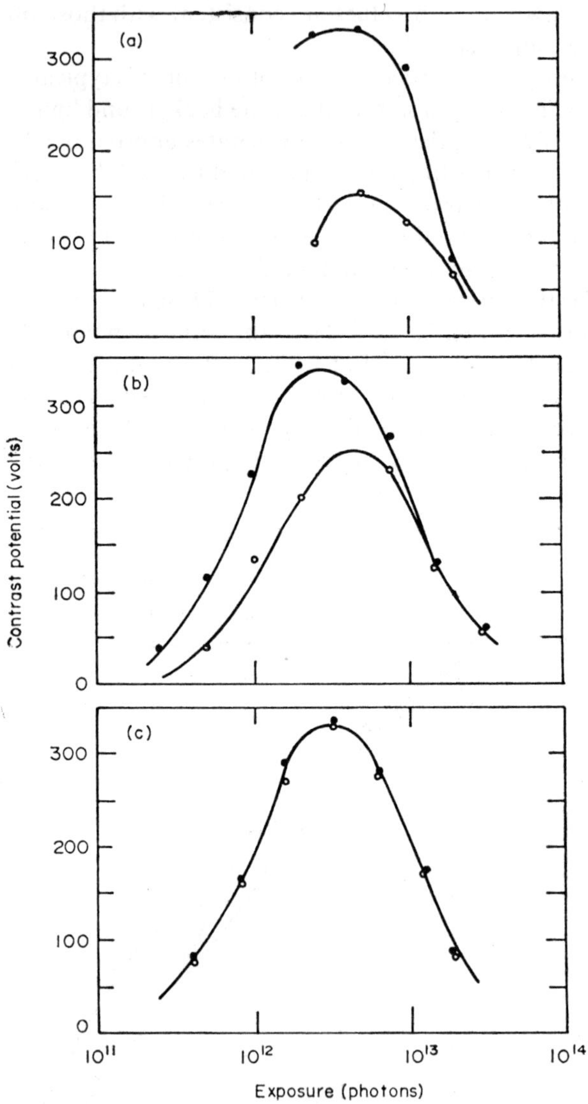

FIG. 20.7. Contrast potential versus exposure curves for an optical density of unity at different white background fluxes. The initial surface potential was 820 V in each case. The open circles denote the instantaneous contrast potential and the black circles rep-represent the maximum contrast potential. White background flux (a) $3{\cdot}1\times10^{14}$ photons cm^{-2} sec^{-1}, (b) $3{\cdot}1\times10^{13}$ photons cm^{-2} sec^{-1} and (c) $3{\cdot}1\times10^{12}$ photons cm^{-2} sec^{-1}. (After Mort *et al.*, Ref. 13.)

ways. In the first, the contrast potential attained immediately at the termination of the exposure was measured, while in the second the maximum contrast potential attainable was measured. In some cases this was obtained only after times longer than the exposure time. The exposures given in the abscissae of Fig. 20.7 equal the product of the white background flux and the exposure time. The exposure times therefore varied from 8 msec. for the smallest exposure in Fig. 20.7(a) to 6 sec. for the largest exposure in Fig. 20.7(c). From Figs. 20.7(a), (b) and (c) it can be clearly seen that the contrast potential measured by both techniques agree at the low flux but that increasingly at higher fluxes the contrast potential measured immediately on the termination of the exposure is smaller than the maximum ultimately attainable. This is because in these cases a CV of charge is already in transit in the sample and continues to flow even after the exposure is terminated. The development of the contrast potential continues to grow after the exposure. This effect is therefore a direct result of the TFSCL discharge.

20.3. SUMMARY

The goals which motivated the preparation of this contribution on xerography were two-fold. The first was to demonstrate the depth of understanding of the physical mechanisms operative in the xerographic process as they relate the photogeneration and transport properties of amorphous materials to the latent electrostatic image and hence ultimately the copy quality. The second was to indicate the power of the xerographic discharge itself, divorced from its technological ramifications, as a mature experimental technique for investigating the electrical properties of insulating amorphous materials.

ACKNOWLEDGEMENTS

The author is grateful to Dr. I. Chen for many stimulating discussions concerning the theory of xerographic discharge.

REFERENCES

1. Dessauer, J. H. and Clark, H. E., "Xerography and Related Processes". London, The Focal Press, 1965.
2. Schaffert, R. M., "Electrophotography", London, The Focal Press, 1965.
3. Chen, I. and Mort, J., *J. Appl. Phys.* **43,** 1164 (1972).
4. Chen, I., Mort, J. and Tabak, M. D., I.E.E.E. Trans. on Elec. Devices—Special Issue on Electrophotography, April (1972).
5. Tabak, M. D. and Warter, P. J., Jr., *Phys. Rev.* **173,** 899 (1968).
6. Tabak, M. D., Ing, S. W. and Scharfe, M. E., I.E.E.E. Trans. on Elec. Devices—Special Issue on Amorphous Materials, 1972.

7. Scharfe, M. E., *Phys. Rev.* **B2,** 5025 (1970).
8. Batra, I. P., Kanazawa, K. Keiji, and Seki, H., *J. Appl. Phys.* **41,** 3416 (1970).
9. Chen, I., *J. Appl. Phys.* **43,** 1137 (1972).
10. Spear, W. E., *Proc. Phys. Soc.* **B76,** 826 (1960).
11. Regensburger, P. J., *Photochem. Photobiol.* **8,** 429 (1968).
12. Mort, J., *Phys. Rev.* **B5,** 3329 (1972).
13. Mort, J., Chen, I., Emerald, R. L. and Sharp, J. H., *J. Appl. Phys.* **43,** 2285 (1972).

AUTHOR INDEX

Numbers in *italics* refer to the pages where references are listed in full.

C

D

I

J

N

O

P

T

Y

Z

SUBJECT INDEX

P

Q

R

S

T

U

V

W

X